T0180525

Advances in Intelligent Systems and Computing

Volume 1020

The series "Advances in Intelligent Systems and Computing" contains publications on theory, applications, and design methods of Intelligent Systems and Intelligent Computing. Virtually all disciplines such as engineering, natural sciences, computer and information science, ICT, economics, business, e-commerce, environment, healthcare, life science are covered. The list of topics spans all the areas of modern intelligent systems and computing such as: computational intelligence, soft computing including neural networks, fuzzy systems, evolutionary computing and the fusion of these paradigms, social intelligence, ambient intelligence, computational neuroscience, artificial life, virtual worlds and society, cognitive science and systems, Perception and Vision, DNA and immune based systems, self-organizing and adaptive systems, e-Learning and teaching, human-centered and human-centric computing, recommender systems, intelligent control, robotics and mechatronics including human-machine teaming, knowledge-based paradigms, learning paradigms, machine ethics, intelligent data analysis, knowledge management, intelligent agents, intelligent decision making and support, intelligent network security, trust management, interactive entertainment, Web intelligence and multimedia.

The publications within "Advances in Intelligent Systems and Computing" are primarily proceedings of important conferences, symposia and congresses. They cover significant recent developments in the field, both of a foundational and applicable character. An important characteristic feature of the series is the short publication time and world-wide distribution. This permits a rapid and broad dissemination of research results.

** Indexing: The books of this series are submitted to ISI Proceedings, EI-Compendex, DBLP, SCOPUS, Google Scholar and Springerlink **

More information about this series at http://www.springer.com/series/11156

Volodymyr Lytvynenko ·
Sergii Babichev · Waldemar Wójcik ·
Olena Vynokurova · Svetlana Vyshemyrskaya ·
Svetlana Radetskaya
Editors

Lecture Notes in Computational Intelligence and Decision Making

Proceedings of the XV International Scientific Conference "Intellectual Systems of Decision Making and Problems of Computational Intelligence" (ISDMCI'2019), Ukraine, May 21–25, 2019

 Springer

Editors
Volodymyr Lytvynenko
Kherson National Technical University
Kherson, Ukraine

Waldemar Wójcik
Lublin University of Technology
Lublin, Poland

Svetlana Vyshemyrskaya
Kherson National Technical University
Kherson, Ukraine

Sergii Babichev
Jan Evangelista Purkyně University in Ústi
and Labem
Usti and Labem, Czech Republic

Olena Vynokurova
IT Step University
Lviv, Ukraine

Svetlana Radetskaya
Kherson National Technical University
Kherson, Ukraine

ISSN 2194-5357   ISSN 2194-5365   (electronic)
Advances in Intelligent Systems and Computing
ISBN 978-3-030-26473-4   ISBN 978-3-030-26474-1   (eBook)
https://doi.org/10.1007/978-3-030-26474-1

This Springer imprint is published by the registered company Springer Nature Switzerland AG
The registered company address is: Gewerbestrasse 11, 6330 Cham, Switzerland

Preface

Collecting, analysis, and processing information are one of the current directions of modern computer science. Many areas of current existence generate a wealth of information which should be stored in a structured manner, analyzed, and processed appropriately in order to gain the knowledge concerning investigated process or object. Creating new modern information and computer technologies for data analysis and processing in various fields of data mining and machine learning create the conditions for increasing effectiveness of the information processing by both the decrease of time and the increase of accuracy of the data processing.

The international scientific conference "Intellectual Decision-Making Systems and Problems of Computational Intelligence" is a series of conferences performed in East Europe. They are very important for this geographic region since the topics of the conference cover the modern directions in the field of artificial and computational intelligence, data mining, machine learning, and decision making.

The conference is dedicated to the memory of Professor, Academician of the National Academy of Sciences of Ukraine Yuriy Kryvonos (April 12, 1939–February 12, 2017). Professor Y. Kryvonos was a well-known specialist in informatics, mathematical modeling, and artificial intelligence. Under his leadership, the fundamentals of the perturbation theory of pseudo-inverse and projection operators, the theory of analysis, and synthesis of high-quality clustering systems, recognition, and prediction of information were developed. In his studies, tools were identified for the optimal synthesis of linear and nonlinear recursive data classification systems (pattern recognition) and methods for analyzing and synthesizing voice speech information were proposed. He has developed a unified approach to solving problems of modeling wave and fast physical and technological processes and a new approach to the synthesis of active artificial media with desired properties. The distributed information technologies based on the concept of an electronic document have been created under his leadership too. Yuriy Kryvonos was one of the founders of this conference, the chairman of the program, and international committees.

The aim of the conference is the reflection of the most recent developments in the fields of artificial and computational intelligence used for solving problems in variety areas of scientific researches related to data mining, machine learning, and decision making.

The 15th International ISDMCI Scientific Conference (ISDMCI'2019) held in Zaliznyi Port, Kherson region, Ukraine, from May 21 to 25, 2019, was a continuation of the highly successful ISDMCI conference series started in 2004. For many years, ISDMCI has been attracting hundreds or even thousands of researchers and professionals working in the field of artificial intelligence and decision making. This volume consists of 49 carefully selected papers that are assigned to three thematic sections:

Section 1. **Analysis and Modeling of Complex Systems and Processes**:

- Methods and means of system modeling in the conditions of uncertainty
- Problems of identification of complex system models and processes
- Modeling of operated complex systems
- Modeling of various nature dynamic objects
- Time series forecasting and modeling
- Information technology in education

Section 2. **Theoretical and Applied Aspects of Decision-Making Systems**:

- Decision-making methods
- Multicriteria models of decision making in the conditions of uncertainty
- Expert systems of decision making
- Methods of artificial intelligence in decision-making systems
- Software and instrumental means for synthesis of decision-making systems
- Applied systems of decision-making support

Section 3. **Computational Intelligence and Inductive Modeling**:

- Inductive methods of model synthesis
- Computational linguistics
- Data mining
- Multiagent systems
- Neural networks and fuzzy systems
- Evolutionary algorithm and artificial immune systems
- Bayesian networks
- Hybrid systems and models
- Fractals and problems of synergetics
- Images' recognition and cluster analysis

We hope that the broad scope of topics related to the fields of artificial intelligence and decision making covered in this proceedings volume will help the reader to understand that the methods of data mining and machine learning have become an important element of modern computer science.

June 2019

Oleh Mashkov
Volodymyr Stepashko
Yuri Krak
Yuriy Bardachov
Volodymyr Lytvynenko

Organization

ISDMCI'2019 is organized by the Department of Informatics and Computer Science, Kherson National Technical University, Ukraine, in cooperation with:

Black Sea Scientific Research Society, Ukraine
IT Step University, Ukraine
Jan Evangelista Purkyne University in Usti nad Labem, Czech Republic
Lublin University of Technology, Poland
Taras Shevchenko National University of Kyiv, Ukraine
Glushkov Institute of Cybernetic of NAS of Ukraine, Ukraine
International Centre for Information Technologies and Systems of the National Academy of Sciences of Ukraine, Ukraine

Program Committee

Chairman

Oleh Mashkov	State Ecological Academy of Postgraduate Education and Natural Resources Management of Ukraine, Kyiv, Ukraine

Vice-chairmen

Volodymyr Stepashko	International Centre for Information Technologies and Systems of the National Academy of Sciences of Ukraine, Kyiv, Ukraine
Yuri Krak	Taras Shevchenko National University of Kyiv, Ukraine

Members

Igor Aizenberg	Manhattan College, New York, USA
Tetiana Aksenova	Grenoble University, France
Mikhail Alexandrov	Autonomous University of Barcelona, Spain
Svitlana Antoshchuk	Odessa National Polytechnic University, Ukraine
Sergii Babichev	Jan Evangelista Purkyne University in Usti nad Labem, Czech Republic
Peter Bidyuk	National Technical University of Ukraine "Igor Sikorsky Kyiv Polytechnic Institute", Ukraine
Yevgeniy Bodyanskiy	Kharkiv National University of Radio Electronics, Ukraine
Vitaliy Boyun	Glushkov Institute of Cybernetic of NAS of Ukraine, Ukraine
Yevhen Burov	Lviv Polytechnic National University, Ukraine
Mykola Dyvak	Ternopil National Economic University, Ukraine
Aleksandr Gozhyj	Petro Mohyla Black Sea National University, Ukraine
Larysa Globa	National Technical University of Ukraine "Igor Sikorsky Kyiv Polytechnic Institute", Ukraine
Volodymyr Hnatushenko	Oles Honchar Dnipro National University, Ukraine
Ivan Izonin	Lviv Polytechnic National University, Ukraine
Maksat Kalimoldayev	Institute of Information and Computational Technologies, Kazakhstan
Bekir Karlik	Neurosurgical Simulation Research and Training Center, Canada
Alexandr Khimich	Glushkov Institute of Cybernetic of NAS of Ukraine, Ukraine
Pavel Kordik	Czech Technical University in Prague, Czech Republic
Andrzej Kotyra	Lublin University of Technology, Poland
Yuri Krak	Taras Shevchenko National University of Kyiv, Ukraine
Victor Krylov	Odessa National Polytechnic University, Ukraine
Roman Kuc	Yale University, USA
Frank Lemke	KnowledgeMiner Software, Germany
Vitaly Levashenko	Zilinska univerzita v Ziline, Slovakia
Volodymyr Lytvynenko	Kherson National Technical University, Ukraine
Vasyl Lytvyn	Lviv Polytechnic National University, Ukraine
Leonid Lyubchyk	National Technical University "Kharkiv Polytechnic Institute", Ukraine
Igor Malets	Lviv State University of Life Safety, Ukraine
Viktor Mashkov	Jan Evangelista Purkyne University in Usti nad Labem, Czech Republic

Mykola Malyar	Uzhhorod National University, Ukraine
Sergii Mashtalir	Kharkiv National University of Radio Electronics, Ukraine
Volodymyr Mashtalir	Kharkiv National University of Radio Electronics, Ukraine
Oleksandr Mikhalyov	National Metallurgical Academy of Ukraine, Ukraine
Vadim Mukhin	National Technical University of Ukraine "Igor Sikorsky Kyiv Polytechnic Institute", Ukraine
Sergii Olszewski	Taras Shevchenko National University of Kyiv, Ukraine
Volodymyr Osypenko	Kyiv National University of Technologies and Design, Ukraine
Nataliya Pankratova	National Technical University of Ukraine "Igor Sikorsky Kyiv Polytechnic Institute", Ukraine
Dmytro Peleshko	IT Step University, Ukraine
Eduard Petlenkov	Tallinn University of Technology, Estonia
Hanna Rudakova	Kherson National Technical University, Ukraine
Yuriy Rashkevych	Ministry of Education and Science of Ukraine
Yuriy Romanyshyn	Lviv Polytechnic National University, Ukraine
Anatoliy Sachenko	Ternopil National Economic University, Ukraine
Natalia Savina	National University of Water and Environmental Engineering, Ukraine
Galina Setlak	Rzeszow University of Technology, Poland
Natalya Shakhovska	Lviv Polytechnic National University, Ukraine
Ihor Shelevytsky	Kryvyi Rih Institute of Economics, Ukraine
Volodimir Sherstyuk	Kherson National Technical University, Ukraine
Galyna Shcherbakova	Odessa National Polytechnic University, Ukraine
Andrzej Smolarz	Lublin University of Technology, Poland
Miroslav Snorek	Czech Technical University in Prague, Czech Republic
Volodymyr Stepashko	International Centre for Information Technologies and Systems of the National Academy of Sciences of Ukraine
Sergey Subbotin	Zaporizhzhia National Technical University, Ukraine
Roman Tkachenko	Lviv Polytechnic National University, Ukraine
Ivan Tsmots	Lviv Polytechnic National University, Ukraine
Oleksii Tyshchenko	Institute for Research and Applications of Fuzzy Modeling, CE IT4Innovations, University of Ostrava, Czech Republic
Kristina Vassiljeva	Tallinn University of Technology, Estonia
Alex Voloshin	Taras Shevchenko National University of Kyiv, Ukraine
Viktor Voloshyn	IT Step University, Ukraine

Olena Vynokurova	IT Step University, Ukraine
Waldemar Wojcik	Lublin University of Technology, Poland
Mykhaylo Yatsymirskyy	Institute of Information Technology, Lodz University of Technology, Poland
Elena Zaitseva	Zilinska univerzita v Ziline, Slovakia
Jan Zizka	Mendel University in Brno, Czech Republic
Taras Rak	IT Step University, Ukraine
Olexandr Barmak	Khmelnitsky National University, Ukraine
Volodymyr Buriachok	Borys Grinchenko Kyiv University, Ukraine
Oleksandr Khimich	Glushkov Institute of Cybernetic of NAS of Ukraine, Ukraine
Arkadij Chikrii	Glushkov Institute of Cybernetic of NAS of Ukraine, Ukraine
Sergiy Gnatyuk	National Aviation University, Ukraine
Volodymyr Hrytsyk	Lviv Polytechnic National University, Ukraine
Leonid Hulianytskyi	Glushkov Institute of Cybernetic of NAS of Ukraine, Ukraine
Volodymyr Khandetskyi	Oles Honchar Dnipro National University, Ukraine
Mykola Korablyov	Kharkiv National University of Radio Electronics, Ukraine
Viktor Morozov	Taras Shevchenko National University of Kyiv, Ukraine
Sergiy Pavlov	Vinnytsia National Technical University, Ukraine
Petro Stetsyuk	Glushkov Institute of Cybernetic of NAS of Ukraine, Ukraine
Vasyl Trysnyuk	Institute of Telecommunications and Global Information Space, Ukraine
Valeriy Zadiraka	Glushkov Institute of Cybernetic of NAS of Ukraine, Ukraine
Yurii Yaremchuk	Vinnytsia National Technical University, Ukraine
Fedir Geche	Uzhhorod National University, Ukraine

Organization Committee

Chairman

| Yuriy Bardachov | Kherson National Technical University, Ukraine |

Vice-chairmen

| Volodymyr Lytvynenko | Kherson National Technical University, Ukraine |
| Yuriy Rozov | Kherson National Technical University, Ukraine |

Members

Igor Baklan	National Technical University of Ukraine "Igor Sikorsky Kyiv Polytechnic Institute", Ukraine
Anatoliy Batyuk	Lviv Polytechnic National University, Ukraine
Oleg Boskin	Kherson National Technical University, Ukraine
Liliya Chyrun	Lviv Polytechnic National University, Ukraine
Oleksiy Didyk	Kherson National Technical University, Ukraine
Nataliya Kornilovska	Kherson National Technical University, Ukraine
Yurii Lebedenko	Kherson National Technical University, Ukraine
Olena Liashenko	Kherson National Technical University, Ukraine
Irina Lurje	Kherson National Technical University, Ukraine
Anton Omelchuk	Kherson National Technical University, Ukraine
Oksana Ohnieva	Kherson National Technical University, Ukraine
Viktor Peredery	Kherson National Technical University, Ukraine
Svetlana Radetskaya	Kherson National Technical University, Ukraine
Victoria Vysotska	Lviv Polytechnic National University, Ukraine
Svetlana Vyshemyrskaya	Kherson National Technical University, Ukraine
Maryna Zharikova	Kherson National Technical University, Ukraine
Andriy Kogut	IT Step University, Ukraine
Natalia Axak	Kharkiv National University of Radio Electronics, Ukraine
Oleksandr Melnychenko	Kherson National Technical University, Ukraine
Iryna Mukha	National Technical University of Ukraine "Igor Sikorsky Kyiv Polytechnic Institute", Ukraine
Pavlo Mulesa	Uzhhorod National University, Ukraine

Additional Reviewers

Fabio Bracci	Scientific Researcher at the Institute for Robotics and Mechatronics, German Aerospace Center (DLR), Germany
Shruti Jain	Jaypee University of Information Technology, India
Aigul Kaskina	University of Fribourg, Switzerland
Swietlana Kasuba	University of Economics, Bydgoszcz, Poland
Anoop Kumar Sahu	Madanapalle Institute of Technology and Science, India
Hakan Kutucu	Karabuk University, Turkey
Kevin Li	University of Windsor, Canada
Stan Lipovetsky	GfK Custom Research North America, USA
Zbigniew Omiotek	Lublin University of Technology, Poland
Scott Overmyer	Baker College, Flint Township, USA
Marek Pawełczyk	Silesian University of Technology, Poland
Ali Rekik	Higher Institute of Computer Sciences, Tunisia

Contents

Theoretical and Applied Aspects of Decision-Making Systems

Computational Intelligence and Inductive Modeling

Analysis and Modeling of Complex Systems and Processes

Soft Filtering of Acoustic Emission Signals Based on the Complex Use of Huang Transform and Wavelet Analysis

Sergii Babichev[1,2]([⊠]) [ID], Oleksandr Sharko[3] [ID], Artem Sharko[4],
and Oleksandr Mikhalyov[5]

[1] Jan Evangelista Purkyně University in Ústí nad Labem,
Ústí nad Labem, Czech Republic
sergii.babichev@ujep.cz
[2] Ukrainian Academy of Printing, Lviv, Ukraine
[3] Kherson State Maritime Academy, Kherson, Ukraine
[4] Kherson National Technical University, Kherson, Ukraine
[5] National Metallurgical Academy of Ukraine, Dnipro, Ukraine

Abstract. The paper presents the results of the research concerning development of acoustic emission signals soft filtering model based on the complex use of Huang transform and wavelet analysis. The acoustic emission signals which were generated during crack progression from initiation to final failure with several distinct phases have been used as the experimental signals during the simulation process. The families of biorthogonal wavelets were used during the filtering process. The Shannon entropy criterion which was calculated with the use of James-Stein estimator was used as the main criterion to estimate the filtering process quality. The optimal parameters of the wavelet filter (type of wavelet, level of wavelet decomposition, value of the thresholding coefficient) were determined based on the minimum value of the Shannon entropies ratio which were calculated for filtered signal and for allocated noise component.

Keywords: Acoustic emission signal · Filtering · Wavelet analysis ·
Huang transform · Shannon entropy

1 Introduction

Acoustic emission (AE) technique is one of the current directions of structural state monitoring methods which are developed as an alternative of nondestructive testing methods. Implementation of this technique allows us to perform both the continuous or on-demand diagnostics and discovering defects using permanently installed sensors [1–4]. The main advantages of the AE technique are high level of availability and low maintenance costs. Identification of a defect location is performed by evaluation of the time difference of AE signals arrival to the sensors which are allocated at the different places of the object [5,6]. High

© Springer Nature Switzerland AG 2020
V. Lytvynenko et al. (Eds.): ISDMCI 2019, AISC 1020, pp. 3–19, 2020.
https://doi.org/10.1007/978-3-030-26474-1_1

level of noise component which appears at the stages of signal generation, propagation and detection is one of the main reasons which complicates the successful application of this technique. Thus, the filtering of initial AE signal in order to remove the noise component is the one of the necessary conditions of the AE signals processing technique successful implementation.

A lot of techniques for different types of signals filtering exist nowadays. So, in [7–9] the authors presented the signal processing methods based on smoothing the signal by the use of both the extrapolation technique and minimizing the mean square error between the estimated random and the desired processes. The main disadvantage of these techniques is their low effectiveness in the case of processing of non-stationary and non-linear signals with local particularities. Implementation of these techniques in these cases promotes to the loss of the large amount of useful information. The current methods of non-stationary and non-linear signals processing are based on decomposition of the signal with allocation of its components and the following processing of these components in order to remove the noise. The paper [10] presents the results of the research concerning the use of fast Fourier transform for evaluation of the anisotropic relaxation of composites and nonwovens. Implementation of the fast Fourier transforms technique for time-frequency analysis of pressure pulsation signal is presented in [11]. The frequency spectrum including frequency-domain structure and approximate frequency-scope was obtained during the simulation process. However, it should be noted, that fast Fourier transform technique is effective in the case of stationary signals processing. In the case of non-stationary and non-linear signal processing the effectiveness of this technique decreases.

An alternative and logical continuation of the fast Fourier transforms technique is wavelet analysis [12–15]. Implementation of this technique involves wavelet-decomposition on levels from 1 to N with calculation of both the approximation coefficients on N-th level and the detail coefficients on levels from 1 to N. In the most cases the detail coefficients contain the noise component, thus these coefficients should be processed during the filtering process. Reconstruction of the signal is performed with the use of both the approximation coefficients and the processed detail coefficients. The effectiveness of this technique implementation depends on the choice of the type of the used wavelet, level of the wavelet decomposition and the thresholding coefficient value to process of the detail coefficients. It should be noted that effective techniques for these parameters objective determining are absent nowadays. Moreover, the direct implementation of this technique for signals processing increases the requirements to the wavelet filter parameters determination. In this case more effective can be techniques which are based on decomposition of the signal into components with the further allocation and wavelet-processing of the noised components.

In [16,17] the authors proposed the use of the empirical mode decomposition (EMD) method based on complex use of both the Huang transform and Hilbert spectrum for non-stationary and non-linear signals analysis and processing. The main concept of this method consists in decomposition of the initial signal into mutually independent intrinsic mode functions (IMFs) based on

Huang transform. Then, the Hilbert spectrum is formed by applying the Hilbert transform to the obtained modes (IMFs). The analysis of the Hilbert spectrum for the allocated modes allows us to receive the detail information concerning particularities of the investigated signal. Nowadays the Hilbert-Huang technique has been implemented in various fields of scientific research. So, the paper [18] presents the technique to decompose the multicomponent micro–Doppler signals based on the complex use of Hilbert-Huang transform and analytical mode decomposition (HHT-AMD). The approach concerning the implementation of the Hilbert-Huang transform (HHT) for detection, diagnostic and prediction of the degradation in the ball bearing is proposed in [19]. The papers [20–22] present the results of the research concerning implementation of the HHT for analysis of the vibration signals from different objects. In the paper [23] the authors present the results of the research concerning the use of HHT for the processing and analysing of ECG signal in order to diagnose the brain functionality abnormalities. The results of the research concerning the implementation of the HHT for the analysis of the non-stationary financial time series and the acoustic wave frequency spectrum characteristics of rock mass under blasting damage are presented in the papers [24,25]. However, it should be noted that in spite of the achievements in this subject area the problem of denoising of the non-stationary and non-linear signals has no effective solution nowadays. This problem can be solved based on the complex use of modern techniques of both the data mining and machine learning which are applied successfully in different areas of the scientific research nowadays [26–29]. In this paper we propose the technique of acoustic emission signals filtering based on the complex use of both the Huang empirical mode decomposition method and wavelet analysis. The optimal parameters of the wavelet filter for each of the intrinsic modes are determined on the basis of the quantitative criterion minimum value which is calculated as the ratio of Shannon entropies for the filtered data and for the allocated noise component.

The aim of the research is the development of technique of acoustic emission signals filtering based on the complex use of Huang transform and wavelet analysis.

2   Materials and Methods

The Huang transform technique involves that initial signal is a complex one and it can be decomposed into intrinsic mode functions (IMFs) [16]:

$$y(x) = \sum_{i=1}^{n} f_i(x) + r_n(x) \tag{1}$$

where n is the number of the IMFs functions, $f_i(x)$ is the IMFs function on i-th level of the signal decomposition, $r_n(x)$ is the residual function, which represent the average trend of the investigated signal. Implementation of the Huang empirical mode decomposition technique (EMD) intendes the following conditions:

- the number of each of the IMFs functions extrema and the number of zero crossing should be equal or not differ by more than one;
- in any point of the IMFs function the mean value of the envelope defined by local maximums and local minimums should be zero.

The signal decomposition process is stopped if one of the following conditions is performed:

- the residual function $r_n(x)$ does not contain more than 2–3 extrema points;
- the residual function $r_n(x)$ in whole interval of x change is insignificant in comparison with appropriate values of the IMFs functions.

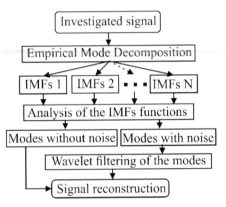

Fig. 1. A structural block chart of step-by-step process of the signal filtering

A structural block-chart of the step-by-step process of the signal filtering based on the complex use of Huang empirical mode decomposition technique and wavelet analysis is presented in Fig. 1. As it can be seen, the result of the Huang transform is selection of the IMFs functions which contain the noise component for purpose of their further filtering using discrete wavelet transform. Figure 2 presents the main idea of the discrete wavelet decomposition process. Implementation of this process involves calculation of both the approximation coefficients at N-th level and the detail coefficients at levels from 1 to N using the low frequency (LF) and high frequency (HF) filters:

$$y(x) \rightarrow \{CA(N), CD(N), ..., CD(2), CD(1)\} \tag{2}$$

The noise component in the most cases is contained in detail coefficients therefore these coefficients should be processed during the signal processing. To process the detail coefficients we propose to use the soft thresholding in accordance with the following conditions:

$$\begin{cases} d = 0, & if\ d \leq \tau \\ d = d - \tau, & if\ d > \tau \end{cases} \tag{3}$$

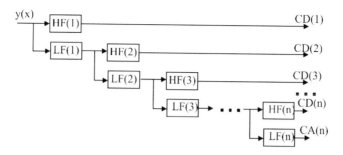

Fig. 2. A structural block chart of the discrete wavelet decomposition process

where d is the detail coefficient and τ is the thresholding coefficient value. It is obvious, that quality of wavelet filtering process depends on type of the used wavelet, level of the wavelet decomposition and thresholding coefficient value to process the detail coefficients. In [15] the authors proposed the technology to determane the optimal parameters of the wavelet filter based on the use of the Shannon entropy criterion which is calculated on the basis of James-Stein estimator [30]. This method is based on the complex use of two different models: a high-dimensional model with low bias and high variance, and a lower dimensional model with larger bias but smaller variance. Evaluation of the values distribution probability in a cell in accordance with the James-Stein shrinkage method is calculated in the following way:

$$p_i^{Srink} = \lambda p_i + (1 - \lambda)p_i^{ML} \qquad (4)$$

where p_i^{ML} is the probability of the values distribution in the i-th cell, which is calculated by the maximum likelihood method; $p_i = \dfrac{1}{n_i}$ is the maximum entropy target in the i-th cell; n_i is the number of the features in this cell. It is obvious, that p_i^{ML} corresponds to the high-dimensional model with low bias and high variance and p_i is the estimation with higher bias and lower variance of the features distribution. Intensity parameter λ in the proposed model is calculated as follows:

$$\lambda = \frac{1 - \sum_{i=1}^{k}(p_i^{ML})^2}{(n-1)\sum_{i=1}^{k}(p_i - p_i^{ML})^2} \qquad (5)$$

where n is the number of the features in the vector. The value of Shannon entropy is calculated with the use of standard formula taking into account the method of the probability estimation:

$$H^{Shrink} = -\sum_{i=1}^{k} p_i^{Shrink} \log_2 p_i^{Shrink} \qquad (6)$$

In this paper we propose the technique of the wavelet filter optimal parameters determination based on the use of the ratio of Shannon entropies which are calculated for both the filtered signal and the allocated noise component:

$$RH = \frac{H(filtered\ signal)}{H(noise\ component)} \tag{7}$$

It is obvious that the optimal parameters of the wavelet filter corresponds to the minimum value of the Shannon entropy for filtered signal and the maximum value of this criterion for the allocated noise component. In this case the value of the relative criterion (7) should be achieved the minimum one. The structural block chart of the process of this criterion calculation within the framework of the proposed technique is presented in Fig. 3. Figure 4 shows the structural block chart of the algorithm to determine the wavelet filter optimal parameters. The stages of this algorithm implementation are the following:

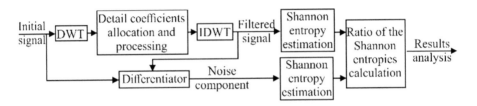

Fig. 3. A structural block chart of the technique to calculate the Shannon entropies ratio

Stage I. Signal loading and Huang transform performing.

1. Loading of the investigated signal. Application of Huang transform to the signal. Empirical mode decomposition of the signal.
2. Visualization and analysis of the obtained modes. Allocation of the noised modes for their following processing.

Stage II. Wavelet filtering of the selected modes.
3. Setup the ranges and the steps of the wavelet filter parameters change.
   3.1 Formation of the vector of different types of wavelets for the appropriate mother wavelet.
   3.2 Calculation of the thresholding coefficients initial value to process the detail coefficients:

$$\tau_0 = \sigma\sqrt{2\ln k}$$

where k is the length of the investigated signal; σ is the median absolute deviation for the allocated detail coefficients:

$$\sigma = \delta \cdot (|CD(i) - median(CD(i))|)$$

where $i = 1, ..., n$ is the wavelet decomposition level, coefficient δ is determined empirically during the simulation process.

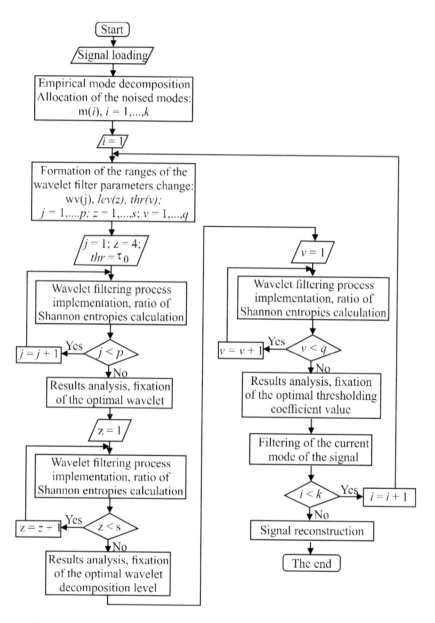

Fig. 4. A structural block chart of the algorithm to determine the wavelet filter optimal parameters

3.3 Formation of the range and the step of the thresholding parameter value change:

$$\tau_{min} = 0.1\tau; \ \tau_{max} = 5\tau \ d\tau = 0.02 \cdot (\tau_{max} - \tau_{min})$$

3.4 Evaluation of the wavelet decomposition maximum level.

4. Determination of the optimal type of the wavelet.

4.1 Initialization of the counter, which corresponds to the first wavelet in the appropriate sequence ($j = 1$). Setup of the initial values of both the wavelet decomposition level ($N = 3$) and the thresholding coefficient value ($\tau = \tau_0$).

4.2 Discrete wavelet decomposition of the signal with calculation of both the approximation coefficients at the N-th decomposition level and the detail coefficients at the levels from 1 to N.

4.3 Soft thresholding of the detail coefficients using the conditions (3).

4.4 Reconstruction of the signal based on both the approximation coefficients and the processed detail coefficients.

5. Calculation of the data processing quality criteria.

5.1 Extraction of the noise component as the difference of both the initial and filtered signals.

5.2 Calculation of the Shannon entropies for the filtered signal and for the allocated noise component by the formula (6). Calculation of their ratio by the formula (7).

5.3 If the counter value is maximal, the results analysis and fixation of the optimal type of wavelet which corresponds to the minimum value of the criterion (7). Otherwise, increment of the counter value and go to the step 4.2 of this procedure.

6. Determination of the optimal wavelet decomposition level.

6.1 Initialization of the counter, which corresponds to the first level of wavelet decomposition ($z = 1$).

6.2 Repetition of the steps from 4.2 to 5.3 of this procedure.

6.3 If the counter value is maximal, the results analysis and fixation of the optimal wavelet decomposition level. Otherwise, increment of the counter value and go to the step 6.2 of this procedure.

7. Determination of the thresholding coefficient optimal value.

7.1 Initialization of the counter ($v = 1$), which corresponds to the minimum value of the thresholding coefficient: ($\tau = \tau_{min}$).

7.2 Repetition of the steps from 4.2 to 5.3 of this procedure.

7.3 If the counter value is maximal, the results analysis and fixation of the thresholding coefficient optimal value. Otherwise, increment of the counter value and go to the step 7.2 of this procedure.

8. Filtering of the current IMFs function with the use of the wavelet filter optimal parameters.

9. Repetition of the stage 2 for other of the allocated IMFs functions.

Stage III. Reconstruction of the signal.

10. Reconstruction of the signal with the use of both the processed and non-processed components of the signal.

3 Experiment

The experimental device which were used to fixation of the acoustic emission signals for the different levels of mechanical loading of the tested material is shown in Fig. 5. The experimental device contains three main mechanisms: the deformation, the force-fixation and the AE signal fixation mechanisms. The samples for four-point bend test were cutted out from steel flat in the size $300 \times 20 \times 4$ mm. The simulation process involved the fixation of both the AE signals and the level of the sample deformation for different levels of the tested sample loading. The broadband sensors for acoustic emission instrument AF15 with a bandwidthes of 0.2–2.0 was used as the measurement device. The artificial load was step-by-step increased from 150 N to 400 N with fixation of the AE signals for different values of the sample deformation. The examples of the AE signals which were obtained during the simulation process are shown in Fig. 6. As it can be seen, the shape of the signals is changed during the load increase. However, the existence the noise component complicates the obtained results interpretation. Thus, at the first step it is necessary to decrease the level of the noise component with saving useful information concerning state of the investigated sample. The familiy of biorthogonal wavelets were used during the simulation process within the framework of the hereinbefore technique.

Fig. 5. Four-point bend test device: (1) test sample; (2) support; (3) deformation indicator; (4) AE signal indicator

4 Results and Discussion

Figure 7 presents the result of the Huang empirical modes decomposition implementation for the signal which is shown in Fig. 6g. The same results were obtained for other signals. The analysis of the IMFs functions allows us to conclude that the first and the second modes contain the noise component, thus these modes should be processed for the signal denoise. Figure 8 presents the results of the simulation

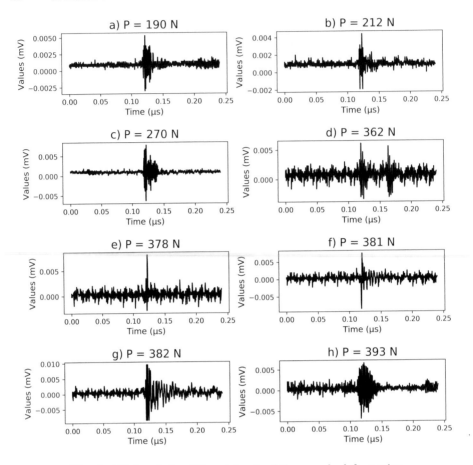

Fig. 6. AE signals for different levels of the sample deformation

concerning determination of the optimal wavelet from the family of the biorthogonal ones in terms of the minimum value of the criterion (7). The results of the simulation has shown that the biorthogonal wavelet *bior*1.1 is the optimal one for processing both the IMFs 1 and IMFs 2 functions since the criterion values of Shannon entropies ratio which have been calculated by the formula (7) achieved the minima values in these cases. Figure 9 shows the results of the simulation concerning determination of the optimal wavelet decomposition level in the cases of the use of both the IMFs 1 and IMFs 2 functions. The analysis of the obtained charts allows us to conclude that the wavelet decomposition levels 7 and 8 are the optimal in terms of the criterion (7) minima for the functions IMFs 1 and IMFs 2 respectively. Figure 10 presents the same results in the case of the thresholding coefficient optimal values determination. The range and the step of the thresholding coefficient value change was determined in accordance with the steps 3.2 and 3.3 of the hereinbefore described algorithm. The value of multiplier δ was taken as 0.5. The optimal value of the thresholding coefficient was determined as the first achieved of

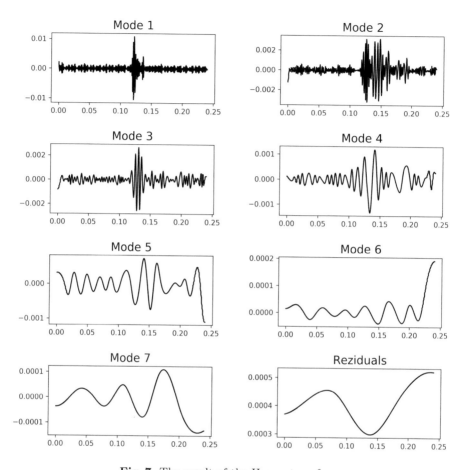

Fig. 7. The result of the Huang transform

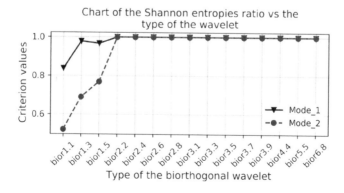

Fig. 8. Results of the simulation concerning determination of the biorthogonal wavelet optimal type

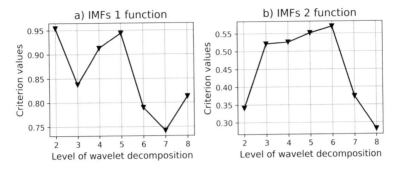

Fig. 9. Charts of the Shannon entropies ratio *vs* the wavelet decomposition level

the global minimum within the range of this parameter change. The thresholding coefficient values $\tau_1 = 1.47 \cdot 10^{-3}$ and $\tau_2 = 1.16 \cdot 10^{-3}$ were determined for the IMFs 1 and the IMFs 2 functions respectively as the result of this stage implementation. These values were used to process the detail coefficients within the framework of the proposed technique.

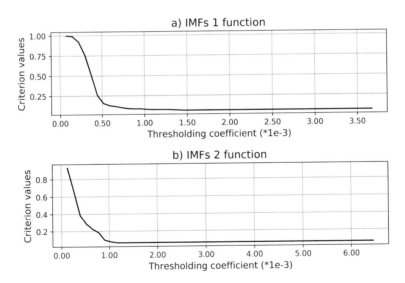

Fig. 10. Results of the simulation to determine the thresholding coefficient optimal values

The final stage of the hereinbefore described procedure is the reconstruction of the signal with the use of both the processed and unprocessed IMFs functions. Figure 11 presents the results of the simulation concerning filtration of the AE signal using parameters of the wavelet filter which were determined within the framework of the proposed technique. The analysis of the obtained results indicates the high effectiveness of the proposed technique, since the level of noise in the signal in Fig. 11b is significantly less in comparison with the level of noise in the initial signal (Fig. 11a). Table 1 presents the values of wavelet filter optimal parameters which were determined within the framework of the proposed technique for the AE signals which have been shown in Fig. 6. In all cases the biorthogonal wavelet *bior*1.1 was determined as an optimal one in terms of the minimum value of the criterion (7). The results of the simulation concerning filtration of the AE signals which have been shown in Fig. 6 are presented in Fig. 12. The analysis of the obtained results allows us to conclude about high efficiency of the proposed technique. The level of noise component in signals in Fig. 12 is significantly less to compare with the noise level in appropriate initial signals which are shown in Fig. 6. It should be noted, that local particularities of the signals have not been changed during the signals processing. Moreover, the filter parameters are adapted to the filtered component. In all cases the parameters are determined empirically based on the minimum value of the quality filtration criterion which takes into account both the useful signal maximum informativity and noise component minimum informativity.

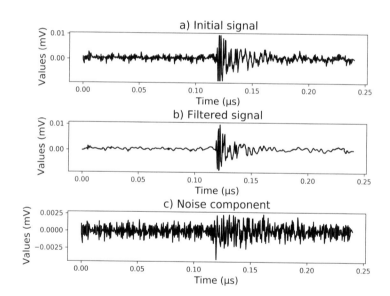

Fig. 11. Results of the AE signal filtering

Table 1. Values of the wavelet filter parameters for the allocated IMFs functions processing

Weight, N	190		212		270		362		378		381		382		393	
Modes	M1	M2	M1	M2	M1	M2	M1	M2	M1	M2	M1	M2	M1	M2	M1	M2
Level	8	8	8	2	7	2	8	2	8	2	3	3	7	8	2	3
THR, 10^{-3}	0.66	0.31	0.37	0.24	0.64	0.25	0.8	0.71	0.95	0.55	0.75	0.82	1.47	1.16	0.89	0.47

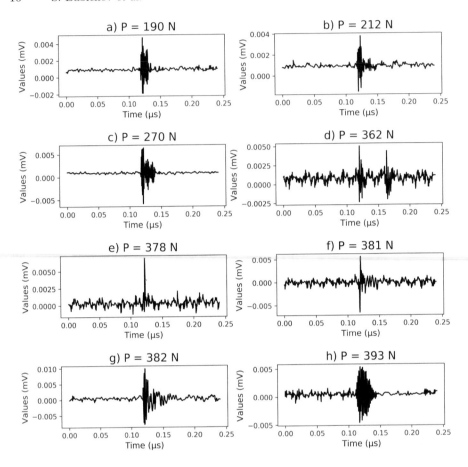

Fig. 12. The final results of the AE signals filtering

5   Conclusions

The technique of acoustic emission (AE) signal filtering based on the complex use empirical mode decomposition (EMD) method (Huang transform) and wavelet analysis has been presented in the paper. Implementation of this technique involves the following stages. Firstly, the Huang transform is performed to decompose the initial signal into the components (modes). The modes with noise are allocated at this stage. Then, the optimal parameters of the wavelet filter are determined for each of the selected modes. The wavelet filtering of the allocated modes is performed as the result of this stage implementation. Finally, the reconstruction of the signal is performed with the use of both the processed and non-processed IMFs functions. The ratio of Shannon entropies which are calculated for both the filtered signal and the allocated noise component has been used as the main criterion to determine the wavelet filter optimal parameters. The effectiveness of the proposed technique was estimated with the use of the

AE signal obtained as the result of the experiment concerning identification of the structural particularities of the mechanism of the materials deforming based on the AE signals analysis. The family of biorthogonal wavelets were used during the simulation process. The results of the simulation have shown that the first two modes from 8 contain the noise component, wavelet *bior*1.1 is the optimal one for both modes in all cases. The optimal wavelet decomposition level and the thresholding coefficient value for each of the allocated IMFs functions were determined during the simulation process. The practical implementation of the proposed technique for the AE signals filtering has shown its high effectiveness since the level of noise component in the filtered signals was significantly less to compare with the level of the noise in appropriate initial signals. Moreother, the local particularities of the signals have not been changed during the signals processing.

References

1. Gong K, Hu J (2019) Online detection and evaluation of tank bottom corrosion based on acoustic emission. In: Springer series in geomechanics and geoengineering, (216039), pp 1284–1291. https://doi.org/10.1007/978-981-10-7560-5_118
2. Ridgley KE, Abouhussien AA, Hassan AAA, Colbourne B (2019) Characterisation of damage due to abrasion in SCC by acoustic emission analysis. Mag Concr Res 71(2):85–94. https://doi.org/10.1680/jmacr.17.00445
3. Berte R, Della Picca F, Poblet M, Li Y, Cortés E, Craster RV, Maier SA, Bragas AV (2018) Acoustic far-field hypersonic surface wave detection with single plasmonic nanoantennas. Phys Rev Lett 121(25). https://doi.org/10.1103/PhysRevLett.121.253902. Article no 253902
4. Zhang X, Zou Z, Wang K, Hao Q, Wang Y, Shen Y, Hu H (2018) A new rail crack detection method using LSTM network for actual application based on AE technology. Appl Acoust 142:78–86. https://doi.org/10.1016/j.apacoust.2018.08.020
5. Xu J, Shu S, Han Q, Liu C (2018) Experimental research on bond behavior of reinforced recycled aggregate concrete based on the acoustic emission technique. Constr Build Mater 191:1230–1241. https://doi.org/10.1016/j.conbuildmat.2018.10.054
6. Su F, Li T, Pan X, Miao M (2018) Acoustic emission responses of three typical metals during plastic and creep deformations. Exp Tech 42(6):685–691. https://doi.org/10.1007/s40799-018-0274-x
7. Kalman RE (1960) A new approach to linear filtering and prediction problems. J Basic Eng 82(1):35–45
8. Wiener N (1949) Extrapolation, interpolation, and smoothing of stationary time series. Wiley, New York
9. Izonin I, Trostianchyn A, Duriagina Z, Tkachenko R, Tepla T, Lotoshynska N (2018) The combined use of the wiener polynomial and SVM for material classification task in medical implants production. Int J Intell Syst Appl 10(9):40–47. https://doi.org/10.5815/ijisa.2018.09.05
10. Staub S, Andrä H, Kabel M (2018) Fast FFT based solver for rate-dependent deformations of composites and nonwovens. Int J Solids Struct 154:33–42. https://doi.org/10.1016/j.ijsolstr.2016.12.014

11. Cui L, Ma F, Gu Q, Cai T (2018) Time-frequency analysis of pressure pulsation signal in the chamber of self-resonating jet nozzle. Int J Pattern Recogn Artif Intell 32(11). https://doi.org/10.1142/S0218001418580065. Article no 1858006
12. Zalik RA (2019) On orthonormal wavelet bases. J Comput Anal Appl 27(5):790–797
13. Riabova S (2018) Application of wavelet analysis to the analysis of geomagnetic field variations. J Phys Conf Ser 1141(1). https://doi.org/10.1088/1742-6596/1141/1/012146. Article no 012146
14. Bodyanskiy Y, Perova I, Vynokurova O, Izonin I (2018) Adaptive wavelet diagnostic neuro-fuzzy network for biomedical tasks. In: 14th international conference on advanced trends in radioelectronics, telecommunications and computer engineering, TCSET 2018 - proceedings, April 2018, pp 711–715. https://doi.org/10.1109/TCSET.2018.8336299
15. Babichev S, Škvor J, Fišer J, Lytvynenko V (2018) Technology of gene expression profiles filtering based on wavelet analysis. Int J Intell Syst Appl 10(4):1–7. https://doi.org/10.5815/ijisa.2018.04.01
16. Huang N, Shen Z, Long S, Wu M, Shih H, Zheng N, Yen N, Tung C, Liu H (1998) The empirical mode decomposition and Hilbert spectrum for nonlinear and nonstationary time series analysis. Proc Math Phys Eng Sci 454:903–995
17. Huang NE, Wu Z (2008) A review on Hilbert-Huang transform: method and its applications to geophysical studies. Rev Geophys 46(2). https://doi.org/10.1029/2007RG000228. Article no RG2006
18. Li W, Kuang G, Xiong B (2018) Decomposition of multicomponent micro-Doppler signals based on HHT-AMD. Appl Sci (Switz) 8(10). https://doi.org/10.3390/app8101801. Article no 1801
19. Soualhi A, Medjaher K, Zerhouni N (2015) Bearing health monitoring based on Hilbert-Huang transform, support vector machine, and regression. IEEE Trans Instrum Meas 64(1):52–62. https://doi.org/10.1109/TIM.2014.2330494. Article no 6847199
20. Susanto A, Liu C-H, Yamada K, Hwang Y-R, Tanaka R, Sekiya K (2018) Application of Hilbert-Huang transform for vibration signal analysis in end-milling. Precis Eng 53:263–277. https://doi.org/10.1016/j.precisioneng.2018.04.008
21. Susanto A, Liu C-H, Yamada K, Hwang Y-R, Tanaka R, Sekiya K (2018) Milling process monitoring based on vibration analysis using Hilbert-Huang transform. Int J Autom Tech 12(5):688–698. https://doi.org/10.20965/ijat.2018.p0688
22. Trusiak M, Styk A, Patorski K (2018) Hilbert-Huang transform based advanced Bessel fringe generation and demodulation for full-field vibration studies of specular reflection micro-objects. Opt Lasers Eng 110:100–112. https://doi.org/10.1016/j.optlaseng.2018.05.021
23. Oweis RJ, Abdulhay EW (2011) Seizure classification in EEG signals utilizing Hilbert-Huang transform. BioMed Eng Online, 10. https://doi.org/10.1186/1475-925X-10-38. Article no 38
24. Huang NE, Wu M-L, Qu W, Long SR, Shen SSP (2003) Applications of Hilbert-Huang transform to non-stationary financial time series analysis. Appl Stoch Models Bus Ind 19(3):245–268. https://doi.org/10.1002/asmb.501
25. Yuan H, Liu X, Liu Y, Bian H, Chen W, Wang Y (2018) Analysis of acoustic wave frequency spectrum characters of rock mass under blasting damage based on the HHT method. Adv Civ Eng 2018. https://doi.org/10.1155/2018/9207476. Article no 9207476

26. Babichev S, Lytvynenko V, Osypenko V (2017) Implementation of the objective clustering inductive technology based on DBSCAN clustering algorithm. In: Proceedings of the 12th international scientific and technical conference on computer sciences and information technologies, CSIT 2017, vol 1, pp 479–484. https://doi.org/10.1109/STC-CSIT.2017.8098832. Article no 8098832

27. Babichev S, Lytvynenko V, Gozhyj A, Korobchynskyi M, Voronenko M (2019) A fuzzy model for gene expression profiles reducing based on the complex use of statistical criteria and Shannon entropy. Adv Intell Syst Comput 754:545–554. https://doi.org/10.1007/978-3-319-91008-6_55

28. Bidyuk P, Gozhyj A, Kalinina I, Gozhyj V (2017) Methods for processing uncertainties in in solving dynamic planning problems. In: Proceedings of the 12th international scientific and technical conference on computer sciences and information technologies, CSIT 2017, vol 1, pp 151–155. https://doi.org/10.1109/STC-CSIT.2017.8098757. Article no 8098757

29. Bidyuk P, Gozhyj A, Kalinina I, Gozhyj V (2018) Analysis of uncertainty types for model building and forecasting dynamic processes. Adv Intell Syst Comput 689:66–78. https://doi.org/10.1007/978-3-319-70581-1_5

30. Hausser J, Strimmer K (2009) Entropy inference and the james-stein estimator, with application to nonlinear gene association networks. J Mach Learn Res 10:1469–1484

Some Features of the Numerical Deconvolution of Mixed Molecular Spectra

Serge Olszewski[1] , Paweł Komada[2] , Andrzej Smolarz[2] ,
Volodymyr Lytvynenko[3] , Nataliia Savina[4] , Mariia Voronenko[3(✉)] ,
Svitlana Vyshemyrska[3] , Anton Omelchuk[3] , and Iryna Lurie[3]

[1] Taras Shevchenko National University of Kyiv, Kyiv, Ukraine
olszewski.serge@gmail.com
[2] Lublin University of Technology, Lublin, Poland
{p.komada,a.smolarz}@pollub.pl
[3] Department of Informatics and Computing Technology,
Kherson National Technical University, Kherson, Ukraine
immun56@gmail.com, mary_voronenko@i.ua, printvvs@gmail.com,
tareon@ukr.net, lurieira@gmail.com
[4] National University of Water and Environmental Engineering, Rivne, Ukraine
n.b.savina@nuwm.edu.ua

Abstract. The direct method features of finding the weight coefficients of the mixed molecular spectrum components on the basis of their reference samples are considered in this paper. It has been established that the presence of additive noise in the output mixed spectrum generates a noise component with an unidentified probability distribution law in the found weight coefficients. The power generated by the noise can be several orders of magnitude higher than the power of the output signal additive noise. It is shown that the use of numerical methods for suppressing this noise, which is not based on its statistical characteristics, in particular, the median filtration, expands the limits of SNR, in which the proposed method maintains efficiency.

Keywords: Problem of deconvolution · Median filtration ·
Molecular spectra · Additive noise

1 Introduction

The main problem of spectral analysis tasks is the problem of establishing the quantitative composition of the mixture based components on the experimentally recorded spectra, in which the overlapping of the spectral bands belonging to different components takes place.

The urgency of removing this problem is dictated by a wide range of activities in which spectral analysis is used, as the leading method of controlling the process flow models. This is a technological control of combustion modes by spectral

V. Lytvynenko et al. (Eds.): ISDMCI 2019, AISC 1020, pp. 20–34, 2020.
https://doi.org/10.1007/978-3-030-26474-1_2

analysis of exhaust gases and smoke. This is the control of active pharmaceutical industry reactors, environmental monitoring of the environment, and so on.

Typically, methods for the establishment of such components are based on the replacement of the output signals by commonly used approximations, which contain, in their basis, the explicit form of the form of spectral bands obtained from approximate theoretical representations. Such an approach involves the risk of obtaining uncontrolled errors in the experimental data processing, which, for example, is pharmacologically inadmissible.

In the paper, the direct finding method of weight coefficients of the mixed spectrum components is considered at the condition of the availability of a basis of reference spectra for all investigated mixtures components and the features, conditions, and limits of the proposed method application.

2   Review of the Literature

Observed infrared spectra and spectra of combination scattering (Raman spectra) of radiation mixtures centres or absorption with a high degree of probability can have a strong overlap between the bands of individual components. In addition, any real spectrum always overlaps with random noise. In practice, infrared and Raman molecular spectra are used for qualitative and quantitative analysis in many chemical and biochemical studies. In particular, the analysis of infrared and Raman spectra makes it possible to characterize the composition of unknown chemical mixtures, for example - red wine [1], coal [2], amorphous carbon [3], zirconium [4], and the like. To interpret spectra, it is necessary to determine the parameters of the studied mixtures individual component's molecular bands. Spectral data points can be divided into three different types. The set of areas points without absorption in a wave number having the same intensity with adjacent points is the baseline. The set of points that relate to a strong absorption band and strongly differ from adjacent points are points of a useful signal. The set of noise points is a point whose values are between the points of the baseline and points of the useful signal and do not correlate with each other.

The reasons for the mandatory presence of noise and self-overlapping of individual molecular bands are due to such natural effects as the Doppler effect, the expansion of the structural elements of the band due to collisions, etc. The separation of the experimental signal can be accomplished by using mathematical methods of noise suppression. However, the separation of the molecular bands of the mixture's individual components by methods remains an urgent task. Two classical methods are known which are effective and attract considerable research effort [3–14]. This is the fitting of the curve and the spectral deconvolution.

The curve fit is a special case of data modelling and parameter estimation. Experimental spectral data is modelled by the amount of bandwidth to obtain the result as a bandwidth parameter estimate. The easiest method for fitting the curve is to estimate the spectral bands overlapping by the optimization method for the smallest squares [5]. In addition to the least squares method, a number of other effective methods are proposed in a number of papers, such as the fitography of the Voigt profile [6], polynomial fit [7], the frequencies identified by the second derivative of the spectrum [8], and parametric fit, for example, fitting by the Gaussian profile, or the Lorentz and Voigt profile [9]. Lorenz-Fonfria and Padros [10] analyzed the stability of the curve fitting method and constructed a more precise model for processing errors generated by noise. However, it should be emphasized that the method of fitting the curve is effective only in the event of a slight degradation of spectra due to obstacles. For strongly overlapped spectra, it does not provide an adequate coincidence of model and spectral maxima, especially at a low signal-to-noise ratio (SNR).

Deconvolution has become one of the most useful methods for solving the problem of separation of spectral bands with overlap on component components. The simplest method of deconvolution is the processing of spectral data by the Fourier method, the FourierSelf-Deconvolution (FSD) [11,12].

Recently, Preser performed a combination spectral deconvolution of cybacillus using the method of weighted least squares [4]. Lorenzo-Fonfriya and Padroz proposed an effective method of deconvolution with entropy expression, which allowed negative values in the expression of entropy [13]. Then they successfully evaluated the superposition of strips with strong overlap, using only the Fourier transform module and the derivative of the first order of the spectrum [14]. The results of both methods mentioned earlier are used to interpret the spectra with overlapping.

Crylli [15] analyzed several iterative deconvolution algorithms and came to the conclusion that the Janson method [16] yields the best results. To overcome the restriction of conventional methods based on filters, we propose methods of blind deconvolution without any previous filters, for example, homomorphic filtering [17], homomorphic deconvolution, van Sitter's iteration [18] and high order statistics [19–21]. However, the results obtained were rather noisy.

The peculiarities of the considered methods of analysis of mixed spectra are that their accuracy depends essentially on the choice of the form of the spectral band characteristic. That is, the achievement of the spectra adequate deconvolution and the successful extraction of the individual component's features of the characteristics of the mixed spectral curve is determined primarily by the quality of the fitting of the extrapolation curves for the reference spectra of the individual components and the mixed spectrum. Simultaneous extrapolation of parameters and restoration of the overlapping spectrum with noise was presented in [22,23].

As an alternative to the methods considered in this paper, consideration is given to the method for the exact solution of the problem of mixed molecular spectra deconvolution, which is direct and not based on the fitting of approximation curves.

3   Problem Statement

In the problems of spectral analysis, there is always a need to establish the quantitative content of components in mixed spectra.

One of the directions of this problem solution is the direct finding of the weight coefficients of the spectrum components of each individual substance in the mixture based on the Bouguer-Lambert-Ber's law, which is called the exact solution of deconvolution problem. The strong sensitivity ofthe deconvolution problem exact solutionto the noise component, which is always present in the observed signal, requires further utilization of the mathematical methods of the exact solution noise component inhibition (see Fig. 1).

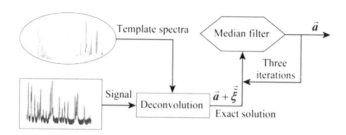

Fig. 1. Solving the problem of molecular spectra deconvolution by an exact method

The peculiarities of finding the exact solution of deconvolution problem are that in finding the solution of inhomogeneous linear equations systems with respect to the weight coefficients of the mixed spectrum components with the additive noise, the additive noise in the cleavage itself is rapidly increasing. At the same time, the structure of the exact solution does not make it possible to establish an explicit form of the distribution law of the solution noise component probability. This means that in order to oppress this component, it is necessary to use methods of noise clearing that do not rest on the statistical characteristics of the deconvolution problem exact solution noise.

4 Materials and Methods

4.1 General Solving Problems of Deconvolution

For optical emission or absorption spectra, in accordance with Bouguer-Lumbert-Beer law, the spectrum of a mixture of heterogeneous refraction or absorption centres $F(\lambda)$ is a superposition M of the spectra of components of a mixture $f_i(\lambda)$:

$$F(\lambda) = \sum_{i=0}^{M} a_i f_i(\lambda),\tag{1}$$

where a_i is the unknown weight coefficient of each component of the mixture, which in the future we will call the weight coefficient of deconvolution. For uniformly discriminated spectra, the same look:

$$F[n] = \sum_{i=0}^{M} a_i f_i[n].\tag{2}$$

As a rule, when registering optical spectra by modern devices, the total number of samples of a sample signal N significantly exceeds the number M of components of the mixture. Since the unknown coefficients a_i that are the subject of the solution of the deconvolution problem are the same for $\forall n$, there is a finite set $\mathfrak{M}$ of systems of non-homogeneous linear algebraic equations with M unknowns, which binds between the numerical values of the reference spectra $f_i[n]$, the spectrum of the mixture $F[n]$ and the weight coefficients of deconvolution a_i :

$$\begin{pmatrix} f_1[n+1] & \cdots & f_i[n+1] & \cdots & f_M[n+1] \\ f_1[n+2] & \cdots & f_i[n+2] & \cdots & f_M[n+2] \\ \cdots & \cdots & \cdots & \cdots & \cdots \\ f_1[n+M] & \cdots & f_i[n+M] & \cdots & f_M[n+M] \end{pmatrix} \times \begin{pmatrix} a_1 \\ a_2 \\ \cdots \\ a_M \end{pmatrix} = \begin{pmatrix} F[n+1] \\ F[n+2] \\ \cdots \\ F[n+M] \end{pmatrix}. \quad (3)$$

For each of these systems of equations, the total number of which $L = \dfrac{N!}{M}$, the vector of solutions $\vec{a} = (a_1, a_2, \ldots, a_i, \ldots, a_M)^T$ is the same. We call this solution a solution to the problem of deconvolution.

Thus, the general task of finding the weight coefficients of the mixture of radiation or absorption centres or the weight coefficients of deconvolution, based on the known reference true spectra of each individual component and the true spectrum of the mixture, has an exact solution.

It should be emphasized that the application of the above approach for practical purposes has a fundamental complication.

One of them is that the experimentally registered spectra do not coincide with the true ones due to the influence of the hardware function of the response of the spectral devices. So the observed optical spectrum $Q(\lambda)$ is associated with the true spectrum $F(\lambda)$ of the correlation:

$$Q(\lambda) = \int_0^\infty G(\lambda - \lambda') F(\lambda') d\lambda', \quad (4)$$

where $G(\lambda)$ is the hardware function of the device response. Using the Bouguer-Lambert-Beer law for the observed spectrum gives a relation:

$$Q(\lambda) = \int_0^\infty G(\lambda - \lambda') \sum_{i=0}^{M} a_i f_i(\lambda') d\lambda'$$

$$= \sum_{i=0}^{M} a_i \int_0^\infty G(\lambda - \lambda') f_i(\lambda') d\lambda' = \sum_{i=0}^{M} a_i q_i(\lambda), \quad (5)$$

where $q_i(\lambda)$ is the observed spectra of individual components of a mixture of heterogeneous radiation or absorption centres. That is, the registration of the studied and reference spectra by devices with the same hardware function leaves the exact solution of the deconvolution problem to be correct.

Another complication, due to the fact that any real signal contains an integral component of the additive noise, is more fundamental in nature, since arithmetic operations over realizations of random variables are not reflected in similar operations over their probability distributions.

Thus, the nature of the noise component in fluence on the solution of the deconvolution problem requires a more detailed study.

4.2   Influence of Additive Gaussian Noise

The mathematical model of the maximally close to the reality of the observed optical spectrum of the mixture of radiation or absorption centres will take the form:

$$Q(\lambda) + \zeta(\lambda) = \sum_{i=0}^{M} a_i (q_i(\lambda) + \zeta_i(\lambda)), \tag{6}$$

where $\zeta(\lambda)$ is the random variable corresponding to the noise component of the spectral signal of the mixture and $\zeta_i(\lambda)$ is the noise components of spectral signals component. In general $\zeta(\lambda)$ and $\zeta_i(\lambda)$ are random processes depending on the wavelength. For most of the experimental conditions, they are sufficiently considered as ergodic random processes.

Accordingly, for a uniformly sampled signal:

$$Q[n] + \zeta[n] = \sum_{i=0}^{M} a_i (q_i[n] + \zeta_i[n]). \tag{7}$$

The given modification of the model of the observed signal causes the vector of solutions of the set of systems of nonhomogeneous linear algebraic equations (3) $- \overrightarrow{a}$ into a multidimensional random variable $\overrightarrow{\xi} = (\widetilde{a}_1, \widetilde{a}_2, \ldots, \widetilde{a}_i, \ldots, \widetilde{a}_M,)^T$ with an unknown M-dimensional probability distribution $P\left(\overrightarrow{\xi}\right)$.

Hypothetically, the mathematical expectation of this magnitude $\overrightarrow{\xi}$ should be directed to the vector of solutions:

$$M\left\{\overrightarrow{\xi}\right\} = \lim_{K \to \infty} \sum_{k=1}^{K} \cdots \sum_{k=1}^{K} \overrightarrow{\xi}_k P\left(\overrightarrow{\xi}_k\right) = \overrightarrow{a}. \tag{8}$$

Here and below the index k denotes the solutions of a particular system of multiplication equations $\mathfrak{M}$ generated by the relation (7).

To simulate the influence of additive noise on statistical characteristics $\overrightarrow{\xi}$, the relation (7) was modified in relation:

$$\sum_{i=0}^{M} a_i q_i[n] = Q[n] + \theta[n], \tag{9}$$

where $\theta[n]$ are samples of the discrete random process, given by expression:

$$\theta[n] = \zeta[n] - \sum_{i=0}^{M} a_i \varsigma_i[n]. \tag{10}$$

As the size of the deconvolution problem solution's sample increases with the averaging of the array of solutions to the ensemble, the mathematical expectation of the noise component must go to zero, since the output noise was symmetric. However, the confirmation of the above hypothesis can be obtained only from the analysis of the exact solution structure of the deconvolution problem of the signal with noise. In addition, such an analysis will allow us to draw conclusions about the law of the distribution of the boiling value, which is the solution of the deconvolution problem of the signal with noise.

4.3   Structure of the Precise Problem Solving of the Deconvolution of the Signal with Noise

According to the model described in Eq. (8), the elements of the vector of free members of the inhomogeneous linear equations will include a random variable with an unknown distribution function. The matrix form of the system of equilibrium for finding the weight coefficients of deconvolution will take the form:

$$
\begin{pmatrix}
q_1\,[n+1] & \cdots & q_i\,[n+1] & \cdots & q_M\,[n+1] \\
q_1\,[n+2] & \cdots & q_i\,[n+2] & \cdots & q_M\,[n+2] \\
\cdots & \cdots & \cdots & \cdots & \cdots \\
q_1\,[n+M] & \cdots & q_i\,[n+M] & \cdots & q_M\,[n+M]
\end{pmatrix}
\times
\begin{pmatrix}
a_1 \\ a_2 \\ \cdots \\ a_M
\end{pmatrix}
$$
$$
=
\begin{pmatrix}
Q\,[n+1] + \theta_G\,[n+1] \\
Q\,[n+2] + \theta_G\,[n+2] \\
\cdots \\
Q\,[n+M] + \theta_G\,[n+M]
\end{pmatrix}.
\tag{11}
$$

Solutions of the system of linear equations can be found by the formula Cramer:
$\widetilde{a_i} = \dfrac{\widetilde{\Delta_i}}{\Delta}$, where $\Delta = \det\left(q_i\,[n]\right)$ are the determinants of the matrix of coefficients. According to the model, it is a constant value. $\widetilde{\Delta_i}$ are the determinants of matrices obtained from the matrix of coefficients by replacing the i-th column with the vector of free members:

$$
\widetilde{\Delta_i} =
\begin{vmatrix}
q_1\,[n+1] & \cdots & Q\,[n+1] + \theta_G\,[n+1] & \cdots & q_m\,[n+1] \\
q_1\,[n+2] & \cdots & Q\,[n+2] + \theta_G\,[n+2] & \cdots & q_m\,[n+2] \\
\cdots & \cdots & \cdots & \cdots & \cdots \\
q_1\,[n+M] & \cdots & Q\,[n+M] + \theta_G\,[n+M] & \cdots & q_m\,[n+M]
\end{vmatrix}.
\tag{12}
$$

According to the properties of the determinant $\widetilde{\Delta_i} = \Delta_i + \Xi_i$, where $\dfrac{\Delta_i}{\Delta} =$

$\dfrac{1}{\Delta}
\begin{vmatrix}
q_1\,[n+1] & \cdots & Q\,[n+1] & \cdots & q_M\,[n+1] \\
\cdots & \cdots & \cdots & \cdots & \cdots \\
q_1\,[n+M] & \cdots & Q\,[n+M] & \cdots & q_M\,[n+M]
\end{vmatrix} = a_i$ are the elements of the vector

of the exact solution of the back deconvolution for a signal without a noise component, and $\dfrac{\Xi_i}{\Delta} = \dfrac{1}{\Delta}
\begin{vmatrix}
q_1\,[n+1] & \cdots & \theta_G\,[n+1] & \cdots & q_M\,[n+1] \\
\cdots & \cdots & \cdots & \cdots & \cdots \\
q_1\,[n+M] & \cdots & \theta_G\,[n+M] & \cdots & q_M\,[n+M]
\end{vmatrix}$ is the noise

component of the vector of solutions of the deconvolution problem for a signal with additive noise. Decomposition of the determinant Ξ_i by the column of random variables gives the elements of the solution of the deconvolution problem:

$$\widetilde{a}_i = a_i + \sum_{j=1}^{M} \frac{A_{ij}}{\Delta} \theta_G [n + j], \tag{13}$$

where A_{ij} are the algebraic supplements of realizations of a random variable $\theta_G [n + j]$. According to the chosen model A_{ij} have fixed values. Denote the stochastic component of the i-th element of the vector of the solution of the problem of the deconvolution of the signal with the noise as $\vartheta_i = \sum_{j=1}^{M} \frac{A_{ij}}{\Delta} \theta_G [n + j]$. Thus, the solution of the problem of deconvolution of a signal with additive noise is the sum of the solution of the problem of deconvolution of a signal without noise and additive noise, which depends on the form and the additive noise of the signal itself and the form of reference spectra. According to the rules of mathematical operations over random variables, and based on the ergodicity of the discrete process $\theta_G [n]$, the variance of the value $\widetilde{a}_i$ is given by the expression:

$$D\{\widetilde{a}_i\} = D\{a_i + \vartheta_i\} = D\{\vartheta_i\} = D\left\{ \sum_{j=1}^{M} \frac{A_{ij}}{\Delta} \theta_G \right\}$$
$$= \sum_{j=1}^{M} D\{\theta_G\} \left(\frac{A_{ij}}{\Delta} \right)^2. \tag{14}$$

The order of quantities $\left(\frac{A_{ij}}{\Delta} \right)^2$ can be arbitrary since the physical task of the problem generates only a limitation $\sum_{j=1}^{M} \frac{A_{ij}}{\Delta} Q_i [n + j] = const$, where $Q[n]$ denotes thesignal samples of the mixed spectrum in the absence of noise.

The resulting structure of the solution can confirm the hypothesis (8). So for a sound with additive noise:

$$\overrightarrow{\xi} = (\widetilde{a}_1, \widetilde{a}_2, \ldots, \widetilde{a}_i, \ldots, \widetilde{a}_M,)^T$$
$$= (a_1 + \vartheta_1, a_2 + \vartheta_2, \ldots, a_i + \vartheta_i, \ldots, a_M + \vartheta_M)^T \tag{15}$$
$$= (a_1, a_2, \ldots, a_i, \ldots, a_M)^T + (\vartheta_1, \vartheta_2, \ldots, \vartheta_i, \ldots, \vartheta_M)^T = \overrightarrow{a} + \overrightarrow{\vartheta}.$$

Hence, proceeding from (8), and the properties of the mathematical expectation:

$$M\left(\overrightarrow{\xi} \right) = M\left\{ \overrightarrow{a} + \overrightarrow{\vartheta} \right\} = M\{\overrightarrow{a}\} + M\{\overrightarrow{\vartheta}\} \tag{16}$$

Since $\overrightarrow{a}$ is a value that is the same for all plural systems $\mathfrak{M}$, its mathematical expectation coincides with it itself: $M\{\overrightarrow{a}\} = \overrightarrow{a}$. By virtue of symmetry $\overrightarrow{\vartheta}$, its mathematical expectation $M\{\overrightarrow{\vartheta}\} = 0$. Thus, the mathematical expectation of a set of solutions of the problem of deconvolution of a signal with an additive semitric noise is the solution of the problem of deconvolution of a signal in the absence of noise $M\left\{\overrightarrow{\xi}\right\} = \overrightarrow{a}$. However, the problem of finding $M\left\{\overrightarrow{\xi}\right\}$ lies in the fact that it is necessary to know the explicit form of the distribution of the probability of a random variable $\overrightarrow{\vartheta}$, which, in general, is unknown. The situation does not improve and is proposed in the chosen model of the mixed spectrum of assumptions regarding the normal law of distribution of the probability of quantities θ_G, since the composition of the laws of their distributions will also be a normal law, but its parameters are determined by the values of algebraic additions A_{ij} (13).

Thus, in order to determine the weight factors of the deconvolution of a real signal, it is necessary to apply methods of digital processing of the array of solutions aimed at the silencing of noise. One such method, which does not require accurate knowledge of the probability distribution function of a noise component, can be a digital media filter.

4.4   De-Noise of Weight Deconvolution Factors by Median Filtration

The medians have long been used and studied in statistics as an alternative to the arithmetic average of the reference values in the evaluation of sample mean values. The median $med\{\ldots\}$ of numerical sequence $x[1], x[2], \ldots, x[n], \ldots, x[N]$ with an odd one N is the mean value of a term in a row, which is obtained by arranging this sequence in ascending order (or decrease). For even medians, the mean value of two ordered sequences is usually defined.

The median filter is a window filter that slides sequentially across the waveform and returns at one step the median value of the array of elements that fall into the window (aperture) of the filter. The output signal $y[k]$ of the sliding median filter is the width of the window $2n + 1$ for the current count is formed from the input row $\ldots, x[k-1], x[k], x[k+1], \ldots$ according to the formula:

$$y[k] = med\{x[k-n], x[k-n+1], \ldots,$$
$$\ldots, x[k], \ldots, x[k+n-1], x[k+n]\}, \tag{17}$$

where $med\{x[1], \ldots, x[m], \ldots, x[2n+1]\} = x[n+1]$ and $x[m]$ are the elements of the variation series, that is, ranks in order of increasing values.

Thus, the median filtration replaces the values of the readings in the centre of the median filter aperture by the value of the output samples in the middle of the aperture filter. In practice, the aperture of the filter to simplify the data processing algorithms is usually set with an odd number of readings.

Median filtering was implemented as local processing of the counters in a sliding window, which includes a certain number of signal counts. For each position of the window, selected in it countdowns are ranked ascending or descending values. The average position in its ranked list is called the median of this set of readings. this count is replaced by the central countdown in the window for the processed signal. Thus, the median filter is one of the nonlinear filters, which replaces the anomalous points and emission values of the median, regardless of their amplitude values. It is stable by definition and is able to cancel even infinitely large counts.

The median filtration algorithm has a clearly expressed selectivity for elements of an array from a nonmonotonic component of a sequence of numbers within the aperture and most effectively excludes from signals negative and positive single releases that fall on the edge of a ranked list. In this way, media filters are well suppressed by noise and interference, which extends to less than half the window. A stable point is a sequence that does not change with median filtration. In the one-dimensional case, stable points of median filters are "locally-monotone" sequences that the median filter leaves unchanged. An exception is some periodic binary sequences.

Due to this feature, media filters at optimally chosen aperture inhibit uncorrelated and slightly correlated interferences without relying on their probability distribution law.

5   Experiment

The study of the influence of additive noise on the deconvolution problem solution was carried out on experimental infrared spectra of carbon monoxide, carbon dioxide and water. Examples of reference spectra are shown in Fig. 2. The graph (a) corresponds to the normalized maximum carbon monoxide spectrum, graph (b) - normalized carbon dioxide spectrum and graph (c) - reference water spectrum.

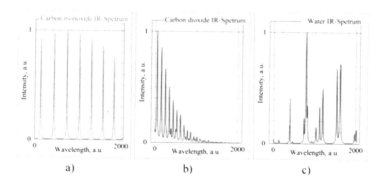

Fig. 2. Examples of normalized to maximum reference infrared spectra

To find the weight coefficients of deconvolution, a synthesized signal was used $Q_s[n]$. This signal was obtained as the sum of a linear combination of reference samples and random variables with a normal probability distribution:

$$Q_s[n] = 0.5713 \times q_{CO}[n] + 1.0000 \times q_{CO_2}[n] + 1.0000 \times q_{H_2O}[n] + \theta_G[n], \quad (18)$$

here $q_{CO}[n]$ is the normalized to the maximum sampled in frared spectrum of carbon monoxide, $q_{CO_2}[n]$ – a similar spectrum of carbon dioxide and $q_{H_2O}[n]$ is the spectrum of water. Additionally $\theta_G[n]$ are realizations of a random variable with a normal probability distribution. The selected synthesized signal satisfies all the requirements of the model of a continuous Gaussian transmission of information, which adequately reproduces the properties of most spectral devices.

The general set of solutions of the deconvolution problem was obtained by sequential solving of systems of linear equations from the set $\mathfrak{M}$ by Gauss method. To do this, a two-dimensional array of reference samples and a synthesized signal $\{q_{CO}\,[n]\,;\; q_{CO_2}\,[n]\,;\; q_{H_2O}\,[n]\,;\; Q_s\,[n]\}$ were split into fragments of M lines ($M = 3$), thus obtaining the combined matrices for systems of linear equations. After exhausting the array of samples, the process of forming the general set of solutions was completed. The described procedure was performed for synthesized spectra with different ratios of signal power without noise to noise power. On Fig. 3 examples of the influence of the ratio of the signal noise on the scattering of the solutions of the deconvolution problem are given. The graphs (a), (b), (c) show the comparison of the synthesized spectrum in the absence of noise - curves 1, and the spectrum with additive Gaussian noise with different dispersion - curves 2. The graph (a) corresponds to the signal-to-noise ratio of 57.07 dB. Chart (b) and (c) illustrate cases for 54.57 and 3.04 dB, respectively.

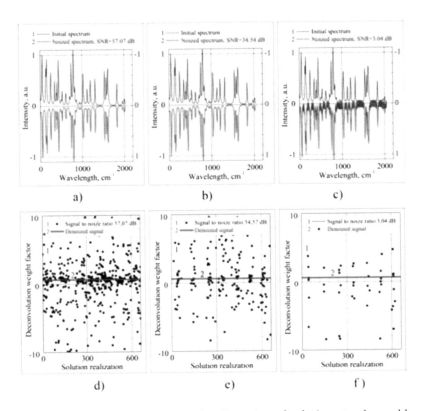

Fig. 3. Influence of additive noise on the dispersion of solutions to the problem of deconvolution

The graphs (d), (e), (f) represent the general set of solutions of the deconvolution problem for different cases of the signal-to-noise ratio. Curves 2 in all graphs illustrate the solution of the deconvolution problem for the synthesized signal in the absence of noise. Curves 1 correspond to the values of the general set of solutions of the deconvolution problem for cases of perturbated by the additive noise of a signal with different signal t noise ratios. Graphs (d), (e) and (f) illustrate cases for SNR values of 57.07, 54.57 and 3.04 dB, respectively.

To suppress the noise of weight deconvolution coefficients, a median filter was used, the window of which was 89% of the size of the sample of solutions. Such a choice is dictated by the fact that the solution of all systems of equations from the set must be the same, so any changes in the countdown in the trend are caused by a pure noise component. The noise suppression procedure was repeated three times, successively passing the pre-silent signal through the median filtration.

6   Results and Discussion

As the study even showed a slight signal disturbance by a noise component, when the signal strength is almost three orders of magnitude higher than the noise power, it leads to strong scattering of the deconvolution problem solutions and an increase in the error of weight coefficients determination. In addition, an increase in noise power by several dB results in an increase in the mean square error of determining the deconvolution coefficients by several orders of magnitude.

The full dependence of deconvolution weight coefficients dispersion on the ratio of signal to noise is presented on rice Fig. 4.

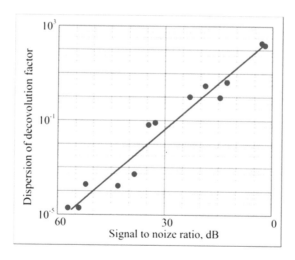

Fig. 4. The dependence of the weight coefficients of the deconvolution of the mixed molecular spectrum on the signal-to-noise ratio

As studies have shown with a decrease in the signal-to-noise ratio from 60 to 0 dB, there is an increase in the mean-square deviation of the deconvolution coefficients of the noise signal from the deconvolution coefficients of the signal without noise by seven orders of magnitude. Consequently, the leading problem of using the exact solution of the deconvolution problem for real signals is the superstrong increase in the power of the noise component of the solution with a slight perturbation of the output signal as a component of the low-power additive noise.

In Fig. 5. the results of processing the solution of the deconvolution problem for the synthesized signal of the mixture of carbon monoxide, carbon dioxide and water vapour at various noise ratio ratios are presented. The graph illustrates the dependence of the relative error of determining the weight coefficients of deconvolution on the ratio of the noise signal of the mixed molecular spectrum.

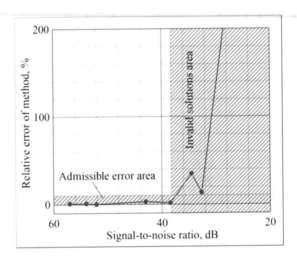

Fig. 5. Dependence of the relative error of the method of determining the deconvolution coefficients from SNR

Studies have shown that when the signal-to-noise ratio is reduced from 60 to 37 dB, the relative error of determining the deconvolution weight coefficients varies little and is within 10%, which is the area of permissible efficiency of the method. Exceeding the signal-to-noise ratio of 37 dB results in a rapid increase in relative error to unacceptably high values of 200% at 30 dB. That is, the region of SNR values >37 dB is the area of themethod disability.

7   Conclusions

A direct method for finding the exact solution of the problem of deconvolution of molecular mixed spectra is proposed. It is shown that the presence of additive noise in the output mixed spectrum makes the exact solution of the deconvolution problem to a random variable with an unidentified probability distribution law whose dispersion is an order of magnitude higher than the noise dispersion of the output signal. Reducing the SNR of the output signal from 60 to 3 dB causes an increase in the dispersion of this random variable by 7 orders of magnitude.

The structure of the deconvolution problem exact solution of the molecular spectrum with the additive noise is found. It is shown that the mathematical expectation of the random variable representing this solution equals the exact solution of the problem of deconvolution of the mixed spectrum in the absence of noise.

The non-established law for the distribution of the exact solution of deconvolution problem of the molecular spectrum with the additive noise does not allow the use of the mathematical expectation operation to find exact values of the spectrum components weight coefficients. To suppress the noise component of the solution it is necessary to use numerical methods that do not rely on the statistical characteristics of the noise. In particular, the iterative use of the median filter showed that the direct deconvolution method maintains efficiency within the SNR of the output signal from 60 to 37 dB.

References

1. Gallego AL, Guesalagu AR, Bordeu E, Gonza X, Lez AS (2011) Rapid measurement of phenolics compounds in red wine using Raman spectroscopy. IEEE Trans Instrum Meas 60(2):507–512
2. Potgieter-Vermaak S, Maledi N, Wagner N, Van Heerden J, Van Grieken R, Potgieter J (2011) Raman spectroscopy for the analysis of coal. Raman Spectrosc 42(2):123–129
3. Wang Q, Allred D, Knight I (1995) Deconvolution of the Raman spectrum of amorphous carbon. Raman Spectrosc 26(12):1039–1043
4. Presser V (2009) Metamictization in zircon, Part I: Riman investigation following a Rietveld approach: profile line dtconvolution technique. Raman Spectrosc 40(5):491–498
5. Fraser R, Suzuki E (1996) Resolution of overlapping absorption bands by least squares procedures. Anal Chem 38(12):1770–1773
6. Sundius T (1996) Computer fitting of Voigt profiles to Raman lines. Raman Spectrosc 1(5):471–488
7. Singh RK, Singh SN, Ashtana BP, Pathak CM (1994) Deconvolution of Lorentzian Raman linewidth: techniques of polynomial fitting and extrapolation. Raman Spectrosc 25(6):423–428
8. Talulian SA (2003) Attenuatid total reflection Fourier transform infrared spectroscopy: a method of choice for studying membrane proteins and lipids. Biochemistry 42(41):11898–11907
9. Alsmeyer F, Marquardt W (2004) Automatic generation of peak-shaped models. Appl Spectrosc 58(8):986–994
10. Lorenz-Fonfria V, Padros E (2004) Curve-fitting overlapped bands: quantification and improvement of curve-fitting robustness in the presence of errors in the model and in the data. Analyst 129(12):1243–1250

11. Kauppinen JK, Moffatt DJ, Muntsch HH, Cameron DG (1981) Fourier transforms in the computation of self-deconvoluted and firstorder derivative spectra of over-lapped band contours. Anal Chem 53(9):1454–1457
12. Kauppinen JK, Moffatt DJ, Mantsch HH, Cameron DG (1981) Fourier self-deconvolution: a method for resolving intrinsically overlapped bands. Appl Spectrosc 35(3):271–276
13. Lorenz-Fonfria VA, Padros DE (2005) Maximum entropy deconvolution of infrared spectra: use of a novel entropy expression without sign restriction. Appl Spectrosc 59(4):474–486
14. Lorenz-Fonfria VA, Padros DE (2008) Method for the estimation of the mean Lorentzian bandwidth in spectra composed of an unknown number of highly over-lapped bands. Appl Spectrosc 62(6):689–700
15. Crilly PB (1991) A quantitative evaluation of various iterative deconvolution algo-rithms. IEEE Trans Instrum Meas 40(3):558–562
16. Jansson PA (1984) Deconvolution: with applications in spectroscopy. Academic, New York
17. Senga Y, Minami K, Kawata S, Minami S (1984) Estimation of spectral slit width and blind deconvolution of spectroscopic data by homomorphic filtering. Appl Opt 23(10):1601–1608
18. Sarkar SC, Dutta PK, Roy NC (1998) A blind-deconvolution approach for chro-matographic and spectroscopic peak restoration. IEEE Trans Instrum Meas 47(4):941–947
19. Yuan J, Hu Z, Sun J (2005) High-order cumulant-based blind deconvolution of Raman spectra. Appl Opt 44(35):7595–7601
20. Yuan J (2009) Blind deconvolution of x-ray diffraction profiles by using highorder statistics. Opt Eng 48(7):076501–076505
21. Babichev S, Korobchynskyi M, Lahodynskyi O, Korchomnyi O, Basanets V, Borynskyi V (2018) Development of a technique for the reconstruction and valida-tion of gene network models based on gene expression profiles. East-Eur J Enterp Technol 1(4–91):19–32
22. Morawski RZ, Miekina A, Barwicz A (1996) Combined use of Tikhonov decon-volution and curve fitting for spectrogram interpretation. Instrum Sci Technol 24(3):155–167
23. Miekina A, Morawski R, Barwicz A (1997) The use of deconvolution and iter-ative optimization for spectrogram interpretation. IEEE Trans Instrum Meas 46(4):1049–1053

Approach to Piecewise-Linear Classification in a Multi-dimensional Space of Features Based on Plane Visualization

Iurii Krak[1,2](✉)📳, Olexander Barmak[3]📳, Eduard Manziuk[3]📳,
and Hrygorii Kudin[1]📳

[1] Taras Shevchenko National University of Kyiv, Kyiv, Ukraine
krak@univ.kiev.ua
[2] Glushkov Cybernetics Institute, 40 Glushkov avenue, Kyiv 03187, Ukraine
krak@nas.gov.ua
[3] National University of Khmelnytskyi, Khmelnytskyi, Ukraine
barmakov@khnu.km.ua

Abstract. Information technology allowing to implement problems of classification, clustering, researching of topology of the information component of the data is proposed. The multidimensional feature space is reduced to the visual presentation space to determine the information content of the data. Optimized reduction of the space dimension to two-dimensional one applying multidimensional scaling methods has been used. The use of the piecewise linear constraints allows us to implement projecting into original multidimensional feature space. Visual construction of restrictive separators makes possible consideration of tolerance fields of changing of features parameters, measure separation of classes, nonlinearity of data grouping. Thus, restrictive areas of hyperspace for the necessary categories of classes are formed. At the same time, visualization of the classification processes in hyperspace has been provided. Information technology is the multidimensional space projecting into visual (two-dimensional) space, constructing of piecewise linear constraints of studied areas, subsequent constraints projecting into multidimensional space. Thus, the information technology enables to synthesize separating hyperplanes limiting categories of classes in multidimensional space. The technology application successive stages and example for fingerspelling alphabet recognition have been described.

Keywords: Classification · Data visualization ·
Information technology

1 Introduction

Investigations which is presented in this paper are correlated with studies that use the Group Method of Data Handling (GMDH) [1], Support Vector Machine

V. Lytvynenko et al. (Eds.): ISDMCI 2019, AISC 1020, pp. 35–47, 2020.
https://doi.org/10.1007/978-3-030-26474-1_3

(SVM) [2,3], optimal synthesis of linear and nonlinear transformations [4], regression analysis, and the like. Despite the fact that there are quite a lot of results of modern research for designing information classification systems [1–3], the problem of improving the quality of classification in image recognition problems, analysis of text, voice and visual contents and others remains actual. In contrast to these important and well-known methods, this paper proposes an approach implemented in the form of a certain combination of computational and informational procedures, which allows visualizing the objects in a two-dimensional space in order to estimate their distribution. Further, in the resulting two-dimensional space lines are built empirically, dividing objects into the necessary classes. Hyperplanes-analogues in the input n-dimensional space are built along the given lines, which are further used for classification.

2   Problem Statement and Information Model of Classification

Let's $x = (x_1, x_2, ..., x_n)^T$ is a vector of features characterized the object which should be classified, where T is the transposition symbol. Let us accept the hypothesis of identity or similarity of objects according to the distance between them in the space of characteristic features. Take a sequence of objects $\Omega_x = \{x : x(1), ..., x(m)\}$, where $x(j) = (x_{j1}, x_{j2}, ..., x_{jn})^T$, $j = \overline{1, m}$ in the features space. We assume that the vectors correspond to the different states of the objects in the domain that is being studied. Without loss of generality, will consider only the division into two classes. From here, we formulate the classification problem as follows: it is necessary to find such a hyperplane $w^T x + b = 0$, $x \in R^n$, so that $x_j \in \Omega_x(1)$ than $w^T x + b > 0$, and for $x_j \in \Omega_x(2)$ there is inequality $w^T x + b < 0$ correspondly. There are: $\Omega_x(1) \subset \Omega$ – subset of the features vectors, that corresponds to first class objects,$\Omega_x(2)$ – subset of the features vectors, that corresponds to second class objects, $\Omega_x = \Omega_x(1) \bigcup \Omega_x(2)$, $w = (w_1, ..., w_n)^T$ - vector of the coefficients, b – some scalar parameter.

For practical implementation of the problems that are formalized by proposed above method, the process of validation is very important. Therefore, at the first stage of the information technology implementation it is proposed to visualize the training sequence Ω_x to ensure its ability to present data into two separate classes.

Visualization requires understanding such a way of representing vectors of a training sequence of the n-dimensional features space in the 2-dimensional space, in which the basic patterns inherent in the initial distribution are qualitatively displayed - its cluster structure, topological features, information about the location of data in the original space, etc. As a criterion of similarity/difference, we take the distance between the vectors in the space: we assume that two vectors are identical if the distance between them is zero, and the vectors are similar, if the distance between them is less than some predetermined value $\varepsilon > 0$, and different if the distance is large ε. The proposed method of mapping the n-dimensional space of features into a space of smaller dimension correlates with the method of multidimensional scaling (Multidimensional scaling (MDS)) [5,6].

A visual representation of the structure of a multidimensional data set can be mapped into one-, two-, and three-dimensional mapping spaces. Note that the most informative for human perception of geometric structures for information extraction and analysis is a two-dimensional space.

Further, the data visualization is used as a way of displaying a multidimensional distribution of the data on the two-dimensional plane, in which, the basic patterns inherent in the original distribution are qualitatively displayed. At the same time, it is necessary to minimize the loss of information content and its manifestations in the cluster structure, topological features, and dependencies between the features of the location of the data in the original space. With a small amount of data, visual display allows you to determine the existence of information links, which are weakly manifested when using methods in combination. In this case, informational links are difficult to determine with approaches that use a different nature of the model formation [7–9].

The initial information is not presented in the form of a "object-feature" type table, but in the form of a square symmetric matrix of mutual distances of objects from each other. At the intersection of the i-th row of the j-th column in the matrix is the value of the distance from the i-th object to the j-th object.

Thus, at the first, each object is assigned coordinates in a multidimensional space. The task of multidimensional scaling is to construct a data set in the usual three-dimensional space or on a plane so that the distances between objects most closely correspond to the distances specified in the matrix [5,6]. The input coordinate axes can be interpreted as some implicit factors, the values of which determine the differences between objects. If we provide each object with a pair of coordinates, then the result is an image of the data visualization.

3   Algorithm of the Multidimensional Scaling Information Data

To build scaling algorithm we consider input information as elements of the matrix of pairwise distances between classified objects determined by multidimensional points $X_i = \{x_{i1}, x_{i2}, ..., x_{in}\}$, $i = \overline{1, n}$. Denote the distance between the objects i and j as $\delta_{ij} = d(X_i, X_j)$, which is calculated as follows:

$$d(X_i, X_j) = \sqrt{\sum_{k=1}^{n}(x_{ik} - x_{jk})^2} \tag{1}$$

Denote by $d(Y_i, Y_j)$ the distance between points in space of smaller dimension (reduced space with dimension m, $m < n$) which will be found by formula similar to (1).

Scaling purpose is finding the points $Y_i = \{y_{i1}, y_{i2}, ..., y_{in}\}$, $i = \overline{1, n}$ of space so that the distance between points in the space of lower dimension would have been the closest to the distance in multidimensional source space.

The proximity measure is defined as the stress-function σ, that can be found by least square function [8,9]:

$$\sigma = \sum_{i<j} w_{ij}(d(Y_i, Y_j) - \delta_{ij})^2, \tag{2}$$

where w_{ij} is non-negative weights. When normalized, stress function is determined as follows:

$$\sigma = \frac{\sum_{i<j} w_{ij}(d(Y_i, Y_j) - \delta_{ij})^2}{\sum_{i<j} w_{ij}\delta_{ij}^2}, \tag{3}$$

Note that operation of normalization (3) allows us to reduce dependency from number of objects and their location, and thereby to improve interpretation of objects' visualization quality on plane.

With scaled (smaller dimension) matrix of pairwise distances D in multidimensional space m let us perform the following transformations:

1. Matrix double centering by one of known methods;
2. For the obtained matrix we define eigenvectors $e_1, e_2, ..., e_m$
3. Calculate the matrix $X = E_m \Lambda_m^{0.5}$, where E_m - is matrix of eigenvectors $e_1, e_2, ..., e_m$, Λ_m - is a diagonal matrix of eigenvalues.

After the transformations the coordinate matrix which is used in multidimensional scaling we obtain by eigenvector decomposition of the matrix $B = XX^T$.

Stress functions σ in various data reduction variants are quite numerous and are based on the multidimensional scaling interpretations and optimization algorithms. Note that for multidimensional scaling non-metric methods not quantitative measures of objects' similarity are used but only their relative order. Stress function σ minimization corresponds to finding the most optimal matching between initial distances matrix and resulting distances matrix.

4   Algorithm of Majorizing for Stress-Function Minimization

To minimize the stress function σ proximity matrix finding will be used iteratively the SMACOF (Scaling by Majorizing a Complicated Function) algorithm [10,11]. The $SMACOF$ algorithm is based on a strategy that provides good model convergence, minimizing data influence. When using the majorizing method, for the objective function $f(x)$ a simpler and more controlled function $g(x, y)$ is sought to majorize function $f(x)$, i.e. for all x, $g(x, y) \geq f(x)$, where y - fixed value of reference point. Wherein reference point is the contact point of the surface $g(y, y) = f(y)$, and the minimizing point x_* satisfies the inequality $f(x_*) \leq g(x_*, y) \leq g(y, y) = f(y)$, forming thus a layered structure.

Majorizing algorithm is an iterative procedure which consists of sequence of steps:

1. Reference point determination $y = y_0$;
2. Calculation x_*, based on condition $g(x_*, y) \le g(y, y)$;
3. Transition to previous step with setting $y = x_*$, if the condition $f(y) - f(x_*) \le \varepsilon$ is not reached.

Note that proposed approach can be transferred to multidimensional spaces while respecting inequality and can be used to minimize the objective function.

Let's formulate a number of conditions that are imposed on majorizing function $g(x, y)$ and which determine the advantage of its using:

1) majorizing function $g(x, y)$ is easier to minimize than the function $f(x)$;
2) in source field it must be no less than the original function $f(x) \le g(x, y)$;
3) majorizing function $g(x, y)$ must be tangent to the function $f(x)$ at the reference point $f(y) = g(y, y)$.

The set $Y = \{Y_1, Y_2, ..., Y_m\}$ with m points is iteratively calculated using Gutman's transformations [11]:

$$Y_{k+1} = V^+ B(Y_k) Y_k, \tag{4}$$

where k is the iteration number; V^+- pseudo inverse matrix for weights matrix V with elements:

$$v_{ij} = -w_{ij}, \; i \ne j; \; v_{ij} = \sum_{i=1, i \ne j}^{m} w_{ij}. \tag{5}$$

$B(Y_k)$ - the matrix with elements:

$$b_{ij} = \begin{cases} -\dfrac{w_{ij} \delta_{ij}}{d(Y_i Y_j)}, \; i \ne j \; \& \; d(Y_i Y_j) \ne 0; \\ 0, \; i \ne j \; \& \; d(Y_i Y_j) = 0. \end{cases} \tag{6}$$

$$b_{ij} = -\sum_{i=1, i \ne j}^{m} b_{ij} \tag{7}$$

If in the formula (4) the weights $w_{ij} = 1$, we get:

$$Y_{k+1} = \frac{1}{m} B(Y_k) Y_k. \tag{8}$$

Then a majorizing algorithm consists in performing the following actions:

1) set the initial reduced space value Y_0;
2) write down stress function $\sigma = \sum_{i<j} w_{ij}(d(Y_i Y_j) - \delta_{ij})^2$;
3) find values $Y_{k+1} = V^+ B(Y_k) Y_k$;
4) calculate stress function for $\sigma(Y_{k+1})$;
5) set iteration increment $k + +$;

6) verify conditions of convergence $\sigma(Y_{k-1}) - \sigma(Y_k) < \varepsilon$, otherwise go to the stress function.

Thus, at this stage the process of information technology creating consists of the following steps:

1) formation of pairwise distances matrix based on the input data;
2) finding the square of distances for matrix of distances;
3) using of matrix double centering;
4) determining matrix eigenvalues and matrix eigenvectors;
5) optimization with SMACOF algorithm.

As the result, we obtain set of objects with a pair of coordinates that can be displayed in two-dimensional space (on a plane). These objects are mappings of marked data from multidimensional space and, since the objects are marked, it is necessary to mark classes on resulting plane to form a decision tree using a linear classifier.

5   Stress-Function Minimization Based on the Majorization Algorithm

For a linear discriminant function, the classifier $d(\bar{x})$ is determined by the ratio:

$$d(\bar{x}) = \bar{W}^T \bar{x} + w_n, \tag{9}$$

where $\bar{x} = (x_0, x_1, ..., x_{n-1})^T$ - is the feature vector characterizing the object being classified; $\bar{W} = (w_0, w_1, ..., w_{n-1})^T$ - vector of classifier weight coefficients; w_n - threshold value.

Belonging to one of two classes $\Omega(1)$, $\Omega(2)$ is determined by the rule:

if $d(\bar{x}) = \sum_{i=0}^{n-1} w_i x_i + w_n < 0$, then the object belongs to the first class $\Omega(1)$,

if $d(\bar{x}) = \sum_{i=0}^{n-1} w_i x_i + w_n > 0$, then the object belongs to the second class $\Omega(2)$.

Note, that to find only the vector of coefficients $\bar{W}$ and the threshold value w_n for separation into two classes using only a linear classifier is rather difficult, and it is not possible to separate the data by a curve or a broken line. To solve givens problem, it is proposed to use a combination of linear classifiers by finding a set of a sequence of piecewise linear classifiers. Given approach has the advantage of visualization in the reduced space and allows to demonstrate the controllability of data classification. When using a linear classifier in a multidimensional space, a hyperplane is sought, which will be the dividing criterion for compliance with the class. Next, the y_{new} is searched for a new element represented by a point x_{new} and for some boundary value b from the conditions:

$$y_{new} = \begin{cases} +1, & \bar{W}\bar{x}_{new} > b; \\ 0, & \bar{W}\bar{x}_{new} = b; \\ -1, & \bar{W}\bar{x}_{new} < b; \end{cases} \tag{10}$$

Equation (10) with equality to zero describes the hyperplane. It is known that the vector $\bar{W}$ is perpendicular to the desired separation line with the corresponding properties: the best separation line is as far as possible from the points of separation classes nearest to it. Note that the distance between these points specifies the separation strip, which corresponds to the condition $-1 < \bar{W}\bar{x}_{new} - b < 1$, and there are no element points within the strip boundaries, and the width of the separation strip is equal to $2/|\bar{W}|$ [12].

Note that when separation classes with the help of a dividing strip, only boundary points are important, since the strip consists of parallel lines running along the class boundaries. These lines no longer represent the division of classes (this function takes the separation strip), but denote the boundaries of the classes, since these lines limit them. Thus, the problem is reduced to finding the boundaries of the classes that represent the lines of restriction. From here, the hyperspace is divided into some limited hyper-volumes within which classes are represented.

Note that with the existence of several classes and the construction of a linear classifier, intersection of lines and the construction of segments occur, forming a piecewise linear structure, which in the general case is non-linear. Using restriction lines of the classes, we obtain some geometric structure that restricting class. The element that locate into this restriction belongs to this class.

To improve the linear classification problem solving, an increase (expansion) of dimension of the space is used. The space is expanded by the mapping function $\phi(x)$ to the new space. To expand the two-dimensional space in three-dimensional, the transformation function is represented as follows:

$$\phi(x) = \phi(x_1, x_2) = (x_1^2, x_2^2, \sqrt{2}x_1x_2) \qquad (11)$$

Increasing the dimension of space allows us to find a hyperplane of linear classification. Thus, the hyperplane allows us to linearly distinguish separable classes, which is possible under the condition that the target function is convex. Further, with the inverse decrease of space, the line of separation of the classes can be described in an acceptable way in a piecewise linear way by restriction lines. This makes it possible to use a multiple increase in the dimensionality of space under reverse projection to define hyperplanes. As a consequence of this fact, using the approaches of space reduction by scaling methods, it is possible to determine the class boundaries by visualization methods with their subsequent projection into multidimensional space. In this case, the function of mapping into n-dimensional space will be the following:

$$\phi(x) = \phi(x_1, ..., x_n) \qquad (12)$$

An approach to the construction of a piecewise linear classifier is proposed using a decision tree based on a data visualization system. A decision tree is a way of displaying rules in a hierarchical, sequential structure, where each object corresponds to a single node that provides a solution [13]. A rule is a logical construction represented as "if ... then ...". Rules are defined by curves that separate groups of objects on the plane of the visualization system. Curves are defined as a piecewise linear structure with different required level of discretization. Thus forming curves in the first approximation. As a result, when adding a new object, can clearly indicate to which class it belongs.

Let define several situations from the set of possible situations T when building a decision tree.

1. The set T contains elements that belong to the same class. In this case, the decision tree defines the class. If the set T contains no elements, the decision tree determines the branch and the class associated with this branch is retrieved from another set other than T, for example, an ancestor node.
2. The set T contains elements that belong to different classes, the set is divided into subsets. For this, a feature is defined that contains more than two distinct values $O_1, O_2, ..., O_n$. The set T is divided into subsets, with each subset T_i containing elements O_i that are relevant to the selected trait. The process is recursive, the final condition of which is the formation of subsets, which consist of elements of one class.

When a tree construction on each internal node, it is necessary to find the condition for dividing the set on this node into subsets. As a condition, one of the attributes is accepted. The general rule is as follows: the attribute divides the set in such a way that the resulting subsets consist of objects that belong to the same class or are maximized by this feature. To find the attributes, the algorithm $C4.5$ [14] is used, where the attribute $Gain(\Theta)$ of the set Θ is selected by the following criteria:

$$Gain(\Theta) = Info(T) - Info_x(T) Gain(\Theta) = Info(T) - Info_x(T), \qquad (13)$$

where $Info(T)$ - is the entropy of the set T:

$$Info_x(T) = \sum_{i=1}^{n} \frac{|T_i|}{|T|} Info(T_i). \qquad (14)$$

Subsets $T_1, T_2, ..., T_n$ are obtained from the original set T when checking the set Θ. An attribute is selected that gives the maximum value behind the criterion (14). At the same time, in order to reduce the number of subsets, it is necessary to minimize the number of nodes and branches.

6 Algorithm for the Piecewise Linear Classifier Constructing

Note that initially the piecewise linear classifier works in a multidimensional space dimension n. In order to create the separating constraints in this space it is necessary to transform the constraints (lines) of the piecewise linear classifier of the reduced space into the restrictions of the hyperplanes of the multidimensional space. For this purpose it is necessary to expand dimension of the reduced space to the original one. After the space expansion and the formation of hyperplanes, their equations are determined. For construction hyperplanes in n-dimensional space need corresponding n points. Given points was obtained by adding additional $n-2$ points on linear segments. Thus, a system of linear equations is obtain as follows:

$$\begin{cases} wX_1 + b = 0; \\ \qquad\ldots\ldots\ldots; \\ wX_n + b = 0. \end{cases} \qquad (15)$$

Here W is the unknown vector of the coefficients of hyperplanes. Note that in the general case the system (15) is solved by Gauss method.

A class in the multidimensional space is determined by bounding hyperplanes. For new data classification, their locations are determined in the multidimensional space by determining their position relative to hyperplanes. Substituting the coordinates of the data into the equation of the hyperplane, we determine their relative position from the set $\{-1, 0, 1\}$. If the result is less than zero, the element is arbitrarily "right" relative to the plane, if the result is greater than zero - the element is on the "left" plane and, accordingly, if it is zero, the element is on the separating plane.

The formation of rules for nonlinear classification it is carried out by the following sequence of actions:

1) formation of piecewise-linear visual limitations of the class in the reduced space;
2) calculation of support point-rules for the class;
3) transformation of point-rules into a multidimensional space;
4) construction of hyperplanes in a multidimensional space based on transformed points;
5) formation of rules for a class in multidimensional space on the basis of boundary hyperplanes.

The piecewise linear restrictive rules define the set of positions of the classes and allows us to determine visually the need of the increasing or limitting the area of the class, which is important for the boundary data. This allows us to ensure good interpretability of the classification results and controllability of the class boundary domain and, in fact, the interactivity of the classification system. Ensuring the presence of a visual component in the classification is particularly important in comparison with other approaches, especially in the context of the boundary complex conditions of classification. At the same time, an additional information component is provided with the help of visual interactive means of determining class constraints. This allows providing the toolkit for obtaining additional information by the system and controllability of the classification process.

The results of the system operation well understood and managed due to the visual representation and interactivity of the restrictive rules. The scope of the restriction is provided by the minimum necessary visual boundaries, which, if it is necessary, can be redefined. In this case, the limitation lines are transformed into a multidimensional space and represented in it by hyperplanes, thus forming limiting regions. The classification of new data takes place in a multidimensional space based on the calculated data and their corresponding position with respect to the bounding hyperplanes. The spatial position of the new element is determined in relation to all hyperplanes, thus determining its position in the categories of classes of bounded volumes. This process is controlled because the results are scaled to the reduced space and have a visual representation of the new classified data elements. As a consequence, the result of classification in multidimensional space is presented and evaluated visually, providing the possibility of interpreting and analyzing the correspondence of new data elements with respect to class categories. In multi-classes conditions, if an element has the necessary informational content, according to which it belongs to several classes, the problem of interpretation and qualitative conformity is determined by the visual aspect.

7   Example of Proposed Information Technology Realization

Effectiveness of proposed approach to synthesis of separating hyperplanes was tested on a problem of character recognition of sign language dactyl alphabet (dactylemes or fingerspelling alphabet) [12,15]. Note that in fingerspelling alphabet each letter of it is shown by some configuration of fingers of a person's hand. To show dactylemes a right hand is used as a rule. The difficulty of dactyl information recognition is connected with some features of the geometric-topological characteristics of each person's hands, as well as differences in dactylemes demonstration by different people (see Fig. 1). Therefore for effective classification, dactylemes classes centers obtained when different people showed the

Fig. 1. Examples of sizes and geometrical characteristics of human hands, which were used when showing fingerspelling alphabet

same dactylemes were calculated. In experimental researches dactylemes photographic images were used to demonstrate the alphabet of Ukrainian sign language [12,15]. From the images various characteristic features were extracted according to which further classification was carried out [12]. It is worth to note that some dactylemes can be recognized easy enough by hand and fingers spatial position. However, there is a group dactylemes that are similar due to minor changes in fingers configuration and it is difficult to classify them in spaces of characteristic features used. For Ukrainian sign language such dactylemes are E, S, H, Ch, Ia, V, Ie, L, U, Y, K, N, R, A, I, M, F, Iu, they will be called difficult separable dactylemes.

Approach proposed in the paper was applied to divide the group of dactylemes into several classes. Each tested dactylemes' characteristic features were obtained using contour analysis descriptors [12] representing a set of points in four-dimensional space. Proposed information technology of synthesis of separating hyperplanes allowed identify hyperplanes with the help of which the study group of difficult separable dactylemes steadily divided into 4 classes (see Fig. 2).

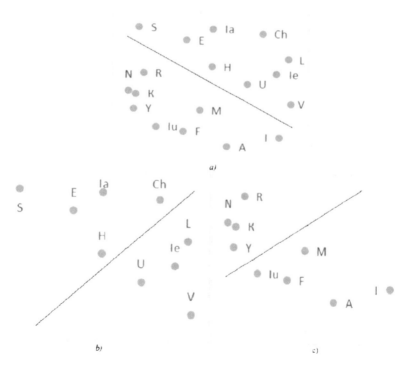

Fig. 2. Dactylemes analogs on the plane from four-dimensional feature space

With the approach proposed such showing made possible to define hyperplanes in four-dimensional space using lines in two-dimensional space. Dactylemes separation into two classes is presented on Fig. 2 a): (1) E, S, H, Ch, Ia, V, U, Ie, L; and (2) F, Iu, Y, K, N, R, A, I, M. Dactylemes separation from the first class into two classes is presented on Fig. 2 b): (1.1) E, S, H, Ch, Ia; (1.2) V, U, Ie, L. The result of proposed approach to separate into two classes dactylemes from the set (2) is presented on Fig. 2 c): (2.1) Y, K, N, R; (2.2) F, Iu, A, I, M. Thus stable rules are obtained according to which dactylemes are separated into four classes with a small number of objects in each. Finally exact object identity in each particular class is obtained easy enough by analyzing features of hand and fingers spatial arrangement [15].

8   Conclusions

The proposed information technology makes possible to train a classification system using a visual definition of classes in a reduced space based on multidimensional scaling. This allows us to determine the necessary parameters of permissible changes in the boundaries of the signs of classes, as well as the spatial nonlinearity of their grouping. The visual representation of the hyperspace makes it possible to most effectively determine the information content of the data. Based on this, the data sets that form the classes are determined. Data relationship visualization is a major factor in the definition and training of a classification system. Visually defined class boundaries are projected into the original multidimensional space defining class boundaries. At the same time, the maximum information content of the distribution of space objects and the controllability of classification processes are preserved.

References

1. Davison M (1988) Multidimensional scaling. Finance and Statistics
2. Zhambyu M (1988) (StatIerarkhicheskii klaster-analiz i sootvetstviya) Hierarchical cluster analysis and matching. Fin Stat 342 p.
3. Vapnik VN (1998) Statistical learning theory. Wiley, New York
4. Kirichenko MF, Krak YV, Polishchuk AA (2004) Pseudo inverse and projective matrices in problems of synthesis of functional transformers. Kibernetika i Sistemnyj Analiz 40(3):116–129
5. Cox TF, Cox MAA (2001) Multidimensional scaling. Chapman and Hall/CRC, Boca Raton
6. Vorontsov KV. Lectures on algorithms of clustering and multidimensional scaling. http://www.ccas.ru/voron/download/Clustering.pdf. Accessed 10 Mar 2019
7. Manziuk EA, Barmak OV, Krak IV, Kasianiuk VS (2018) Definition of information core for documents classification. J Autom Inf Sci 50(4):25–34
8. Mair P, Borg I, Rusch T (2016) Goodness-of-fit assessment in multidimensional scaling and unfolding. Multivar Behav Res 51(6):772–789
9. Stojkoska BR (2016) A taxonomy of localization techniques based on multidimensional scaling. In: 39th international convention on information and communication technology, electronics and microelectronics, Opatija, 30 May-3 June, pp 649–654

10. de Leeuw J, Mair P (2009) Multidimensional scaling using majorization: SMACOF in R. J Stat Softw 31(3):1–30
11. Guttman L (1968) A general nonmetric technique for finding the smallest coordinate space for a configuration of points. Psychometrics 33(4):469–506
12. Krak IV, Kryvonos IG, Barmak OV, Ternov AS (2016) An approach to the determination of efficient features and synthesis of an optimal band-separating classifier of dactyl elements of sign language. Cybern Syst Anal 52(2):173–180
13. Kruskal JB, Wish M (1978) Multidimensional scaling. Sage University Paper
14. Quinlan JR (1993) C4.5: programs for machine learning. Morgan Kaufmann Publishers Inc., Burlington
15. Kryvonos IG, Krak IV, Barmak OV, Shkilniuk DV (2013) Construction and identification of elements of sign communication. Cybern Syst Anal 49(2):163–172

Model of the Operator Dynamic Process of Acoustic Emission Occurrence While of Materials Deforming

Volodymyr Marasanov⬡, Artem Sharko⬡, and Dmitry Stepanchikov$^{(\boxtimes)}$⬡

Kherson National Technical University, Kherson, Ukraine
volodymyr.marasanov@gmail.com, sharko_artem@ukr.net, dmitro_step75@ukr.net

Abstract. The task of identifying and predicting the processes of mechanical properties changes during the loading of metal structures according to acoustic emission signals is considered. A model and algorithm for finding the operator of the dynamic process of the output signal appearance in the acoustic emission source are proposed. To find the dynamic process operator, the experimental results of the occurrence of AE signals when testing on four-point bending of specimens of St3sp steel were used. To restore the original analogue signal AE from the spectrum of the diagnosing system output signal, sampling theorem was used. It is proposed to represent the polynomial approximation of the AE signal envelope as a model of a potential function in the Schrödinger equation for the anharmonic oscillator as a structural unit of the solid state vibrational dynamics. A quantitative estimate of the potential energy of the anharmonic oscillator is obtained. The adequacy and accuracy of the considered operator model of the dynamic process of the AE signals appearance is established.

Keywords: Information diagnostics · Prediction ·
Mechanical properties · Operator · Wave function · Acoustic emission

1 Introduction

Acoustic emission (AE) is the phenomenon of the appearance and propagation of elastic vibrations during the deformation of a stressed material. The accumulation of materials damage under load and their identification allows us to trace the dynamics of their development. Using AE information diagnostics materials with the AE phenomenon, it is possible to determine the degree of material performance under changing external conditions [1,2], measure the level of stresses and deformations [3,4], detect developing defects [5,6], observe the growth of cracks [7,8] and other discontinuities caused by the influence of aggressive media [9,10], determine their coordinates, assess the degree of their danger. Informational diagnostics is a targeted scientific and practical activity on the assessment of the structures state based on a dynamic information model with predetermined similarity criteria.

© Springer Nature Switzerland AG 2020
V. Lytvynenko et al. (Eds.): ISDMCI 2019, AISC 1020, pp. 48–64, 2020.
https://doi.org/10.1007/978-3-030-26474-1_4

The problem issues of information diagnostics studying using the phenomenon of AE are covered in researches [11–13]. The theoretical aspects of informational diagnostics from the standpoint of localization of AE sources were studied in research [14]. Such issues as the registration and processing of AE signals [15–17], the possibility of technical implementation under dynamic and static loads are justified, but attention is not focused on the quantum-mechanical issues of information diagnostics.

The main advantage of the AE method is associated with the ability to diagnose the entire object as a whole. The study of the mechanism of AE occurrence is of great practical interest since the destruction of the material of structures under cyclic loading conditions is especially dangerous, since it occurs under the action of stresses of much smaller strength and yield strengths. Since the destruction of materials is related to the state space of the object, the nature of loading, which is unknown a priori, uncertainty arises in measurements of monitoring of the material state, which is caused by insufficient reliability and limited information obtained from the tests.

The most difficult stage in the study of materials deformation under load is the identification of AE signals. The flaw detection of materials using the AE method solves the direct problem of elastic waves receiving from a developing defect and establishing the location of the AE source using a diagnostic system. The inverse problem of information diagnostics is to develop a dynamic model for determining the mechanical properties of materials based on experimental data, that allow to calculate the technical characteristics of the material at the source of the AE signal at the moment of material loading. This determines the relevance of the work.

The aim of the research is to develop a predictive model of the operator of AE signals dynamic process while deforming the structure of materials, assessing its adequacy according to four Gauss-Markov criteria and its accuracy.

2   Materials and Methods

The transition from considering oscillations of atoms in a discrete structure of a material under load to a set of propagating waves can be performed on the basis that each operation on the functions of a discrete argument can be compared to operations on their images – functions of a continuous argument [18].

The developed model belongs to the class of predictive multifactor models with incomplete initial information [19].

The operator, including spatial and temporal variables characterizing the occurrence and propagation of emission waves, can be represented by a partial differential equation of a hyperbolic type. Its solution is the wave function Ψ. It is postulated by the method of analogies on the basis of experimental data. Such experimental data are the AE signals obtained by the diagnostic system when the material is loaded.

The main element of the oscillatory dynamics of a solid is the anharmonic oscillator and its behavior under adiabatic mechanical loading. It can be represented by the stationary Schrödinger equation:

$$\Delta\Psi + \frac{2m}{\hbar^2}\left(E - U\right)\Psi = 0 \tag{1}$$

where $\hbar$ is the Dirac constant, Ψ is wave function, E is the total energy of the particle, U is a potential function, m is particle mass.

The cubic parabola can be chosen as a potential function of the oscillator [20]

$$U(x) = \frac{f}{2}x^2 + \frac{g}{3}x^3 \tag{2}$$

where f is elastic force constant, g is the anharmonicity coefficient, x is oscillator displacement from equilibrium position.

This expression describes the change in potential energy near the bottom of the potential well.

The initial stage of the dynamic process of the changing of material structure under load can be represented as follows. A time-dependent mechanical force $F(t)$, which is assumed to increase in time from 0 to some final value $F_0 \ll F_m$ is applied to the oscillator. This force is applied to the oscillator adiabatically that means the characteristic time of its change is much longer than the oscillator oscillation period. Then the total energy of the oscillator in the force field $F(t)$ is equal to:

$$E = E_k + U(x) - F(t)x \tag{3}$$

where E_k is kinetic energy of the oscillator.

By introducing a dimensionless unit of length $\xi = x\sqrt{\dfrac{m\omega}{\hbar}}$, where $\omega = \sqrt{\dfrac{f}{m}}$ is the oscillation frequency of an equivalent harmonic oscillator, the Schrödinger equation can be converted to a dimensionless form [20]:

$$-\frac{1}{2}\frac{\partial^2}{\partial x^2}\Psi + \left(\frac{1}{2}\xi^2 - \frac{\xi^3}{3\xi_0} - \frac{1}{4}P\xi\xi_0\right) = \lambda\Psi \tag{4}$$

In this ratio $P = \dfrac{4g}{f^2}F$ is the external force, measured in units of the ultimate strength of the oscillator, $\xi_0 = \sqrt{\dfrac{6U}{\hbar\omega}}$ is dimensionless length equal to the distance from the bottom of the pit to the top of the barrier, $U = \dfrac{f^3}{6g^2}$ is the magnitude of the energy barrier for the considered oscillator, $\lambda = \dfrac{E}{\hbar\omega}$ is the energy eigenvalue measured in relative units.

The wave function of the ground state of the anharmonic oscillator can be approximated by a Gaussian dependence [20]:

$$\Psi(\xi) = \frac{A}{\beta\sqrt{\frac{\pi}{2}}}exp\left(-2\frac{(\xi - \xi_c)^2}{\beta^2}\right) \tag{5}$$

where A is amplitude, ξ_c is displacement, β is dispersion.

The boundary conditions of the Schrödinger equation have the form [20]:

$$\Psi\left(-\frac{\xi_0}{2}\right) = \Psi\left(\frac{\xi_0}{2}\right) = 0 \tag{6}$$

When an external force field is applied to a quantum oscillator, the energy of the system changes with a change in this field: the vibrational energy is a result of the dynamic component of the deformation, and the elastic energy is a result of the static component. For a solid body, the above formulas must be extended to the three-dimensional case and generalized for an ensemble of quantum anharmonic oscillators.

3   Experiment

The identification of the structural features of the deformation mechanisms [21] according to the AE data was carried out when the samples were deformed for four-point bending at a facility designed on the basis of the MIP-10 spring testing machine operating on the principle of a given deformation. The general scheme of the experimental setup is shown in Fig. 1.

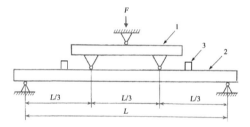

Fig. 1. Test scheme for four-point bending: 1 – loading device, 2 – test sample, 3 – acoustic emission sensor

Structurally, the installation contains two main mechanisms: the deformation mechanism and the force meter mechanism. The deformation mechanism includes: an engine, a worm gearbox, a loading screw, a carriage with an upper plate and a screw for switching off the compensation spring. Nonius is fixed on the plate. The mechanism of the weighing device includes: a lower plate and a lever-type transmission mechanism located inside the body. To implement the four-point bending, punches in the form of a load-bearing indenter and a support were additionally built in the installation. The supports were fixed in place to allow the sample to be accurately centered. In this case, the longitudinal axis of the sample was parallel to the lateral surface of the crosshead, and the center of the sample symmetry coincided with the axis of the applied load. The AE sensors were fastened to the sample using screw clamps. The tests were carried out with the registration of the load and the corresponding deflection of the sample with simultaneous fixation of the moments of AE signals occurrence.

The measurement setup used broadband sensors for acoustic emission instrument AF15 with a bandwidth of $0.2 \div 0.5$ MHz and $0.2 \div 2.0$ MHz. The informational measuring system used in the experiment provided an indication, registration and preprocessing of AE signals with its further storage in the computer memory using a storage oscilloscope (RIGOL DS1052E Digital oscilloscope).

4   Results and Discussion

The nature of the AE is determined by the interaction of the internal vibrational degrees of freedom with an external mechanical field (3). As a result of this interaction, a multiplicative term appears in the system energy, which is proportional to both the external force and the internal vibrational energy, which can be interpreted as the work of internal dynamic forces on the displacement created by the external field. This term determines the change in the vibrational energy of a solid. The mentioned work of internal dynamic forces can be associated with the area under the envelope of the AE signal, and thus obtain its experimental estimate. To do this, it is necessary to find the operator of the dynamic process of the material structure changing under load.

Finding the operator of the dynamic process of the material structure changing under load is part of the algorithm of the multifactor model of information diagnostics, shown in Fig. 2.

To find the operator of the dynamic process of material structure changing under load, as the analyzed signal of the AE a fragment of the results of the occurrence of AE signals at different loads on steel samples St3sp was selected, in which the time of occurrence, duration and ending of the AE signal was selected as the analyzed parameter [21].

The pulse shape is determined by the transfer functions of the elements of the acoustic path and the frequency band of the receiving transducer. Since the attenuation increases with the distance traveled and increases greatly with increasing frequency, low frequency components associated with its energy spectrum and containing characteristic local extremes dominate in the recorded AE signal. In the character of the AE signals presented, it is possible to distinguish high-frequency components associated with oscillations, i.e. filling with electrical signals.

The shape of the recorded pulse is distorted in proportion to the distance between the source and receiver of the AE signal. It depends on the design of acoustic sensors, the location of transducers, the configuration of the surface and the material of the product. Impulse AE has a wide frequency spectrum.

The experimentally obtained signal and its envelope, conducted through local amplitudes U_{max} in the range of the signal existence, are shown in Figs. 3 and 4.

Since the signal has a symmetrical shape the Fig. 4 shows only the upper part of the AE signal envelope, corresponding to positive amplitudes.

The dynamic characteristics of the experimental AE signal are presented in Table 1.

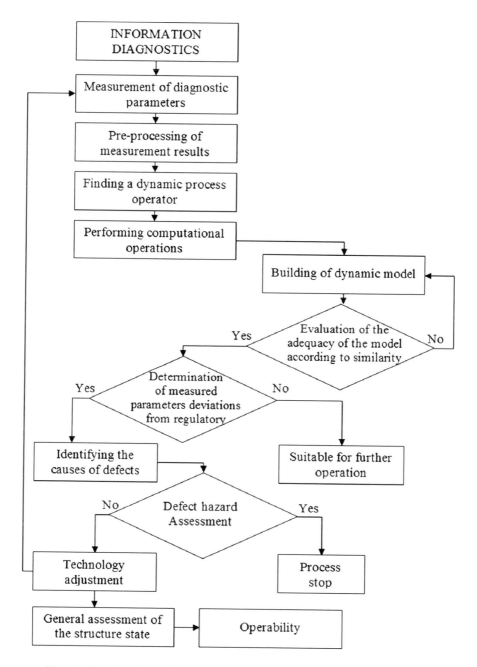

Fig. 2. The algorithm of the multifactor model of information diagnostic

During the construction and spectral analysis of the operator of dynamic deformation of material structure under load, the known value of the initial function y is determined from the known experimental output signal $y(t)$.

To restore the original analog signal AE from the spectrum of the output signal of the diagnosing system, sampling theorem was used, which postulating that any periodic function consisting of oscillation frequencies from zero to f_1 can be continuously transmitted with any given accuracy using samples following each other after $\frac{1}{2}f_1$ s. But for fragments of true non-periodic signals, which are AE signals (Fig. 3), this sum becomes infinite. Therefore, the AE signal in the frequency domain can be represented only with a certain limited accuracy. In addition, real and calculated AE signals have an infinite spectrum and cannot formally be sampled according to the sampling theorem, but due to the fact that the analyzed signal has a damped spectrum, the high-frequency components neglect what can be performed by their filtering. This difficulty is overcome by increasing the sampling rate.

It is necessary to select discrete points at intervals of half the duration of this period. Thus, if the analogue AE signal has a limited spectrum, it can be

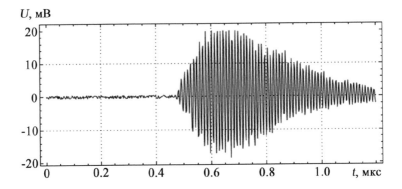

Fig. 3. Experimental acoustic emission signal

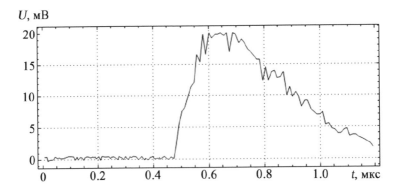

Fig. 4. Acoustic emission signal envelope shape

Table 1. Dynamic characteristics of the experimental AE signal

t, ms	U, mV	t, ms	U, mV	t, ms	U, mV	t, ms	U, mV
0.456	0.04	0.634	19.8	0.816	12.4	1.010	7.4
0.463	0.05	0.644	20.0	0.828	13.8	1.020	5.4
0.474	0.08	0.656	19.6	0.838	14.0	1.032	5.6
0.482	2.0	0.666	20.0	0.848	12.8	1.042	4.8
0.492	5.4	0.676	17.0	0.860	13.0	1.054	4.6
0.504	7.6	0.688	20.0	0.870	13.8	1.064	4.0
0.514	8.2	0.698	19.8	0.882	10.0	1.074	4.0
0.526	10.0	0.710	18.4	0.892	11.4	1.086	4.6
0.536	11.4	0.720	19.0	0.902	9.8	1.096	4.8
0.548	12.2	0.752	17.0	0.914	10.6	1.106	3.6
0.558	16.6	0.730	18.4	0.924	9.8	1.118	3.8
0.570	15.4	0.742	17.4	0.934	8.2	1.128	3.8
0.580	19.8	0.752	17.0	0.946	9.2	1.140	3.4
0.590	16.6	0.774	15.8	0.956	9.2	1.150	3.2
0.602	20.0	0.784	15.8	0.968	7.8	1.160	3.0
0.612	19.2	0.796	12.4	0.978	7.4	1.182	2.4
0.624	19.8	0.806	14.6	0.988	7.0	1.192	1.8

reconstructed from information obtained from an experimental spectrum with a frequency twice as large as doubled upper frequency.

With this in mind, the sampling of the input information of the experimentally obtained signal is performed with a time step equal to 0.02 ms and presented in the first two columns of the Table 2.

The calculated values of the voltage taken from the AE sensor, presented in Table 2, were found by linear interpolation, the envelope between two adjacent maxima $U_1 = U(t_1)$ and $U_2 = U(t_2)$:

$$U(t) = U_1 + \frac{(U_2 - U_1)(t - t_2)}{t_2 - t_1} \qquad (7)$$

where U_1, U_2 are the envelope values at time t_1 and t_2, respectively, $t_1 < t < t_2$.

To simplify the computational operations, the averaging over three counts, the third and fourth columns of Table 2, was held.

The output signal $y(t)$ is described by the m-th order differential equation with constant coefficients [19]:

$$\frac{d^m y(t)}{dt^m} + \sum_{m=0}^{m-1} a_m \frac{d^m y(t)}{dt^m} = 0 \qquad (8)$$

where m is order of the highest derivative of the differential operator.

Table 2. The results of the experimental signal preprocessing

Discretization of experimental data		Averaging over three counts		Difference rows			
t, ms	U, mV	t, ms	y, mV	$u^{(1)}$	$u^{(2)}$	$u^{(3)}$	$u^{(4)}$
0.478	1.063						
0.498	6.500	0.498	5.454	10.230	−6.498	2.343	1.004
0.518	8.800						
0.538	11.533						
0.558	16.600	0.558	15.684	3.732	−4.155	3.347	−5.312
0.578	18.920						
0.598	18.867						
0.618	19.500	0.618	19.416	−0.423	−0.808	−1.965	7.729
0.638	19.880						
0.658	19.680						
0.678	17.500	0.678	18.993	−1.231	−2.773	5.764	−10.736
0.698	19.800						
0.718	18.880						
0.738	17.733	0.738	17.762	−4.004	2.991	−4.972	7.985
0.758	16.673						
0.778	15.800						
0.798	12.840	0.798	13.758	−1.013	−1.981	3.013	−4.779
0.818	12.633						
0.838	14.000						
0.858	12.967	0.858	12.745	−2.994	1.032	−1.766	4.290
0.878	11.267						
0.898	10.440						
0.918	10.280	0.918	9.751	−1.962	−0.734	2.524	−4.590
0.938	8.533						
0.958	8.967						
0.978	7.400	0.978	7.789	−2.696	1.790	−2.066	
0.998	7.000						
1.018	5.800						
1.038	5.120	1.038	5.093	−0.906	−0.276		
1.058	4.360						
1.078	4.200						
1.098	4.560	1.098	4.187	−1.182			
1.118	3.800						
1.138	3.467						
1.158	3.040	1.158	3.005				
1.178	2.509						

The solution of this equation can be represented as

$$y(t) = \sum_{i=1}^{m} C_i e^{-\gamma_i t} \tag{9}$$

The characteristic polynomial of Eq. (8) has the form

$$a_m r^m + a_{m-1} r^{m-1} + \ldots + a_1 r + a_0 = 0 \tag{10}$$

where r_i are roots of the characteristic Eq. (10).

Equation (10) displays the structure of the linear operator (8) and establishes the relationship between the set of roots r_i and the coefficient vector $(a_0, a_1, \ldots, a_m)$ of the trend equation. On the other hand, for each set of roots, we have m equalities:

$$\sum_{i=1}^{m} C_i r_i^{(m)} = y^{(m)} \tag{11}$$

which uniquely links the vectors $(C_1, C_2, \ldots, C_m)$ and $(y^{(0)}, y^{(1)}, \ldots, y^{(m-1)})$ [18, 19].

To obtain estimates of the characteristic polynomial (10), the finite difference method is used. This is explained by the fact that according to Table 2, the time series levels consist of two components: a trend and a random component. The trend is fairly smooth and can be approximated by a m-th degree polynomial. The order of the polynomial is determined using experimental information processed by the least squares method for finite differences in Table 2. The equation of this polynomial represents a model of a potential function in the Schrödinger equation (1, 4), the area under which the envelope determines the potential energy of an oscillator.

The algorithm for finding the input signal operator in dynamic processes of changing the structure of a material is realized through the inverse problem of selecting a model that adequately describes the experimentally received signal from a developing defect.

The initial stage of the proposed algorithm implementation is the calculation of the voltage average values from the output of the diagnostic system at regular intervals. Then the difference series are compiled:

1st differences $u^{(1)} = y_i - y_{i-1}$

2nd differences $u^{(2)} = u_i^{(1)} - u_{i-1}^{(1)}$

3rd differences $u^{(3)} = u_i^{(2)} - u_{i-1}^{(2)}$

4th differences $u^{(4)} = u_i^{(3)} - u_{i-1}^{(3)}$

The results of the differences series calculation are given in Table 2. For the initial difference series, the dispersion is:

$$\sigma_0^2 = \frac{\sum_{i=1}^{n} \left(y_i - \frac{1}{n} \sum_{i=1}^{n} y_i \right)^2}{n-1} \tag{12}$$

For k-th order difference series, the dispersion is:

$$\sigma_k^2 = \frac{\sum_{i=1}^{n-k+1} \left(u_i^{(k)}\right)^2}{n-k} \tag{13}$$

The dispersion of the original σ_0^2 series is used to determine the accuracy of the model, the dispersion of the difference series of the k-order – to determine the order of the polynomial model of operator of the dynamic process of the experimental AE signal occurrence. This order is found by determining the minimum values of the dispersion differences $\left|\sigma_2^2 - \sigma_1^2\right|$, $\left|\sigma_3^2 - \sigma_2^2\right|$, $\left|\sigma_4^2 - \sigma_3^2\right|$, from which determine the order of the trend model by the least squares method. The numerical calculations showed that $m = 3$.

The regression equation for the characteristic polynomial (10) and the differential operator (8) is written in the form:

$$\widehat{y} = a_0 + a_1 t + \ldots + a_m t^m \tag{14}$$

The coefficients of the characteristic polynomial $a_0, a_1, \ldots, a_m$ are determined by solving the system of normal equations by the least squares method, which for $m = 3$ has the form:

$$
\begin{aligned}
a_0 n \quad + a_1 \sum t + a_2 \sum t^2 + a_3 \sum t^3 &= a_1 \sum y \\
a_0 \sum t + a_1 \sum t^2 + a_2 \sum t^3 + a_3 \sum t^4 &= a_1 \sum yt \\
a_0 \sum t^2 + a_1 \sum t^3 + a_2 \sum t^4 + a_3 \sum t^5 &= a_1 \sum yt^2 \\
a_0 \sum t^3 + a_1 \sum t^4 + a_2 \sum t^5 + a_3 \sum t^6 &= a_1 \sum yt^3
\end{aligned}
\tag{15}
$$

$$a_0 = \frac{\Delta_0}{\Delta}; \; a_1 = \frac{\Delta_1}{\Delta}; \; a_2 = \frac{\Delta_2}{\Delta}; \; a_3 = \frac{\Delta_3}{\Delta} \tag{16}$$

where Δ is the determinant of system (15), Δ_0, Δ_1, Δ_2, Δ_3 are the main minors of this determinant.

The characteristic equation of the differential operator (8) for $m = 3$ has the form:

$$a_3 r^3 + a_2 r^2 + a_1 r + a_0 = 0 \tag{17}$$

Solving Eq. (17), we can find the roots r_1, r_2, r_3 of the characteristic equation of the differential operator (8) and then obtain its solutions as a function (9).

The results of calculations performed using the appropriate batch programs are presented in the form of a trend equation and an integral of the AE signal existence region.

$$\widehat{y} = -236.859 + 942.794t - 1116.167t^2 + 415.990t^3 \tag{18}$$

The integral of Eq. (18) determines the potential energy of an oscillator:

$$U = I \int_{t_1}^{t_2} \widehat{y} dt \tag{19}$$

where I is the current intensity of the output signal AE, its value can be estimated using the resistance value of the piezosensor of the AE signal (≥ 60 MOhm) and the average value of the signal voltage (11.14 mV), t_1, t_2 is start and end time of the AE signal.

The estimated calculation of integral (19) gives the result of $U \leq 1.43 \cdot 10^{-12}$ J, which can be considered the upper limit of the potential energy of the anharmonic oscillators ensemble.

A comparison of the polynomial approximation with the experimental data is given in Table 3 and in Fig. 5.

Table 3. Polynomial approximation and adequacy of the AE signal model

t, ms	y_{exp}, mV	y_{calc}, mV	$y_{exp} - y_{calc}$, mV	Turning points
0.498	5.454	7.216	−1.762	−
0.558	15.684	13.960	1.724	0
0.618	19.416	17.682	1.734	1
0.678	18.993	18.921	0.072	0
0.738	17.762	18.215	−0.453	0
0.798	13.758	16.104	−2.346	1
0.858	12.744	13.127	−0.383	0
0.918	9.751	9.854	−0.073	0
0.978	7.789	6.732	1.057	1
1.038	5.093	4.392	0.701	1
1.098	4.187	3.342	0.845	1
1.158	3.005	4.121	−1.116	−

The question of the possibility of using trend models for analyzing and predicting the mechanical properties of a material directly at the source of AE can only be resolved after determining the adequacy of the model for the process under study.

The trend model $\widehat{y}$ of a specific time series y is considered adequate if it correctly reflects the systematic components of the time series, i.e. its residual component $\epsilon_t = y_{exp} - y_{calc}$ satisfies the properties of randomness ϵ_t.

The verification of the randomness of the character of the time series levels deviations from the trend was carried out using the criterion of turning points. The sequence level ϵ_t is considered a maximum if it is more than two adjacent levels $\epsilon_{t-1} < \epsilon_t > \epsilon_{t+1}$ and the minimum if it is less than two adjacent levels $\epsilon_{t-1} > \epsilon_t < \epsilon_{t+1}$. In both cases, ϵ_t is considered a turning point.

The presence and distribution of such turning points is given in Table 3.

The criterion of randomness with a confidence level of 95% is the inequality

$$P > \left| \overline{P} - 1.96\sqrt{\sigma_p^2} \right| \tag{20}$$

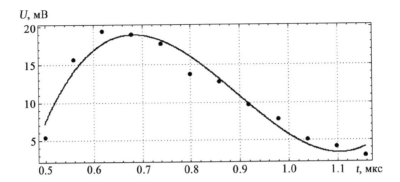

Fig. 5. Polynomial approximation of the AE signal experimental data (the points are the experimental data, the line is the polynomial approximation)

where P is the total number of turning points (for Table 3; $P = 5$), $\overline{P}$ is expectation of the number of turning points, σ_p^2 is dispersion of a number of turning points.

The $\overline{P}$ and σ_p^2 values entering into formula (20) were calculated by the formulas

$$P = \frac{2}{3}(n-2)\,;\, \sigma_p^2 = \frac{16n-29}{90} \tag{21}$$

where n is the total number of elements of the calculated sample (we have $n = 12$)

The calculations showed that $\overline{P} = 6.666$ and $\sigma_p^2 = 1.811$, i.e. inequality (20) is satisfied, which indicates the adequacy of the model according to this criterion.

The second condition for checking the adequacy of the model is to check the requirement of normality of distribution ϵ_t for which the indicators of asymmetry $\hat{\nu}_1$ and excess $\hat{\nu}_2$ and their dispersion are calculated

$$\hat{\nu}_1 = \frac{\frac{1}{n}\sum_{i=1}^{n}\epsilon_t^3}{\sqrt{\left(\frac{1}{n}\sum_{i=1}^{n}\epsilon_t^2\right)^3}} \tag{22}$$

$$\hat{\nu}_2 = \frac{\frac{1}{n}\sum_{i=1}^{n}\epsilon_t^4}{\left(\frac{1}{n}\sum_{i=1}^{n}\epsilon_t^2\right)^2} - 3 \tag{23}$$

$$\sigma_{\hat{\nu}_1} = \sqrt{\frac{6(n-2)}{(n+1)(n+3)}} \tag{24}$$

$$\sigma_{\hat{\nu}_2} = \sqrt{\frac{24n(n-2)(n-3)}{(n+1)^2(n+3)(n+5)}} \tag{25}$$

If the following inequalities are fulfilled simultaneously $|\hat{\nu}_1| < 1.5\sigma_{\hat{\nu}_1}$ and $\left|\hat{\nu}_2 + \frac{6}{n+1}\right| < 1.5\sigma_{\hat{\nu}_2}$, then the third order model is adequate.

The next verification criterion is the verification of the equality of the mathematical expectation of the random component ϵ_t to zero by the Student criterion. The estimated value of this criterion is given by

$$t = \frac{\bar{\epsilon} - 0}{S_\epsilon} \sqrt{n} \qquad (26)$$

where $\bar{\epsilon}$ is the arithmetic mean of the residual sequence levels ϵ_t, S_ϵ is the standard root-mean-square deviation for this sequence.

If the calculated value t is less than the table value of t_α in Student statistics with a given level of significance α and the number of degrees of freedom $n - 1$, then the hypothesis that the expected value of the random sequence is zero for the random sequence is accepted. Calculated test data by Student criterion are given in Table 4.

The absence of significant autocorrelation in the residual sequence is checked using the Durbin-Watson model.

$$d = \frac{\sum_{i=2}^{n} (\epsilon_i - \epsilon_{i-1})^2}{\sum_{i=1}^{n} \epsilon_i^2} \qquad (27)$$

where $\epsilon_i = y_{i_{exp}} - y_{i_{calc}}$.

The calculated value of criterion d is compared with the upper d_2 and lower d_1 critical values of the Durbin-Watson statistics. If $d > d_2$, then the hypothesis of the autocorrelation absence in the residual sequence is accepted and the model is considered adequate. If $d < d_1$, then this hypothesis is rejected and the model is inadequate.

Calculated by the formulas (22–27) values are given in Table 4. The values of the Durbin-Watson statistics d_1 and d_2 are given for a confidence level of 99%, sample size $n = 12$ and the number $k = 3$ determined parameters of the model without a_0.

For adequate models it makes sense to set the task of assessing their accuracy. The accuracy of the model is characterized by the deviation of the model output from the real value of the variable being modeled. For an indicator represented by a time series, accuracy is defined as the difference between the value of the actual level of the time series and its estimate obtained by calculation using the model, with the following being used as statistical indicators of accuracy:
root-mean-square deviation

$$\sigma_\epsilon = \sqrt{\frac{1}{n-k} \sum_{i=1}^{n} (y_i - \widehat{y}_i)^2} \qquad (28)$$

mean relative approximation error

$$\bar{\epsilon} = \frac{1}{n} \sum_{i=1}^{n} \left| \frac{y_i - \widehat{y}_i}{y_i} \right| \qquad (29)$$

coefficient of convergence

$$\phi^2 = \frac{\sum_{i=1}^{n}(y_i - \widehat{y}_i)^2}{\sum_{i=1}^{n}(y_i - \overline{y}_i)^2} \tag{30}$$

where n is number of row levels (sample size), k is the number of determined parameters of the model, $\widehat{y}$ is evaluation of row levels by model, $\overline{y}$ is the arithmetic mean of the values of the levels of the series.

The results of the model accuracy assessment of the operator of the dynamic process of the acoustic emission signals appearance when the structure of materials is deformed are given in Table 4.

Table 4. The results of the model adequacy and accuracy assessment of the operator dynamic process of AE signals appearance

Evaluation of the model adequacy								
$\widehat{\nu}_1$	$\widehat{\nu}_2$	$\sigma_{\widehat{\nu}_1}$	$\sigma_{\widehat{\nu}_2}$	t	t_α	d	d_1	d_2
-0.325	-0.850	0.555	0.775	$4\cdot10^{-13}$	2.23	1.511	0.449	1.373
$|\widehat{\nu}_1| < 1.5\sigma_{\widehat{\nu}_1}$			$\left|\widehat{\nu}_2 + \frac{6}{n+1}\right| < 1.5\sigma_{\widehat{\nu}_2}$			$d > d_1$		
$0.325 < 0.832$			$0.388 < 1.162$			$d > d_2$		
Model accuracy assessment								
σ_ϵ			$\overline{\epsilon}$			ϕ^2		
1.43			0.13			0.05		

5   Conclusions

Predicting the structural states of materials based on establishing the relationship between the evolution of the defective structure and the kinetics of damage accumulation at different stages of plastic deformation and destruction is not only a scientific, but also a technical problem. The problem of identifying and predicting the processes of changes in mechanical properties during the loading of metal structures was solved on the basis of an estimate of the potential function of the anharmonic oscillator from an experimental AE signal. For the first time, a model and algorithm for finding the operator of the dynamic process of the output signal occurrence in the acoustic emission source are proposed. The algorithm for finding the operator is implemented through the inverse problem of selecting a model that adequately describes the experimentally received signal from a developing defect. It is shown that the dynamic process operator of the AE output signal can be represented by a partial differential equation of hyperbolic type. When solving the problem, the experimental results of the appearance of AE signals when testing for four-point bending of specimens from St3sp steel were used. To restore the original analog AE signal from the spectrum of the output signal of the diagnostic system, sampling theorem was used.

The results of the calculations are presented in the form of an envelope equation and an integral over the region of AE signal existence. It is proposed to represent the polynomial approximation of the AE signal envelope as a model of a potential function in the Schrödinger equation, where the area under the envelope allows to quantify the potential energy of the anharmonic oscillator as a structural unit of the vibrational dynamics of a solid. The adequacy and accuracy of the considered model of the operator of the dynamic process of the AE signals appearance is established. The calculations allow to establish the upper limit of the potential energy of the anharmonic oscillators ensemble $U \leq 1.43 \cdot 10^{-12}$ J. By the known value of the potential function, the equation of the spherical potential of atoms interaction in the crystal lattice of the metal, through which the interaction constants and strength metal properties in the area of the AE source can be get.

References

1. Srickij V, Bogdevicius M, Junevicius R (2016) Diagnostic features for the condition monitoring of hypoid gear utilizing the wavelet transform. Appl Acoust 106:51–62. https://doi.org/10.1016/j.apacoust.2015.12.018
2. Li C, Sanchez RV, Zurita G, Cerrada M, Cabrera D, Vasquez RE (2016) Gearbox fault diagnosis based on deep random forest fusion of acoustic and vibratory signals. Mech Syst Sig Process 76(77):283–293. https://doi.org/10.1016/j.ymssp.2016.02.007
3. Hase A, Wada M, Koga T, Michina H (2014) The relationship between acoustic emission via piezoelectric actuator wave control. Int J Adv Manuf Technol 70:947–955. https://doi.org/10.1007/s00170-013-5335-9
4. Kumar J, Sarmah R, Ananthakrishna G (2015) General framework for acoustic emission during plastic deformation. Phys Rev B 92(14):1441. https://doi.org/10.1103/PhysRevB.92.144109
5. Madarshahian R, Ziehl P, Caicedo JM (2019) Acoustic emission Bayesian source location: onset time challenge. Mech Syst Sig Process 123:483–495. https://doi.org/10.1016/j.ymssp.2019.01.021
6. Bohmann T, Schlamp M, Ehrlich I (2018) Acoustic emission of material damages in glass fibre-reinforced plastics. Compos Part B Eng 155:444–451. https://doi.org/10.1016/j.compositesb.2018.09.018
7. Liu S, Li X, Li Z, Chen P, Yang X, Liu Y (2019) Energy distribution and fractal characterization of acoustic emission (AE) during coal deformation and fracturing. Meas J Int Meas Confederation 136:122–131. https://doi.org/10.1016/j.measurement.2018.12.049
8. Wang K, Zhang X, Hao Q, Wang Y, Shen Y (2019) Application of improved least-square generative adversarial networks for rail crack detection by AE technique. Neurocomputing 332:236–248. https://doi.org/10.1016/j.neucom.2018.12.057
9. Cho H, Shoji N, Ito H (2018) Acoustic emission generation behavior in A7075–T651 and A6061–T6 aluminum alloys with and without cathodic hydrogen charging under cyclic loading. J Nondestr Eval 37:83. https://doi.org/10.1007/s10921-018-0536-7

10. Babichev S, Korobchynskyi M, Lahodynskyi O, Korchomnyi O, Basanets V, Borynskyi V (2018) Development of a technique for the reconstruction and validation of gene network models based on gene expression profiles. Eastern-Eur J Enterp Technol 1(4–91):19–32

11. Marasanov V, Sharko A (2017) Energy spectrum of acoustic emission signals of nanoscale objects. J Nano- Electron Phys 9(2):02012-1–02012-4. https://doi.org/10.21272/jnep.9(2).02012

12. Marasanov V, Sharko A (2017) Energy spectrum of acoustic emission signals in complex media. J Nano- Electron Phys 9(4):04024-1–04024-5. https://doi.org/10.21272/jnep.9(4).04024

13. Huang C, Ju S, He M, Zheng Q, He Y, Xiao J, Zhang J, Jiang D (2018) Identification of failure modes of composite thin-ply laminates containing circular hole under tension by acoustic emission signals. Compos Struct 206:70–79. https://doi.org/10.1016/j.compstruct.2018.08.019

14. Marasanov V, Sharko A (2018) Information-structural modeling of the forerunners of origin of acoustic emission signals in nanoscale objects. In: IEEE 38th international conference on electronics and nanotechnology (ELNANO), pp 494–498. https://doi.org/10.1109/ELNANO.2018.8477473

15. Marasanov V, Sharko A (2017) Discrete models characteristics of the acoustic emission signal origin forerunners. In: IEEE first Ukraine conference on electrical and computer engineering (UKRCON), pp 680–683. https://doi.org/10.1109/UKRCON.2017.8100329

16. Mi Y, Chen Z, Wu D (2018) Acoustic emission study of effect of fiber weaving on properties of composite materials. In: IEEE international ultrasonics symposium, IUS, October 2018, art. no. 8579807. https://doi.org/10.1109/ULTSYM.2018.8579807

17. Wang SG, Liu YR, Tao ZF, Zhang Y, Zhong DN, Wu ZS, Lin C, Yang Q (2018) Geomechanical model test for failure and stability analysis of high arch dam based on acoustic emission technique. Int J Rock Mech Min Sci 112:95–107. https://doi.org/10.1016/j.ijrmms.2018.10.018

18. Himmelblau D (1970) Process analysis by statistical methods. Wiley, New York

19. Hudzenko L (1969) Some questions about the structure of the object on the steady-state signal. Proc Lebedev Phys Inst 45:110–133

20. Gilyarov V, Slutsker A (2010) Energy loaded quantum anharmonic oscillator. FTT 52(3):540–544

21. Aleksenko VL, Sharko AA, Sharko AV, Stepanchikov DM, Yurenin KY (2019) Identification by acoustic emission method of structural features of deformation mechanisms at bending. Tech Diagn Non-destructive Test 1:32–39. https://doi.org/10.15407/tdnk2019.01.04

Euclidean Combinatorial Configurations: Continuous Representations and Convex Extensions

Oksana Pichugina$^{(\boxtimes)}$ ⓘ and Sergiy Yakovlev ⓘ

National Aerospace University "Kharkiv Aviation Institute",
17 Chkalova Street, Kharkiv 61070, Ukraine
oksanapichugina1@gmail.com, svsyak7@gmail.com

Abstract. This paper presents general approaches to continuous functional representations (f-representations) of sets of Euclidean combinatorial configurations underlying the possibility of applying continuous programming to optimization on them. A concept of Euclidean combinatorial configurations (e-configurations) is offered. Applications of f-representations in optimization and reformulations of extreme combinatorial problems as global optimization problems are outlined. A typology of f-representations is presented and approaches to construct them for classes that were singled out are described and applied in forming a number of polynomial f-representations of basic sets of e-configurations related to permutation and Boolean configurations. The paper's results can be applied in solving numerous real-world problems formulated as permutation-based, binary or Boolean.

Keywords: Combinatorial optimization · Continuous optimization ·
Euclidean combinatorial set · Vertex-located set · Surfaced-located set

1 Introduction

A special class of mathematical models of extreme combinatorial problems (COPs) is a collection of their Euclidean statements, where an admissible domain is a finite point configuration in Euclidean space. Their advantage is the possibility of applying to their solving: properties of Euclidean space; a transition from a discrete set to considering continual ones, such as a convex hull or a circumsurface; combining discrete and nonlinear optimization techniques with properties of the admissible domain. By now, a Euclidean formulation is known for a narrow class of combinatorial problems, and for those for which it is available, such a formulation is defined ambiguously due to the discreteness of the domain. Respectively, there is a choice of the mathematical model and, therefore, there is an alternative of methods to solve the optimization problem.

Regarding combining discrete and continuous optimization methods, Euclidean formulations of extreme combinatorial problems in algebraic form

© Springer Nature Switzerland AG 2020
V. Lytvynenko et al. (Eds.): ISDMCI 2019, AISC 1020, pp. 65–80, 2020.
https://doi.org/10.1007/978-3-030-26474-1_5

(continuous formulations) are of particular interest, where the admissible discrete domain is described by a finite set of functional dependencies, thereby generating its continuous model. Therefore, it is relevant to expand the class of Euclidean statements; to construct their various forms, including continuous ones; to compare the efficiency of solutions of COPs depending on the Euclidean statement form chosen.

In this work, we consider in detail the first two aspects for sets of combinatorial configurations according to Berge [1], whose elements are mappings of a finite set into a finite set of vectors of the same dimension. Such sets often arise in theoretical and practical areas. For instance, in problems of Geometric Design (GD), geometric information on objects is often given in this form; classic geometric configurations on a plane are represented by ordered samples of Boolean vectors, etc.

2   Euclidean Combinatorial Configurations

We consider configurations according to [1]. This implies that any configuration is a mapping ψ of some initial set B of certain elements into a resulting finite abstract set $A = \{a_1, ..., a_k\}$ of certain elements having specific structure, if a given set of constraints Λ is true, i.e.,

$$\psi : B \to A. \tag{1}$$

Suppose B is a finite set $B = \{b_1, ..., b_n\}$, then the result of mapping (1) is an ordered set π of elements from A:

$$\pi = \begin{pmatrix} b_1 & \cdots & b_n \\ a_{j_1} & \cdots & a_{j_n} \end{pmatrix} = \langle a_{j_1} a_{j_2} ... a_{j_n} \rangle, \tag{2}$$

where $\{j_1, ..., j_n\} \subseteq J_k = \{1, ..., k\}$, further referred to as a combinatorial configuration or a c-configuration.

Now, introduce a set Π of c-configurations, induced by various tuples (2) for the given A, B, and constraints Λ (further referred to as $\mathcal{E}_c$-set).

Then we single out the following class of c-configurations. Let

$$\mathbf{A} = \{\mathbf{a}_1, ..., \mathbf{a}_k\} \tag{3}$$

be a collection of vectors of the same dimension m, i.e.,

$$\mathbf{a}_l = (a_{1l}, ..., a_{ml})^T \in \mathbb{R}^m, \quad l \in J_k. \tag{4}$$

Form a multiset of coordinates of vectors (4) and single out its ground set $\mathcal{A} = \{e_1, ..., e_K\} = S(\{a_{ij}\}_{i \in J_m, j \in J_k})$. Let $\mathcal{A}$ be called a generated set (further referred to as $(E.\mathrm{GS})$) of $\mathcal{C}$-set E. Consider (3) as a resulting set for configurations formation, i.e., in (1)

$$A = \mathbf{A}.$$

According to (2), c-configuration π is an ordered set of vectors from $\mathbf{A}$, i.e.,

$$\pi = \langle \mathbf{a}_{j_1}, \mathbf{a}_{j_2}, ..., \mathbf{a}_{j_n} \rangle . \tag{5}$$

With every π configuration (5), let us associate a multiset

$$\tilde{A}(\pi) = \{\alpha_i(\pi)\}_{i \in J_N} = \{a_{1j_1}, ..., a_{1j_n}, ..., a_{mj_1}, ..., a_{mj_n}\},$$

and a point

$$x = (x_1, ..., x_N) \in \mathbb{R}^N, \ N = m \cdot n : \ \{x_1, ..., x_N\} = \tilde{A}(\pi).$$

Point x is a result of a mapping φ such that:

$$x = \varphi(\pi), \ \pi = \varphi^{-1}(x). \tag{6}$$

Definition 1. *Euclidean combinatorial configuration (e-configuration) is a mapping as follows:*

$$\varphi : (\psi, \mathbf{A}, \Theta) \to \mathbb{R}^N, \tag{7}$$

where $\mathbf{A}$ *is a resultant set* (3), (4), ψ *is a mapping* $\psi : J_n \to \mathbf{A}$, *and* Θ *is a collection of constraints on* φ, ψ.

The Euclidean combinatorial configuration is completely determined by a tuple

$$\langle \varphi, \psi, \mathbf{A}, \Theta \rangle . \tag{8}$$

It is an image of combinatorial configuration (4) in Euclidean space $\mathbb{R}^N$ with the given φ, ψ, being also the x-vector (6) of the dimension N.

By selecting the mapping φ and a way of multiset $\tilde{A}(\pi)$ ordering, we can obtain various e-configurations, e.g.,

$$x = vec((\pi)) = (a_{1j_1}, ..., a_{1j_n}, ..., a_{mj_1}, ..., a_{mj_n})^T$$
$$x = vec((\pi)^T) = (a_{1j_1}, ..., a_{mj_1}, ..., a_{1j_n}, ..., a_{mj_n})^T, \tag{9}$$

where $vec((\pi))$ is a vectorization of matrix $(\pi) = (a_{lj_i})_{l,i} \in \mathbb{R}^{m \times n}$ that corresponds to c-configuration π.

Further, we shall use mapping φ with reference to (9) with no other constraints to its form. As a result, we shall have $\Theta = \Lambda$ that fashions formula (7) to:

$$\varphi : (\psi, \mathbf{A}, \Lambda) \to \mathbb{R}^N,$$

and formula (8) looks as:

$$\langle \varphi, \psi, \mathbf{A}, \Lambda \rangle . \tag{10}$$

The mapping of $\mathcal{E}_c$-set Π into $\mathbb{R}^N$ brings about set

$$E = \varphi(\Pi) \subset \mathbb{R}^N$$

of all e-configurations (10) (further referred to as $\mathcal{C}$-set) such that $\Pi = \varphi^{-1}(E)$.

We can single out a special class of $\mathcal{C}$-sets when vectors (4) have a single coordinate, i.e., $m = 1$, with the resultant set expected to be numerical. Thus, $K = k$, $N = n$, $\mathbf{a}_j = a_j \in \mathbb{R}^1$, $j \in J_k$. Hence, (9) can be expressed as follows:

$$x = vec((\pi)) = (a_{j_1}, ..., a_{j_n})^T.$$

Depending on, whether $m = 1$ or $m > 1$ the class of $\mathcal{C}$-sets is divided into two subclasses of numerical and vector $\mathcal{C}$-sets.

Suppose $\mathcal{C}$-set E is an FPC in $\mathbb{R}^n$:

$$E = \{x^i\}_{i \in J_{n_E}} \subseteq \mathbb{R}^N, \ x^i = (x_{ij})^T_{j \in J_N}, \ i \in J_{n_E}, \tag{11}$$

$n_E > 1$. Then multiset $G = \bigcup_{x \in E} \{x\}$ is called a multiset that induces E (an induced multiset, $(E.\mathrm{IM})$). Its ground set $S(G)$ coincides with $(E.\mathrm{GS})$, i.e., $S(G) = \mathcal{A}$. One can express $(E.\mathrm{IM})$ and $(E.\mathrm{GS})$ as follows:

$$G = \{g_1, ..., g_\eta\}, \ g_i \leq g_{i+1}, \ i \in J_{\eta-1},$$
$$\mathcal{A} = \{e_1, ..., e_K\}, \ e_i < e_{i+1}, \ i \in J_{K-1},$$

or $G = \{e_i^{\eta_i}\}_{i \in J_K}$, where $[G] = (\eta_i)_{i \in J_K}$ is a primary specification of G, and $\eta_i = \mu_G(e_i)$ is a multiplicity of e_i in G, $i \in J_K$.

We shall solve different problems of Combinatorial Analysis on $\mathcal{C}$-sets analyzing them in two independent ways as follows [16–19]:

1. Analysis of $\mathcal{A}, G, n$ (a constructive analysis, C-A);
2. Analysis of algebraic and topological properties of E (a geometric analysis, G-A).

We restrict our consideration to the case $m = 1$, that together with (11) means that $K = k$, wherefrom

$$E = \{x^i\}_{i \in J_{n_E}} \subseteq \mathbb{R}^n, \ x^i = (x_{ij})^T_{j \in J_n}, \ i \in J_{n_E},$$
$$\mathcal{A} = \{e_1, ..., e_k\}, \ e_i < e_{i+1}, \ i \in J_{k-1}, \ k, n, n_E > 1.$$

In order to single out a specific set E among other $\mathcal{C}$-sets of the same combinatorial type that are induced by $G = E.\mathrm{IM}$, we shall associate it with a basic $\mathcal{C}$-set (further referred to as $\mathcal{C}_b$-set [35–37]) of the same combinatorial type as E, that has the same parameters $\mathcal{A}, G, n$ and unites $\mathcal{C}$-sets of the type and parameters. Hence, the general $\mathcal{C}_b$-set of permutations induced by n-element multiset $\mathcal{A}$ will be expressed as follows:

$$E_{nk}(G) = \{x \in \mathbb{R}^n : \{x\} = \{x_1, ..., x_n\} = G\}; \tag{12}$$

$\mathcal{C}_b$-set of permutations without repetitions induced by G is

$$E_n(G) = \{x \in \mathbb{R}^n : \{x\} = \{x_1, ..., x_n\} = \mathcal{A}\};$$

a binary $\mathcal{C}_b$-set -
  $B'_n = \{x \in \mathbb{R}^n : x_i \in \{-1, 1\}, i \in J_n\}$,
  $B'^+_n = \{x \in B'_n : x \text{ has an even number of minus ones}\}$,
  $B'^-_n = \{x \in B'_n : x \text{ has an odd number of minus ones}\}$, where $J^0_m = J_m \cup \{0\}$, etc.

3 Optimization on $\mathcal{C}$-sets

Let us formulate the problem of optimization of some functional $h : \Pi \to \mathbb{R}^1$, defined on $\mathcal{E}_c$-set Π, as follows: we need to find $\pi^* \in \tilde{\Pi}$ such that

$$h(\pi^*) = \min_{\pi \in \tilde{\Pi}} h(\pi), \tag{13}$$

where $\tilde{\Pi} \subseteq \Pi$ is a feasible domain determined by the given set of constraints.

As a result of mapping of $\mathcal{E}_c$-set Π into Euclidean space, we obtain an equivalent problem of finding such e-configuration $x^* \in \tilde{E} \subseteq E$ that

$$f(x^*) = \min_{x \in \tilde{E}} f(x),$$

where $E = \varphi(\Pi)$, $\tilde{E} = \varphi(\tilde{\Pi})$, and function $f : E \to \mathbb{R}^1$ is such that $h(\pi) = f(x)$ for any $x = \varphi^{-1}(\pi)$, where $\pi \in \Pi$.

Without loss of generality, we can assume that E is some $\mathcal{C}_b$-set. As a result, we get a discrete optimization problem over $\mathcal{C}$-set $\tilde{E}$, that is a subset of the $\mathcal{C}_b$-set E, which can be formalized as an optimization problem:

$$f(x) \to \min, \ x \in \tilde{E}, \tag{14}$$

where the feasible domain $\tilde{E}$ is given by a collection of constraints

$$x \in E, \tag{15}$$

$$\psi_i(x) \le 0, \ i \in J_{\mu'}; \psi_i(x) = 0, \ i \in J_\mu \backslash J_{\mu'}, \tag{16}$$

and functions $f(x)$, $\psi_i(x)$, $i \in J_\mu$, are defined on E.

Solving the problem (14)–(16) assumes finding such a point $x^* \in \tilde{E}$ that

$$f(x^*) = \min_{x \in \tilde{E}} f(x).$$

Remark 1. We can assume that these functions are continuous on a predefined set $K \supset E$, otherwise, they can be continuously extended from E onto K in accordance with [25, 26, 29, 30, 32].

The goal of the paper is to present approaches to constructing continuous equivalent formulations of the problem (14)–(16), which means that the condition (15) is replaced by relationships evolving continuous functions on K only. After that, the original extreme combinatorial problem (13) can be solved by means on continuous optimization [2, 20, 27].

4 Continuous Functional Representations of $\mathcal{C}$-sets

Suppose

$$\mathcal{F} = \{f_j(x)\}_{j \in J_m}, \tag{17}$$

where $f_j : E \to \mathbb{R}^1$ are continuous functions for $j \in J_m$.

Definition 2. *The representation of C-set E with the help of functional dependencies expressed as follows:*

$$f_j(x) = 0, \ j \in J_{m'}, \tag{18}$$

$$f_j(x) \le 0, \ j \in J_m \backslash J_{m'}. \tag{19}$$

will be called a continuous functional representation (a f-representation) of E.

In f-representation (18), (19): (a) (18) is a strict part; (b) (19) is a nonstrict part; (c) m is an order; (d) m', $m'' = m - m'$ is an order of the strict and nonstrict parts, respectively.

Introducing notations

$$S_j = \{x \in \mathbb{R}^n : f_j(x) = 0\}, \ j \in J_{m'}, \tag{20}$$

$$C_j = \{x \in \mathbb{R}^n : f_{j+m'}(x) \le 0\}, \ j \in J_{m''}, \tag{21}$$

for geometric locused, determined by expressions (18), (19), E can be represented as follows: $E = \left(\underset{j \in J_{m'}}{\cap} S_j\right) \cap \left(\underset{i \in J_{m''}}{\cap} C_j\right)$.

For instance, if dimension of varieties (20) is $n-1$, and $f_j(x), \ j \in J_m \backslash J_{m'}$ are convex, C-set E is formed as an intersection of hypersurfaces (20) with convex bodies (21).

We shall provide the classification of f-representations in several ways according to: (a) the type of functions (17) including linear, nonlinear, differentiable, smooth, convex, polynomial, trigonometrical, etc.; (b) the combination of parameters m, m', m''.

Definition 3. *System* (18), (19) *is called:*

- *a strict f-representation E (further referred to as (E.SR)), if it contains only a strict part $m' = m$, $m'' = 0$;*
- *a nonstrict (further referred to as (E.NR)), if a f-representation contains only a nonstrict part $m' = 0$, $m'' = m$;*
- *a mixed (further referred to as (E.MR)), if it contains both strict and nonstrict parts, i.e., $m'(m - m') > 0$.*

System of constraints (18), (19) will be called an irredundant f-representation of E-set (E.IR), if the exclusion of any of its constraints assumes forming its proper superset:

$$\forall j \in J_{m'} \ E \backslash S_j \supset E;$$

$$\forall i \in J_{m''} \ E \backslash C_i \supset E.$$

Finally, a bi-component strict f-representation of E will be called tangential (E.TR), if set E coincides with a set of tangential points between S_1 and S_2 surfaces that are tangent to each other; a n-component irredundant strict f-representation will be called intersected (E.IIR).

Among the mixed f-representations we can single out a class of polyhedral-surfaced representations (further referred to as (E.PSR)), that consist of equation of a strictly convex surface S, circumscribed around E, and an H-representation of P. The outlook of the surface S allows classifying these f-representations, specifically, singling out polyhedral-spherical (E.PSpR), polyhedral-supersphe-rical (E.PSsR), polyhedral-ellipsoidal (E.PER) ones, etc.

Remark 2. If there is a constraint that the family of functions (17) consists only of polynomials, one can use tools of Real Algebraic Geometry [4,8,28]. For instance, (E.SR) can be formed in the way of finding of a base of an ideal generated by algebraic set E. And when we determine (E.NR) or (E.MR), we bear in mind that E is a semi-algebraic set.

   In the next section, we outline approaches to the construction of f-representa-tions of $\mathcal{C}$-sets that are based on their C-A and G-A.

5   Approaches to Construction of f-Representations of $\mathcal{C}$-set

Nonstrict and mixed f-representations of $\mathcal{C}$-sets can be found in the following ways:

1. find the equation of strictly convex surface S and H-representation of P, if the existence of (E.PSR) is substantiated;
2. extract a family of functions, for which the range of value changes over E is known, form (18) and (19) in view of the above and check:

$$x \in E \Longleftrightarrow x \text{ satisfies } (18), (19). \tag{22}$$

3. construct a strict f-representation of $\mathcal{C}$-set $E' \supset E$ and define E as E' subject to some constraints including some inequalities. To exemplify, we can use as E' the grid $E' = \mathcal{A}^n$.

   For some VLSs, PSRs, which are non-strict f-representation, can be found by deriving H-representation of P and a strictly convex circumsurface for E.

Example 1. Taking into account that $E_{nk}(G)$, B'_n are PSpSs, and their convex hulls are a generalized permutohedron and a hypercube correspondingly [38], we obtain:

$$(B'_n.\text{PSR}): \quad -1 \le x_i \le 1, i \in J_n; \ \sum_{i=1}^{n} x_i^2 = n;$$

$$(E_{nk}(G).\text{PSR}): \sum_{j\in\omega} x_j \ge \sum_{j=1}^{|\omega|} g_j, \ \omega \subset J_n; \ \sum_{i=1}^{n} x_i^l = \sum_{i=1}^{n} g_i^l, \ l = 1, 2;$$

For SRs, the relationship (22) is simplified to the following: the system (18) is (E.SR), if

$$x \in E \Longleftrightarrow x \text{ satisfies } (18).$$

   Let us outline some approaches to constructing SPSs, which use some tools and terminology of Real Algebraic Geometry generalized from the ring of poly-nomials to the ring of continuous functions.

Algorithm 1. We can single out a family of functions $\Phi(E)$ that takes zero value on E sufficient for single outing $\mathcal{F}$ and use it to form (18). Selecting subfamily (18) from $\Phi(E)$, we are to justify that: (a) it yields a finite point configuration (FPC) [7]; (b) the FPC contains no other points except for E.

Theorem 1. $\Phi(E_{nk}(G)) = \Phi^{sym}(G)$, where $\Phi^{sym}(G)$ is a set of symmetric functions that are zero-valued at a point $g = (g_1, ..., g_n)$.

Theorem 2. $\Phi(B'_n) = \Phi^{even}(\{-1^n, 1^n\})$, where $\Phi(B'_n)$ is a set of functions even on every coordinate that are zero-valued at a point $e = (1, ..., 1)$.

Theorems 1 and 2 imply that strict f-representations of $E_{nk}(G)$, B'_n are constructed only by functions of families $\Phi^{sym}(G)$ and $\Phi(B'_n)$, respectively, whereas the construction of their proper subsets requires the application of other functions as well. In this relation, we need to find such functions that are zero-valued on $\mathcal{C}$-set E and nonzero-valued on $E' \backslash E$, where E' is the correspondent $\mathcal{C}_b$-set.

Example 2. $(B'_n.\text{SR})$: $x_i^2 - 1 = 0$, $i \in J_n$.

Algorithm 1 was used to construct strict f-representations of the general $\mathcal{C}_b$-set of permutations and its special classes [16,18,35].

Theorem 3. *If* (18) *is a strict f-representation of $\mathcal{C}$-set $E' \supset E$,*

$$f \in \overline{\Phi}(E') = \Phi(E) \backslash \Phi(E'), \tag{23}$$

$$f(x) \underset{E' \backslash E}{\neq} 0, \tag{24}$$

then (18), $f(x) = 0$ *is (E.SR).*

The given theorem specifies conditions necessary to single out one $\mathcal{C}$-set from another using the only equality constraint.

Corollary 1. *If* (18) *and* (19) *are f-representations of $\mathcal{C}$-set $E' \supset E$, and function f satisfies* (23) *and* (24), *then* (18), (19), *and* $f(x) = 0$ *form (E.FR).*

Example 3. Function

$$f(x) = \prod_{i=1}^n x_i \underset{B'_n}{\in} \{-1, 1\}, \tag{25}$$

enables splitting of B'_n into two subsets – B'^-_n and B'^+_n, because $f(x)$ takes the value of -1 and 1 correspondingly. According to Theorem 3, $(B'_n.\text{SR})$ with $f(x) = 1$ is $(B'^+_n.\text{SR})$, whilst $(B'_n.\text{SR})$ and $f(x) = -1$ form $(B'^-_n.\text{SR})$. Taking into account that $B'_n = 2B_n - 1$, these f-representations allow obtaining strict f-representations of $\mathcal{C}_b$-sets of even and odd Boolean vectors [35].

According to Corollary 1, $(B'_n.\text{PSpR})$ together with $f(x) = 1$ forms a mixed f-representation B'^+_n (further referred to as $(B'^+_n.\text{MR1})$) of the order of $2n + 2$, and $(B'_n.\text{PSpR})$ together with $f(x) = 1$ forms f-representation B'^-_n (further referred to as $(B'^-_n.\text{MR1})$).

Algorithm 2. We can express constraints on $\mathcal{A}, G, \psi, \Omega$ in terms of coordinates of e-configurations. This is much more convenient for constructing of the above-mentioned strict f-representations of $\mathcal{C}_b$-sets as for the way of their constructions. We shall illustrate this with a $\mathcal{C}_b$-set of permutations.

Example 4. Suppose $E = E_n(G)$. The constraint $\{x_1, ..., x_n\} = \mathcal{A}$ can be expressed in its turn in two ways as follows: (1) $x_i \in \mathcal{A}$, $i \in J_n$; (2) $x_i \neq x_j$, $i, j \in J_n, i \neq j$. Consequently, we obtain the following $(E_n(G).SR)$:

$$\prod_{j=1}^{n} (x_i - e_j) = 0, \; i \in J_n; \; (x_i - x_j)^2 \geq \delta^2, \; 1 \leq i < j \leq n, \; \text{where}$$

$\delta = \min_{i \in J_{n-1}} \{e_{i+1} - e_i\}$.

To construct a strict f-representation of $E_{nk}(G)$, we can simply express the following condition, given in (12), in Cartesian variables:

$$\{x_1, ..., x_n\} = \{g_1, ..., g_n\} \tag{26}$$

It can be expressed as shown below:

$$(x - g_1) \cdot ... \cdot (x - g_n) = 0, \; x \in \mathbb{R}^1. \tag{27}$$

Indeed, to solve Eq. (27) with respect to x, we need to find the collection $x_1, ..., x_n$ of its roots, that exactly coincides with the multiset G, thus making condition (26) true.

Now we shall rewrite (27) in terms of its roots according to Vieta formula:
$$x^n - (g_1 + ... + g_n) \, x^{n-1} + (g_1 g_2 + ... + g_{n-1} g_n) \, x^{n-2} + ... + (-1)^n g_1 \cdot g_2 \cdot ... g_n = 0.$$
As a result we obtain a system of n equations:

$$\sum_{\omega \subseteq J_n, |\omega|=j} \prod_{i \in \omega} x_i = \sum_{\omega \subseteq J_n, |\omega|=j} \prod_{i \in \omega} g_i, \; j \in J_n, \tag{28}$$

whose solution is no other set except for the set of n real numbers $x_1, ..., x_n$ the same as in (26). On the other hand, dealing with every equation of the system (28) as with the equation of some variety in $\mathbb{R}^n$, we come up with the fact that a complete solution to these nonlinear system is exactly $\mathcal{C}$-set of permutations $E_{nk}(G)$. Thus, (28) is a strict polynomial representation of this set (further referred to as $(E_{nk}(G).SR1)$), whose degree and order coincide with the dimension of the Euclidean space and are equal to n.

Let us denote elementary symmetric polynomials as follows:

$$u_j(x) = \sum_{\omega \subseteq J_n, |\omega|=j} \prod_{i \in \omega} x_i, \; j \in J_n, \tag{29}$$

and rewrite $(E_{nk}(G).SR1)$ as:

$$u_j(x) = u_j(g), \; j \in J_n. \tag{30}$$

We should note that the use of $(E_{nk}(G).SR1)$ for large dimensions is problematic because it is quite difficult to evaluate functions (29).

We shall construct another functional representation of a $\mathcal{C}$-set $E_{nk}(G)$ based on $(E_{nk}(G).\mathrm{SR1})$ that relies on the given interrelations between elementary symmetric polynomials (29) with power sums:

$$q_j(x) = \sum_{i=1}^{n} x_i^j, \ j \in J_n^0, \tag{31}$$

that are reflected in the Newton-Girard identities [3]:

$$q_j(x) = j \cdot (-1)^{-j+1} h_j(x) + \sum_{i=1}^{j-1} (-1)^{i-j+1} q_i(x) \cdot h_{j-i}(x), \ j \in J_n. \tag{32}$$

Applying the recurrent formula (32) to both parts of the Eq. (30), we obtain $q_j(x) = q_j(g), \ j \in J_n$, or, considering (31),

$$\sum_{i=1}^{n} x_i^j = \sum_{i=1}^{n} g_i^j, \ j \in J_n. \tag{33}$$

Similar to (28), the system of Eq. (33) can be considered from two points of view: first, as a system used to determine a set of solutions of Eq. (27); second, as a system that defines a set of varieties in $\mathbb{R}^n$, whose intersection is exactly the set $E_{nk}(G)$. Thus, we found another f-representation (33) of $E_{nk}(G)$ (further referred to as $(E_{nk}(G).\mathrm{SR2})$). Similar to $(E_{nk}(G).\mathrm{SR1})$, it is strict, polynomial, with its degree and order being equal to n. At the same time, it has apparent benefits over $(E_{nk}(G).\mathrm{SR1})$, namely, the simplicity of the functions involved, and its convexity in $\mathbb{R}^n_{\geq e_1}$.

The use of the concept of Euclidean combinational configurations and property (26) of e-configuration of permutations allowed us to offer a new, much simpler proof of the following theorem, stated in [18].

Theorem 4. *Each of the systems of Eqs. (28), (33) defines a strict continuous functional representation of $E_{nk}(G)$.*

6 Approaches to Construction of Tangential f-Representations of $\mathcal{C}$-sets

In the class of convex f-representations, tangential ones have a minimal number of components. The minimum degree of polynomial tangential f-representation is three. In view of the fact that the minimum degree of a polynomial f-representation is two, it becomes apparent that $(E.\mathrm{TR})$ has benefits as for an order and a degree.

We shall describe some ways of constructing these f-representations. Let us outline a general construction scheme for tangential f-representations of $\mathcal{C}$-set E

that rests on properties of differentiable functions over E (further referred to as (TR.Scheme1)).

The tangential representation will be constructed as shown below:

$$f_1(x) = 0, \tag{34}$$
$$f_2(x) = 0. \tag{35}$$

First, we should select the differentiable functions $f_1(x), f_2(x) \in \Phi(E)$, $f_1(x) \neq f_2(x)$ that have no singularities. This means that, in $\mathbb{R}^n$,

$$f_1(x) \underset{E}{=} 0, \; f_2(x) \underset{E}{=} 0 \tag{36}$$

determine surfaces

$$S_j = \{x \in \mathbb{R}^n : f_j(x) = 0\}, \; j = 1, 2. \tag{37}$$

Second, we should check if the condition below is true:

$$\forall x \in E \quad \exists k(x) \neq 0 : \nabla f_2(x) \underset{E}{=} k(x) \cdot \nabla f_1(x).$$

Next, we should choose $j \in J_2$ and solve the optimization problem using the method of Lagrange multipliers:

$$f_j(x) \underset{S_{3-j}}{\to} extr, \tag{38}$$

Suppose

$$X^{j\,\min} = \underset{S_{3-j}}{Argmin} \; f_j(x), \; Z^{j\,\min} \underset{X^{j\,\min}}{=} f_j(x);$$

$$X^{j\,\max} = \underset{S_{3-j}}{Argmax} \; f_j(x), \; Z^{j\,\max} \underset{X^{j\,\max}}{=} f_j(x) -$$

is a complete solution of the problem (37).

Then the system of Eqs. (34), (35) will be a tangential f-representation of E, if one of the conditions below is true: $X^{j\,\min} = E$, $Z^{j\,\min} = 0$ or $X^{j\,\max} = E$, $Z^{j\,\max} = 0$.

(TR.Scheme1) can be applied to quite a narrow class of functions $f_1(x)$, $f_2(x)$ that allow solving problem (37), (38) explicitly (ref. e.g., [16,19] and Theorem 7).

Next we shall introduce another method (further referred to as (TR.Scheme2)) of analytic foundation of a tangential f-representation of two-level sets [7], i.e., sets that can be decomposed exactly along two parallel hyperplanes towards the normal vectors to facets of the correspondent polytope $P = conv\,E$.

Theorem 5. *If E is two-level and fulldimensional, then*

$$S^2 : f_0(x, 2) = \sum_{F \in \mathbf{F}} (\overline{n}_F'^T x - a_F')^2 - |\mathbf{F}| = 0;$$

$$S^4 : f_0(x, 4) = \sum_{F \in \mathbf{F}} (\overline{n}_F'^T x - a_F')^4 - |\mathbf{F}| = 0 -$$

is its tangential representation (further referred to as (E (2-level). TR)), where $\mathbf{F}$ is a set of P-facets; $\overline{n}_F'^T \in \mathbb{R}^n$, $a_F' \in \mathbb{R}^1$, $\overline{n}_F'^T = \frac{\overline{n}_F}{\delta_F}$, $a_F' = \frac{a_F}{\delta_F}$ for all $F \in \mathbf{F}$.

Here is one more method for tangential f-representation construction (further referred to as (TR.Scheme3)) that is used in case if Eqs. (34), (35) can be expressed in terms of some norm:

$$\exists \|.\|_{(\alpha)}, \exists \alpha_1, \alpha_2 \in \mathbb{R}^1_+ : f_1(x) = \|x\|_{(\alpha_1)} - 1; \ f_2(x) = \|x\|_{(\alpha_2)} - 1. \qquad (39)$$

Correspondingly, surfaces (37) are spheres in a norm space equipped with a norm $\|.\|_{(\alpha_i)}$ (further referred to as $\|.\|_{(\alpha_i)}$-spheres, $i = 1, 2$).

The proof of the fact that this norm is strictly monotonous with respect to α, i.e., $\forall \alpha_1, \alpha_2 \in \mathbb{R}^1_+$, $\alpha_1 \neq \alpha_2$ one of the conditions is true:

$$\forall\, x \in \mathbb{R}^n : \|x\|_{(\alpha_1)} \geq \|x\|_{(\alpha_2)}, \qquad (40)$$

$$\forall\, x \in \mathbb{R}^n : \|x\|_{(\alpha_1)} \leq \|x\|_{(\alpha_2)}, \qquad (41)$$

substantiates that (34), (35) is $(E.\text{SR})$. Besides, if this norm is a differentiable function in domain $K \supset E$, then (34), (35) will be $(E.\text{TR})$.

In turn, it means that in case (40) true, then $S_1 \subseteq C_2$, with $E = S_1 \cap \partial C_2$. And in case (41) we have $E = S_2 \cap \partial C_1$, where $C_i = conv\, S_i$, $i = 1, 2$. In other words, in case (40), S_1 is inscribed into S_2, and in case (41), S_1 is circumscribed around S_2. At that, in terms of norm $\|.\|_{(\alpha)}$, (36), (39) are expressed as: $\|x\|_{(\alpha_1)} \overset{=}{E}$ $\|x\|_{(\alpha_2)} \overset{=}{E} 1$.

Consequently, the following theorem is proved.

Theorem 6. *If there is function $\|.\|_{(\alpha)}$ that is strictly monotonous with respect to α and such that functions $f_1(x), f_2(x)$ in (39) satisfy the conditions of (36), (39), then the pair of Eqs. (34), (35) form $(E.\text{TR})$.*

Example 5. (TR.Scheme3) can be used to substantiate the existence of $(B'_n.\text{TR})$ expressed as:

$$(B'_n.\text{TR}(\alpha_1, \alpha_2)) : \sum_{i=1}^n x_i^{\alpha_1} = n, \ \sum_{i=1}^n x_i^{\alpha_2} = n,$$

where $1 \leq \alpha_1 < \alpha_2 < \infty$, because the scaled l_p-norm $\|x\|_{\langle p \rangle} = \frac{1}{n}\|x\|_p$ is monotonous [6,13].

(TR.Scheme2) is applied to both B_n, B'_n, and further two-level $\mathcal{C}_b$-sets, that belong to class $\mathcal{SS}$s, such as $E_{n2}(G)$, $E^n_{n+1,2}(G)$. When using Theorem 6 in this case, we should note that polytopes $conv E_{n2}(G)$, $conv E^n_{n+1,2}(G)$ are hypercubes with maximum two additional constraints.

Finally, (TR.Scheme1) can be used to substantiate the existence of $(B'_n.\text{TR}(\alpha_1, \alpha_2))$ for even α_1, α_2 [19], and to substantiate the existence of cubic $(E_{n2}(G).\text{TR})$ for the cases $G = \{e_1, e_2^{n-1}\}$ and $G = \{e_1^{n-1}, e_2\}$ [19].

We should also mention the fact, we come up with using (TR.Scheme1).

Theorem 7. *(a)* $\sum_{i=1}^{n} x_i^2 = n, \prod_{i=1}^{n} x_i = 1$ *is* $(B_n'^+.TR)$;

(b) $\sum_{i=1}^{n} x_i^2 = n, \prod_{i=1}^{n} x_i = -1$ *is* $(B_n'^-.TR)$.

Corollary 2. *Other mixed f-representations of* $B_n'^+$, $B_n'^-$ *are:*

$$(B_n'^+.MR2): -1 \le x_i \le 1,\ i \in J_n;\ \prod_{i=1}^{n} x_i = 1;$$

$$(B_n'^-.MR2): -1 \le x_i \le 1,\ i \in J_n;\ \prod_{i=1}^{n} x_i = -1.$$

Corollary 2 proves that $(B_n'^+.MR1)$, $(B_n'^-.MR1)$ are redundant. If we extract the circumscribed hypersphere equation out of them or the H-representation of the polytope, they become irredundant.

We should also note that notwithstanding $(B_n'^+.MR2)$, $(B_n'^-.MR2)$ consist of an equation of a circumscribed surface and a H-representation of the polytope, they are not polyhedral-surfaced, as function (25) is not convex. However, since $B_n'^+$, $B_n'^-$ are PSpSs, they also enable PSR. Indeed, according to [26], there is $M > 0$ such that $\forall \mu > M\ F(x, \mu) = \prod_{i=1}^{n} x_i + \mu(\sum_{i=1}^{n} x_i^2 - n)$ – is strongly convex. Correspondingly, there are PSRs of these sets that involve $F(x, \mu) = 0$ and H-representation of the hypercube.

7 Conclusion and Further Research

The theory of Euclidean Combinatorial Optimization on $\mathcal{C}$-sets is supposed to be further developed in the following directions. Theoretically, it is of interest to develop new approaches to the construction of convex extensions of functions defined on the corresponding $\mathcal{C}_b$-sets. Primarily the focus is on smooth convex extensions since most effective optimization methods assume at least the continuity of the second partial derivatives. At that, it is natural to single out various special classes of $\mathcal{C}$-sets, such as sets of e-configurations of permutation matrices, even, cyclic, or signed permutations, and so on. Considering these sets as new types of $\mathcal{C}_b$-sets, it is of considerable interest to single out their special subclasses and explore properties of the classes of $\mathcal{C}_b$-sets both in general and particular cases. Expectedly, this will provide new approaches to the construction of the required convex extensions. On the other hand, we need to conduct a comparative analysis of various continuous functional representation of $\mathcal{C}$-sets, since this greatly affects the efficiency of applied methods of Nonlinear Optimization that use the representations. Naturally, both of these directions should be considered integrally.

Further, we intend to proceed to the study of genetic algorithms for optimization problems on $\mathcal{C}$-sets, multiobjective combinatorial optimization, problems of Network Analysis, etc. in view of previous research [10–12, 14, 34].

We also expect to focus further research on the development of methods for solving the problems of packing, layout, and covering [5,9,15,21–24]. Historically, it was this class of problems, also called the problems of geometric design that laid the foundation for Euclidean Combinatorial optimization. The problems of geometric design focus on spatial objects having a shape, metric parameters and placement parameters that characterize their mutual position in space. Taken together, these characteristics determine geometric information that induces the configurational space of geometric objects [24,31]. The generalized variables of this space are considered as components of a vector represented a Euclidean combinatorial configuration [24]. In turn, selecting the combinatorial structure in problems of placement of geometric objects [33] we can consider this class of optimization problems as DO problems on $\mathcal{C}$-sets.

References

1. Berge C (1971) Principles of combinatorics. Academic Press, New York
2. Bertsekas D, Nedic A, Ozdaglar A, Nedic A (2003) Convex analysis and optimization. Athena Scientific, Belmont
3. Blum-Smith B, Coskey S (2017) The fundamental theorem on symmetric polynomials: history's first whiff of Galois theory. Coll Math J 48:18–29. https://doi.org/10.4169/college.math.j.48.1.18
4. Bochnak J, Coste M, Roy M-F (1998) Real algebraic geometry. Springer, Berlin, New York
5. Gerasin SN, Shlyakhov VV, Yakovlev SV (2008) Set coverings and tolerance relations. Cybern Syst Anal 44:333–340. https://doi.org/10.1007/s10559-008-9007-y
6. Gotoh J, Uryasev S, Gotoh J, Uryasev S (2016) Two pairs of families of polyhedral norms versus ℓ_p-norms: proximity and applications in optimization. Math Program 156:391–431. https://doi.org/10.1007/s10107-015-0899-9
7. Grande F (2015) On k-level matroids: geometry and combinatorics. http://www.diss.fu-berlin.de/diss/receive/FUDISS_thesis_000000100434
8. Hartshorne R (1983) Algebraic geometry. Springer, New York
9. Grebennik IV, Kovalenko AA, Romanova TE, Urniaieva IA, Shekhovtsov SB (2018) Combinatorial configurations in balance layout optimization problems. Cybern Syst Anal 54:221–231. https://doi.org/10.1007/s10559-018-0023-2
10. Koliechkina LM, Dvirna OA (2017) Solving extremum problems with linear fractional objective functions on the combinatorial configuration of permutations under multicriteriality. Cybern Syst Anal 53(4):590–599. https://doi.org/10.1007/s10559-017-9961-3
11. Koliechkina LN, Dvernaya OA, Nagornaya AN (2014) Modified coordinate method to solve multicriteria optimization problems on combinatorial configurations. Cybern Syst Anal 50(4):620–626. https://doi.org/10.1007/s10559-014-9650-4
12. Koliechkina L, Pichugina O (2018) Multiobjective optimization on permutations with applications. In: DEStech transactions on computer science and engineering. Supplementary volume OPTIMA 2018, pp 61–75. https://doi.org/10.12783/dtcse/optim2018/27922
13. Pavlikov K, Uryasev S (2014) CVaR norm and applications in optimization. Optim Lett 8:1999–2020. https://doi.org/10.1007/s11590-013-0713-7
14. Pichugina O, Farzad B (2016) A human communication network model. In: CEUR workshop proceedings. KNU, Kyiv, pp 33–40

15. Pichugina O (2017) Placement problems in chip design: modeling and optimization. In: 2017 4th international scientific-practical conference problems of infocommunications. Science and technology (PIC S&T), pp 465–473. https://doi.org/10.1109/INFOCOMMST.2017.8246440

16. Pichugina O, Yakovlev S (2016) Convex extensions and continuous functional representations in optimization, with their applications. J Coupled Syst Multiscale Dyn 4(2):129–152. https://doi.org/10.1166/jcsmd.2016.1103

17. Pichugina OS, Yakovlev SV (2016) Continuous representations and functional extensions in combinatorial optimization. Cybern Syst Anal 52(6):921–930. https://doi.org/10.1007/s10559-016-9894-2

18. Pichugina OS, Yakovlev SV (2016) Functional and analytic representations of the general permutation. Eastern-Eur J Enterp Technol 79(4):27–38. https://doi.org/10.15587/1729-4061.2016.58550

19. Pichugina O, Yakovlev S (2017) Optimization on polyhedral-spherical sets: theory and applications. In: 2017 IEEE 1st Ukraine conference on electrical and computer engineering, UKRCON 2017 - Proceedings. KPI, Kiev, pp 1167–1174. https://doi.org/10.1109/UKRCON.2017.8100436

20. Rockafellar RT (1996) Convex analysis. Princeton University Press, Princeton

21. Shekhovtsov SB, Yakovlev SV (1989) Formalization and solution of a class of covering problems in the design of monitoring and control systems. Avtomat i Telemekh 5:160–168

22. Stoyan Y, Romanova T (2013) Mathematical models of placement optimisation: two- and three-dimensional problems and applications. In: Fasano G, Pintér J (eds) Modeling and optimization in space engineering. Springer, New York, pp 363–388

23. Stoyan YG, Semkin VV, Chugay AM (2014) Optimization of 3D objects layout into a multiply connected domain with account for shortest distances. Cybern Syst Anal 50:374–385. https://doi.org/10.1007/s10559-014-9626-4

24. Stoyan YG, Yakovlev SV (2018) Configuration space of geometric objects. Cybern Syst Anal 716–726. https://doi.org/10.1007/s10559-018-0073-5

25. Stoyan YG, Yakovlev SV, Parshin OV (1991) Quadratic optimization on combinatorial sets in R^n. Cybern Syst Anal 27:561–567. https://doi.org/10.1007/BF01130367

26. Stoyan YG, Yakovlev SV, Emets OA, Valuĭskaya OA (1998) Construction of convex continuations for functions defined on a hypersphere. Cybern Syst Anal 34:27–36. https://doi.org/10.1007/BF02742066

27. Tuy H (2016) Convex analysis and global optimization. Springer, New York

28. Vinzant C (2011) Real algebraic geometry in convex optimization

29. Yakovlev SV (1989) Bounds on the minimum of convex functions on Euclidean combinatorial sets. Cybernetics 25:385–391. https://doi.org/10.1007/BF01069996

30. Yakovlev SV (1994) The theory of convex continuations of functions on vertices of convex polyhedra. Comp Math Math Phys 34:1112–1119

31. Yakovlev SV (2018) On some classes of spatial configurations of geometric objects and their formalization. J Autom Inform Sci 50:38–50. https://doi.org/10.1615/JAutomatInfScien.v50.i9.30

32. Yakovlev S (2017) Convex extensions in combinatorial optimization and their applications. In: Optimization methods and applications. Springer, Cham, pp 567–584. https://doi.org/10.1007/978-3-319-68640-0_27

33. Yakovlev SV (2017) The method of artificial space dilation in problems of optimal packing of geometric objects. Cybern Syst Anal 53:725–731. https://doi.org/10.1007/s10559-017-9974-y

34. Yakovlev S, Kartashov O, Yarovaya O (2018) On class of genetic algorithms in optimization problems on combinatorial configurations. In: 2018 IEEE 13th international scientific and technical conference on computer sciences and information technologies (CSIT), pp 374–377. https://doi.org/10.1109/STC-CSIT.2018.8526746
35. Yakovlev SV, Pichugina OS (2018) Properties of combinatorial optimization problems over polyhedral-spherical sets. Cybern Syst Anal 54(1):99–109. https://doi.org/10.1007/s10559-018-0011-6
36. Yakovlev S, Pichugina O, Yarovaya O (2018) On optimization problems on the polyhedral-spherical configurations with their properties. In: 2018 IEEE first international conference on system analysis intelligent computing (SAIC), pp 94–100. https://doi.org/10.1109/SAIC.2018.8516801
37. Yakovlev S, Pichugina O, Yarovaya O (2019) Polyhedral spherical configuration in discrete optimization. J Autom Inf Sci 51(1):38–50
38. Ziegler GM (1995) Lectures on polytopes. Springer, New York

Unconventional Approach to Unit Self-diagnosis

Viktor Mashkov[1(✉)] [ID], Josef Bicanek[1] [ID], Yuriy Bardachov[2] [ID],
and Mariia Voronenko[2] [ID]

[1] Jan Evangelista Purkyně University in Ústí nad Labem,
Ústí nad Labem, Czech Republic
{viktor.mashkov,josef.bicanek}@ujep.cz
[2] Department of Informatics and Computing Technology,
Kherson National Technical University, Kherson, Ukraine
y.bardachov@kntu.net.ua, mary_voronenko@i.ua

Abstract. Unit self-diagnosis is considered at system level. As distinct from system level self-diagnosis based on units mutual tests, we have researched the method based on the tests which a unit performs on other system units. Taking into account the obtained test results, a unit evaluates its own state. In our research, we have considered different faulty assumptions and testing procedures. Diagnosis model was developed and analyzed. Computer simulation is performed by using the web application developed for this research. Results of simulation were analyzed and assessed. Some recommendations were made for achieving better diagnosis results.

Keywords: Unit self-diagnosis · Testing procedure ·
Faulty assumptions · Intermittent fault · Hybrid-fault situation ·
Computer simulation

1 Introduction

Usually, system level self-diagnosis is based on the mutual tests among the system units [1,2]. In Fig. 1a, state of unit u_i is diagnosed based on the results of tests τ_1 and τ_2. Whereas, in case of Fig. 1b, unit u_i diagnoses its own state based on the results of tests τ_1 and τ_2.

Thus, in the second case, other system units play the role of testing environment for the unit being diagnosed. This environment allows unit u_i to expose (manifest) its faulty behavior, if any. It is expected that the behavior of a faulty unit will be different from the behavior of a fault-free unit. Such difference will be more evident if we observe the unit behavior for a long period of time. The suggested by us unconventional approach to unit self-diagnosis tackles two main problems. First problem concerns the issue of how to describe and evaluate the unit behavior. Second problem concerns the issue of how long the unit that diagnoses its own state should perform testing. The key idea of diagnosis suggested

© Springer Nature Switzerland AG 2020
V. Lytvynenko et al. (Eds.): ISDMCI 2019, AISC 1020, pp. 81–96, 2020.
https://doi.org/10.1007/978-3-030-26474-1_6

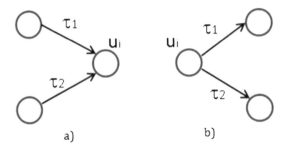

Fig. 1. Different approaches to the diagnosis of unit u_i.

by us consists in forcing the fault of a unit, if any, to exhibit itself during the testing which a unit performs on other units. It is assumed that unit faults can affect the results of tests. For each particular type of systems (e.g., sensor network, multi-agent system, multi-robot system etc.) assessment of such affect can be carried out on the basis of additional investigation. This task is not considered in our research. We only assume that a fault in the unit will affected the results of the tests which the unit performs on other units. We also assume that a unit has a fault-tolerant part that allows him to perform diagnosis algorithm correctly.

Implementation of the proposed self-diagnosis requires also unit capability to perform tests on other units (i.e., connection among the units, tests for each tested unit).

The suggested self-diagnosis is time consuming. That is why, much attention were paid to the issue of determining the time required for achieving correct result of unit self-diagnosis. Particularly, we have determined how many times a unit should test other units to achieve highly credible result of self-diagnosis. The proposed self-diagnosis was first introduced in [3] and then was modified and extended in [4]. In this research, we have generalized this approach and considered the corresponding diagnosis process. We also have considered computer simulation of unit self-diagnosis in more detail. Presented results are structured as follows. In Sect. 3, related work is presented. Section 4 describes the generalized approach to unit self-diagnosis. Computer simulation and its results are considered in Sect. 5 and Sect. 6. Conclusion summarizes the advances of the proposed approach to unit self-diagnosis and evaluates the results obtained in this research.

2   Problem Statement

Unit can have a fault which is initially dormant. A fault, when active, can produce an error. A fault can be activated by either external or internal facilities.

An error can be understood as invalid state of a unit. For example, the state of a computational unit is expressed by values of variables. Then, invalid values of variables can be detected with the help of tests performed by unit on other units. In the presented research, the problem of how to activate a fault is not considered. Determining location of fault after error detection is a separate problem which is not considered by us either.

The subject of this research is only the problem of how to detect the error caused by the activated fault.

3   Review of the Literature

Comparison-based system level diagnosis has been researched in many papers, and analyzed and structured in many surveys. This diagnosis was considered for possible implementation in wireless sensor networks, in multi-robot systems, in many-core processors and other areas.

In the case of wireless sensor networks, researchers take into account the network topology (changed or fixed), types of possible node and communication faults (permanent or transient/soft/intermittent), protocol of communication among the nodes (one-to-one, one-to-many, one-to-all). Also, different comparing-based models were proposed (e.g., generalized, broadcast, probabilistic).

Application of comparison-based model in sensor networks allows a sensor node to identify its own status based on the information received from the neighbors [5–8]. Alternatively, node state can be determined by the other system nodes [9,10]. In the case of multiple or related faults of sensor nodes the unit diagnosis may be incorrect.

The developed of comprehensive mutual diagnosis [11] is built on top of self-tests and presents a combination of self-testing and comparing approach. This diagnosis applied to multicore arrays implies that each core first executes its self-tests and generates a test signature. Then, it sends this signature to its four neighbors. Each core compares its own signature with those received from its neighbors. Comparing of signatures allows the core to detect faulty neighbors. The proposed in [11] mutual diagnosis also depends considerably on the state of the unit neighbors. The majority of comparison-based models impose an upper bound on the number of faulty units in the system. Some researchers consider faulty situations in which multiple and/or related faults take place. Related faults or common faults mean that some system units simultaneously become faulty due to the same reason. Xu in [12] proposes to combine comparison testing with $t/(n-1)$ variant programming. It is worth noting that the adjudicator (i.e., diagnostor) in this scheme intends to detect only the correct variant. However, in practice it is also important to detect the faulty units (variants). In [13], the adjudicator with extended functionality was presented. This adjudicator allows to detect not only the correct units but also all the faulty ones. Both these papers consider independent and related faults. It is worth noting that these schemes were designed only for software systems to provide their fault-tolerance.

Mostly, the self-diagnosis performed on the basis of comparison-based model is used to detect permanently faulty system units [8,14].

Chessa and Santi [9] have considered the problem of fault identification in ad-hoc networks and presented a comparison-based diagnostic model based on the one-to-many communication paradigm. In this paper, the authors show how both hard and soft faults can be detected. Elhadef et al. [15] have presented distributed comparison-based self-diagnosis protocol for wireless ad hoc networks. The proposed protocol identifies hard and soft faults. Khilar [16] proposed a distributed fault diagnosis algorithm for wireless sensor networks to diagnose intermittently faulty sensor nodes.

It is worth noting that these researches do not consider a model to describe the behavior of soft/intermittent faults. However, the different behavior of such faults can have different impact on the system. A more detailed consideration of system faults which are different from hard faults will allow to provide a more effective system recovery.

Using the Bayesian approach for system diagnosis is a common practice. Krishnamahari and Iyengar [17] presented Bayesian fault recognition algorithm to solve the fault-event disambignation problem in sensor networks. Luo et al. [18] have proposed a fault-tolerant energy-efficient event detection paradigm for wireless sensor networks and presented Bayesian detection method.

In our research, we also use this well-known approach to diagnose the state of a system unit. However, in our case the event that a unit is faulty is split into three events. Explanation of such splitting is given in the next sections.

4 Generalized Approach to Unit Self-diagnosis

Proposed approach is based on the assumption that fault in a unit can affect the results of test which it performs on other units. Results of tests are used as input to diagnosis algorithm.

The diagnosis process includes the following tasks:

- test generation;
- test execution;
- test evaluation;
- performing diagnosis algorithm.

In Fig. 2, the parts of a unit that perform the above mentioned tasks are shown. The following denotations are used:

P_1 - test generation part;
P_2 - part of unit engaged in test execution;
P_3 - test evaluation part that forms the test result;
P_4 - fault-tolerant part that performs a diagnosis algorithm and delivers its result to the unit's environment.

At the beginning, it is assumed that tested unit is fault-free and it always replies to the test correctly. It means that its reply matches the one which is expected when no errors occur. Fault in the unit diagnosing its own state can be either in part P_1 or part P_2 or part P_3 or outside these parts. If fault is in one of these parts, then test result is "1". Here we accept the presentation of test result which is used in system level self-diagnosis [19]. Particularly, correct result is denoted as "0" and incorrect one as "1".

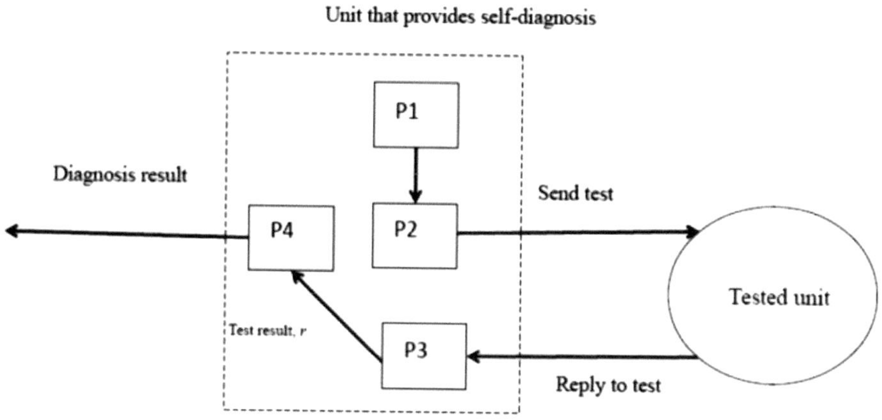

Unit that provides self-diagnosis

Fig. 2. Diagnosis process as a sequence of tasks.

For increasing fault coverage of the unit diagnosing its own state parts P_1, P_2 and P_3 may change. Also, it is assumed that fault in part P_4 cannot influence the diagnosis algorithm. Any fault will be tolerated (e.g., it can be masked).

The proposed approach is based on the repeated execution of tests. If the fault of one of the parts P_1, P_2 and P_3 is a permanent fault, then test result r obtained after each test execution will be the same (i.e., "1"). In case when test execution is repeated m times, $m = 1, 2, ...N$, the resulting tuple T is:

$$T = (r_1, r_2, ..., r_m) = (1, 1, ..., 1) \tag{1}$$

If the parts of a unit change, then some of test executions may produce the correct result (i.e., "0"). It can be explained by the fact that after changes of parts P_1, P_2 and P_3 the fault may be outside these parts. In the given case, the resulting tuple T may contain one or more zeros.

There is also a probability (albeit, very small) that fault in part P_3 will constantly affect the test result r so that it will be constantly equal to "0". Now let us assume that unit diagnosing its own state is fault-free, and tested unit is permanently faulty. Also, we assume that test always detects any error in tested unit caused by any fault (i.e., fault coverage for tested unit is equal to 100%). In the given case, the resulting tuple will be $T = (1, 1, ..., 1)$. When both unit diagnosing its own state and tested unit are faulty, the resulting tuple will be, except the case when faulty part P_3 constantly forms $r = 0$. We can summarize this consideration in the form of table (see Table 1).

We denote the tuple $T = (0, 0, ..., 0)$ by T_0 and the tuple $T = (1, 1, ..., 1)$ by T_1. From Table 1 it follows that it is not possible to discriminate between the fault-free and faulty states of unit under diagnosis when tuple is obtained. This situation can be resolved by way of using several tested units (see Fig. 3). Further we assume that parts of unit that provide self-diagnosis may change.

When k tested units are used, we can expect, with great probability, that at least one of the tested units will be fault-free. This probability is equal to

Table 1. Possible resulting tuples

State of unit under diagnosis	Parts change	State of tested unit	Resulting tuple
Fault-free	Doesn't affect	Fault-free	$T = (0, 0, ..., 0)$
		Faulty	$T = (1, 1, ..., 1)$
Faulty	No	Fault-free	$T = (1, 1, ..., 1)$
		Faulty	$T = (1, 1, ..., 1)$
	Yes	Fault-free	$T = (1, 0, 1, 0, ..., 1)$
		Faulty	$T = (1, 1, ..., 1)$

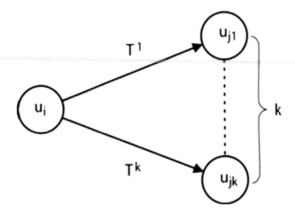

Fig. 3. Repeated testing of k units

$1 - q^k$. Where q is the probability of unit fault. Since a faulty testing unit after testing a fault free unit will, with great probability (denoted as P_E), produce the tuple different from T_1 and T_0, we can use this fact as the basis for diagnosis algorithm. Probability P_E can be computed as:

$$P_E = 1 - P_S^m - (1 - P_S)^m, \tag{2}$$

where P_S is the probability that a faulty unit will produce the test result equal to 1 when a tested unit is fault-free.

If testing unit is fault-free, all of the obtained tuples will equal either T_0 or T_1 depending on the states of tested units. Thus, diagnosis algorithm is as follows. If each of the obtained tuples is either T_0 or T_1, then testing unit is fault-free. Otherwise, testing unit is faulty.

Credibility of such diagnosis can be evaluated on the basis of probability P_E. Diagnosis will be incorrect if testing unit is faulty and event $A_i, i \in [1, 2, ..., k]$, takes place. Here A_i is the event when i tested units are fault-free, and each of the tuples is either of type T_0 or T_1. Probability of event A_i is determined as:

$$P(A_i) = C_i^k q^{k-i} (1 - q)^i P_d^i \tag{3}$$

where probability $P_d = 1 - P_E$.

Probability of incorrect diagnosis, P_{ID}, is equal to:

$$P_{ID} = \sum_{i=0}^{k} P(A_i) = \sum_{i=0}^{k} C_i^k q^{k-i} (1-q)^i P_d^i \tag{4}$$

Thus,

$$D = 1 - P_{ID} = 1 - \sum_{i=0}^{k} C_i^k q^{k-i} (1-q)^i P_d^i \tag{5}$$

It is worth noting that probability of incorrect diagnosis, P_{ID}, is computed under the condition that testing unit is faulty. Figures 4 and 5 give the idea of how parameters m, q, k and P_S influence the credibility of proposed diagnosis.

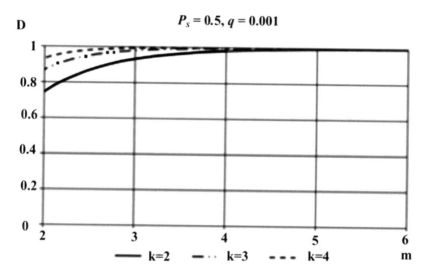

Fig. 4. Functional dependence $D = f(m)$ for the best case.

Figure 4 presents the results for the case of $P_S = 0.5$. In the given case, credibility of diagnosis is greater than the credibility obtained for $P_s \neq 0.5$. Figure 5 depicts the credibility of diagnosis for $P_S = 0.1$. Even for this worse case, the number of test repetitions doesn't exceed a few dozens. For example, for $k > 2$ it is sufficient to repeat the tests 14 times to obtain highly credible result of diagnosis.

Unit that provides self-diagnosis can also have intermittent faults. Detailed consideration of intermittent faults was performed by us in [4]. Intermittent faults can be defined as the faults whose presence is bounded in time. More precisely, a unit can possess an intermittent fault but effect of this fault is present only part of time.

In our research, we accepted the approach to describing behavior of intermittent faults presented in [20]. In the given case, an intermittent fault has two

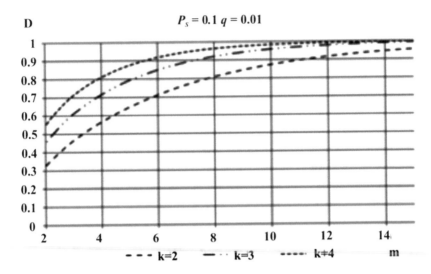

Fig. 5. Functional dependence $D = f(m)$ for the case of $P_S = 0.1$

states active (AS) and passive (PS). When an intermittent fault is in AS, the effect of intermittent fault is present. Whereas when an intermittent fault is in PS, its effect is not present. Transfers from one state to the other one are described with the corresponding intensities (rate of transition) λ and μ.

It is assumed that if a testing unit possesses an intermittent fault which at the moment of testing is in the AS, then the result of the test will be affected by this fault, and such intermittent fault can be detected. In the sequel, this fault can be identified on the basis of the affected test result(s). It is evident that depending on the time allocated to testing procedure, it is possible to detect different types of intermittent faults. Here, we consider two classes of intermittent faults.

To the first class C_1, we refer the intermittent faults which can frequently (more than once) appear in AS during testing procedure. To the second class C_2, we refer the intermittent faults which can rarely (not more than once) appear in AS during testing procedure. The introduced classification of intermittent faults is relative and depends considerably on the parameters of testing procedure. Such classification of intermittent faults is important for diagnosis based on limited time of testing procedure.

In [4], the diagnosis algorithm allowing to discriminate among permanent and intermittent faults was introduced. Essentiality of algorithm can be expressed by the following table (see Table 2).

In the table, the Chebyshev's inequality for the case when is denoted as $Chi\{m/2\}$.

When a fault-free unit tests a fault-free unit (suppose m times), the resulting tuple will contain the total number of "1" equal to zero (in the table, see the intersection of the first row and the first column). In case when a tested unit is permanently faulty, the resulting tuple will contain the total number of "1"

Table 2. Total number of "1" in the tuple

Total number "1" in the resulting, tuple, S	Tested unit of			
Testing unit	A	P	C_2	C_2
A	0	m	$1 \leq m$	$0 \cup 1$
P	$Chi\{m/2\}$	$Chi\{m/2\}$	$Chi\{m/2\}$	$Chi\{m/2\}$
C_1	$0 \leq s < m$	$0 \leq s < m$	$0 \leq s < m$	$0 \leq s < m$
C_2	$0 \cup 1$	$(m-1) \cup m$	$0 \leq s < m$	–

equal to m. When a tested unit possesses an intermittent fault of class C_1, the resulting tuple will contain, with "very high" probability, the total number of "1" which satisfies the expression $1 \leq s < m$. If a tested unit possesses an intermittent fault of class C_2, then the resulting tuple will contain, with "very high" probability, the total number of "1" equal to either zero or one (in Table, it is denoted as $0 \cup 1$).

Results produced by a permanently faulty unit are similar to tossing of fair coin. When a permanently faulty unit can produce result either 0 or 1 with equal probability, the total number of "1" in m tests (which are considered as Bernoulli trials) will follow the Chebyshev's inequality

$$p\{|s - mp| > \varepsilon\} < \frac{var(s)}{\varepsilon^2} \tag{6}$$

where var(s) is the variance of s, i.e. $var(s) = mp(1-p)$

Let's set the deviation of s from mp as three or more standard deviations, i.e.

$$\varepsilon = 3\sqrt{mp(1-p)} \tag{7}$$

Thus, Chebyshev's inequality will take the form:

$$p\left\{|s - mp| > 3\sqrt{mp(1-p)}\right\} \leq \frac{1}{9} \tag{8}$$

For example, for $m = 100$ we have

$$p\{|s - 50| < 15\} > \frac{8}{9} \tag{9}$$

In other words, with probability greater than or equal to 0.888 we can expect that the total number of "1" in 100 tests performed by a permanently faulty unit will be within the range $35 < s < 65$.

As distinct from permanently faulty units, a unit possessing an intermittent fault of class C_1 doesn't have such consistent pattern in producing tests results as a permanently faulty unit can produce. This fact allows us to discriminate between permanent and intermittent faults. In the given case, achieving correct diagnosis is not very important. As a rule, incorrect diagnosis results in that a permanently faulty unit is diagnosed as a unit with an intermittent fault whose behavior is very similar to that of a permanently faulty unit.

We assume that the probability that both testing and tested units possess an intermittent fault of class C_2 and both these intermittent faults are in AS during the same testing procedure is so small that it can be neglected. In the Table 2, this situation is depicted as "-".

5   Computer Simulation

In order to perform the computer simulation the web application was developed [21]. Application allows us the following:

- to set the testing procedure (i.e., set the number of tested units k, number of tests repetitions m and duration of single test t_{test});
- to set the states of all units (both testing and tested units). There are several options for each unit: fault-free; permanently faulty; intermittent fault of type C_1; intermittent fault of type C_2.
- to input parameters which will be used in diagnosis algorithm, such as probabilities of unit states, threshold for decision making and z-score used in Chebyshev's inequality.
- performing diagnosis algorithm;
- comparing the result produced by diagnosis algorithm with the actual unit state set by us.

The main characteristics of testing procedure can be set on the main web page (homepage) of the application. See Fig. 6.

At the first, the parameters k and m are set. After that, the application redirects a user to the page allowing setting the state of each unit (see Fig. 7). Different types of fault are denoted as follows:

- A means fault-free state;
- P means permanent fault;
- C_1 denotes an intermittent fault which has high intensity (rate of transition) λ (i.e., very often appears in active state);
- C_2 denotes an intermittent fault which has low intensity λ (i.e., very rarely appears in active state). Most of the time this fault is in the passive state and doesn't affect the test result.

The form for setting the total number of repetitions of the whole testing procedure (needed for processing of statistics) is located at the bottom of blue rectangle (see Fig. 7). In the figure, this number is denoted as "number of cycles". The application permits to set the following values: 100, 1000, 10000 and 100000.

States of intermittent fault are modeled by Stochastic Timed Petri Nets (STPN) [22]. In Fig. 8, the fragment of STPN is shown. This fragment models only one test which testing unit performs on tested unit. Both these units can have an intermittent fault. This fact is modeled by two states: active state (AS) and passive state (PS) of intermittent fault. When intermittent fault is in PS,

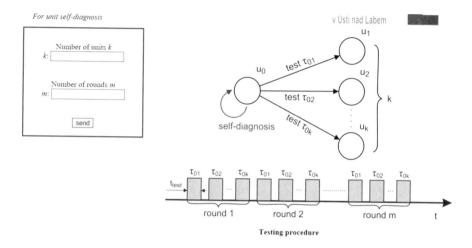

Fig. 6. Settings of the testing procedure.

the unit which has such fault behaves like a fault-free unit. If intermittent fault is in AS, then unit behaves like a permanently faulty unit. Transitions from AS to PS and vice versa are modeled by timed transition T_1 and T_2 for testing unit and T_3 and T_4 for tested unit. Test is modeled by four timed transitions: T_5, T_6, T_7 and T_8. Each of these timed transitions models the test for each particular combination of possible states of testing and tested units (e.g., timed transition T_6 models the test when testing unit has intermittent fault in AS and tested unit in PS).

Duration of test is modeled by firing time of transitions T_5, T_6, T_7 and T_8. Firing of one of these transitions removes tokens from the corresponding places and puts a token to the corresponding output place. For example, firing of transition T_6 removes tokens from the place denoted as AS (testing unit) and from the place denoted as PS (tested unit) and puts a token to output place denoted as "1 or 0".

When states of all units are set, user can change the default values used in the application (see Fig. 9).

On the left hand side of the Fig. 9 there is the form. It allows a user to assign the concrete values to the variables which are used in diagnosis algorithm.

For the unit performing self-diagnosis a user can set the following probabilities of it states:

P_a – the probability of fault-free state;
P_p – the probability of permanent fault;
PC_1 – the probability of intermittent fault of type C_1;
PC_2 – the probability of intermittent fault of type C_2;

The default value for test duration is set to 1 ms. It is assumed that for each particular unit this value should be known. User can also change the default values λ and μ (intensities transitions between AS and PS).

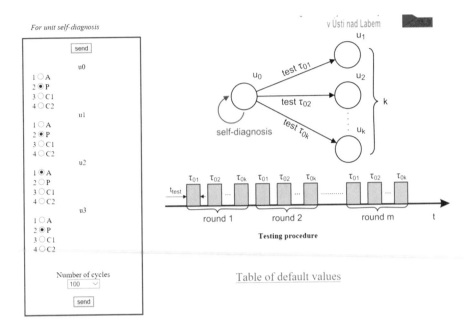

Fig. 7. Settings of the units states

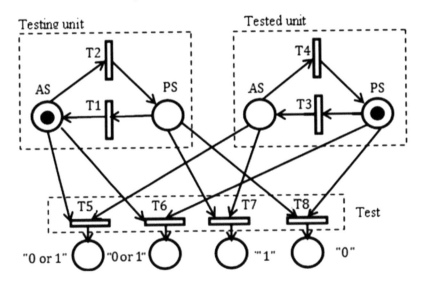

Fig. 8. Petri Net which models test performed by testing unit on tested unit

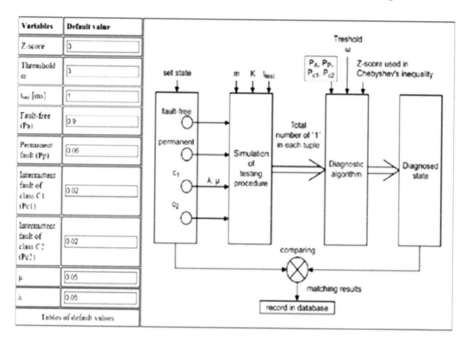

Fig. 9. Main parts of application

Threshold ω is used for making the decision about the unit's state. Decision about the state of the unit is made by using the following likelihood ratio:

$$\varkappa = \frac{P(H_1/R)}{P(H_2/R)} \qquad (10)$$

where R is a set of all test results (i.e., syndrome); H_1 is the hypothesis that unit is fault-free; H_2 is the hypothesis that unit is faulty;

It is also necessary to choose the threshold ω. The value of ω can be determined on the basis of Bayesian discriminant analysis, so as to minimize the average misjudgment cost [23]. Thus, if $\varkappa \geq \omega$, then hypothesis H_1 is accepted. Otherwise, additional measures should be undertaken to provide correct diagnosis.

To discriminate between permanent and intermittent fault the z-score is used. In the given case, z-score means standardized value indicating the number of standard deviations above or below the mean. It is expressed as

$$z = \frac{s - E[s]}{\sigma(s)}, \qquad (11)$$

where s is the total number of "1" in the resulting tuple (random value); $E[s]$ is the mean of s; $\sigma(s)$ is standard deviation. In the given case, the mean is equal to $m/2$.

In the form, user can input certain value for variable z (e.g., 3). This value will be used by the diagnosis algorithm. Particularly, this value is used for calculating the permissible range of random variable s when unit is permanently faulty (e.g., $40 < s < 60$). If the obtained value of s is outside of this range (e.g., $s = 26$), then the result of self-diagnosis concludes that unit has an intermittent fault.

6   Results and Discussion

We introduce two probabilities:

  P_1 – probability of correct diagnosis;

  P_2 – probability of fault detection (without discrimination between permanent and intermittent fault).

The whole diagnosis process is repeated N times, where value of N (number of cycles) is set on the second page of web application. The first column of the table contains the total number of correct diagnosis, Nc. For example, $N = 100$ and $Nc = 98$. Thus, $P_1 = 0.98$. Statistical data can be saved in a database.

Computer simulation of the proposed unit self-diagnosis showed the correctness of the diagnosis algorithm. Diagnosis results are evaluated by the values of probability P_1 and probability P_2. In several cases, the value of probability P_1 may be less than P_2. It can be explained by several reasons: not accurate input data, limited number of cycles N, incorrectly chosen threshold ω and z-score.

It should be noted that the proposed unit self-diagnosis allows increasing the probability of correct diagnosis P_1 if statistical characteristics of possible intermittent faults are known and taken into account.

7   Conclusions

The paper presents a novel approach to system-level fault diagnosis. Traditionally, system level self-diagnosis uses the results of tests performed on a unit for its diagnosis. Unlike this traditional approach to unit diagnosis, we proposed to use for it the results of tests which the unit performs on other units.

The proposed by us unit self-diagnosis has several advantages as compared to unit diagnosis based on mutual tests. Mutual tests are used self-diagnosis at system level. One of the main problems in the theory of system level self-diagnosis is the characterization problem. This problem consists in finding necessary and sufficient conditions for a system testing assignment which should be satisfied to achieve a given level of diagnosability given a fault model and an allowable family of fault sets. This problem is important when the predefined testing assignment will be used. It may be very difficult to implement the determined testing assignment. For example, it will be difficult to implement the requirement that each system unit must be tested by at least t distinct other units. In case when tests are performed randomly, the diagnosability problem becomes very important. There can be obtained arbitrary structure of tests among system units. For the obtained structure it is necessary to determine the family of fault sets that can be diagnosed. Solving this problem for the systems with large number of units may be very difficult. The proposed unit self-diagnosis doesn't tackle these problems.

System level self-diagnosis is considered only for homogeneous systems. System units should have similar connection facilities in order to perform the tests on the assigned units. Each system unit must be able to perform tests on other units. These conditions restrict its applicability. Proposed unit self-diagnosis can cope with these problems.

It is very difficult to diagnose intermittently faulty units based on the results of mutual tests. There are not many researches in which this issue was considered. In our research, we devoted much attention to this issue and suggested the method of how to discriminate between permanent and intermittent faults.

System level self-diagnosis deals with diagnosis nucleus. Diagnosis nucleus is a system unit or subset of units which provides the following tasks: processing of the syndrome; making decision about the system state; delivering information about the system state to the system environment. In the proposed unit self-diagnosis, there is no task to determine the diagnosis nucleus. In the given case, unit that provides self-diagnosis is at the same time diagnosis nucleus.

The proposed diagnosis can deal with different types of test. Test may present complete assessment of a tested unit. In this case, the result of such test indicates if tested unit is fault-free or faulty. The other extreme is the "empty" test. Such test implies that tested unit receives some data and sends it back without any processing.

The main drawback of the proposed unit self-diagnosis is that it is time consuming, which may restrict its applicability. When time is crucial factor, the suggested diagnosis could be used as a supplementary facility for the traditional system-level diagnosis, and could be performed in background mode (i.e., as a daemon).

Carried out computer simulation testified that the proposed unit self-diagnosis allows to achieve high credibility of diagnosis results.

Finally, the proposed unit self-diagnosis is intended to be applicable to complex systems such as, for example sensor networks, multi-robot systems, many-core processors, multi-agent systems [24, 25] and possibly, in other type of systems.

References

1. Preparata T, Metze G, Chien R (1967) On the connection assignment problem of diagnosable system. IEEE Trans Electron Comput EC–16(12):848–854
2. Mashkov V, Barabash O (1998) Self-checking and self-diagnosis of module systems on the principle of walking diagnostic kernel. Eng Simul 15:43–51
3. Mashkov V (2011) New approach to system level self-diagnosis. In: Proceedings of IEEE 11th international conference on computer and information technology, CIT 2011, Cyprus, pp 579–584
4. Mashkov V, Lytvynenko V (2019) Method for unit self-diagnosis at system level. Int J Intell Syst Appl (IJISA) 11(1):1–12
5. Chen J, Kher S, Somani A (2006) Distributed fault detection of wireless sensor network. In: Proceedings of the international conference on mobile computing and networking, New York, USA, pp 65–72
6. Jiang P (2009) A new method for fault detection in wireless sensor neworks. In: Proceeding, Hangzhou Dianzi Unversity, ISSN 1424-8220

7. Jangale S, Hadsul D (2013) Detection of faulty sensor nodes in wireless sensor network. Comput Technol Appl 4(1):150–154
8. Lee MH, Choi YH (2008) Fault detection on wireless sensor networks. Comput Commun 31. https://doi.org/10.1016/j.comcom.2008.06.014
9. Chessa S, Santi P (2001) Comparison-based system-level fault diagnosis in ad hoc network. In: 20th symposium on reliable distributed systems, pp 257–266
10. Albini L, Duarte J, Ziwich R (2005) A generalized model for distributed comparison-based system-level diagnosis. J Brazil Comput Soc 10(3):44–56
11. Collet J, Zajac P, Psarakis M, Gizopoulos D (2011) Chip self-organization and fault-tolerance in massively defective multicore arrays. IEEE Trans Dependable Secure Comput 8(2):207–217
12. Xu J (1991) The t/(n-1) diagnosability and its application to fault tolerance. Technical report, series No. 340, University of Newcastle upon Tyne
13. Mashkov V, Pokorny J (2007) Scheme for comparing results of diverse software versions. In: Proceedings of ICSOFT Conference, Barcelona, Spain, pp 341–344
14. Ding M, Chen D, Xing K, Cheng X (2005) Localized fault-tolerant event boundary detection in sensor networks. In: IEEE Infocom, pp 902–913
15. Elhadef M, Boukerche A, Elkadiki H (2006) Performance analysis of a distributed comparison-based self-diagnosis protocol for wireless ad hoc networks. In: Proceedings of the 9th ACM international symposium on modeling analysis and simulation of wireless and mobile systems, pp 165–172
16. Khilar PM (2010) Performance analysis of distributed intermittent fault diagnosis in wireless networks using clustering. In: Proceedings of 5th international conference on industrial and information systems, ICIIS, pp 13–18
17. Krishnamachari B, Iyengar S (2004) Distributed Bayesian algorithms for fault-tolerant event region detection in wireless sensor networks. IEEE Trans Comput 53(3):241–250
18. Luo X, Dong M, Huang Y (2006) On distributed fault-tolerant detection in wireless sensor networks. IEEE Trans Comput 55(1):58–70
19. Blount ML (1977) Probabilistic treatment of diagnosis in digital systems. In: 7th IEEE international symposium on fault-tolerant computing, pp 72–77
20. Mallela S, Masson G (1978) Diagnosable systems for intermittent faults. IEEE Trans Comput C–27(6):560–566
21. PNsimulator. http://vtan.ujep.cz/PNsimulator
22. Ciardo G, Muppala J, Trivedi K (1989) SPNP: Stochastic Petri Net Package. In: Proceedings of 3rd international workshop on Petri Nets and performance models, Japan, pp 142–150
23. Wang Z, Zhang J, Zhang Y (2012) Bayes-based fault discrimination in wide area backup protection. Adv Electr Comput Eng 12(1):91–96. https://doi.org/10.4316/AECE.2012.01015
24. Mashkov V (2005) Task allocation among agents of restricted alliance. In: Proceedings of IASTED ISC 2005 conference, Cambridge, MA, USA, pp 13–18
25. Mashkov V (2004) Restricted alliance and coalitions formation. In: Proceedings of IEEE WICACM international conference on intelligent agent technology, Beijing, China, pp 329–332

Digital Acoustic Signal Processing Methods for Diagnosing Electromechanical Systems

Oksana Polyvoda$^{(\boxtimes)}$ (ID), Hanna Rudakova (ID), Inna Kondratieva (ID),
Yuriy Rozov (ID), and Yurii Lebedenko (ID)

Kherson National Technical University, Kherson, Ukraine
pov81@ukr.net

Abstract. An effective means of preventing accidents, identifying critical operating modes, and diagnosing equipment failures of electromechanical systems are the methods of functional diagnostics. A number of problems of diagnostics of electromechanical complexes at the present time can be realized by acoustic methods, by analyzing the signals received from working assemblies. The actual scientific task is the formation of a procedure for analyzing acoustic signals generated by working equipment of electromechanical complexes, based on the use of modern methods of digital time series processing in real time. The article analyzes the acoustic signals obtained as a result of an experiment when operating electromechanical equipment. At the first stage of processing, the signals are passed through a low-pass filter and a band-pass filter. The spectra of the amplitudes of the signals before and after filtering, as well as the dynamics of signals in the phase space, are studied. For signals before and after processing, the autoregressive moving average models were calculated and their standard deviations were analyzed. The application of the procedure for analyzing acoustic signals and standard methods for their digital processing will allow real-time decision-making support systems to be implemented with automatic detection (diagnosis at the rate of measurement of diagnostic signals) of machinery equipment malfunctions, their degree of danger and the formation of a list of compensating measures.

Keywords: Functional diagnostics · Acoustic signal · Filtering · Autoregressive model

1 Introduction

Modern electromechanical industrial complexes, as a rule, consist of a large number of interacting elements. The relative displacements of these elements generate vibrations that can critically affect the operation of precision mechatronic systems. This can lead to limiting operating conditions of the equipment, and in some cases even to its failure [1].

© Springer Nature Switzerland AG 2020
V. Lytvynenko et al. (Eds.): ISDMCI 2019, AISC 1020, pp. 97–109, 2020.
https://doi.org/10.1007/978-3-030-26474-1_7

An effective means of preventing accidents, identifying critical modes of operation, diagnosing faults in equipment of electromechanical systems (EMS) are the methods of functional diagnostics [2]. Recently, there has been a particular interest in the creation of methods and techniques for diagnosing the technical state of electromechanical systems based on the study of vibrational and acoustic processes in them.

2   Review of the Literature

The essence of the problem of functional diagnostics is the development and practical implementation of algorithms for estimating the parameters of the technical states of electromechanical assemblies, without disassembling them according to the characteristics of the vibration processes accompanying their functioning. Modern computing technology allows us to improve the technology of checking the parameters of EMS by automating the measurement process and the use of diagnostic software (Fig. 1). Measurement and analysis of signals in the systems of vibroacoustic diagnostics of electromechanical systems are most often performed using equipment adapted for work in industrial conditions. At the same time, these operations can also be carried out with the help of a computer, at the input of which devices are installed that supply measuring transducers, amplifying electrical signals and converting signals into digital form.

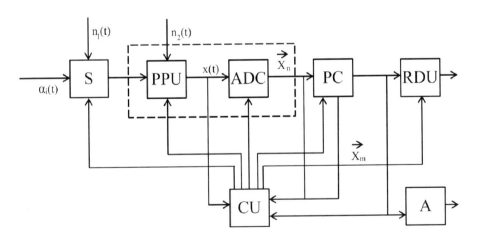

Fig. 1. Block diagram of the technical diagnostics system.

In Fig. 1 the following notation is used: S - Sensor; PPU - preprocessing unit; ADC - analog-to-digital converter; PC - computer; CU - control unit of parameters and algorithms; RDU - information registration and display unit; A is the actuator, X_n is the working sample, X_m is the training sample (n, m are the sample volumes).

In the process of technical diagnostics information about the behavior of the object of control is recorded by the sensor (S). PPU and ADC receivers realize the functions of preliminary analog processing of a mixture of signals and interference, ensuring the amplification of weak signals against the background of self noise; band-pass, low-frequency and notch filtering of external and internal interference; normalization of output processes in intensity. The converted signals arrive at the computer, where they undergo statistical, mathematical processing. The CU implements the control functions of the signal handling process in the receivers and the processor, and also sets the parameters for the actuator (A).

Currently, the main methods for processing the measurement results are the Fourier and Laplace transforms, classical methods for analyzing time series, multiple-scale, wavelet analysis, etc. [3–6]. The disadvantage of the approaches used is the computational complexity in the implementation of discrete transformations and the need for large amounts of memory to store reference values.

Methods of functional diagnostics in real time require a large number of calculations, mathematical modeling of the object, processing a large amount of information, which leads to the mandatory use of computing equipment. Different methods of processing acoustic signals allow you to select a wide range of characteristics, both static and dynamic. For digital processing of acoustic signals from EMC, a number of methods are used [7–12]:

- methods of linear filtering, which allow selection of signals in the required frequency range;
- spectral analysis, by which various types of signals are processed;
- time-frequency analysis, which is used to find deviations in sensitive receivers of signals;
- adaptive filtering, allowing recognition of sound samples based on previously identified patterns, as well as muffle noises;
- non-linear processing for calculating correlations;
- high-speed processing - interpolation (increase) and decimation (decrease) of the sampling rate.

The actual scientific task is the development of efficient algorithms for functional diagnostics implemented on a computer, which fully take into account technical and economic requirements, constraints in optimization, the stochastic nature of external influences, etc. When building functional diagnostics algorithms, it is necessary to take into account the specifics of the problem being solved, to use modern techniques and optimization methods, to apply the accumulated experience in related areas.

3   Problem Statement

Typical noise signals of electromechanical equipment have periodic and non-periodic components. The parameters of the noise signals change in time - in defect-free devices it is slow, and in equipment that approaches the state of destruction, very quickly. It is assumed that, within the device's entire service life, its characteristic noise signal is an interval-stationary process, provided that the observation intervals are selected for each type of device, and the signals are considered to be realizations of a random process with a normal distribution.

The purpose of diagnosis is to identify the development of failure before the area of most intense wear or destruction. Therefore, it is necessary to have a time dependence of the work of defective and defect-free objects of control, and to ensure reliable results, apply statistical modeling methods. In accordance with the probabilistic approach, all deviations from the norm are considered as random variables, and the main requirement is the minimum permissible probability of failure.

A number of problems of diagnostics of electromechanical complexes can now be realized by acoustic methods, by analyzing signals received from working assemblies in real time.

Noises from equipment and machines also characterize both the general properties of systems and the properties of their parts. The experience of using acoustic methods shows that, in a state of normal functioning, the noise energy is mainly concentrated in the low-frequency region, and the energy corresponding to defects is located at higher frequencies. This circumstance is used for the timely detection of nascent violations. The disadvantage of most acoustic methods is the need to have in the memory of the monitoring system a set of signal realizations or statistical characteristics of the signals of all functioning states [2].

The aim of the research is to formulate a procedure for analyzing acoustic signals generated by working equipment of electromechanical complexes based on the use of modern methods for digital processing of time series in real time.

4   Materials and Methods

As a rule, stationary random processes and, therefore, stationary changes in the parameters of time series correspond to the normal state or mode of functioning of the objects under control. Objects whose informative parameters depend on the conditions of their operation are non-stationary in nature and are described by the nonstationarity function [1].

In case of violation of the operating modes of the monitored equipment, the type of non-stationarity function changes. For the analysis of non-stationary discrete signals, an autoregressive moving average model is often chosen as a model within a moving window [3,4] as expression

$$y\,[k] = A_0 y\,[k-1] + A_1 y\,[k-2] + A_2 y\,[k-3] + A_3 y\,[k-4]\,, \tag{1}$$

where coefficients are determined by the least squares method.

Building models of the form (1) for acoustic signals recorded during the operation of electromechanical equipment with different rotational frequencies and degrees of loading showed the presence of a significant dependence of the model coefficients on changes in the operating conditions [4]. However, the mean square recovery error values of the model were in the range from 3.2% to 25.2%, which indicates a low quality of a models.

The main energy of the acoustic signal recorded during the operation of electromechanical equipment is concentrated in the medium frequency range of 200–7000 Hz. For the analysis of the signal in engineering practice, two channels are used: the first channel with a frequency range of 200–500 Hz, the second - 1000–2500 Hz. The first channel is characterized by the absence of fictitious high-energy components. In the second range, the energy of the useful signals significantly (approximately 2 times) exceeds the energy of the noise components of the signal components. In fact, the frequency bands of the channels were chosen so that the "noise" components (signals that do not carry useful information) were as small as possible.

To identify the frequency range traditionally used filtering procedures. The best characteristics among low-pass filters (LPF) are Butterworth filters [5] with a transfer function of the form:

$$H_B(p) = \frac{1}{B_n(p)}, \qquad (2)$$

where $B_n(p)$ is the n-th order Butterworth polynomial. Lower order polynomials of Butterworth are:

$$B_1(p) = p + 1, \qquad (3)$$

$$B_2(p) = p^2 + 1.414136p + 1, \qquad (4)$$

$$B_3(p) = p^3 + 2p^2 + 2p + 1, \qquad (5)$$

$$B_4(p) = p^4 + 2.6131259p^3 + 3.41421362p^2 + 2.6131259p + 1, \qquad (6)$$

$$B_5(p) = p^5 + 3.236068p^4 + 5.236068p^3 + 5.236068p^2 + 3.236068p + 1. \qquad (7)$$

An increase in the filtering range ω_c of the normalized low-pass filter is achieved by substituting into $H(p)$

$$p \to \frac{p}{\omega_c}. \qquad (8)$$

The bandpass filter can be obtained on the basis of a low-pass filter using standard transforms [9]. To convert a normalized low-pass filter to a bandpass with a medium frequency ω_0 rad/s and bandwidth $\Delta\omega$ rad/s, it is necessary to make a replacement in the transfer function $H(p)$ of the normalized filter:

$$p \to \frac{p^2 + \omega_0{}^2}{p\Delta\omega}. \qquad (9)$$

The average frequency of the passband is defined as the geometric average of the frequencies of the lower ω_1 and upper ω_2 bandwidths of the bandpass filter, that is:

$$\omega_0 = \sqrt{\omega_1\omega_2}. \qquad (10)$$

The bandwidth is defined as:

$$\Delta\omega = \omega_2 - \omega_1. \qquad (11)$$

5   Experiment, Results and Discussion

Figure 2 shows the acoustic signal obtained as a result of the experiment during the work of the electromechanical equipment. A number of characteristic zones can be distinguished on the recorded signal:

- zone 1 is the initial section corresponding to the acceleration of the engine to the nominal speed;
- zone 2 is the section corresponding to the normal operating mode of the engine;
- zone 3 is the section with additional perturbations characterizing the presence of disturbances in the engine operation (3a - occurrence of failures in work, 3b - established violations).

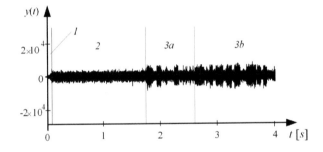

Fig. 2. Acoustic signal obtained as a result of an experiment.

The results of the experiments are saved with the help of sound recording equipment in the format *.wav, with a sampling frequency of 44 kHz and 16-bit depth. For the subsequent processing of the received time signal, to analyze the effect of changing the load on the characteristics of the acoustic signal, a fragment from zone 2 is cut out. If it is necessary to identify the faulty operation modes and diagnose the cause of the emergency state, then a fragment of 3 zones (fragments 3a or 3b) must be selected for processing. Figure 3 shows the signal fragments selected from zone 2 (see Fig. 3a) and zone 3 (see Fig. 3b).

At the first stage of processing, the signals were passed through a low frequency filter with a cut-off frequency of 100 Hz and a band-pass filter with a frequency range of 1500–2500 Hz. Transfer functions of 5th order filters

- low frequencies filter

$$H_1\left(p\right) = \frac{1}{p^5 + 3.236068\,p^4 + 5.236068\,p^3 + 5.236068\,p^2 + 3.236068\,p + 1}.$$
$$(12)$$

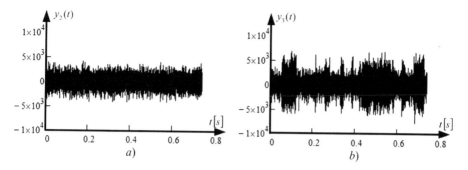

Fig. 3. Fragments of an acoustic signal: (a) a signal selected from zone 2, (b) a signal selected from zone 3.

– bandpass filter

$$H_2\,(p) = 1.52 \cdot 10^{13} p^5 \big/ \big(0.002 p^{10} + 9.71 p^9 + 4.86 \cdot 10^4 p^8 + 1.32 \cdot 10^8 p^7 +$$
$$+\, 3.3 \cdot 10^{11} p^6 + 5.6 \cdot 10^{14} p^5 + 8.4 \cdot 10^{17} p^4 + 8.3 \cdot 10^{20} p^3 +$$
$$+\, 7.6 \cdot 10^{23} p^2 + 3.8 \cdot 10^{26} p + 1.9 \cdot 10^{29}\big)\,.$$

$$(13)$$

Original signals and the signals obtained after filtering are shown in Fig. 4: (a), (b) signals selected from the zones 2 and 3, respectively, and the signals obtained after filtering with LPF, (c), (d) the selected signals from the zones 2 and 3, respectively, and signals obtained after filtering with band-pass filter. The spectra of the amplitudes of the signals before and after processing are shown in Fig. 5. Figure 6 shows the dynamics of signals in the phase space.

For signals before and after processing, autoregressive moving average models were calculated, which have the form:

$$y_2\,[k] = 2.04 y\,[k-1] - 2.05 y\,[k-2] + 1.66 y\,[k-3] - 1.09 y\,[k-4] + 0.36 y\,[k-5]\,,$$

$$y_{2l}\,[k] = 4.85 y\,[k-1] - 9.49 y\,[k-2] + 9.39 y\,[k-3] - 4.69 y\,[k-4] + 0.95 y\,[k-5]\,,$$

$$y_{2b}\,[k] = 4.61 y\,[k-1] - 8.78 y\,[k-2] + 8.63 y\,[k-3] - 4.38 y\,[k-4] + 0.92 y\,[k-5]\,,$$

$$y_3\,[k] = 1.42 y\,[k-1] - 1.61 y\,[k-2] + 1.49 y\,[k-3] - 0.71 y\,[k-4] + 0.22 y\,[k-5]\,,$$

$$y_{3l}\,[k] = 4.74 y\,[k-1] - 9.07 y\,[k-2] + 8.77 y\,[k-3] - 4.28 y\,[k-4] + 0.84 y\,[k-5]\,,$$

$$y_{3b}\,[k] = 4.6 y\,[k-1] - 8.74 y\,[k-2] + 8.57 y\,[k-3] - 4.33 y\,[k-4] + 0.9 y\,[k-5]\,.$$

Figure 7 shows a fragment of an acoustic signal from zone 2 and zone 3 (denoted by a solid line), and its model (shown by dots).

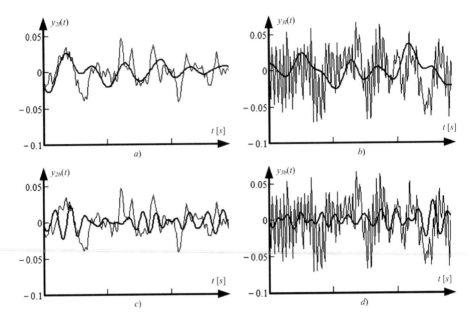

Fig. 4. Original signals and signals obtained after filtering.

To assess the performance of the obtained model can use the standard error of the mathematical models of signal (σ), calculated according to the formula

$$\sigma = \sqrt{\frac{\sum_{i=1}^{N} \left(y_i - y_k\right)^2}{\sum_{i=1}^{N} y_i^2}}.$$

The calculation results of the error recovery mathematical models of signal from zone 2 and from zone 3 are given in Table 1.

Table 1. Calculation results of the standard error of the mathematical models of signal

Zone/Filter type	Before filtering	After the low pass filter	After bandpass filter
For signal from zone 2	23.7%	0.01634%	0.08563%
For signal from zone 3	45.1%	0.03056%	0.09039%

A number of problems of diagnostics of electromechanical complexes can now be realized by acoustic methods, by analyzing the signals received from working assemblies in real time. Acoustic signals underwent a preprocessing procedure using the transfer functions found using expressions (12) and (13). As a result, an array of information data was obtained for further study of the properties

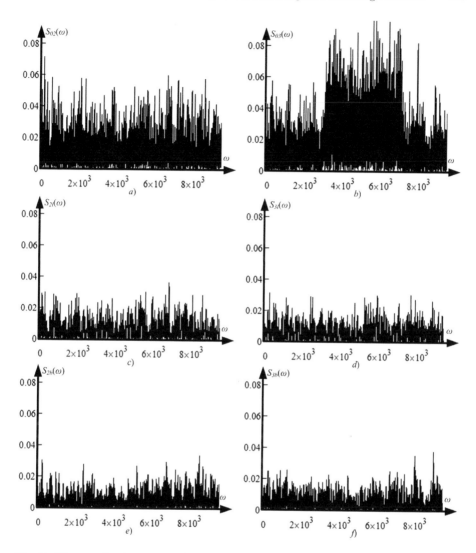

Fig. 5. The amplitude spectra of the signals before and after processing: (a), (b) the amplitude spectra of the signals selected from zones 2 and 3, respectively; (c), (d) spectra of amplitudes of signals selected from zones 2 and 3, respectively, after low-pass filtering; (e), (f) spectra of amplitudes of signals selected from zones 2 and 3, respectively, after filtering with a band-pass filter.

of signals. The separation of the acoustic signal into frequency ranges increases the adequacy of the models. The auto-regression model of the moving average adequately describes the time sequence corresponding to the acoustic signals generated by the operating equipment.

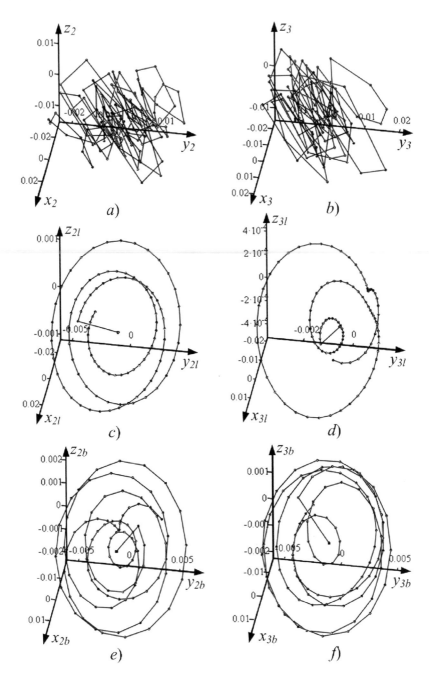

Fig. 6. The dynamics of signals in the phase space: (a), (b) selected from zones 2 and 3, respectively, before filtering; (c), (d) selected from zones 2 and 3, respectively, after low-frequency filtering; (e), (f) selected from zones 2 and 3, respectively, after filtering with a band-pass filter.

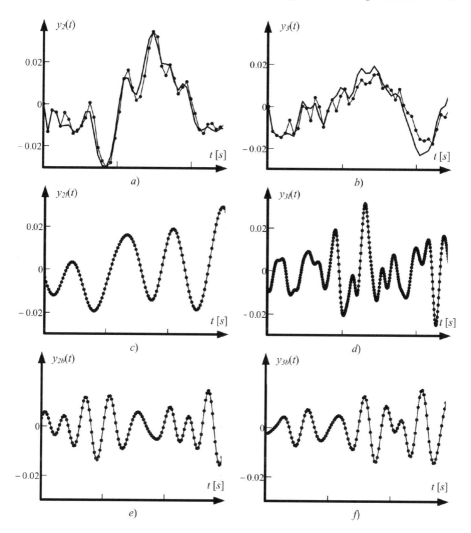

Fig. 7. Fragment of an acoustic signal from zone 2 and zone 3, and its model: (a), (b) from zones 2 and 3, respectively, before filtering; (c), (d) from zones 2 and 3, respectively, after low-pass filtering; (e), (f) from zones 2 and 3, respectively, after filtering with a band-pass filter.

It is enough to use the model up to 5 orders. The characteristics of the acoustic signal (model parameters) really depend on the level of loading, which makes it possible to estimate not the nominal and pre-emergency operating modes of the engine.

6   Conclusion

Methods of digital processing of acoustic signals of electromechanical systems are a reliable diagnostic method. After the formation of two channels from the original signal, in which all useful information is concentrated, it is possible to use standard methods of digital signal processing to extract informational signals.

This will allow real-time decision-making support systems to be implemented with automatic detection (diagnosis at the rate of measurement of diagnostic signals) of malfunctions of machinery components, the degree of their danger and the formation of a list of compensating measures.

To monitor the operating modes of the drive in mobile objects, you can use acoustic emission signals generated by operating electromechanical equipment. The application of the method of digital processing of registered acoustic signals opens the possibility of both was controlling the degree of loading of the electric drive and finding the causes of malfunctions in the operation of the equipment.

The use of the recursive least squares method for adaptive model identification makes it possible to respond to monitoring subsystems for monitoring and diagnostics of the drive in real time, which is essential for timely detection of overloads and prevention of extreme operating conditions. All of the above will help ensure the reliability and safety of mobile objects.

References

1. Kuznetsov YuM, Dmitriev DO, Dinevich GYu (2009) Compounding of machine tools with mechanisms of parallel structure. PP Vyshemirsky V.S., Kherson (in Ukrainian)
2. Balitsky FYa, Ivanova MA, Sokolov AG, Khomyakov EI (1984) Vibroacoustic diagnostics of incipient defects. Science, Moscow (in Russian)
3. Malaychuk VP, Mozgovoy AV (2005) Mathematical flaw detection. System technologies, Dnepropetrovsk (in Russian)
4. Kondratieva IU, Rudakova HV, Polyvoda OV (2018) Using acoustic methods for monitoring the operating modes of the electric drive in mobile objects. In: 2018 IEEE 5th international conference, methods and systems of navigation and motion control (MSNMC), pp 218–221
5. Voloshchuk YuI (2005) Signals and processes in radio engineering. LLC "Company SMIT", Kharkov (in Ukrainian)
6. Malaychuk VP, Petrenko OM, Rozhkovsky VF (2010) Processing of measurements and signals of non-destructive control. DNU, Dnipropetrovsk (in Ukrainian)
7. Oppenheim AV, Schafer RW, Buck JR (1999) Discrete-time signal processing, 2nd edn. Prentice Hall, Upper Saddle River
8. Mikhalev AI, Vinokurova EA, Sotnik SL (2014) Computer methods of intellectual data processing. System technologies, Dnipropetrovsk (in Russian)
9. Babichev S, Korobchynskyi M, Lahodynskyi O, Korchomnyi O, Basanets V, Borynskyi V (2018) Development of a technique for the reconstruction and validation of gene network models based on gene expression profiles. Eastern-Eur J Enterp Technol 1(4–91):19–32
10. Sergienko AB (2002) Digital signal processing. Peter, St. Petersburg (in Russian)

11. Solonina AI, Ulakhovich DA, Arbuzov SM, Solov'eva YeB (2005) Fundamentals of digital signal processing. BHV-Petersburg, St. Petersburg (in Russian)
12. Bezvesilna AN, Larin VYu, Chichikalo NI, Fedorov EE, Dobrzhanskyy OO (2011) Transforming devices of the equipment. Technological measurements and devices, ZhDTU, Zhytomyr (in Ukrainian)

Construction of the Job Duration Distribution in Network Models for a Set of Fuzzy Expert Estimates

Yuri Samokhvalov[✉] [ID]

Taras Shevchenko National University of Kyiv, Kyiv, Ukraine
yu1953@ukr.net

Abstract. This article presents the mechanisms for processing fuzzy expert estimates in network models for project management. The distribution of the probability function of the execution time is proposed. This function allows us to build a β-distribution of a random variable over the entire domain of its definition for a combination of approximate points and interval expert estimates. This article describes also the approximation of linguistic estimates by using fuzzy triangular and trapezoidal numbers. It is based on the construction of the membership function of a fuzzy triangular number, accounting for its scale. The proposed approach makes it possible to obtain more accurate prediction estimates for making informed decisions during informational uncertainty.

Keywords: Project management · Duration of the work ·
β-distribution · Expert estimates · Fuzzy estimates · Fuzzy numbers ·
Membership function

1 Introduction

The project approach has recently become a new trend in both the business and government structures. The use of project management technology allows organizations to reduce project duration time and total cost. One of the main tasks in project management is estimation of the time to complete the work. The PERT (Program Evaluation and Review Technique) method is widely used to solve this problem. This method is based on the assumption of the beta distribution of a random variable that describes the duration of operation.

In [1,2] approaches are considered to select the β-distribution for each duration in the design model. In this case, three expert estimates of the duration of work are used: optimistic, pessimistic, and most probable time. However, special difficulties are normally caused by the estimations for projects where each operation is unique (for example, scientific, innovative, organizational, and other projects). In such situations there is no complete and accurate information about the duration of each operation – that is called informational uncertainty. Therefore, it is easier for a person to estimate the time to perform the work by using fuzzy categories.

© Springer Nature Switzerland AG 2020
V. Lytvynenko et al. (Eds.): ISDMCI 2019, AISC 1020, pp. 110–121, 2020.
https://doi.org/10.1007/978-3-030-26474-1_8

The generalization of the PERT method to use fuzzy estimates for evaluating the duration of operation was considered in many researches [3–10]. These papers proposed various project management techniques in which work duration is expressed by trapezoidal fuzzy numbers. There are also algorithms which were provided for constructing a critical path based on the trapezoidal fuzzy intervals of events occurrences.

The use of fuzzy estimates for the duration of operation reduces the psychological uncertainty for a person while assessing the time of occurrence of an event. However, approximation of the most probable time, even by fuzzy linguistic constructions, is difficult. In this regard, numerous probabilistic network models were created [11,12]. One of the simplest and most successful is the probabilistic model Golenko [13]. This model involves the assignment of only two estimates—an optimistic and pessimistic estimate of the duration of work.

In [14] the mechanism based on this model was proposed for constructing a continuous distribution of the duration of work over a set of interval estimates. This article is a further development of this research. It discovers an approach to build a continuous distribution of the duration of work for the set of the fuzzy estimates.

2   Model Distribution of the Duration of Work with Two Estimates

For better understanding of the essence of the proposed approach we consider it in the context of article [14]. This paper proposes mechanisms for estimating the most probable time to solve problems using the Golenko model and interval expert estimates of the duration of work.

The Golenko model is based on β-distribution in the interval $[a, b]$ with density:

$$f(t|a, b) = \frac{12}{(b-a)^4} \cdot (t-a)(b-t)^2$$

and distribution function with parameters $\alpha = 2$ and $\beta = 3$:

$$F(t) = \frac{1}{B(2,3)} \int_0^x y \cdot (1-y)^2 dy,$$

where $B(2,3) = \int_0^1 y \cdot (1-y)^2 dy$ is a beta function, $x = \dfrac{t-a}{b-a}$ is a scaling variable $(0 \leq x \leq 1)$. Since $B(2,3) = \dfrac{1}{12}$, finally have: $F(t) = 12 \int_0^x y \cdot (1-y)^2 dy$.

As can be noted, this model involves only two estimations: optimistic and pessimistic. Optimistic (minimum) estimation t_{min} estimates the duration of work in extremely favorable conditions. In this case, it is considered that the probability of performing work in a time less than t_{min} will be less than 0.01. The pessimistic (maximum) estimation t_{max} is the estimation of the duration of work in the most adverse conditions. For such estimation it is considered that the probability of performing work in a time bigger than t_{max} will be less than

0.01 [15]. Therefore, in terms of practice, the interval $[t_{min}, t_{max}]$ is well suited for β-distribution which is given as the points t_{min} and t_{max}, the beta function in this case takes values of 0 and 1 respectively.

Let $[t', t'']$ is an estimated duration of work s. Then according to [14] the probability $p(t|t', t'')$ that the work s will be completed by the time $t' \leq t \leq t''$ can be calculated by the formula:

$$p(t|t', t'') = \begin{cases} 0, & t < t' \\ F(t), & t' \leq t \leq t'' \\ 1, & t > t'' \end{cases} \tag{1}$$

The considered β-distribution is simple and convenient for predicting likely execution time of work in the exact interval. We will use this model as a basis for building the distribution of the duration of work and under conditions of fuzzy time estimates.

3 Approximation of Linguistic Estimates of Fuzzy Numbers

Let's consider the cases when the duration of the work is given by the approximate point or the interval estimate. Such estimates are formulated by statements like "duration approximately equal T" and "duration is located approximately in interval from T_1 to T_2" and approximated by triangular and trapezoidal fuzzy numbers respectively.

3.1 Approximation of Approximate Point Estimates

Let $FN = (T, \alpha, \beta)$ is the triangular fuzzy number, where T is a mode of this number, α and β are the fuzziness coefficients of these numbers. These coefficients are the values of the transition points of the membership function (MF) of a fuzzy number FN. These values will be considered as the estimates of t_{min} and t_{max} of execution time.

To find the interval $[\alpha, \beta]$ the membership function of "numbers approximately equal to the number T" should be built. Moreover, such a function should take into account the scale of number T.

There is a large number of types of membership functions of fuzzy numbers, that have their own pros and cons [16]. The biggest interest for ensuring the flexibility of models is the MF in the form of a generalized Gaussian number [17]. Taking this into account we use as the MF the Gauss function [18], which constructively takes into account the scale of the exact values of numbers. This function has the form:

$$\mu_T(x) = exp(-a(x - T)^2), \tag{2}$$

where a depends on the required degree of fuzziness $\mu_T(x)$ and it is determined from the expression:

$$a = -\frac{4ln0.5}{b^2},$$

where b is a distance between the transition points of the function $\mu_T(x)$, that is the points at which function (2) takes the value 0.5.

Thus, the task of constructing a function $\mu_T(x)$ is scoped to finding the parameter b. To determine the parameter b we use the following algorithm [18]. The purpose of these studies is finding out the following: how the experts represent the boundaries of the classes of "numbers approximately equal to the number T". The respondents were offered to come up with such numbers $k_1(T)$ and $k_2(T)$ what, in their opinion, separate the numbers approximately equal to the given T from the numbers that are not existed. Numbers $k_1(T)$ and $k_2(T)$ can be viewed as the function transition points $\mu_T(x)$. The obtained processed results are shown in Table 1.

Table 1. The distance between the transition points

Number x	The distance between the transition points $b(x)$
$1, 2, 3, 4, 6, 7, 8, 9$	$0.46x$
$10, 20, 30, 40, 60, 70, 80, 90$	$(0.357 - 0.00163)x$
$35, 45, 55, 65, 75, 85, 95$	$(0.213 - 0.00067x)x$
5	2.8
15	6.48
25	6.75
50	24
Other two-digit numbers	$\frac{1}{2}(b([\frac{x}{10}] \cdot 10 + 5) + b(x - [\frac{x}{10}] \cdot 10))$

Let's consider the positive integer number T. Let his least significant digit is the order q. We divide the possible values of q into classes of residues modulo 3 and enter the variable d, the values of which will be representatives of these classes $0, 1, 2$. As a result, we obtain equivalence classes: $M_d : \{d = 0, 1, 2\}, d = q \bmod(3)$.

Let x is an integer variable. If $x \in [1, 99]$, then the values in $b(x)$ depending on x and are located according to Table 1, in which [...] is an integer part of the number.

The value of $b(x)$ depends also on class, in which the M_d number T belongs to one. Let us to denote by r_q a digit which standing in q digit of the number T. Then:

1. By $T \in M_0$ (for example, 300, 300000 and etc.) $b(T)$ depends only on the least significant digit of the number T, i.e. from r_q : $x = r_q \cdot 10$; $b(T) = b(x) \cdot 10^{q-2}$, takes $b(x)$ from Table 1.
2. By $T \in M_1$ (for example, 101, 202000, 15000 and etc.) two options are possible:
   (a) if $r_{q+1} = 0$, then $b(T)$ depends only on r_q : $x = r_q$; $b(T) = b(x) \cdot 10^{q-1}$;

(b) if $r_{q+1} \neq 0$, then $b(T)$ depends only on the least significant digit of the number T : $x = r_{q+1} \cdot 10 + r_q$; $b(T) = b(x) \cdot 10^{q-1}$.
3. By $T \in M_2$ (for example, 2030, 2140 and etc.) also two options are possible:
   (a) if $r_{q+1} = 0$, then $x = r_q \cdot 10$; $b(T) = b(x) \cdot 10^{q-2}$;
   (b) if $r_{q+1} \neq 0$, then $x = r_{q+1} \cdot 10 + r_q$; $b(T) = b(x) \cdot 10^{q-1}$.

After, the value which is found for the number T we construct the membership function $\mu_T(x)$ for $x \in R$ using formula (2). Then the points k_1 and k_2 of the transition for the function $\mu_T(x)$ are found from the following relations:

$$k_1 = T - \frac{b(T)}{2} \quad and \quad k_2 = T + \frac{b(T)}{2}.$$

These values define the limits of the confidence interval of numbers approximately equal to a given T and are coefficients α, β of the fuzziness of a fuzzy number FN.

To better understanding the essence of such calculations we consider the example of constructing the membership function for an approximate point estimate "duration of work which approximately equal to 121".

According to the hereinbefore algorithm, $T = 121$. Next, we determine the values of the variables q, r_q, r_{q+1} and d. The least significant digit of the number 121 is in the unit one, i.e. $q = 1$, $r_q = r_1 = 1$, $r_{q+1} = r_2 = 2$ - a digit whose order is one higher than the order of the least significant digit of 121. Since when dividing q by 3 in the remainder we get 1, therefore, the number 121 belongs to the equivalence class M_1, therefore $d = 1$.

Insofar as $r_{q+1} \neq 0$, then according to example 2b, $x = r_{q+1} \cdot 10 + r_q = r_2 \cdot 10 + r_1 = 21$ and $b(T) = b(x)$, where $b(x) = b(21)$ is determined from the Table 1 according to the formula:

$$b(21) = \frac{1}{2}(b([\frac{21}{10}] \cdot 10 + 5) + b(21 - [\frac{21}{10}] \cdot 10)) = \frac{1}{2}(b(25) + b(1)).$$

According to the Table 1 $b(25) = 6.75$ and $b(1) = 0.46$, so, $b(121) = \frac{1}{2}(6.75 + 0.46) = 3.6$ and $a = -\frac{4ln0.5}{b^2} = -\frac{4ln0.5}{3.6^2} = 0.21$. As a result, we obtain the following membership function: $\mu_T(t) = exp(-0.21(t - 121)^2)$. In this case $k_1 = T - \frac{b(T)}{2} = 121 - \frac{3.6}{2} = 119.2$, $k_2 = T + \frac{b(T)}{2} = 121 + \frac{3.6}{2} = 122.8$ and fuzzy number FN takes the form $FN = (121, 119.2, 122.8)$.

Let $FN_j = (T_j, t'_j, t''_j)$ is a fuzzy approximation-number of a linguistic approximate point estimate of the execution time of the work s, given by an expert j. Then, the probability $p(t|t'_j, t''_j)$ that the work s will be performed by the time $t'_j \leq t \leq t''_j$ will be calculated according to (1).

3.2   Approximation of Approximate Interval Estimate

Let $FN = (T_1, T_2, \alpha, \beta)$ is a trapezoid fuzzy number, where T_1 and T_2 are the limits of tolerance of the number FN, and α and β are the coefficients of the fuzziness of this number.

To find the interval $[\alpha, \beta]$ it is necessary to construct the membership function of "numbers listed approximately in the interval from T_1 to T_2". This function is constructed as follows:

$$\mu_{(T_1,T_2)}(x) = \begin{cases} x < T_1, & \mu_{T_1}(x) \\ T_1 \le x \le T_2, & 1 \\ x > T_2, & \mu_{T_2}(x), \end{cases}$$

where $\mu_{(T_1,T_2)}(x)$ is the membership function of the fuzzy interval (T_1, T_2); $\mu_{T_1}(x)$ and $\mu_{T_2}(x)$ are the membership functions for fuzzy sets of numbers which approximately equal to T_1 and T_2 respectively. These functions are constructed similarly to the above algorithm.

In this case, the points k_1 and k_2 of the transition for the function $\mu_{(T_1,T_2)}(x)$ are found from the relations: $k_1 = T_1 - \dfrac{b(T_1)}{2}$, $k_2 = T_2 + \dfrac{b(T_2)}{2}$. These values are taken as the coefficients α and β of a fuzzy number FN.

Let $FN_j = (T'_j, T''_j, t'_j, t''_j)$ is a fuzzy estimate of the time to complete the work s, given by the expert j. Then, the probability $p(t|t'_j, t"_j)$ that the work s will be performed by the time $t'_j \le t \le t''_j$ will be calculated according to (1).

Let FN_j is a fuzzy estimate of the duration of work s, given by the expert j. Then, the probability $p(t)$ that the work s will be done for the time t, we will calculate by the formula:

$$p(t) = \sum_{j=1}^{m} r_j p(t|t'_j, t''_j),$$

where r_j is a weight coefficient of the expert j ($\sum_{j=1}^{m} r_j = 1$), m is a number of the experts, which evaluate the work s.

4   Model of the β-Distribution of the Duration of Conditional Work

We have considered the use of fuzzy time estimates in the construction β-distribution of the work duration on the execution of which do not impose certain conditions (restrictions). Let us to generalize the considered approach to the case of conditional works. By conditional work is understood one, for the fulfillment of which it is necessary to fulfill some condition. Let's consider this model in the context of solving the following tasks [19].

Let it will be required to estimate the probable lead time $s_1, s_2, ..., s_m$ which will be called *major*. Firstly, this list of works is supplemented by *intermediate* works $s_{m+1}, ..., s_{m+n}$, which may be necessary or useful for carrying out major works.

Then, for each work s_i ($i = \overline{1, m+n}$), experts formulate the conditions for its achievement $s_{i1}, s_{i2}, ..., s_{ik}$ and give estimates of the time t_i of execution of work after the fulfillment of the condition. As a result, for each work s_i several such conditions will be received in accordance with the number of experts involved.

According to (1), we introduce the probability $p(t|t'_j, t''_j)$ that by the time $t'_j \le t \le t''_j$ point the work s_i will be completed, where $t'_{ij}(t_{ij})$ is an estimate of expert j of working time s_i. Then the probability $p_i(t)$ that the work will be completed by the time point, taking into account [14], we will calculate by the formula:

$$p(t) = \sum_{j=1}^{m_i} r_{ij} p_{ij}(t|t'_{ij}, t''_{ij}) p_{ij1}(t) p_{ij2}(t)...p_{ijk_j}(t), \qquad (3)$$

where $ij_1, ij_2, ..., ijk_j$ are the numbers of intermediate works, determined by the expert j as a condition for the performance of the work s_i; r_{ij} is an expert weight. Thus, $\sum_{j=1}^{m_i} r_{ij} = 1$; t'_{ij}, t''_{ij} are the boundaries of the "shifted" estimate of the time to complete the work s_i after the conditions formulated by the expert are fulfilled; m_i is number of experts evaluating the work s_i.

This formula gives the average probability estimate, taking into account the weights of the experts. Moreover, for unconditional works, composition: $r_{ij} p_{ij}(t|t'_{ij}, t''_{ij}) p_{ij1}(t)...p_{ijk_j}(t)$ in (3) changing to $r_{ij} p_{ij}(t|t'_{ij}, t''_{ij})$.

In order to use this formula to consistently find functions $p_i(t)$ for all jobs s_i, $(i = \overline{1, m+n})$, it is necessary to carry out the separation of jobs into nonintersecting sets $M_0, M_1, ..., M_p$. The set M_0 should consist of jobs with only unconditional estimates of execution time. And for job in any of the sets M_i as conditions may protrude only jobs from the sets $M_0, M_1, ..., M_{i-1}$ $(i = \overline{1, p})$. If initially such a stratification is not possible, then by introducing new auxiliary jobs, such a stratification can be achieved.

In addition, for each job s_i $(i = \overline{1, m+n})$ a "shifted" estimate of its duration is calculated relative to the time of the condition and the maximum time to complete. The shifted estimates are calculated as follows.

Case 1. Let s_i be an unconditional job, and $t_{ik} = [t'_{ik}, t''_{ik}]$ $(k = \overline{1, m_i})$ is an estimate of the time of its execution. Then these estimates will be shifted estimates of the job time s_i, and $t_i^{max} = \max_k t''_{ik}$ is a maximum duration.

Case 2. Let $s_{i1}, s_{i2}, ..., s_{ik_i}$ be jobs (operations), which are the conditions of the work s_i, $t_{ik} = [t'_{ik}, t''_{ik}]$ $(k = \overline{1, m_i})$ of its duration after the fulfillment of this condition, and t_{ij}^{max} is the maximum job duration $s_{ij}(j = \overline{1, k_i})$. Then the boundaries of the shifted estimates of the job execution time are calculated as follows:

$$t'_{ik} = t'_{ik} + t_{iu} \quad and \quad t''_{ik} = t''_{ik} + t_{iu},$$

where t_{iu} is the maximum time to fulfill the condition $s_{i1}, s_{i2}, ..., s_{ik_i}$.

At the same time, the fulfillment of this condition depends on the logic of the job s_{ij} sequence. If the jobs are arranged in a sequence of the "end-start" type, then the maximum time to fulfill the condition $s_{i1}, s_{i2}, ..., s_{ik_i}$ is calculated as the sum of the values t_{ij}^{max}. If the works are performed in parallel, then the maximum value is taken as the maximum time for the fulfillment t_{ij}^{max} of the condition $s_{i1}, s_{i2}, ..., s_{ik_i}$. In other cases, the time t_{iu} is calculated according to the corresponding graph work s_{ij}.

For a better understanding of the nature of such calculations, consider another example. Let two main jobs are given s_1, s_2, and one intermediate s_3.

Let also each work be evaluated by two experts. Table 2 shows the results of the assessment (separation of works, conditions, initial and shifted estimates of time, taking into account the parallel execution of job conditions and the maximum duration of work).

Table 2. The examination results

Works	s1		s2		s3	
Layers	M2		M1		M0	
Experts	1	2	1	2	1	2
Weights	0.6	0.4	0.3	0.7	0.4	0.6
Conditions	(s2, s3)	s2	s3	-	-	-
Time estimates	[1, 3]	[2, 4]	[1, 4]	[2, 3]	[1, 3]	[2, 3]
Shifted time estimates	[8, 10]	[9, 11]	[4, 7]	[2, 3]	[1, 3]	[2, 3]
Max time	11		7		3	

In this table the interval time estimates is changed within the range $[\alpha, \beta]$, where α and β are the coefficients of fuzziness numbers, which are approximated by linguistic estimates of the duration of jobs s_1, s_2 and s_3.

Based on the data in the table and taking into account (1), for each job we obtain the following probability distribution functions:

$$p_3(t) = 0.4 \cdot p_{31}(t|1,3) + 0.6 \cdot p_{32}(t|2,3),$$

$$p_2(t) = 0.3 \cdot p_{21}(t|4,7) \cdot p_3(t) + 0.7 \cdot p_{22}(t|2,3),$$

$$p_1(t) = 0.6 \cdot p_{11}(t|8,10) \cdot p_2(t) \cdot p_3(t) + 0.4 \cdot p_{12}(t|9,11) \cdot p_2(t).$$

These functions can be used to predict the most likely time to complete each job. As such the time considered to be a distribution median, i.e. time t, for which the probability $p_i(t)$ equals 0.5: $t = p_i^{-1}(0.5)$. Given the type of function $p_i(t)$ (3), it is difficult to get an analytical expression for a function $p_i^{-1}(x)$. Therefore, in this case it is more constructive to find the median of the distribution by a numerical method. In this case, the problem is to find t for which, for example the following inequality becomes true: $|p_i(t) - 0.5| \leq \varepsilon$, where ε is a calculation error.

In addition, for these distributions, you can get the probability density using the formula:

$$f_i(t|1, t_i^{max}) = p_i'(t),$$

where $[1, t_i^{max}]$ is an interval of the maximum duration of work s_i.

So, for the density $p_2(t)$ distribution is described by the function:

$$f_2(t|1,7) = p_2'(t) = 0.3(p_{21}'(t|4,7) \cdot p_3(t) + p_{21}(t|4,7) \cdot p_3'(t)) + 0.7 \cdot p_{22}'(t|2,3).$$

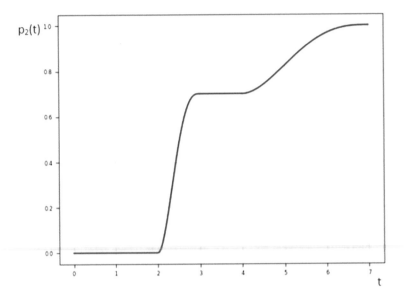

Fig. 1. Integral distribution function $p_2(t)$

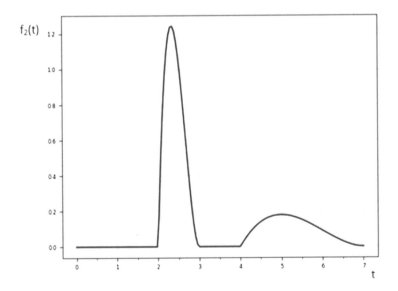

Fig. 2. Differential distribution function $f_2(t)$

Knowing the density function, you can get a prediction of the expected execution time. The mode of the corresponding distribution acts as such a time. Figures 1 and 2 show the graphs of the differential and integral distribution functions $p_2(t)$.

5   Refinement of the Estimates

With collective expertise, differences in expert estimates are inevitable. The results of the examination can be considered sufficiently reliable only if there is good consistency in the assessments of individual specialists.

The issues of harmonization of the group assessments were considered by many researchers, specifically [14, 20]. For example, [14] presents the mechanism for coordinating estimates of the execution time for conditional work.

At the same time, the time estimates are refined in two stages. At the first stage, estimates of the duration of unconditional works are refined, and at the second, estimates of conditional goals are specified. The reason for the refinement of estimates is rather low degree of consistency they have. The coefficient of variation is used as a measurement for consistency of estimates.

At the first stage, the coefficient of variation is determined separately for the left and right borders of time intervals using the formula $V = s/\bar{x}$, where s is a standard deviation of estimates; $\bar{x}$ is its mean value.

Let $[a_{i1}, b_{i1}], ..., [a_{ik_i}, b_{ik_i}]$ are the estimates of the time for performing unconditional work, which are given by k_i experts. Then the coefficients of variation of the boundaries of these intervals will be determined in the following way:

for left limit by formula $V_{iL} = s_{iL}/\bar{x}_{iL}$, where

$$S_{iL} = \sqrt{\frac{1}{k_i - 1} \sum_{j=1}^{k_i}(a_{ij} - \bar{x}_{iL})^2 r_{ij}}, \quad \bar{x}_{iL} = \sum_{j=1}^{k_i} a_{ij} r_{ij},$$

for right limit by formula $V_{iR} = s_{iR}/\bar{x}_{iR}$, where

$$S_{iR} = \sqrt{\frac{1}{k_i - 1} \sum_{j=1}^{k_i}(b_{ij} - \bar{x}_{iR})^2 r_{ij}}, \quad \bar{x}_{iR} = \sum_{j=1}^{k_i} b_{ij} r_{ij}.$$

In these formulas r_{ij} is a weight coefficient expert j, who has evaluated the work s_i, thus $(\sum_{j=1}^{k_i} r_{ij} = 1)$.

In the second stage, the relation $V = \sigma/m_t$ is used as the coefficient of the estimates variation, where σ is a standard deviation of random value t; m_t is its math expectation.

Let $[t_i', t_i'']$ is the shifted estimate of the execution time of the conditional work s_i. Next, let $t_{i1}, t_{i2}, ..., t_{in_i}$ are the arbitrary points of this interval, and $p_{ik}(t_{ik})$ are the probabilities that the work s_i will be completed by the time t_{ik}. These probabilities are calculated by the formula (1). Then

$$\sigma = \sqrt{\sum_{j=1}^{n_i}(t_{ij} - m_t)^2 p_{ij}}, \quad m_t = \sum_{j=1}^{n_i} t_{ij} p_{ij}.$$

In practice of applying methods for expert estimates, there are various conditional classifications of sample variability based on the coefficient of variation.

So, the results of the examination can be considered satisfactory, if $0.2 \leq V \leq 0.3$ and better, if $V < 0.2$. These conditions can be used as a criterion for the consistency of expert assessments.

To clarify the estimates, the examination is made to be continuous. Each time an expert changes one or another opinion, the degree of consistency of the time for performing unconditional work is determined and recalculation of probabilities $p_{ik}(t_{ik})$ is performed. At the same time, jobs for the refinement of their ratings are selected from the sets M_k, starting with a smaller number. As a result of refining the estimates for each of the main job s_i, a graph of forecast ways of its execution will be obtained in accordance with the number of conditions formulated.

If it is necessary to translate the main jobs into a real plan, then the task arises of choosing ways to perform these works. According to [19], the selection criteria are: the degree of confidence that this path will lead to the performance of the main job; expected time and the expected cost of the job in this way. In [14] the mechanisms for calculating relevant indicators were considered and the use of an improved method for analyzing hierarchies [21] as a method of choice was proposed.

6   Conclusion

The use of fuzzy expert estimates in network models of project management of the duration of operations allows, under conditions of uncertainty, to increase the objectivity of the results. The distribution function of the probability of duration of work is proposed, which allows to build the β-distribution of a random variable over the entire range of its definition from a set of approximate point and interval expert estimates. This makes it possible to create flexible distributions of the duration of work and to obtain more accurate estimates of the forecast problem-solving paths. There is considered approximation of linguistic estimates by fuzzy triangular and trapezoidal numbers. It is based on the growth of the membership function of a fuzzy triangular number, accounting for its scale.

In general, the considered approach makes it possible under the condition of information uncertainty, to obtain more accurate performance prediction estimates for making informed decisions.

References

1. Oleynikov S, Kirilov A (2011) Numerical estimation of beta distribution parameters. Bull Voronezh State Tech Univ 7(7):209–212
2. Davis R (2008) Teaching note teaching project simulation in excel using PERT-beta distributions. INFORMS Trans Educ 8(3):139–148
3. Dubois D, Prades A (1990) Theory of opportunities. Applications to the representation of knowledge in computer science. Network (calculation by the PERT method and the sum of fuzzy trapezoidal numbers with fuzzy durations of works). Radio and communication, p 288

4. Shushura A, Yakimova Yu (2012) The fuzzy critical path method for project management based on fuzzy interval estimates. Artrificial Intalagence 3:332–337
5. Yang M, Chou Y, Lo M, Tseng W (2014) Applying fuzzy time distribution in the PERT model. In: International multi-conference 2014 of engineers and computer scientists, Hong Kong
6. Akimov V, Balashov V, Zalozhnev A (2003) The method of fuzzy critical path. In: Management of large systems, vol 3, pp 5–10
7. Głądysz B (2017) Fuzzy-probabilistic PERT. Ann Oper Res 258(2):437–452
8. Samman T, Ramadan M, Brahemi R (2014) Fuzzy PERT for project management. Int J Adv Eng Technol 7(4):1150–1160
9. Habibi F, Birgani O, Koppelaar H, Radenović S (2018) Using J Project Manage 3(4):183–196
10. Sushanta Kumer Roy S, Miah M, Uddin M (2016) Alternative approach of project scheduling using fuzzy logic and probability distribution. J Comput Math Sci 7(3):130–143
11. Barishpolets V (2011) Network modeling of stochastic processes for performing a complex of interrelated operations. RENSIT 3(2):49–73
12. Gelrud Yu (2010) Generalized stochastic network models for managing complex projects. Vestnik, NSU 36–51
13. Golenko-Ginzburg D (2010) Stochastic network models of planning and development management. In: Scientific book, Voronezh, p 284
14. Samokhvalov Yu (2018) Development of the prediction graph method under incomplete and inaccurate expert estimates. Cybern Syst Anal 54(1):75–82
15. Presnyakov V. Internet university of information technologies, project management basics lecture 5: risk management. In: E-book - basics of project management. LNCS, p 132. http://www.intuit.ru/studies/courses/2194/272/info
16. Pegat A (2009) Fuzzy modeling and control. Laboratory of Knowledge, p 798
17. Ledeneva T, Chermenev D (2015) Fuzzy model of the project with the duration of work in the form of generalized Gaussian numbers. VSU, Series: system analysis and information technology, vol 2, pp 72–81
18. Borisov A, Krumberg O, Fedorov I (1990) Decision making based on fuzzy models: examples of use. Knowledge, Riga, 184 p
19. Glushkov V (1969) About forecasting based on expert assessments. In: Cybernetics, vol 2, pp 8–17
20. Samokhvalov Yu (2002) Matching of expert estimates in preference relation matrices. Upravlyayushchie Sistemy i Mashiny 182(6):49–54
21. Samokhvalov Yu (2004) Distinctive features of using the method of analysis of hierarchies in estimating problems on the basis of metric criteria. Kibernetika i Sistemnyj Analiz 40(5):15–19

Configuration Spaces of Geometric Objects with Their Applications in Packing, Layout and Covering Problems

Sergiy Yakovlev[✉] [iD]

National Aerospace University "Kharkiv Aviation Institute", Kharkiv, Ukraine
svsyak7@gmail.com

Abstract. The concept of spatial configuration is introduced as a mapping of a finite set of geometric objects into a partially ordered set of a certain structure subject to a given constraints. Depending on the choice of generalized variables, various classes of spatial configurations are investigated. New approaches have been proposed for an analytical description of the main constraints in the packing, layout and covering optimization problems. This approach is based on the use of a special class of functions defined on a set of generalized variables of the configuration space.

Keywords: Geometric object · Configuration space · Optimization · Packing · Layout · Covering

1 Introduction

The problems of synthesizing spatial configurations of geometric objects are of constant interest to scientists and practitioners. Currently, the focus is on important classes problems of packing [1–8], layout [9–14], covering [15–18], partitioning [19,20]. At the same time, a variety of practical productions forces us to consider more and more complex spatial configurations, which takes us beyond the framework of classical formulations. Complex spatial configurations assume that objects can be located in different relationships both in relation to each other, and to different collections of objects. In this case, it is necessary to take into account specific conditions of behavior. Such spatial configurations must simultaneously take into account the specifics of the packing, layout and covering problems for the various groups of objects that form the configuration. Therefore, at present it is impossible to offer a unified approach to the study of problems of optimization of spatial configurations, and the development of new methods for their solution is an urgent task.

The purpose of this paper is to generalize the known results related to the problems of placing, layout, and covering in the case of variable metrical characteristics of objects. Formalization and the solution of such problems very little attention is paid due to their high complexity.

© Springer Nature Switzerland AG 2020
V. Lytvynenko et al. (Eds.): ISDMCI 2019, AISC 1020, pp. 122–132, 2020.
https://doi.org/10.1007/978-3-030-26474-1_9

2   Problem Statement

Let us consider the problem of synthesizing optimal spatial configurations of geometric objects. Each object is characterized by its shape, sizes and location in the space. Together, these characteristics determine the geometric information about the geometric object. In addition, the geometric objects are in a certain relationship with each other. In general, the problem lies in the determination of such an acceptable mutual location of objects, at which some quality criterion reaches an extremum.

A feature of the approach to the formalization of this class of problems in this paper is a new view at the description of their structure by introducing the concept of the configuration space of geometric objects. The peculiarity of the proposed results is the possibility of their generalization to arbitrary placement problems, when groups of objects are in different relations at the same time. Moreover, the rotation of objects may be taken into account.

3   Materials and Methods

3.1   Basic Concepts and Definitions

Consider a geometric object S in the space R^n as a geometric placement of points $P \in R^n$ that satisfy the inequality $f(P) \geq 0$. In this case, the equation $f(P) = 0$ describes the boundary of the object S, which we denote $fr\ S$. It is clear that the boundary equation $f(P) = 0$ determines the shape of the object S. Fundamental studies on the problem of constructing a boundary equation for a given geometric object are described in [21].

The same geometric object can be defined in countless ways. Therefore, in practice, additional conditions are introduced that the function $f(P)$ must satisfy. In the general case, the equation of the object S boundary includes constants $m = (m_1, ..., m_\alpha)$ that characterize its topological and metric properties (sizes), i.e. $f(P, m) = 0$, $P \in R^n$. The components of the vector $m = (m_1, ..., m_\alpha)$ are called the metric parameters of the object. The object S having metric parameters m will be denoted $S(m)$. In the following, we confine ourselves to considering geometrical objects in the spaces R^3 and R^2, putting $P = (x, y, z) \in R^3$ and $P = (x, y) \in R^2$ respectively.

Let a function $f(P, m)$ be defined and continuous by $P \in R^3$ for any $m \in D$, where D is a admissible set of the metrical parameters m. Then a geometric object $S \subset R^3$ will be a set of points $P \in R^3$ subject to:

$$f(P, m) = 0,\ if\ P \in fr\ S(m);$$

$$f(P, m) > 0,\ if\ P \in int\ S(m);$$

$$f(P, m) < 0,\ if\ P \in R^3 \setminus cl\ S(m),$$

for any fixed $m \in D$. Here and in the future $int\ S$ and $cl\ S$ topological interior and closure of the object S.

In the papers of Yu.G. Stoyan the class of so-called φ-objects is distinguished, which in the spaces R^2 and R^3 are the models of real material bodies existing in nature. We note in particular the fact that the φ-objects are Lebesgue measurable, and the functions $f(P, m)$ defining their boundaries are also measurable and continuous by $P \in R^3$ for any $m \in D$. Further we will consider the φ-objects of space R^3 as geometric objects, and the symbol φ will be omitted.

We define the coordinate system $Oxyz$ in space R^3, which called stationary, and associate our own (moving) coordinate system with some point, which we call the pole of the object S. To characterize the relative position of these coordinate systems, we introduce the so-called placement parameters $p = (p_1, ..., p_\beta) = (v, \theta)$, where v is the vector of the coordinates of the poles of the object S in a fixed coordinate system, as well as the vector θ of angular parameters that determine the relative position of the axis of the eigen and fixed coordinate systems.

For $S \subset R^3$ we have $p = (p_1, ..., p_\beta) = (v, \theta)$. In general $\beta = 6$, $v = (v_1, v_2, v_3)$, $\theta = (\theta_1, \theta_2, \theta_3)$. Depending on the shape of the object, the number of its placement parameters may be less. In particular, $\beta = 3$ for a centrally symmetric object (sphere). In addition, according to their formulation, some placement parameters can be fixed.

The position of the object S relative to a stationary coordinate system $Oxyz$ is given by the general equation, which has the form

$$F(P, m, v, \theta) = f(A(P - v), m) = 0, \qquad (1)$$

where A is an orthogonal operator, expressed through angular parameters θ.

3.2 Configuration Spaces and Generalized Variables of Geometric Objects

To formalize the spatial configurations optimization problem, we use the concept of constructing configuration spaces of geometric objects proposed in [22,23]. In the general case, the configuration space is defined by a set of generalized variables that specify the position in space of a certain system and its parts both relative to each other and relative to this system.

Let the equation of the general position of the object S has the view (1). We form its configuration space $\Xi(S)$ by choosing metrical and placement parameters $m = (m_1, ..., m_\alpha)$, $p = (p_1, ..., p_\beta) = (v, \theta)$ as generalized variables. Then a parameterized geometric object $S(g) \subset R^3$ will correspond to the point $g = (g_1, ..., g_\gamma) = (m, p)$, $\gamma = \alpha + \beta$ of the configuration space $\Xi(S)$, such that

$$F(P, m, p) = 0, \; if \; P \in fr \; S(g)$$

$$F(P, m, p) > 0, \; if \; P \in int \; S(g)$$

$$F(P, m, p) < 0, \; if \; P \in R^3 \setminus cl \; S(g).$$

Denote $J_n = \{1, 2, ..., n\}$. Let $\Sigma = \{S_1, ..., S_n\}$ be the initial set of objects $S_i \subset R^3$, $i \in J_n$. By analogy with the above arguments, we introduce their metrical parameters $m^i = (m_1^i, ..., m_{\alpha_i}^i)$ and placement parameters $p^i = (p_1^i, ..., p_\beta^i) = (v^i, \theta^i)$, $i \in \mathbf{J}_n$, $\beta \leq 6$. We form a configuration spaces $\Xi(S_i)$ of a objects S_i with generalized variables $g^i = (m^i, p^i)$, $i \in J_n$. Each point $g^i \in \Xi(S_i)$ will correspond to a parametrized geometric object $S_i(g^i) \subset R^3$ defined by its general equation $F_i(P, m^i, v^i, \theta^i) = 0$. Then

$$F_i(P, m^i, p^i) = 0, \; if \; P \in fr \; S_i(g^i)$$

$$F_i(P, m^i, p^i) > 0, \; if \; P \in int \; S_i(g^i)$$

$$F_i(P, m^i, p^i) < 0, \; if \; P \in R^3 \setminus cl \; S_i(g^i).$$

Let us to form a configuration space:

$$\Xi(\Sigma) = \Xi(S_1) \times \Xi(S_2) \times ... \times \Xi(S_n)$$

with generalized variables $g = (g^1, ..., g^n)$. In accordance with [24], under the configuration we mean the mapping ξ of some initial set Σ of elements of an arbitrary nature into an abstract set Ω of a certain structure subject to a given set of constraints Λ, i.e. $\xi \colon \Sigma \to \Omega$. We choose $\Sigma = \{S_1, ..., S_n\}$ as a finite set of geometrical objects.

- *The mapping $\xi \colon \Sigma \to \Xi(\Sigma)$ of a set Σ of geometric objects into a configuration space $\Xi(\Sigma)$ that satisfies a given set Λ of constraints defines a spatial configuration of objects S_i, $i \in J_n$*

Thus, the spatial configuration defines in space R^3 a set of parametrized geometric objects $S_i(g^i) \subset R^3$, $i \in J_n$, that allow to build a complex object of a certain structure. In this regard, consider the mapping $\chi \colon \Sigma \to S_\chi$ that gives the so-called complex object

$$S_\chi = \chi(S_1, ..., S_n). \tag{2}$$

The mapping χ defines the structure of a complex object S_χ and forms a parameterized object

$$S_\chi(g) = S_\chi(g^1, ..., g^n) = \chi(S_1(g^1), ..., S_n(g^n)) \tag{3}$$

in the configuration space $\Xi(\Sigma)$. We will say that the formation of the structure of a complex object in combination with the choice of generalized variables sets the configuration structure.

3.3 Typology and Modeling of Spatial Configurations

The generalized variables $g^1, ..., g^n$ of the configuration space $\Xi(\Sigma)$ can be imposed restrictions Λ that allow you to proposed different classes of spatial configurations [25]. Such restrictions follow from the fact that the objects forming the spatial configuration are in various relations, both binary and n-ary in general case. The typology of spatial configurations based on the formation of configuration space $\Xi(\Sigma)$, the choice of the corresponding generalized variables and the formation of the structure of a complex object. To form a system of constraints Λ, let us define binary relations on the set of geometric objects:

- *non-overlapping* {∗}, *assuming* S′ ∗ S″, if int S′ ∩ int S″ = ∅;

- *inclusion* {∘}, *assuming* S′∘S″, if int S″ ⊂ S′.

We indicate some classes of spatial configurations.

- *A mapping* $\xi: \Sigma \rightarrow \Xi(\Sigma)$ *specifies a packing configuration, if* $S_i(g^i) *$ $S_j(g^j) \forall i,j \in J_n, i < j$.

We introduce an additional object S_0, called a container, and denote its configuration space $\Xi(S_0)$ with generalized variables $g^0 = (m^0, p^0)$. Let us $\Sigma^0 = \{S_0, S_1, ..., S_n\} = \Sigma \cup S_0$ and build a configuration space:

$$\Xi(\Sigma^0) = \Xi(S_0) \times \Xi(S_1) \times ... \times \Xi(S_n)$$

- *A mapping* $\xi: \Sigma^0 \rightarrow \Xi(\Sigma^0)$ *specifies a layout configuration, if* $S_0(g^0) \circ S_j(g^j)$, $S_i(g^i) * S_j(g^j) \ \forall i,j \in J_n, i < j$.

Let us to assume $p^0 = (0, ..., 0)$. Note that relations $S_0(g^0) \circ S_j(g^j)$ and $cS_0(g^0) * S_j(g^j) \ \forall j \in J_n$ are equivalent. Therefore, the layout configuration can be considered as packing configuration of objects cS_0, $S_j, j \in J_n$.

In the study of configurations, additional constraints are imposed on the minimum and maximum allowable distances between objects. It is this feature that forms the basis for the distinction between the classes of configurations of packing and layout in [13, 14]. When geometric objects S_i, $i \in J_n$ are solid bodies of a given mass, the condition of system balance (unbalance constrains) [26–28] is imposed as additional restrictions on their mutual position. This configuration is called the balanced packing configuration.

Let us to form a complex object:

$$S_{\check{\chi}} = \check{\chi}(S_1, ..., S_n) = \bigcup_{i=1}^{n} S_i$$

The mapping $\xi: \Sigma^0 \rightarrow \Xi(\Sigma^0)$ specifies the covering configuration, if

$$\check{\chi}(S_1(g^1), ..., S_n(g^n)) \circ S_0(g^0).$$

An object S_0 is called a coverage domain, and an objects $S_1, ..., S_n$ are covering ones. Some formulations of covering problems are given, for example, in [15–18].

When metrical and placement parameters of geometric objects are fixed, we have a class of assignment problems, which is the subject of research in the field of discrete optimization [29,30]. On the other hand, as shown in [8], the synthesis of spatial configurations implies makes the isolation of a combinatorial structure of problems. This makes it possible to consider geometric objects as combinatorial ones [31] and to involve for their research the results of the theory of Euclidean combinatorial configurations [32–35].

4   Results and Discussion

4.1   On Φ-function for Geometric Objects with Variable Metrical Parameters

Currently, the focus is on packing, layout and covering problems with fixed metrical parameters. Variables could only be sizes of container or covering domain. The use of configuration spaces of geometric objects allows us to significantly expand the class of formalized and solvable problems. In this case, both metrical and placement parameters are considered as generalized variables, some of which can be fixed.

The formation of various spatial configurations (packing, layout, covering) involves the choosing of generalized variables $g^0, g^1, ..., g^n$ of the configuration space $\Xi(\Sigma^0)$ and their admissible set description in the form of functional constraints system. To formalize the conditions of non-overlapping and inclusion Yu.G. Stoyan introduced the concept of Φ-functions, the properties of which were elucidated in [36,37], and an analytical description was obtained for basic 2D and 3D objects. In this case, it is assumed that only own congruent transformations are performed over objects, that is, the form and metric parameters of objects are fixed. Studies of Φ-functions of complex shape objects with variable metric and angular parameters are at the initial stage.

The study of spatial objects configurations requires a generalization of the notion of a Φ-function taking into account the generalized variables of the configuration space of geometric objects. Consider geometric objects S′ and S″, which have generalized variables $g' = (m', p')$, $g'' = (m'', p'')$, where $m' = (m'_1, ..., m'_\gamma)$, $m'' = (m''_1, ..., m''_\nu)$, $p' = (p'_1, ..., p'_\alpha)$, $p'' = (p''_1, ..., p''_\alpha)$. Let us create a configuration space $\hat{\Xi} = \Xi(S') \times \Xi(S'')$. Suppose that the admissible domains $D' \subseteq R^\gamma$ and $D'' \subseteq R^\nu$ of metrical parameters m' and m'', respectively.

- *A continuous function $\Phi^{S'S''}(g', g'')$ defined everywhere on $R^{2\alpha} \times D' \times D''$ is called a generalized Φ-function of geometric objects S′, S″ with generalized variables $g' = (m', p')$, $g'' = (m'', p'')$, if it satisfies the following conditions:*

$$\Phi^{S'S''}(g', g'') > 0, \; if \; cl \; S'(g') \bigcap cl \; S''(g'') = \emptyset,$$

$$\Phi^{S'S''}(g', g'') = 0, \; if \; cl \; S'(g') \bigcap int \; S''(g'') = \emptyset,$$

$$fr \; S'(g') \bigcap fr \; S''(g'') \neq \emptyset,$$

$$\Phi^{S'S''}(g', g'') < 0, \; if \; int \; S'(g') \bigcap int \; S''(g'') \neq \emptyset$$

- *A generalized Φ-function is called normalized if its value for any fixed metrical parameters $\hat{m}' \in D'$ and $\hat{m}'' \in D''$ is equal to the Euclidean distance between the objects $S'(g')$ and $S''(g'')$, subject to*

$$\tilde{G} = \{(g'g'')|int \; S'(g') \bigcap int \; S''(g'') \neq \emptyset\}$$

where $(g', g'') \in \tilde{G}$, $g' = \left(\hat{m}', p'\right)$, $g'' = \left(\hat{m}'', p''\right)$.

An analysis of existing methods for constructing the Φ-function of basic 2D and 3D objects with fixed metric parameters allows us to conclude that it is possible to naturally generalize these results on the case of variable metrical parameters.

The theory of Φ-functions is primarily effective for the formalization of constraints with the binary relations between geometrical objects. Indeed, the condition of non-overlapping of parameterized geometric objects $S'(g')$ and $S''(g'')$ is given by the inequality:

$$\Phi^{S'S''}(g', g'') > 0$$

Constraints on the minimum and maximum allowable distances (respectively, d^{min} and d^{max}) between objects S' and S'' will be

$$d^{min} \leq \Phi^{S'S''}(g', g'') \leq d^{max}$$

where $\Phi^{S'S''}$ is the normalized Φ-function.

However, the use of Φ-function for n-ary relations of geometrical objects is associated with great computational difficulties. In particular, we are talking about a covering problem.

4.2   A New Class of Functions for Formalization of Mutual Position Relations of Geometric Objects

We propose an approach related to the introduction of a new class of functions based on the calculation of the measure (area, volume) of a parametrized complex object of a given structure [38]. Consider a complex structure object (2). In the configuration space $\Xi(\Sigma)$ with generalized variables $g^1, ..., g^n$, it corresponds to a parameterized object $S_\chi(g)$ of the form (3). Define the function:

$$\omega_\chi(g) = \omega_\chi(g^1, ..., g^n) = \mu(S_\chi(g^1, ..., g^n)),$$

where $\mu(S)$ is Lebesgue measure of a set S.

- *A function $\omega_\chi : \Xi(\Sigma) \to R^1$ is called a ω-function of a parameterized geometric object $S_\chi(g)$.*

To formalize a ω-function, we introduce the characteristic function:

$$\lambda_\chi(P, g) = \begin{cases} 1, \ if \ P \in S_\chi(g); \\ 0, \ if \ P \notin S_\chi(g) \end{cases}$$

Then

$$\omega_\chi(g) = \iiint \lambda_\chi(P, g) \, dP.$$

Note that if $P \in R^2$, then

$$\omega_\chi(g) = \iint \lambda_\chi(P, g) \, dP.$$

We use the above results for a complex object that has the structure:

$$\tilde{\chi}_0(S_0, S_1, ..., S_n) = S_0 \cap \bigcup_{i=1}^{n} S_i.$$

As the result, we have:

$$\omega_{\tilde{\chi}_0}(g) = \mu \left(S_0(g^0) \cap \bigcup_{i=1}^{n} S_i(g^i) \right) = \iiint \lambda_{\tilde{\chi}_0}(P, g) dP =$$

$$= \iiint \lambda_{\chi_0}(P, g^0) \left(1 - \prod_{i=1}^{n} \left(1 - \lambda_{\chi_i}(P, g^i) \right) \right) dP =$$

$$= \iiint_{S_0(g^0)} \left(1 - \prod_{i=1}^{n} \left(1 - \lambda_{\chi_i}(P, g^i) \right) \right) dP.$$

In a class of φ-objects, a covering of a domain S_0 by covering sets $S_1, ..., S_n$ holds if and only if

$$\omega_{\tilde{\chi}_0}(g) = \mu \left(S_0(g^0) \right).$$

Using the proposed class of ω-functions, one can formalize the conditions of non- overlapping and inclusion of objects S_i and S_j. Let

$$S_{\chi_{ij}} = \chi_{ij}(S_1, ..., S_n) = S_i \cap S_j$$

Then

$$\omega_{\chi_{ij}}(g^i, g^j) = \mu \left(S_i(g^i) \cap S_j(g^j) \right) = \iiint \lambda_{\chi_i}(P, g^i) \lambda_{\chi_j}(P, g^j) dP.$$

Since objects S_i and S_j should not have common points, conditions of non-overlapping are satisfied when:

$$\omega_{\chi_{ij}}(g) = 0, \ i, j \in J_n, \ i < j.$$

Conditions for the inclusion of objects $S_j, j \in J_n$ in the container S_0 will be

$$\omega_{\chi_{0j}}(g^0, g) = 0, \ j \in J_n.$$

Thus, the new class of functions proposed in this section allows one to naturally formalize the conditions for the mutual location of geometric objects by analyzing the areas (volumes) of geometric objects and their parts. From a computational point of view, in many cases it is much easier than calculating the distances between objects. For example, it is possible to use pixel image processing.

5 Conclusion

The concept of configuration spaces has significantly expanded the scope of application of existing methods for the synthesis of optimal spatial configurations of geometric objects. The proposed approaches to the formalization of the basic constraints of the problems of packing, layout and covering are naturally integrated with the known approaches by additionally taking into account generalized variables and the structure of spatial configurations. The extension of the concept of Φ-functions and a new class of ω-functions allow us to describe the characteristic constraints of various classes of spatial configurations. Further research is the optimization of spatial configurations in accordance with specified quality criteria. This will allow us to propose new mathematical models and methods for optimizing spatial configurations by highlighting the corresponding classes of geometric objects and choosing their generalized variables.

References

1. Fasano G (2015) A modeling-based approach for non-standard packing problems. Optim Pack Appl 105:67–85
2. Fasano G (2013) A global optimization point of view for non-standard packing problems. J Global Optim 155(2):279–299
3. Sriramya P, Parvatha BV (2012) A state-of-the-art review of bin packing techniques. Eur J Sci Res 86(3):360–364
4. Hifi M, M'Hallah R (2009) A literature review on circle and sphere packing problems: model and methodologies. Adv Optim Res 2009:1–22
5. Wascher G et al (2007) An improved typology of cutting and packing problems. Eur J Oper Res 183:1109–1130
6. Bortfeldt A, Wascher G (2013) Constraints in container loading: a state-of-the-art review. Eur J Oper Res 229(1):1–20
7. Fadel GM, Wiecek MM (2015) Packing optimization of free-form objects in engineering design. Optim Pack Appl 105:37–66
8. Yakovlev SV (2017) The method of artificial space dilation in problems of optimal packing of geometric objects. Cybern Syst Anal 53(5):725–732
9. Sun Z-G, Teng H-F (2003) Optimal layout design of a satellite module. Eng Optim 35(5):513–529
10. Coggan J, Shimada K, Yin S (2002) A survey of computational approaches to three-dimensional layout problems. CAD Comput Aided Des 34(8):597–611
11. Tian T et al (2016) The multiple container loading problem with preference. Eur J Oper Res 248(1):84–94
12. Drira A, Pierreval H, Hajri-Gabouj S (2007) Facility layout problems: a survey. Ann Rev Control 31(2):255–267
13. Stoyan YG, Semkin VV, Chugay AM (2014) Optimization of 3D objects layout into a multiply connected domain with account for shortest distances. Cybern Syst Anal 50(3):374–385
14. Grebennik IV et al (2018) Combinatorial configurations in balance layout optimization problems. Cybern Syst Anal 54(2):221–231

15. Yakovlev SV (1999) On a class of problems on covering of a bounded set. Acta Mathematica Hungarica 53(3):253–262
16. Stoyan YG, Patsuk VM (2014) Covering a convex 3D polytope by a minimal number of congruent spheres. Int J Comput Math 91(9):2010–2020
17. Shekhovtsov SB, Yakovlev SV (1989) Formalization and solution of one class of covering problem in design of control and monitoring systems. Avtomatika i Telemekhanika 5:160–168
18. Kiseleva EM, Lozovskaya LI, Timoshenko EV (2009) Solution of continuous problems of optimal covering with spheres using optimal set-partition theory. Cybern Syst Anal 45(3):421–437
19. Kiseleva EM, Koriashkina LS (2015) Theory of continuous optimal set partitioning problems as a universal mathematical formalism for constructing Voronoi diagrams and their generalizations. Cybern Syst Anal 51(3):325–335
20. Kiseleva EM, Prytomanova OM, Zhuravel SV (2018) Algorithm for solving a continuous problem of optimal partitioning with neurolinguistic identification of functions in target functional. J Autom Inform Sci 50(3):102–112
21. Rvachov VL (1982) Theory R-function and its applications. Nauk. Dumka, Kiev
22. Stoyan YG, Yakovlev SV (2018) Configuration space of geometric objects. Cybern Syst Anal 54(5):716–726
23. Yakovlev SV (2018) On some classes of spatial configurations of geometric objects and their formalization. J Autom Inf Sci 50(9):38–50
24. Berge C (1968) Principes de combinatoire. Dunod, Paris
25. Yakovlev S, Kartashov O (2018) System analysis and classification of spatial configurations. In: Proceedings of 2018 IEEE first international conference on system analysis and intelligent computing, SAIC 2018, Kyiv, pp 90–93
26. Kovalenko AA et al (2015) Balance layout problem for 3D-objects: mathematical model and solution methods. Cybern Syst Anal 51(4):556–565
27. Stoyan YuG, Sokolovskii VZ, Yakovlev SV (1982) Method of balancing rotating discretely distributed masses. Energomashinostroenie 2:4–5
28. Stoyan Yu, Romanova T, Pankratov A, Kovalenko A, Stetsyuk P (2016) Balance layout problems: mathematical modeling and nonlinear optimization. In: Space engineering. Modeling and optimization with case studies, vol 114, pp 369-400
29. Korte B, Vygen J (2018) Combinatorial optimization: theory and algorithms, 6th edn. Springer, New York
30. Burkard RE (2013) Quadratic assignment problems. In: Handbook of combinatorial optimization, vol 5, no 1, pp 2741–2814
31. Hulianytskyi L, Riasna I (2017) Formalization and classification of combinatorial optimization problems. In: Optimization and its applications, vol 130, pp 239–250. Springer
32. Yakovlev S (2017) Convex extensions in combinatorial optimization and their applications. In: Optimization and its applications, vol 130, pp 567–584. Springer
33. Yakovlev SV (1989) Bounds on the minimum of convex functions on Euclidean combinatorial sets. Cybernetics 25(3):385–391
34. Yakovlev SV, Pichugina OS (2018) Properties of combinatorial optimization problems over polyhedral-spherical sets. Cybern Syst Anal 54(1):99–109

35. Yakovlev SV, Pichugina OS, Yarovaya OV (2019) Polyhedral spherical configuration in discrete optimization. J Autom Inf Sci 51(1):38–50
36. Stoyan Y, Romanova T (2013) Mathematical models of placement optimization: two- and three-dimensional problems and applications. Model Optim Space Eng 73:363–388
37. Bennell J et al (2010) Tools of mathematical modelling of arbitrary object packing problems. J Ann Oper Res 179(1):343–368
38. Yakovlev SV (2019) Formalization of spatial configuration optimization problems with a special function class. Cybern Syst Anal 55(4):512–523

Model of the Internet Traffic Filtering System to Ensure Safe Web Surfing

Vitaliy Serdechnyi(ID), Olesia Barkovska(✉)(ID), Dmytro Rosinskiy(ID),
Natalia Axak(ID), and Mykola Korablyov(ID)

Kharkiv National University of Radio Electronics,
Nauky Avenue 14, Kharkiv 61166, Ukraine
{vitalii.serdechnyi,olesia.barkovska,dmytro.rosinskyi,
nataliia.axak,mykola.korablyov}@nure.ua

Abstract. The paper proposes a generalized model of network traffic filtering system that includes three modules: network operations' module (low-level drivers needed to capture and modify network traffic); initialization and control module (initiates classification operations, preparing the received content for transfer to the classification module; manages the network operations' module; keeps statistics on the detection of unwanted resources); web resource classification module (determines whether information provided in HTML format refers to one of three thematic categories that are banned for children). The system provides safe Internet surfing of the child and excludes access to prohibited categories with a probability of 99.53%. Classification is subject to HTTP request and HTTP response. Studies have been performed on the work of SVM classifiers and Naive Bayes classifier depending on the response threshold. The effect of stop words removal on classification accuracy of input data is also analyzed; it is able to increase the classification accuracy up to 6%.

Keywords: Parental control system · Traffic classifier · Stop words · Black list · White list · Internet · Naive Bayes · SVM

1 Introduction

The ongoing rapid development of information and telecommunication technologies supports as well as disrupts the spiritual and emotional bond among people and their development. On the one hand, easier and more convenient storage of information, time cutting, increase in the accuracy of calculations made within economic, medical and scientific research as well as learning capabilities enhancement are guaranteed [1,2].

On the other hand, dependence on technologies, electronic educational and information resources as well as computer games increases. A common consequence is information neurosis diagnosed by psychiatrists form many countries of the world. It entails deterioration of the qualitative and quantitative values

V. Lytvynenko et al. (Eds.): ISDMCI 2019, AISC 1020, pp. 133–147, 2020.
https://doi.org/10.1007/978-3-030-26474-1_10

of the human life; it is caused by an enormous amount of information and a lack of time for logical processing of its extreme volumes. Emergence of social networks increased the number of people exposed to mind virtualization with the inability of adequate evaluation of the world and self-evaluation as a biological and social being, which is, namely, the negative aspect of information technology development. A particular problem is integration of the oncoming generation, namely, children whose psycho is weak, thus, more exposed to the destructive impact, into the world of information technologies [3]. Predominance of information technologies in children's development does not always have a positive impact on their development, often thwarting their individual personality progress. Among the information sources, which influence the child's progress, the following ones are distinguished (Fig. 1): desktop and mobile applications, cartoons, electronic learning games etc.

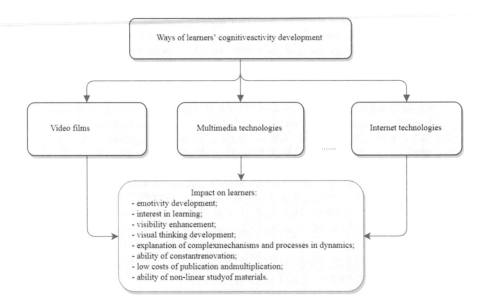

Fig. 1. Information sources' influence on the child's progress

Among the child's development factors influenced by information technologies and information provided by them, mental, physical and social factors can be distinguished (Fig. 2).

Despite the fact that a child's psycho adapts to mechanical devices, information provided by the abovementioned devices may cause shift in a child's mindset and perception, result in a child's psychic trauma and affect their adult life. This information includes resources containing materials with sexual content, descriptions and graphic violence, drugs as well as resources propagating suicide, various radical movements and pages containing coarse language.

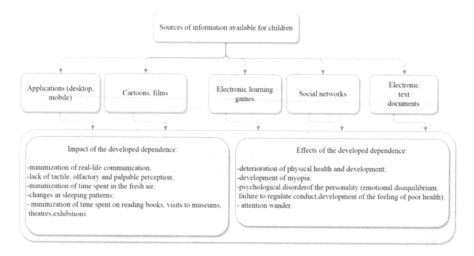

Fig. 2. The child's development factors influenced by information technologies and information

All the abovementioned facts justify the necessity of control and analysis of information streams transmitted via Internet channels by means of creating a secure environment for the use of the Internet as the basic source of information which a child can access as well as prove relevance and necessity for the development of models and methods aiming at classifying traffic provided on receipt of HTTP-response from the server and further definition of unsolicited resources.

A large number of studies devoted to the application of machine learning techniques for text classification [4,5]. Classic examples of the use of automatic text classification is the task of cataloging sites, the fight against spam, recognizing emotional color of the text, advertising personalization, automatic annotation and so on. Statement of the problem in general form is formulated as "to build a classifier which assigns a selected document to one of several predefined categories based on the presence of document attributes that are relevant to this category correlated with the signs of the category, and the lack of irrelevant signs and symptoms not correlated with signs in this category". An example of classification methods based on machine learning are probabilistic, metric, logical, linear, and also methods based on artificial neural networks. The results of studies on various text classification methods show the advantage of the linear support vector method over other machine learning methods when evaluating such an indicator as the learning quality [6], and Naive Bayes method in estimating such parameters as speed, simple software implementation of the algorithm, easy interpretability of the algorithm results [7].

2 Review of the Literature

The existing solutions for the set task are Internet traffic filters known as "parental control" systems, which provide for the Internet use "safe" for the child [8].

The given software tools enable parents to control or limit what their child may see or do by classifying the obtained content into two groups – white and black list as well as limiting the time for these actions. Currently, this control is exercised predominantly in two ways: special-purpose web applications and correspondent modules within antivirus software suites. There are also independent software solutions, however, they have not become widely used in comparison with antivirus suites (Fig. 3).

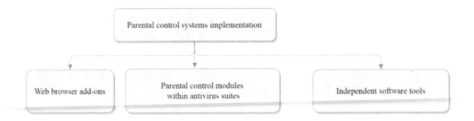

Fig. 3. Provision of child safety in the Net

Web browser add-ons are free, however, the features of such add-ons are quite poor and are often limited to the black list. Moreover, if two or more browsers are installed in the system, it is necessary to install applications designed specifically for a certain web browser. Meanwhile, it is worth noting such drawbacks as ease of deleting such applications while it would be difficult for a child to delete an antivirus solution because access to its settings is usually password protected. The leading antivirus suites such as ESET Smart Security and Kaspersky Internet Security contain integrated modules with the parental control features. These are rather powerful tools, which effectively block potentially unwanted web resources. These suites are not free any longer and are provided with paid subscription.

In order to evaluate the efficiency of the existing software solutions a special set of tests (analyses), the outcomes of which are provided below, was elaborated.

Test 1. Pages contain English text and "18+" rated pictures. A simple HTML page, which includes language offencive for children, was written for the analysis. The written site was hosted on https://ru.000webhost.com.

The test findings are provided in Table 1. Mobicip software passed the test for 100%, having blocked all the three pages. ESS Premium did not block any of the pages, thus, becoming an outsider of the research. The remaining software passed the test for 41.64% on average.

Test 2. Pages contain English text and "18+" rated pictures. A simple HTML page, which includes language and images offencive for children, was written for the analysis. The specificity of the page was the absence of any hints on the unwanted thematic contents in the title and meta. The second HTML page differed from the first one by the fact that in $<img\,alt =$ "..."$/>$ "Nude young girl #i", $i = [1, 4]$ was written. Therefore, the second page clearly but not rudely

Table 1. Test 1 findings

Software	indexMetaTitle	indexMeta	indexNoMeta	Total
ESS	−	−	−	0/3
Kaspersky TS	+	−	−	1/3
Child WG	+	+	−	2/3
HT PC	+	−	−	1/3
Mobicip	+	+	+	3/3
Qustodio	+	−	−	1/3

pointed at the erotic content of the pictures. At the moment of the analysis, the page was hosted on the same site as during test 1. The test findings are provided in Table 2. None of the software solutions passed the test. All the software solutions under scrutiny did not block any of the pages.

Table 2. Test 2 findings

Software	indexClean	indexAlt
ESS	−	−
Kaspersky TS	−	−
Child Web Guardian	−	−
Mobicip	−	−

Table 3. Test 3 findings

Software	indexClean	indexTitle
ESS	−	−
Kaspersky TS	+	+
Child Web Guardian	−	−
Mobicip	−	−

Test 3. Pages contain Russian text rated as "18+" and "coarse language". A simple HTML page with inappropriate content was written for the analysis. In the title tag, there was a reference to the subject scope of the page. The pages were hosted on the same site as for tests 1 and 2. The test findings are provided in Table 3. Kaspersky Total Security 2017 software product fully passed the given test. Its heuristic analysis enabled to distinguish the coarce language on the page and its subject scope. All the other software products fully failed the test. The conducted tests did not distinguish any software products, which could fully pass the tests. The test findings are very heterogenious for all the software solutions.

None of the software products were able to block the proposed pages in test 2, whereas, tests 1 and 3 revealed definite leaders Kaspersky Total Security 2017 (3) and Mobicip (1). Table 4 provides the eventual findings for basic software products used in tests.

Table 4. Test 3 findings

Software	indexClean	indexTitle
ESS	−	−
Kaspersky TS	+	+
Child Web Guardian	−	−
Mobicip	−	−

The obtained results for the existing software products gives a clear understanding that, currently, there is no effective solution of the task to protect children from the unwanted information in the Internet. Therefore, limitations to children's access to the given kinds of information along with the creation of additional information filtering tools in compliance with the established rules, are considered to be a relevant task.

3   Problem Statement

The tasks of the paper are:

- development of a generalized model of system for filtering Internet traffic;
- analysis of the impact of text processing methods on the accuracy of web page content classification;
- study the SVM classifier performance depending on the learning method;
- study the Naive Bayes classifier performance for the task of classifying text depending on the threshold.

The aim of the paper is the research into the efficiency of network traffic classification methods with regard to their application to the systems providing children's security in the Internet.

4   Materials and Methods

A generalized model of the proposed traffic filtering system is shown in Fig. 4, where the developed system's boundaries are represented by a dotted line. The model is divided into three working together modules, which will guarantee the effective traffic filtering:

- network operations' module (NOpM);
- initialization and control module (InitCM);
- classification module (ClassM).

The work of the system can be devided into the input traffic processing T_{in} and the output traffic processing T_{out}. Input traffic T_{in} is intercepted by NOpM, which holds its progress through the network driver stack and checks for the page URI in a whitelist, that can be created by the user, or can be filled in programmatically after visiting great number of webpages. If a match is found, the traffic is allowed to display without going further down the stack. If the site's address is not in the whitelist, NOpM sends unknown packets D to the module InitCM for further content pre-processing (in case of text content classification, D is the HTML page of the open resource, in the case of image classification, D is an image). This module, after preliminary preparation of the page (D^* - modified data intercepted by the network operations module, in most cases $D^* = D$), sends it to the web resource classification module, which in turn generates a plurality of thematic categories $K = \{K_1, K_2, ..., K_n\}$ to which data are classified. The total number of outgoing thematic categories is $0 \leq n \leq m$. If the set K is empty, the web resource is not blocked. If at least one category has been determined, the web resource is blocked and its URI address is entered in the blacklist of the system. After processing, the modified incoming network traffic T_{in}^* is formed. In case of the web-page locking - $T_{in}^* = 0$. In case of the web-page permission $T_{in}^* = T_{in}$ and InitCM generates a signal M that affects the behavior of the NOpM module (Fig. 4).

Output traffic T_{out} is intercepted by the NOpM module when it path from the client program to the network driver stack before it gets into the NIC driver. Traffic is delayed and NOpM verifies the presence of the URI in the system blacklist. If a match has been found, traffic movement is blocked ($T_{out}^* = 0$, where T_{out}^* - modified outgoing network traffic). Therefore, the request does not get to the Internet. If the URIs are not blacklisted, the traffic moves freely further down the network driver stack ($T_{out}^* = T_{out}$).

Network operations' module (NOpM) is a set of low-level drivers for interception and modifying network traffic. It performs the following tasks:

- input/output network traffic monitoring;
- network traffic blocking for the purpose of inadmissibility of receiving it by the client program or sending to the network;
- operations with black/white lists of the system;
- generating certain network traffic (for example, a message about blocking access to the Internet resource for the browser trying to open it) or modifying it;
- low-level network traffic operations to ensure correct and stable operation of the system

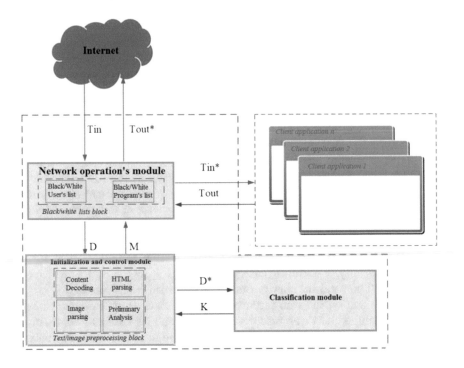

Fig. 4. A generalized model of network traffic filtering system

All information, required for the parent system, consists of two parts:

– requested host URI;
– received HTML web page code from a remote server (Fig. 5).

Data transmitted over a secure channel can not be parsed using direct reading. Therefore, in order to allocate the necessary module data, one must use invasion to TLS traffic.

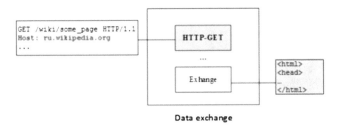

Fig. 5. Receiving's HTML workflow

The InitCM module is the central module of the system. It connects two other modules and initiates their work at the right time points. The module performs the following tasks:

- initiates the classification operation, performing input content preliminary processing for transfering it to the classification module;
- manages the network operations' module;
- keeps track of unwanted resources detection statistics.

One of the InitCM's component is the text/image preprocessing block, which converts the data into a understood form for the classifier.

The process of the web page text content preprocessing is divided into several stages:

- useful text information extracting;
- unnecessary characters removal;
- text language definition and, if necessary, its translation;
- lematization;
- stop words removal;
- weighing of terms.

These steps are reasonable, since the text normalization before further classification using machine learning methods, aims to remove grammatical information from the source text (cases, numbers, verb types and tenses, voices of participles, gender, etc.), leaving the semantic component and, thus, reducing the amount of information to be processed. The works [9–11] show that in Russian the use of stemming does not improve the accuracy, and the use of the Bernoulli model for the naive Bayes classifier markedly affects the accuracy. Thus, the work excluded consideration of the stemming effect on the accuracy of training. As the stages of text preprocessing, the stop words removal is considered, as well as preliminary processing influence on the accuracy of the classifiers is analyzed. The scheme of the text preliminary processing which is supposed to be used in the system is shown in Fig. 6.

The classification module (ClassM) is the intellectual part of the system. It determines whether a webpage corresponds to any prohibited thematic category or not. The answer is returned to the InitCM module. The classifiers were considered and analyzed: the reference vector (SVM) method and the naive Bayesian classifier (NB).

At the stage of creating training and test set for analyzing HTTP responses and requests, following categories were regarded:

- materials for adults (18+);
- alcohol, tobacco, drugs;
- means of anonymous access to the network.

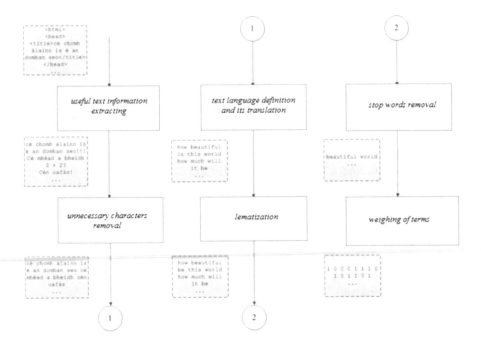

Fig. 6. Text preprocessing workflow schematic representation

5   Experiment

Purpose of the module classifying web resources is to resolve classification task for textual information presented in HTML format in real time mode. Each text may belong to one of the prohibited categories of content for children. If this text does not belong to any of them, it is considered safe for children and allow to display.

This system involves three thematic categories: materials for adults (18+); alcohol, tobacco, drugs; means of anonymous network access (this category includes web resources that contain information about tools that circumvent access restrictions to certain resources that are prohibited in the country or region. Includes VPN, TOR, proxy servers, anonymizers, etc).

Input text to be classified, is an HTML-page. Therefore, first of all, it is necessary to remove useful text information from certain HTML-tags and their attributes. There are certain tags which can contain useful text information:

- Title (often the title itself indicates thematic focus of the page);
- Meta (can contain keywords which indicate a thematic orientation of page or page description);
- p, h1, h2, h3, h4, h5, h6, a, and span tags contain a lot of useful text information;
- Img (its alt attribute contains text describing the image, which is also the input for classifiers).

To investigate an effect of preprocessing stages on the accuracy of recognition using reference vectors and Naive Bayes, the implementation of proposed system within the Windows OS was performed in accordance with Fig. 7.

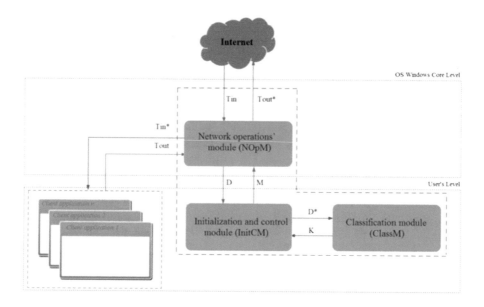

Fig. 7. Parent control system model implementation on Windows base

Developed prototype has no GUI and works in console mode. When user opens Web resource which is subject to inspection a message will be displayed in the console window. As result of classification, if a web resource is subject to blocking, the corresponding message with indicated reason (prohibited categories) of the lock is displayed in console window. The browser also appears with a message as shown in Fig. 8.

6 Results and Discussion

In this paper we investigate the effectiveness of SVM classifier and the Naive Bayes classifier for task of textual information classification. The efficiency of these tools was evaluated relative to the percentage errors of the first and second kind, allowed in the classification process, and depending on performed steps of HTTP-request and HTTP-response content preprocessing. SVM classifier study was divided into two parts, depending on the method of training: Sequential Minimal Optimization (SMO) and Stochastic Gradient Descent (SGD). To investigate the Naive Bayes classifier there were used different thresholds, ranging from classic threshold of a posteriori probability of 0.5 to 0.99 in steps of 0.1.

Fig. 8. Real time locking the opening a web resource using developed prototype

Table 5. The stop words removal stage impact on the classification accuracy

	Classification accuracy with the removal of stop words	Classification accuracy without removing stop words
	Adults (18+)	
SVM	95.3%	89.77%
Naive Bayes	99.53%	94.88%
	Alcohol, tobacco, drugs	
SVM	92.08%	88.12%
Naive Bayes	97.03%	97.03%
	Anonymizers	
SVM	96.05%	88.16%
Naive Bayes	94.74%	97.37%

6.1   Study of the Effect of Preprocessing Steps (Removing Stop Words) on Classification Accuracy

The results of study the impact of stop words removal stage on the classification accuracy are shown in Table 5. For SVM classifier, stop words removal stage significantly affects the classification quality. For the Naive Bayes classifier, presence of stop words removal stage is not critical but desirable. For all classifiers, the effect of stop words removal stage varies depending on thematic category. The obtained results show the need to use stop words removal stage in the preparation of text for classification by each of the classifiers under study increasing classification accuracy of up to 6%.

6.2   Accuracy of SVM and Naive Bayes Classifiers

For the reasons stated above, SVM and Naive Bayes classifiers conducted another study to determine the percentage of false positives, depending on size of the training set. Due to the nature of binary SVM classifier, testing took place on the training set which consist of examples and anti-examples of equal ratio, the total size of 200 to 815 in increments of 150 items. When training multiclass Naive Bayes classifier, there was no splitting the training set examples and anti-examples. The size of test sample was 100 items that are not included in the training set. The average value of impact of training set size on classification accuracy is shown in Fig. 9. The timetable is the basis for further research on the training sample with largest number of items in each category. ClassM provides functional support for several classifiers. One of the main classifier choosing criteria is its precision and ability to work close to in real time mode.

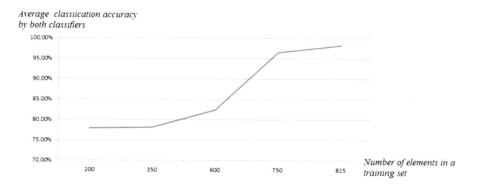

Fig. 9. The impact of training set size on recognition accuracy

Table 6 shows the results of experimental studies for the SVM classifier, depending on SVM (SVO) and SVM (SGD) optimization algorithms. Research of Naive Bayes was performed according to threshold (0.5–0.99). Naive Bayes classifier showed stable results. Maximum classifier accuracy at the threshold of 0.5 was 98.78%. At the maximum threshold of 0.99 classifier accuracy drops to 98.31%. Results of SVM classifier using an optimized algorithm SGD is also quite high with an average accuracy 97.29%. However, the same SVM classifier, using the SMO optimizing method, showed significantly worse average accuracy at 90.12%. It should be noted that at the stage of study, the size of sample for training classifiers was 815 pages for the category "Adult (18+)", page 750 for the category "Alcohol, tobacco, drugs" and 684 pages for the category "Anonymizers".

Research the number of false positives (i.e. locking frequency of secure web pages) are listed in Table 7. The Naive Bayes classifier has twice the number of false positives than the SVM (SGD) classifier and four times higher than the SVM (SMO) classifier.

Table 6. Classification accuracy depending on the optimization algorithms for SVM classifier and on Naive Bayes classifier threshold

Classifier	Adults (18+)	Alcohol, tobacco, drugs	Anonymizers	Average accuracy
SVM (SMO)	87.86%	87.68%	94.81%	90.12
SVM (SGD)	98.55%	98.52%	94.81%	98.78
Naive Bayes (0.50)	99.53%	98.08%	98.73%	98.78
Naive Bayes (0.60)	99.53%	98.08%	98.73%	98.78
Naive Bayes (0.70)	99.53%	98.08%	98.73%	98.78
Naive Bayes (0.80)	99.53%	98.08%	98.73%	98.78
Naive Bayes (0.90)	99.53%	98.08%	94.47%	98.36
Naive Bayes (0.99)	99.08%	97.11%	98.73%	98.31

Table 7. Locking frequency of secure web pages

Classifier	Average percentage of false positives for all categories studied
SVM (SMO)	1.16%
SVM (SGD)	2.32%
Naive Bayes (0.50)	12.79%
Naive Bayes (0.60)	11.63%
Naive Bayes (0.70)	6.98%
Naive Bayes (0.80)	6.98%
Naive Bayes (0.90)	6.98%
Naive Bayes (0.99)	4.65%

7   Conclusions

In the paper the efficiency of network traffic classification methods with regard to their application to the systems providing children's security in the Internet was researched.

The paper presents a generalized model of Internet traffik filtering system which will permit safe Internet serfing for child and eliminate the probability of 99.53% to the access to one of the three considered prohibited categories "Adult (18+)", "Alcohol, tobacco, drugs", "Means of anonymous access to the network". The proposed model includes three interacting modules: network operations' module (NOpM); initialization and control module (InitCM); classification module (ClassM).

Analysis of the effect of text preprocessing, namely, removal of stop words, prior to actual classification by the considered methods, on the classification accuracy of the web page content showed the need to use the stop word removal stage, since it increases classification accuracy up to 6%.

The paper investigates the performance of SVM classifiers and Naive Bayes classifier for the task of classifying text information in the parental control system by the values of first and second kind errors allowed in classification process. Studies evaluating locking frequency of secure web pages showed the best result 1.16% when classifying with the support vectors method and learning algorithm SMO. However, the Naive Bayes classifier showed the highest classification accuracy, reaching an accuracy of 99.53% at the aposteriori threshold of 0.5.

References

1. Jackson L, Witt EA, Eye AV, Fitzgerald HE, Zhao Y (2010) Children's information technology (IT) use and their physical, cognitive, social and psychological well-being. In: 2010 fourth international conference on digital society, St. Maarten, pp 198–203
2. Gonzales L. Best practices around social media safety. http://www.go.galegroup.com/ps/i.do?p=AONE&sw=w&u=mcc_pv&v=2.1&id=GALE%7CA488965839&it=r&asid=bc418b8d53bb2b1d8222a3713289b79d. Accessed 05 Apr 2019
3. Yakovin A, Myltasova O. The impact of information technology on the development of cognitive activity of children. http://www.elar.urfu.ru/bitstream/10995/46681/1/klo_2017_158.pdf. Accessed 28 Apr 2019
4. Dai X, Bikdash M, Meyer B (2017) From social media to public health surveillance: word embedding based clustering method for twitter classification. In: Southeast-Con 2017, Charlotte, NC, pp 1–7
5. Liu K (2009) Text classification based on KNN algorithm. Sci Technol Ecnony Market 06:12–13 (in Chinese)
6. Liu Z, Lv X, Liu K, Shi S (2010) Study on SVM compared with the other text classification methods. In: 2010 second international workshop on education technology and computer science, Wuhan, pp 219–222
7. Kim H, Kim J, Kim J (2016) Semantic text classification with tensor space model-based Naïve Bayes. In: 2016 IEEE international conference on systems, man, and cybernetics (SMC), Budapest, pp 004206–004210
8. Barkovska O, Axak N, Rosinskiy D, Liashenko S (2018) Application of mydriasis identification methods in parental control systems. In: 2018 IEEE 9th international conference on dependable systems, services and technologies, Kiev, pp 459–463
9. Hartmann J, Huppertz J, Schamp C, Heitmann M (2019) Comparing automated text classification methods. Int J Res Mark 36(1):20–38
10. Haryanto AW, Mawardi EK, Muljono (2018) Influence of word normalization and chi-squared feature selection on support vector machine (SVM) text classification. In: 2018 international seminar on application for technology of information and communication, Semarang, pp 229–233
11. Huosong X, Jian L (2011) The research of feature selection of text classification based on integrated learning algorithm. In: 2011 10th international symposium on distributed computing and applications to business, engineering and science, Wuxi, pp 20–22

Analysis of Objects Classification Approaches Using Vectors of Inflection Points

Diana Zahorodnia[1,2], Pavlo Bykovyy[1,2(✉)], Anatoliy Sachenko[2,3],
Viktor Krylov[2,4], Galina Shcherbakova[4], Ivan Kit[1,2], Andriy Kaniovskyi[1,2],
and Mykola Dacko[1,2]

[1] Ternopil National Economic University, Ternopil, Ukraine
pb@tneu.edu.ua
[2] Research Institute for Intelligent Computer Systems, Ternopil, Ukraine
[3] Department of Computer Science,
Kazimierz Pulaski University of Technology and Humanities, Radom, Poland
[4] Odessa National Polytechnic University, Odessa, Ukraine

Abstract. An increase of the automated video surveillance system operational performance was proposed by reducing the amount of data using the contour inflection points for object identification. The authors developed a structural statistic method to identify facial images. It allowed obtaining a set of vectors that consist of inflection points of every hierarchical layer of the processed object. The research described in this paper shows the comparison of such vectors identifying similarities and referring them to a particular class. Three objects classification approaches were proposed and analyzed: approaches based on centroid, sum and maximum. Two alternative schemes for the calculation of distance between inflection point vectors were also proposed and analyzed: the classical Euclidean meter and the modified one, based on distances from the coordinate origins.

Keywords: Image processing · Contour inflection points ·
Objects classification · Cluster analysis

1 Introduction

1.1 Classification Problem

Nowadays the number of cameras in automated video surveillance systems is growing fast, which requires a large number of human operators [1–4]. Therefore, intelligent video surveillance systems become more popular; however, it requires significant computing resources.

Tasks that appear during intelligent video surveillance systems construction can be attributed to several main areas [5]. The first one is connected with the output data submission received from measurement results of the object

© Springer Nature Switzerland AG 2020
V. Lytvynenko et al. (Eds.): ISDMCI 2019, AISC 1020, pp. 148–157, 2020.
https://doi.org/10.1007/978-3-030-26474-1_11

being under identification. Each measured value has some characteristics of the object or image. Image vectors contain all the measurable information about the images. The measurement process, where objects are being referred to a certain class of images, can be considered as encoding process. It is good to consider image vectors as points of an n-dimensional Euclidean space.

The second task is image recognition using allocation of specific features or properties from the obtained data which decreases dimension of the image vectors. This task is often defined as image pre-processing and feature selection. The image class features are common properties for all images of this class. Features that characterize differences between individual classes can be interpreted as interclass features. Interclass features are common to all considered classes and do not provide useful information for object recognition and may not be taken into account during the recognition process. Feature selection is one of the main tasks of subsystems recognition.

The third task is finding the best procedures for identification and classification processes. After collecting the data from recognized object, which are represented as points or identification vectors within the image space, the system recognizes to which class of images this data is being referred.

To increase operational performance of such systems, reduction of the data amount was proposed by using contours and inflection points for object identification [6–15] and hierarchical classification [16–19].

Authors developed a structural statistic method for identifying facial images by contour inflection points [20–22] and the usage of contour segmentation structural-hierarchical principle [23–25]. These approaches allowed obtaining a set of vectors that consist of inflection points for each hierarchical layer of the processed object. The next step is the comparison of such data vectors to identify similarities and refer them to a particular class – performance of the clustering process.

1.2   Cluster Analysis of the Identification Vectors

The problem of classification (partitioning, clustering) has been solved efficiently by methods of cluster analysis, where a set of objects is arranged in a way that all of them represent one class (cluster or group) and are more similar to each other rather than similar to other clusters [26]. Let $\mathbf{a}_{ij}$ is a vector of the inflection points of j-th image of the i-th person, aggregated by all detail levels:

$$\mathbf{a}_{i,j} = (a_{i,j,1}, a_{i,j,2}, \ldots, a_{i,j,N}), \tag{1}$$

where $a_{i,j,k}$ is k-th inflection points of j-th image of i-th person, N is the number of inflection points. In back, $a_{i,j,k}$ is also the vector which describes Cartesian ordinates of the inflection points:

$$a_{i,j,k} = (ax_{i,j,k}, ay_{i,j,k}), \tag{2}$$

where $ax_{i,j,k}$ is the abscise of the k-th inflection points of j-th image of i-th person, $ay_{i,j,k}$ is the ordinate of the k-th inflection points of j-th image of i-th person.

If the research requires not to use all detail levels but only some of them, then appropriate subset of aggregated inflection points set is used as follows:

$$\mathbf{a}_{i,j} = (a_{i,j,1}, a_{i,j,2}, \dots, a_{i,j,N}),\tag{3}$$

To ensure invariance of classification procedure in affine transformations of scale and shift it is preferable to number inflection points according to their centroid and geometrical size of the object:

$$na_{i,j,k} = s_{i,j}(ax_{i,j,k} - cx_{i,j}, ay_{i,j,k} - cy_{i,j}),\tag{4}$$

where $na_{i,j,k}$ is k-th normalized characteristic point of j-th image of i-th person, $s_{i,j}$ is the scaling coefficient of j-th image of i-th person, $cx_{i,j}$ is the abscissa of the centroid of j-th image of i-th person, $cy_{i,j}$ is the ordinate of the centroid of the j-th image of the i-th person.

Scaling coefficient $s_{i,j}$ and coordinates of the centroid $(cx_{i,j}, cy_{i,j})$ could be calculated in the following way:

$$s_{i,j} = \frac{1}{(\max_k ax_{i,j,k} - \min_k ax_{i,j,k})}\tag{5}$$

$$cx_{i,j} = \frac{1}{N}\sum_{k=1}^{N} ax_{i,j,k}, \qquad cy_{i,j} = \frac{1}{N}\sum_{k=1}^{N} ay_{i,j,k}.\tag{6}$$

Let $\mathbf{b}$ is a new vector of characteristic points of a new image:

$$\mathbf{b} = (b_1, b_2, \dots, b_N).\tag{7}$$

For the classification of the inflection points vector in a new picture $\mathbf{b}$ a cluster analysis can be used. Thus, each person is described by the set of inflection points from all of their images, which represent one cluster:

$$A_i = (\mathbf{a}_{i,1}, \mathbf{a}_{i,2}, \dots),\tag{8}$$

where A_i is a set of inflection points if i-th cluster.

The distance between any two vectors of inflection points $\mathbf{a} = (a_1, a_2, \dots, a_N)$. and $\mathbf{b} = (b_1, b_2, \dots, b_N)$. can be estimated as the Euclidean norm:

$$\rho_1(\mathbf{a}, \mathbf{b}) = \|\mathbf{a} - \mathbf{b}\|_1 = \sqrt{\sum_{k=1}^{N}(ax_k - bx_k)^2 + (ay_k - by_k)^2},\tag{9}$$

where ax_k, ay_k are the abscissa and ordinate of the k-th inflection points of the $\mathbf{a}$ vector, bx_k, by_k are the abscissa and ordinate of the k-th inflection points of the $\mathbf{b}$ vector.

Alternatively, to provide invariance of classification procedures in order to affine rotation transformations, the formula of the modified Euclidean distance can be used. It describes the difference between the distances of separately taken characteristic points and the point of origin:

$$\rho_2(\mathbf{a}, \mathbf{b}) = \|\mathbf{a} - \mathbf{b}\|_2 = \sqrt{\sum_{k=1}^{N}(ax_k^2 + ay_k^2 - bx_k^2 - by_k^2)^2}. \tag{10}$$

Classification of inflection points of a new image $\mathbf{b}$ can be carried out in a different way. This way it would differ by calculating the degree of the involvement, in other words - similarity $similarity(\mathbf{b}, A_i)$ of vector $\mathbf{b}$ to the cluster A_i:

$$0 < similarity(\mathbf{b}, A_i) \leq 1 \tag{11}$$

The greater the value of the similarity function $similarity(\mathbf{b}, A_i)$, the higher the probability that the inflection points of the image $\mathbf{b}$ belong to the cluster A_i. When the value of the similarity function exceeds a certain threshold, then we can assume that the picture $\mathbf{b}$ belongs to the corresponding cluster.

1.3 Alternative Approaches

Alternative approaches to calculate $similarity(\mathbf{b}, A_i)$ were considered. There are the following methods of cluster analysis without preliminary training [27,28]:
Hierarchical methods:

- close communication method;
- method of middle link of King;
- Ward's method.

Iterative methods of grouping:

- MacQueen k-means method.

Algorithms of graph cutting type:

- method of correlation galaxies Terentyev;
- Wroclaw taxonomy method.

Taking into account the specificity of the problem of adding new items to one of the existing clusters, it is recommended to use the k-means method, where the values in clusters and centroids are known by default.

The MacQueen k-means method [27,28] is a method of cluster analysis, which aims on the division of m observations (objects) into k clusters, where each observation refers to the same cluster which center (centroid) the closest.

Similarity function by k-means method with fixed centers is calculated as follows:

$$similarity(\mathbf{b}, A_i) = \frac{1}{1 + \rho(\mathbf{b}, \mu(A_i))}, \tag{12}$$

where $\rho(\mathbf{b}, \mu(A_i))$ is the function of the distance between the vectors of inflection points, calculated by the formula (9) or (10), $\mu(A_i)$ is the centroid of cluster A_i, which is calculated by the formula:

$$\mu(A_i) = \frac{1}{|A_i|} \sum_{\mathbf{a} \in A_i} \mathbf{a}, \tag{13}$$

where $|A_i|$ is the power of the i-th cluster or the number of vectors of inflection points that belonging to this cluster.

The second approach is to calculate the similarity function using the sum of distances of inflection points vector $\mathbf{b}$ to all vectors of the cluster A_i:

$$similarity_2(\mathbf{b}, A_i) = \frac{1}{1 + \sum_{\mathbf{a} \in A_i} \rho(\mathbf{a}, \mathbf{b})}, \tag{14}$$

where $\rho(\mathbf{a}, \mathbf{b})$ is the function of the distance between the inflection points vectors, calculated by the formula (9) or (10).

The third and last approach is to calculate the membership function as the maximum of similarity across all vectors of the cluster A_i:

$$similarity_3(\mathbf{b}, A_i) = \max_{\mathbf{a} \in A_i} \frac{1}{1 + \rho(\mathbf{a}, \mathbf{b})}. \tag{15}$$

After entering the inflection points of the object into clusters, the condition of minimization the total quadratic deviation of cluster points from centroids of these clusters is shown above:

$$\min[\sum_{i=1}^{k} \sum_{\mathbf{a} \in A_i} \|\mathbf{a} - \mu(A_i)\|^2] \tag{16}$$

where $\|\mathbf{a} - \mu(A_i)\|$ is the distance between inflection point vectors calculated by one of the formulas (9) or (10), $\mu(A_i)$ is the centroid in the cluster A_i.

Using the described method causes difficulties in the coordination between the results of different levels of detail. To overcome these difficulties it's useful to expedient the usage of parity vote algorithm.

The essence of the algorithm is to find one head element in the majority. The main one is that it is repeated more than half times for the existing number of elements. If there is no majority, the algorithm will not detect this fact, and still provide one of the elements. The algorithm may not find the item that has the most repeated value if the number of repetitions is not big.

2   Software Implementation

An information technology for structural and statistical identification of hierarchical objects [24] and automated video surveillance system based on such hierarchical object identification [25]. The proposed methods were implemented in the software system.

Figure 1 represents the main window of the system, where image can be uploaded and system automatically allocate faces, presenting the percentage of similarity of available faces in the database. Selecting an object allows you to view its image in the form on the right. In current sample there were no images from the loaded picture in the database, therefore the similarity percentage is low.

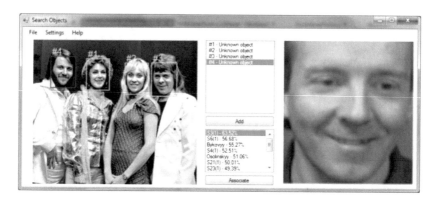

Fig. 1. The main window of the system

Unknown objects could be added to the system, see Fig. 2.

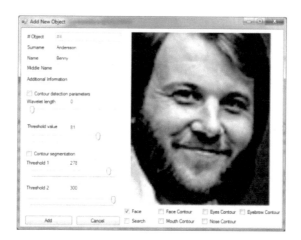

Fig. 2. Window of adding new object to the database

After the object was entered into the database, the upload of different image of the same objects gives 63.97% of similarity with the existing in the database (Fig. 3).

Fig. 3. Identification of available in the database image

Selecting the menu "Settings $\longrightarrow$ View objects" will display a window with available in the database objects and its vector clusters (Fig. 4).

Fig. 4. Display of available in the database objects

3 The Effectiveness of Classification Methods

The image faces database ORL [29] of AT&T laboratory, the Cambridge university, were used to test the system. Table 1 presents the results of proposed methods of distance estimation between characteristic point vectors (MDE) (based on centroid, sum and the maximum) and schemes to calculate the distance between characteristic point vectors (classical Euclidean metric and modified based on the distance from the origin point) on a sample of 40 people, where 4 considered as suspicious.

To evaluate the values of errors in classification, two objects were selected: the first shows the face of investigated (wanted) person, and the second – image of face that is not wanted (regular person).

Table 1. Comparison of the classification methods

Detecting Object 1 in a frame of 4 people				Detecting Object 2 in a frame of 4 people			
	Classification method				Classification method		
MDE	Centroid (12)	Sum (14)	Max (15)	MDE	Centroid (12)	Sum (14)	Max (15)
Classic (9)	66.3%	56.1%	73.0%	Classic (9)	12.5%	20.6%	10.1%
Modified (10)	65.5%	57.2%	78.4%	Modified (10)	9.3%	15.6%	3.4%
Detecting Object 1 in a frame of 20 people				Detecting Object 2 in a frame of 20 people			
	Classification method				Classification method		
MDE	Centroid (12)	Sum (14)	Max (15)	MDE	Centroid (12)	Sum (14)	Max (15)
Classic (9)	58.2%	43.8%	69.8%	Classic (9)	17.3%	31.2%	22.4%
Modified (10)	61.0%	49.0%	65.3%	Modified (10)	14.5%	16.5%	5.3%
Detecting Object 1 in a frame of 40 people				Detecting Object 2 in a frame of 40 people			
	Classification method				Classification method		
MDE	Centroid (12)	Sum (14)	Max (15)	MDE	Centroid (12)	Sum (14)	Max (15)
Classic (9)	43.3%	35.6%	51.1%	Classic (9)	23.3%	33.4%	30.1%
Modified (10)	65.5%	57.2%	58.2%	Modified (10)	19.4%	22.3%	17.1%

Classification errors are divided into the first (I) and the second (II) type
[30]. The error of the first type is the failure of the wanted person classification
(object 1) on the image, where a priori it is known that it exists there. The
error of second type - false classification of a person as being wanted, when a
priori it is known that this is not true (objects 2). From the table it is clear that
with the increase of the number of people in the frame worse the severity of the
second type errors (ordinary people referred to suspicious). If the number of in
the frame people is increased up to 8–10 times, the automated video surveillance
system makes a decision, but it is not always reliable.

Such result can be explained by a low resolution of images that contains large
amount of faces, in this case, the human operator makes the final decision.

4 Conclusions

Use of structural and hierarchical contour segmentation allows to improve the
performance of automated video surveillance systems by reducing the amount
of information required for processing at each level of the hierarchy for several
times. Three approaches for objects classification were proposed and analyzed:
based on centroid, sum and maximum; and two alternative schemes for the
calculation of distance between inflection points vectors: the classical Euclidean
meter and the modified, based on distances from the origin of coordinates.

On a sample of 40 people, including 4 suspicious the classification authenticity
by second type of errors was reduced 5–10%.

According to the Table 1, it can be concluded that the highest quality results
of classification ratio provides methods for evaluating the distance between characteristic point vectors by modified Euclidean metric (10) and the maximum
method (15).

References

1. Fleck S, Strasser W (2008) Smart camera based monitoring system and its application to assisted living. Proc IEEE 96(10):1698–1714
2. Patrick R, Bourbakis N (2009) Surveillance systems for smart homes: a comparative survey. In: 21st IEEE international conference on tools with artificial intelligence, pp 248–252
3. Brezovan M, Badica C (2013) A review on vision surveillance techniques in smart home environments. In: Proceedings of the 19th international conference on control systems and computer science, pp 471–478
4. Kale PV, Sharma SD (2014) Review of securing home using video surveillance. Int J Sci Res (IJSR) 3(5):1150–1154
5. Caputo AC (2014) Digital video surveillance and security, 2nd edn. Butterworth-Heinemann, Oxford, 440 p
6. Schimid C, Mohr R, Bauckhane C (2000) Evaluation of interest point detectors. Int J Comput Vis 37:151–172 2nd edn
7. Rodehorst V, Koschan A (2006) Comparison and evaluation of feature point detectors. In: Proceedings of 5th International Symposium Turkish-German Joint Geodetic Days "Geodesy and Geoinformation in the Service of our Daily Life". Technical University of Berlin, Germany, p 8, March 2006
8. Nain N, Laxmi V, Bhadviya B, Deepak BM, Mushtaq A (2008) Fast feature point detector. In: IEEE international conference on signal image technology and internet based systems, pp 301–306
9. Tuytelaars T, Mikolajczyk K (2008) Local invariant feature detectors: a survey. Found Trends Comput Graph Vis 3(3):177–280
10. Dornaika F, Chakik F (2010) Efficient object detection and matching using feature classification. In: 20th international conference on pattern recognition, Istanbul, Turkey, 23–26 August, pp 3073–3076
11. Jiang D, Yiy J (2012) Comparison and study of classic feature point detection algorithm. In: Proceedings of the international conference on computer science and service system, pp 2307–2309
12. Kurt Z, Özkan K (2013) Description of contour with meaningful points. In: 21st signal processing and communications applications conference (SIU), Haspolat, Turkey, 24–26 April, pp 1–4
13. Liang J, Zhang Y, Maybank S, Zhang X (2014) Salient feature point detection for image matching. In: IEEE China summit & international conference on signal and information processing (ChinaSIP), pp 485–489
14. Liang C-W, Juang C-F (2015) Moving object classification using a combination of static appearance features and spatial and temporal entropy values of optical flows. IEEE Trans Intell Trans Syst 16(6):3453–3464
15. Zhu R, Dornaika F, Ruichek Y (2018) Flexible and discriminative non-linear embedding with feature selection for image classification. In: 24th international conference on pattern recognition (ICPR), Beijing, China, 20–24 August, pp 3192–3197
16. Tarabalka Y, Tilton JC (2012) Improved hierarchical optimization-based classification of hyperspectral images using shape analysis. In: IEEE international geoscience and remote sensing symposium, pp 1409–1412
17. Brik Y, Zerrouki N, Bouchaffra D (2013) Combining pixel- and object-based approaches for multispectral image classification using Dempster-Shafer theory. In: International conference on signal-image technology & internet-based systems, Kyoto, Japan, 2–5 December, pp 448–453

18. Kim J, Kim D (2014) Static region classification using hierarchical finite state machine. In: IEEE international conference on image processing (ICIP), Paris, France, 27–30 October, pp 2358–2362

19. Banerjee P, Patel A, Das S, Seraogi B, Roy R, Majumder H et al (2018) A robust system for visual pattern recognition in engineering drawing documents. In: TENCON 2018 - 2018 IEEE region 10 conference, Jeju, Korea, 28–31 October, pp 2050–2055

20. Paliy I, Dovgan V, Boumbarov O, Panev S, Sachenko A, Kurylyak Y, Zagorodnya D (2011) Fast and robust face detection and tracking framework. In: Proceedings of the 6th IEEE international conference on intelligent data acquisition and advanced computing systems: technology and applications (IDAACS 2011), vol 1, Prague, Czech Republic, 15–17 September. IEEE, pp 430–434

21. Zahorodnia D, Pigovsky Y, Bykovyy P, Krylov V, Paliy I, Dobrotvor I (2015) Structural statistic method identifying facial images by contour inflection points. In: Proceedings of the 8th IEEE international conference on intelligent data acquisition and advanced computing systems: technology and applications (IDAACS 2015), vol 1, Warsaw, Poland, 24–26 September, pp 293–297

22. Zahorodnia D, Pigovsky Y, Bykovyy P, Krylov V, Rusyn B, Koval V (2017) Criteria to estimate quality of methods selecting contour inflection points. In: Proceedings of the 9th IEEE international conference on intelligent data acquisition and advanced computing systems: technology and applications (IDAACS 2017), vol 2, Bucharest, Romania, 21–23 September. IEEE, pp 969–973

23. Bykovyy P, Kochan V, Sachenko A, Markowsky G (2007) Genetic algorithm implementation for perimeter security systems CAD. In: 4th IEEE workshop on intelligent data acquisition and advanced computing systems: technology and applications (IDAACS 2007), Dortmund, Germany, 06–08 September, pp 634–638

24. Zahorodnia D, Pigovsky Y, Bykovyy P, Krylov V, Sachenko A (2018) Information technology for structural and statistical identification of hierarchical objects. In: Proceedings of the 14th international conference on advanced trends in radioelecrtronics, telecommunications and computer engineering (TCSET), Lviv-Slavske, Ukraine, 20–24 February, pp 272–275

25. Zahorodnia D, Pigovsky Y, Bykovyy P, Krylov V, Sachenko A, Molga A (2018) Automated video surveillance system based on hierarchical object identification. In: 14th international conference on development and application systems (DAS), Suceava, Romania, 24–26 May, pp 194–199

26. Starck J-L, Murtagh FD, Bijaoui A (1998) Image processing and data analysis: the multiscale approach. Cambridge University Press, Cambridge, 287 p

27. Gonzalez RC, Woods RE (2007) Digital image processing, 3rd edn. Pearson, London, 976 p

28. Pratt WK (2013) Introduction to digital image processing. CRC Press, Boca Raton, 750 p

29. Image database: AT&T Laboratories Cambridge. http://www.cl.cam.ac.uk/research/dtg/attarchive/facesataglance.html

30. Mansfield AJ, Wayman JL (2002) Best practices in testing and reporting performance of biometric devices. NPL Report CMSC 14/02, 32 p, August 2002

Regression Spline-Model in Machine Learning for Signal Prediction and Parameterization

Ihor Shelevytsky[1]($\boxtimes$) (iD), Victoriya Shelevytska[2]($\boxtimes$) (iD), Kseniia Semenova[3]($\boxtimes$) (iD), and Ievgen Bykov[1]($\boxtimes$) (iD)

[1] Kryvyi Rih Economic Institute of Kyiv National Economic University, 16, Medychna Street, Kryvyi Rih, Ukraine
sheleviv@gmail.com, bykov.evgen@gmail.com
[2] Dnipropetrovsk Medical Academy, 9, Vernadsky Street, Dnipro, Ukraine
shelevika@gmail.com
[3] National Aviation University, 1, Komarova Street, Kyiv, Ukraine
XenaShelev@gmail.com

Abstract. Traditionally, polynomial models are used to analyze and simulate time series. The parameters of these models are estimated using the ordinary least squares (OLS) method. The complex form of the time series requires the complication of the polynomial and the growth of its order. This leads to problems with the conditionality of the equations and the inadequacy of the model. Applying splines allows you to partially solve these problems. In this paper it is proposed to use cubic Hermite splines with infinite first and last fragments to analyze and predict time series, which allows using spline to predict. To estimate the spline parameters, the least squares method is used. For placement of nodes (docking points) algorithms of coordinate optimization with constraints and algorithm of sequential buildup of fragments are offered. In order to avoid over-training or under-training of the spline model, randomness control of residues is proposed. The results are applied for parameterization of heart sounds.

Keywords: Spline · Regression · Algorithm · OLS · Approximation · Machine learning · Time series · Heart tone

1 Introduction

Process forecasting (including time series forecasting) is a typical machine learning problem. Traditionally a regression analysis with algebraic polynomials models is used to solve such problems. The training goal for such models is to choose an order of a polynomial and to find the values of polynomial coefficients. The problem of choosing an order of a polynomial lies in finding a compromise between undertraining (with a low order) and overtraining (with a high order).

© Springer Nature Switzerland AG 2020
V. Lytvynenko et al. (Eds.): ISDMCI 2019, AISC 1020, pp. 158–174, 2020.
https://doi.org/10.1007/978-3-030-26474-1_12

When increasing the order of a polynomial, problems arise in connection with the poor conditionality of the matrices in the ordinary least squares (OLS) method. From a point of view of forecast statistical reliability it is desirable to have more values, but expanding the training interval leads to undertraining. This has to be compensated by the increasing the order. It lowers the reliability of the forecast and can lead to poor condition. Applying a spline as a model allows to avoid these difficulties.

The use of splines is justified in cases where traditional polynomial models cannot adequately describe the existing series and signals. Usually these are series and signals of complex shape. Under the complex shape the authors consider the presence of more than 4–6 non-periodical extremes. This paper will study using splines to analyze time series and signals as an OLS model. The issue of using splines was studied in many papers, however often the barrier for this usage is presence of boundary conditions for last splines fragments, which distort the forecast. However, this paper proposes a method to solve this problem.

The boundary conditions for last splines fragments problem can be evaded using the splines with the last fragment of infinite length. The paper studies such splines. The advantage of splines over classical polynomials is that increasing the order of a polynomial spline does not change the complexity of its basic functions. In addition, the spline models have additional parameters for model training: quantity and layout of nodes of the spline (merging points of the fragments). The paper describes the algorithms for placing nodes (training a spline model) using coordinate optimization with constraints, as well as an algorithm for sequential construction of the fragments.

2   Problem Statement

Let us consider the basic assumptions of the OLS. The numerical series is represented by discrete samples $y_i = f_i + \xi_i$, which are the sum of samples of non-random function $f_i = f(t_i)$ and random samples ξ_i. Random samples have zero mean $M[\xi] = 0$ and uncorrelated $V[\xi] = \sigma^2 \mathbf{I}$. Let's describe the samples y_i with the help of a function $g(\mathbf{A}, t)$ that is linearly dependent on the vector of numerical parameters $\mathbf{A}$ of length r. The function $g(\mathbf{A}, t)$ is a linear combination of basis functions - splines: $g(\mathbf{A}, t) = \sum_{j=1}^{r} a_j B_j(t)$. Values $\mathbf{A}$ need to satisfy minimum value of the standard deviation of the model function $g(\mathbf{A}, t)$ from the original samples: $\sum_i (y_i - g(\mathbf{A}, t_i))^2 \rightarrow \min$. The minimization problem in a matrix form $(\mathbf{Y} - \mathbf{XA})^{\mathbf{T}}(\mathbf{Y} - \mathbf{XA}) \rightarrow \min$ can be described as the system of linear algebraic equations $\mathbf{A} = (\mathbf{X}^{\mathbf{T}}\mathbf{X})^{-1}(\mathbf{X}^{\mathbf{T}}\mathbf{Y})$ of r order. This system is called ordinary. The matrix $\mathbf{X}$ (of the dimension of n rows (number of samples) by r columns) contains the samples of the basic spline functions $x_{i,j} = B_j(t_i)$ in the columns. In case of a random component, the resulting values of the parameters are random numbers $\widehat{\mathbf{A}}$—estimates of the true values $\widehat{\mathbf{A}}$ (obtained by the absence of random samples ξ_i). The linearity of the model with respect to the estimated parameters allows building accurate confidence intervals for the estimates if ξ_i have a normal distribution. The covariance estimation matrix is $V\left[\widehat{\mathbf{A}}\right] = \sigma^2 \left(\mathbf{X}^T\mathbf{X}\right)^{-1}$. Since σ is unknown, we use $\widehat{\sigma}^2 = s^2$ as the estimate with $s^2 = \left(\mathbf{Y} - \mathbf{X}\widehat{\mathbf{A}}\right)^T \left(\mathbf{Y} - \mathbf{X}\widehat{\mathbf{A}}\right) / (n - r - 1)$. The confidence interval of probability α for $\widehat{a}_j$ is $\pm t\left(\alpha, v\right) s\sqrt{g_{i,j}}$, where $t\left(\alpha, v\right)$ is the two-sided $\dfrac{\alpha + 1}{2}$ quantile of the Student's distribution with the $v = n - r - 1$ degree of freedom, $g_{j,j}$ is the diagonal element of the matrix $\mathbf{G} = \left(\mathbf{X^T X}\right)^{-1}$. Thus, the spline model is an ordinary polynomial and does not fundamentally change the OLS.

Let's consider the OLS from a slightly different point of view, as a toll for solving the interpolation problems. In the matrix form $\mathbf{S} = \mathbf{X}\mathbf{A}$, where $\mathbf{X}$ is the matrix with basic spline functions as columns, $\mathbf{A}$ - interpolation coefficients. We assume that the samples of a non-random function are the result of interpolation. Then the OLS can be considered an inverse problem of interpolation. If the spline basis is an interpolation pulse, it is necessary to estimate the values of the spline at the interpolation nodes using known (with an error) interpolated values. We can also agree that the interpolation spline is a model in the OLS.

Therefore we need to construct an interpolation spline, which is most convenient for us to solve forecasting problems in machine learning algorithms. We need to demonstrate the differences between the spline model and the classical algebraic polynomials and build learning algorithms that take into account this specificity. We need to find the best location of the spline nodes. We also use these algorithms to parameterize such complex signals as heart sounds for their classification.

3   Literature Review

Usually the spline is determined as described in [1]. A function $S(t)$ defined and continuous on a segment $[a, b]$ is called a polynomial spline of order m with nodes $x_j \in (a = x_0 < x_1 < \cdots < x_n = b)$ if on each of the segments $[x_{j-1}x_j)$, $j = \overline{1, n}$. $S(t)$ is an algebraic polynomial of m order, at each x_j point some derivative may have a discontinuity. If functions $S(t), \ldots, S^{(m-k_i)}(t)$, are continuous at a point x_j, and the derivative $S^{(m-k_i+1)}(t)$ at a point x_j has a discontinuity, the number $k = \min_{0 \leq i \leq n} k_i$ is called a spline defect.

Let's consider the most simple and well-known signal interpolation problem. A detailed historical review on interpolation from antiquity to modern era is contained in article [2]. The history of the splines begins with the work of Schoenberg [3] published in 1946 and is closely connected with the development of computer technology. An electronic database of spline bibliography is located here www.math.auckland.ac.nz/~waldron/Bib/. Mathematical aspects of the splines were studied in the works of many mathematicians: Alberg [4], Schumaker [5,6], Korniychuk, Ligun [7] and others.

The method of least squares (LS) is the oldest, multiuse and versatile method in the practice of processing of numerical series and signals. Specialists of different fields use different names for the solutions using the OLS and models with linear parameters. Those are a regression analysis, parameter estimation, smoothing, filtering, econometrics, biometrics. In machine learning, getting regression model estimates is called model training. As a result, regardless of the initial content of the problem and the interpretation of the results, we have identical models and computational algorithms for solving systems of linear equations. The OLS is rather fully described in the monograph [8]. Computational aspects of OLS are considered in detail in [9, 10].

The use of splines in machine learning algorithms is discussed in [11], where the advantages of spline approximation of functions are shown in comparison with iterative methods. In [12], it is proposed to use cubic interpolation splines for radial basis function (RBF) approximation in perceptrons and to ensure the continuity of input data. The input data is approximated with smoothing cubic splines. In [13], authors use multivariate adaptive regression splines (MARS) for approximation. In [14], MARS is used for approximation in the methods of reference vectors. The MARS algorithm uses Hinge functions to divide the approximation region into fragments in [15]. This approach creates a good theory, but it is not very convenient in practice.

4 Materials and Methods

4.1 Construction of Interpolation Splines

Let's $B_i(x) = x^i$ are algebraic basis functions. The first derivative is $B_i'(x) = ix^{i-1}$.

For a cubic polynomial $m = 2$, $n = 4$:

$$P_4(x) = a_0 + a_1 x + a_2 x^2 + a_3 x^3 \tag{1}$$

The system of interpolation Eq. (1) for a separate segment of a cubic polynomial is:

$$\begin{bmatrix} 1 & x_1 & x_1^2 & x_1^3 \\ 1 & x_2 & x_2^2 & x_2^2 \\ 0 & 1 & 2x_1 & 3x_1^2 \\ 0 & 1 & 2x_2 & 3x_2^2 \end{bmatrix} \cdot \begin{bmatrix} a_0 \\ a_1 \\ a_2 \\ a_3 \end{bmatrix} = \begin{bmatrix} f_1 \\ f_2 \\ f_1' \\ f_2' \end{bmatrix}.$$

Finding the solution, substituting in (1) and performing the transformations, we get:

$$P_4(x) = \left[f(x_1)\widetilde{X}_0(x) + f(x_2)\widetilde{X}_1(x) + f'(x_0)\widetilde{\widetilde{X}}_0(x) + f'(x_1)\widetilde{\widetilde{X}}_1(x) \right] \tag{2}$$

where

$$\widetilde{X}_0(x) = \frac{2x^3 - 3x^2(x_1 + x_2) + 6x_1x_2x - x_2^2(3x_1 - x_2)}{(x_1^2 - 2x_1x_2 + x_2^2)(x_2 - x_1)},$$

$$\widetilde{X}_1(x) = \frac{(x - x_1)^2(2x + x_1 - 3x_2)}{(x_1 - x_2)(x_1^2 - 2x_1x_2 + x_2^2)},$$

$$\widetilde{\widetilde{X}}_0(x) = \frac{(x - x_1)(x^2 - 2x_2x + x_2^2)}{(x_1 - x_2)^2},$$

$$\widetilde{\widetilde{X}}_1(x) = \frac{(x - x_1)^2(x - x_2)}{(x_1 - x_2)^2}.$$

A general view of the basic functions is easy to obtain by joining the fragments. Formula (2) can be used if the values of the derived numerical series are known.

For the case when only the values of the series are known, we define the derivatives in terms of the central differences:

$$f_1' = \frac{x_2 - x_1}{x_2 - x_0}\frac{f_1 - f_0}{x_1 - x_0} + \left(1 - \frac{x_2 - x_1}{x_2 - x_0}\right)\frac{f_2 - f_1}{x_2 - x_1}$$

$$f_2' = \frac{x_3 - x_2}{x_3 - x_1}\frac{f_2 - f_1}{x_2 - x_1} + \left(1 - \frac{x_3 - x_2}{x_3 - x_1}\right)\frac{f_3 - f_2}{x_3 - x_2}.$$

Substituting in (2) and performing a series of transformations, we finally have the following expressions for determining the value of the spline at the point belonging to the j-th fragment of the spline:

$$h_{j-1} = x_j - x_{j-1}, h_j = x_{j+1} - x_j, h_{j+1} = x_{j+2} - x_{j+1},$$

$$p_1 = \frac{h_j}{h_{j-1}}, p_2 = \frac{h_j}{(h_{j+1} + h_j)}, p_3 = \frac{h_j}{h_{j+1}}, p_4 = \frac{h_j}{(h_{j-1} + h_j)},$$

$$a = \frac{(x - x_j)}{h_j}, b = 1 - a, c = ab^2, d = a^2b,$$

$$X_0(x) = -p_1p_4b, X_1(x) = p_1c - p_2d + b,$$

$$X_2(x) = -p_4c - p_3d + a, X_3(x) = -p_3p_2d.$$

(3)

Therefore the spline is

$$S(x) = \sum_{j=0}^{R} f(x_j)\widetilde{X}_j(x),$$

where $\widetilde{X}_j(x)$ is the local basic spline function consisting of four fragments:

$$\widetilde{X}_j(x) = \begin{cases} X_{0,j-1}(x), & x \in [x_{j-1}, x_j) \\ X_{1,j}(x), & x \in [x_j, x_{j+1}) \\ X_{2,j+1}(x), & x \in [x_{j+1}, x_{j+2}) \\ X_{3,j+2}(x), & x \in [x_{j+2}, x_{j+3}) \\ 0, & x \notin [x_{j-1}, x_{j+3}) \end{cases}.$$

Finally, to calculate the value of the spline at a point on the j-th fragment, we have:

$$S(x) = f(x_{j-1})X_{0,j-1}(x) + f(x_j)X_{1,j}(x)$$
$$+ f(x_{j+1})X_{2,j+1}(x) + f(x_{j+2})X_{3,j+2}(x). \quad (4)$$

Note that with such interpolation there is no need to solve a system of interpolation equations. Interpolated values are obtained with minimal operations, in contrast to two-stage interpolation using the cspline and interp functions in the MathCad package, and rather complicated calculations of the spline function in MATLAB. The values on a separate fragment are fully defined by the nearest four nodes. The boundary conditions are obtained in a natural way. To do this, we consider the extreme fragments tend to infinity. The corresponding values p become zero. This spline is quite suitable for extrapolation. Note that in the case of $c = 0$, $d = 0$, we get a linear spline - a broken line.

The function of calculating the four components of the basic spline functions in the Matlab language (Octave) is shown below.

```
function [n, x] = spl1xu(tu,t,r)
% tu abscissa of nodes spline, t abscis of interpolation points,
% r number of spline nodes
% n number of spline fragment, x vector of values of 4 basic functions
%
for i = 1: r-1% spline fragment search
    if((tu(i)<=t)&(t < tu(i+1)))
        break;
    end
end
if   i>1      %  if the fragment is not the first
    tu1 = tu(i-1) ;
end
tu2 = tu(i);        tu3 = tu(i+1);
if   i<r-1 % if the fragment is not the last
    tu4 = tu(i+2);
end
    hn = tu3-tu2;
if   i>1
    hp = tu2-tu1;
else
    hp = hn;
end
if   i<r-1
    hb = tu4-tu3;
else
    hb = hn;
end
if   i>1
    p1 = hn/hp;   p4 = hn/(hp+hn);
else        % if the fragment is the first
    p1 = 0 ;   p4 = 0 ;
```

```
end
if    i<r-1
      p2 = hn/(hb+hn);    p3 = hn/hb;
else        % if the fragment is the last
      p2 = 0 ;   p3 = 0 ;
end
a = ( t-tu2 )/hn ;
b = 1 - a        ;
c = a * b * b ; d = a * a * b ;
x(1) = -p1 * p4 * c  ;
x(2) = p1 * c - p2 * d + b ;
x(3) = -p4 * c + p3 * d + a ;
x(4) = -p3 * p2 * d ;
n=i;
```

To calculate an interpolation spline on a definite fragment n, it is sufficient to multiply the value of the basis functions to the corresponding spline values in the nodes:

```
tu=[0,1,2,3,4,5];   a=[0,10,8,12,6,3]; t=2.5;
[n, x] = spl1xu(tu,t,5);
s=a(n-1)*x(1)+ a(n)*x(2)+a(n+1)*x(3)+a(n+2)*x(4);
```

4.2   Construction of Interpolation Hermite Splines

LS with a spline model is not fundamentally different from one used for the classical models. Nevertheless, a number of features of spline bases allows to get rid of the problems of poor conditionality and effectively organize the computational process. We use the interpolation model approach. Let us apply the cubic spline obtained above. The difference from the cubic spline of the minimum defect (most studied in theory) is that the continuity of the second derivative is not guaranteed (visually it is not noticeable), but the first derivative is not subject to the discontinuity of the second order. In addition, the weight coefficients of the basis splines (estimated LS parameters) are the values of the spline at the node points. That feature provides the convenience of interpreting the result. The use of other splines does not fundamentally change what was just stated, but it is necessary that the base splines consist of four fragments.

Let's consider an interpolation spline built on a set of nodes $T = \{tu_1, tu_1, \ldots, tu_R\}$. We need to find the values of the spline S on the set N of $t = \{t_1, t_1, \ldots, t_N\}$ points, and $N >> R$. The system of equations is:

$$S(t_i) = \sum_{j=1}^{R} f_j X(t_i), i = \overline{1, N},$$

or in matrix form $\mathbf{S} = \mathbf{XA}$.

In this system of equations, the form of the matrix $\mathbf{X}$ is important. The values of the j-th column are the values of the j-th basis spline. Due to the local properties of the basic splines, the matrix $\mathbf{X}$ is block-diagonal:

$$\begin{bmatrix} \cdots & \cdots & \cdots & \cdots & \cdots & \cdots & \cdots \\ \cdots G1_{j-1} & 0 & 0 & 0 & 0 & \cdots \\ \cdots G2_{j-1} & G1_j & 0 & 0 & 0 & \cdots \\ \cdots G3_{j-1} & G2_j & G1_{j+1} & 0 & 0 & \cdots \\ \cdots G4_{j-1} & G3j & G2_{j+1} & G1_{j+2} & 0 & \cdots \\ \cdots & 0 & G4_j & G3_{j+1} & G2_{j+2} & G1_{j+3} \cdots \\ \cdots & \cdots & \cdots & \cdots & \cdots & \cdots & \cdots \end{bmatrix}, \tag{5}$$

where $G1_j, G2_j, G3_j, G4_j$ are matrix-columns, which consist of samples of the basis on the corresponding adjacent fragments. The specific shape of the matrix $\mathbf{X}$ determines the seven-diagonal symmetric shape of the matrix $\mathbf{C} = \mathbf{X}'\mathbf{X}$. Often when solving the LS there is no need to store the matrix $\mathbf{X}$. It is much easier to directly calculate the matrix $\mathbf{C}$ and vector $\mathbf{B} = \mathbf{X}'\mathbf{Y}$. The matrix $\mathbf{C}$ can also be stored as four diagonals.

The peculiarity of the matrix $\mathbf{C}$ distinguishes it from similar ones for classical models. An increase in the number of parameters (spline nodes) does not significantly affect the matrix order. This is due to the fact that an increase in the number of estimated parameters leads to an increase in the number of blocks in the planning matrix $\mathbf{X}$. The elements of the planning matrix for a spline have the same order (as nodes of a spline basis).

Since the changes in the norm of the matrix $\mathbf{C}$ are insignificant, and taking into account the good initial conditionality of the system of equations, we should expect a very slight change in the conditionality of the extended system of equations. This conclusion is confirmed empirically by numerous examples of calculations.

Let us show an example of calculations for constructing a spline using least squares.

```
% MODELING INPUT DATA
n = 100; % NUMBER OF INITIAL DATA
sigma = 0.1; % NOISE
t = [-1:2/(n-1):1]'; % ABCISS OF THE DATA
% MODEL VALUES WITHOUT INTERFERENCE


f = 1./sqrt(1+25*t.^2); % EXAMPLE
% f = sin(1*pi*t);
eps = normrnd(0,sigma,size(f)); % MODELING INTERFERENCE
y = f+eps;
% LS SOLUTION
r = 5; %ORDER OF LS MODEL
tu = [-1 -0.5 0 0.5 1]; % nodes
X=creat_ps(tu,t,100,r); % PLANNING MATRIX
```

```
C=X'*X;
G=C^(-1);     % INVERSE MATRIX
B=X'*y;
A=G*B;        % LS ESTIMATES OF MODEL PARAMETERS
Y=X*A;        % FORECAST LS ESTIMATES
e=y-Y;          % REMAINS
s=sqrt((e'*e)/(n-r-1));      % DEVIATION OF REMAINS
dA=tinv((0.9+1)/2,n-r-1)*s*sqrt(diag(G)); % CONFIDENCE INTERVALS
dY=tinv((0.9+1)/2,n-r-1)*s*sqrt(diag(X*G*X'));
cond(C)       % NUMBER OF CONDITION
figure(1); plot(X); title('Base functions');
figure(2); spy(X);
figure(3); plot(tu,A,'+',t,f,t,y,'.',t,Y,t,Y-dY,t,Y+dY);
function [p]=creat_ps(tu,x,n,r)
%         p  - planning matrix
%         tu  abscissa of the spline nodes
%         x   abscissa of the data
%         n   number of data
%         r   number of nodes

    for  i=1:n
        [z,xx]= spl1xu(tu,x(i),r);
      if (z>1)
        p(i,z-1) = xx(1);
      end
        p(i,z)   = xx(2);   p(i,z+1) = xx(3);
      if (z<r-1)
        p(i,z+2) = xx(4)  ;
      end
    end
```

4.3   Layout and Optimization of the Spline Node Grid

Unlike classical polynomials, the spline has additional parameters that significantly affect the quality of the model. This is the number and location of the
spline nodes. LS allows to find the optimal values of the spline in the nodes for
a fixed grid of nodes. The number of estimated parameters of the spline is equal
to the number of nodes of the spline. With two nodes we get the usual case with
least squares for one segment. The maximum number of nodes is equal to the
number of source data and the task is reduced to the repetition of the source data
(interpolation). If the quality of a model with one fragment is not satisfactory,
then in the case of the classical model the polynomial is complicated. For an
algebraic polynomial, the complication means the increasing of the order, which
causes deterioration of conditionality and a significant increase in computation.
For a spline, the complication of a polynomial means dividing a segment into
fragments. This process does not worsen the conditionality and the growth of

calculations is not so significant. But questions arise with the position of the breakpoints and their optimal number. Perhaps the difficulty with determining the successful placement of nodes is a significant factor of limited of the LS with spline models. After trying one or two schemes of nodes randomly (as a rule on equal fragments), the practicing researcher does not obtain a satisfactory result and refuses the tool. If, however, we managed to get a good result, then it may be necessary to explain the choice of this particular scheme. Using the optimization algorithms that can be considered as machine learning algorithms can help to place the nodes the best way.

In the most general case, for a given number of nodes, we have the classical nonlinear optimization problem with constraints. Such problems are solved numerically using gradient methods. The practice has shown that a method of coordinatewise descent is quite acceptable in terms of the simplicity of the algorithm and the minimum number of settings. The method makes it very easy to take into account spline limitations. The essence of these restrictions is that the nodes must form a strictly increasing sequence and at least one sample must be present on each fragment of the spline. The algorithm is implemented in the MATLAB environment, where the function of searching for the minimum of the function of one variable over a given interval is used for optimization along the coordinate: fminbnd. The program cycles through the optimization of the position of the nodes from the second to the last one (arbitrary order is possible):

$$\min_{tu_j \in [tu_{j-1}+h, tu_j - h]} \Psi\left(tu_j, \{tu_1, \ldots, tu_{j-1}, tu_{j+1}, \ldots, tu_r\}\right), j = \overline{2, r-1},$$

where $\Psi()$ is the objective function (the sum of the squares of the deviation between the model and the data); h - the maximum gap between the data.

The limited scope of the search area $[tu_{j-1} + h, tu_{j+1} - h]$ takes into account the limitations on the ordering of the grid of nodes and the presence of data on each segment. The accuracy of determining the position of the extremum is limited to a certain value ($h/10$), in order to reduce the amount of calculations.

Numerous experiments with optimization of spline nodes have shown that the optimization procedure is quite stable. Optimization tends to move nodes to areas with features (faster changes and breaks). Although it cannot be guaranteed that different starting positions of the nodes and different sequences of their optimization will give the same result, in practice these results are quite similar. This is probably due to the presence of the only one optimal node allocation scheme for the deterministic case. The method works with a fixed number of nodes.

The question of the correct choice of the parameters of the algorithm requires the choice of a specific criterion for the quality of the regression model. In classical methods, the coefficient of determination and the corrected coefficient of determination or simply the sum of the squares of the error are used. The disadvantage of the coefficient of determination is that it formally shows the relationship between the variance of the input data and the residuals, the real value of which is unknown.

It is proposed to use the statistical criterion for the probability of randomness of residuals, defined on the nearest fragments. It is based on the hypothesis of the presence of an additive data model as the sum of the deterministic component and uncorrelated noise. If there is a random component in the data, the analysis of the quality of the model is associated with the analysis of residuals:

$$e_i = y_i - S(t_i, A), i = \overline{1, N}.$$

Under the condition that the random component is uncorrelated (the conditions of the Gauss – Markov theorem), the residuals of the qualitative model should also be random. The qualitative model should approximate only the deterministic component of the data and the criterion should reflect the probability of their randomness. This criterion is convenient for a number of reasons. The application of the criterion to the input data makes it possible to priori determine the feasibility of further regression analysis (if the data are not random). With an excessive increase in the number of estimated parameters (decrease in degrees of freedom), the criterion shows a deterioration of the model. The criterion has non-parametric solutions and requires a minimum of a priori information.

It is possible to divide the data into deterministic and random components knowing the flexibility of the spline model, which depends on the width of the fragments. The problem can be formulated as follows: it is necessary to find such grid of spline nodes that with the minimum number of nodes ensures the absence of a deterministic component in the residuals. The absence of a deterministic component in the residuals should be understood in a statistical sense. That is, a given level of error of the first kind should be guaranteed when testing the null hypothesis of the random nature of the residuals. There are a number of statistical criteria for checking the random nature of the data. The most convenient are the non-parametric criteria for series and inversions. These criteria require a minimum of a priori information and allow the effective implementation.

The approach described above allows sequential construction of the spline. At the beginning of construction, it is enough to have at least four data values to build one fragment of a spline. A spline fragment can include these points and have a minimum length. Upon receiving of the next data reference, the right boundary of the fragment moves so that the new observation point belongs to the spline. If necessary, the fragment length can take into account the forecast interval. After the LS calculation of the spline, the inversion criterion calculates and compares with the critical values. In the absence of a deterministic component, the process of the interval increasing continues. If the residual randomness hypothesis cannot be accepted, the last node returns to its previous position and a new node and a new fragment of the spline are added, it includes the new data node. To reduce the amount of calculations, the number of samples for which the criterion is calculated, is limited to 20–40 (if a sufficient amount of data is available). It should be noted that the addition of fragments affects the estimates of the next three to four nodes. If the changes lead to the appearance of a deterministic component in the residuals, then they are compensated in the process of building the next fragments by reducing the subsequent fragments.

This process can lead to the formation of a whole series (3–8) of close nodes and fragments with one node. Such a process is capable of forming an excessive number of nodes and reduces the degrees of freedom in statistical evaluations. To reduce this phenomenon, two restrictions are introduced: the criterion is calculated only on the last fragment and when more than four samples are received. The fragmentation of fragments largely depends on the specifics of the data and the level of error of the first kind.

5   Experiments

To test the results of the proposed algorithms a test time series is synthesized as a sum of a sinusoid fragment and a white noise with SD of 0.1. $t = 0 : 0.02 : 6$; $s = sin(t)$; $eps = normrnd(0, 0.1, size(s))$; $y = s + eps$; The spline model is built using a sequential extension algorithm for spline fragments with a type I error set to 0.01, criteria calculation interval of 20 samples and a forecast interval of 10 samples. The result is shown in Fig. 1. The spline nodes are shown with "plus"-signs, the confidence intervals in the nodes are shown with "star"-signs, the confidence intervals between the nodes are shown with lines.

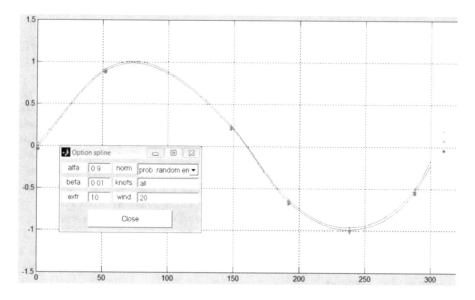

Fig. 1. The spline model with sequential construction of the fragments with a type I error set to 0.01

Increasing a type I error setting leads to excessive spline fragmentation and multiplication of nodes near extremum points. This is similar to the process of overtraining the model, when the model describes not only the deterministic dependence but also the random fluctuations. The result obtained with an

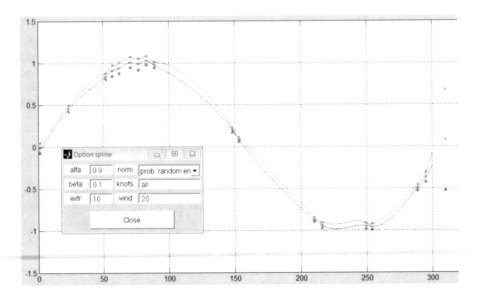

Fig. 2. The spline model with sequential construction of the fragments with a type I error set to 0.1

increase in the level of significance to 0.1 is shown in Fig. 2. Decreasing a type I error setting leads to decrease spline fragmentation and may lead to under-training of the spline model. The result obtained with an increase in the level of significance to 0.001 is shown in Fig. 3.

The method is applied to other data from the real task of processing and classifying heart sounds. The heart sounds were recorded in newborn babies to form a training sample of sounds of a healthy heart and heart that has medical problems. Such records do not contain significant random noise and the problem is reduced to the spline approximation of the complex shape curve. The heart tones are located in the in the sound series. [16] The preliminary data processing task is parametrization of tones and reduction of polynomial order for further classification and preservation in the database. The sounds are recorded at a frequency of 44.1 kHz. Figure 4 shows parametrization of the second tone with the help of 14 nodes spline (13 fragments) with the nodes placement optimization based on a uniform grid. The algorithm allow describing the tone by the model with only 14 extremums.

6   Results and Discussion

Cubic Hermite spline with infinite last fragments is demonstrated as an OLS model in time series processing. Unlike conventional polynomial models with unlimited basic functions, spline splint models have additional parameters that provide flexibility and allow to complete additional study. The essence of this

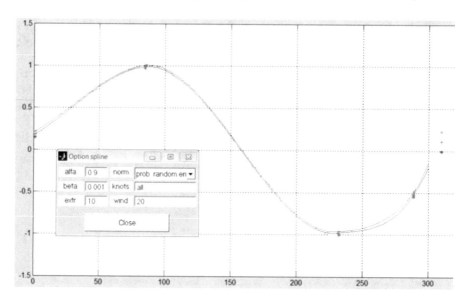

Fig. 3. The spline model with sequential construction of the fragments with a type *I* error set to 0.001

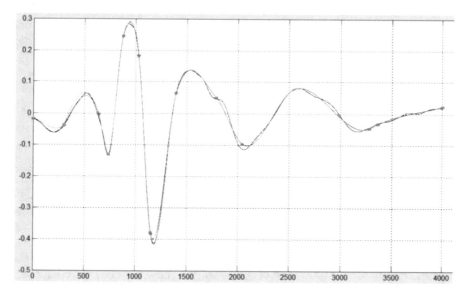

Fig. 4. Parametrization and reduction of the dimension of the spline data for the second tone of the sound of the heart after OLS method

study is to find the optimal in a certain sense positioning of the spline fragments in parallel with OLS values optimization at the nodes.

The local nature of the basis spline functions provides a good condition for the OLS method equations. The increase in an order of the spline polynomial is reduced to adding a spline fragment (increasing the number of fragments) without complication of the basic functions and calculations. If some simple conditions are guaranteed (for example, the presence of at least one point of the input data on each fragment of the spline) then the growth of order of the model does not worsen the condition. Table 1 shows how the condition number of the OLS equations changes with an increase in the number of parameters (with the same number of data on each fragment). In practice, we are talking about almost perfect condition.

Table 1. OLS matrix condition number.

r	5	20	25
cond(C)	1.9691	1.8547	1.8300

The application of statistical test of randomness of the residues deserves a particular attention. It is expedient to use when there is a significant level of additive uncorrelated noise in the data. The test does not solve the problem of undertraining or overtraining, but it allows to manage this process with one clear parameter - type I error. High value of type I error can lead to overtraining and low value to undertraining. Perhaps for additive models with uncorrelated noise the p-value of the probability of the assumption that the null hypothesis is correct can be an indicator of the level of training. This requires additional research.

7   Conclusion

The cubic Hermite spline is used as an ordinary least squares (OLS) model for time series processing. Hermite splines interpolation has been demonstrated to allow extending the last fragments to infinity. In this case the boundary conditions do not distort the forecast. This spline option allows its usage for time series forecasting.

The application of the OLS spline model does not fundamentally change the expressions. However, the growth of spline polynomial order does not complicate calculations and does not lead to an increase in condition number. This approach gives additional to OLS parameters for optimizing the model. These are abscissas of spline nodes and quantity of the fragments.

Two algorithms for spline nodes placement optimization have been proposed: successive coordinate optimization with constraints and successive buildup of spline fragments. The experiments with synthesized model data and real data of sound tones demonstrate stability of the algorithms and possibilities for machine learning.

The statistical test of randomness of the residues has been proposed for spline approximation as one of the criteria for optimizing the location of spline nodes. It is expedient to apply this criterion to the series, which are the sums of a deterministic function and an uncorrelated noise.

This test allows successively building a spline so that the residues of the approximation remain random. This way we avoid overtraining of the model. However, high significance threshold values of the randomness test lead to overtraining. This creates the significance threshold parameter that can be predictably regulated.

The application of the discussed algorithms to the processing of the synthesized time series and the parameterization and order reduction of the real data series (the second heart sound tone parametrisation) has been shown.

References

1. Mathématicien EU, De Boor C (1978) A practical guide to splines. Springer, New York, vol 27, 325 p
2. Mejering E (2002) A chronology of interpolation: from ancient astronomy to modern signal and image processing. In: Proceedings of the IEEE conference, vol 90, no 3, pp 329–342
3. Schoenberg I (1946) Contribution to the problem of approximation of equidistant data by analytic functions. Q Appl Math 4:45–99, 112–141
4. Ahlberg JH, Nilson EN, Walsh JL (1967) The theory of splines and their applications. Academic Press, New York
5. Schumaker LL (2007) Spline functions: basic theory, 3rd edn. (with supplement), Cambridge University Press, Cambridge, 600 p
6. Schumaker LL (2015) Spline functions: computational methods. SIAM, Philadelphia, 422 p
7. Korneichuk NP, Ligun AA (1981) Error bound of spline interpolation in an integral metric. Ukrainian Math J 33(3):301–303
8. Rao CR, Toutenburg H et al (2008) Linear models: least squares and alternatives. Springer series in statistics, 3rd edn. Springer, Berlin, 571 p
9. Hamming RW (1987) Numerical methods for scientists and engineers (Dover books on mathematics), 2nd revised edn. Edition Dover Publications, 752 p
10. Lawson CL, Hanson RJ (1995) Solving least squares problems, vol 15, 336 p. https://doi.org/10.1137/1.9781611971217
11. Richardson J, Reiner P, Wilamowski BM (2015) Cubic spline as an alternative to methods of machine learning. In: IEEE 13th international conference on industrial informatics (INDIN), Cambridge, pp 110–115. https://doi.org/10.1109/INDIN.2015.7281719
12. Marsland S (2015) Machine learning: an algorithmic perspective. Machine learning & pattern recognition series, 2nd edn. Chapman and Hall/CRC, 430 p
13. Sekhar Roy S, Roy R, Balas VE (2018) Estimating heating load in buildings using multivariate adaptive regression splines, extreme learning machine, a hybrid model of MARS and ELM. Renew Sustain Energy Rev 82:4256–4268. https://doi.org/10.1016/j.rser.2017.05.249
14. Lee T-S, Chiu C-C, Chou Y-C, Lu C-J (2006) Mining the customer credit using classification and regression tree and multivariate adaptive regression splines. Comput Stat Data Anal 50(4):1113–1130. https://doi.org/10.1016/j.csda.2004.11.006

15. Friedman JH (1991) Multivariate adaptive regression splines. Ann Stat 19(1):1–67
16. Shelevytsky I, Shelevytska V, Golovko V, Semenov B (2018) Segmentation and parametrization of the phonocardiogram for the heart conditions classification in newborns. In: 2018 IEEE second international conference on data stream mining & processing (DSMP). https://doi.org/10.1109/dsmp.2018.8478495

Intelligent Agent-Based Simulation of HIV Epidemic Process

Dmytro Chumachenko[1]([envelope]) [iD] and Tetyana Chumachenko[2] [iD]

[1] National Aerospace University "Kharkiv Aviation Institute", Kharkiv, Ukraine
dichumachenko@gmail.com
[2] Kharkiv National Medical University, Kharkiv, Ukraine

Abstract. The research deals with the problems of predicting the epidemic process of HIV infection. To solve this problem, an intelligent agent-based model has been developed, which allows to take into account factors affecting the incidence, distribution by age categories and risk groups. The model uses statistical data on the real incidence of HIV infection in the Kharkiv region from 2010 to 2018. The intelligent agent-based model is implemented in the Visual Studio 2013 development environment with use of the C# programming language. The developed software package shows high accuracy of the forecast.

Keywords: Expert system · Knowledge assessment ·
Infections related to medical care

1 Introduction

The problem of spreading of infectious diseases has always been extremely important [1]. Urbanization, the constant increase in the population of the Earth, the decrease in the total number and pollution of fresh water on the planet are all factors that to one degree or another contribute to the emergence of ever newer, more resistant to known, methods of treating infections [2]. Moreover, the growth rate only increases with time [3]. This situation leads to the fact that the modern epidemiology of infectious diseases is one of the fastest growing and most important sciences of our time, on the scale of all mankind [4].

The epidemiology of infectious diseases as a science about the patterns of development and termination of the epidemic process, ways of its limitation and elimination has accumulated a wealth of theoretical, practical and experimental material, which is the basis for the development and implementation of scientifically based effective measures to prevent and combat infectious diseases [5].

2 Background on HIV Infection and Simulation of HIV Epidemic Process

HIV infection is a global threat to the public health and social development of mankind [6]. According to the World Health Organization (WHO), at the end of

© Springer Nature Switzerland AG 2020
V. Lytvynenko et al. (Eds.): ISDMCI 2019, AISC 1020, pp. 175–188, 2020.
https://doi.org/10.1007/978-3-030-26474-1_13

2016, there were approximately 36.7 million people living with HIV in the world, and 1.8 million people became infected with HIV in 2018. In 2018, 1.0 million people worldwide died of HIV-related causes [7].

Key population groups are those at increased risk of HIV infection, regardless of the type of epidemic or local conditions. Such groups include: men who have sex with men, people who inject drugs, people who are in prisons and other isolated conditions, sex workers and their clients, as well as transgender people [8].

Behavioral characteristics make these groups of people vulnerable to HIV and make it difficult to access testing and treatment programs [9]. The use of antiretroviral drugs for the treatment of HIV-infected people can increase life expectancy, improve its quality and reduce the risk of transmission of infection with risky behavior [10].

According to WHO, over the period from 2000 to 2018, 13.1 million lives were saved in the world due to antiretroviral therapy (ART) [11]. However, despite the effectiveness of ART for the long-term preservation of life of HIV-infected people, this disease is still incurable for life, which dictates the need to improve preventive measures [12].

During the ten months of 2018 in Ukraine, according to the Ukrainian Center for Control over Socially Dangerous Diseases of the Ministry of Health of Ukraine [13], 13,381 new cases of HIV infection were registered (of which 2,349 children under 14 years old). Since 1987, a total of 293,739 new cases of HIV have been officially registered in Ukraine, during which time 40,816 people have died of AIDS. The regions most affected by HIV are the Dnipro, Donetsk, Kyiv, Mykolaiv and Odessa regions, as well as Kiev [14].

The lack of effective treatment, vaccination currently does not allow the authorities and health care institutions of Ukraine to adequately carry out epidemiological surveillance, to carry out the full range of measures to combat the HIV epidemic [15]. The current direction of the strategy to combat HIV infection is determined by a scientifically-based system for evaluating the epidemiological situation of HIV infection using mathematical models that are likely to reveal patterns of the epidemiological process and predict the expected incidence rate [16]. At present, the World Health Organization uses three mathematical models to obtain the estimated data "Software package for estimation and forecasting" [17], "Workbook" [18], and the Spectrum program [19]. These models provide, to some extent, rough estimates of changes in the rate of HIV infection over time, the number of people living with HIV, new infections and deaths due to AIDS, children orphaned by AIDS, and treatment needs. However, the parameters that form the basis of these models, significantly need to be adjusted, taking into account epidemiological and sociological studies in specific territories.

In order to develop effective prevention programs, it is necessary to identify the leading risk factors, primarily behavioral, that contribute to the spread of the disease. At the same time, different factors may prevail in different regions of the world and in different territories, depending on the socio-economic, cultural, religious, and other historically established behaviors of the population of a particular region or country. Predicting the dynamics of the spread of morbidity will identify the leading risk factors for the spread of HIV infection in specific conditions [20]. The most effective tool is a simulation mathematical model, the initial data for the construction of which should be data on the incidence and behavioral characteristics of risk groups of the territory for which the preventive program is being developed [21].

A large number of information products for healthcare [22–24], as well as approaches to modeling the epidemic process of various infectious diseases [25–27] show their effectiveness.

3   Problem Statement

Thus, this research proposes an intelligent agent-based model for the spread of HIV infection, the result of which is a prediction of the dynamics of morbidity. The research used data from the statistics on the incidence of HIV infection in the Kharkiv region and the results of a survey of people at risk for the period from 2010 to 2018.

4   Materials and Methods

4.1   Intelligent Agent-Based Simulation as an Approach to Forecasting Epidemic Processes

Speaking about mathematical modeling, one cannot but pay attention to the evolutionary process of changing the paradigms of modeling, which is characteristic of many disciplinary areas in which methods of control theory are used [28]. This process began to be considered in modern scientific research, as a "change of generations" of mathematical models, most recently. Nevertheless, now we can talk about three such generations. At the first stages, we most often talk about the mathematical recording of separate fragments of observations of real objects [29]. They are characterized by simplicity of descriptions, typical linearity of equations and small dimension (often only one or two variables are reproduced). Analysis methods are mainly related to obtaining analytical solutions and graphical viewing on the phase plane. The next stage is the models describing the object in its entirety – in them the object is presented as a system – the model reflects its structure and the laws according to which it functions [30]. Models become essentially non-linear, a purely mathematical apparatus is supplemented with a logical-semantic one. Dimension increases, reaching several dozen. Such models are called "complex", "large", and the computational experiment becomes the working tool at this stage. Currently, the actual subject of research is the transition to the third generation of mathematical models – the models of the virtual world [31]. Virtual modeling can be defined as the reproduction of the three-dimensional world by computer means. The volume of information processed and reproduced sharply increases (for example, the number of "parts" visualized reaches several thousand). In the task of mathematical modeling, in addition to the object of modeling and the model, the subject of modeling, the person is necessarily present, by the efforts and in whose interests the model is implemented [32]. The role of the subject of modeling is decisive, because it is his goals, interests and preferences that form the model.

It was the specific needs of the subject of modeling that led to the fact that such an area of research of objects and building models as simulation modeling separated from mathematical modeling. In the aforementioned "complex" models of virtual processes, there are often many components that are similar in nature but different in behavior [33]. Similarity means that components have a number of properties that play the role of control parameters in modeling. Difference means the same individuality in behavior, that is, depending on the values of certain characteristics of objects, each component of which has its own unique line of conduct [34]. In such models, as a rule, the subject of modeling is the task of predicting the behavior of a group of such objects with different sets of input (control) parameters.

The type of modeling described above is called agent-based modeling [35]. We give the same definition. Agent modeling is a simulation method that investigates the behavior of the entire system as a whole [36]. Unlike system dynamics, the analyst determines the behavior of agents at the individual level, and global behavior arises as a result of the activities of a multitude of agents [37].

An agent is a certain entity that has activity, autonomous behavior, can act in accordance with a certain set of rules, interact with the environment and other agents, and, in the course of operation, can change its behavior and take into account changes in the external environment [38].

An agent-based model that provides for the interaction of a large number of particles and contains a discrete event model in its basis is called a multi-agent model. Multi-agent simulation modeling of active systems is a new concept of intelligent information technologies [39]. It is focused on the joint use of models and methods of natural and artificial intelligence for virtual research, identification and prediction of the state and behavior of active systems in a given environment. The principal difference in the new concept of modeling is the introduction and formalization of sensory connections (variables) between the interacting active elements of the dynamic system. At an early stage of execution, the multi-agent model is a set of state variables describing the model, as well as a list of planned events that should occur in the future.

4.2 Intelligent Agent-Based Model of HIV Epidemic Process Simulation

To simulate the dynamics of the spread of infection, model was developed containing all the principles of agent-based modeling. HIV infection was chosen as a disease, the main mechanism of transmission of which is contact (sexual contact, contact with blood and biological fluids of a person, blood transfusion, etc.). This is an anthroponotic infection, i.e. virus transmission is from one person to another. The main idea of the method is the principle of decentralization of agent-based models. The behavior of agents is set at the individual level, and the dynamics of the system is defined as the result of the interaction of multiple agents. Based on the agent-based approach, a simulation model of the spread of infection has been developed and the rules for the interaction of agents have been defined. The model assumes the following basic assumptions: there is a probability of interaction of any agent with any other of the system; the unit of time is one iteration, the countdown starts from zero iterations, the time step is equal to one; all agents are divided into types, the system of rules is determined for each type of object.

In the model, the following agent states are defined: Susceptible (S), Infected (I), Curing (C), Recovered (R), Dead (D). The transition between states is shown in Fig. 1.

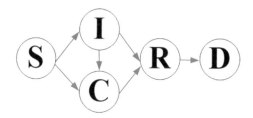

Fig. 1. Scheme of agent's transition between states

Agent properties are the following:

- type of agent (prudent and unreasonable);
- time of the next event: t;
- agent age: t_old;
- location in which the agent is located: current_location;
- type of the nearest event (death, transition to the infected zone, transition to a safe zone or birth);
- position of the agent in the location;
- status of the agent relative to the disease (0 – susceptible, 1 – infected, 2 – curing).

Agents in the system are objects with a common set of properties, but with individual numerical values. They differ in type, each of which has its own line of conduct. Rules of behavior are given by initial probabilities, which remain unchanged until the end of the program.

Locations are one-dimensional dynamic arrays, the size of each is given and can be maximally equal to the total number of agents in the system. Zones contain the number of agents that reside in them during a given period of time, their numbers, cells (array indices) with infected fields, as well as the "busy" identifier of the cell. In one place of a location there can be no more than one agent.

The system is a dynamic array of agents. It sets all the numerical character-istics of locations, such as the geometric dimensions (array length), also creates each of the agents with their own unique sets of properties.

System properties are follows:

- average age of the agent: MaxAge;
- number of agents processed per iteration;
- probabilities (different for each type of agent);
- global system time.

Input data is divided into two types: initial data for system management and data describing the individual properties of agents.

Parameters for the operation of the system: the total number of agents in the system, the number of agents processed per iteration, the number of infected zones, the maximum age of the agent.

In this case, the unique properties of agents are influenced by the probabilities of transition from one location to another, according to its type:

- HealthyInRisk: being healthy go to the risk zone;
- SickInRisk: being sick, go to the risk zone;
- SickInHosp: consent to hospitalization;
- HealthOnHosp: a favorable outcome of treatment.

The initial position of the agent *current_location* is 0, that is, all agents are in the safe zone. When the system starts up, a type of the nearest event is randomly assigned to each of the agents, according to the entered probabilities and its time, in the interval from 0 to *MaxAge*. The main idea that implements the agent-based approach is an event modeling apparatus. This is achieved by implementing a tournament algorithm or a heap algorithm [40]. It contains the numbers (array indices) of the agents - *ind_treat*. After randomizing the time and type of the nearest event, the initial construction of the heap is conducted. Hip can only be built for 2^m objects. If the total number of agents does not constitute any power of two, the array is supplemented with fictitious elements, the time of which is obviously large. The event structure, in this case, serves only to track global time. At the top of the heap there is an agent with a minimum (that is, with the closest) event time. Next, the event is handled for the *ind_treat* agent, that is, the transition from the current location to the one provided for by the event in question. In turn, the current location (*current_location*) becomes the location to which the agent has moved. This agent is assigned a new type of the nearest event and a new, random time. Then the hip is rebuilt. This processing is carried out using the *switch()* block and conditional statements. The probabilities are implemented using *random()* random number generation. Since randomization has a uniform distribution, the required probability is achieved by generating a random number ranging from 0 to $1/P\_i$, where P_i is the particular probability entered.

These transitions may entail an agent status change. For example, if an agent has moved to an unsafe area on an infected area, its status will change from 0 to 1, of course, if he has not been sick before or has no immunity.

Infected agents, depending on the type, may be admitted to hospital. If this condition is fulfilled, then with a certain probability the patient can recover, respectively, his condition will change from 2 to 3.

The algorithm realizes the ability of agents to interact with each other. Every time the *ind_treat* agent hits the new *current_location*, it will be placed a randomly selected index within the dimension of the array representing the location. Here we are not talking about the vacancy of this cell array. If the array index is free, then simply insert the current agent number there. Otherwise, the interaction of the agent *ind_treat* with the agent located on the resulting index in *current_location* occurs, we call it *find_agent*.

After the agents interact, the *ind_treat* agent receives a new index for the room, and the transition processing to the specified location begins anew.

The model simulates the interaction of agents with each other and the external environment, until an external command is issued that means the process is stopped.

4.3   Interpolation of Simulation Results

Interpolation methods were used to apply the results obtained in the simulation to practice. This is necessary because it is necessary to study the dynamics of the spread of HIV, and it is very important to know the number of people infected during any arbitrary period of time. Consequently, it is necessary to operate on the values of the function that imitates the number of infected individuals at intermediate points (not nodal), that is, at the points in time at which the simulation was performed, since the model uses a discrete time representation. This is the definition of the interpolation operation of the objective function, in this case, the function of the number of infected (healthy or dead people). For this case, it was necessary to apply interpolation polynomials for the function of two variables. According to the HIV incidence model, Newton interpolation polynomial [41] was taken.

Newton's formula for functions of two variables, otherwise the Newton interpolation polynomial, is a class of methods based on the approximation of functions of a nth degree polynomial. This formula, as well as in the case of bilinear interpolation, can be obtained directly from the Newton polynomial for the one-dimensional case.

For a function defined by a double table (matrix) z, you can define partial finite differences:

$$\Delta_x z_{i,j} = z_{i+1,j} - z_{i,j} \quad and \quad \Delta_x z_{i,j} = z_{i,j+1} - z_{i,j} \tag{1}$$

Re-applying these operations, we obtain the double differences of higher orders:

$$\Delta^{m+n} z_{i,j} = \Delta^{m+n}_{x^m,y^n} = \Delta^m_{x^m}(\Delta^n_{y^n} z_{i,j}) = \Delta^n_{y^n}(\Delta^m_{x^m} z_{i,j}) \tag{2}$$

Using the difference function of two variables $z = f(x,y)$, one can construct an interpolation polynomial similar to the Newton interpolation polynomial.

For ease of calculation, variables are usually introduced:

$$\frac{x - x_0}{h} = p \qquad \frac{y - y_0}{k} = q \tag{3}$$

Hence, Newton's formula takes the form:

$$z \approx z_0 + (p\Delta^{1+0}z_{0,0} + q\Delta^{0+1}z_{0,0}) + \\ + \frac{1}{2!}[p(p-1)\Delta^{2+0}z_{0,0} + 2pq\Delta^{1+1}z_{0,0} + q(q-1)\Delta^{0+2}z_{0,0}] + \ldots \tag{4}$$

With real problems, the matrix of the values of the objective function can reach a dimension of several hundreds, or even thousands, and the exact factorial calculation of such numbers is not possible. For this, approximate formulas for calculating factorial are used, for example, the Stirling formula, but this leads to an increase in the error.

5   Program Realization of Agent-Based Model

The software implementation of the task is performed in the Visual Studio 2013 development environment in the C# programming language. The developed software makes it possible to predict the dynamics of the spread of an infectious disease for a short period of time, based on the initial data entered by the user.

When starting the program, the input data entry window will open. The top panel contains the program navigation menu, which has three tabs: "Initial data", "System operation" and "Schedule". The user can independently switch from one menu tab to another at any time, but it will be possible to work with the system only after correct input of initial parameters. If data is entered incorrectly, the corresponding icon will appear; when you hover over it, you can find out the cause and type of error. Some controls may be unavailable at one time or another - this means that the user has not made the necessary preparatory actions, the type of which will be reported in the output field of numerical statistics of the system, in the tab "System operation". Each element of the program interface has a signature denoting the purpose of a particular control component. The "Initial data" navigation menu tab contains fields for entering numerical values of initial parameters. "System operation" is a tab for displaying the disease in real time. Also there are components of manual control of the modeling process. Finally, in the last menu option there is a field for displaying the simulation result as a graph of the function of the number of agents' relative to their state dependent on time.

Figure 2 shows the initial window of developed software product. Input blocks specify the appropriate numerical parameters for the model to work, according to user input. The control saves all previously entered parameters, but if an error was made in any field, an error warning icon will appear next to it. The control also sets default parameters, after clicking you can immediately see their values in the corresponding fields. The default values for numeric parameters cannot be edited. This data was specially selected for the most rapid and visual demonstration of the program.

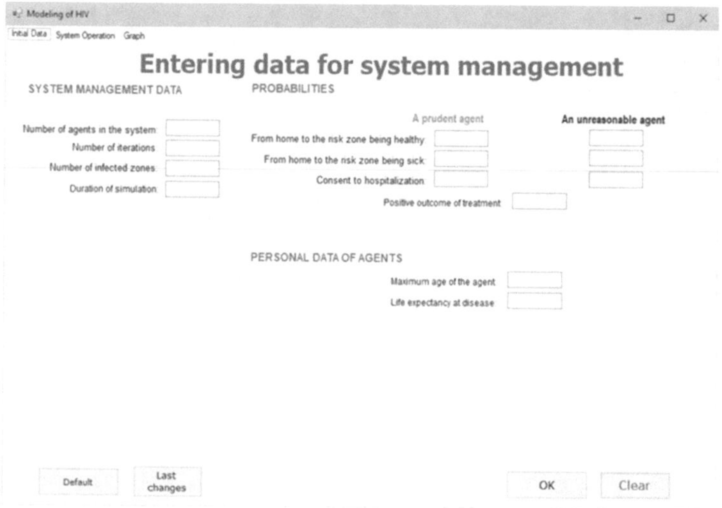

Fig. 2. The initial window of the software package

Figure 3 shows the main window of the system.

In order to avoid errors when the program is running, restrictions have been introduced on some controls. More precisely, all the buttons at one time or another may not be available for use. This is due to the fact that the user could incorrectly or not enter the initial data at all. It may also be due to the fact that the current one was not stopped when starting a new simulation. More information about the cause of this situation can be found in the error field.

Using this software, the user can simulate and see in real time the spread of not only HIV infection, but any arbitrary anthroponotic infectious disease, which tends to indulge from person to person upon contact (the type of contact is not specified) for a short period of time, in accordance with the initial parameters set manually.

6   Results and Discussions

To test this software, several options of input data were checked. With each set of parameters, the program showed good temporal indicators. It was noticed that two factors influence the speed of work: the total number of agents in the system and the number of agents processed in one iteration. The maximum value of these two parameters when testing reached 100,000:

- the total number of agents in the system is 20000;
- the number of agents processed per iteration is 10000;
- the number of infected zones is 18000;
- the maximum age of agents is 60 (years).

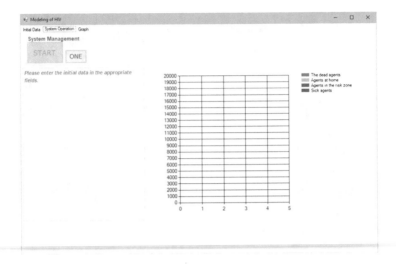

Fig. 3. The main window of the software package

The initial probabilities for agents' behavior have been found out by experimental way. The main results obtained in the modeling process are shown in Fig. 4.

The model demonstrates the behavior of an infectious disease, the main property of which is transmission from one person to another upon contact.

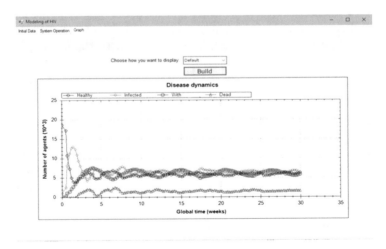

Fig. 4. The results of simulation

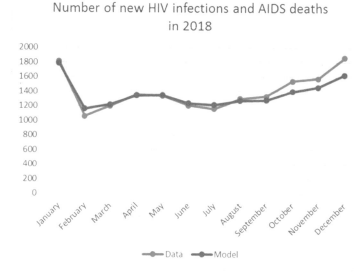

Fig. 5. Comparison of statistical data on the incidence of HIV infection for 2018 and the forecasted incidence, based on data from 2010 to 2017

The forecast is made for a short time, approximately 4 months (15 weeks). Figure 5 shows the "spikes" in the number of diseased agents, after which some of them died. These waves tend to repeat at approximately equal intervals of time, which is also manifested in real life. If we continue the simulation for a longer period, then, despite the increasing error, we can observe a tendency towards a decrease in the number of cases in each epidemic period.

Empirical data were taken from the State Center of Public Health of the Ministry of Health of Ukraine. As can be seen from the graph presented in Fig. 5, the model predicts data very well, dated to August inclusive, that is, approximately for a period of 32 weeks. Further, there are more significant discrepancies. This is explained by the fact that this implementation of the model is applicable for forecasting for a short period of time of several months, although in some cases it can give a result with acceptable accuracy and for more than half a year.

7   Conclusions

The intelligent agent-based approach proposed in the research for simulation of the epidemic process of HIV infection allows to take into account the factors affecting the incidence, the distribution of the population by age categories and risk groups. To check the adequacy of the developed intellectual multi-agent model, statistics on real morbidity in the Kharkiv region 2010–2018 were used.

Based on the built multi-agent model, a program complex was developed in the C# programming language in the Visual Studio 2013 environment, which makes it possible to calculate the prognosis of HIV infection for a short period (15 weeks). The approach proposed in the article will make it possible to identify the main factors influencing the epidemic process of HIV infection, as well as to develop effective preventive measures to reduce the incidence rate.

References

1. Polyvianna Y, Chumachenko D, Chumachenko T (2019) Computer aided system of time series analysis methods for forecasting the epidemics outbreaks. In: 2019 15th international conference on the experience of designing and application of CAD systems (CADSM), pp 7.1–7.4
2. Schlecht HP, Schellhorn S, Dezube BJ, Jacobson JM (2008) New approaches in the treatment of HIV/AIDS - focus on maraviroc and other CCR5 antagonists. Ther Clin Risk Manag 4(2):473–485
3. Ford N et al (2018) Managing advanced HIV disease in a public health approach. Clin Infect Dis: Off publ Infect Dis Soc Am 66(2):S106–SS110
4. Prentice RL, Huang Y (2018) Nutritional epidemiology methods and related statistical challenges and opportunities. Stat Theor Relat Fields 2(1):2–10
5. Brownson R et al (2018) Getting the word out: new approaches for disseminating public health science. J Publ Health Manag Pract 24(2):102–111
6. Hill AL, Rosenbloom DIS, Nowak MA, Siliciano RF (2018) Insight into treatment of HIV infection from viral dynamics models. Immunol Rev 285(1):9–25
7. Global HIV & AIDS statistics – 2018 fact sheet. (2019) UNAIDS report
8. Yu F et al (2018) Evolution of HIV-1 quasispecies within one couple: a follow-up study based on next-generation sequencing. Sci Rep 8:1404
9. Molina JM et al (2017) Efficacy, safety, and effect on sexual behaviour of on-demand pre-exposure prophylaxis for HIV in men who have sex with men: an observational cohort study. Lancet HIV 4(9):e402–e410
10. Saag MS et al (2018) Antiretroviral drugs for treatment and prevention of HIV infection in adults: 2018 recommendations of the international antiviral society-USA panel. J Am Med Assoc 320(4):379–396
11. Van Zyl G, Bale MJ, Kearney MF (2018) HIV evolution and diversity in ART-treated patients. Retrovirology 15(1):14
12. Simonetti FR, Kearney MF (2015) Review: influence of ART on HIV genetics. Curr Opin HIV AIDS 10(1):49–54
13. Global AIDS Monitoring 2018 (2019) Ukraine Summary. UNAIDS report, 11 p
14. Holt E (2018) Conflict in Ukraine and a ticking bomb of HIV. Lancet HIV 5(6):e273–e274
15. Booth R et al (2016) HIV incidence among people who inject drugs (PWID) in Ukraine: results from a clustered randomized trial. Lancet HIV 3(10):e482–e489
16. Mazorchuck M, Dobriak V, Chumachenko D (2018) Web-application development for tasks of prediction in medical domain. In: 2018 IEEE 13th international scientific and technical conference on computer sciences and information technologies (CSIT), Lviv, pp 5–8
17. Scheffler RM, Liu JX, Kinfu Y, Dal Poz MR (2008) Forecasting the global shortage of physicians: an economic and needs-based approach. Bull World Health Organ 86:516–523

18. Hughes BB et al (2011) Projections of global health outcomes from 2005 to 2060 using the international futures integrated forecasting model. Bull World Health Organ 89:478–486
19. Chin J, Mann J (1989) Global surveillance and forecasting of AIDS. Bull World Health Organ 67(1):1–7
20. Funk S et al (2018) Real-time forecasting of infectious disease dynamics with a stochastic semi-mechanistic model. Epidemics 22:56–61
21. Verma M et al (2018) Google search trends predicting disease outbreaks: an analysis from India. Healthc Inf Res 24(4):300–308
22. Krak I (2019) Computer technologies for gestures communication systems construction. In: Communications in computer and information science, pp 135–144
23. Krak Y, Barmak O, Mazurets O (2018) The practice implementation of the information technology for automated definition of semantic terms sets in the content of educational materials. In: CEUR workshops proceeding, vol 2139, pp 245–254
24. Krak I, Kondratiuk S (2017) Cross-platform software for the development of sign communication system: dactyl language modelling. In: Proceedings of the 12th international scientific and technical conference on computer sciences and information technologies, CSIT 2017, pp 167–170
25. Chumachenko D, et al (2018) On agent-based approach to influenza and acute respiratory virus infection simulation. In: Proceedings of the 14th international conference on advanced trends in radioelectronics, telecommunications and computer engineering, TCSET 2018, pp 192–195
26. Chumachenko D (2018) On intelligent multiagent approach to viral Hepatitis B epidemic processes simulation. In: Proceedings of the 2018 IEEE 2nd international conference on data stream mining and processing, DSMP 2018, pp 415–419
27. Meniailov D et al (2019) Using the K-means method for diagnosing cancer stage using the pandas library. In: CEUR workshop proceeding, vol 2386, pp. 107–116
28. Liu L, Hu J (2012) The practice of MATLAB simulation in modern control theory course teaching. In: 2012 IEEE fifth international conference on advanced computational intelligence (ICACI), Nanjing, pp 896–899
29. Moss R, et al (2017) Epidemic forecasts as a tool for public health: interpretation and (re)calibration. Infect Commun Dis 42(1):69–76
30. Hosseinichimeh N, Rahmandad H, Jalali MS, Wittenborn AK (2016) Estimating the parameters of system dynamics models using indirect inference. Syst Dyn Rev 32(2):156–180
31. Wolf P et al (2017) Learning how to drive in a real world simulation with deep Q-networks. In: 2017 IEEE intelligent vehicles symposium (IV), Los Angeles, CA, pp 244–250
32. Bazilevych K et al (2018) Stochastic modelling of cash flow for personal insurance fund using the cloud data storage. Int J Comput 17(3):153–162
33. Chumachenko D, Chumachenko K, Yakovlev S (2019) Intelligent simulation of network worm propagation using the code red as an example. Telecommun Radio Eng 78(5):443–464
34. Mashtalir VP, Shlyakhov VV, Yakovlev SV (2014) Group structures on quotient sets in classification problems. Cybern Syst Anal 50(4):507–518
35. Chumachenko D, Yakovlev S (2019) On intelligent agent-based simulation of network worms propagation. In: 2019 15th international conference on the experience of designing and application of CAD systems (CADSM), pp 3.11–3.13
36. Badham J et al (2018) Developing agent-based models of complex health behavior. Health Place 54:170–177

37. Prokhorov OV, Prokhorov VP, Matiushko AO, Kuznetsova YA (2016) Regional resources management by agent-based simulation. Naukoviy Visnyk Natsionalnogo Hirnychoho Universytetu 4:107–114
38. Bora S, Emek S (2019) Agent-based modeling and simulation of biological systems. In: Modeling and computer simulation, pp 121–132
39. Roses R, Kadar C, Gerritsen C, Rouly C (2018) Agent-based simulation of offender mobility: integrating activity nodes from location-based social networks. In: Proceedings of the 17th international conference on autonomous agents and multiagent systems (AAMAS 2018), pp 804–812
40. Marszalek Z (2017) Performance test on triple heap sort algorithm. Tech Sci 20(1):49–61
41. Yang Y, Gordon SP (2016) Visualizing and understanding the components of Lagrange and Newton interpolation. Probl Res Issues Math Undergraduate Stud 26(1):39–52

On the Computational Complexity of Learning Bithreshold Neural Units and Networks

Vladyslav Kotsovsky[1]([✉])(iD), Fedir Geche[2](iD), and Anatoliy Batyuk[3](iD)

[1] IMST Department, Uzhhorod National University, Uzhhorod, Ukraine
kotsavlad@gmail.com
[2] Department of Cybernetics and Applied Mathematics, Uzhhorod National
University, Uzhhorod, Ukraine
fgeche@hotmail.com
[3] ACS Department, Lviv Polytechnic National University, Lviv, Ukraine
abatyuk@gmail.com

Abstract. We study the questions concerning the properties and capabilities of computational bithreshold real-weighted neural-like units. We give and justify the two sufficient conditions ensuring the possibility of separation of two sets in n-dimensional vector space by means of one bithreshold neuron. Our approach is based on application of convex and affine hulls of sets and is feasible in the case when one of the two sets is a compact and the second one is finite. We also correct and refine some previous results concerning bithreshold separability. Then the hardness of the learning bithreshold neurons is considered. We examine the complexity of the problem of checking whether the given Boolean function of n variables can be realizable by single bithreshold unit. Our main result is that the problem of verifying the bithreshold separability is NP-complete. The same is true for neural networks consisting of such computational units. We propose some continuous modifications of the bithreshold activation function to smooth away these difficulties and to make possible the application of modern paradigms and learning techniques for such networks.

Keywords: Bithreshold neuron · Neural network · Machine learning · Computational complexity

1 Introduction

Bithreshold neurons (or units) and multithreshold ones were introduced in the early 1960s. They have capabilities to recognize more complicated pattern than ordinary threshold units (e.g. bithreshold neuron is capable to solve XOR problem). But unlike the threshold units bithreshold ones are not widely used. We know no examples of modern effective system of computational intelligence developed on the base of multithreshold devices similar to the ones proposed in [1–3].

© Springer Nature Switzerland AG 2020
V. Lytvynenko et al. (Eds.): ISDMCI 2019, AISC 1020, pp. 189–202, 2020.
https://doi.org/10.1007/978-3-030-26474-1_14

The hybrid multidimensional neuro-system [4] or the special problem-oriented packages [5] providing the facilities based on bithreshold paradigm are still missing. We know little about properties of multithreshold units. The effective learning techniques for them are also unknown. We can say that bithreshold and multithreshold units remain a "blank spot" in the theory of artificial neural networks.

2 Review of the Literature

Let us consider some significant papers devoted to multithreshold units, their study and applications. Bithreshold neurons have been studied in [6] by means of tolerance matrices. But this technique is rather complicated and applicable in practice only in the case of small number of input features. The adaptive synthesis of multithreshold devices based on cumbersome version of random search has been proposed in [7]. The problem of the estimation of the capacity of multithreshold elements has been considered in [8,9]. The spectral approach in study of bithreshold and multithreshold neurons has been introduced in [10,11] and developed in [12]. However, very little attention was payed to practical aspect of learning in [6–11]. Deolalikar [13] studied bithreshold neurons and two-layer neural networks, which hidden layer consists of ordinary threshold units, and the single neuron in the output layer is the bithreshold one. The main result stated in [13] is the following proposition: the network of above-mentioned architecture is capable to recognize an arbitrary sets which are the intersection at most $m-1$ convex polytopes in R^n, where m is the number of neurons in hidden layer. This result has been based on Theorem 1 from [13] concerning some properties of bithreshold neuron. In Subsect. 4.2 we will give the counterexample denying the statement of Theorem 1. The complex angle-like generalizations of bithreshold activation function [14] are also worth mentioning.

Our paper is motivated by the lack of knowledge about bithreshold units. We study both the conditions ensuring separability by means of bithreshold devices and the "hardness" of learning such units and networks consisting of them. The paper has the following structure. First, we give necessary definitions, models, describe the problem domain and our methods. Then we state our result concerning the separation in real vector space by means of one bithreshold neuron. Finally, we examine the complexity of learning bithreshold neurons to realize given Boolean function or more generally to classify patterns in n-dimensional real vector space. Our main observation is that the problem of checking bithreshold separability is NP-hard. We show that these inconveniences can be partly overcome by using smoothed generalizations of bithreshold activation functions.

3 Problem Statement

3.1 Model of the Bithreshold Neuron

Let $\mathbf{w} = (w_1, \ldots, w_n) \in \mathrm{R}^n$ be n-dimensional real vector. For an arbitrary vector $\mathbf{x} \in (x_1, \ldots, x_n) \in \mathrm{R}^n$ the value of the inner product $(\mathbf{w}, \mathbf{x}) = w_1 x_1 + \cdots + w_n x_n$ is said to be a weighted sum corresponding to the vector $\mathbf{x}$.

A computation unit with n inputs $x_1, x_2, \ldots, x_n$ and one output y is said to be a linear bithreshold unit (LBU) with the weight vector $\mathbf{w} = (w_1, \ldots, w_n) \in \mathrm{R}^n$ and thresholds $t_1, t_2 \in \mathrm{R}\, (t_1 < t_2)$ if $y = f_{t_1,t_2}\,((\mathbf{w}, \mathbf{x}))$, where the bithreshold activation function $f_{t_1,t_2}(x) : \mathrm{R} \to \{-1, 1\}$ is defined in the following way:

$$f_{t_1,t_2}(x) = \begin{cases} -1, & \text{if } t_1 < x < t_2, \\ 1, & \text{otherwise.} \end{cases}$$

The graph of the activation function $f_{t_1,t_2}(x)$ is shown in Fig. 1.

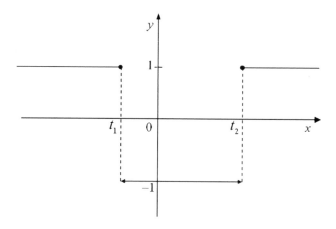

Fig. 1. A graph of bithreshold activation function

The LBU is completely defined by the ordered triplet $(\mathbf{w}, t_1, t_2)$. We call this triplet the structure of LBU. LBU with the structure $(\mathbf{w}, t_1, t_2)$ divides the space R^n into two subsets

$$R^{n+} = \{\mathbf{x} \in \mathrm{R}^n | (\mathbf{w}, \mathbf{x}) \le t_1\} \cup \{\mathbf{x} \in \mathrm{R}^n | (\mathbf{w}, \mathbf{x}) \ge t_2\},$$
$$R^{n-} = \{\mathbf{x} \in \mathrm{R}^n | t_1 < (\mathbf{w}, \mathbf{x}) < t_2\}.$$

3.2 B-Separable Sets

We call two sets $A^+ \subset \mathrm{R}^n$ and $A^- \subset \mathrm{R}^n$ *bithreshold separable* (B-separable) if there exists such LBU that $A^+ \subset \mathrm{R}^{n+}$ and $A^- \subset \mathrm{R}^{n-}$. In this case the LBU with structure $(\mathbf{w}, t_1, t_2)$ induces the dichotomy (A^+, A^-) of the set A in two disjoint sets A^+ and A^- (i.e. $A = A^+ \cup A^-$, $A = A^+ \cap A^- = \varnothing$). If we have LBU with the structure triplet $(\mathbf{w}, t_1, t_2)$ and A is an arbitrary set in the space R^n, then sets $A^+ = A \cap \mathrm{R}^{n+}$, $A^- = A \cap \mathrm{R}^{n-}$ are B-separable. The notion of the B-separability is the generalization of the notion of linear separability. It easy to verify that there exists very simple B-separable sets which are not linear separable even in one-dimensional space (e.g. $A^- = \{b\}$, $A^+ = \{a, c\}$, where $a < b < c$).

The set of all weight vectors providing the given linear dichotomy (A^+, A^-) is convex. The weights of bithreshold units do not have the previous property. Let $A^+ = \{-1, 1\}^2$, $A^- = \varnothing$. The dichotomy (A^+, A^-) can be obtained using LBUs with structures $(\mathbf{w}^1, -1, 1)$ and $(\mathbf{w}^2, -1, 1)$, respectively, where $\mathbf{w}^1 = (2, 0)$, $\mathbf{w}^2 = (0, 2)$. But this dichotomy is not realizable by any LBU with the weight vector $\frac{1}{2}(\mathbf{w}^1 + \mathbf{w}^2)$.

The main goal of our paper is to establish necessary and sufficient conditions ensuring the B-separability of two bounded set in n-dimensional vector space and to investigate the existence of effective learning algorithms for bithreshold neural units and networks.

4   Materials and Methods

Throughout the text we deal with the properties of bithreshold neurons. We use models and methods of linear algebra, convex analysis and the theory of linear vector spaces to formulate and to establish conditions of B-separability. The methods of complexity theory are used to study the computational complexity of learning bithreshold units and networks. We also use machine learning methods to train the networks consisting of generalized bithreshold units.

4.1   Necessary Condition of B-Separability

It is well known that there exists strong dependence between the linear separability of two sets in $\mathbf{R}^n$ and the properties of their convex hulls: two finite sets A and B are linearly separable if and only if $\text{Conv}(A) \cap \text{Conv}(B) = \varnothing$. In the case of LBU the relation between B-separability and convex hulls is more complicated.

Lemma 1. *If LBU with the structure triplet* $(\mathbf{w}, t_1, t_2)$ *performs dichotomy* (A^+, A^-), *then*

$$A^+ \cap \text{Conv}(A^-) = \varnothing \tag{1}$$

and sets A^+ *and* $\text{Conv}(A^-)$ *are B-separable.*

Proof. Let $\mathbf{y}$ be an arbitrary elements of $\text{Conv}(A^-)$. Then by the definition of the convex hull

$$\mathbf{y} = \sum_{i=1}^{m} \lambda_i \mathbf{x}^i, \lambda_i \in [0, 1], \sum_{i=1}^{m} \lambda_i = 1, \mathbf{x}^i \in A^-, i = 1, \ldots, m.$$

Thus

$$t_1 = \sum_{i=1}^{m} \lambda_i t_1 < \sum_{i=1}^{m} \lambda_i (\mathbf{w}, \mathbf{x}^i) = (\mathbf{w}, \mathbf{y}) < \sum_{i=1}^{m} \lambda_i t_2 = t_2.$$

The last equation implies that $\mathbf{y} \notin A^+$ and LBU with the structure $(\mathbf{w}, t_1, t_2)$ provides B-dichotomy of the sets $A^+ \cup \text{Conv}(A^-)$ on A^+ and $\text{Conv}(A^-)$.

The statement inverse to Lemma 1 is not true in the case $n \geq 2$. In Fig. 2 we can see the example of two sets $A^+ = \{a, b, c\}$ and $A^- = \{x, y, z\}$ satisfying Eq. 1, but it is impossible to separate these sets using single LBU (the convex hull of the set A^- is shown in grey). The same is true for $A^+ = \{e, d\}$ and $A^- = \{x, y, z\}$.

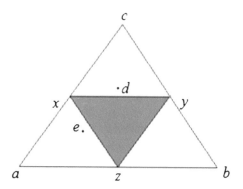

Fig. 2. The example of two non B-separable sets satisfying Eq. 1

4.2   Sufficient Conditions of B-Separability

It is interesting enough to find additional conditions that along with Eq. 1 ensure the B-separability of sets A^+ and A^-. The first steps in this direction were made in [13], but the proof of Theorem 1 in this paper contains some logical errors. Moreover, further we will show that the main statement of this theorem is not correct.

The vector $\mathbf{y} \in \mathbb{R}^n$ is called an affine combination of vectors $\mathbf{x}^1, \ldots, \mathbf{x}^k$ if

$$\mathbf{y} = \alpha_1 \mathbf{x}^1 + \cdots + \alpha_k \mathbf{x}^k, \quad \sum_{i=1}^{k} \alpha_i = 1, \ \alpha_i \in \mathbb{R}, \ i = 1, \ldots, k. \tag{2}$$

We denote by $\mathrm{Aff}(X)$ the affine hull of the set X (the set of all possible affine combination of some elements of X).

Proposition 2. *Let A^+ be a finite or countable compact in the space $\mathbb{R}^n$ and $A^- = \{\mathbf{x}^1, \ldots, \mathbf{x}^k\}$ is the set of linearly independent vectors, $k \leq n$. If there exists such index l $(1 \leq l \leq k)$ that for every $\mathbf{y} \in A^+ \cap \mathrm{Aff}(A^-)$ its coefficient α_l in Eq. 2 satisfies the following condition*

$$\alpha_l \in (-\infty, 0) \cup (1, +\infty), \tag{3}$$

then the sets A^+ and A^- are B-separable.

Proof. Let us show that there exists a hyperplane $H_{\mathbf{w}} = \{\mathbf{x} \in \mathbb{R}^n \mid (\mathbf{w}, \mathbf{x}) = 1\}$ such that A^- belongs to $H_{\mathbf{w}}$ and for all $\mathbf{y} \in A^+$ from $\mathbf{y} \in H_{\mathbf{w}} \cap A^+$ it follows that $\mathbf{y} \in \langle A^- \rangle$, where $\langle A^- \rangle$ is the linear hull of the set A^-. In the case $k = n$ the existence of such hyperplane is evident, because $\langle A^- \rangle = \mathbb{R}^n$. Let $k < n$, $A^+ \setminus \langle A^- \rangle = \{\mathbf{y}^1, \mathbf{y}^2, \dots\}$, $D = \max\{\|\mathbf{y}\| \mid \mathbf{y} \in A^+\} \neq 0$ (we suppose that A^+ contains only nonzero elements; this condition can be always obtained by the appropriate translation of both the sets A^+ and A^- without loss of their B-separability). Let us build the convergent sequence of vectors $\{\mathbf{w}^m\}$ and the numerical sequence $\{d_m\}$ satisfying the following conditions:

$$\forall \mathbf{x} \in A^- \; (\mathbf{w}^m, \mathbf{x}) = 1, \left|(\mathbf{w}^m, \mathbf{y}^j) - 1\right| \geq d_j, j = 1, \dots, m - 1. \tag{4}$$

Let $\mathbf{w}^1$ be the normal vector of an arbitrary hyperplane $H_1 = \{\mathbf{x} \in \mathbb{R}^n \mid (\mathbf{w}^1, \mathbf{x}) = 1\}$ such that $A^- \subset H_1$. The existence of H_1 follows from the linear independence of the elements of A^-. Let us suppose that the vector $\mathbf{w}^m$ and numbers d_j, $1 \leq j < m$ are already defined. We shall demonstrate how we can use them to obtain $\mathbf{w}^{m+1}$ and d_m. If $(\mathbf{w}^m, \mathbf{y}^m) \neq 1$, then let $\mathbf{w}^{m+1} = \mathbf{w}^m$. If $(\mathbf{w}^m, \mathbf{y}^m) = 1$, then we can choose such vector $\mathbf{v}^m$ that $(\mathbf{v}^m, \mathbf{x}^j) = 0$, $1 \leq j \leq k$, $(\mathbf{v}^m, \mathbf{y}^m) = 1$. Let us consider $\mathbf{w}^{m+1} = \mathbf{w}^m + \beta_m \mathbf{v}^m$, where $\beta_m = \left(2^{m+1} \cdot \max\{\|\mathbf{v}^m\|, 1\} \cdot D\right)^{-1} \cdot \min\limits_{1 \leq j < m} d_i$ (if $m = 1$, then we suppose that $\min\limits_{1 \leq j < m} d_i = 1$), $d_m = \dfrac{1}{2}|(\mathbf{w}^m, \mathbf{y}^m) - 1|$. Then the vector $\mathbf{w}^{m+1}$ satisfies (4). So long as $\|\mathbf{w}^{m+p} - \mathbf{w}^m\| < 2^{-m} d_1 D$, then the fundamental sequence $\{\mathbf{w}^m\}$ converges to such $\mathbf{w} \in \mathbb{R}^n$ that $\forall \mathbf{x} \in A^- \; (\mathbf{w}, \mathbf{x}) = 1$, $\forall j \in \mathbb{N} \left|(\mathbf{w}, \mathbf{y}^j) - 1\right| \geq d_j$. Note that in the case of finite $A^+ \setminus \langle A^- \rangle$ it suffices to assume $\mathbf{w} = \mathbf{w}^{m+1}$, where m is the power of $A^+ \setminus \langle A^- \rangle$. Therefore, the hyperplane $H_{\mathbf{w}}$ is sought. Then we have two possible cases for $H_{\mathbf{w}}$ and A^+:

Case 1. $A^+ \cap H_{\mathbf{w}} = \varnothing$. According to the Weierstrass theorem the continuous function $f(\mathbf{x}) = |(\mathbf{w}, \mathbf{x}) - 1|$ possess its minimum δ on A^+ and $\delta > 0$. Then LBU with the triplet $(\mathbf{w}, 1 - \delta, 1 + \delta)$ performs dichotomy (A^+, A^-).

Case 2. $A_H^+ = A^+ \cap H_{\mathbf{w}} \neq \varnothing$. We show that in this case the set A_H^+ contains only those elements of A^+, which belong to $\mathrm{Aff}(A^-)$. Let $\mathbf{y} \in A_H^+$. By the rule of choice of $H_{\mathbf{w}}$ we can conclude that $\mathbf{y} = \alpha_1 \mathbf{x}^1 + \dots + \alpha_k \mathbf{x}^k$. Thus, $\mathbf{y}$ belongs to the affine hull of the set A^-, because

$$\sum_{i=1}^{k} \alpha_i = \sum_{i=1}^{k} \alpha_i (\mathbf{w}, \mathbf{x}^i) = (\mathbf{w}, \mathbf{y}) = 1.$$

In addition, it is easy to see that at least one coefficients α_i in (2) is negative, because from (3) it follows that $\mathbf{y} \notin \mathrm{Conv}(A^-)$. Let us complete the set A^- up to the basis of the space $\mathbb{R}^n$. Let $\{\mathbf{x}^1, \dots, \mathbf{x}^k, \mathbf{x}^{k+1}, \dots, \mathbf{x}^n\}$ be an obtained basis. Then every $\mathbf{y} \in \mathbb{R}^n$ can be represented as follows:

$$y = \alpha_1 \mathbf{x}^1 + \dots + \alpha_l \mathbf{x}^l + \dots + \alpha_n \mathbf{x}^n.$$

From the boundedness of the set A^+ we can infer that there exists such $\alpha_{\max}$ that for all $\mathbf{y} \in A^+$ $|\alpha_j| < \alpha_{\max}$, $j = 1, \dots, n$. Let $A_l^+ = \{\mathbf{y} \in A^+ \mid \alpha_l \in [0, 1]\}$.

It is easy to verify that the set A_l^+ is closed and $A_l^+ \cap H_{\mathbf{w}} = \varnothing$. Furthermore, for all $\mathbf{z} \in A_l^+$ $(\mathbf{w}, \mathbf{z}) \neq 1$. Therefore, there exists such an open ball $B(\mathbf{z}, r_{\mathbf{z}})$ with the center $\mathbf{z}$ and radius $r_{\mathbf{z}}$ that its closure $B[\mathbf{z}, r]$ has an empty intersection with the hyperplane $H_{\mathbf{w}}$. From the Borel lemma it follows that the open cover of the closed set A_l^+ contains the finite subcover $\{B(\mathbf{z}^1, r_1), \ldots, B(\mathbf{z}^s, r_s)\}$ and the value

$$\delta = \frac{1}{2} \min_{1 \leq j \leq s} \{|(\mathbf{w}, \mathbf{z}) - 1| \, |\mathbf{z} \in B[\mathbf{z}^j, r_j]\}$$

is strictly positive. Let us show that there exists an $\varepsilon > 0$ such that

$$\text{for all } \mathbf{y} \in A^+ \left|\alpha_l - \frac{1}{2}\right| \geq \frac{1}{2} + \varepsilon \text{ or } |(\mathbf{w}, \mathbf{y}) - 1| \geq \delta. \tag{5}$$

Let us suppose the contrary. Then it is possible to build the sequence $\{\mathbf{y}^m\}$ satisfying the following condition:

$$\text{for all } m \in \mathrm{N} \left|\alpha_{ml} - \frac{1}{2}\right| < \frac{1}{2} + \frac{1}{m} \text{ and } |(\mathbf{w}, \mathbf{y}^m) - 1| < \delta.$$

From the first inequality we can infer that the sequence $\{\mathbf{y}^m\}$ contains the subsequence that converges to some $\mathbf{y}^* \in A_l^+$ such that $|(\mathbf{w}, \mathbf{y}^*) - 1| \leq \delta$. But it contradicts to the condition that for all $\mathbf{y} \in A_l^+$ $|(\mathbf{w}, \mathbf{y}) - 1| \geq 2\delta$. Let us select a vector $\mathbf{u} \in \mathrm{R}^n$ in the way $(\mathbf{u}, \mathbf{x}^j) = 0$ when $j \in \{1, \ldots, n\}, j \neq l$ and $(\mathbf{u}, \mathbf{x}^l) = 1$. Let us consider $\widetilde{\mathbf{w}} = \mathbf{w} + \gamma\mathbf{u}$, where

$$\gamma < \frac{\delta}{\max\{\alpha_{\max}, \varepsilon + 1\}}.$$

For all $\mathbf{x} \in A^-$ $(\widetilde{\mathbf{w}}, \mathbf{x}) \in \{1, 1 + \gamma\}$. If $\mathbf{y} \in A^+$, then Eq. 5 implies $|(\mathbf{w}, \mathbf{y}) - 1| \geq \delta$ or $\alpha_l \in (-\infty, -\varepsilon] \cup [1 + \varepsilon, +\infty)$. In the first case we have $|(\widetilde{\mathbf{w}}, \mathbf{y}) - 1| \geq \delta \geq \gamma(1 + \varepsilon)$. In the second case it easy to see that $(\widetilde{\mathbf{w}}, \mathbf{y}) \leq 1 - \gamma\varepsilon$ or $(\widetilde{\mathbf{w}}, \mathbf{y}) \geq 1 + \gamma(1 + \varepsilon)$. Thus, LBU with the structure $(\widetilde{\mathbf{w}}, 1 - \gamma\varepsilon, 1 + \gamma(1 + \varepsilon))$ separates the sets A^+ and A^-. The proof is complete.

It should be noted the following direct consequence of Proposition 2: the sets A^+ and $\mathrm{Conv}(A^-)$ are also B-separable.

Remark. Let us show it on the following example. Suppose that

$$A^+ = \{(1, 1, 1, 0, 0), (1, -1, -1, 0, 0), (-1, 1, -1, 0, 0), (-1, -1, 1, 0, 0)\},$$
$$A^- = \{(1, 1, -1, 0, 0), (1, -1, 1, 0, 0), (-1, 1, 1, 0, 0), (-1, -1, -1, 0, 0)\}.$$

It is well known [10], that two sets consisting of non-adjacent vertices of the unit cube $\{-1, 1\}^3$ in R^3 aren't B-separable. Therefore, the sets A^+ and A^- having the same first three coordinates aren't B-separable too. This example is the counterexample to the statement of Theorem 1 in [13]. This theorem states that for B-separability of two finite sets A^+ and A^- in the space R^n it suffices that $\mathrm{Card} A^- \leq n - 1$ and $A^+ \cap \mathrm{Conv}(A^-) = \varnothing$. Two last conditions hold in our example, but the sets A^+ and A^- are not B-separable. Thus, the linear independence of the elements of the set A^- is significant.

The importance of Eq. 3 can be confirmed by the example shown in the Fig. 2. It is easy to verify that we have the following equations for the points a, b, c: $a = x - y + z$, $b = -x + y + z$, $c = x + y - z$. The corresponding coefficients do not satisfy Eq. 3 (for the complete correspondence to the conditions of Proposition 2 we can consider the Fig. 2 as the surface of the plane $z = 1$ in space R^3—the assumption ensuring the linear independence of $A^- = \{x, y, z\}$). Similarly, for both points d and e two of their coordinates lie in the interval $(0, 1)$ and one is negative. Thus, the non-separable dichotomy $(\{d, e\}, \{x, y, z\})$ does not satisfies Eq. 3.

In the previous proof we have used the constraint on the power of the set A^+ (namely, we have required A^+ to be finite or countable) only for building $\mathbf{w}$—the normal vector of the hyperplane. We can relax this constraint by restricting conditions imposed on the coordinates in Eq. 3.

Proposition 3. *Let A^+ be a compact in R^n and $A^- = \{\mathbf{x}^1, \ldots, \mathbf{x}^k\}$, $(k \leq n)$ be the set of linearly independent vectors. If there exists an index l and a completion $\{\mathbf{x}^1, \ldots, \mathbf{x}^k, \mathbf{x}^{k+1}, \ldots, \mathbf{x}^n\}$ of the set A^- up to the base of the space R^n such that $1 \leq l \leq k$ and for all $\mathbf{y} \in A^+$ its coordinate α_l in this basis satisfies Eq. 3, then the sets A^+ and $\mathrm{Conv}(A^-)$ are B-separable.*

Proof. By using the equivalence of the convergence in R^n to the coordinatewise convergence similarly to the proof of Proposition 2 we can prove the existence of such $\varepsilon > 0$ that

$$\text{for all } \mathbf{y} \in A^+ \ \alpha_l \in (-\infty, -2\varepsilon] \cup [1 + 2\varepsilon, +\infty).$$

Since the set A^+ is bounded, there exists such $\alpha_{\max}$ that for all $\mathbf{y} \in A^+$ $|\alpha_j| < \alpha_{\max}$, $j = 1, \ldots, n$. Let δ be a positive number satisfying the inequalities $\delta \leq \dfrac{\varepsilon}{n\alpha_{\max}}$. Then we can select $\mathbf{w} \in R^n$ satisfying following conditions: $(\mathbf{w}, \mathbf{x}^l) = 1$, $(\mathbf{w}, \mathbf{x}^j) = \delta$, $j \in \{1, \ldots, l-1, l+1, \ldots, k\}$, $(\mathbf{w}, \mathbf{x}^j) = 0$, $k < j \leq n$. Thus, for all $\mathbf{y} \in A^+$ $(\mathbf{w}, \mathbf{y}) \leq -\varepsilon$ or $(\mathbf{w}, \mathbf{y}) \geq 1 + \varepsilon$. So, LBU with the structure $(\mathbf{w}, 0, 1 + \varepsilon)$ separates the sets A^+ and A^-. Finally, for the purpose of obtaining the complete proof, we can apply Lemma 1.

Let us to show the geometrical sense of Eq. 3 on the following example. In Fig. 3 is shown the surface of the plane $z = c$, $c \neq 0$ in R^3, containing the set $A^- = \{a, b, c\}$. The filled area consists of the set of points satisfying (3) with respect to the point a. All other points (including the two horizontal lines) do not satisfy (3). It easy to see that if A^+ is a compact subset of filled area in Fig. 3, then the sets A^+ and A^- are B-separable. It is also evidently from Fig. 3 that the compactness of the set A^+ is significant. In Proposition 3 B-separability follows from Eq. 3 for coordinates of the elements of A^+ in their expansions on some basis containing the set A^-. It is possible to demonstrate that the similar result holds when expansion is made on the base of elements of the set A^+.

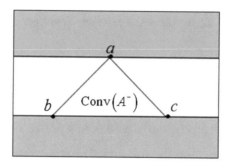

Fig. 3. Graphic illustration of the importance of conditions imposed in Eq. 3

4.3 Learning Complexity of Bithreshold Neurons

In this section we consider the problem of the learning LBU to recognize given Boolean function. In agreement with [15] we consider the learning algorithm to be "effective" if its time complexity is $O\left(s^{r}\right)$, where s is the input size and r is some fixed positive integer number (i.e. the algorithm has polynomial complexity). In the case of learning Boolean functions an instance of the problem is disjunctive normal formula in n variables corresponding to the given function f or the set of Boolean vectors which are true points of the function f.

In the next two subsections we consider Boolean functions in the binary base $Z_2 = \{0,1\}$. We can always use linear transformations $y = \dfrac{1 \pm x}{2}$, to move from the base $\{-1, 1\}$ to Z_2 and vice versa.

Complexity of Learning Boolean Functions Represented by Disjunctive Normal Form. Now we prove that if "NP $\neq$ P conjecture" is true, that there is no polynomial time learning algorithm for LBU. In order to prove this statement, let us consider the well-known *Membership*(C) *problem*. Suppose that $C = \{C_n\}_{n \geq 1}$ is a class of Boolean functions, that is, $C_n \subset \{f | f : Z_2^n \to Z_2\}$, $n \in \mathrm{N}$. For the class C of functions the *Membership Problem* is as follows.
  Membership(C)
  Input: A disjunctive normal form formula φ in n variables for the function f.
  Output: 1, if f belongs to the class C_n, and -1 otherwise.

Proposition 4. *The learning of the LBU for a given Boolean function represented by a disjunctive normal form formula φ in n variables is NP-hard.*

Proof. We use the following result from [15]. *Membership*(C) is NP-hard if the class C has the projection property:

1. for all $f \in C_n$ and for every $i \in \{1, \ldots, n\}$ functions $f(x_1, \ldots, x_{i-1}, 1, x_{i+1}, \ldots, x_n)$ and $f(x_1, \ldots, x_{i-1}, 0, x_{i+1}, \ldots, x_n)$ belong to C_{n-1};
2. for every $n \in \mathrm{N}$ the identically-1 function belongs to C_n;

3. there exists $k \in \mathbb{N}$ such that some Boolean function on Z_2^k does not belong to C_k.

Let LBT_n be the set all Boolean function in n variables realizable by some LBU. Let us show that the class $LBT = \{LBT_n\}_{n \geq 1}$ satisfies conditions 1–3. The first condition follows from Shannon expansion formula

$$f(x_1, \ldots, x_n) = f(x_1, \ldots, x_{n-1}, 0) \wedge \overline{x}_n \vee f(x_1, \ldots, x_{n-1}, 1) \wedge x_n.$$

If function $f(x_1, \ldots, x_n)$ can be realized on LBU with the structure

$$(\mathbf{w} = (w_1, \ldots, w_{n-1}, w_n), t_1, t_2),$$

then both functions of $n - 1$ variables $f(x_1, \ldots, x_{n-1}, 1)$ and $f(x_1, \ldots, x_{n-1}, 0)$ are realizable by bithreshold units with structures $((w_1, \ldots, w_{n-1}), t_1 - w_n, t_2 - w_n)$ and $((w_1, \ldots, w_{n-1}), t_1, t_2)$, respectively [6]. The second condition is almost evident. The third condition follows from the existence of non-bithreshold functions (for example see one in remark to Proposition 2 or in [10]). Thus, *Membership*(LBT) is NP-hard.

It should also be mentioned that the complexity of the learning problem can be related with the disjunctive form of the representation of Boolean function. It is possible to demonstrate that the corresponding problem is NP-hard even in the case where the function is defined by 3-DNF (each disjunction consists of exactly 3 variables).

Complexity of Checking B-Separability. Now we demonstrate that the complexity of the learning follows from the shape of the activation function. Let us give some intuitive reasons in the case of on-line learning. Suppose that the new learning sample $\mathbf{x} \in A^-$ is presented on some step of the learning process and the current LBU with the structure $(\mathbf{w}, t_1, t_2)$ misclassifies it. Then $(\mathbf{w}, \mathbf{x}) \geq t_2$ or $(\mathbf{w}, \mathbf{x}) \leq t_1$. In both cases the learner knows how it has to change weight vector (or its thresholds) in order to satisfy the inequalities $t_1' < (\mathbf{w}', \mathbf{x}) < t_2'$, where $(\mathbf{w}', t_1', t_2')$ is a corrected structure. Suppose that LBU misclassified $\mathbf{x} \in A^+$. In this case the learner does not know how it has to correct the current structure. Must it change $\mathbf{w}$ so that $(\mathbf{w}', \mathbf{x}) > t_2'$ or maybe so that $(\mathbf{w}', \mathbf{x}) < t_1'$? Therefore, in the worst case for misclassified samples from A^+ the learner may be forced to "blindly" try the both ways of correction. Thus, the total number of correction can grow exponentially. So, in the worst case we have exponential learning time. NP-hardness of the on-line learning algorithms results from it. Our previous arguments are not sufficiently exhaustive and well-reasoned. The strict justification of the complexity of the B-separability verification is given in the following proposition.

Proposition 5. *The problem of checking the B-separability of finite sets A^+ and A^- is NP-hard even in the case $A^+ \cup A^- \subset \{a, b\}^n$, where $a \in \mathbb{R}$, $b \in \mathbb{R}$ $(a \neq b)$ and the absolute values of the weight coefficients of LBU can possess only two different values.*

Proof. Blum and Rivest showed [16] that the problem of learning a neural network consisting of two neurons in the hidden layer, which weight vectors are opposite and belong to $\{-1, 1\}^n$, and one output neural unit realizing the conjunction of its two inputs is NP-complete in the cases, when the network inputs are binary. The NP-hardness of this task implies from the fact that well-known Set-Splitting problem [15] can be reduced to it.

It is easy to verify that an arbitrary dichotomy (A^+, A^-) is B-separable if and only if it is realizable by the neural network having the above-mentioned shape. Indeed, $\mathbf{x} \in A^- \Leftrightarrow (\mathbf{w}, \mathbf{x}) \leq t_2$ and $(-\mathbf{w}, \mathbf{x}) \leq -t_1$, and the transformation from the base $\{a, b\}^n$ to the base $\{-1, 1\}^n$ can be accomplished by means of the linear transformation of variables (the same is true for the coefficients of weight vectors).

Consequence. *Learning a neural network consisting of bithreshold neuron is NP-complete.*

5   Experiment

We have already mentioned that there are serious deficiencies in the suitable learning methods for bithreshold units and networks. The results stated in Proposition 5 and its consequence suggest "hardness" of the learning with such frameworks. It seems that main origin for this intractability is the shape of the activation function. For the purpose to preserve the advantage of bithreshold paradigm over single threshold one and to obtain appropriate learning techniques we shall use the "smoothed" modification of bithreshold activation functions.

Let us consider the following two continuous bithreshold-like activation functions:

$$f_{p,q}(x) = 1 - 2e^{-p(x-q)^2}, \tag{6}$$

$$f_{p,q,r}(x) = \frac{2}{1 + e^{-p(x-q-r)}} - \frac{2}{1 + e^{-p(x-q+r)}} + 1. \tag{7}$$

The parameters p and q in Eqs. 6 and 7 are responsible for the slope of the activation function and the bias, respectively. The parameter r in Eq. 7 allows us to regulate the width of the "ridge" on the curve of activation function (it can be roughly estimated as $2r$ and is similar to the distance between thresholds t_1 and t_2 on x-axis in Fig. 1).

Further we shall use activation functions with $p = 1$, $q = 0$ in Eq. 6 and $p = 10$, $q = 0$, $r = 1$ in Eq. 7.

In computer simulation we train feedforward neural network with one hidden layer and one output neuron. The training error on kth iteration in the case of on-line learning protocol is

$$E\left(\mathbf{w}^k\right) = \frac{1}{2}\left(\text{out}\left(\mathbf{x}_k\right) - t_k\right)^2,$$

where out $(\mathbf{x}_k)$ is the output of the network at the kth iteration step, $\mathbf{x}_k$ is the input vector, t_k is the target (desired output) and the vector $\mathbf{w}^k$ represents all the weights in the network at this step.

We use the backpropagation learning rule based on gradient descent:

$$\mathbf{w}^{k+1} = \mathbf{w}^k + \Delta \mathbf{w}^k, \Delta \mathbf{w}^k = -\eta_k \ \mathrm{grad} \ E\left(\mathbf{w}^k\right),$$

where η_k is the learning rate and $\Delta \mathbf{w}^k$ is the correction vector on kth iteration.

We use the backpropagation algorithm for solving the XOR problem in the case of 100 variables. We compared the performance of the network consisting of neurons with sigmoidal and bithreshold-like activation functions for different combinations of network parameters and various sizes of the training sample. For example, the following Table 1 contains results concerning the performance of the three-layer networks with 50 neurons in hidden layer for the corresponding kind of activation function after 1000000 steps of learning. The size of the training sample is 1000 and the learning rate η_k is 0.1.

Table 1. Learning results for XOR problem.

Activation function	Maximum network error on training sample
Modified logistic function	1.994454
Hyperbolic tangent	1.999761
Activation function (6)	0.478229
Activation function (7)	0.246016

It should be mentioned that for the data set described in [3] the networks with sigmoidal activation functions show mainly better results in the solving of time series prediction problems.

6   Results and Discussion

We proved the two sufficient conditions ensuring B-separability in R^n. They deal with conditions on coefficients in affine combination of vectors and are not always convenient in practice. It seems promising to design more suitable test of B-separability. The questions concerning the utilization of the spectral coefficients of dichotomy [17] in the study of bithreshold neurons is still open. The same is true for the bounds on size of their integer weights [18].

The complexity of learning bithreshold neurons has been also studied. We have established that the verification of B-separability is NP-hard. So the development of the systems of computational intelligence on the base of bithreshold neurons and their practical applications are somewhat problematic.

The last conclusion suggests to design some modified bithreshold-like model with better learning capability. For example, neural networks consisting of neurons with continuous "smoothed" bithreshold-like activation functions similar to

(6) or (7) may be used in the context of backpropagation learning framework similar to one described in [19]. The results of the computer simulation (see Table 1) demonstrate that such networks have sometimes better performance on the hard model problem than networks built on the base of neurons with classical sigmoidal activation functions. It should be mentioned that the parameters p, q, r in Eqs. 6 and 7 can be used for "tuning" the network capability to deal with specific training data, etc.

7   Conclusions

The necessary and sufficient conditions of B-separability in R^n have been proposed in the paper. The complexity of the learning bithreshold neurons has also been established. The main result is that even the simple form of the problem of checking B-separability is NP-complete. The possible way out of this situation consists in the use of continuous modifications of the bithreshold activation function. We obtained some empirical results confirming that network on the base of such neurons can be useful in computational intelligence and decision making.

References

1. Tkachenko R, Izonin I (2019) Model and principles for the implementation of neural-like structures based on geometric data transformations. Adv Intell Syst Comput 754:578–587
2. Bodyanskiy Y, Dolotov A, Peleshko D, Rashkevych Y, Vynokurova O (2019) Associative probabilistic Neuro-Fuzzy system for data classification under short training set conditions. Adv Intell Syst Comput 761:56–63
3. Geche F, Kotsovsky V, Batyuk A, Geche S, Vashkeba M (2015) Synthesis of time series forecasting scheme based on forecasting models system. In: CEUR workshop proceedings, Lviv, Ukraine, pp 121–136
4. Vynokurova O, Peleshko D, Borzov Y, Oskerko S, Voloshyn V (2018) Hybrid multidimensional wavelet-neuro-system and its learning using cross entropy cost function in pattern recognition. In: Proceedings of the 2018 IEEE 2nd international conference on data stream mining and processing, DSMP 2018, Lviv, Ukraine, pp 305–309
5. Bovdi V, Laver V (2019) Thelma, a package on threshold elements, Version 1.02. https://gap-packages.github.io/Thelma
6. Geche F (1999) The realization of Boolean functions by bithreshold neural units. Her Uzhhorod Univ Ser Math 4:17–25. (in Ukrainian)
7. Rastrigin LA (1981) Adaption of the complex systems: methods and applications. Zinatne, Riga. (in Russian)
8. Olafsson S, Abu-Mostafa YS (1988) The capacity of multilevel threshold function. IEEE Trans Pattern Anal Mach Intell 10(2):277–281
9. Takiyama R (1985) The separating capacity of multithreshold threshold element. IEEE Trans Pattern Anal Mach Intell PAMI-7(1):112–116
10. Geche F, Batyuk A, Kotsovsky V (2001) Properties of Boolean functions realizable by bithreshold units. Her Lviv Polytech Natl Univ Ser Comput Sci Inf Technol 438:22–25. (in Ukrainian)

11. Geche F, Kotsovsky V (2001) Representation of finite domain predicates using multithreshold neural elements. Her Uzhhorod Univ Ser Math Inform 6:32–37. (in Ukrainian)

12. Tsmots I, Medykovskyi M, Andriietskyi B, Skorokhoda O (2015) Architecture of neural network complex for forecasting and analysis of time series based on the neural network spectral analysis. In: Proceedings of 13th international conference: the experience of designing and application of CAD systems in microelectronics, CADSM 2015, Lviv-Polyana, Ukraine, pp 236–238

13. Deolalikar V (2002) A two-layer paradigm capable of forming arbitrary decision regions in input space. IEEE Trans Neural Netw 13(1):15–21

14. Kotsovsky V, Geche F, Batyuk A (2015) Artificial complex neurons with half-plane-like and angle-like activation function. In: Proceedings of the international conference on computer sciences and information technologies, CSIT 2015, Lviv, Ukraine, pp 57–59

15. Anthony M (2001) Discrete mathematics of neural networks: selected topics. SIAM, Philadelphia

16. Blum A, Rivest R (1992) Training a 3-node neural network is NP-Complete. Neural Netw 5(1):117–127

17. Kotsovsky V, Geche F, Batyuk A (2018) Finite generalization of the offline spectral learning. In: Proceedings of the 2018 IEEE 2nd international conference on data stream mining and processing, DSMP 2018, Lviv, Ukraine, pp 356–360

18. Geche F, Kotsovsky V, Batyuk A (2015) Synthesis of the integer neural elements. In: Proceedings of the international conference on computer sciences and information technologies, CSIT 2015, Lviv, Ukraine, pp. 63–66

19. Tsmots I, Teslyuk V, Teslyuk T, Ihnatyev I (2018) Basic components of neuronetworks with parallel vertical group data real-time processing. Adv Intell Syst Comput 689:558–576

Implementation of FPGA-Based Barker's-Like Codes

Ivan Tsmots[1], Oleg Riznyk[1]([✉]), Vasyl Rabyk[2], Yurii Kynash[1],
Natalya Kustra[1], and Mykola Logoida[1]

[1] Lviv Polytechnic National University, Lviv, Ukraine
`ivan.tsmots@gmail.com`, `riznykoleg@gmail.com`, `yuk.itvs@gmail.com`,
`kno1935@ukr.net`, `mykola.m.lohoida@lpnu.ua`
[2] Ivan Franko National University of Lviv, Lviv, Ukraine
`rabykv@ukr.net`

Abstract. Pseudorandom code sequences with a low level of side lobe autocorrelation function have been used in radars, communication systems and information security. The results of the studies have shown that there are no Barker's signals with a length greater than 13, for which the value of the side lobe autocorrelation function doesn't exceed one. Consequently, in many cases, code sequences with length greater than 13 are used with the minimum possible value of the side petals level.

The article provides an overview of the synthesis of Barker-like sequence with the arbitrary length using the ideal ring bundles and an example of implementation by means of FPGA, their testing, and evaluation of FPGA hardware resources that are required for this. FPGA EP3C16F484 of the Cyclone III family has been used in the implementation process. The development has been carried out in the VHDL hardware programming language in the development environment of Quartus II with the using of libraries of the development environment.

Keywords: Autocorrelation function · Barker's signal ·
Barker-like sequence · FPGA · Quasi-barker code · Ideal ring bundle ·
Noise-like code

1 Introduction

The increase of strong cryptography, impedance protection and secrecy in onboard systems of protection and real-time data transmission, which comply with requirements for mass-overall parameters, energy consumption and price is a burning problem. For this purpose, noise-like signals are used which have such well-known advantages as high impedance in relation to narrow-band interference of high power, the possibility of separating subscribers by code, the secrecy of transmission, high protection to multi-beam propagation, and even high resolution in radar and navigational location measurements.

The development of methods and means of noiseproof coding using noise-like codes based on Barker-like sequences are shown in the papers of famous scientists, in particular: M. Kelman and F. Rivest real-time encoding and decoding

© Springer Nature Switzerland AG 2020
V. Lytvynenko et al. (Eds.): ISDMCI 2019, AISC 1020, pp. 203–214, 2020.
https://doi.org/10.1007/978-3-030-26474-1_15

algorithm using Barker sequences [1]; Kim and Jang the study of noisy codes based on the sequences of Goley and Barker [2]; Konig and Schmidt [3] wireless data protection and real-time data transmission systems with given parameters; Omar and Kassem methods of solving the problem of ambiguity, based on Barker sequences [4]; Matsuyuki and Tsuneda application in control systems and communication codes with a minimum auto-correlation function [5]. From these papers, it's known that attempts to find Barker codes for lengths more than 13 don't have any solutions. In addition, the development of regular methods for constructing Barker-like sequences of any length with the minimum possible value of the level of side lobes of the autocorrelation function is the actual unresolved problem. Thus, the known Barker codes can be used just only for signals with a relatively small base.

Therefore, the problem of finding barker-like sequences with a length greater than 13 with the minimum possible value of the side lobes arises in many cases, making the actual task of the proposed regular method for constructing barter-like codes based on ideal ring bundles. With the use of noise-like codes on the basis of Barker-like sequences of arbitrary length with the minimum possible value of the level of side lobes, the implementation of hardware and software components for the purpose of synthesis of small-sized transmission systems in real time, which are characterized by high impedance, the possibility of separating subscribers by code sign, secrecy of transmission and high resistance to multi-beam dissemination.

2   Review of the Literature

The impedance protection is one of the most important characteristics of modern information transmission systems. The possibility of its further increase at fixed broadcast speed is a very topical problem.

The aim of the present work is to develop a method of noise-like coding for enhancing of the impedance protection of the information transmission system. The object of research in this paper is the noise-like coding which is based on Barker-like sequences [6].

The subject of the study is the development of an advanced method for constructing Barker-like sequences. Objectives of the study are to identify the possibility of applying Barker-like sequences for the impedance protection coding [7–9].

The practical value is to find Barker-like sequences, where the value of the ratio of the main petal to the side lobes is better rather than at the known Barker codes. It will allow use them to solve tasks of impedance protection encoding [10, 11].

Such known advantages of noise-like signals as high impedance protection against narrowband impediments of high power, the possibility of separating subscribers by code, secrecy of the transfer, high resistance to multi-beam propagation and even high resolution for radar and navigation measurements, have led to their use in various communication systems and location determination [1, 12–15].

Barker sequence is the object of our research. Barker sequence is a series which consist with N elements a_i for $0 \leq j \leq N$, which have values $+1$ and -1, and alternate in order to the condition is fulfilled:

$$\left| \sum_{j}^{N-v} a_j a_{j-v} \right| \leq 1, \tag{1}$$

where $1 \leq v \leq N$.

The autocorrelation function (ACF) of a signal $s(t)$, localized in time and finite in energy is a quantitative integral characteristic of the waveform, and is determined by the integral of the product of two copies of the signal $s(t)$, shifted relative to each other for a while t:

$$B_s(\tau) = \int_{-\infty}^{+\infty} s(t)s(t+\tau)dt. \tag{2}$$

AFC is the scalar product of a signal and its copy in functionality depending on the variable value of the shift value t. Signals, the base of which varies according to the Barker code, is a unique phase-manipulated signal. The module of their autocorrelation function has the lowest achievable level of side petals [4,5,16,17].

3   Problem Statement

The choice of the pseudorandom code sequence in the radio engineering system of information transmission is very significant. Since, the enhancement of system processing, its impedance protection, and sensitivity depend on the parameters of the code sequence. On condition that the code sequence will have the same length, system parameters can be different from each other by the level of interference protection, transmission speed, and so on. To use the noise-like signals, code combinations must be endowed with certain mathematical characteristics, the main of which is autocorrelation [14,18–20].

Orthogonality is a concept that is a generalization of perpendicularity for linear spaces with a scalar product introduced. If the scalar product of two space elements equal to zero, then they are called orthogonal to each other. An important feature of the notion is its binding to the particular scalar product which is used. While product changing, the orthogonal elements might become non-orthogonal and vice versa.

Depending on the method of formation and statistical properties, orthogonal code sequences are divided into orthogonal and quasi-orthogonal. An earmark feature of the sequence is the correlation coefficient ρ_{ij}, which in general case varies from -1 to $+1$.

In the coding theory has proved that the maximum achievable value of the coefficient of mutual correlation is determined from the condition [21,22]:

$$\rho_{ij} = \begin{cases} -1/N, & \text{where } N \text{ is odd number} \\ -1/(N-1), & \text{where } N \text{ is ever number} \end{cases} \tag{3}$$

The minimum value of the correlation coefficient provides codes in which the correlation coefficients for any pair of sequences are negative (trans-orthogonal codes). The coefficient of mutual correlation of orthogonal sequences is equal to zero by definition. Thus $\rho_{ij} = 0$. For large one's values of N, the difference between the coefficients of correlation of orthogonal and trans-orthogonal codes can be almost neglected. Unfortunately, orthogonal codes do not always exist to all lengths. Therefore, the problem is reduced to the search for synthesis methods of quasi-orthogonal codes. I mean case when the coefficient of mutual correlation is minimal $\rho_{ij} \longrightarrow min$.

That is to say that Barker's codes are in line with these requirements. It's known that attempts to find Barker's codes with the number of elements $m > 13$ does not have any decision. So, Barker's codes can be used just only for signals with a relatively small base. Therefore, the development of the algorithm for constructing Barker-like codes with the number of elements $m > 13$ based on the ideal ring bundles (IRB) is an urgent task [2,23]. Ideal ring bundle is called the circular sequence $L_N = (l_1, l_2, ..., l_N)$ of numbers N in which all possible circular sums exhaust R-times of values of numbers within natural numbers from 1 to $S_N^R = N(N-1)/(R+1)$.

4   The Method of Constructing on the Basis of IRB Families

The method of constructing on the basis of IRB families on the criterion of the minimum value of the function of autocorrelation of the discrete signal is as follows:

- choose a variant of the IRB of a given order of N the required length L_N of multiplicity R;
- build L_N-positioned code μ_i, $i = 1, 2, ..., L_N$ with one-level periodic autocorrelation function based on choose variant of IRB $(k_1, k_2, .., k_l, ..., k_N)$, where on N positions of code with ordinal numbers x_l, $l = 1, 2, ..., N$ which determined with next formula:

$$x_l = 1 + \sum_{i=1}^{l} k_i (mod\, L_N), \qquad (4)$$

where place symbols "1", and on the rest $L_N - N$ positions - symbols "−1".

To construct the Barker sequences, we choose from the different types of IRB those sequences where the ratio is $N/R \approx 2$ [24].

Let's consider an example of constructing of Barker sequences based on IRB according to the given methodology, where $N = 12$, $L_N = 28$, $R = 5$. Out of two existing variants of the simplest shortest IRB order $N = 12$, constructed by the algorithm of selective displacements, for instance, we choose the first IRB type:

$$(1, 1, 3, 1, 1, 7, 2, 2, 3, 3, 3, 1).$$

Let us create a sequence with the length of the code $L_N = 28$; We place the characters "1" by the formula (4) in twelve positions ($N = 12$), whereas remained positions fill with characters "-1":

$$1, 1, 1, -1, -1, 1, 1, 1, -1, -1, -1, -1, -1, -1, 1, -1, 1, -1, 1, -1,$$
$$-1, 1, -1, -1, 1, -1, -1, 1.$$

The resulting sequence is a Barker-like sequence with the property "no more matches" for which the value of the autocorrelation function doesn't overdraw 2 (apart from the main petal):

$$28, -1, 0, 1, -2, 1, 2, 1, -2, 1, -2, 1, -2, 1, 2, -1, 2, -1, -2, -1, -2,$$
$$-1, 2, -1, -2, -1, 0, 1.$$

In that case, the ratio of the main petal to others is 14 and this is more than in the classical Barker sequences. We can similarly obtain variants of Barker-like sequences with the indicated properties by choosing others types of IRB. Consequently, the use of IRBs for the synthesis of Barker-like sequences makes it possible to simplify their construction, thanks to the application of the results of constructing different types of IRBs. It also makes possible to simplify of finding the complete families of these configurations, searching among them the Barker-like sequences with the best characteristics.

In addition to the widely known Singer IRB families, several families of IRB are known [25]. Let us list them with a short description (Fig. 1).

Type Q. A pair of differences of quadratic residue in $GF(p^r)$, $p_r = 3(mod\,4)$. Parameters:

$$S_N^R = p^r = 4t - 1, \ N = 2t - 1, \ R = t - 1. \tag{5}$$

Type H_6. Primitive root r modulo p, such that $Ind_r(3) \equiv 1(mod\,6)$, where p is prime number in the form $p = 4x^2 + 27$. Calculation $a_1, ..., a_{\frac{p-1}{2}}\,(mod\,p)$,

such that $Ind_r(a_i) \equiv 0.1$ or $3(mod\,6)$. Then, the pair of differences of these calculations form the IRB plural with parameters:

$$S_N^R = p = 4t - 1, \ N = 2t - 1, \ R = t - 1. \tag{6}$$

Type T. Among of $(p-1)(q-1)$ residue modulo pq, mutually prime with pq, where p and $q = p+2$ is prime numbers. Let $a_1, ..., a_m$, where $m = (p-1)(q-1)/2$ are the residue, for which $(\frac{a_i}{p}) = (\frac{a_j}{q})$; we will also denote $0, q, 2q, ..., (p-1)q$ by $a_{m+1}, ..., a_{m+p}$. Then $m + p = (pq - 1)/2 = N$. Differences between pairs of residues $a_1, ..., a_N$ form IRB modulo S_N^R with parameters $S_N^R = pq$, $N = (pq - 1)/2$, $R = (pq - 3)/4$, as always: $pq \equiv -1(mod\,4)$. Then:

$$S_N^R = 4t - 1, \ N = 2t - 1, \ R = t - 1. \tag{7}$$

Similarly, IRB is determined for $GF(p^r)$ and $GF(q^s)$, if $q^s = p^r + 2$.

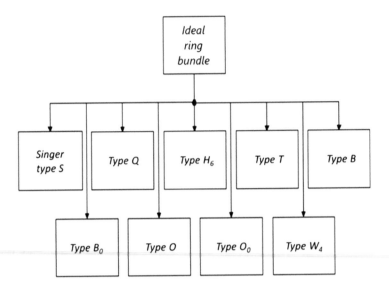

Fig. 1. Types of ideal ring bundle

Type B. The difference of the pairs of two-quadratic residues of prime numbers of the form $p = 4x^2 + 1$, where x is odd number, where

$$S_N^R = p = 4x^2 + 1, \ N = x^2, \ R = \frac{x^2 - 1}{4}. \tag{8}$$

Type B_0. The difference of the pairs of two-quadratic residues of prime numbers of the form $p = 4x^2 + 9$, where x is odd number, where

$$S_N^R = p = 4x^2 + 9, \ N = x^2 + 3, \ R = \frac{x^2 + 3}{4}. \tag{9}$$

Type O. The difference between pairs of octal residues of prime numbers of the form $p = 8a^2 + 1 = 64b^2 + 9$, where a, b are odd numbers, where

$$S_N^R = p, \ N = a^2, \ R = b^2. \tag{10}$$

Type O_0. The difference between pairs of octal residues modulo of prime numbers of the form $p = 8a^2 + 49 = 64b^2 + 441$, where a is odd number and b is even number, where

$$S_N^R = p, \ N = a^2 + 6, \ R = b^2 + 7. \tag{11}$$

Type W_1. The difference of the pairs of two-quadratic residues. Here p and q are two prime numbers, such that $d = (p - 1, q - 1)/4$ is integer. If the number g is a primitive root for p and q, then IRB consists of pairs of differences $(g - 1), (g^2 - g)...., (g^{d-1} - g^{d-2}), (0 - g^{d-1}), (g - 0), (2g - g), ..., ((p - 1)q - (p - 2)q)$ modulo pq, where

$$S_N^R = pq, \ N = (S_N^R - 1)/4, \ R = (S_N^R - 5)/16. \tag{12}$$

Then, the equality must be fulfilled $q = 3p + 2$, and $(S_N^R - 1)/4$ should be an odd square.

All these IRB families can be used to construct Barker's codes and Barker's-like codes of high order.

5   Implementation of Barker's-Like Codes

Figure 2 presents the scheme of the generator of a pseudorandom sequence code which has been implemented with dimensionality $N = 28$. The main component of it is a sequential parallel shift register (REG_S_PAR). This register saves the value 0x8F1112D (1000111100010001000100101101B) as the initial value for the offset. The data have been taken from work [25], where all code sequences with the maximum level of side lobes PSL $= 2$ and the maximum length of the code $N = 28$ have shown.

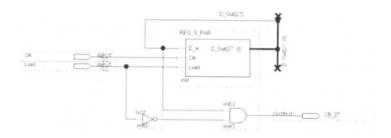

Fig. 2. The scheme of the generator of a Barker's-like code with dimensionality $N = 28$

The inputs of the generator are Clk - the input of clock pulses of a sequence-parallel shift; Load (active signal level "1") - the signal of loading the initial data into the register. The output of the generator is a Barker's-like code with dimension. During each Clk synchronization impulse, a one-bit offset to the left of the data array D_Out [27...0] is done and record the input value D_In register REG_S_PAR into junior class D_Out [0]. Figure 3 shows the results of the simulation of generator operation. It should be noted that the sequence at the generator output has repeated after 28 impulses.

The generator of quasi-barker codes with another dimension N can be implemented in the same way. So, as a result, the initial value of the shift register for a generator with a dimension (PSL $= 4$) is 0x025863ABC266F (00 0010 0101 1000 0110 0011 1010 1011 1100 0010 0110 0110 1111). Under those circumstances, the hardware resources which are needed for their realization will change. The implementation of the generator with dimension requires 29 logical elements (out of 15,408), 28 registers, and 3 (out of 347) FPGA EP3C16F484 outputs, and for a generator with a dimension of $N = 50$, 50 logical elements, 50 registers, and 3 FPGA outputs are required.

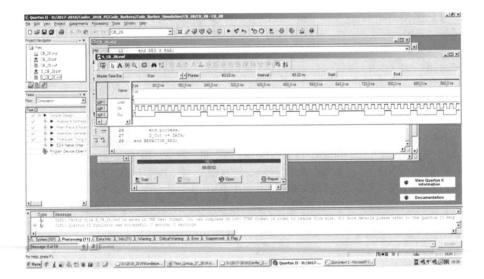

Fig. 3. The time charts of generator operation of the Barker's-like code with dimensionality $N = 28$

Barker's-like codes often use in extended-spectrum communication systems (DS-SS). To extend the spectrum of a narrowband signal in DS-SS systems, each transmitting information bit embeds a sequence of chips called a pseudorandom code sequence. Consider expanding the spectrum of input data using Barker's-like codes on the FPGA.

6   Results and Discussions

Suppose that the clock frequency of chips of Barker's-like code consists of $F_c = 200$ kHz (Tc = 5 us) and pulse frequency $F_{Clk} = 50$ MHz. Under those circumstances, counter with a credit factor of $K_c = 250$ is needed to get the clock frequency F_c. The period of data bits determines from the equation:

$$T_d \geq \frac{N}{F_c}, \tag{13}$$

where N is the length of the Barker's-like code. Thus, the period of data bits for $N = 14$ should be $T_d \geq 70us$. The Principle scheme of expanding the spectrum of input data is shown in Fig. 4, and modeling its operation - in Fig. 5.

Input signals for this scheme: Clk – signal of clock frequency (FClk = 50 MHz); Data_In – 8-bit input data; Reset – signal of reset register of REG_PAR_SER to zero (active signal level "1"); Load – the signal of loading the initial data into the Barker's-like code generator (active signal level "1"). Output signals: Clk_N1 – sync-impulses of chips of Barker's-like code; Clk_N2 – sync-impulses following up bits of input data; Clk_N3 – sync-impulses loading

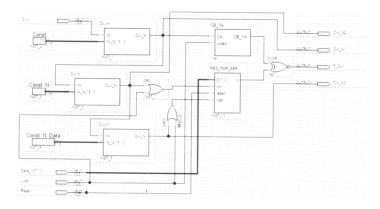

Fig. 4. The scheme for expanding the spectrum of input data by Barker's-like codes with dimensionality $N = 14$

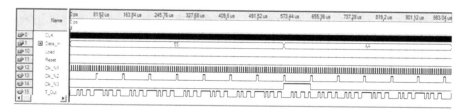

Fig. 5. Time charts operation of the spectrum expansion scheme by the sequence of Barker's-like code $N = 14$

of input data Data_In; T_Out – signal of the extended range of input data. The appearance of these signals is shown in Fig. 5.

Synchronization signals Clk_N1 are obtained from the signal clock frequency Clk by dividing it by the coefficient N1 = 250 using a counter with a variable division factor INST_1. The dividing factor N1 is given by the mega function Const (INST_2). Clock impulses Clk_N2 are obtained from Clk_N1 with the division of the latest into a coefficient N2 = 14 by the counter INST_5. The dividing factor N2 is given by the mega function Const (INST_3). The synchronization signal Clk_N3 is obtained from the signal Clk_N2 by dividing it by the coefficient N3 = 8. The counter INST_6 uses to this purpose. The dividing factor N3 is given by the mega function Const (INST_4). Entrances the counters with a variable coefficient of account: Clk – signal of clock frequency which divides by the coefficient N; IN_N – 16-bit data that specifies the dividing factor for the counter. The maximum dividing factor for them is 216-1.

Clock impulses Clk_N1 and signal "Load" are served on the entrance of the generator of the Barker's-like codes CB_14 (INST). A sequence of Barker's-like codes of length $N = 14$, which are repeated with the clock pulse Clk_N2 we obtain at its output. These impulses go to one of the inputs of the XNOR (INST_11) element. Impulses from the output of the register REG_PAR_SER

(INST_10) come on its second input. At the output of the XOR (T_Out) element we get input with an expanded spectrum with dimension $N = 14$. The REG_PAR_SER register performs the conversion of the parallel Data_In input code into the consistent code. His work synchronizes with clock pulses generated by combining signals Clk_N2 and Load (INST_7). The hardware resources which are required to implement this scheme are 99 logical elements, 71 registers, and 15 FPGA outputs.

7 Conclusions

With the help of the developed software complex, a code search with a number of discretes greater than 13. A unique barker-like code was found for phase-manipulated signals longer than 13 periods that have, for this number of discretions, the minimum achievable level of side-lobe correlation function. In this way, the possibility of constructing barter-like codes with the help of IRB models, creation of effective algorithms for their construction is shown. Gotten code sequences are sequences with property no more than $R-$coincidences and by the minimum value of autocorrelation function. Choosing other variant of IRB models with such parameters can get other Barker's-like sequences with property no more R-coincidences and by the minimum value of autocorrelation function.

The generator of Barker's-like codes and scheme expansion of the spectrum by these codes have been developed in the Quartus II environment. Furthermore, the hardware resources which are necessary for their implementation have been estimated.

References

1. Kellman M, Rivest F, Pechacek A, Sohn L, Lustig M (2017) Barker-coded node-pore resistive pulse sensing with built-in coincidence correction. In: 2017 IEEE international conference on acoustics, speech and signal processing (ICASSP), New Orleans, LA, pp 1053–1057. https://doi.org/10.1109/ICASSP.2017.7952317
2. Kim P, Jung E, Bae S, Kim K, Song T (2016) Barker-sequence-modulated golay coded excitation technique for ultrasound imaging. In: 2016 IEEE international ultrasonics symposium (IUS), Tours, pp 1–4. https://doi.org/10.1109/ULTSYM.2016.7728737
3. Konig S, Schmidt M, Hoene C (2010) Precise time of flight measurements in IEEE 802.11 networks by cross-correlating the sampled signal with a continuous Barker code. In: The 7th IEEE international conference on mobile ad-hoc and sensor systems (IEEE MASS 2010), San Francisco, CA, pp 642–649. https://doi.org/10.1109/MASS.2010.5663785
4. Omar SM, Kassem F, Mitri R, Hijazi H, Saleh M (2015) A novel barker code algorithm for resolving range ambiguity in high PRF radars. In: 2015 European radar conference (EuRAD), Paris, pp 81–84. https://doi.org/10.1109/EuRAD.2015.7346242
5. Matsuyuki S, Tsuneda A (2018) A study on aperiodic auto-correlation properties of concatenated codes by barker sequences and NFSR sequences. In: 2018 international conference on information and communication technology convergence (ICTC), Jeju, pp 664–666. https://doi.org/10.1109/ICTC.2018.8539367

6. Riznyk O, Povshuk O, Noga Y, Kynash Y (2018) Transformation of information based on noisy codes. In: 2018 IEEE second international conference on data stream mining & processing (DSMP), Lviv, Ukraine, pp 162–165. https://doi.org/10.1109/DSMP.2018.8478509
7. Ahmad J, Akula A, Mulaveesala R, Sardana HK (2019) Barker-coded thermal wave imaging for non-destructive testing and evaluation of steel material. IEEE Sens J 19(2):735–742. https://doi.org/10.1109/JSEN.2018.2877726
8. Fu J, Ning G (2018) Barker coded excitation using pseudo chirp carrier with pulse compression filter for ultrasound imaging. In: BIBE 2018, international conference on biological information and biomedical engineering, Shanghai, China, pp 1–5
9. Dua G, Mulaveesala R (2013) Applications of barker coded infrared imaging method for characterisation of glass fibre reinforced plastic materials. Electron Lett 49(17):1071–1073. https://doi.org/10.1049/el.2013.1661
10. Arasu KT, Arya D, Kedia D (2016) Comparative analysis of punctured sequence pairs for frame synchronization applications. In: 2016 international conference on computational techniques in information and communication technologies (ICCTICT), New Delhi, pp 470–475. https://doi.org/10.1109/ICCTICT.2016.7514626
11. Nazarkevych M, Yavourivskiy B, Klyuynyk I (2015) Editing raster images and digital rating with software. In: The experience of designing and application of CAD systems in microelectronics, Lviv, Ukraine, pp 439–441. https://doi.org/10.1109/CADSM.2015.7230897
12. Kellman MR, Rivest FR, Pechacek A, Sohn LL, Lustig M (2018) Node-pore coded coincidence correction: coulter counters, code design, and sparse deconvolution. IEEE Sens J 18(8):3068–3079. https://doi.org/10.1109/JSEN.2018.2805865
13. Wang M, Cong S, Zhang S (2018) Pseudo chirp-barker-golay coded excitation in ultrasound imaging. In: 2018 Chinese control and decision conference (CCDC), Shenyang, pp 4035–4039. https://doi.org/10.1109/CCDC.2018.8407824
14. Chunhong Y, Zengli L (2013) The superiority analysis of linear frequency modulation and barker code composite radar signal. In: 2013 ninth international conference on computational intelligence and security, Leshan, pp 182–184. https://doi.org/10.1109/CIS.2013.45
15. Nilawar RC, Bhalerao DM (2015) Reduction of SFD bits of WiFi OFDM frame using wobbulation echo signal and barker code. In: 2015 international conference on pervasive computing (ICPC), Pune, pp 1–3. https://doi.org/10.1109/PERVASIVE.2015.7087095
16. Riznyk O, Povshuk O, Kynash Y, Yurchak I (2017) Composing method of anti-interference codes based on non-equidistant structures. In: 2017 XIIIth international conference on perspective technologies and methods in MEMS design (MEMSTECH), Lviv, Ukraine, pp 15–17. https://doi.org/10.1109/MEMSTECH.2017.7937522
17. Wang S, He P (2018) Research on low intercepting radar waveform based on LFM and barker code composite modulation. In: 2018 international conference on sensor networks and signal processing (SNSP), Xi'an, China, pp 297–301. https://doi.org/10.1109/SNSP.2018.00064
18. Xia S, Li Z, Jiang C, Wang S, Wang K (2018) Application of pulse compression technology in electromagnetic ultrasonic thickness measurement. In: 2018 IEEE far east NDT new technology & application forum (FENDT), Xiamen, China, pp 37–41. https://doi.org/10.1109/FENDT.2018.8681975
19. Banket V, Manakov S (2018) Composite walsh-barker sequences. In: 2018 9th international conference on ultrawideband and ultrashort impulse signals (UWBUSIS), Odessa, pp 343–347. https://doi.org/10.1109/UWBUSIS.2018.8520220

20. Riznyk O, Parubchak V, Skybajlo-Leskiv D (2007) Information encoding method of combinatorial configuration. In: 2007 9th international conference the experience of designing and applications of CAD systems in microelectronics, Lviv-Polyana, Ukraine, p. 370. https://doi.org/10.1109/CADSM.2007.4297583
21. Sekhar S, Pillai SS (2016) Comparative analysis of offset estimation capabilities in mathematical sequences for WLAN. In: 2016 international conference on communication systems and networks (ComNet), Thiruvananthapuram, pp 127–131. https://doi.org/10.1109/CSN.2016.7824000
22. Riznyk O, Kynash Y, Povshuk O, Noga Y (2018) The method of encoding information in the images using numerical line bundles. In: 2018 IEEE 13th international scientific and technical conference on computer sciences and information technologies (CSIT), Lviv, Ukraine, pp 80–83. https://doi.org/10.1109/STC-CSIT.2018.8526751
23. Riznyk O, Povshuk O, Kynash Y, Nazarkevich M, Yurchak I (2018) Synthesis of non-equidistant location of sensors in sensor network. In: 2018 XIV-th international conference on perspective technologies and methods in MEMS design (MEMSTECH), Lviv, Ukraine, pp 204–208. https://doi.org/10.1109/MEMSTECH.2018.8365734
24. Oleg R, Yurii K, Oleksandr P, Bohdan B (2017) Information technologies of optimization of structures of the systems are on the basis of combinatorics methods. In: 2017 12th international scientific and technical conference on computer sciences and information technologies (CSIT), Lviv, Ukraine, pp 232–235. https://doi.org/10.1109/STC-CSIT.2017.8098776
25. Li J, Kijsanayotin T, Buckwalter JF (2016) A 3-Gb/s radar signal processor using an IF-correlation technique in 90-nm CMOS. IEEE Trans Microw Theory Tech 64(7):2171–2183. https://doi.org/10.1109/TMTT.2016.2574983

Controlled Spline of Third Degree: Approximation Properties and Practical Application

Oleg Stelia$^{(\boxtimes)}$ ⓘ, Iurii Krak ⓘ, and Leonid Potapenko ⓘ

Taras Shevchenko National University of Kyiv, Kiev, Ukraine
oleg.stelya@gmail.com

Abstract. The interpolation properties of a Hermitian cubic spline of C^1 smoothness controlled by a broken line, are studied. The angles of inclination of the polygonal line links and linear combinations of values at the vertices of the polygonal line determine the derivatives and spline values in multiple interpolation nodes. Spline nodes are selected to coincide with the abscissas of the vertices of the polygonal line. To calculate the polynomial coefficients of a piecewise polynomial curve, a system of algebraic equations with a three-diagonal matrix is solved. The position of the interpolation nodes plays the role of the shape parameters. When you select spline nodes that match the interpolation nodes, you get a local version of the spline. For the global spline, theoretical estimates of the interpolation error are obtained. The interpolation properties of the considered spline variants are illustrated with various examples. The influence of shape parameters on the behavior of the constructed curve is shown. Some differences between the local and global spline options are shown. An example of using a spline to approximate the electrocardiogram data is given.

Keywords: Cubic controlled spline · Approximation error · Local spline

1 Introduction

This publication is a continuation of the study of spline polynomial curves with special properties. In [1], a construction method is presented and the conditions for the existence and uniqueness of a piecewise polynomial curve of the third degree are formulated, which is controlled using control points. Examples of using such a curve as a Bezier curve are considered. In this paper, we study the curve from the point of view of a cubic interpolation spline of C^1 smoothness. Local version of the curve is built. Using numerical examples, the interpolation properties of the considered spline are illustrated. The recommendations on its use for solving practical problems are given.

© Springer Nature Switzerland AG 2020
V. Lytvynenko et al. (Eds.): ISDMCI 2019, AISC 1020, pp. 215–224, 2020.
https://doi.org/10.1007/978-3-030-26474-1_16

2   Review of the Literature

Cubic interpolation splines of C^1 smoothness are one of the most well-known constructions for piecewise polynomial interpolation of grid functions. Algorithms for their construction, justification, as well as illustration of various properties are given in numerous works, see for example [2–4]. Despite of such splines unconditional popularity, they have several disadvantages, which will be emphasized in this publication. One of the major drawbacks of the cubic splines of C^2 smoothness is that they generally do not preserve the monotony of the original data sets, resulting in oscillation of the constructed functions. Splines can also make uncontrolled outliers of the constructed functions values. Thus, the construction of piecewise polynomial functions that are not subject to oscillation and outliers is relevant.

One way to eliminate the drawbacks of global cubic splines is to use local splines. Akima in the paper [5] proposed a cubic piecewise polynomial interpolation spline of C^1 smoothness. The polynomial for the selected segment is constructed from the given values of the function and the first derivatives at its ends. The article introduces the concept of curve slope, and derivatives at the ends of a segment are equal to the found slope. Unlike the global cubic interpolation spline of C^2 smoothness, Akima's spline is less affected by outliers and allows building functions with almost no oscillation. The local Catmull-Rom spline, which is a special case of the cubic splines family, was proposed in the paper [6]. The first derivatives at the ends of the segment are constructed as difference derivatives obtained from the values of the interpolating grid function. A subclass of the Catmull-Rom splines was built in [7], which contains the shape parameters. These parameters are used to change the shape of the curve, regardless of the interpolated function values. In the paper [8], an analysis of various ways of the Catmull-Rom cubic curve parametrization was carried out. It is shown that the choice of parameterization is essential in applications. In particular, an example of the robot arm animation is given. The use of shape parameters in Catmull-Rom splines is also the subject of work [9]. A class of cubic basis functions of the Catmull-Rom spline with a shape parameter, which is called a cubic α-Catmull-Rom spline, is constructed. On the basis of the introduced basic functions, the corresponding cubic spline curves are generated.

Thus, as follows from the analysis of existing interpolation methods, expressions for calculating derivatives are obtained on the basis of the combination of the interpolated functions values. In the proposed approach, the values of the interpolated function and differential derivatives, obtained on the basis of the values at the control points, are used.

3   Problem Statement and Solution Method

Let a polygonal line is set on the interval $[a, b]$. The vertices abscissas of this line are given in the grid nodes:

$$\Delta_\tau: \quad a = \tau_1 < \tau_2 < ... < \tau_N = b.$$

The vertices ordinates denoted F_i (control points in terms of Bezier curves). We construct an interpolation spline that satisfies the following conditions: the angles of inclination of the polygonal line links determine the values of the first derivatives of the spline at the nodes x_i of the grid Δ_x : $\tau_1 = x_1 < x_2 < \ldots < x_{N+1} = \tau_N$, such that $\tau_{i-1} < x_i < \tau_i$, $i = \overline{2, N}$, $f_i' = \dfrac{(F_i - F_{i-1})}{h_i}$, here $h_i = \tau_i - \tau_{i-1}$, $i = \overline{2, N}$.

The values of the interpolated function in the nodes of grid Δ_x are determined as follows: $f_i = \dfrac{[F_{i-1}(h_i - \mu_i) + F_i\mu_i]}{h_i}$, where $\mu_i = \tau_i - x_i$. In this case, the parameters μ_i play the role of the parameters of the curve shape. In this notation, the relations $x_i - \tau_{i-1} = h_i - \mu_i$ are satisfied.

In [1], an algorithm for constructing is presented and the conditions for the existence and uniqueness of a cubic spline curve $S(x)$ of the defect 2 in the interval $[a, b]$ are obtained that meet the following conditions:

$$S(x_i) = f_i \tag{1}$$

$$S'(x_i) = f_i', \ i = \overline{2, N}. \tag{2}$$

The points τ_i are nodes of the spline, and the points x_i are multiple nodes of interpolation. Denoting by ϕ_i, $i = \overline{1, N}$ the unknown values of the spline at points τ_i, the system of equations for their definition is written in the form:

$$A_i\phi_{i-1} - (B_i^{(1)} + B_i^{(2)})\phi_i + C_i\phi_{i+1} = \Phi_i, \ i = \overline{2, N-1}, \tag{3}$$

where:

$$A_i = \frac{\mu_i^2}{(h_i - \mu_i)^2 h_i}, \quad B_i^{(1)} = \frac{2h_i + \mu_i}{\mu_i h_i}, \quad B_i^{(2)} = \frac{3h_{i+1} - \mu_{i+1}}{(h_{i+1} - \mu_{i+1})h_{i+1}},$$
$$C_i = \frac{(h_{i+1} - \mu_{i+1})^2}{h_{i+1}\mu_{i+1}^2}. \tag{4}$$

$$\Phi_i = f_i \frac{h_i}{(h_i - \mu_i)\mu_i} + f_i \frac{(h_i - 2\mu_i)h_i}{(h_i - \mu_i)^2\mu_i} + \frac{F_i - F_{i-1}}{(h_i - \mu_i)}$$
$$+ f_{i+1}\frac{h_i}{(h_{i+1} - \mu_{i+1})\mu_{i+1}} - f_{i+1}\frac{(h_{i+1} - 2\mu_{i+1})h_{i+1}}{(h_{i+1} - \mu_{i+1})\mu_{i+1}^2} - \frac{F_{i+1} - F_i}{\mu_{i+1}}. \tag{5}$$

To close the system of equations it's necessary to add the conditions:

$$\phi_1 = F_1, \quad \phi_N = F_N.$$

In this case, the spline is written as:

$$S(x) = \phi_{i-1}\frac{(x - x_i)(x - \tau_i)}{(\tau_{i-1} - x_i)(\tau_{i-1} - \tau_i)} + \phi_i \frac{(x - x_i)(x - \tau_{i-1})}{(\tau_i - x_i)(\tau_i - \tau_{i-1})} +$$
$$f_i \frac{(x - \tau_i)(x - \tau_{i-1})}{(x_i - \tau_i)(x_i - \tau_{i-1})} + Q(x - \tau_i)(x - \tau_{i-1})(x - x_i),$$

$$Q = -\phi_{i-1}\frac{1}{(h_i - \mu_i)^2 h_i} + \phi_i \frac{1}{h_i \mu_i^2} - f_i \frac{h_i - 2\mu_i}{(h_i - \mu_i)^2 \mu_i^2} - \frac{F_i - F_{i-1}}{h_i \mu_i (h_i - \mu_i)}$$

for $x \in [\tau_{i-1}, \tau_i], \quad i = \overline{2, N}$.

To simplify further transformations, we'll consider uniform grids Δ_τ and Δ_x. Let $h_i = h$, $i = 2, \ldots, N-1$, $\tau_i = x_{i+1/2} = (x_{i+1} + x_i)/2$, $i = 1, \ldots N-1$. In this case, the spline takes the form:

$$S(x) = -\phi_{i-1}\frac{4}{h^3}(x - x_i)^2(x - x_{i+1/2}) - f_i \frac{4}{h^2}(x - x_{i-1/2})(x - x_{i+1/2})$$
$$+ \phi_i \frac{4}{h^3}(x - x_i)^2(x - x_{i-1/2}) - f_i' \frac{4}{h^2}(x - x_i)(x - x_{i-1/2})(x - x_{i+1/2}). \tag{6}$$

The system of equations for the definition of unknowns ϕ_i is written as:

$$\phi_1 = F_1,$$
$$\phi_{i-1} - 10\phi_i + \phi_{i+1} = 8F_i, \quad i = \overline{2, N-1},$$
$$\phi_N = F_N.$$

4   Interpolation Error

Consider the interpolation error of the considered spline. For obtaining estimates, the approach proposed in the book [10] is used. Earlier in the paper [11], this approach was used to study a parabolic spline.

Theorem. *Let the spline of the defect 1 interpolate the function $f(x) \in C^2[a, b]$ at the nodes of the grid Δ_x. Then the following estimates are made:*

$$\left\| S^{(r)}(x) - f^{(r)}(x) \right\|_{C[a,b]} \leq K_r \ h^{2-r}\omega(h, f''), \quad r = 0, 1, \quad K_0 = \frac{1}{4}, \quad K_1 = \frac{29}{8}.$$

Proof. Consider the difference $|S(x) - f(x)|$ for $x \in [x_{i-1/2}, x_{i+1/2}]$, $i = 2, \ldots, N-1$.
Add and subtract the terms $f_{i-1/2}\frac{4}{h^3}(x - x_i)^2(x - x_{i+1/2})$, $f_{i+1/2}\frac{4}{h^3}(x - x_i)^2(x - x_{i-1/2})$ on the right-hand side of (6), and consider for $i = 2, \ldots, N-1$ the expression:

$$|S(x) - f(x)| = \left| -\left(\phi_{i-1} - f_{i-1/2}\right)\frac{4}{h^3}(x - x_i)^2(x - x_{i+1/2}) \right.$$
$$+ \left(\phi_i - f_{i+1/2}\right)\frac{4}{h^3}(x - x_i)^2(x - x_{i-1/2}) - f_{i-1/2}\frac{4}{h^3}(x - x_i)^2(x - x_{i+1/2})$$
$$+ f_{i+1/2}\frac{2}{h^2}(x - x_i)(x - x_{i-1/2}) \tag{7}$$
$$\left. - \frac{4}{h^2}(f_i + (x - x_i)f_i')(x - x_{i-1/2})(x - x_{i+1/2}) - f(x) \right|.$$

Using Taylor expansions with a residual Lagrange term and, substituting them in the right-hand side of (7), we obtain:

$$|S(x) - f(x)| = \left| -\left(\phi_{i-1} - f_{i-1/2}\right)\frac{4}{h^3}(x - x_i)^2(x - x_{i+1/2}) \right.$$

$$+ \left(\phi_i - f_{i+1/2}\right)\frac{4}{h^3}(x - x_i)^2(x - x_{i-1/2}) - \left(f_i - \frac{h}{2}f'_i + \frac{h^2}{8}f''(\xi_2)\right)$$

$$\times \frac{4}{h^3}(x - x_i)^2(x - x_{i+1/2}) + \left(f_i + \frac{h}{2}f'_i + \frac{h^2}{8}f''(\xi_3)\right) \tag{8}$$

$$\times \frac{4}{h^3}(x - x_i)^2(x - x_{i-1/2}) - \frac{4}{h^2}(f_i + (x - x_i)f'_i)(x - x_{i-1/2})(x - x_{i+1/2})$$

$$\left. - \left(f_i + (x - x_i)f'_i + \frac{(x - x_i)^2}{2}f''(\xi_1)\right)\right|,$$

where $\xi_1 \in (x, x_i)$, $\xi_2 \in (x_{i-1/2}, x_i)$, $\xi_3 \in (x_i, x_{i+1/2})$.

Consider and transform the terms containing the values of function f_i and its derivatives f'_i, f''_i:

$$-\frac{4}{h^3}f_i(x - x_i)^2(x - x_{i+1/2}) + \frac{4}{h^3}f_i(x - x_i)^2(x - x_{i-1/2})$$

$$- \frac{4}{h^2}f_i(x - x_{i-1/2})(x - x_{i+1/2}) - f_i$$

$$= -\frac{4}{h^3}f_i(x - x_i)^2(x - x_i - \frac{h}{2}) + \frac{4}{h^3}f_i(x - x_i)^2(x - x_i + \frac{h}{2})$$

$$- \frac{4}{h^2}f_i(x - x_i + \frac{h}{2})(x - x_i - \frac{h}{2}) - f_i$$

$$= \frac{4}{h^3}f_i[-(x - x_i^3) + \frac{h}{2}(x - x_i^2) + (x - x_i^3) + \frac{h}{2}(x - x_i^2) - h(x - x_i)^2 + \frac{h^3}{4}] - f_i = 0.$$

Consider and transform the terms containing f'_i:

$$-\frac{2}{h^2}f'_i(x - x_i)^2(2x - x_i - \frac{h}{2} - x_i + \frac{h}{2})$$

$$- \frac{4}{h^2}f'_i(x - x_i)(x - x_i + \frac{h}{2})(x - x_i - \frac{h}{2}) - (x - x_i)f'_i$$

$$= \frac{4}{h^2}f'_i(x - x_i)^3 - \frac{4}{h^2}f'_i(x - x_i)((x - x_i)^2 - \frac{h^2}{4}) - (x - x_i)f'_i = 0.$$

Consider and transform the terms containing f''_i:

$$-\frac{1}{2h}f''(\xi_2)(x - x_i)^2(x - x_{i+1/2}) + \frac{1}{2h}f''(\xi_3)(x - x_i)^2(x - x_{i-1/2})$$

$$- \frac{(x - x_i)^2}{2}f''(\xi_1) = -\frac{1}{2h}f''(\xi_2)(x - x_i)^2\left(x - x_i - \frac{h}{2}\right)$$

$$+ \frac{1}{2h}f''(\xi_3)(x - x_i)^2\left(x - x_i + \frac{h}{2}\right) - \frac{(x - x_i)^2}{2}f''(\xi_1)$$

$$= -\frac{1}{2h}f''(\xi_2)(x - x_i)^3 + \frac{1}{4}f''(\xi_2)(x - x_i)^2 + \frac{1}{2h}f''(\xi_3)(x - x_i)^3$$

$$+ \frac{1}{4}f''(\xi_3)(x - x_i)^2 - \frac{(x - x_i)^2}{2}f''(\xi_1)$$

$$= \frac{1}{2h}(f''(\xi_3) - f''(\xi_2))(x - x_i)^3 + \frac{1}{4}(f''(\xi_2) - f''(\xi_1))(x - x_i)^2$$

$$+ \frac{1}{4}(f''(\xi_3) - f''(\xi_1))(x - x_i)^2.$$

Since the coefficients for the terms containing f_i and f_i' equal to zero, we can write:

$$|S(x) - f(x)| = \left| -\frac{4}{h^3} \left(\phi_{i-1} - f_{i-1/2} \right) (x - x_i)^2 (x - x_{i+1/2}) \right.$$
$$+ \frac{4}{h^3} \left(\phi_i - f_{i+1/2} \right) (x - x_i)^2 (x - x_{i-1/2}) + \frac{1}{2h} \left(f''(\xi_3) - f''(\xi_2) \right) (x - x_i)^3$$
$$\left. + \frac{1}{4} \left(f''(\xi_2) - f''(\xi_1) \right) (x - x_i)^2 + \frac{1}{4} \left(f''(\xi_3) - f''(\xi_1) \right) (x - x_i)^2 \right|$$
$$\leq \frac{4}{h^3} \left| \left(\phi_{i-1} - f_{i-1/2} \right) (x - x_i)^2 (x - x_{i+1/2}) \right.$$
$$\left. + \left(\phi_i - f_{i+1/2} \right) (x - x_i)^2 (x - x_{i-1/2}) \right|$$
$$+ \left| \frac{1}{2h} \left(f''(\xi_3) - f''(\xi_2) \right) (x - x_i)^3 \right| + \left| \frac{1}{4} \left(f''(\xi_2) - f''(\xi_1) \right) (x - x_i)^2 \right|$$
$$+ \left| \frac{1}{4} \left(f''(\xi_3) - f''(\xi_1) \right) (x - x_i)^2 \right|.$$

Evaluating expressions $R_1(x) = (x - x_i)^2 \left(x - x_{i+1/2} \right)$ and
$R_2(x) = (x - x_i)^2 \left(x - x_{i-1/2} \right)$, one can have: $|R_1| \leq \dfrac{h^3}{32}$ and $|R_2| \leq \dfrac{h^3}{32}$.
Using the obtained estimates, it's got:

$$|S(x) - f(x)| \leq \frac{1}{8} \left| \phi_{i-1} - f_{i-1/2} \right| + \frac{1}{8} \left| \phi_i - f_{i+1/2} \right|$$
$$+ \frac{h^2}{16} \left(|f''(\xi_3) - f''(\xi_2)| + |f''(\xi_2) - f''(\xi_1)| + |f''(\xi_3) - f''(\xi_1)| \right) \qquad (9)$$
$$\leq \frac{1}{4} \max_{2 \leq i \leq N} \left| \phi_{i-1} - f_{i-1/2} \right| + \frac{3}{16} h^2 \omega(h, f'').$$

Estimating $\max\limits_{2 \leq i \leq N} \left| \phi_{i-1} - f_{i-1/2} \right|$ in the same way as it was done in [10, 11], we can obtain:

$$\max_{2 \leq i \leq N} \left| \phi_{i-1} - f_{i-1/2} \right| \leq \frac{1}{4} h^2 \omega(h, f''), \qquad (10)$$

where ω is the modulus of continuity.
Taking into account (9) and (10), we can obtain the estimation:

$$|S(x) - f(x)| \leq \frac{1}{4} h^2 \omega(h, f''). \qquad (11)$$

Consider $|S'(x) - f'(x)|$ for $x \in \left[x_{i-1/2}, x_{i+1/2} \right]$, $i = 2, \ldots, N - 1$. Differentiating the expression, one can get

$$S'(x) = -\phi_{i-1} \frac{12}{h^3} (x - x_i)^2 + \phi_{i-1} \frac{4}{h^2} (x - x_i) + \phi_i \frac{12}{h^3} (x - x_i)^2$$
$$+ \phi_i \frac{4}{h^2} (x - x_i) - f_i \frac{8}{h^2} (x - x_i) - f_i' \frac{12}{h^2} (x - x_i)^2 + f_i'.$$

Estimate $|S'(x) - f'(x)|$:

$$|S'(x) - f'(x)| = \left| -(\phi_{i-1} - f_{i-1/2}) \frac{12}{h^3} (x - x_i)^2 \right.$$
$$+ (\phi_{i-1} - f_{i-1/2}) \frac{4}{h^2} (x - x_i) + (\phi_i - f_{i+1/2}) \frac{12}{h^3} (x - x_i)^2$$
$$+ (\phi_i - f_{i+1/2}) \frac{4}{h^2} (x - x_i) - f_i \frac{8}{h^2} (x - x_i) \qquad (12)$$
$$- f_{i+1/2} \frac{12}{h^3} (x - x_i)^2 + f_{i+1/2} \frac{4}{h^2} (x - x_i) - f_{i-1/2} \frac{12}{h^3} (x - x_i)^2$$
$$\left. + f_{i-1/2} \frac{4}{h^2} (x - x_i) - f_i' \frac{12}{h^2} (x - x_i)^2 + f' - f'(x) \right|.$$

Using in the formula (12) the Taylor expansion, one can get

$$
\begin{aligned}
|S'(x) - f'(x)| = \Big| &-(\phi_{i-1} - f_{i-1/2})\frac{12}{h^3}(x - x_i)^2 + (\phi_{i-1} - f_{i-1/2})\frac{4}{h^2}(x - x_i) \\
&+ (\phi_i - f_{i+1/2})\frac{12}{h^3}(x - x_i)^2 + (\phi_i - f_{i+1/2})\frac{4}{h^2}(x - x_i) - f_i\frac{8}{h^2}(x - x_i) \\
&- \left(f_i + \frac{h}{2}f_i' + \frac{h^2}{8}f_i''(\xi_2)\right)\frac{12}{h^3}(x - x_i)^2 + \left(f_i + \frac{h}{2}f_i' + \frac{h^2}{8}f_i''(\xi_2)\right)\frac{4}{h^2}(x - x_i) \\
&- \left(f_i - \frac{h}{2}f_i' + \frac{h^2}{8}f_i''(\xi_1)\right)\frac{12}{h^3}(x - x_i)^2 + \left(f_i - \frac{h}{2}f_i' + \frac{h^2}{8}f_i''(\xi_1)\right)\frac{4}{h^2}(x - x_i) \\
&- f_i'\frac{12}{h^2}(x - x_i)^2 + f' - f_i' - (x - x_i)f_i''(\xi_3)\Big|.
\end{aligned}
$$

Consider the coefficients at f_i and f_i'. It is easy to see that these coefficients are zero, which allows us to write

$$
\begin{aligned}
|S'(x) - f'(x)| = \Big| &-(\phi_{i-1} - f_{i-1/2})\frac{12}{h^3}(x - x_i)^2 + (\phi_{i-1} - f_{i-1/2})\frac{4}{h^2}(x - x_i) \\
&+ (\phi_i - f_{i+1/2})\frac{12}{h^3}(x - x_i)^2 + (\phi_i - f_{i+1/2})\frac{4}{h^2}(x - x_i) \\
&- \frac{3}{2h}(f''(\xi_2) - f''(\xi_1))(x - x_i)^2 + \frac{1}{2}(f''(\xi_1) - f''(\xi_3))(x - x_i) \\
&+ \frac{1}{2}(f''(\xi_2) - f''(\xi_3))(x - x_i)\Big| \leq \frac{3}{h}|\phi_{i-1} - f_{i-1/2}| + \frac{2}{h}|\phi_{i-1} - f_{i-1/2}| \\
&+ \frac{3}{h}|\phi_i - f_{i+1/2}| + \frac{2}{h}|\phi_i - f_{i+1/2}| + \frac{3h}{8}|(f''(\xi_2) - f''(\xi_1))| \\
&+ \frac{h}{4}|(f''(\xi_1) - f''(\xi_3))| + \frac{h}{4}|(f''(\xi_2) - f''(\xi_3))| + \frac{1}{2}(f''(\xi_2) - f''(\xi_3))(x - x_i)\Big| \\
&\leq \frac{10}{h}\max_{2\leq i \leq N}|\phi_{i-1}' - f_{i-1/2}| + \frac{9}{8}h\omega(h, f'') = \frac{29}{8}h\omega(h, f'').
\end{aligned}
$$

Thus, the theorem is proved.

5   Local Controlled Spline Option

Previously, a curve composed of polynomials was considered whose coefficients were determined by solving a system of linear algebraic equations. By slightly changing the condition of the problem, it is possible to determine the coefficients of polynomials locally, thereby avoiding the solution of the system. If we build polynomials on segments $x \in [x_i, x_{i+1}]$, $i = \overline{1, N - 1}$, then there is no need to find ϕ_i from the solution of a system of equations. In this case, a local spline is obtained, which is written as the following:

$$
\begin{aligned}
S(x) &= f_{i+1}(x - x_i)/h - f_i(x - x_{i+1})/h + (x - x_i)(x - x_{i+1})(ax + b), \\
a &= (f_i' + f_{i+1}')/h^2 - 2(f_{i+1} - f_i)/h^3, \\
b &= (f_{i+1}' - f_i')/2h - (f_i' + f_{i+1}')(x_i + x_{i+1})/2h^2 + (f_{i+1} - f_i)(x_i + x_{i+1})/h^3
\end{aligned}
$$

for $x \in [x_i, x_{i+1}]$, $i = \overline{1, N - 1}$.

The differences in the behavior of the curves obtained using the global and local variants of the curve construction are considered in the examples. The choice of the method of constructing a curve is determined by the user and depends on the problem [11–13].

6   Examples of the Spline Curves Construction

Let the grid function is given on an interval $-3 \leq x \leq 3$. The values of the grid function are given in Table 1. The results of the calculations for variable parameters μ_i are shown in the Fig. 1.

Let the grid function take the values presented in Table 2.

The results of the calculations are presented in Fig. 2. The Fig. 2a shows the curve obtained using with a system of equations to determine unknown parameters ϕ_i. In this case, the parameters μ_i assumed a constant value $1/2$. As can be seen from the graph, the curve gives a slight deviation from the general behavior of the control polygonal line (dashed line in the graphs) in the vicinity of the abscissa values 40 and 80. As shown in the previous example, this fluctuation can be reduced by choosing parameters μ_i. The Fig. 2b shows a local spline curve. The oscillations of the constructed spline are not observed. In this case, the general form of the curve is somewhat different from the previous case.

7   Construction of Spline Curves for ECG Data Approximation

The effectiveness of the third-order controlled spline proposed in this paper will be shown on solving the problem of approximation of data obtained from an

Table 1. Grid function

τ_i	-3	-2	-1	0	1	2	3
F_i	0	0	0	5	5	5	5

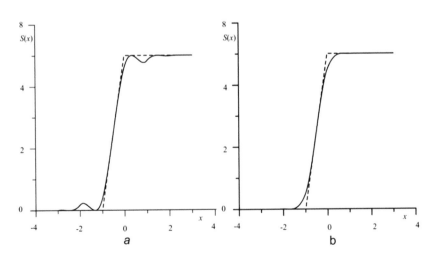

Fig. 1. The results of the calculation with a parameters: (a) $\mu_2 = 0.5$, $\mu_3 = 0.33$, $\mu_4 = 0.5$, $\mu_5 = 0.67$, $\mu_6 = 0.5$, $\mu_7 = 0.5$; (b) $\mu_2 = 0.5$, $\mu_3 = 0.71$, $\mu_4 = 0.5$, $\mu_5 = 0.29$, $\mu_6 = 0.5$, $\mu_7 = 0.5$

Table 2. Grid function

τ_i	10	20	30	40	50	60	70	80	90	100	110
F_i	1	2	12	17	15	200	15	17	12	2	1

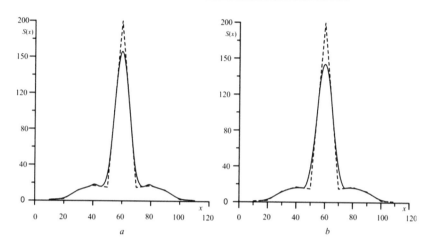

Fig. 2. The results of the calculations: (a) Global spline; (b) Local spline

electrocardiogram (ECG) of human cardiac activity. Note that ECG processing for compression of data [14] and the mining of characteristic features for further classification and clustering of information for the purpose of decoding and statement of the diagnosis is very actual. The proposed spline makes it possible to adapt its form to a variable, non-stationary behavior of real ECG signals. In the Fig. 3a shows of the real signal approximation by the proposed spline are shown. The approximation error is shown in the Fig. 3b. The relative error does not exceed 6%.

8   Conclusion

According to the results of the conducted research, it can be concluded that the global (with additional selection of form parameters) and the local version of the constructed spline curve retain the monotony of the initial data set, do not give oscillation and outliers. Due to the choice of form parameters it is possible to adjust the resulting function to the user's requirements. The above examples demonstrate good approximation properties of curves that can be used in the construction and approximation of the medical data, the trajectories of moving objects, the construction of contours of the human face and body [15,16], in robotics and the control of mechanical processes. Thus, the combination of the interpolation spline construction algorithm with interactive control allows controlling the behavior of the curves on the computer monitor screen. The disadvantages of the presented spline include only the C^1 smoothness, but for most practical applications such smoothness is sufficient. Also, the limitations of the proposed algorithm can include the presence of conditions on the shape parameters that must be observed when constructing the curve.

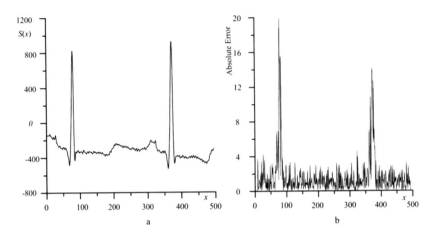

Fig. 3. The results of the approximations: (a) ECG data approximation; (b) Absolute Error

References

1. Stelia O, Potapenko L, Sirenko I (2018) Application of piecewise-cubic functions for constructing a Bezier type curve of smoothness. East Eur J Enterp Technol 2(4–92):46–52
2. de Boor C (2001) A practical guide to splines, Revised edn. Springer, Heidelberg
3. Zavyalov YuS, Kvasov BI, Miroshnichenko VL (1980) Methods of spline functions. Nauka, Moscow (in Russian)
4. Ahlberg JH, Nilson EN, Walsh JL (1967) The theory of splines and their applications. Academic Press, New York
5. Akima H (1970) A new method of interpolation and smooth curve fitting based on local procedures. Boulder J ACM (JACM) 17(4):589–602
6. Catmull E, Rom R (1974) A Class of local interpolation splines. In: Barnhill RE, Riesenfled RF (eds) Computer aided geometric design. Academic Press, pp 317–326
7. Dtrose TD, Barsky BA (1988) Geometric continuity, shape parameters, and geometric constructions for Catmull-Rom splines. ACM Trans Graph 7(1):1–41
8. Yuksel C, Schaefer S, Keyser J (2011) Parameterization and applications of Catmull-Rom curves. J Comput-Aided Des 43(7):747–755
9. Li J, Chen S (2016) The Cubic α-Catmull-Rom spline. Math Comput Appl 21(3):21–33
10. Stechkin SB, Subbotin JuN (1976) Splines in computational mathematics. Nauka (in Russian)
11. Kivva SL, Stelya OB (2002) A parabolic spline on a nonuniform grid. J Math Sci 109(4):1715–1725
12. Samarsky AA (1977) Theory of difference schemes. Nauka (in Russian)
13. Piegl L, Tiller W (1996) The NURBS book, 2nd edn. Springer, Heidelberg
14. Karczewicz M, Gabboujb M (1997) ECG data compression by spline approximation. Sig Process 59(1):43–59
15. Kryvonos IG, Krak IV (2011) Modeling human hand movements, facial expressions, and articulation to synthesize and visualize gesture information. Cybern Syst Anal 47(4):501–505
16. Krak YV, Barmak AV, Baraban EM (2011) Usage of NURBS-approximation for construction of spatial model of human face. J Autom Inf Sci 43(2):71–81

Automatic Search Method of Efficiency Extremum for a Multi-stage Processing of Raw Materials

Igor Konokh$^{(\boxtimes)}$ ⓘ, Iryna Oksanych ⓘ, and Nataliia Istomina ⓘ

Kremenchuk Mykhailo Ostrohradskiy National University, Kremenchuk, Ukraine
icegun.ik@gmail.com

Abstract. The article deals with the theoretical and practical problem of optimal control of raw material processing processes with the continuous consumption of raw materials and energy. Based on the positions of the efficiency theory, the choice of a universal efficiency factor for a selected class of processes has been made, with taking into account resource efficiency, quality indicators of the finished product, productivity of the technological system. The production system model is described including the transporting part, the processing part and implementation modules of control functions. The automatic search method of the extremum of the resource efficiency at the multi-stage processing with continuous supply of raw materials on the basis of the process calculation model is developed. Applying an efficiency factor as an optimization criterion allows making up optimization in a global sense. Within the method borders, the calculation algorithm of the system resources efficiency factor and the extremum search algorithm are proposed. The method efficiency is confirmed by the results of modeling the technological station's operation with three sequence heating stages of the raw material. Each stage is characterized by consumption cost of equipment resources, energy and increment of the output product cost. The efficiency maximum search has a satisfactory timetable, is stable and can be used at real time control of the technological stations.

Keywords: Multi-stage process · Optimal control ·
Resource efficiency · Global extremum

1 Introduction

At present, there are no methods and models for determining the master control of technological regulators on the basis of economic efficiency requirements [1]. This complicates the process of designing optimal control systems. The efficiency of the technological process is determined by a consumption cost decreasing, an increasing of the output product and a reduction of operation time [2,3]. Applying an efficiency factor as an optimization criterion allows make optimization in a global sense [4]. Thereby, the results of the technological system operation

ⓒ Springer Nature Switzerland AG 2020
V. Lytvynenko et al. (Eds.): ISDMCI 2019, AISC 1020, pp. 225–241, 2020.
https://doi.org/10.1007/978-3-030-26474-1_17

will increasingly accords to owner's purpose. The development of effectiveness evaluation's models for multi-stage continuous processing of raw materials will allow defining and comparing possible technological regimes in viewpoint of effectiveness. Thereby, a transition from the general economic requirements of technological equipment operation to regulators tuning became possible. Increasing the technological regimes efficiency by several percent will greatly increase the enterprise profitability [5].

On the basis of analysis, we can conclude that there is no general methodology of optimality criterion formulation in the analytic form for a wide class of processes. The approach based on forming an additive integral expression for each individual case using empirical weighting factors for forming a compromise requirement for the production quality, total productivity and consumption, for example as shown in [6,7].

2   Review of the Literature

In economic theory, the "balanced indicators" of KPI and BSC are known. At the same time, in the analytical review [3], more than 700 varied packages of balanced KPI indicators are offered, which are proposed as a solution to the problem of determining the effectiveness of operational processes. Also, as a criterion for optimization they try to use technical indicators: the minimum energy consumption [8], the maximum speed [9], the minimum trajectory [10], the minimum error [11], the fuel consumption [12], etc. It is proved that high accuracy is the confirmation of optimality [13].

Papers [14–16] are concern optimization of technological processes of drying with simultaneous increase of energy efficiency. In [17] the management of a pumping units group and the reduction of specific energy consumption are researched.

The reviewed researches use traditional approaches to the formation of optimality criterion [18]. They provide a set description of the process qualitative requirements and the proposed restrictions in the form of weighted sums forming an additive criterion [19,20]. For example, the frequently used integral quality criterion may include requirements for a minimum deviation of the controllable value from the set value, the minimum values of the controlled amount derivatives and the minimum fuel consumption:

$$I = \int_{t_1=t_0}^{t_2=t_k} \left(a_1 y^2 + a_2 y^2 + \cdots + a_m \overset{(n)}{y}{}^2 + a_p u^2 \right) dt,$$

where t_0, t_k – corresponding initial and final time of observation interval; a_1, a_2, a_m, a_p – weight factors of the criteria components, determining the degree of their contribution (importance) to the final value; y – the value of the controllable value; u – the value of the control action, which is directly proportional to the fuel consumption.

The disadvantage of this approach is absence of rigorous mathematical justification for weight factors. This can lead to unjustified understatement of system capacity and higher costs.

It can also be noted that in the papers [17, 21, 22] the described models are aimed at increasing the observability of the technological system and improving the control quality, but do not solve the problem of maximizing the profit from the operation under reducing the material costs.

In work on optimal control of continuous technological processes of firing pellets, drying loose products, ore enrichment [23, 24] the multicriterial optimization is used. However, in none of the works, it is not justified how to combine the criteria to ensure the highest conformity of the equipment operation results with the enterprise owner's objectives. As a result, situations with unreasonably reduced capacity with a high demand for a product, or overrate quality indicators under overrun of energy source, resource, and time of equipment operation may arise. All this in general reduces the profitability of enterprises.

The sources [25, 26] prove the condition for achieving the global optimum of the operation, determined by the efficiency factors. The maximum efficiency of the operation is achieved by minimizing the cost of raw materials and other input costs, the minimum costs of the resource and energy, the minimum operating time and the maximum cost of the finished product [26, 27].

An overview of the sources suggests that at present, methods and models for forming operation modes with extreme efficiency for continuous technological processes are not sufficiently developed.

3   Problem Statement

The research aim is develop and automate the search of effective modes for technological stations with the continuous multi-stage processing of raw materials. To achieve this aim we need to solve the following tasks:

- formulation the optimality criterion of the control system operation based on the efficiency factor;
- development the simulation model of the technological station;
- development the method for optimal mode finding using the simulation model;
- implementation the computer model of the control system of the technological process;
- development of case study and researching the search dynamics of the optimality criterion extremum.

4   Materials and Methods

In works [25–27], the valid and verified efficiency factor for the product portion processing is shown as follows:

$$E = \frac{(PE - RE)^2 \, T_1^2}{PE \cdot RE \cdot TO^2} \qquad (1)$$

where PE is the value of total income; RE is the value of total consumption; T_1 is the dimension coefficient; TO is the operation time.

Flow processing, unlike portion, requires modification of the factor, since the process is continuous in time and the operation must be consistent with the joint ones.

It is also necessary to take into account the product quality limitations and the fact that the technological requirements, performance, qualitative characteristics and cost of the original product may change over a long time period. According to [28] the efficiency factor takes the form (2).

$$E = sign\,(PE \cdot Yr - RE)\,\frac{(PE \cdot Yr - RE)^2\,T_1^2}{PE \cdot Yr \cdot RE \cdot TO^2} \tag{2}$$

where Yr is the correction factor of product value.

Under using formula (2) to evaluation the continuous process efficiency, we assume that this process undergoes a conditional time quantization. In this case, the efficiency of the portion processing of raw material with mass equal to the mass of smallest filled sections of the technological station will be estimated, and the operation time is determined as sum of the operating time of all sections. Then the operation time TO is calculated by the formula (3):

$$TO = \frac{M}{F} \tag{3}$$

where M is the mass of processed portion of raw materials; F is the flow of raw materials.

The value of total consumption RE is calculated by (4):

$$\begin{cases} RE = RE_{in} + RE_{e1} + RE_{e2} + RE_{res1} + RE_{res2} + RE_{add}, \\ RE_{in} = C_{in} \cdot F \cdot TO, \\ RE_{e1} = C_{e1} \sum_{i=1}^{n} (P_{e1i} \cdot TO_i), \\ RE_{e2} = C_{e2} \cdot P_{e2} \cdot TO, \\ RE_{res1} = C_{res1} \sum_{i=1}^{n} (f_{\nu1}(P_{e1i}) \cdot TO_i), \\ RE_{res2} = C_{res2} \cdot f_{\nu2}(P_{e2}) \cdot TO, \\ RE_{add} = C_{add} \cdot TO \end{cases} \tag{4}$$

where RE_{in}, RE_{e1}, RE_{e2}, RE_{res1}, RE_{res2}, RE_{add} – the value of the input raw material, the energy of the processing part, the energy of the transporting part, the processing part resource wear, the transport part resource wear, the additional costs (including the cost of staff time, tax, lease payments, etc.); C_{in}, C_{e1}, C_{e2}, C_{res1}, C_{res2}, C_{add} – the cost of input raw materials, the energy of the processing part, the energy of the transporting part, the processing part resource wear, the transport part resource wear, the additional costs; i – section/stage number; n – amount of sections/stages; P_{e1i}, TO_i – a power consumption and time of i-th section; P_{e2} – power of the transporting part; $f_{\nu1}(P_{e1i})$, $f_{\nu2}(P_{e2})$ – a determining functions of resource wear rate for the processing and transporting parts of the technological station, depending on the power consumption.

The value of total income in this case is determined by the cost of the output product (5):

$$PE = C_m \cdot F \cdot TO, \tag{5}$$

where C_m is the product cost.

At calculating income, the product cost is the function of its qualitative characteristic Q. The graph of the function is shown in Fig. 1. The function can be specified as an interpolation based on a known values array of Q and C_m.

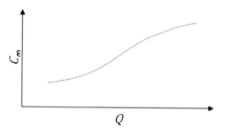

Fig. 1. Product cost graph

The correction factor of product value Yr is a function of its qualitative product characteristics. It can be expressed using the Gaussian function, as shown in Fig. 2.

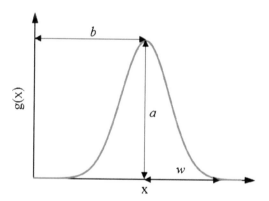

Fig. 2. Gaussian function's graph

The Gaussian function is described by formula (6):

$$g(x) = ae^{\frac{-(x-b)^2}{2c^2}}, \tag{6}$$

where a is the parameter defining the graph maximum height; b is the parameter determining the maximum displacement from 0 along the OX axis; c is the parameter related to the bell width w:

$$w = 2c\sqrt{2\ln(2)}. \tag{7}$$

Thereby, correction factor of product value Yr determined by the formula (8):

$$Yr\left(Q_{set}, Q_{act}\right) = 0.3 + 0.7e^{\frac{-(Q_{act}-Q_{set})^2}{2c^2}}, c = \frac{\Delta Q}{2\sqrt{2\ln{(2)}}},\qquad(8)$$

where Q_{act} is the actual value of product qualitative characteristic; Q_{set} is the setting value of product qualitative characteristic; ΔQ is the acceptable setting value deviation of product qualitative characteristic.

Due to the influence of the correction factor on the finished product cost (Fig. 3), the factor (2) becomes more sensitive to a given range of qualitative characteristics.

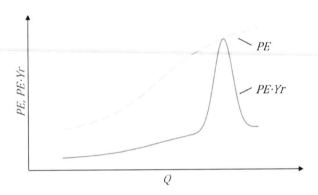

Fig. 3. Correction factor effect on product cost

5 Experiment

Based on formulas (2)–(5), (8), the model of technological station and calculation of the efficiency factor can be constructed.

In general, basing on the works [25–28], the flow processes can be described by the scheme shown in Fig. 4.

On the basis of the conceptual model a block diagram of the technological station system and its control system are developed (Fig. 5).

The maximum point of the efficiency factor (2) determines the optimal operation mode of the technological station and the corresponding input actions. For the most part, the technological station has more than two input actions. Therefor the problem of extremum searching for function of many variables appears. In this case, it is appropriate to use the simplex searching method of Nelder-Mead method [29]. This is a direct search method that does not use multiple or analytic gradient values.

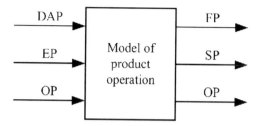

Fig. 4. Conceptual model of the flow processing of raw materials: DAP – directed action product; EP – energy product; OP – operating product; FP – finished product; SP – secondary product

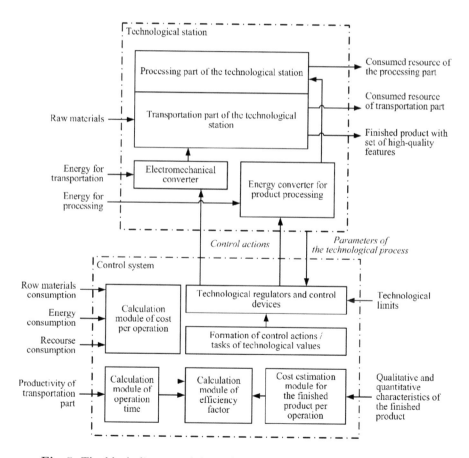

Fig. 5. The block diagram of the technological station and control system

6   Results and Discussion

For a function with n-th number of variables, in the n-dimensional space, the simplex is characterized by $n + 1$ by different vectors. These vectors are simplex apexes. At each step of the search, a new point or current simplex is generated. The value of a function in a new point is compared with the functions values at simplex apexes, and, as a rule, one of the apexes is replaced by a new point creating a new simplex. This step is repeated until the diameter of the simplex is less than the given accuracy ε. The scheme of the algorithm for simplex search is shown in Fig. 6.

In addition, empirical researches have shown that factor (2) has local extremum (Fig. 7) [5].

In this case, the extremum search should be carried out several times from several starting points, distributed in the space of technological limitations on input actions. In particular, the local extremum functions are observed only along the axis F, while for the P axis the function has a smooth dome-shaped form (Fig. 8).

Therefore, F should be tricky step by step, and the search for the optimal P_i can be done less accurately, but by the quickest method. Thus, the algorithm for finding the optimal operation mode of the technological station has the form as shown in Fig. 9.

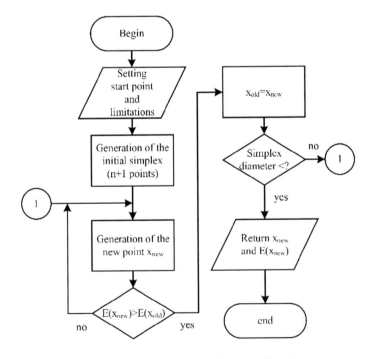

Fig. 6. Algorithm for the simplex search of the function extremum

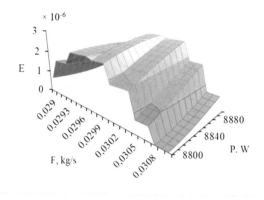

Fig. 7. Dependence E on F and P (1 stage system)

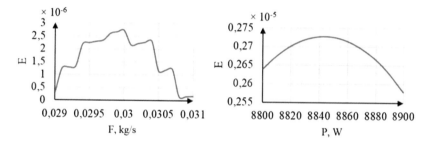

Fig. 8. Dependences E on $F(P = \text{const} = 8840\,\text{W})$ and E on $P(F = \text{const} = 0.3\,\text{kg/c})$

Let there is a technological station of raw materials heating. The station has 3 identical length sections with independent supplying for each heater. Control actions for the process are:

– flow rate of raw materials;
– power of three sections heaters.

It is necessary to determine the optimal operation mode of the station according to the factor (2) with the given limitations (Table 1).

Necessary data (energy cost, equipment resources wear, dependence of technological station physical parameters, physical constans) are taken from reference sources.

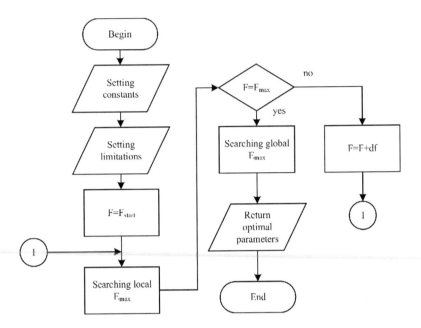

Fig. 9. Algorithm for finding the optimal operation mode of the technological station

In software application Matlab R2014a the computational model of techno-logical station and calculation of efficiency factor was build. Model consists of files:

set_vars.m – initialization of variables and constants;
getE1.m – calculation function of efficiency factor;
getRE.m – calculation function of total consumption;

Table 1. Limitations of operation mode of technological station.

Parameter	Value
Range of flow rate of raw materials F, kg/s	from 0.02 to 0.06
Tolerance of flow rate of raw materials dF, kg/s	0.01
Range of power of each section heating part P, kW	from 1 to 9
Tolerance of power of each section heating part dP, kW	1
Initial temperature of raw materials T_start, °C	18
Optimal final temperature of raw materials T_opt, °C	85 ± 6
Mass of drum loading mass, tone	2.6
Environment temperature T_env, °C	18

getT.m – calculation function of final temperature of the raw material;
sim_temp.slx – Simulink-model of heating section of raw materials;
getPE.m – calculation function of total income.

In Fig. 10 shows the simulink model of raw material heating section "sim_temp.slx".

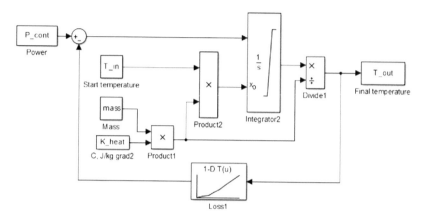

Fig. 10. Simulink model of raw material heating section

In Fig. 11 shows an algorithm for calculating the efficiency factor reflected the interaction of computational model files.

A simplex search is performed using the $fminsearch()$ function from Matlab's Optimization Toolbox.

The research results are given in Table 2.

Table 2. Research results.

Parameter	Value
Efficiency factor	2.9871e−05
Final temperature of raw materials T, °C	85.0255
Power of 1 section $P1$, W	7647.3809
Power of 2 section $P2$, W	8999.7178
Power of 3 section $P3$, W	9000.0221
Summary power of sections P, W	25647.1208
Flow F, kg/s	0.03
Search time, s	103.7563

The average calculation time of efficiency factor is 0.01 c. If you carry out a rigorous search of all combinations of arguments, the search will take $5 \cdot 10^7$ c.

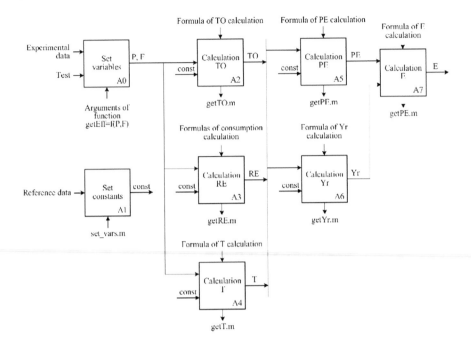

Fig. 11. Scheme of the calculation algorithm for efficiency factor

Thus, the obtained data shows the effectiveness of simplex search method application: the search time is reduced by 99.9998%.

The graph of the dynamics of simplex search the optimal operation mode is shown in Fig. 12.

The graphs show conditionally allocated 5 stages corresponding to 5 values of the flow F. At each stage, the power $P1$–$P3$ is selected, starting from $1\,\mathrm{kW}$ each and at iteration reducing the step. The temperature graph T at each stage is close to the specified level of $85\,^{\circ}\mathrm{C}$. Characteristic T contains overshoot, as a consequence of the large initial steps $P1$–$P3$. The efficiency factor E reaches its highest value by the end of each stage and forms an array of local maximums (Fig. 13).

After the passage of all stages of flow change search program finds the highest value of the effectiveness of local maximums. The search program accepts highest of local maximums as a global maximum and returns to the screen of operator corresponding optimal parameters of the system.

The dynamics of raw material heating is shown in Fig. 14. A solid line depicts a temperature graph with optimal control effects, dashed – at changing one of the actions in one direction or another.

Comparing graphs in Fig. 14 shows that in optimal operation mode the raw material reaches the required temperature evenly, whereas the change of at least one of the control actions leads to graph curvature and deviation from the optimum of the final raw material temperature.

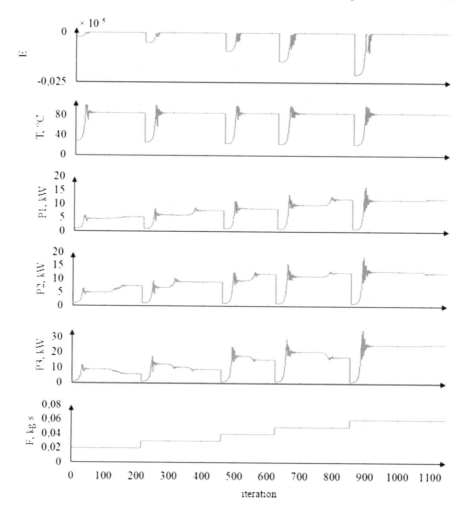

Fig. 12. Dynamic of simplex search of optimal operation mode

The accuracy of simulation results could have affected:

- step of the change control network;
- rounding of the operation time;
- linear interpolation of determining the process characteristics (cost dependence on the quality indicator, etc.).

To improve the accuracy of the results, the simplex search accuracy can be changed by adjusting the optional parameters of the $fminsearch()$ function by setting the fixed value of the changing step of argument or function. The setting of process characteristics by cubic interpolation, or interpolation by splines, will give a more precise characteristic definition at the operating point.

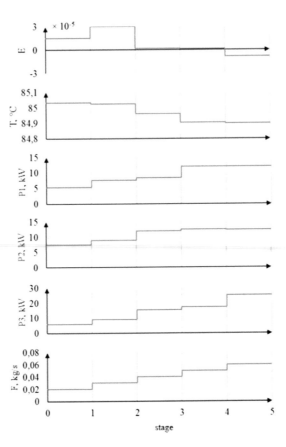

Fig. 13. Graph of local optimal modes

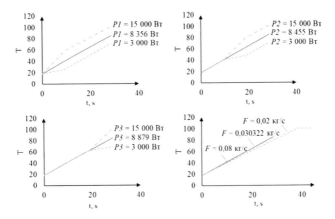

Fig. 14. Dynamics of raw material heating

7   Conclusions

Using the efficiency factor as an optimization criterion allows us to optimize the processes of processing raw materials with continuous material and energy flows in the global sense, which implies compliance with the goal of the enterprise owner. The transition from the economic and qualitative requirements to the choice of master controls for the automatic regulation systems forming a operation mode is carried out.

The production systems including the transporting subsystem are considered. For such systems gross productivity determined as chain of successive processing sections with a "rigid" control of processing energy supply.

The development of specialized models for effectiveness evaluation of a multi-stage continuous process of raw material processing combined with models of dynamic transforming energy processes and raw materials into the final product allows selecting and comparing possible technological regimes. It becomes possible to organize a directed search for the global optimum of the production system functioning.

The optimality criterion of control system operation based on the efficiency factor is formulated taking into account:

– cost estimates for raw materials, energy and equipment resources;
– cost estimation of the finished product, taking into account the current received quality indicator and boundary requirements for its quantity and qualitative indicators;
– operation time.

The paper proposes a method for the automatic search of the efficiency extremum for a multi-stage process of raw material processing, differed the simplex search function of the extremum using the Nelder-Mead method on the basis of a computational model with evaluation modules of the efficiency factors. Developed efficiency extremum search method allows to determine the master control of regulators and to form the most efficient operation mode. The simulation allowed to approve perspective of chosen approach for model designing of the extreme control system of multistage processes with a continuous flow of raw materials. The extremum search is performed in automatic mode, the process is stable and carried out at a given accuracy.

References

1. Bellman RE (2003) Dynamic programming. Princeton University Press, Princeton, 401 p
2. Encinar M (2016) Evolutionary efficiency in economic systems. Cuadernos de Economía 39(110):93–98. https://doi.org/10.1016/j.cesjef.2015.11.001
3. Vinícius P, Daniela CA, Tim C (2016) Process-related key performance indicators for measuring sustainability performance of ecodesign implementation into product development. J Clean Prod 39:416–428. https://doi.org/10.1016/j.jclepro.2016.08.046

4. Thomas AW (2011) Optimal control theory with applications in economics. The MIT Press, Cambridge, 362 p
5. Kulej M (2010) Operations research. Bus Inf Syst 70
6. Gong S, Shao C, Zhu L (2017) Energy efficiency evaluation in ethylene production process with respect to operation classification. Energy 118:1370–1379
7. Arena L (2014) Measure guideline: condensing boilers - optimizing efficiency and response time during setback operation. Technical report DOE/GO-102014-4369 KNDJ-0-40342-03. https://doi.org/10.2172/1126299
8. Gregory J, Olivares A (2012) Energy-optimal trajectory planning for the Pendubot and the Acrobot. Optim Control Appl Methods 34(3):275–295. https://doi.org/10.1002/oca.2020
9. Bing-Feng J, Xiaolong B, Ju JC, Yaozheng G (2013) Design of optimal fast scanning trajectory for the mechanical scanner of measurement instruments. Scanning 36(2):185–193. https://doi.org/10.1002/sca.21084
10. Gasparetto A, Zanotto V (2010) Optimal trajectory planning for industrial robots. Adv Eng Softw 41(4):548–556. https://doi.org/10.1016/j.advengsoft.2009.11.001
11. Wang H, Tian Y, Vasseur C (2015) Non-affine nonlinear systems adaptive optimal trajectory tracking controller design and application. Stud Inform Control 24(1):05–12. https://doi.org/10.24846/v24i1y201501
12. Burmistrova ON, Korol SA (2013) Opredelenie optimalnyih skorostey dvizheniya lesovoznyih avtopoezdov iz usloviy minimizatsii rashoda topliva. Lesnoy vestnik 1:25–28
13. Vega-Alvarado EA, Portilla-Flores EA, Mezura-Montes E, Flores-Pulido L, Calva-Yáñez MB (2016) Memetic algorithm applied to the optimal design of a planar mechanism for trajectory tracking. Polibits 53:83–90
14. Dinçer İ, Zamfirescu C (2015) Optimization of drying processes and systems. In: Drying phenomena, pp 349-380. https://doi.org/10.1002/9781118534892.ch9
15. Weigler F, Scaar H, Franke G, Mellmann J (2016) Optimization of mixed flow dryers to increase energy efficiency. Drying Technol 35(8):985–993. https://doi.org/10.1080/07373937.2016.1230627
16. Paramonov AM (2017) The technological raw material heating furnaces operation efficiency improving issue. In: AIP conference proceedings, vol 1876, no 1. https://doi.org/10.1063/1.4998868
17. Zdor GN, Sinitsyn AV, Avrutin OA (2017) Pump group automatic control for reducing its energy consumption. Energ Proc CIS High Educ InstS Power Eng Assoc 60(1):54–66. https://doi.org/10.21122/1029-7448-2017-60-1-54-66
18. Grad S (2016) Duality for multiobjective semidefinite optimization problems. Oper Res Proc 189–195. https://doi.org/10.1007/978-3-319-28697-6_27
19. Löber J (2017) Analytical approximations for optimal trajectory tracking. Bull Volcanol 78(10):119–193. https://doi.org/10.1007/978-3-319-46574-6_4
20. Kasperski A, Zieliński P (2016) Robust discrete optimization problems with the WOWA criterion. In: Criterion operations research proceedings, pp 271–277
21. Zagirnyak M, Kovalchuk V, Korenkova T (2015) Power model of an electrohydraulic complex with periodic nonlinear processes in the pipeline network. In: 2015 international conference on electrical drives and power electronics (EDPE), pp 345–352. https://doi.org/10.1109/EDPE.2015.7325318
22. Ha QP, Vakiloroaya V (2015) Modeling and optimal control of an energy-efficient hybrid solar air conditioning system. Autom Constr 49:262–270. https://doi.org/10.1016/j.autcon.2014.06.004

23. Liakhomskii AV, Vakhrushev SV, Petrov MG (2006) Modelirovanie poverkhnosti pokazatelei kachestva energoeffektivnosti obogatitelnykh proizvodstv gornykh predpriiatii. Min Inf Anal Bull 10:313–316

24. Carrington CG, Sun ZF, Sun Q, Bannister P, Chen G (2000) Optimizing efficiency and productivity of a dehumidifier batch dryer. Part 1: capacity and airflow. Int J Energy Res 24(3). https://doi.org/10.1081/DRT-120014056

25. Lutsenko I (2014) Identification of target system operations. 1. Determination of the time of the actual completion of the target operation. East-Eur J Enterp Technol 6(2(72)):42–47

26. Lutsenko I (2015) Identification of target system operations. 2. Determination of the value of the complex costs of the target operation. East-Eur J Enterp Technol 1(2(73)):31–36 (2015)

27. Lutsenko I, Fomovskaya E, Vihrova E, Serdiuk O (2016) Development of system operations models hierarchy on the aggregating sign of system mechanisms. East-Eur J Enterp Technol 3(2(81)):39–46

28. Konokh I. (2018) Synthesis of the structure for the optimal system of flow treatment of raw materials. East-Eur J Enterp Technol 5(2(95)):57–65

29. Lagarias JC, Reeds JA, Wright MH, Wright PE (1998) Convergence properties of the Nelder-Mead simplex method in low dimensions. SIAM J Optim 9(1):112–147

Intellectual Information Technology of Analysis of Weakly-Structured Multi-Dimensional Data of Sociological Research

Olena Arsirii$^{(\boxtimes)}$ ⓘ, Svitlana Antoshchuk ⓘ, Oksana Babilunha ⓘ,
Olha Manikaeva ⓘ, and Anatolii Nikolenko$^{(\boxtimes)}$ ⓘ

Odessa National Polytechnic University, Odessa, Ukraine
e.arsiriy@gmail.com, anatolyn@ukr.net

Abstract. To automate the process of obtaining knowledge (metadata) about the respondents of sociological surveys or questionnaires, an intelligent information technology has been developed for analyzing weakly-structured multi-dimensional medical-social data. The technology is based on the developed methods of presenting sociological data in the spaces of primary and secondary features and the neural network classification of respondents based on cluster analysis of aggregated sociological data in the space of secondary features. Approbation of the developed technology on real data of sociological surveys showed to increase the reliability of making classification decisions on the respondents' lifestyle in comparison with the sociologist-analyst and their own definition of respondents.

Keywords: Data Mining · Information technology · Cluster analysis · Social data

1 Introduction

Conducting of various sociological studies in the form of surveys, questioning or interviewing with the subsequent analysis of answers containing heterogeneous information allows the analyst "to extract" valuable knowledge about the target (analyzed) audience in the form of quantitative metrics or qualitative assessments of behavioral, social and demographic, geographical, psychophysical, medical or any other characteristics. Such kind of knowledge, presented in a convenient visual form, allows specialists in the subject area to make strategic decisions to improve the interaction between their business and the studied sociological environment (the target audience). However, the computer form for representing heterogeneous empirical data of the sociological studies contains information which is necessary for expert assessment the target audience state only in an implicit form. To extract it for the purpose of making a management decision, it is necessary to use special methods for data analysis. For example,

© Springer Nature Switzerland AG 2020
V. Lytvynenko et al. (Eds.): ISDMCI 2019, AISC 1020, pp. 242–258, 2020.
https://doi.org/10.1007/978-3-030-26474-1_18

in order to make decisions about the organization of a specialized library and information services, an expert analyst needs to assess the interests, requirements and preferences of the target audience. For the formation of such knowledge the following personal data is used: (gender, age, educational back-ground), activity (frequency and purpose of library visits, the basis for literature choice), literary preferences (entertaining, professional, etc.) [1]. A complete knowledge examination of the graduate of an educational institution with the subsequent formation of recommendations on the specifics of employment is impossible without evaluating data on the qualitative composition of teachers, the adequacy of curricula and practice technologies for the requirements of practice, the preliminary level of training of the trainees, etc. [2].

As a result, a class of tasks associated with Data Mining has been formed - an approach that combines methods, using of which in a certain method allows to detect previously unknown, non-trivial, practically useful and accessible interpretations of knowledge necessary for decision making regarding the target audience [3]. Modern level of development of information technologies allows to automate the process of conducting of intellectual analysis of empirical quantitative and categorical data of sociological surveys for building various data models in the source and feature spaces and for their visual representation in the decision space [4,5].

2   Review of the Publications and Formulation of the Research Problem

In the general case, new knowledge about the target audience is extracted on the basis of the analysis of empirical data of the sociological survey, and according to [6,7] this process is presented in the form of a cyclic sequence consisting of the following steps:

– awareness of the theoretical or practical insufficiency of the existing knowledge of the target audience;
– formulation of the problem and the hypothesis (in qualitative research hypotheses are usually presented at the last stages of research);
– collecting of empirical material on the basis of which hypothetical assumptions can be confirmed or refuted;
– analysis of empirical data using various methods, strategies, research programs and models;
– interpretation of the processed data and decision making, explanation of the social phenomenon with the use of them;
– redefinition and clarification of a problem or hypothesis leading to a new re-search cycle (return to stage 3).

On the other hand, recently, due to the advent of modern software, Data Mining methods are gradually becoming the most popular tools for the sociologist - analyst. From the point of view of the existing standards describing the organization of the Data Mining process and the development of Data Mining systems, the most popular and common methodology is CRISP-DM (The Cross Industry

Standard Process for Data Mining) [8,9]. In accordance with the CRISP-DM standard, Data Mining is a continuous process with many cycles and feedbacks and includes the following six steps: Business understanding, Data Understanding, Data Preparation, Modeling, Evaluation, Deployment.

The seventh step is sometimes added to this sequence of stages - Control; it ends the circle. Using the CRISP-DM Data Mining methodology, it becomes a business process during which Data Mining technology focuses on solving specific business problems.

Consider some of the problems of organizing the Data Mining process of the sociological survey data for a sociologist-analyst using the CRISP-DM standard.

Thus, according to [6,7] at the stage of understanding business, the sociologist-analyst on the basis of his knowledge or lack of knowledge about the target audience, solves the problem of hypothesizing, which must be confirmed or refuted as a result of a sociological research. According to [7] such a hypothesis is a partially substantiated pattern of knowledge, serving either for the connection between various empirical facts or for explaining a fact or a group of facts. For example, as a formalized goal of the sociological survey there exists a hypothesis that the dependent variable (lifestyle) varies depending on some reasons (quality of food, alcohol consumption, playing sports, etc.) that are independent variables. However, this variable is not initially dependent or independent. It becomes such at the stage of understanding the business.

At the stage of data understanding, the sociological expert solves the problems connected with collecting sociological (or empirical) data. Such data can be defined as primary information of any kind, obtained as a result of one of the many types of sociological information collecting [10]. As a rule, to conduct deep analytical studies, data are collected through questionnaires and interviews with a "complex" structure. Any empirical data is always structured. Depending on the degree of structuredness, the data can be subdivided into the following types:

- unstructured Data–text-type data, obtained in the process of conducting different types of interviews (narrative, leitmotiv, etc.), the texts of answers to open questions with an unlimited search field for answers and any other texts or documents to which the sociologist could address [9,10];
- strongly-structured data– data, existing in the form of matrices of any type (for example, tables), obtained through the part of the questionnaire (interview) according to the scheme of closed questions;
- weakly-structured data of an intermediate type, not only quantitative but also categorical (nominal and ordinal) types, as well as existing in textual form, however, while being specially organized. Examples of such kind of data – data with a limited number of unique values or categories. As a rule, they contain answers to open-ended questions of an interview (interview) with a limited field of search for answers, either obtained by the method of incomplete sentences (test for sentence completion), or by using the method of repertory grids (G. Kelly's theory of personality constructs) [11,12].

At the stage of data preparation, the sociologist-analyst faces the tasks associated with the organization of a multidimensional attribute space—the formalization and structuring of sociological survey data. When transforming unstructured data into feature space, special text recognition and analysis methods (OCR and Text Mining) are used for the further modeling. Rigidly structured data is a quantitative data type that is susceptible to the formalization and automatically transformed feature space. For the structuring and formalization of poorly structured data, various methods of preliminary data processing are used, for example, cleaning, filtering, rating and coding, etc. As a result of this preliminary processing, the weakly structured sociological survey data is transformed into a multidimensional feature space.

To conduct the sociological survey data modeling stage in the multidimensional attribute space, the sociologist-analyst solves the problems associated with the choice of methods and models for conducting Data Mining. Moreover, he has at his disposal the entire arsenal of regression, discriminant, dispersive, cluster, correlation, factor analysis methods [12, 13].

At the stages of results evaluating and implementing, the tasks of interpreting the so-called sociological survey metadata or extracted knowledge obtained after the simulation, solving the characteristics of the target audience and explaining the studied social phenomenon with their help are solved [6, 7].

Thus, the complexity of developing Data Mining technologies and systems to solve practical problems of extracting knowledge about the target audience is associated with the need of using sociological survey data for the analysis and modeling which are poorly structured because they are collected through surveys, questionnaires and interviews with structure, interpreted by different and not always related scales and are often contradictory. And the expert decision on attributing the respondent to a certain class based on the analysis of the multidimensional attributes of such kind of data is ambiguous and depends on the qualifications of the sociologist-analyst.

To eliminate the above-mentioned deficiencies of the sociological survey of data processing systems and to facilitate the work of the sociologist-analyst, a problem-oriented Data Mining system has been developed. The system consists of subsystems of initial data preparing, modeling and decision-making support, interconnected with the help of interface tools for a decision maker - sociologist-analyst [14–16]. The creation of such problem-oriented intellectual systems is always based on previously developed models, methods, methods and information technologies (IT).

The goal and tasks of the research. The goal of this study is to develop IT-intelligence analysis of weakly structured multidimensional data from sociological surveys or questionnaires to automate the extraction of knowledge about the target audience, which will improve the accuracy of the classification of respondents.

To achieve this goal, the following tasks are solved:

1. In terms of the tasks of classifying/clustering sociological survey respondents, a formalization of sociological survey hypothesis has been suggested and detailed and aggregated data models have been developed.
2. A method for presenting detailed and aggregated sociological survey data in the spaces of primary and secondary attributes has been developed.
3. The method of neural network classification of respondents has been developed on the basis of cluster analysis of aggregated sociological survey data in the space of secondary features.
4. IT of the intellectual analysis of poorly structured multidimensional data of sociological survey data has been developed and tested on data "Ukraine - lifestyle"

3 Materials and Methods

The main study material includes a consistent presentation of the results of solving the formulated tasks within the framework of the unified methodical Data Mining approach.

3.1 Models of Detailed and Aggregated Data of Sociological Surveys

The task of extracting knowledge about the target audience (obtaining metadata) as a result of sociological survey data analyzing can be reduced to the classification task in terms of discriminant analysis or machine learning [13]. Let's consider there are many respondents belonging to the target audience (for example, students), which are hypothetically divided into groups - classes (for example, "healthy", "not healthy, not sick" and "sick"). A finite set of respondents is given. They know (from the point of view of an expert analyst) which classes they belong to. This set is called a training set. Class affiliation other respondents is unknown. It is required to build an algorithm capable of attributing (classifying) any respondent from the target audience to one of the hypothetical classes.

The task of classifying respondents that belong to the target audience being analyzed can be formally presented as follows.

Let's consider that Target Audience consists of many Respondents:

$$TA = \{r_1, r_2, \ldots, r_i, \ldots, r_n\}, \tag{1}$$

where r_i – survey respondent, n – number of respondents.

Each respondent is characterized by a multidimensional set of variables that denote its properties:

$$R_i = \{x_1, x_2, \ldots, x_j, \ldots, x_m, y_k\}, \tag{2}$$

where x_j – independent variables denoting the properties of the respondent, m – the number of properties of the respondent which are significantly less than the analyzed respondents themselves $m < n$, y_k – dependent variable, belongs to a set of class values $Y = \{y_1, \ldots, y_k, \ldots, y_l\}$, l – number of classes; subset of respondents is divided into them (1). The value y_k is determined by analyzing a set of independent variables $X = \{x_j\}$.

A multidimensional set of independent variables (2) in conducting sociological research is a set of poorly structured data, the type and format of which depend on the method of obtaining initial information Thus, for example, data about the respondent can be obtained not only with the help of filled survey forms, questionnaires and interviews but also extracted from electronic mail messages, business cards, pdf and txt files, instant messages in various messengers (WhatsApp, Viber, Facebook Messenger, Skype, ICQ, Google Hangouts, etc.), documents, web pages, invoices, audio/video, checks and contracts, pictures, etc. This Extraction from various sources is called Detailed Data.

In solving the problem of respondent classifying (R) that belong to an analyzed TA (1) taking into account representation (2), a detailed data model D_{R_i} of respondent R_i in m-dimensional property space is suggested. It is formally defined using a tuple

$$D_{R_i} = \langle\, V_j, T_j, F_j, S_j, Q_j, Mt_j \,\rangle, j = \overline{1, m}, \tag{3}$$

where $V_j, T_j, F_j, S_j, Q_j, Mt_j$ – value, type, format, source, quality assessment, data transformation method of the j – property.

To increase the reliability of the classification decision y_k detailed data D_{R_i} are aggregated (merged) and using the method Mt_j are transformed (Transformation) into feature space: $X_j = \bigcup_\phi Mt_j(V_j)$. As a function ϕ of detailed data of aggregation of D_{R_i} can be used min, max, Σ, etc.

Then the aggregated data model A_{R_i} of respondent R_i in q - dimensional feature space, taking into account the representation (2), is specified using a tuple:

$$A_{R_i} = \langle\, x_1, \ldots, x_j, \ldots, x_q, y_k \,\rangle, \tag{4}$$

where q – dimension of feature space $q < m$. Taking into the consideration (4) and (1) data set is a set of aggregated respondent data R_i of target audience TA:

$$A_{TA} = \left\langle \begin{bmatrix} x_{11} & \ldots & x_{1q} \\ \ldots & \ldots & \ldots \\ x_{n1} & \ldots & x_{nq} \end{bmatrix}, \begin{bmatrix} y_1 \\ \ldots \\ y_k \end{bmatrix} \right\rangle, \tag{5}$$

where matrix X – a set of respondent descriptions R_i in feature space, and vector Y – finite set of numbers (names, labels) classes. There exists also unknown target dependency - $y^* : X \to Y$, values of which are known only for a finite number of respondents R_i from the target audience TA (of the training set) $R_i^r = (x_1, y_1), \ldots, (x_r, y_r), r < n$. There is a need to build an algorithm $a : X \to Y$ that could classify an arbitrary respondent $R_i \in TA$.

In its turn, the accuracy of algorithm construction $a : X \to Y$ directly depends on the previously presented hypothesis about the power of the set of classes – Y, the adoption of which is ambiguous and is determined by the qualifications of the sociologist-analyst.

To determine the power of $|Y|$ classes Data Clustering is used, within the framework of which the task of splitting the target audience TA is solved on disjoint subsets, that are called clusters in the way that each cluster consisted of similar respondents, and respondents of different clusters differed significantly.

In this case, the task of clustering of target audience respondents (5) on a set of aggregated data A_{R_i} consists in building a set Implementation of the Huang empirical mode decomposition technique (EMD) intends the following conditions:

$$Y = y_1, \ldots, y_k, \ldots, y_l, \tag{6}$$

where y_k – cluster containing similar A_{R_i} from a set (5):

$$y_k = \{A_{R_i}, A_{R_j} | A_{R_i} \in A_{TA}, A_{R_j} \in A_{TA}, d(A_{R_i}, A_{R_j}) < \sigma\}, \tag{7}$$

where σ – the value that determines the measure of proximity for inclusion in one cluster, and $d(A_{R_i}, A_{R_j})$ – proximity measure between objects, called distance. Data set of aggregated data A_{TA} (5) is loading in the data warehouse and is used to build a classification algorithm $a : X \rightarrow Y$.

Thus, for obtaining, storing and managing weakly structured multidimensional data sociological surveys use a specialized data preparation subsystem built on the principle of ETL-systems (Extraction, Transformation Loading) [17]. Development of a method for presenting detailed and aggregated sociological survey data in the spaces of primary and secondary attributes.

3.2 Development of a Method for Presenting Detailed and Aggregated Sociological Survey Data

To get and load aggregate data sets into the repository A_{TA} (5) has been developed a method for presenting detailed and aggregated data of sociological surveys in spaces of primary and secondary signs consisting of the following stages:

1. Preliminary processing of detailed data of respondents R_i taking into account the suggested model D_{R_i} in m-dimensional property space (3), in order to clean and filter them.
2. Transformation of the cleared data D_{R_i} in q-dimensional space of primary signs $q < m$, based on aggregated data model A_{R_i} and formalized research objectives for data sampling (5).
3. The use of technique of nonlinear reduction of the dimension of the q-dimensional space of primary features for visualization of aggregated data A_{R_i} and for the purpose of the subsequent clustering and classification of respondents R_i.

At the first stage, preliminary processing of detailed data of sociological survey requires the following steps:

– representation of text data D_{R_i} and their names in the standard research language (for example, English using Google translate) and in one of the open text formats for structured hierarchical data, for example – csv (comma-separated values) format;

– data checking of D_{R_i} matching the range of permissible values of each variable (element or column of data), removing false values;
– the definition of variables V_j, in the data model of D_{R_i}, that contain a large number of missing values, the removal of such variables if they are not informative, or a decrease in the number of missing values of variables by replacing gaps with averages or the most frequent values.

An example of statistics on the number of variables that contain the passed values is shown in Fig. 1 (here in after, a dataset is used to present the results of processing which contains the answers of 1143 respondents to 114 questions of the sociological survey "Ukraine – lifestyle" [18]. Results in the form of minimum, maximum, lower and upper quartiles, average value and standard deviation are displayed in the form of diagrams—box plot before and after the deletion of variables.

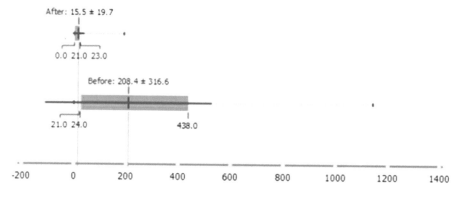

Fig. 1. Boxplot of an average number of missing values in the dataset "Ukraine - life-style" per variable before and after variables imputation

As a result of preliminary processing, the amount of empirical data of D_{R_i} that are subject to further processing, decreases on average by 25%.

At the second stage of the suggested method for the transformation of the purified detailed data D_{R_i} (3) primary feature space based on aggregated data model A_{R_i} (4) to get a sample of data (5) three steps are required:

– to formalize the goal of sociological survey, a hypothesis of dependent variable $y_k(1)$ changing is presented. It depends on some reasons which are independent variables $\{x_j\}$ (1). Thus, for example, for a dataset [16], the expert sociologist has put forward the hypothesis "lifestyle", according to which the number of classes (meta-classes) $l = 3$, $y_1 = 1$ - "good", $y_2 = 2$ - "average" and $y_3 = 3$ - "bad". A set $X = \{x_j\}$, $j = \overline{1,114}$ contains the respondents' answers to the questionnaire in which it is suggested to determine, for example, the quality of their food, alcohol use, sports, etc.;

- determination of independent variables $\{x_j\}$, which are relevant formalized goals of sociological survey and elimination of irrelevant variables from the space of primary features; From the point of view of a sociological expert, relevant goals are variables that directly or indirectly affect the respondent's lifestyle. Thus, for example, dataset [16] variables were removed from the attribute space, which connect information sources and taste preferences, as well as motives for playing sports, etc.;
- development of procedures for rationing, binarization and coding of variables $\{x_j\}$, relevant goals of sociological survey. For the subsequent use of Data Mining methods, the values of the studied variables should be presented in the same scales. Continuous values of attributes are reduced to the same ranges, for example, the range from -1 to 1 or from 0 to 1. The values of "yes" and "no" or "male" and "female" are binarized, the values of categorical variables, if necessary, are translated into the Likert Scale and then converted using One-Hot Encoding [14]. In particular, a characteristic that characterizes playing sports, can take on 5 different meanings: "I do not do it at all", "I rarely train and rather not regularly", "I train several times a month", "I train several times a week", "I train every day and regularly". In this case, after coding such kind of the feature using One Hot Encoding, 5 binary features will be created, all values of which are zero, except for one. The position corresponding to the numerical value of the attribute is placed 1.

Figure 2 shows fragments of sociological survey data of first and second stages of processing. At the third stage of the suggested method, nonlinear reduction techniques of the attribute space and visualization of multidimensional data of sociological survey are used. When choosing a method for reducing the dimensionality of the attribute space, the possibilities of multiple learning signs have been compared (stochastic neighbor embedding, SNE) and multidimensional scaling (Multidimensional scaling MDS). Reduction of the attribute space makes visual presentation of sociological survey data possible and understandable. They are subject to further clustering and classification.

In general, the problem of reducing the dimension can be formulated in the following way: there exist data of sociological survey that are described by a multidimensional variable $X = \{x_j\}$, $j = \overline{1, m}$ with the dimension of space substantially more than three $- m \gg 3$. It is necessary to obtain a new variable that exists in two-dimensional or three-dimensional space, as it were, of "secondary" features, which, to the maximum extent, would retain the structure and patterns in the original data of the sociological survey. The task of MDS is as follows [19]. There is a training selection of objects $X^l = \{x_1, \ldots, x_l\} \in X$, for which only pairwise distances are known $R_{ij} = \rho(x_i, x_j)$ for some pairs of learning objects $(i, j) \in D$. Required for each object $x_i \in X^l$ build its trait description – vector $x_i = (x_i^1, \ldots, x_i^n)$ in Euclidean space $\mathbb{R}^n$ so that Euclidean distances d_{ij} between objects x_i and x_j

$$d_{ij}^2 = \sum_{d=1}^{n} (x_i^d - x_j^d)^2 \tag{8}$$

	health_assessment	id	meal_frequency_breakfast	meal_frequency_second_breakfast	meal_frequency_lunch	meal_frequency_second_lunch
1	Хорошее	4	Иногда	Иногда	Несколько раз в неделю	Ежедневно
2	Среднее	5	Ежедневно	Ежедневно	Ежедневно	Ежедневно
3	Хорошее	8	Ежедневно	Несколько раз в неделю	Ежедневно	Несколько раз в неделю
4	Среднее	9	Ежедневно	Иногда	Иногда	Не принимаю
5	Хорошее	11	Ежедневно	Иногда	Ежедневно	Иногда
6	Среднее	12	Ежедневно	Несколько раз в неделю	Иногда	Иногда
7	Среднее	13	Ежедневно	Не принимаю	Ежедневно	Не принимаю
8	Среднее	14	Ежедневно	Не принимаю	Ежедневно	Иногда
9	Среднее	15	Несколько раз в неделю	Не принимаю	Ежедневно	Иногда
10	Хорошее	16	Ежедневно	Иногда	Несколько раз в неделю	Иногда
11	Хорошее	17	Несколько раз в неделю	Не принимаю	Ежедневно	Иногда
12	Хорошее	18	Ежедневно	Иногда	Ежедневно	Не принимаю
13	Хорошее	19	Ежедневно	Не принимаю	Ежедневно	Не принимаю
14	Среднее	20	Ежедневно	Не принимаю	Ежедневно	Не принимаю
15	Среднее	21	Ежедневно	Не принимаю	Ежедневно	Не принимаю
16	Хорошее	23	Ежедневно	Несколько раз в неделю	Ежедневно	Несколько раз в неделю
17	Среднее	24	Ежедневно	Иногда	Ежедневно	Ежедневно
18	Хорошее	25	Ежедневно	Не принимаю	Ежедневно	Иногда
19	Хорошее	28	Ежедневно	Не принимаю	Ежедневно	Ежедневно
20	Плохое	29	Не принимаю	Не принимаю	Не принимаю	Не принимаю
21	Хорошее	30	Ежедневно	Несколько раз в неделю	Ежедневно	Ежедневно
22	Хорошее	32	Не принимаю	Иногда	Ежедневно	Иногда
23	Среднее	33	Несколько раз в неделю	Не принимаю	Ежедневно	Иногда
24	Хорошее	34	Ежедневно	Иногда	Ежедневно	Иногда
25	Хорошее	35	Ежедневно	Несколько раз в неделю	Ежедневно	Ежедневно
26	Плохое	36	Ежедневно	Ежедневно	Ежедневно	Ежедневно
27	Хорошее	37	Ежедневно	Иногда	Ежедневно	Иногда
28	Плохое	38	Несколько раз в неделю	Иногда	Ежедневно	Иногда

a

	id	breakfast	lunch	dinner	cereals	milk	vegetables	meat	poultry	eggs	fast food	tea	energetic
1	4.000	0.000	0.000	0.000	0.000	0.000	0.000	0.000	0.000	0.000	0.000	0.000	0.000
2	5.000	0.000	0.000	0.000	0.000	0.000	0.000	0.000	0.000	0.000	0.000	0.000	0.000
3	8.000	0.000	0.000	0.000	0.000	0.000	0.000	0.000	0.000	0.000	0.000	0.000	0.000
4	9.000	0.000	0.000	0.000	0.000	0.000	0.000	0.000	0.000	0.000	0.000	0.000	0.000
5	11.000	0.000	0.000	0.000	0.000	0.000	0.000	0.000	0.000	0.000	0.000	0.000	0.000
6	12.000	0.000	0.000	1.000	0.000	0.000	1.000	0.000	1.000	0.000	0.000	0.000	0.000
7	13.000	0.000	0.000	0.000	0.000	0.000	0.000	0.000	0.000	0.000	0.000	0.000	0.000
8	14.000	0.000	0.000	0.000	0.000	0.000	0.000	0.000	0.000	0.000	0.000	0.000	0.000
9	15.000	0.000	0.000	0.000	0.000	0.000	0.000	0.000	0.000	0.000	0.000	0.000	0.000
10	16.000	0.000	0.000	0.000	0.000	0.000	0.000	0.000	0.000	0.000	0.000	0.000	0.000
11	17.000	0.000	0.000	0.000	0.000	0.000	0.000	0.000	0.000	0.000	0.000	0.000	0.000
12	18.000	0.000	0.000	0.000	0.000	0.000	0.000	0.000	0.000	0.000	0.000	0.000	0.000
13	19.000	0.000	0.000	1.000	0.000	0.000	0.000	0.000	0.000	0.000	0.000	0.000	0.000
15	21.000	0.000	0.000	0.000	0.000	0.000	0.000	0.000	0.000	0.000	0.000	0.000	0.000
16	23.000	0.000	0.000	0.000	0.000	0.000	0.000	0.000	0.000	0.000	0.000	0.000	0.000
17	24.000	0.000	0.000	0.000	0.000	0.000	0.000	0.000	0.000	0.000	0.000	0.000	0.000
18	25.000	0.000	0.000	0.000	0.000	0.000	0.000	0.000	0.000	0.000	0.000	0.000	0.000
19	28.000	0.000	0.000	0.000	0.000	0.000	0.000	0.000	0.000	0.000	0.000	0.000	0.000
21	30.000	0.000	0.000	0.000	0.000	0.000	0.000	0.000	0.000	0.000	0.000	0.000	0.000
22	32.000	1.000	0.000	0.000	0.000	0.000	0.000	0.000	0.000	0.000	0.000	0.000	0.000
24	34.000	0.000	0.000	0.000	0.000	0.000	0.000	0.000	0.000	0.000	0.000	0.000	0.000
25	35.000	0.000	0.000	1.000	0.000	0.000	0.000	0.000	0.000	0.000	0.000	0.000	0.000
26	36.000	0.000	0.000	0.000	0.000	0.000	0.000	0.000	0.000	0.000	1.000	0.000	0.000
27	37.000	0.000	0.000	0.000	0.000	0.000	0.000	0.000	0.000	0.000	0.000	0.000	0.000

b

Fig. 2. Fragments of the sociological survey data: a - after preprocessing; b - after transformation into a primary feature space

could approximate the original distances as close as possible R_{ij} for all $(i, j) \in D$. This requirement can be formalized in different ways; one of the most common ways is by minimizing the functional which is called stress:

$$S(X^l) = \sum_{(i,j) \in D} w_{ij}(d_{ij} - R_{ij})^2 \to min, \qquad (9)$$

where the minimum is taken cumulatively ln variables $(x_i^d)_{i=1,l}^{d=1,n}$. The dimension n is usually small. In particular, with $n = 2$ multidimensional scaling allows you to display a sample as a scatter plot. The flat representation is usually distorted $(S > 0)$, but generally reflects the main structural features of multidimensional sampling, in particular, its cluster structure. The weights w_{ij} are set on the basis of scaling goals. Usually $w_{ij} = (R_{ij})^\gamma$ is taken. When $\gamma < 0$ priority is given to more accurate approximation of small distances; when $\gamma > 0$ – of large distances. The most appropriate value is $\gamma = -2$, when the stress functional acquires the physical meaning of the potential energy of a system l material points connected by elastic bonds, and the minimization problem acquires a clear physical meaning of search for a stable equilibrium [19].

Stress function $S(X^l)$ in a complex way depends on ln of the variables, has a huge number of local minima, and its calculation is time-consuming. This is the reason why many MDS algorithms are based on the iterative placement of objects one by one. At each iteration, the Euclidean coordinates of only one of the objects are optimized, while the coordinates of the other objects calculated at the previous iterations are assumed to be fixed.

MDS allows to calculate the coordinates of objects in Euclidean space, usually of low dimension, from the known pairwise distances between them. In the current implementation (dataset [16]) using the MDS algorithm using (6) and (7) for aggregated data of sociological survey $X = \{x_i\}$, $i = \overline{1,23}$, originally presented in a 23-dimensional attribute space, a space of secondary features is formed $(x_i^d)_{i=1,23}^{d=1,2}$ – the space of the so-called MDS projections, which are defined by the coordinates on the plane w_{ij}. The scatter plot in the form of a dataset respondent similarity map [16] is shown in Fig. 3. On the similarity map, not the coordinates of points have a meaningful interpretation but only their relative position. Studies have shown that the use of MDS algorithm is more preferable if objects in the initial attribute space are specified using binary vectors.

3.3   Development of the Method of Neural Network Classification of Sociological Survey Respondents

For cluster analysis, the set of values of the primary features of the aggregated data of sociological survey $X = \{x_i\}, i = \overline{[1,l]}$ has been transformed into the space of secondary signs $(x_i^d)_{i=1,l}^{d=1,n}$ with the use of the following four options:

1. Using radius values centered at the origin (Fig. 3).
2. Using the values of the radius of "variant 1" and the angle of inclination between each point and the axis X.
3. Using the values of the coordinates in the sector, which divides the plane into 4 parts according to 90°.

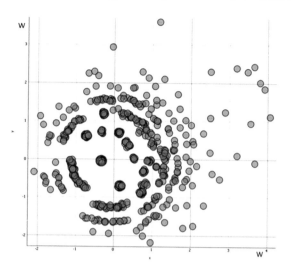

Fig. 3. Map of respondents similarity, built using MDS projections

4. Using the combination of the values of "variant 1" and "variant 2".

Next, in the created spaces of secondary features, clustering is carried out using the methods of K-means or the methods of hierarchical clustering [20,21]. For hierarchical clustering is recommended to use the binding of variables of one cluster by the Method of Ward [21]. The visualization of the results of hierarchical clustering for 4th described variants of converting aggregated data of the sociological survey (dataset [18]) into the space of MDS projections is shown in Fig. 4.

To select the algorithm for classifying variables in the space of secondary signs, a comparative analysis of the following classifiers is carried out [22]: neural network classification using Multilayer Perceptron (MLP), classification using Support Vector Method (SVM), Naive Bayes, classification based on logistic regression (Logistic regression). As a criterion for choosing a classification algorithm, the classification accuracy has been used (classification accuracy, CA) (Table 1):

$$CA = \frac{TP + TN}{TP + TN + FP + FN},\qquad(10)$$

where TP – true positive; TN – true negative; FP – false positive; FN – false negative decisions. Thus, in accordance with the criterion $CA = 0,955$ (Table 1), taking into account (5) and target dependence $y^* : X \rightarrow Y$, to build a classifier $a : X \rightarrow Y$ it is recommended to use MLP based classification with the following parameters: Neurons in hidden layer = 100; neuron activation function (Activation = ReLu); optimization method when learning by BPL (Solver = L-BFGS-G); maximum number of learning epochs (Max iterations = 300).

Table 1. Results of testing classification algorithms

The name of the classifier	CA			
	a	b	c	d
MLP	0,955	0,931	0,952	0,944
SVM	0,915	0,902	0,891	0,873
Naive Bayes	0,788	0,830	0,910	0,918
Logistic regression	0,767	0,822	0,923	0,923

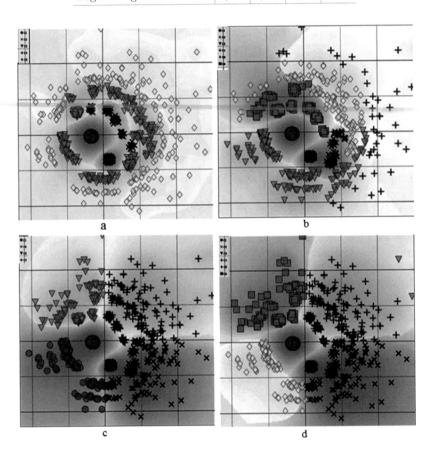

Fig. 4. Examples of hierarchical clustering in MDS projection spaces: a – clusters; b – 6 clusters; c – 4 clusters; d – 6 clusters

4   Approbation of Intelligent Information Technology Using the Example of "Ukraine - Lifestyle" Data

To create an intelligent information technology (IIT) analysis of weakly structured multidimensional sociological survey data, the following fixed as sets were used: an interpreted, interactive, object-oriented programming language Python; framework Orange3, including software for Data Mining based on components and used as a module for Python; library Scikit-Learn, implementing machine learning for Python, as well as interacting with numerical and scientific libraries Python NumPy and SciPy. Moreover, the development of IIT data analysis took into account the provisions of the methodology of CRISP-DM.

For testing of the developed IIT, data from the sociological survey "Ukraine - Lifestyle" has been used (dataset [16]). The purpose of the analysis is to determine if 1143 respondents lead a healthy lifestyle based on their answers to 114 questions. In accordance with the decision of the respondent and the decision of the expert-sociologist, three meta-classes have been defined - "good", "average" and "bad". The standard methodology of the expert sociologist in the distribution of the surveyed respondents to meta classes based on an analysis of their responses to the sociological survey "Ukraine - Lifestyle", included the following steps:

- taking into account the formalized goal of analyzing the target audience, 3 meta-classes are distinguished;
- of 114 questions, 12 main relevant features are determined for the following classification of respondents;
- the response space for each relevant attribute is divided into categories according to the given number of meta-classes;
- when a respondent chooses a definite answer from a category to the specific question, he gets 1 point in the corresponding category;
- respondent belongs to the meta-class corresponding to the category for which he scored the maximum amount of points;
- if the maximum cannot be determined, the respondent belongs to an average class.

Analysis of the effectiveness of the data analysis developed by IIT of sociological survey was carried out using the sociological survey classification accuracy calculation (10). As a feature vector, data from two-dimensional spaces of values of radius R centered at the origin after MDS processing have been used. For two-dimensional spaces of secondary features were used (Fig. 5):

- option 1 – 77-dimensional data presented in the original space after the first stage of processing in accordance with the proposed methods of presenting detailed and aggregated data of sociological survey (Fig. 5, a);
- option 2 – 23-dimensional data presented in the space of primary signs after the second stage of processing in accordance with the proposed methods of presenting detailed and aggregated data of sociological survey (Fig. 5, b);
- option 3 – 36-dimensional data of sociological survey, obtained in accordance with the standard methodology of an expert sociologist (Fig. 5, c).

When visualizing the results of the classification of respondents in Fig. 5, the following classification decisions were used:

– option 1 – respondents' decision about their belonging to one of three meta-classes (Fig. 5, a);
– option 2 – automated decision on whether the respondents belong to one of 6 clusters which has been made by the system Data Mining (Fig. 5, b);
– option 3 – the decision on whether the respondent belongs to one of the three meta-classes which has been taken by the expert based on his analysis using his own method (Fig. 5, c).

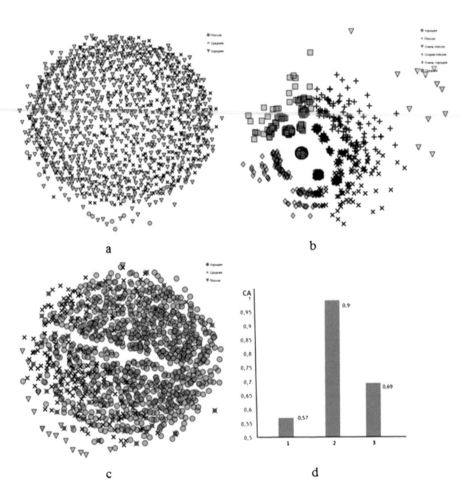

Fig. 5. Visual presentation of the results of the respondents' classification on the basis of the sociological surveys: a – in accordance with their own decision, 3 classes; b – in accordance with an automated solution, with the help of IIT, 6 classes; c – in accordance with the decision of an expert-sociologist, 3 classes; d – value of sociological survey in accordance with the suggested processing options for two meta-classes

To quantify the reliability of the sociological survey value from the original data set, a new data set has been formed from the respondents' answers which are "guaranteed" from the point of view of the respondents; the expert belongs to two polar meta-classes - "bad" and "good". 144 respondents were assigned to the "bad" class and 375 - to the "good" class.

Thus, the use of the sociological survey data analysis developed by IIT allowed to increase the accuracy of classification decision making decisions about the respondents' lifestyle for two artificially formed polar classes of respondents by 30% compared to a professional sociologist and by 43% compared to their own definition of respondents.

5 Conclusions

To automate the process of obtaining metadata about respondents to sociological surveys or questionnaires, an intelligent information technology has been developed for analyzing weakly structured multidimensional social research data. The basis of technological solutions for the implementation of IIT data preparation of sociological survey lay developed models of detailed and aggregated data of sociological survey and methods of their spaces of primary and secondary features and neural network classification based on preliminary cluster analysis.

As studies have shown, data analysis developed by IIT, the most influencing decision making are procedures related to obtaining spaces of primary and secondary features. Testing of the developed IT as part of Data Mining on the data of "Ukraine - Lifestyle", allowed to increase the accuracy of classification decisions on the respondents' lifestyle by 30%, compared to the sociologist of the qualification level "Master", and by 43%, compared with the definition of respondents.

Developed IIT of data analysis of sociological survey may be recommended for use for the analysis of multidimensional weakly-structured data obtained as a result of questioning or interviewing. Presented in a visual form convenient for experts, the knowledge gained about the target audience under study makes it easy to take them into account in order to make informed decisions.

References

1. Arsirii E, Sayenko A (2011) Neural network pattern recognition of the public library readers to provide specialized information services. In: Proceedings of the Odessa Polytechnic University, vol 1(35), pp 118–124 (in Russian)
2. Arsiry E, Zhylenko E (2009) Neural network formation of integrated professional characteristics in the system of distance learning MOODLE. In: Proceedings of the Odessa Polytechnic University, vol 2(32), pp 161–166 (in Russian)
3. Fayyad U, Piatetsky-Shapiro G, Smyth P (1996) From data mining to knowledge discovery in data bases. AI Mag 17(3):37–54. https://doi.org/10.1609/aimag.v17i3.1230

4. Arsirii O, Babilunha O, Manikaeva O, Rudenko O (2018) Automation of the preparation process weakly-structured multi-dimensional data of sociological surveys in the data mining system. In: Herald of Advanced Information Technology, vol 01(01), pp 9–17 (in Russian). https://doi.org/10.15276/hait.01.2018.1
5. Rudenko A, Arsirii E (2018) Method of intellectual analysis of weakly structured multidimensional data of sociological surveys. In: Proceedings of the VIIth international conference "Modern Information Technology" (MIT), Odessa, Ukraine, pp 168–169 (in Russian)
6. Semenov V (2009) Analysis and interpretation of data in sociology. Vladimir State University, Vladimir, Russia (in Russian)
7. Kislova O (2005) Data mining: opportunities and prospects for application in sociological research. In: Methodology, theory and practice of sociological analysis of modern society, Harkiv, pp 237–243
8. Chapman P, Clinton J, Kerber R, Khabaza T, Reinartz T, Shearer C, Wirth R (2000) CRISP-DM 1.0 step - by-step data mining guide. SPSS.: Copenhagen, Denmark
9. Wirth R, Hipp J (2000) CRISP-DM: towards a standard process model for data mining. In: Proceedings of the 4th international conference on the practical applications of knowledge discovery and data mining, Manchester, UK, pp 29–30
10. Praveen S, Chandra U (2017) Influence of structured, semi structured, unstructured data on various data models. Int J Sci Eng Res 8(12):67–69
11. Corbetta P (2011) Social research: theory, methods and techniques. Sage, London. https://doi.org/10.4135/9781849209922
12. Hunter MG (2002) The repertory grid technique: a method for the study of cognition in information systems. MIS Q 26(1):39–57
13. Kim DzhO, Myuller ChU (1989) Factor, discriminant and cluster analysis. Finance and Statistics: Moscow, Russia (in Russian)
14. Hardle W, Simar L (2012) Applied multivariate statistical analysis free preview. Springer, Berlin
15. Merkert J, Mueller M, Hubl MA (2015) Survey of the application of machine learning in decision support systems. In: Proceedings of the twenty-third European conference on information systems (ECIS 2015), Münster, Germany, pp 1–15
16. Arsirii E, Manikaeva O, Vasilevskaja O (2015) Development of the decision support subsystem in the systems of neural network pattern recognition by statistical information. Eastern-Euro J Enterp Technol. Math Cybern Appl Aspects, 4(78), 6:4–12
17. Barsegyan AA, Kupriyanov MS, Stepanenko VV, Holod II (2004) Data analysis methods and modeles: OLAP and data mining. St. Petersburg, Russia, BHV-Peterburg (in Russian)
18. Sociological survey data "Ukraine - life style" (in Russian). http://edukacjainauka.pl/limesurvey/index.php
19. Buja A, Swayne DF, Littman ML, Dean N, Hofmann H, Chen L (2008) Data visualization with multidimensional scaling. J Comput Graph Stat 17(2):444–472
20. Hartigan JA, Wong MA (1979) A k-means clustering algorithm AS 136. Appl Stat 28(1):100–108
21. Jain A, Murty M, Flynn P (1999) Data clustering: a review. ACM Comput Surv 31(3):264–323
22. Haykin S (2009) Neural networks and learning machines. McMaster University, Ontario, Canada

Theoretical and Applied Aspects of Decision-Making Systems

Investigation of Forecasting Methods of the State of Complex IT-Projects with the Use of Deep Learning Neural Networks

Viktor Morozov$^{(\boxtimes)}$ (ID), Olena Kalnichenko$^{(\boxtimes)}$ (ID), Maksym Proskurin$^{(\boxtimes)}$ (ID), and Olga Mezentseva$^{(\boxtimes)}$ (ID)

Management Technology Department, Taras Shevchenko National University of Kyiv, Bohdan Gavrilishin Str., 24, Kyiv 01601, Ukraine
knumvv@gmail.com, kv_vl@ukr.net, proskuryn69@gmail.com,
olga.mezentseva.fit@gmail.com

Abstract. Issues related to the development and implementation of distributed information systems, which today are the most important decision in the IT market, are considered in this article. The tendencies of negative completion of such projects are analyzed. The conclusion is made on the need to take into account the environmental impacts of such projects and the use of proactive project management technologies in these processes, which are currently quite developed and effective. Attention is paid to the development of a conceptual model for the interaction of complex IT projects with their external environment. The authors propose a model "cube of interaction processes" to determine the types of influences on the system (projects) and its reaction. The method of forecasting the status of projects from the effect of changes on the basis of impacts from the project environment is proposed. The basis of the method is the processes of deep learning of neural networks. Initiation and implementation of change management is envisaged on the basis of accumulated knowledge, principles of system analysis and mechanisms of convergence of methodologies for the creation of such systems. The mathematical description of this method is proposed in the article, which defines the interaction of elements and characteristics with the impact of changes in projects for the creation of distributed information systems. Attention is paid to the experimental study of the processes of interaction of elements of such systems with the use of modern neural networks. As variables parameters were chosen the amount of resources that are provided to the work of projects for their implementation.

Keywords: Distributed information systems · IT-projects · Proactive management · Information impacts · Influences · Neural networks

© Springer Nature Switzerland AG 2020
V. Lytvynenko et al. (Eds.): ISDMCI 2019, AISC 1020, pp. 261–280, 2020.
https://doi.org/10.1007/978-3-030-26474-1_19

1   Introduction

The current state of development of cloud technologies in the IT field is characterized by further latest developments and the creation of modern distributed information systems (DIS). This especially concerns in the field of information communications, where the development and use of such systems shows high efficiency and is one of the most promising directions of the development.

This is facilitated by the rapid development of cloud computing, such as distributed data processing technologies, opening horizons and prospects for creating new cloud services [1,2]. Recent trends in this area show that the concept of distributed information technologies is useful and relevant. It is considered as an effective means for solving modern problems and problems arising from the trends of rapid development, globalization, complication of technologies and increased turbulence of the environment.

Today, large corporate volumes are actively shifting to cloud-based solutions. According to estimates, in the next five years 40% to 50% of corporate load will be concentrated in the clouds, and now – by 15%. This indicates an increase in the demand for cloud services and a change in the information policy paradigm at enterprises [3,4].

In addition to cloud technologies, experts also note other important aspects of improving and developing information technologies, for example, in developing technologies for analysing large volumes of data and integrating mobile devices and social networking technologies into the corporate environment [5,6].

Distributed information systems now are increasingly used, particularly in banking, distance learning, telecommunications and others. This is due to the urgent need to process large volumes of data (Big Data). It is known that for processing large amounts of data need to attract additional computing power using hardware supercomputers, distributed systems (compute nodes connected lines), cluster, Cloud-technologies. However, the use of distributed information systems associated with the development of new algorithms (e.g., routing algorithms), new protocols and so on.

The processes of creation of DIS relate to the execution of IT projects, which are characterized not only by functional and constructive complexity, but also by a large number of changes at the planning stages, and especially at the stages of project implementation. In addition, the DIS itself from the point of view of projects and project management processes is the *products* of such projects.

However, sufficient constructive and technological complexity of the processes of creating distributed information systems requires the use of a project approach [7,8], where projects in the design of such systems are applied methods and information technology project management. Over the past 20 years, the use of such tools has proven effective [9].

The use of such technologies for manage communication projects can lead to a change in many rules and principles. At the same time the experience shows that the use of artificial intelligence methods will have a significant impact on the effectiveness of the results of the projects themselves.

The priority task when creating modern integrated telecommunication IT with intellectual support for project management is to optimize costs [8,9] for such development and integration, transforming IT from serving to the element of value generation.

Finding the best options for distributing resources in telecommunication projects can reduce the time frame for project tasks and, consequently, greatly reduce their cost. At the same time, predicting the state [10] of such projects is a multidimensional problem, which can be solved using modern neural network technologies [11]. In addition, numerical changes that often have a negative impact on the various parameters of communication projects and have a significant impact on the results of their implementation should be taken into account.

Thus, consideration of the possibilities of experimental use of neural networks in the research of the interaction of numerical changes on the parameters of information communication projects with the definition of their optimal states is an urgent task.

2   Analysis of Recent Research and Publications

The issue of cloud technologies application in the creation of modern IT was considered in [4–6]. For example, [4] describes the roles and characteristics of distributed computing processes that are used in the creation of modern distributed information systems (DIS). However, the description of project processes is completely absent here. [8] shows 48 project management processes. However, it is unspecified, which of them and whether it should be implemented in full when using modern IT. In [5] describes the features of the methods and technologies used in creating distributed IT projects. But the description of the design approach is also missing here. Herewith, it is noted that the target use of such technologies is to consider the processes of IT functioning in organizations. A substantial degree of financial savings is achieved through integrating the functions of various IT systems, which gives the spatio-temporal balance of business processes.

Support for project coordination was studied in publications of Ukrainian and foreign scientists such as Bushuyev [9], Bushuyeva [10], Gogunskiy [12], Morozov [13], Teslia [14], Biloshchytsky [15], Sutherland [16], Turner [17], Milosevic [18], DeMarco [19] and others.

In particular, of the scope of project management, changes, as well as the synthesis of design product configuration in various subject areas, in particular, in distributed projects are studied deeply. However, the problem of choosing the optimal set of monitored elements of the project affected by the changes has not been thoroughly investigated to offer its practical solutions, including in DIS development and implementation projects.

3   Setting the Objective

Unresolved earlier parts of general problem. Unresolved earlier parts of the general problem. In the above-mentioned works, partially formalized models of DIS elements in distributed projects were partially formalized. However, the impact factors of the IT environment require for such complex projects of further development as to determine the reactions of the management system to change the key parameters of the project and build a mathematical model for experimental research. The question of interaction with the turbulent external environment of organizations and projects, the dynamics of its super-complex influences and management of changes were studied in works [9–19].

At the same time, the analysis of these works shows the range of unresolved problems within the framework of proactive project management and the prospect of further study of this subject area.

Formulating Objectives. The purpose of this research is to develop a method for forecasting the interaction of project elements on the effects of changes. This should be achieved as a result of experimental modelling of changes in the implementation of complex IT projects for the creation of distributed information systems. The research will change the download schemes and volumes of resources for project work. At the same time, the time settings will be fixed. To do this, first of all, it is necessary to determine the peculiarities of the architecture of such projects, to determine the technological topology of creating an IT product, resource parameters and their distribution over tasks and time, to conduct experimental simulation in order to find the optimal option at the output of the system [13].

4   Materials and Methods

4.1   Development of DIS Architecture

It is necessary to distinguish, as a basis, the key components of such systems, architecture (interconnection logical, software and physical structures) systems, as well as the requirements for their functionality in order to understand the complexity of the process of formulating the needs for the configuration of DIS development and implementation projects. This will later identify the features of the project elements, as a second-order configuration. In other words, it is necessary to complete the process of identifying the architecture of such a system.

The procedure for forming a distributed information system can be presented in the form of successive stages: design, project implementation, prototype design, implementation and preparation for use. At the design stage, the structure of the information system is determined, the rules for exchanging information between the different databases that are part of the distributed information system are determined, as well as the rules governing the introduction of changes in such databases.

Thus, we have several groups of components: a remote group of individuals, content developers, including its integration into the central office on the base server, a group of communication with the rules of operation of the system components and a group of affiliated remote centers in other countries with multi-client services network services. Such groups solve all tasks of coordination of processes and operations of delivery of informational content (administrative, interface, terminal, information and computing aspects). The interaction of such components and elements is shown in Fig. 1.

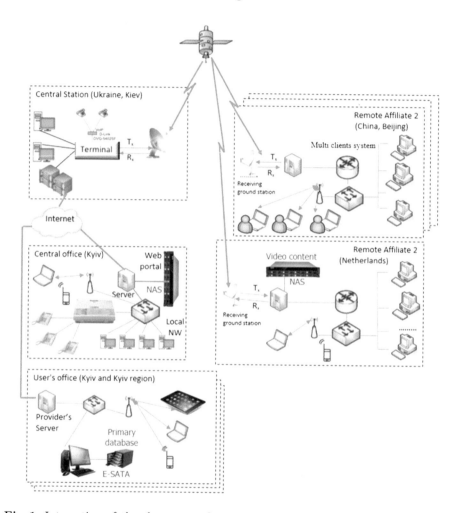

Fig. 1. Interaction of the elements and components of the project with the product when defining the project configuration

4.2 Development of the Model of Interaction with the External Environment

Let's assume that the processes and subjects involved in the project implementation are an integrated "project–product–organization" system (Project – Product – Organization). Then one of the possible models of the "product – project – organization system" (P2O) is a model that consists of [20]:

- a subsystem that includes elements and processes associated with product creation;
- a subsystem that covers the elements and processes of project management;
- a subsystem that describes the elements and processes of the organization implementing the project.

In this case, this system is under constant influence of the external environment. It is assumed [21, 22] that any system can be in a state of rest (the state of normal functioning of the system) until it is exerted on the outside, that is, the system does not receive any information. The system changes its properties and behaviour, and also produces responses to such influences. Let's divide all processes into four categories to understand the situation that occurs during project implementation and interaction with the external environment:

- Processes for managing interactions with the external environment (PMIEE) of the organization and projects.
- Stakeholder Management Processes (SMP).
- Project management processes, which include [8] five groups of processes: initiation, planning, implementation, monitoring and control, closure (PMPs).
- IT product development management processes that comprise [23] of five groups of processes: service strategy management, service design, service development, service exploitation, continuous improvement of services (PDMPs).

Such categorization is considered appropriate on the basis of understanding the interaction in the organization and in the project, as well as to take into account the effects of the environment and the reactions of the organization and the project on these impacts.

If we explore the concept of "external environment" and its impact on the characteristics of the project, then it consists of a neighbor (the environment of the organization implementing the project) and the distant environment of the project (the environment of this organization – political, legal, economic, technological, social and cultural, international, natural and environmental, infrastructure factors).

If we consider the concept of "external environment" from the position of the organization, then the external environment consists of two levels: the level of direct influence on the organization and the level of indirect influence. The level of direct influence includes factors that are also called "business environment" – suppliers, consumers, competitors, investors, customers, intermediaries, shareholders. The level of indirect influence, the so-called "background environment" includes political, legal, economic, technological, social and cultural, international, infrastructure, natural and environmental factors. In fact, this is a distant project environment.

Proceeding from the above, the concept of "external environment" can be represented as a multilayered model (Fig. 2), the processes of which have the following characteristics:

- complexity (the number of factors that make up this model);

- uncertainty (quantity and quality of information);
- mobility (the rate of change occurring in this environment);
- interrelatedness (the level of influence of the components of the environment on each other).

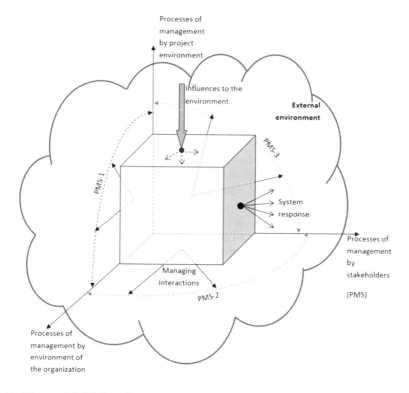

Fig. 2. The model "Cube of interaction processes" with the influence of the environment

Based on the standard PMBOK (Project Management Body of Knowledge) [8], it can be noted that there are external stakeholders of the project. Accordingly, processes related to the activities of stakeholders, their influences and responses to these impacts will have the same characteristics as the external environment to which they belong (complexity, uncertainty, mobility and interconnectivity).

The area "product–project–organization", which is represented by the interaction of the three categories of processes PMPs, PDMPs and SMPs, is shown in Fig. 2. These processes describe a certain system and implement its functions based on the properties and principles of the behaviour of its subsystems and elements. In fact, this is a system of project implementation – P2O.

The external environment impacts on the specified system through certain "entry points" that is capable of perceiving such effects, interpreting them, transmitting through the information communication channels, and "manifesting" the system's response to the environment.

Interesting is the question that represents the "entry point". According to the authors of the article, interested internal parties, who perceive information from the environment, change their perception/vision of the situation and exercise on the basis of this influence on the project in order to adjust its indicators or product parameters. Thus, the impacts are realized through the interaction of external and internal stakeholders through the transfer of information, processing of this information, perception and the formulation of "reaction" (management actions), which in turn is also information generating new perceptions and new reactions.

If, in fact, the impact is the transfer of information and is carried out through the perception of the subject of certain information with subsequent interpretations, we can say that the interaction of external and internal environment (based on the formation of reaction to information) determines further changes and transforms the project, product, organization and external environment. This is due to new data (information) that affect the processes of the system.

Since the elements of the system are processes, it is expedient to consider all these categories of processes in the aggregate and their interconnections. In this context, one can analyze the stability of individual elements or the system as a whole, the sensitivity to effects, the ability to recover, the rate of response to impacts, etc.

Taking into account the above-mentioned, one can conclude that managing changes and solving a plurality of problems that are the result of complex intermittent influences is a very difficult and valuable process. An effective tool for ensuring compliance with a project's stated goals and criteria may be the concept of proactive management, which is to understand the following principles:

– the number of environmental factors as sources of impacts, their interconnectivity and the speed of their changes will only increase;
– the influence and stress of the system will increase;
– it is almost impossible to obtain complete information about all factors, their potential changes and the probability of a certain development of events;
– the complexity of interpreting information will increase;
– increasing internality (the level of subjective control – conscious accountability for all processes, events and results) is the way to effective management in a turbulent environment.

Thus, the concept of proactive management should include the following processes:

– internal and external environment analysis (including event management);
– management of expectations (formation of the desired vision of the stakeholders);

- managing opportunities (creating opportunities);
- management of stability (creation of conditions for increasing the ability of system elements to withstand external influences);
- adaptive management (managing the properties of adaptation to changes in the conditions of existence, self-development, self-organization);
- information management (incoming, outgoing and in-system);
- management of relationships;
- impact management (management of interconnections, perception and interpretation of information by subjects to form their reaction and behavior).

4.3 Development of Mathematical Model of Interaction

Events occurring in the system and events affecting the system from the outside determine its global state at any given time. The objective function is to maintain the optimal state of the system during the project implementation. The optimal state of the system will be called a state in which there is a minimization of the imbalance of elements of the system, maximization of the indicator of significance of the value of the project as a criterion for achieving the objectives, and minimizing the damage (deviations) from the effects of the external environment.

The optimal state of the system can be achieved by activating proactive management processes. This will allow us to move from responding to events to forming the environment that is most suitable for the functioning of the P2O system.

Turning to the formalization and construction of the optimization criteria for the above processes, we introduce the following characteristics of projects [9].

In this case, the input parameters of the project model as the objectives of the project X may also be presented as:

$$X(t) = \{x_1(t),\ x_2(t),\ \ldots, x_i(t),\ \ldots,\ x_n(t)\} \tag{1}$$
$$\text{when } x_i(t) = \left\langle ID_i,\ NM_i,\ \overline{T_i},\ \overline{Rs_i},\ \overline{Rl_i},\ \overline{Vr_i},\ C_i \right\rangle,$$

where $x_i(t)$ is a description of the state of the k task of the project at time t when $i = \overline{(1, n)}$ and $\forall t \in T$;
ID_i is the code of task identification in the project;
NM_i is the text description of the task or its name;
$\overline{T_i}$ is the vector of the time parameters of the current project task;
$\overline{Rs_i}$ is the resource vector foreseen for the task implementation;
$\overline{Rl_i}$ is the vector of technological dependencies of the current problem with other tasks;
$\overline{Cr_i}$ is a vector of cost characteristics of the resource parameters of the task;
C_i is the parameter of the cost of the task;
T is the time of the project implementation;
n the number of tasks (works) of the project.

Since the project needs to have a certain amount of resources, it is necessary to specify a list of such resources for the project in the form of a certain set R.

Usually three types of resources (labour costs, materials and equipment) are used in projects. But from the experience of implementing IT projects, it can be noted that 90–95% of resources are labour costs. Therefore, we can assume that we will use only one type of resources – labour costs or labour resources. Thus we have:

$$R = \{r_j | j = \overline{1,m}\} \tag{2}$$

where m – the number of resources required to complete this project.

In addition, each r_j consists of:

$$r_j = \langle IR, NR, CR, MR, OR \rangle \tag{3}$$

where IR is the identifier of a particular resource;
NR is the name of the resource;
CR is price resource description (price);
MR is maximum possible resource load; OR is characterization of the type of resource.

As described hereinbefore, we now have to download resources R^l for each task $x_i(t)$ of the project. In this case, we obtain the following matrix:

$$R^l = \begin{bmatrix} r^l_{1,1} & \cdots & r^l_{1,m} \\ \vdots & \ddots & \vdots \\ r^l_{n,1} & \cdots & r^l_{n,m} \end{bmatrix} \tag{4}$$

where $r^l_{i,j}$ is the designated volume l of resource j for work i.

The specified download is by expert method. However, as noted in [24], the main approach to obtaining estimates of the time of project execution is the PERT (Program Evaluation and Review Technique) method. It is based on the assumption of beta-distribution of random size, which defines the duration of project tasks. Therefore, conducting the corresponding calculations using the specified method and using the function β we obtain the distribution of resources for each period of time R^d implementation of the project:

$$R^d \xrightarrow{\beta(x)} = \begin{bmatrix} r^d_{1,1} & \cdots & r^d_{t,1} \\ \vdots & \ddots & \vdots \\ r^d_{1,m} & \cdots & r^d_{t,m} \end{bmatrix}, \text{ where } t \in T \tag{5}$$

By obtaining a resource allocation in time and knowing their pricing CR_J, you can apply the function μ of determining the value of the distribution of the project to each project in time.

$$C^d \xrightarrow{\mu(\beta(x))} = \begin{bmatrix} c^d_{1,1} & \cdots & c^d_{t,1} \\ \vdots & \ddots & \vdots \\ c^d_{1,m} & \cdots & c^d_{t,m} \end{bmatrix} \tag{6}$$

Then the planned cost of the project will be:

$$C_p = \sum_{i=1}^{n} \sum_{t=1}^{T} c^d_{i,t} \tag{7}$$

Based on the experimental conditions for using the neural network and conducting its training, we need to get several attempts that will give us the corresponding distributions C_k^d, where $k \in K$ – is the number of simulation attempts (the number of matrices C^d in the simulation process). In this case, the variants R^l will be the variables. Such variants in this case will be considered as a kind of changes that occur under the influence of the external environment.

Thus, at the output of the system we have a family of curves of variants of the distribution of the cost of the project in time. In this case, the averaging distribution can be considered optimal:

$$C^{ia} = \sum_{i=1}^{n} \sum_{t=1}^{T} \frac{max\left(c_{i,t}^d\right) - min\left(c_{i,t}^d\right)}{2} \tag{8}$$

Accordingly, we obtain an averaged distribution:

$$C^a \xrightarrow{avar} = \begin{bmatrix} c_{1,1}^a & \cdots & c_{t,1}^a \\ \vdots & \ddots & \vdots \\ c_{1,m}^a & \cdots & c_{t,m}^a \end{bmatrix}, \tag{9}$$

Thanks to the trained neural network, it is possible to use it to find the optimal schedule of the project cost, corresponding to the optimal load of resources R^{la} for each project work.

In the future, with the use of proactive management of IT projects, for each of the possible variants R^l (simulating changes) we will have, with the help of the developed neural network model, different variants of deviations of the projected cost of the project $\pm \triangle C$ from the optimal variant.

All the main processes and functions that are considered can be presented as a model of the neural network of deep learning (Fig. 3).

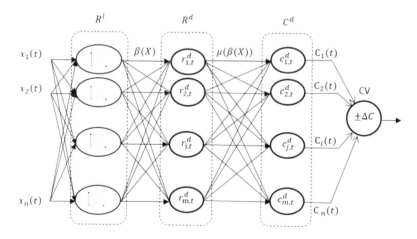

Fig. 3. The architecture of deep neural networks ensemble.

5   Experiments

For conducting experimental studies as inputs, we should determine the list of tasks or works that determine the content of the IT project. Usually such list is from 500 to 1500 tasks, but for the results of these studies should be limited to a fragment of 10 tasks. Description of such tasks is shown in Fig. 4 in the column 3 "Description". Also, as parameters of such tasks, it is possible to determine by certain methods the duration of each task whose values are shown in the column "Original Duration" in Fig. 4.

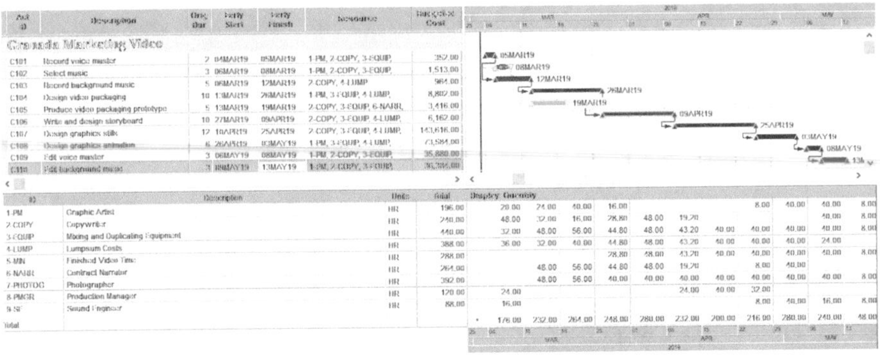

Fig. 4. An example of using software for monitoring IT projects.

Going to the stage of training the neural network, we further define the numerical variants of the loading schemes of resources according to the objectives of the project and the budget volumes of such resources, which in this approach are asked for the entire period of the task. An example of one of the options for a possible boot is shown in Table 1. Such variants should be 10 to 50. These download options will be the generator of changes in the investigated project.

Table 1. A fragment of the schema loading tasks with volumes of resources.

Task ID	Resource ID								
1	4	6	4						
2	1		8	8			1		
3		4		4					
4	1		2	2		4	2		
5						4	8		
6			2	2	4	2			
7		2	3	4	2		2	1	
8	1		2	2	2	2	2		1
9	1	1	2	2	4		1		
10	1	1	1		1		1		4

It is also necessary for each variant of loading tasks (works) resources to have output differentiated and cumulative (ascending sum) distribution of the involved resources in time to train neural networks.

We will use simulation in standard program planning and monitoring projects to obtain such results, for example, the software of Oracle's Primavera [25], an example of which is shown in Fig. 4.

Thus, for each variant of task loading resources, we obtain two corresponding distributions, variants of which are shown in Figs. 5 and 6.

In this case, the differentiated distribution will show us the peak load or failure in the use of resources (Fig. 6). A cumulative diagram will show which of the resources in the time is the most loaded, and which vice versa.

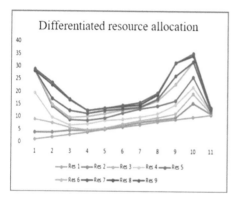
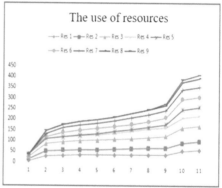

Fig. 5. The option of distributing project resources in a timely manner over the weeks

Fig. 6. The option to calculate resources in increments for weeks

We also load all these graphs in tabular form into the neural network at the learning stage.

The author analyzed the experience of managing IT projects, indicating that as critical source information, in such cases, a project cost schedule or a timetable for project cost sharing is used.

As a result of the simulation of the loading options for the resources for our project to determine their cost, we obtain a family of curves, an example of which is shown in Fig. 7.

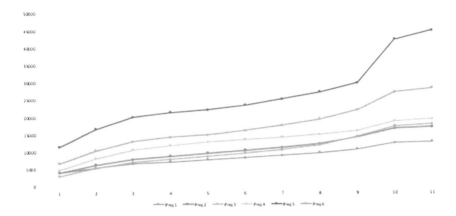

Fig. 7. Results of changes in the initial reactions of the system in the form of the cost of options

We also load all these graphs in tabular form into the neural network at the learning stage.

The author analyzed the experience of managing IT projects, indicating that as critical source information, in such cases, a project cost schedule or a timetable for project cost sharing is used.

As a result of the simulation of the loading options for the resources for our project to determine their cost, we obtain a family of curves, an example of which is shown in Fig. 7.

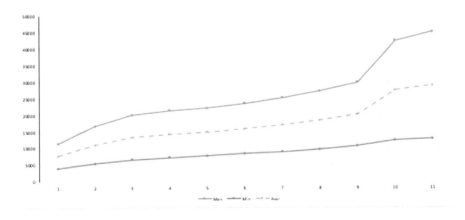

Fig. 8. Determination of averaged results of project cost sharing.

All these obtained data are also transferred to the neural network at the stage of its training. But, analyzing a family of cost curves, one can visually and analytically determine the maximum and minimum distributions. Moreover, the maximum in this case will be formed by two (several) variants of generation of load of tasks with resources, as well as the minimum option.

In practice, from a commercial point of view, this indicates that there may be optimistic and pessimistic options for completing any IT project. But the most interesting is the averaged distribution of project cost over time, which in practice is a compromise between the project customer and its performers. Examples of such distributions are shown in Fig. 8. The average schedule is the optimal result (output) of our system.

Specially designed software was used for conducting experimentation and training of the neural network, the fragment of which is shown in Fig. 9.

```
net.run(tf.global_variables_initializer())
# Setup plot
plt.ion()
fig = plt.figure()
ax1 = fig.add_subplot(111)
line1, = ax1.plot(y_test)
line2, = ax1.plot(y_test * 0.5)
plt.show()


# Fit neural net
batch_size = 256
mse_train = []
mse_test = []


# Run
epochs = 10
for e in range(epochs):
    # Shuffle training data
    shuffle_indices = np.random.permutation(np.arange(len(y_train)))
    X_train = X_train[shuffle_indices]
    y_train = y_train[shuffle_indices]
    # Minibatch training
    for i in range(0, len(y_train) // batch_size):
        start = i * batch_size
        batch_x = X_train[start:start + batch_size]
        batch_y = y_train[start:start + batch_size]
        # Run optimizer with batch
        net.run(opt, feed_dict={X: batch_x, Y: batch_y})
        # Show progress
        if np.mod(i, 50) == 0:
            # MSE train and test
            mse_train.append(net.run(mse, feed_dict={X: X_train, Y: y_train}))
            mse_test.append(net.run(mse, feed_dict={X: X_test, Y: y_test}))
            print('MSE Train: ', mse_train[-1])
            print('MSE Test: ', mse_test[-1])
            # Prediction
            pred = net.run(out, feed_dict={X: X_test})
            line2.set_ydata(pred)
            plt.title('Epoch ' + str(e) + ', Batch ' + str(i))
            plt.pause(0.01)
```

Fig. 9. A fragment of the listings of the training program for the neural network.

6   Results and Discussions

The conducted researches show that for the construction of models of proactive definition of deviations in the system under the influence of the environment of projects on the structural elements of the model of complex projects of the creation of DIS, special attention should be paid to the concept of "event". As the proactive management process consists of trend analysis, forecasting and planning of preventive actions, it is important to identify weak signals (events) that could potentially be a source of future problems. According to the ITIL version is detected: an event is a phenomenon that is relevant to the management of IT infrastructure or IT services.

Typically, events are presented in the form of messages generated by IT services, configuration units, or monitoring tools. In the context of project management, an "event" is any event that can be evaluated as information about a potential problem. In other words, the identification of information signals at the stage of latent development of project problems. At the same time, such events are divided into the following types:

- *information events* (normal implementation of operations) - events that do not require any action, such as completing a planned procedure or obtaining a planned intermediate result;
- *deviations* – events that can be considered as triggers for initiation of incidents, problems, changes. In other words, these are triggers of risky events;
- *warning* (unusual events that are not deviations) – events that cause make worried, some sequence (combination) of which can lead to deviations. It is precisely this type of event that refers to the concept of "weak signals" that is interesting for proactive management. These events can be used in forecasts and for the analysis of trends.

The description of the process of managing complex projects of distributed information systems based on the proactive approach takes into account the classification of the signals received from the turbulent environment of the project. This approach will gradually shift from reactive control to an active predictive situation.

In this case, change management is transformed from the mechanism of responding to incidents into the tool for forming the space (environment) of the project implementation with the following characteristics:

- relevance – changes, according to the latest trends (market, technology);
- flexibility – it is easy to adjust to the effects of changes;
- adequacy – corresponds to the current situation/state of the environment;
- self-adaptability (adaptability) – has the properties of self-development, strives for self-organization and evolution.

In the above description of the conducted studies, the assumption and diminution of the dimension of the research task to carry out numerical experiments on the basis of neural networks; it was assumed that the time parameters of work in the model of the project may remain fixed for some time. The changes concerned only schemas of loading works by certain types of resources. Variables were also the amount of resources for the entire time of each project work.

Activation functions were selected from linear, quadratic, cubic, sigmoid functions due to the fact that the program's operating time is significantly increased as a result of such a selection, implemented in a hidden cycle. We dwelt on the use of sigmoid activation function in the hidden and output layers of the neural network.

Evaluating the results of experiments can be evaluated and made certain conclusions. In particular, it was found that with a sufficiently wide range of changes in the input parameters of the model at the output, we received slight fluctuations in the reaction of the system ($\pm\Delta C$). This is typical only for this configuration of the network model (technology to create a product of the project), time parameters of the project, restrictions, etc.

As can be seen from the following diagram (Fig. 10), the test schedule of reactions to changes in the volume of resources of this project (orange) is closer to the minimum, in the middle between the average and the minimum. This indicates good results and the adequacy of the model with the help of deep artificial neural network training.

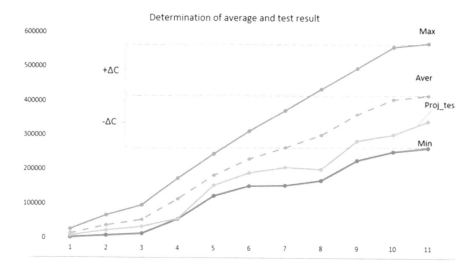

Fig. 10. Diagram of deviation of project cost ($\pm\Delta C$) and determination of system response via neural network

For further research it will be possible to capture the volume of work in the given model and provide a series of changes for the time parameters of the project model works.

In the future, we can also assume that the changes may relate to the technological schemes for creating software products in the form of DIS. But it will be necessary to determine which values of the parameters of the model will be fixed. Otherwise, solving the problem will require quite considerable computing resources.

The obtained results testify about necessity for further in-depth study of the issues of active reaction to environmental impacts of the project. As directions for further research it should be noted that it is necessary to identify the types of processes associated with the creation of complex IT products, project management processes, taking into account the constant dynamics of changes in interaction with the turbulent environment and their mutual influence.

It should be noted that the limitation of this study is the field of research related to the orientation of IT projects, and only when developing distributed information systems. In addition, the planned project cost and the effectiveness of the implementation of the IT project should not exceed the budget allocated for the implementation of the project. Taking into account time limits, in practice the possible over-implementation of the project is in the range of 10–15% of the indicated directive terms.

7   Conclusions

The proposed method of using neural networks with in-depth training to analyze the changes in the reactions of the management system of complex IT projects to the effects of external turbulent environments is based on an integrated approach to the use of information systems to consider the processes of creating complex products of such projects. A distinctive feature of this approach is a coherent presentation and analysis of the environmental impact on a plurality of all elements of the project with strong interacting relationships and influences. In addition, it was possible to identify controls to formalize the development of distributed information systems.

The revealed structural components and interconnections between them formed the basis of the proposed conceptual model "cube of processes of interaction", models for projects for the creation and development of complex IT products. The qualitative characteristic of this model is a platform for the development of DIS with synergistic functions based on knowledge, values, system analysis and convergence.

At the same time, the consideration of the effects of turbulent environments on the design model indicates the existence of various types of impacts that lead to changes in the parameters of the model elements. As a result, the proposed development platform DIS can predict the deviation of the initial reactions of the system. This creates an answer to external influences. However, the analysis of projects shows that such responses are in most cases delayed, which reduces the effectiveness of the project. Therefore, it was proposed to include in the model mechanisms of a proactive approach to minimize possible deviations.

The proposed mathematical model of processes of deep training of artificial neural networks allowed to formalize the processes of realization of projects on the basis of a proactive approach and to define target functions for further study of model behaviour under the influence of turbulent environment. The cost and time of the project implementation were taken as a basis, as the main parameters that determine the effectiveness of the project. In this case, the parameters of time were fixed.

The curves of the planned parameters of the target functions were constructed on the basis of the integration of the three types of information systems. Thus, the indicators simulated in the system indicate slight deviations within 10–15% of the given values under the influence of a wide range of values of the environmental factors and their effects on changes in the volume of resources of the project work for the selected and unchanged technological configuration of the project model. Using proactive managing controls, during re-simulation, it was possible to significantly reduce deviations in costs that do not exceed 10% deviation from optimal values.

References

1. Cloud Terminology – Key Definition. https://www.getfilecloud.com/cloud-terminology-glossary. Accessed 21 Feb 2019
2. Cloud computing and emerging IT platforms: Vision, hype, and reality for delivering computing as the 5th utility. http://www.sciencedirect.com/science/article/pii/S0167739X08001957. Accessed 21 Feb 2019
3. Furht B, Escalante A. Handbook of cloud computing. https://link.springer.com/book/10.1007/978-1-4419-6524-0. Accessed 21 Feb 2019
4. Maimon, O, Rokach L (2005) Data mining and knowledge discovery handbook. http://www.bookmetrix.com/detail/book/ae1ad394-f821-4df2-9cc4-cbf8b93edf40. Accessed 11 Mar 2019
5. Kosyakov M (2014) Introduction to distributed computing. SPb: Public Research Institute IST, 155 p. (in Russian)
6. Proactive Project Management. http://www.itexpert.ru/rus/ITEMS/200810062247/. Accessed 12 Mar 2019
7. International Project Management Association (2015) Individual Competence Baseline Version 4.0. International Project Management Association, 432 p
8. A Guide to the project management body of knowledge (PMBoK guide), 6th Edition. PMI Inc., USA, 537 p (2017)
9. Bushuyev S, Burkov V (2015) Resources management for distributed projects and programs. Monograph. Publisher Torubara V.V., Nikolayev, 338 p
10. Bushuyeva N (2007) Models and methods for proactive management of organizational development programs. Monograph. Scientific World, 199 p
11. Arzamastsev A, Fabrikantov O, Zenkova N, Belousov N (2016) Optimizatsiya formul dlya rascheta IOL [Optimization of formulae FOR intraocular lenses calculating]. Vestnik Tambovskogo universiteta. Seriya Estestvennye i tekhnicheskie nauki - Tambov University Reports. Series: Natural and Technical Sciences, vol 21, no 1, pp 208–213
12. Gogunskiy V, Kolesnikova K, Lukianov D (2016) Lifelong learning is a new paradigm of personnel training in enterprises. Eastern-Eur J Enterp Technol 4/2(82):4–10

13. Morozov V, Kalnichenko O, Bronin S (2018) Development of the model of the proactive approach in creation of distributed information systems. Eastern-Eur J Enterp Technol 43/2(94):6–15

14. Teslia Y, Khlevnyi A, Khlevna Yu (2016) Control of informational impacts on project management. In: Proceedings of the 1st IEEE international conference on data stream mining & processing, Lviv, Ukraine, 23–27 August

15. Biloshchytskyi A, Kuchansky A, Andrashko Yu, Biloshchytska S (2017) A method for the identification of scientists' research areas based on a cluster analysis of scientific publications. Eastern-Eur J Enterp Technol 2(89):4–10. No 5

16. Sutherland J (2017) Scrum: the art of doing twice the work in half the time, 272 p

17. Turner R (2007) Guide to project-based management (trans: from English). Grebennikov Publishing House, Moskow, 552 p. (in Russian)

18. Milosevic D (2006) A set of tools for project management. Dragan Z. Milose-hich (Trans: from English) Mamontova EV (ed.) Unknown S.I. - M.: IT Co., DMK Press, 729 p

19. DeMarco T, Lister T (2005) Waltzing with bears. Risk management in software development projects. - M.: Izd. "Company p.m.Office", 196 p

20. Morozov V, Kalnichenko O (2018) Construction of an integrated model of management processes of IT projects on the basis of a proactive approach. In: Monograph "Moden management: economy and administration". Opole (Poland): The Academy of Management and Administration in Opole, pp 82–89

21. Prigogine I, Nikolis G (2017) Knowledge of the complex. Introduction Per. from English, M.: Lenar, 360 p

22. Garaedagi D (2011) System thinking. How to manage chaos and complex processes. Platform for modeling business architecture, Grevtsov Buks (Grevtsov Publicher), 480 p

23. Free ITIL, v.3. http://www.wikiitil.ru/books/2015_Free_ITIL.pdf. Accessed 10 Mar 2019

24. Samokhvalov Y, (2018) Development of the forecast graph method in conditions of incompleteness and inaccuracy of expert assessments. Mag "Cybern Syst Anal" T. 54(1):84–91

25. Oracle's Primavera P6 Enterprise Project Portfolio Management. https://www.oracle.com/applications/primavera/products/project-portfolio-management/. Accessed 25 Mar 2019

Comparative Analysis of the Methods for Assessing the Probability of Bankruptcy for Ukrainian Enterprises

Oksana Tymoshchuk$^{(\boxtimes)}$, Olena Kirik , and Kseniia Dorundiak

Igor Sikorsky Kyiv Polytechnic Institute, Kyiv 03056, Ukraine
oxana.tim@gmail.com, okirik@ukr.net

Abstract. Various models and methods for analyzing the risk of enterprises bankruptcy using discriminant analysis, artificial neural networks and statistical approaches are presented. The specific problems of the Ukrainian economy are described - the lack of a large number of enterprises in the stock market, the inaccessibility of information about the real financial condition of some enterprises, the vagueness of the definition of bankruptcy. The possibility of using foreign models for Ukrainian enterprises is considered. The advantages, disadvantages and practical significance of the considered models in modern economic conditions are determined. Experimental studies have been carried out to compare statistical models, artificial neural networks of the perceptron type, regression models and binary trees in bankruptcy risk problems. A comparative analysis of the effectiveness of the use of these approaches to assess the financial stability of Ukrainian enterprises has been carried out. The most adequate methods were determined by the example of a specific enterprise actually functioning in Ukraine.

Keywords: Probability of bankruptcy · Discriminant models ·
Neural networks · Regression model · Random forest algorism

1 Introduction

Investigations of preventing bankruptcy of enterprises and the practical experience of leveling its negative consequences remain very significant for the economy of any country. The importance of preventing bankruptcy in advance, at the stages of its inception, led to the emergence of a large number of methods and models for assessing the financial position of companies. But the growing level of bankruptcy of Ukrainian enterprises suggests that there are problems in forecasting and the relevance of developing new modern tools for assessing the financial condition and forecasting the probability of bankruptcy.

There are common and specific problems in predicting the bankruptcy of Ukrainian enterprises. On the one hand, only a part of the indicators of the current financial condition of enterprises is displayed in all mathematical models. The specificity of Ukraine is the absence of a large number of enterprises

© Springer Nature Switzerland AG 2020
V. Lytvynenko et al. (Eds.): ISDMCI 2019, AISC 1020, pp. 281–293, 2020.
https://doi.org/10.1007/978-3-030-26474-1_20

in the stock market and the reluctance of the management of some companies to disclose the real financial condition of their enterprises. In addition, there is no clear definition of the concept of bankruptcy in Ukraine, therefore, part of domestic enterprises continue to operate despite extremely low economic indicators. All these moments complicate or eliminate the option of calculation for many models.

The purpose of the work is to consider and compare the effectiveness of applying various methods of assessing the risk of bankruptcy for Ukrainian enterprises.

The main idea of the work is that the application of not one, but several methods and approaches of bankruptcy assessment to the same enterprise on the basis of a common set of financial indicators will ensure the receipt of rational commercial and investment decisions with the criterion of minimizing the risk of unpredictable bankruptcy.

2   Literature Review

For many years, to predict the risk of bankruptcy, statistical models and one-dimensional classification methods that relate the company to a group of potential bankruptcies or to financially stable with a certain degree of accuracy were used. The most well-known and widely used statistical method is the technique of Prof. Altman [1].

Altmans model was built using the technique of multiplicative discriminant analysis. It allows you to select such indicators, the dispersion of which between the groups would be maximum, and within the group - minimal. As a result of calculations using the Altman model for a particular enterprise, a certain conclusion is made: a very high probability of bankruptcy, a high probability of bankruptcy, possibly bankruptcy, the probability of bankruptcy is extremely small. Since this analysis was suitable only for large enterprises whose shares are listed on the stock exchange, later Altman proposed a new model for enterprises whose shares are not listed on the stock exchange [2].

The approach of Altman based on multidimensional discriminant analysis was further developed by other researchers. But in relation to the economy of Ukraine, the Altman model was not widely used, since certain indicators differ from their values for foreign countries, and information about the financial condition of enterprises is often unreliable.

The universal discriminant function [3] and the discriminant model [4] are the most consistent with the operating conditions of Ukrainian enterprises. However, most researchers who investigated the problem of bankruptcy prediction have come to the conclusion that neural networks are much more accurate for forecast than classic statistical approaches. The advantages of using neural networks include the ability to detect latent nonlinear regularities in the input data, the absence of restrictions on the stationary nature of processes or the invariability of external conditions, rapid self-adaptation with new data without experts.

The mathematical basis of Kaydanovich's model [5] is an artificial neural network constructed using a combination of Kohonen's neuron layer, which can highlight common features in the studied objects using clustering, and the original

Grossberg star, which is characterized by ease of learning and interprets the clustering. The model used the variables selected for the discriminant econometric model A. Matviychuk by checking the financial indicators for multicollinearity.

The article of Debunov [6] showed that the best results were shown by multilayer perceptron networks containing a certain number of neurons in the hidden layer, as well as hyperbolic and logistic activation functions in the neurons of the hidden and output layers.

In [7], the authors carried out a comparative analysis of discriminant models Matviychuk and Tereshchenko with the developed model of an artificial neural network of the perceptron type. Using the example of analyzing the financial condition of several well-known Ukrainian enterprises, the advantages and problems of using discriminant analysis and artificial neural networks are evaluated. Compared with discriminant models, the approach based on the use of an artificial neural network has shown a high accuracy of estimates of the probability of bankruptcy, the ability to identify latent forms of an enterprise crisis, to analyze the financial condition of an enterprise in more detail, relating it to one of the three categories of crisis.

Unlike discriminant analysis, neural networks need extensive statistical data for learning. In article [8], the authors note the imbalance of data as a major problem, i.e. much more solvent data than default data. They propose auto-associative neural network that learns the identity mapping of input. By training the network with only solvent data, they built a bankruptcy predictor with better accuracies than conventional 2-class neural network.

In addition to discriminant analysis, companies sometimes use comparative and qualitative methods for assessing the threat of bankruptcy, because orientation to one selected criterion, even very attractive theoretically, is not always practically justified. The advantages of systems of indicators of possible bankruptcy include systemic and integrated approaches to the problem of assessing financial threats. The disadvantages include the difficulty of making decisions in multicriteria problems, informational and sometimes rather abstract types of indicators and the subjectivity of predictive decisions. Therefore, the estimates obtained are rather recommendatory in nature [9].

Algorithms based on binary classification trees are presented in [10]. There is no clear set of estimated parameters, and this type of model assumes a completely non-linear dependence of the default indicator on regressors. This group includes the random forest algorithm and the actual binary classification tree.

As for models based on behavioral algorithms, this is a relatively new direction in predicting the bankruptcy of industrial enterprises. In the case of these models, such algorithms and their modifications as the genetic algorithm and ant colony optimization algorithm [11, 12] are used. The genetic algorithm is used to optimize decisions made using mechanisms similar to natural selection in nature. The ant algorithm is an algorithm used to find the optimal path that has recently been used to form optimal strategies for managing the risk of bankruptcy of an enterprise. These algorithms were not the subject of research in this work.

3   Problem Statement

The paper is devoted to the analysis of models and methods for forecasting bankruptcies of enterprises of Ukraine, in particular, artificial neural networks, logistic regression, casual forest and decision trees. The advantages, disadvantages and practical significance of the considered models in the current economic conditions are determined, as well as possibilities of using foreign models for domestic enterprises. The purpose of the research is to increase the completeness and adequacy of information support of the commercial and credit decision making process, in particular, to optimize the allocation of investment resources. To achieve the goal, the system approach, modeling, analysis and generalization methods are used.

4   Methods and Algorithms

4.1   Artificial Neural Networks

Artificial neural networks were created to reproduce and simulate the structure of the brain and the ability of the biological nervous systems to learn and correct errors. Each neuron of such a network deals with signals that it receives from time to time, and signals that it periodically sends to others. To train neural networks, two types of algorithms are used: controlled (learning with a teacher) and not controlled (without a teacher) [13, 14].

One part of the neural network bankruptcy prediction models was based on the variables selected for a discriminant econometric model by checking the financial indicators for multicellularity, ensuring a high level of classification of the financial condition of the enterprise. To this set of variables were assigned coefficients of mobility of assets, turnover of accounts payable, turnover of equity, payback of assets, security of own working capital, concentration of attracted capital, debt repayment by equity.

The second group of models on the neural networks was based on a set of the most important indicators for assessing the financial condition of the enterprise, regardless of the existence or absence of linear functional dependencies among them, which are checked for multicellularity: capital efficiency, asset turnover ratios, fast solvency, autonomy, security of own working capital, debt coverage by equity [4].

The advantages of artificial neural networks are the ability to learn without assumptions about the functional dependence of variables. The disadvantage of the neural network is the complexity of the process of creating and optimizing its topology, as well as the fact that it looks like a "black box", so it is problematic to obtain explicit knowledge.

In this paper, an artificial neural network of the perceptron type is constructed and studied, which calculates the output variables through a sequential nonlinear transformation in the neurons of the input layer weighed by the weight coefficients of the interneuronal bonds. The perceptron includes the following three types of elements - the input signals that are passed to the associative elements, and then to the output ones. So, the perceptron allows you to create a certain set of "associations" between the input signals and the required response at the output.

4.2 Logit Model

Logistic regression or logit model is a statistical model used to predict the probability of occurrence of an event, when a dependent variable is categorical, that is, it can only accept two or, more generally, a finite set of values.

One of the logistic regression models is the Ohlson model [15]. The basis of the O-Score model lies in the same approach as in Altman's model, but the author used statistics for two thousand companies, which gives a more precise model with nine variables. According to Ohlson logit model, the index T is calculated using the following formula:

$$T = -1,32 - 0,47X1 + 6,03X2 - 1,43X3 + 0,0757X4 - 2,37X5$$
$$-1,83X6 + 0.285X7 - 1,72X8 - 0,521X9$$

where $X1$ is the natural logarithm of the ratio of aggregate assets to the deflator index of the gross national product; $X2$ - ratio of aggregate liabilities to total assets; $X3$ - ratio of working capital to aggregate assets; $X4$ - ratio of current liabilities to current assets; $X5-1$ if the aggregate liabilities exceed the aggregate assets, if vice versa is equal to 0; $X6$ - the ratio of net profit to aggregate assets; $X7$ - the ratio of proceeds from the main activity to total liabilities; $X8 - 1$ if the net profit was negative for the last two years, if on the contrary the indicator is 0; $X9$ is the ratio of the difference between the net profit in the last reporting period and the net profit in the previous reporting period to the net profit in the last reporting period taken modulo and the profit in the previous reporting period, taken modulo.

The rating coefficient T, which is calculated by Ohlson model, is used to find the probability of bankruptcy following the logistic regression formula:

$$R = \frac{1}{1 + e^T}$$

where R is the probability of occurrence of bankruptcy in shares of one; $e = 2.71828$ is the Euler number.

The main benefit of logistic regression is the quantification of the probability in an explicit form. Moreover, the method of logistic regression is characterized by high stability. The disadvantages of logistic regression are certain inaccuracies in describing the listed variables and its insensitivity to a part of the values of numerical variables.

4.3 Decision Trees

Decision Trees (also called Classification and Regression Trees – CART) used in statistics and in the intelligent analysis of data.

The algorithm is characterized by the fact that binary trees are constructed, that is, each internal node has exactly two output edges. The CART allows users to provide a pre-distribution of probabilities for tree induction. An important feature of the algorithm is its ability to generate regression trees with branching, which minimize the square of the prediction error (less than the quadratic deviation). The evaluation function, used by the CART algorithm, is based on the intuitive idea of reducing the uncertainty (inhomogeneity) in the node. In the CART algorithm, the idea of uncertainty is formalized in the Gini index. If the data set T contains data from N classes, then the Gini index is defined as follows:

$$Gini(T) = 1 - \sum_{i=1}^{n} p_i^2,$$

where p_i is the probability (relative frequency) of class i in T.

The Gini index of a given node of a tree before division, called the probability of mismatching the status of two randomly selected from this node of enterprises. The Gini index of a given tree node after dividing it into two subnodes is the probability of non-matching of the statuses of two randomly selected enterprises if the first is randomly selected from both intermediate nodes and the second is from the same intermediate node as the previous one. The construction of the tree is over when the given difficulty of the tree is reached.

4.4 The Random Forest

The random forest is the machine learning algorithm proposed by Breiman and Kutler, which combines two main ideas: the Breyman's bagging-bootstrap aggregation method [16] and the method of random subspaces proposed by Ho [17]. The algorithm is used for classification, regression and clustering problems.

The basic idea is to use a large ensemble of decision trees, each of which in itself gives a very low quality of classification, but due to their large amount the result turns out to be good. Among the disadvantages is the tendency to retrain the model.

5 Experiment and Results

The training set for all models consisted of 531 observations of 15 financial indicators for Ukrainian enterprises of different industries over the past 5 years, having different financial status for the purpose of their classification into various crisis categories. There were 429 observations to the non-bankruptcy class, and 102 to the class of enterprises with a threat of bankruptcy.

It was important to identify independent and dependent indicators that were used in constructing models. Dependent variable is a class of crisis condition, determined on the basis of financial position of the company.

Two categorical variables indicate the probability of bankruptcy: variable 1 - the company is not in danger, it has good financial stability; variable 2 - enterprises have a high risk of becoming bankrupt, financial difficulties are present.

The input data for assessing the probability of bankruptcy included 15 financial indicators:

$X1$ - total liabilities (balance sheet)/((operating profit + depreciation) * (12/365));

$X2$ - operating profit/sale (sales);

$X3$ - (operating profit - depreciation)/total assets;

$X4$ - (operating profit - depreciation)/sales (profit);

$X5$ - current assets (current assets)/total liabilities;

$X6$ - short-term (current) liabilities/total assets;

$X7$ - (current liabilities * 365)/cost of goods sold);

$X8$ - equity/profit from operating activities;

$X9$ - (sale - cost of sales)/sales;

$X10$ - total costs (net)/total sales (sales);

$X11$ - long-term liabilities/equity;

$X12$ - sale/receivables;

$X13$ - (short-term liabilities * 365)/sales;

$X14$ - sales/short-term liabilities;

$X15$ - sales/operating profit.

The accuracy of the neural network on the training sample was 86.3%. The error matrix is shown in the Table 1.

Table 1. Neural network model error matrix

Class of crisis	1	2
1	419	14
2	59	43

The table shows that the model very accurately recognizes enterprises that are not bankrupt, but more than half of bankrupts did not recognize, which indicates a lack of accuracy in predicting this model.

The classification report is given in Table 2.

The accuracy of the logistic regression model on the training sample was 82.5%.

The error matrix is shown in the Table 3.

Table 2. Classification report for the neural network model

Class of crisis	Precision	Recall	F1-score	Quantity for each class
1-not bankrupt	0.88	0.97	0.92	429
2-bankrupt	0.75	0.42	0.54	102
Average/total	0.85	0.86	0.85	531

Table 3. Error matrix for logistic regression model

Class of crisis	1	2
1	417	12
2	81	21

The table shows that the model also very accurately recognizes non-bankrupt enterprises, but could not recognize most of the bankrupts, which indicates the unreliability of forecasting using this method.

The classification report is shown in Table 4.

The accuracy of the Random Forest model on the training sample was 99%, which indicates a good result, especially compared to the previously considered models.

The error matrix is shown in the Table 5.

The table shows that the model 100% accurately recognized all non-bank businesses, as well as the vast majority of bankrupt enterprises - only 5 enterprises with financial difficulties were mistakenly attributed to the non-bankrupt. Compared to the previous models, this model is characterized by high accuracy and reliability of forecasting.

The classification report is given in Table 6.

The decision tree model on the training sample showed 100% accuracy, so it does not make sense for it to present an error matrix and a classification report.

The generalization of the results of bankruptcy risk prediction for the enterprises under consideration by various methods on the training set is shown in Fig. 1.

To verify the accuracy and quality of the resulted models, the "Obolon" enterprise, which actually operates in Ukraine, was chosen, because its data are

Table 4. Logistic regression model classification report

Class of crisis	Precision	Recall	F1-score	Quantity for each class
1-not bankrupt	0.84	0.97	0.90	429
2-bankrupt	0.64	0.21	0.31	102
Average/total	0.80	0.82	0.79	531

Table 5. Random forest model error matrix

Class of crisis	1	2
1	429	0
2	5	97

Table 6. Random forest model classification report

Class of crisis	Precision	Recall	F1-score	Quantity for each class
1-not bankrupt	0.99	1.00	0.99	429
2-bankrupt	1.00	0.95	0.97	102
Average/total	0.99	0.99	0.99	531

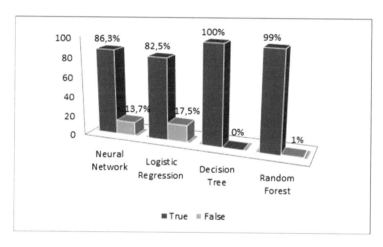

Fig. 1. Results of forecasting the risk of bankruptcy of enterprises by different methods on the training sample

rather informative. For analysis, the data for the financial statements for 2011–2017 were selected from the database smida.gov.ua.

From Fig. 2 it can be seen that the most volatile indicator is $X4$ - "(profit from operating activities - depreciation)/sale" during the entire study period. In the crisis years of 2014–2016, there was a sharp decline in such indicators: $X12$ - "sales/receivables", $X13$ - "(short-term liabilities * 365)/sales, $X15$ - "sales/profits from operating activities". The results of applying various models to the data of the "Obolon" enterprise are given in Table 7.

Table 7. Results of applying the considered models based on the financial indicators of the "Obolon" company for 2011–2017 (1 - not bankrupt, 2 - bankrupt)

Year	Bankruptcy class of the "Obolon" enterprise by different models			
	Neural Network (Perceptron)	Logistic regression	Random Forest	Decision trees
2011	1	1	2	1
2012	1	1	1	1
2013	1	1	2	1
2014	2	1	2	2
2015	2	1	2	2
2016	2	1	2	2
2017	1	1	1	1

6   Discussion

As the result of the analysis of financial indicators of Ukrainian enterprises, 15 key factors were allocated, most accurately reflecting the financial condition of companies.

A comparative analysis of bankruptcy prediction methods in applying to enterprises of Ukraine with the use of different models - artificial neural networks such as perceptron, logistic regression, Random Forest method and Decision Tree using a created software product. The software product was created in the Python programming language in the Jupyter development environment using the library scikit-learn and the functions MLPClassifier, LogisticRegression, RandomForestClassifier and DecisionTreeClassifier.

Analyzing the obtained results, we can say that the model of logistic regression proved to be the worst of all, it had the lowest accuracy even during study (82.5%). The model took all the years of the enterprise financially stable, which is a mistake, because it is known that the company had financial difficulties during the 2014 financial crisis. This is a good example of how external causes of bankruptcy, namely, the economic situation in a country, affect the company's activities and performance.

Although the neural network model did not have the highest accuracy in the training sample (86.3%) compared with other models, it correctly recognized the loss indicators for 2014–2016. This model can manifest latent forms of enterprise crisis and contribute to increasing the reliability of forecasting.

Random Forest model correctly recognized both unprofitable and financially stable years, but it was considered to be unprofitable in 2011, in which the company had only small financial difficulties, which can not be regarded as a significant threat to bankruptcy. Thus, the model is sensitive to very slight changes in coefficients.

Model Decision Trees also correctly recognized as a bankrupt, and not a bankrupt year for the enterprise.

Incorrect classification is the result of errors of type I (by mistake to classify a financially stable company as a bankrupt) and type II errors (mistakenly

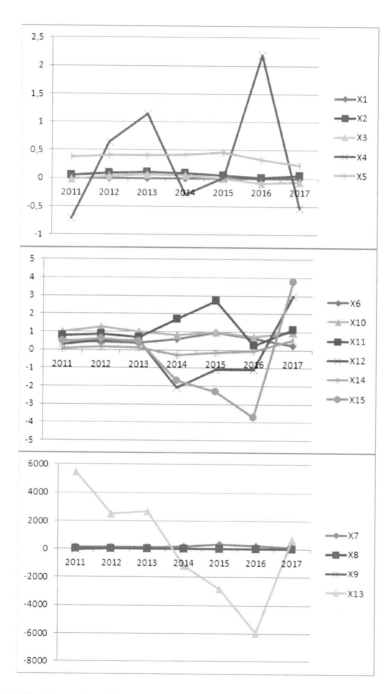

Fig. 2. The Dynamics of changes in the financial ratios of the "Obolon" enterprise

classifying the company's crisis situation as a stable one). Obviously, the consequences of the second type of error are more fatal for the financial sector, so it is advisable for enterprises that are not classified as bankrupt to be further tested by another methodology for more precise classification.

7   Conclusions

Between the models (neural networks of perceptron, logistic regression, Random Forest and decision trees) a comparative analysis was conducted on the example of the Ukrainian "Obolon" enterprise.

The logistic regression model has successfully recognized lucrative businesses, but has had significant bankruptcy difficulties, and its use is unreliable to predict and assess the financial condition of companies.

The models of Random Forest and Decision Trees were the best ones, in which the accuracy of the models on the training sample was 99%, as well as the models correctly recognized the situation at the "Obolon" enterprise. It is worth noting that the Random Forest model is sensitive to small fluctuations in the indicators and may regard them as certain financial difficulties.

Improving the process of forecasting crisis factors for Ukrainian enterprises using decision trees can occur by expanding the information base and improving the procedure for selecting of financial coefficients.

In the future, when creating software products for building an information and analytical systems for optimal allocation of investment resources, it is advisable to use various models and methods for assessing the financial condition of enterprises, in particular, which are able to learn from more financial indicators in order to more accurately recognize the situation in an enterprise with different aspects, as well as take into account the specifics of the industry and the characteristics of the Ukrainian economy.

References

1. Altman E (1968) Financial ratios, discriminant analysis and the prediction of corporate bankruptcy. J Financ 23(4):589–609
2. Altman E, Hotchkiss E (2005) Corporate financial distress and bankruptcy: predict and avoid bankruptcy, analyze and invest in distressed debt, 3rd edn. Wiley, New York
3. Tereshchenko O (2009) Management of financial rehabilitation of enterprises, 2nd edn. KNEU, Kyiv (in Ukrainian)
4. Matviychuk A (2013) Fuzzy, neural network and discriminant models for diagnosing the possibility of bankruptcy of enterprises. Neuro-Fuzzy Simul Technol Econ 2:71–118 (in Ukrainian)
5. Kaidanovich D (2010) Estimation of the risk of bankruptcy of enterprises with the use of counterproliferation neural networks. Sci Notes Natl Univ Ostroh Acad Ser Econ 15:468–474 (in Ukrainian)
6. Debunov L (2017) Application of artificial neural networks in modeling financial sustainability of an enterprise. Bus Inform 9:112–119 (in Ukrainian)

7. Tymoshchuk O, Dorundiak K (2018) Estimation of the probability of bankruptcy of enterprises through discriminant analysis and neural networks. Syst Res Inf Technol 2:22–33 (in Ukrainian)
8. Baek J, Cho S (2003) Bankruptcy prediction for credit risk using an auto-associative neural network in Korean firms. In: Proceedings of the IEEE international conference on computational intelligence for financial engineering, pp 25–29
9. Ostrovska G, Kvasovsky O (2011) Analysis of the practice of using foreign methods (models) for predicting the probability of bankruptcy of enterprises. Galician Econ J 2(31):99–111 (in Ukrainian)
10. Hastie T, Tibshirani R, Friedman J (2009) Random forests. In: The elements of statistical learning: data mining, inference, and prediction, 2nd edn. Springer, Heidelberg. Chap 15
11. Zhang Y, Wu L (2011) Bankruptcy prediction by genetic ant colony algorithm. Adv Mater Res 186:459–463
12. Korol T (2013) Early warning models against bankruptcy risk for Central European and Latin American enterprises. Econ Model 31:22–30
13. Khaykin S (2006) Neural networks: full course, 2nd edn. Publishing House Williams, Montreal Translation from English
14. Nielsen A (2015) Neural networks and deep learning. Determination Press, San Francisco
15. Ohlson J (1980) Financial ratios and the probabilistic prediction of bankruptcy. J Acc Res 18(1):109–131
16. Breiman L (2001) Random forests. Mach Learn 45:1
17. Ho T (1995) Random decision forests. In: Proceedings of the third international conference on document analysis and recognition, vol 1, pp 278–282

Monitoring Subsystem Design Method in the Information Control System

Oksana Polyvoda[1] ⓘ, Dmytro Lytvynchuk[1] ⓘ, and Vladyslav Polyvoda[2](✉) ⓘ

[1] Kherson National Technical University, Kherson 73008, Ukraine
pov81@ukr.net
[2] Kherson State Maritime Academy, Kherson 73000, Ukraine
polivodavv@rambler.ru

Abstract. The article proposes a method for designing and optimizing the monitoring subsystem in the information control system on the example of a grain drying equipment control system. The modern control systems of grain drying equipment provide the presence of an automated system for monitoring the humidity and grain temperature, the development of which should be based on objective data on the grain state at each point of the grain layer. These data can be obtained by analyzing the regularity of temperature and grain dynamics based on heat and mass transfer equations. The article proposes a method for determining the scheme of optimal placement of sensors of temperature and grain moisture, as well as the frequency of polling of sensors for conveyor type grain drying equipment based on the one-dimensional spectral analysis method. For the study, we used data obtained with the help of a mathematical model of the drying process of grain, using the finite difference method. After selecting the characteristic parameters for a specific type of grain and grain drying equipment, the proposed method of optimization of the automated monitoring system can be used to solve many problems, such as determining the static and dynamic characteristics of grain drying equipment, drying parameters selection and optimization, monitoring tasks and process control solution. The developed method for determining the optimal placement of sensors and the frequency of their sampling based on the method of one-dimensional spectral analysis allows to avoid information redundancy and simplify the technical implementation of the monitoring system, the effectiveness of which determines the efficiency of the entire information control system of the grain drying process.

Keywords: Sensor · Moisture · Temperature · Grain · Drying · Spectral analysis · Autocorrelation function

1 Introduction

Modern industrial facilities are complex geographically distributed objects, for the control of which information control systems are used. Their operation is

ⓒ Springer Nature Switzerland AG 2020
V. Lytvynenko et al. (Eds.): ISDMCI 2019, AISC 1020, pp. 294–303, 2020.
https://doi.org/10.1007/978-3-030-26474-1_21

highly dependent on the performance of all its subsystems, especially the monitoring subsystem. Optimization of monitoring subsystems of distributed objects is to determine the optimal layout of sensors, their necessary and sufficient number, the frequency of their sampling in order to reduce information redundancy of measurements and, consequently, reduce the amount of information intended for storage and processing, reduce the requirements for computing hardware and communication channels. The role of the monitoring subsystem is especially important in food production control systems, in particular, in technological processes of storage and processing of grain.

The task of controlling the moisture and temperature of grain arises in the operational control of grain drying equipment in grain receiving and flour milling enterprises. Grain drying gives way to extending the storage period without the possibility of forming self heating center, as well as providing the optimum moisture in grain milling, which will result in the best processing results, the profitability of the flour mill and the competitiveness of the resulting flour in terms of the stability of its baking properties [1]. To prevent overdrying or overmoistening of grain and loss of valuable properties of the product, it is necessary to monitor the moisture and temperature of the grain in real time in order to maintain optimum parameters of the drying process [2,3].

At the present stage of the development of control systems in the flour milling industry, an accurate device is being developed that can measure the grain's moisture content. There are a large number of grain moisture express analyzers based on changing the physical characteristics of the grain when it changes its humidity (electrical conductivity, dielectric constant, absorption and reflection of infrared radiation, etc.), but they can not be used in automated control systems for measuring moisture grain in the flow. More precise today are methods such as measuring grain properties in high and ultrahigh frequency electromagnetic fields, but eventually they lose their accuracy through a series of uncontrolled factors: the change in the geometric dimensions of all components of the sensor with a change in its temperature, the deformation of components of sensitive elements, wear of their surfaces, degradation of the properties of materials used, moisture absorption in their micropores and microcracks.

The promising direction of creating a grain humidity sensor in the flow is the use of optical methods. At the same time, the actual task is to determine the number of sensors and their layout scheme, the frequency of polling of sensors to determine the static and dynamic characteristics of the drying equipment, to select and optimize the drying conditions, to solve the problems of control and process control.

2   Review of the Literature

Traditionally, to dry the grain, tower, modular and shaft driers are used [4,5]. But recently, conveyor dryers have become increasingly widespread. Their main advantage is the ability to dry the grain of high moisture (up to 30%) with a small layer height (from 15 to 35 cm) and a low grain speed (up to 1.2 m/min).

They do not require preliminary purification of grain from the impurity, and also removes from the dryer the remnants of the previous product when changing the kind of grain. They are highly productive and environmentally friendly [6]. The diagram of the technological process of grain drying is presented in Fig. 1. In this figure the following notation is used: 1 - plate conveyor; 2 - rummers; 3 - windows; 4 - grain; 5 - a window for the suction air; 6 - steam heater; 7 - centrifugal fan; 8 - distribution channel; 9 - suction fan; 10 - moisture sensors; 11 - temperature sensors; 12 - the speed sensor of the conveyor belt; 13 - sensor of the rupture of the conveyor belt; 14 - a grain level sensor.

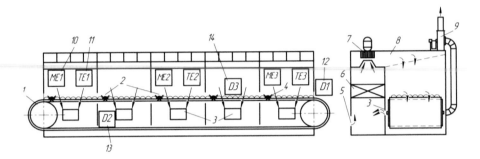

Fig. 1. Scheme of grain drying process using a conveyor dryer.

Each dryer is equipped with a control panel. The operator specifies drying parameters based on statistical data, using personal experience. The dryer is equipped with sensors of grain level, moisture and temperature of the grain, as well as sensors controlling the parameters of the conveyor. These measurements allow the operator of the information control system of the grain drying process to optimize the regime parameters of the process, such as the height of the layer, the speed of the conveyor movement and the temperature of the drying agent. A complete set of operational data is displayed on the monitor of the operator's workstation, so control and change of parameters can be performed directly by the operator from his workstation.

Due to the existence of uncertainties in the assessment of the state of the grain drying facility caused by stochastic changes in the external and internal conditions of the facility, the main task of the monitoring subsystem is to identify the state of the object in real time, which is used to synthesize control in the information control system. The calculation of optimal control is usually performed in specialized software at the dispatcher's workplace, using a Decision Support System (DSS) [7], a block diagram of which is shown in Fig. 2.

Modern systems for controlling grain drying equipment use automated systems for monitoring the moisture and temperature of grain, the development of which should be based on objective data on the state of grain at each point of the grain layer.

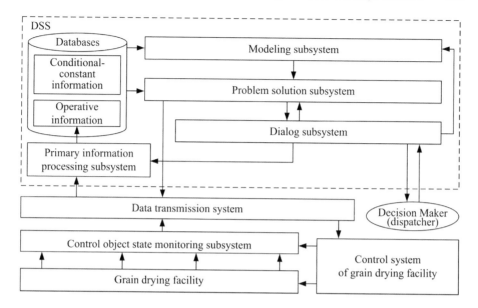

Fig. 2. Block diagram of grain drying facility information control system.

The task of the monitoring subsystem is to periodically measure all current parameters of the functioning of the grain drying facility, environmental parameters, add new information to the databases and take into account changes in the system parameters to enable further prognosis. Thus, the purpose of the monitoring subsystem is to provide the DSS with operational information about the state of the grain drying facility, as well as information about the operating mode of the executive equipment.

In order to avoid information redundancy and loss of time when carrying out unnecessary measurements, it is necessary to determine the scheme of optimal placement of sensors and the frequency of their sampling. For calculations, we can use the methods of spectral analysis of the dynamic model of moisture and grain temperature on the basis of the equations of heat and mass transfer [8] of the form:

$$\frac{\partial W}{\partial \tau} = \frac{\partial}{\partial x}\left(a_m \frac{\partial W}{\partial x} + a_m \delta \frac{\partial t_g}{\partial x}\right), \tag{1}$$

$$c\rho_0 \frac{\partial t_g}{\partial \tau} = \frac{\partial}{\partial x}\left(\gamma \frac{\partial t_g}{\partial x} + \varepsilon r \rho_0 \frac{\partial W}{\partial x}\right), \tag{2}$$

where W is the moisture, t_g is the temperature, ρ_0 is the density, c is the heat capacity, γ is the heat conductivity of grain, a_m is the coefficient of diffusion of moisture; δ is the coefficient of thermo gradient; ε is the criterion of phase transformation; x is the coordinate; τ is the time. To solve Eqs. (1), (2) the finite difference method [9] was used. The obtained model of the dynamics of temperature and grain moisture [10] on the basis of the equations of heat and mass transfer has the form

$$W(i, j+1) = a_m \frac{K}{h^2} [W(i+1, j) - 2W(i, j) + W(i-1, j)]$$
$$+ a_m \delta \frac{K}{h^2} [t_g(i+1, j) - 2t_g(i, j) + t_g(i-1, j)] + W(i, j), \quad (3)$$

$$t_g(i, j+1) = \frac{\gamma \cdot K}{c \cdot \rho_0 \cdot h^2} [t_g(i+1, j) - 2t_g(i, j) + t_g(i-1, j)]$$
$$+ \frac{\varepsilon \cdot r \cdot K}{c \cdot h^2} [W(i+1, j) - 2W(i, j) + W(i-1, j)] + t_g(i, j). \quad (4)$$

Equations (3) and (4) make possible to calculate the moisture and grain temperature at the time τ_{j+1} using the thermophysical and thermodynamic characteristics at the previous moment of time τ_j. Limit values are obtained on the basis of analysis of the structural and technological features of grain drying equipment.

3   Problem Statement

The aim of the research is to determine the scheme of optimal placement of moisture and grain temperature sensors as well as the period of their sampling as essential functional parameters of the monitoring subsystem of the information management system by drying the grain using the methods of spectral analysis of the dynamic model.

4   Materials and Methods

Spectral analysis methods can be used to determine the scheme of optimal placement and the period of sensor sampling. We will use a dynamic model of moisture and grain temperature, which allows analyzing these parameters directly at a given level of grain mass on the conveyor, taking into account the quantitative characteristics of the thermophysical and thermodynamic properties of the grain, which influence the process of heat and mass transfer.

One-dimensional spectral analysis allows us to determine the period of sensor sampling in time. The degree of dependence of the values of the analyzed parameter $x(t)$, namely moisture $W(t)$ or temperature $t_g(t)$, on a given level z of the grain layer, can be estimated in time using the autocorrelation function [11–13] as

$$K_t(\tau) = \frac{1}{(M-n)} \sum_{k=0}^{M-n-1} \frac{\overset{\circ}{x}(t_k) \cdot \overset{\circ}{x}(t_k + \tau)}{D_x}, \quad (5)$$

where $\tau = n\Delta t$, $n = 1, 2, ..., M$ is the number of samples in time.

The centered values of the parameter being analyzed can be found as

$$\overset{\circ}{x}(t_k) = x(t_k) - m_x, \tag{6}$$

where m_x is the mathematical expectation can be calculated as

$$m_x = \frac{1}{M} \sum_{k=0}^{M-1} x(t_k). \tag{7}$$

The dispersion D_x of the process $x(t)$ in (5) can be written as

$$D_x = \frac{1}{M-1} \sum_{k=0}^{M-1} [\overset{\circ}{x}(t_k)]^2. \tag{8}$$

Spectral analysis allows us to estimate the maximum frequencies of the spectrum of random processes, on the basis of which one can determine the discretization step according to the known Nyquist–Shannon–Kotelnikov sampling theorem [14]: $\Delta t \leq \pi/\omega_c$.

To find the cut-off frequency, the Parseval equality is traditionally used:

$$\frac{1}{\pi} \int_0^{\omega_c} [S(\omega)]^2 \, d\omega = \frac{\eta}{\pi} \int_0^{\infty} [S(\omega)]^2 \, d\omega, \tag{9}$$

where η is the coefficient characterizing the reproduction accuracy, $\eta = 0,95$; $S(\omega)$ is the spectral density of the amplitudes of the random process, which can be found according to the Fourier transform of the correlation function [15]:

$$S(\omega) = \frac{2}{\pi} \int_0^{\infty} K_t(\tau) \cos(\omega\tau) \, d\tau. \tag{10}$$

5 Experiment, Results and Discussion

To apply the method of spectral analysis, it is necessary to have information on the dynamics of moisture and grain temperature, depending on various initial conditions and parameters of the drying mode. Sample periods of grain moisture and temperature sensors were determined for three different drying modes at a drying agent temperature $T_{a1} = 100\,°C$, $T_{a2} = 105\,°C$, $T_{a3} = 110\,°C$, based on modelling the dynamics of grain moisture and temperature at initial humidity $W_0 = 20\%$ and initial grain temperature $t_{g1} = 10\,°C$, $t_{g2} = 15\,°C$, $t_{g3} = 20\,°C$, according to Eqs. (3) and (4). Fragments of the simulation results are shown in Figs. 3 and 4.

Examples of the obtained correlation dependencies corresponding to the second series of each experiment (middle graph of (a)–(b) Figs. 3 and 4) are shown in Fig. 5. To find the spectral density of the processes of change in moisture and temperature, the correlation dependencies must be represented in an analytical form, which can be obtained from the graphs using the least squares method.

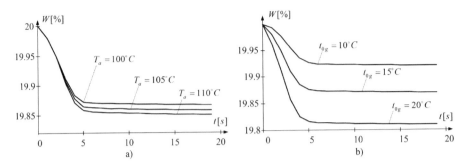

Fig. 3. Dynamics of grain moisture depending on the different initial conditions: (a) dynamics of grain moisture depending on the drying agent temperature at the initial grain temperature $t_{0g} = 15$ °C, (b) dynamics of grain moisture depending on the initial grain temperature at the drying agent temperature $T_a = 105$ °C

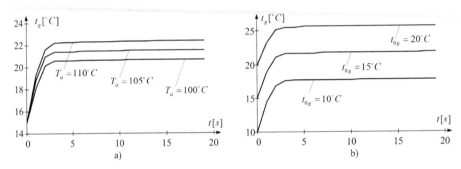

Fig. 4. Dynamics of grain temperature depending on the different initial conditions: (a) Dynamics of grain temperature depending on the drying agent temperature at the initial grain moisture $W_0 = 20\%$, (b) Dynamics of grain temperature depending on grain initial temperature at the drying agent temperature $T_a = 105$ °C

The results of the approximation, as well as the cut-off frequency and estimation of the discretization steps for the considered grain drying modes calculated on the basis of analytical correlation dependencies are given in Fig. 6.

To find the approximation accuracy of the correlation function set $K(\tau)$ one can use the standard error σ [16], calculated according to the formula

$$\sigma = \sqrt{\frac{\sum_{i=1}^{N} (K_i - K_a)^2}{\sum_{i=1}^{N} K_i^2}}, \qquad (11)$$

where K_i is the value calculated with use of (5), K_a is the approximated value.

Calculated values of the standard error σ are shown in Fig. 6 also. Analyzing the information in Fig. 6 gives important results:

(1) All calculated values of the standard errors σ do not exceed 0.05, which proves the possibility of using the proposed type of correlation functions to

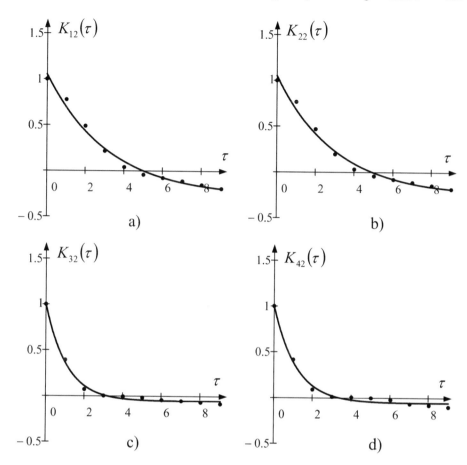

Fig. 5. The results of the correlation dependencies calculations for the three drying modes ($\bullet\bullet\bullet$ – calculated values, — — approximated values).

solve the problem of optimizing the system for monitoring the grain drying process using conveyor type grain drying equipment.

(2) Now we can optimize the sensor placement pattern for particular conveyor dryer model presented in Fig. 1, taking into account the design and technological features of conveyor (the conveyor length is $l = 30\,\text{m}$ and its speed is $v = 1.2\,\text{m/min}$) and the calculated periods of polling of the sensors. So, for this particular conveyor dryer it is advisable to choose the minimum value for the sensor sampling period – to measure grain temperature it will be $\Delta t_t \le 1.72\,\text{min}$ and for moisture it will be $\Delta t_m \le 5.95\,\text{min}$.

(3) When designing a monitoring system for a conveyor grain dryer with the specified design parameters, in order to effectively control the drying process, it is necessary and sufficient to foresee the presence of temperature sensors in the quantity of $l/(v\Delta t_t) = 14\,\text{pcs}$ and moisture sensors in the quantity of $l/(v\Delta t_m) = 4\,\text{pcs}$.

The results of the Fig. 3 (a) data processing by the method of spectral analysis		
$K_{11}(\tau) = -1.323e^{-0.005\tau} \times$ $\times\left(0.206 - e^{-0.321\tau}\right)$ $\sigma_{11} = 0.044$ $\omega_{c11} = 0.437\ \text{min}^{-1}$ $\Delta t_{11} \leq 7.192\ \text{min}$	$K_{12}(\tau) = -1.338e^{-0.005\tau} \times$ $\times\left(0.214 - e^{-0.309\tau}\right)$ $\sigma_{12} = 0.0442$ $\omega_{c12} = 0.414\ \text{min}^{-1}$ $\Delta t_{12} \leq 7.588\ \text{min}$	$K_{13}(\tau) = -1.343e^{-0.005\tau} \times$ $\times\left(0.215 - e^{-0.306\tau}\right)$ $\sigma_{13} = 0.0458$ $\omega_{c13} = 0.412\ \text{min}^{-1}$ $\Delta t_{13} \leq 7.633\ \text{min}$
The results of the Fig. 3 (b) data processing by the method of spectral analysis		
$K_{21}(\tau) = -1.383e^{-0.005\tau} \times$ $\times\left(0.213 - e^{-0.303\tau}\right)$ $\sigma_{21} = 0.05$ $\omega_{c21} = 0.413\ \text{min}^{-1}$ $\Delta t_{21} \leq 7.61\ \text{min}$	$K_{22}(\tau) = -1.316e^{-0.005\tau} \times$ $\times\left(0.202 - e^{-0.325\tau}\right)$ $\sigma_{22} = 0.0432$ $\omega_{c22} = 0.446\ \text{min}^{-1}$ $\Delta t_{22} \leq 7.038\ \text{min}$	$K_{23}(\tau) = -1.268e^{-0.005\tau} \times$ $\times\left(0.177 - e^{-0.365\tau}\right)$ $\sigma_{23} = 0.0404$ $\omega_{c23} = 0.528\ \text{min}^{-1}$ $\Delta t_{23} \leq 5.95\ \text{min}$
The results of the Fig. 4 (a) data processing by the method of spectral analysis		
$K_{31}(\tau) = -1.068e^{-0.005\tau} \times$ $\times\left(0.056 - e^{-0.884\tau}\right)$ $\sigma_{31} = 0.025$ $\omega_{c31} = 1.733\ \text{min}^{-1}$ $\Delta t_{31} \leq 1.813\ \text{min}$	$K_{32}(\tau) = -1.06e^{-0.005\tau} \times$ $\times\left(0.053 - e^{-0.919\tau}\right)$ $\sigma_{32} = 0.0225$ $\omega_{c32} = 1.825\ \text{min}^{-1}$ $\Delta t_{32} \leq 1.721\ \text{min}$	$K_{33}(\tau) = -1.061e^{-0.005\tau} \times$ $\times\left(0.053 - e^{-0.898\tau}\right)$ $\sigma_{33} = 0.0189$ $\omega_{c33} = 1.783\ \text{min}^{-1}$ $\Delta t_{33} \leq 1.762\ \text{min}$
The results of the Fig. 4 (b) data processing by the method of spectral analysis		
$K_{41}(\tau) = -1.066e^{-0.005\tau} \times$ $\times\left(0.056 - e^{-0.862\tau}\right)$ $\sigma_{41} = 0.0184$ $\omega_{c41} = 1.7\ \text{min}^{-1}$ $\Delta t_{41} \leq 1.849\ \text{min}$	$K_{42}(\tau) = -1.062e^{-0.005\tau} \times$ $\times\left(0.055 - e^{-0.885\tau}\right)$ $\sigma_{42} = 0.0298$ $\omega_{c42} = 1.698\ \text{min}^{-1}$ $\Delta t_{42} \leq 1.85\ \text{min}$	$K_{43}(\tau) = -1.071e^{-0.005\tau} \times$ $\times\left(0.063 - e^{-0.846\tau}\right)$ $\sigma_{43} = 0.0227$ $\omega_{c43} = 1.612\ \text{min}^{-1}$ $\Delta t_{32} \leq 1.949\ \text{min}$

Fig. 6. Calculation results of analytical correlation dependencies, cut-off frequencies and estimates of discretized steps.

(4) The proposed method for determining the structure of the monitoring subsystem of the grain drying facility information control system is universal and easily adapts to a specific type of grain drying equipment.

6   Conclusion

The use of an information control system to optimize the operational control of technological processes in modern geographically distributed systems reduces the subjectivity in assessing the current state of the controlled process and provides the ability to control such systems in an automatic mode.

The developed method for determining the optimal placement of sensors and the frequency of their sampling based on the method of one-dimensional spectral analysis allows to avoid information redundancy and simplify the technical implementation of the monitoring system, the effectiveness of which determines the efficiency of the entire information control system.

The proposed method of designing subsystems of monitoring, considered on an example of a technological process of drying grain using conveyor type grain drying equipment, is based on models of dynamics of temperature and moisture of grain. This model must take into account the main features of the design of grain drying equipment, the height of the grain layer and the quantitative characteristics of the thermophysical and thermodynamic properties of the grain.

References

1. Delcour JA, Hoseney RC (2010) Principles of cereal science and technology, 3rd edn. AACC International, St. Paul
2. Hurburgh CR, Bern CJ, Brumm TJ (2013) Managing grain after harvest. Iowa State University
3. Brooker D, Bakker-Arkema FW, Hall CW (1992) The drying and storage of grains and oilseeds. Springer, New York
4. Maier D, McNeill S, Hellevang K, Ambrose K, Ileleji K, Jones C, Purschwitz M (2017) Grain drying, handling, and storage handbook, 3rd edn. MidWest Plan Service, Iowa State University, Ames
5. Pabis S, Jayas DS, Cenkowski S (1998) Grain drying: theory and practice. Wiley-Blackwell, Hoboken
6. Nabok V (2012) The Ukrainian experience of conveyor dryers. "Zerno" J 11(80):154–163 (in Russian)
7. Rudakova GV, Polyvoda OV (2014) Optimization of monitoring subsystems of the distributed objects by methods of the spectral analysis. Syst Technol Regional Interuniversity Collection of Scientific Papers 2(91):98–106 (in Russian)
8. Ostapchuk NV (1977) Mathematical modeling of technological processes of storage and processing of grain. Kolos, Moscow (in Russian)
9. Samarskij AA, Nikolaev ES (1989) Numerical methods for grid equations. Birkhäuser Basel, Boston
10. Lytvynchuk DG, Polyvoda OV, Polyvoda VV, Gavrylenko VO (2018) Mathematical model of grain humidity and temperature dynamics in the drying process. "Bull KNTU" J N3(66):85–90 (in Ukrainian)
11. Wentzel ES (1969) Probability theory. Science, Moscow (in Russian)
12. Broersen PMT (2006) Automatic autocorrelation and spectral analysis. Springer, New York
13. Warner RM (1998) Spectral analysis of time-series data. Guilford Press, New York
14. Dmitriev VI (1989) Applied information theory. Vysshaya Shkola, Moscow (in Russian)
15. Marks RJ II (1991) Introduction to Shannon sampling and interpolation theory. Springer, New York
16. Bidyk PI, Baklan YI, Baklan MY, Korshevnyuk LO, Litvinenko VI, Minin, Petrenko VV, Petrenko O, Selin YM, Fefelov AA: Modeling and forecasting of nonlinear dynamic processes. EKMO, Kyiv (2004) (in Ukrainian)

Information-Entropy Model of Making Management Decisions in the Economic Development of the Enterprises

Marharyta Sharko$^{(\boxtimes)}$ ⓘ, Nataliya Gusarina ⓘ, and Nataliya Petrushenko ⓘ

Kherson National Technical University, Kherson, Ukraine
mvsharko@gmail.com

Abstract. The main problem of making managerial decisions in predicting the innovative development of the enterprises under the influence of the environment is to obtain the required volume of quality information, since its small number reduces the accuracy of forecasts, and the greater is associated with the difficulties of processing the production data. The methodological approach is substantiated and the information-entropy model of the quantitative estimation of necessary input information is proposed when making managerial decisions on the innovative development of enterprises caused by fluctuations of the external environment such as the difference between a priori and a posteriori information of the main factors characterizing production activity. An algorithm for eliminating uncertainty, structuring information needs, and means of providing information management as a strategic resource of the dynamics of the flow of production processes is created. An experimental verification of the representativeness of the input information is carried out and it provides the opportunity to remove uncertainty in the formation of the mechanism of innovation development of enterprises in order to achieve a steady state of production. Determining the quantitative assessment of the required input information provides the opportunity to reduce the amount of stocks to compensate the impact of the environment, which will provide a significant economic effect. Calculations of apriori and aposteriori information and the size of entropy will give us the possibility to regulate the process of accumulation of the required amount of information when making managerial decisions.

Keywords: Innovation development · Uncertainty · Management ·
Informational support · Apriori entropy · Quantitative estimations

1 Introduction

The complication of the tasks of enterprises operation on an innovative basis emergence of new goals, priorities and resource constraints today put research into theoretical and practical problems in the management of innovation development in the category of relevant. At the solution of tasks of management of

© Springer Nature Switzerland AG 2020
V. Lytvynenko et al. (Eds.): ISDMCI 2019, AISC 1020, pp. 304–314, 2020.
https://doi.org/10.1007/978-3-030-26474-1_22

enterprises innovative development in the conditions of uncertainty, one of the main problems is obtaining the necessary volume and quality of incoming information for prediction of financial and economic indicators when changing parameters and processes that occur under the influence of the external environment. The total amount of information, which is needed for analytical substantiation of the managerial decisions to increase the innovation activity of the enterprises, needs a quantitative assessment.

On the indicated problem reveals a constant interest in the regulation of production and development of applied tools of management of innovation activity, taking into account the functional characteristics of the production activity of enterprises.

2   Review of the Literature

The main problem solved by the methodology of management of economic development is the adaptability of the enterprises to unforeseen the influences of the environment. Production processes are to be considered multivariate, due to resource and financial constraints. The resources owned by the enterprise during its operation are not always fully used. As a rule, there are side processes such as consuming stocks of an enterprise without creating useful output [1–3]. Therefore, it is necessary to have some stock of resources the optimal value of which is determined by the uncertainty of the environmental impact [4,5]. Since the optimization of the uncertainty is an almost insoluble task, it is necessary to expand significantly the volume of the unused insurance reserves [6–9]. This fact causes significant economic losses associated with their achievement, storage, warehousing etc. Determining the quantitative impact of the environment will reduce the volume of these stocks while ensuring the reliability of the management decisions and provide significant economic benefits [10–13].

There are no general guidelines for choosing the most preferred solutions for multiplying existing management alternatives in a variety of informational situations and management actions for meeting the goals of achieving economic growth, so any steps in solving this problem are extremely useful.

The purpose of the article is to create a model of substantiation of the volume of information for making managerial decisions on the economic development of the enterprise.

3   Materials and Methods

An assessment of the state of the production facility and its dynamic change is a necessary element in the process of economic development of the enterprise determination [14,15]. The variety of the economic indicators does not allow us to determine unambiguously the trends in the development of the production facilities, because, in addition to quantitative estimates of these indicators, which are characterized by different scales of the measured values, there is also their diversification. It is necessary to create a common indicator, which takes into account the state of the production facility as a whole [16, 17].

Quantitative assessment of the true state of the production facility is the starting point for the diagnosing and forecasting changes in the state of the enterprise under uncertain dynamic impacts of the environment. Since there is no generalized indicator characterizing this state, so it is natural to use the device of synectics, that is, the borrowing of scientific achievements from other areas of knowledge. It is proposed to consider the value of entropy as a measure of the state of production. Entropy is a degree of uncertainty and incompleteness of knowledge about the specific state of the object. In physics, entropy is a spatial and energy interaction, which manifests itself as a measure of the probability of a system being in this state.

In sociology - entropy is a deterioration of self-organization.

In computer science - entropy represents a measure of uncertainty of random events. With the help of entropy, it is possible to decrypt the encoded text, analyzing the probability of the appearance of characters in the text.

In the management of the development of the economic system, the concept of entropy refers to the characteristics of the collection, transmission and processing of the information. The entropy of the production system is the generalizing indicator that characterizes the state of production at a given time.

It is proposed to use the concept of entropy in assessing the economic development of the enterprise as a characteristic of finding the system in this state [2]:

$$H = \frac{I}{n} = -\sum_{i=1}^{m} P_i log_2 P_i$$

where P_i is the prodability of finding a production system in this state; I – volume information; N is the number of the analyzed indicators.

For unambiguous determination of the unit of measurement of entropy it is necessary to specify the number of the states of the object m and the base of the logarithm. The least possible number of possible states of a production facility is characterized by its efficiency and can be estimated by the symbols "suitable" or "not suitable", or "yes" or "no", that is, this number is equal to two. Such a code with the base two is considered binary, since it corresponds to the state of acquisition of information or its absence, that is, $m = 2$. Therefore, as the basis of the logarithm, it is expedient to choose number two. Consequently, the unit of uncertainty is the entropy of equal probability states. This unit is called a *bit*. Turning to the assessment of information, per unit of information can be considered as the amount of information containing messages that reduce the uncertainty of knowledge twice. Comparison of these two studies of message units reveals their general character and manifests itself in removing the uncertainty of the input information or its sufficiency. Entropy is a valid and inseparable value, since for any i the value of P_i varies from 0 to 1.

Entropy H is related to the probability of a W system by the relation:

$$H = k \cdot lnW$$

where k is Bostzmann's conltant.

The entropy of the production system is a logarithmic measure of the inversion of the information source and characterizes the average degree of uncertainty in the assessment of the state of this source of information.

The entropy of the production system is a mathematical expectation of the logarithm of the probability of a system being in a given state m. On the basis of the law of large numbers with a large number of states of the production system as a whole, the arithmetic meaning of these states of one and the same value will acquire a stable value. This allows us to identify the general patterns of self-organization of the functioning of enterprises in the external disturbing actions caused by the fluctuations in the environmental impact. Thus, the principles of the feedback between the production and information system are used. With a small number of observations, this value becomes random. It stabilizes when the number of observations increases and approaches to the mathematical expectation. In this formulation, entropy is a mathematical proof of the states of the production system by partial observation of quantitative information.

The statistical analysis of the properties of the information sources consists in the certainty of its probability $P(a)$ as the ratio of the number of favorable results $N(a)$ to the total number of probable results N:

$$P(a) = \frac{N(a)}{N}$$

Mathematical expectation is one of the most important concepts of mathematical statistics and probability theory, which characterizes the distribution of the random variables. It is possible in the assessment of risk and is used to develop a strategy in the theory of games. Mathematical expectation $M[X]$ is the sum of the multiplication of all probable values of random variables x_i is the probability of these values of p_i. The average value of a discrete random variable is equal to:

$$M[X] = \frac{x_1 p_1 + x_2 p_2 + ... + x_n p_n}{p_1 + p_2 + ... + p_n} = \frac{\sum_{i=1}^{n} x_i p_i}{\sum_{i=1}^{n} p_i}$$

The state uncertainty of the production system depends on the number of probable states and the probability distribution of these states. The degree of state uncertainty of the production system of facilities or information source depends not only on the number of its probable states, but also on the probabilities of these states. The entropy of the source with two states u_1 and u_2 on change in the ratio of their probabilities $P(u_1) = P$ and $P(u_2) = 1 - P$ is determined by the equation:

$$H(u) = -[P log_2 P + (1 - P) log_2 P]$$

Practical interest is not the absolute value of entropy, but its change. At reduction of entropy, the incoming information increases and, conversely, at increase of entropy – decreases. Change in entropy is the main criterion for the effectiveness of innovation. Apriori and a posteriori information are interconnected concepts of informational discourse, which denote the knowledge of previous experience and knowledge that the system gained from the experience of its use [7,9,10]. The information structure of apriori data required for the calculation of the apriori entropy of the production facility under uncertainty should contain the information contained in the apriori information on the conditions of the choice of control alternatives, structural alternatives and diagnostic alternatives. Entropy and the amount of information are interconnected. Therefore, the necessary amount of information to make adequate managerial decisions to increase the innovation support of enterprises can be defined as the difference between apriori and aposteriori entropy. This provision is the basis of the proposed model of quantitative assessment of incoming information, when management decisions are made on the innovative development of enterprises under uncertainty. For events $x_1, x_2, ..., x_n$, which have equal probable state $1/n$, the value of apriori entropy is the following:

$$H_0 = -n \times \frac{1}{n} log_2 \frac{1}{n} = -log_2 1 + log_2 n$$

Given the fact that:

$$P_1 = \frac{N(x_1)}{N}, \qquad P_2 = \frac{N(x_2)}{N}, \qquad ... \qquad , P_n = \frac{N(x_n)}{N}$$

the formula of aposteriori entropy take the form:

$$H_1 = -\sum_{i=1}^{n} [P_i (log_2 N(x_i) - log_2 N)]$$

Calculation of the apriori distribution of states of the external business environment is carried out by the processing of statistical material or analytical methods based on the probabilistic representations.

Aposteriori information about the state of a production object is a residual uncertainty of the state of knowledge after additional messages obtaining. The entropy of the compounding of the statistical independent sources of information about the state of the production facility H equals the sum of these entropies:

$$H = \sum_{i=1}^{n} H_i$$

Establishing the probability of realization of the appearance of quantitative values of the analyzed indicators is one of the important stages of the methodology of management of the innovative provision of the economic development of the enterprise in terms of uncertainty, since the number of implementations depends on the amount of information transmitted.

4   Experiment, Results and Discussion

All economic indicators have different order of numbers and different dimensions. To bring them to dimensionless form and reduce the number of analyzed digits, one should either make their normalization, dividing it into some general index, or compute their relative changes in the form of indices.

The calculations of the apriori H_0 and the aposteriori H_1 entropy and the amount of information I at the given values of the probability of occurrence of the analyzed events of the environment influence on functioning of enterprise in the analysis of the three economic indicators: index of average monthly production t in % of the previous year x_1, index of average monthly output in constant prices x_2, index of average monthly wage pay one working in % to the previous year x_3. These sources of information are considered statistically independent. The data submitted for analysis of economic dynamics of production shall have the following features:

– the properties of economic system during its work remain the same, i.e. new capacities, equipment, technologies are not implemented;
– personnel competence is not improved;
– the following indexes are independent of the previous ones, they determine the actual state of the facility this year;
– the represented indexes are consistent, but not dependent.

In the case of equally probable manifestations of the concerned economic indicators x_1, x_2, x_3 in the general economic status of the production facility and the absence of any restrictions, the source of economic growth shall have the maximum information content. Input information on production activities of the business enterprise marine ship building company "Marine Group" representatives in Table 1.

Table 1. Input information on production activities of the business enterprise

Item	2012	2013	2014	2015	2016	2017
Index of average monthly production as a percentage over the previous year, x_1	1.008	1.010	1.022	1.085	1.167	1.647
Index of average monthly output in constant prices, x_2	1.002	1.031	0.916	0.852	0.814	1.119
Index of average monthly wage pay one working as a percentage over the previous year, x_3	1.043	1.051	1.243	1.070	1.340	1.178

In the case of $P_1 = P_2 = P_3$ and $i = 1, 2, 3$ apriori entropy is equal:

$$H_0 = -[\frac{1}{3}(log_2 1 - log_2 3) + \frac{1}{3}(log_2 1 - log_2 3) + \frac{1}{3}(log_2 1 - log_2 3)]$$
$$= -[log_2 1 - log_2 3] = -[0 - 1.58496] = 1.58496 [bit]$$

For values $P_1 = 0.7$, $P_2 = 0.2$, $P_3 = 0.1$, characterizing the possible degree of influence of the external environment on the operation of industrial objects, represented by indicators $x1$, $x2$, $x3$, the aposterior entropy is:

$$H_1 = -[\frac{7}{10}(log_2 7 - log_2 10) + \frac{1}{5}(log_2 1 - log_2 5) + \frac{1}{10}(log_2 1 - log_2 10)]$$
$$= -[(-0.360206) + (-0.464386) + (-0.332193)] = 1.156785[bit]$$

The substruction of these values characterizes entropy changes:

$$\triangle H = H_0 - H_1 = 1.58496 - 1.156785 = 0.428175[bit]$$

In order to determine the minimum amount of dynamic variables that uniquely describe the observed process of economic growth of production, the practical application of the suggested model is made on the basis of the extended data of the enterprise (Table 2). After representation of respective probability of indexes x_1, x_2, x_3, x_4 through P_1, P_2, P_3, P_4, we obtain the apriori entropy of the state of production in the initial year of the adjusted statistical series, with the equal probabilities of the state of the indicators the values are following:

$$H_0 = -[\frac{1}{4}(log_2 1 - log_2 4) + \frac{1}{4}(log_2 1 - log_2 4) + \frac{1}{4}(log_2 1 - log_2 4)$$
$$+\frac{1}{4}(log_2 1 - log_2 4)] = -[log_2 1 - log_2 4] = 2[bit]$$

Aposterior entropy of condition of the production activities of the business enterprise for the values $P_1 = 0.4$, $P_2 = 0.3$, $P_3 = 0.2$, $P_4 = 0.1$, characterizes the possible degree of influence of the external environment, determined on the basis of an expert evaluation is equal to:

$$H_1 = -[\frac{2}{5}(log_2 2 - log_2 5) + \frac{1}{5}(log_2 1 - log_2 5) + \frac{3}{10}(log_2 3 - log_2 10)$$
$$+\frac{1}{10}(log_2 1 - log_2 10)] = -[\frac{2}{5}(1 - 2.322) + \frac{1}{5}(0 - 2.322) + \frac{3}{10}(1.585 - 3.322)$$
$$+\frac{1}{10}(0 - 3.322)] = 1.9138[bit]$$

Table 2. Extended input information on production activities of the business enterprise

Item	2012	2013	2014	2015	2016	2017
Index of average monthly production as a percentage over the previous year, x_1	1.008	1.010	1.022	1.085	1.167	1.647
Index of average monthly output in constant prices, x_2	1.002	1.031	0.916	0.852	0.814	1.119
Index of average monthly wage pay one working as a percentage over the previous year, x_3	1.043	1.051	1.243	1.070	1.340	1.178
Index of average monthly real wages one working in % to the previous, x_4	1.003	1.005	1.114	0.949	1.096	1.049

The difference between these values characterizes entropy variation:

$$\triangle H = H_0 - H_1 = 2 - 1.9138 = 0.0862[bit]$$

The degree of uncertainty removal is the amount of missing information, $I = \triangle H$ (Fig. 1). The algorithm for finding the amount of necessary information in managing the economic development of enterprises under conditions of uncertainty (Fig. 2). If the acquired information eliminates the uncertainty in full information, i.e. information is equal to the eliminated entropy. Representativeness of information is connected with the accuracy of its selection and formation for proper reflection of the properties of the controlled facility.

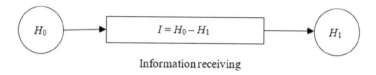

Information receiving

Fig. 1. Uncertainty removing with the informational support of the management system

As can be seen from the above, it was demonstrated that the amount of information that specifies the condition of the production facility at the time of its economic using its innovation provision is proportional to the amount of considered indicators and depends on the probability of their occurrence in general set of implementation of the state of production:

$$I = -n \sum_{i=1}^{m} P_i log_2 P_i$$

where n is a number of the analyzed indicators of production activities of the enterprise.

Information content I depends on the total number of elements of the production system n and the number of states for each element m:

$$I = -n \cdot log_2 m$$

Each system is aimed at an equilibrium state. Non-equilibrium processes in the isolated system are accompanied by increase in the entropy. The stationary state corresponds to the minimum of entropy. The entropy increases under influence of external conditions and due to interference of the equilibrium state of production, if there is no interference the entropy reaches the absolute minimum. If business is not effective then entropy can be characterized, for example, by its state as a difference in the low price in the market and the high cost of production.

In order to implement the proposed methods, it is necessary to perform the functional analysis of the operational procedures of management of innovation provision of economic growth in the context of uncertainty, which shall allow formalization of the management processes of economic growth of enterprises and emphasize basic operations of decision support: assessment of the required amount of information, information situation treatment, choice of an optimization criterion, performing calculations according to the proposed calculation methods.

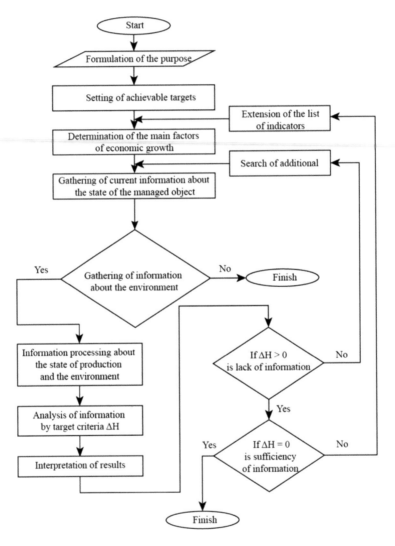

Fig. 2. Algorithm for establishing the required amount of information for making adequate managerial decisions

5 Conclusions

The proposed information and entropy model of the quantitative evaluation of the required input information and algorithms of its realization makes it possible to improve the main principles and rules of organization of information provision for the management process of economic growth of the enterprises. Calculations of apriori and aposteriori information and the entropy value are able to adjust the process of accumulation of the required amount of information when making managerial decisions.

Entropy is used to explain the unforeseen anomalies. Entropy represents the extent of disorganization and state uncertainty of the controlled facility. If the economic system is in order, its entropy is low. The entropy increases in the closed systems. The increase of entropy means the elimination of differences. Entropy is a quantitative measure of state uncertainty.

The entropy of the production system in the context of its dynamic changes due to the influence of the external environment is given as an asymptotic increase in uncertainty. Acquiring of information is a process of disclosing the state uncertainty of a production facility in the current time.

If the managed production system properly describes the total reaction to management actions, the management system takes into account all the main factors that characterize the production activities. If the management system does not take into account all the factors that characterize the management activity, the misrepresentation of the input information occurs and management actions fail to achieve the goal and lead to a number of unintended side effects.

References

1. Martinez-Costa M, Jimenez-Jimenez D, del Pilar Castro-del-Rosario Y (2019) The performance implications of the UNE 166.000 standardised innovation management system. Eur J Innov Manag 22(2):281–301. https://doi.org/10.1108/EJIM-02-2018-0028
2. Sharko MV, Panchenko JV (2014) Formation of the policy of building up intellectual potential. Actual Probl Econ 6(156):30–40
3. Brun EC (2016) Ambidexterity and Ambiguity: the link between ambiguity management and contextual ambidexterity in innovation. Int J Innov Technol Manag 13(04):1650013. https://doi.org/10.1142/S0219877016500139
4. Sharko MV, Zaitceva OI, Gusarina NV (2017) Providing of innovative activity and economic development of enterprise the condition of external environment dynamic changes. Naukovyy visnyk Polissya – Sci Bull Polissia 3(11):65–71 http://ir.stu.cn.ua/123456789/15244
5. Neirotti P, Pesce D (2019) ICT-based innovation and its competitive outcome: the role of information intensity. Eur J Innov Manag 22(2):383–404. https://doi.org/10.1108/EJIM-02-2018-0039
6. Popkov YS (2006) Entropy models of demo-economic dynamics. Trudyi Instituta Sistemnogo Analiza RAN 28:7–47
7. Sharko MV, Gusarina NV (2018) Business analytics ranking of indicators of economic information when making managerial decisions on innovation development of production. Mod Econ 10:146–151. https://doi.org/10.31521/modecon.V10(2018)-24

8. Teece DJ (2017) Towards a capability theory of (innovating) firms: implications for management and policy. Camb J Econ 41(3):693–720. https://doi.org/10.1093/cje/bew063

9. Kravchenko OA (2013) Ensuring the effectiveness and efficiency of the production activity of the enterprise. Econ Realities Time 3(8):29–35 http://economics.opu.ua/files/archive/2013/No3/29-35.pdf

10. Trabukchi D, Buganza T (2019) Innovations driven by data: switching views to large data. Eur J Innov Manag 22(1):23–40. https://doi.org/10.1108/EJIM-01-2018-0017

11. Cerne M, Bortoluzzi G (2019) Micro-foundations of innovation: employee silence, perceived time pressure, flow and innovative work behavior. Eur J Innov Manag 22(1):125–145. https://doi.org/10.1108/EJIM-01-2018-0013

12. Yegorov I, Ryzhkova Y (2018) Innovation policy and implementation of smart specialisation in Ukraine. Ekonomics prognozuvanna 3:48–64. https://doi.org/10.15407/eip2018.03.048

13. Sharko M, Gusarina N, Burenko J (2017) Modeling of management of the information potential of complex economic systems under conditions of risk. Technol Audit Prod Reserves 34:14–19. https://doi.org/10.15587/2312-8372.2017.98275

14. Tavassoli S (2018) The role of product innovation on export behavior of firms: is it innovation input or innovation output that matters? Eur J Innov Manag 21(2):294–314. https://doi.org/10.1108/EJIM-12-2016-0124

15. Vujovic A, Jovanovic J, Krivokapic Y, Pekovic S, Sokovic M, Kramar D (2017) The relationship between innovations and quality management system. Tech Gaz 24(2):551–556. https://doi.org/10.17559/TV-20150528100824

16. Sepúlveda J, Vasquez E (2014) Multicriteria analysis for improving the innovation capability in small and medium enterprises in emerging countries. Am J Ind Bus Manag 4:199–208. https://doi.org/10.4236/ajibm.2014.44027

17. Wang CL, Pervaiz KA (2004) The development and validation of the organisational innovativeness construct using confirmatory factor analysis. Eur J Innov Manag 7(4):303–313. https://doi.org/10.1108/14601060410565056

System Development for Video Stream Data Analyzing

Vasyl Lytvyn[1,2] (ID), Victoria Vysotska[1(✉)] (ID), Vladyslav Mykhailyshyn[1] (ID),
Antonii Rzheuskyi[1] (ID), and Sofiia Semianchuk[1] (ID)

[1] Lviv Polytechnic National University, Lviv, Ukraine
{Vasyl.V.Lytvyn,Victoria.A.Vysotska,antonii.v.rzheuskyi}@lpnu.ua,
vladyslavmykhailyshyn@gmail.com, sonja.semjanzyk@gmail.com
[2] Silesian University of Technology, Gliwice, Poland

Abstract. Information system for analyzing video stream data is developed, intended for conducting various types of video stream analysis. The optimal tools to develop the system are chosen, which allowed us to create a fast, reliable, optimized and user-friendly system with a comfortable web-based interface for users. The optimal hardware and software requirements for complete and uninterrupted operation of the product after development and testing of the system are identified. In addition, all the technical requirements and system parameters were described. Upon completion of the system, a control case analysis is performed, which showed that the system performs all the functions provided and works properly. Despite the fact that the system works correctly, and performs the task, there are still many ways to improve it. It is possible to create a solution that will use several such products on different servers at once to analyze one video stream, which will allow us to perform a better analysis, faster and as the result - to get more data.

Keywords: Video content · Video stream data · Objects analysis ·
Color analysis · Quantitative analysis · People recognition

1 Introduction

The rapid development of information technology in parallel with the increased use of visual forms of communication: digital storage and transmission of images, the availability of digital video technologies and their availability, the spread of visual surveillance technologies, and the transition from text to visual forms of communication, such as the use of such programs as Skype, FaceTime, etc. the approval of visualizations in their various forms is an integral part of modern culture and everyday life [1,2]. Taking into account the rapid pace of computerization of human life and business, the analysis of multimedia is becoming increasingly important, since images, sound, and video are the easiest ways [3] for machines to obtain environmental information in the same form as a person receives [4,5]. Obviously, for the correct processing of data, we need to use a

V. Lytvynenko et al. (Eds.): ISDMCI 2019, AISC 1020, pp. 315–331, 2020.
https://doi.org/10.1007/978-3-030-26474-1_23

variety of algorithms. They can be implemented as software on general-purpose computers, or as hardware in specialized video processing units. Data analysis depends on a good video input, so it's often combined with video quality technologies such as noise reduction, image stabilization, fuzzy masking and super-expansion.

Already, there is a huge number of systems that use these data to perform a significant amount of work without requiring human intervention [6]. Such systems include intelligent cars, security systems based on facial recognition, retina, special graphic codes and other visual identifiers, various visual control systems This analysis is used for entertainment, health care, retail, home automation, etc. [7].

Currently, most of these systems are either paid or not at all available for use by anyone other than the developer [8]. Some companies restrict access to their applications or allow it to be used only with certain hardware, which does not normally allow such systems to be fully used. Existing software is often imperfect, or simply can not be applied to the industry [9]. Some use only 1–2 formats of input data. Also, the form of returning results is often uncomfortable for future use. Most developments are usually created for a single specific purpose, so even minor changes to the method of use may cause completely invalid results. In addition, almost all solutions require significant hardware resources, and have a number of requirements for their installation. Thus, the urgent issue is the creation of a universal system that can handle all popular video data formats, analyze them correctly and present results in a clear way. It will allow users to customize easily the system for performing a specific tasks [10].

The purpose of the work is creation an information system for analyzing video stream data. To achieve the goal we have to solve the following tasks:

- To analyze of literary sources.
- To carry out system analysis of subject area.
- To choose methods and means of implementation.
- Software development.

The object of the study is the process of analyzing the video stream. The subject of the study is image optimization, noise filtering, video stream segmentation, object selection and grouping of results. The relevance of the work is the lack of universal system in the market which makes a multi-dimensional video analysis. So, the development of such software product is necessary and especially for development of the sphere of multimedia analytics.

2 Literature Review

Pre-processing an image, the recorded data must be processed - to limit the errors associated with geometry and adjust the brightness values of the pixels [11]. These errors are corrected with the help of appropriate mathematical models. Improving image quality is achieved by changing the brightness values of the pixels to improve its visual impact [12]. This includes a set of techniques that are used to improve the look of the image or transform the image into a form that is better suited for human or machine interpretation [13]. Sometimes the image obtained with traditional or digital cameras lacks contrast and brightness due to the limitation of the rendering of subsystems and lighting conditions during image capture. Images may have different types of noise [14]. When improving the image quality, the goal is to emphasize certain features of the image for further analysis or to display the image [15]. For example, it could be an increase in contrast, pseudo-coloring, noise filtering, sharpening and magnification [16]. Improving image quality is useful in feature selection, image analysis, and image display. The quality improvement process itself does not increase the content of the data in the data. It just emphasizes some specific image characteristics that are important in this situation. Improvement algorithms, as a rule, are interactive and depend on a particular system. Some of the following methods [17–21]:

1. *Stretching the contrast.* Some images (for example, reservoirs, deserts, dense forests, snow, clouds and foggy conditions) are homogeneous, that is, they have no sharp changes in the color levels. In terms of the presentation of the histogram, they are characterized by the appearance of very narrow peaks. This situation may also occur due to improper lighting of the scene. Ultimately, the images thus obtained are difficult to interpret. This is because there is only a narrow range of gray levels in the image that is used to provide a wider range of such levels. Methods of stretching contrast are designed exclusively for such situations. Different stretching methods have been designed to stretch the narrow area of image for entire available dynamic range.

2. *Noise filtering.* Noise filtering is used to filter out unwanted information from the image. It is also used to remove various types of noise from images. Basically, this feature is interactive. It uses a variety of filters, such as low pass, high pass, middle level, median, etc. [22].

3. *Histogram of modifications.* The histogram is important in improving image quality. It displays the characteristics of the image. By changing the histogram, the image characteristics can be changed. One such example is the alignment of histograms. The alignment of the histograms is nonlinear; the stretching redistributes the pixel values, so that within the range of the range approximately the same number of pixels with each value falls. The result approximates the flat histogram. Thus, the contrast increases in peaks and decreases on the histogram tails [23].

One of the key issues in image processing is segmentation. Image segmentation is a process that divides images into constituent parts or objects. The degree of segmentation may differ depending on the task being solved, that is, segmentation should stop when the objects of interest in the system have been allocated, for example, if we want to identify vehicles on the road, the first step is to segment the road from the image, and then segmentation content of the road, up to potential vehicles [24,25].

Threshold binary image is formed when all pixels of an object have one level of gray and all background pixels have another - usually object pixels are "black" and background "white". The best threshold is one that selects all object pixels and displays them as "black". Different approaches were proposed for automatic threshold selection. The image can be defined as a gray scale map in the binary set [26].

When a histogram has two pronounced maxima that reflect the levels of a gray object and background, we can select one threshold for the entire image. The method is based on this idea and uses the correlation criterion to select the best threshold as described below. Sometimes gray-level histograms have only one maximum. This can be caused, for example, due to the heterogeneous illumination of different areas of the image. In this case, it is not possible to select one value to specify the threshold of the entire image and we must apply a local binary method. But there are no common methods for solving binary problems in homogeneously illuminated images [27].

The results of video analytics are events (messages) that can be passed to the operator of the video surveillance system or recorded in video archives for further search. In addition, the video analyst creates metadata, that is, data structures that describe the content of each frame of the video sequence. Metadata contains information such as location and object identifiers (usually in the form of a disturbing framework), the trajectory and speed of objects, data on the division or merging of objects, data on the occurrence and end of anxiety situations (in security systems) The metadata is recorded in video archives and played with the video.

3   Problem Statement

The main advantage of video analysts before conventional CCTV systems is to automatically select metadata from streaming video without the participation of the operator. The resulting metadata can be used to quickly search for a video archive, send out alert notifications, and collect statistics, monitor events. etc. Compared to "manual video surveillance," video analytics can reduce the cost of video monitoring and the human factor in the detection and response time. Since a significant portion of video data (over 99%) in video surveillance systems has no value for users, video analytics can dramatically reduce the load on communication channels and the system of archiving due to the filtering of unnecessary video data. The main problem of many implementations of video analytics is the high frequency of false positives [18], which greatly reduces the economic effect of technology. The problem is gradually being solved by improving the video analysis algorithms, pre-adjustment of the systems, and a detailed check. Another problem is the significant cost of system integration and the introduction of video analytics. The role of this factor is reduced by the emergence of open standards such as ONVIF [19], simplification of calibration procedures and video analytics settings. Consequently, when developing the video data analysis system, it is necessary to consider the need to process video of different formats, and to implement mechanisms for improving the quality of the image, to improve the accuracy of the analysis. Different data processing algorithms are required.

It will be relevant to use the system for any, or several types of video analytics at once, since there are currently no such programs or there are significant limitations in the application. In addition, it is necessary to provide a convenient type of output analysis results that will be comfortable for any type of video analysis.

4 Materials and Methods

To analyze the purpose of functioning of the system, allocation of the main goals and objectives of the system it is necessary to build a tree of goals. To achieve the aim, we must perform four macro goals, that is, to create the main structural elements of the system, namely (Fig. 1):

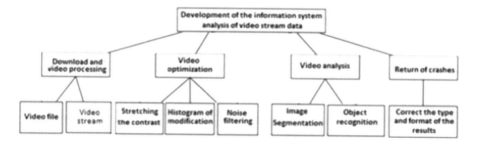

Fig. 1. Tree system goals

1. Uploading and processing video is this subsystem will upload a video file to the system or skip video stream while processing to achieve the format with which the system works. In her turn, there are two ways to download data:
   (a) Video file is uploads data in a finite video file format.
   (b) Video stream is downloading data in a continuous video stream format.
2. Video optimization is Video processing is here for better analysis, that is, throwing extra frames, highlighting key objects, simplifying images, and more. This subsystem in turn includes 3 different ways of optimization:
   (a) Stretching the contrast is providing wider range of gray shades used for analysis.
   (b) Histogram of modifications is redistributed the pixel value in such a way as to get the highest possible uniform image quality.
   (c) Noise filtering is using different filters to remove noise and unwanted information from the image.
3. Video Analysis is a subsystem that actually deals with object recognition on video, and generates master data and has two sub-targets:
   (a) Image Segment is selecting objects in an image, and separating segments with these objects.

(b) Object recognition is the system recognizes each object in the image and stores information about it.

4. Return of results is determining the required data format, which should return the system in accordance with the task and is directly transmitted to the user. Contains one sub-target. Correcting the type and format of the results - determines the best possible data format for the result of the analysis taking into account the results obtained, the type of task, and the subsequent use of the results by the user.

The system accepts input streams with video data, information about previous analyzes, processes them, and returns the results of the analysis that are transmitted to the user and the data obtained during the analysis, which are stored in the database, in order to improve the system in the future (Fig. 2). Now, for more detailed information about the processes occurring inside the system itself, it is necessary to construct a first level decomposition diagram. At this stage, data flows that occur between the main components of the system are presented. Let's perform decomposition for them (Figs. 3, 4, 5 and 6). After decomposition of each of the main processes of the system, it became possible to examine more in detail, and to analyze their subprocesses, as well as data flows that ensure the interaction of all processes and subprocesses, both inside and outside the system.

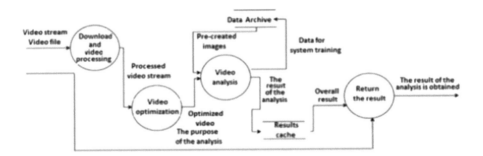

Fig. 2. Chart of the first level of decomposition

It is expedient to construct a hierarchical model to illustrate the structure of this information system, display the levels of the process hierarchy and subprocesses, and to facilitate understanding of the functional relationship between them. The main processes will be located on the first level of the hierarchy, and their subprocesses on the second. Such a diagram will make it possible to uniquely identify the importance of various processes in the system and their hierarchical location (Fig. 7).

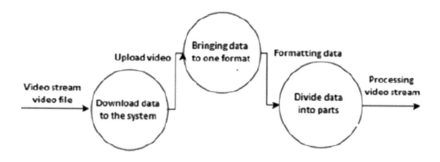

Fig. 3. Decomposition of the process of video uploading and processing

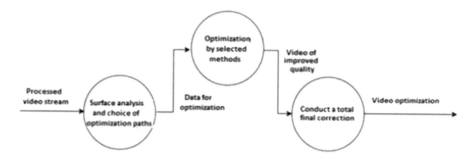

Fig. 4. Decomposition of video optimization process

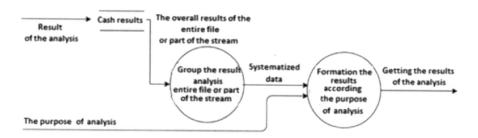

Fig. 5. Decomposition of return process

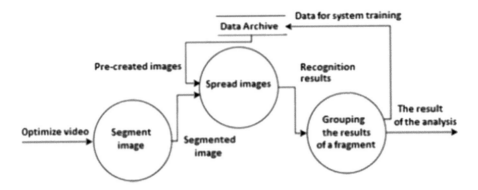

Fig. 6. Decomposition of video analysis process

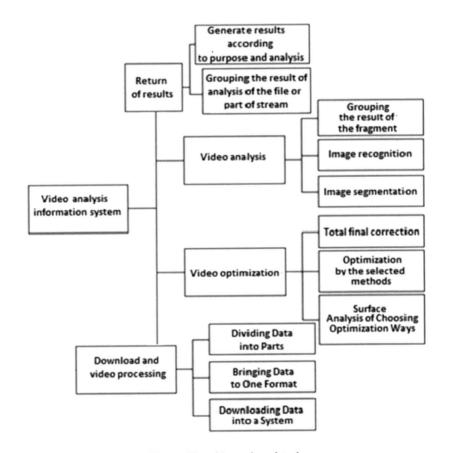

Fig. 7. Tree hierarchy of tasks

5   Experiment

We will select four criteria for choosing an alternative for optimal allocation of hardware resources between processes, this will be the speed of analysis, the accuracy of the analysis, the speed of optimization and its quality. The processes concerning uploading video and returning the results are constant and executed with the help of ready-made embedded technologies, therefore they can not influence the work of the system. We will select for analysis 5 alternatives - A, B, C, D, E, each of which will contain different indicators of the selected 4 criteria. Each alternative will have one criterion with a maximum value, but one will contain the mean for all criteria. Imagine data for analysis in the form of a table, in percentages (100% - the maximum indicator) (Table 1). We structure the task in the form of a hierarchy in accordance with a given target (Fig. 8).

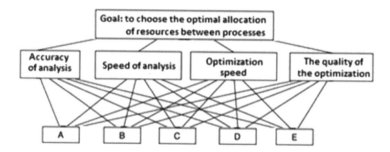

Fig. 8. Structuring a task in the form of a hierarchy

It is necessary to implement the procedure of pairwise comparison of criteria among themselves, and then alternatives. We construct a matrix of pairwise comparison of criteria. Depending on the importance of each process, fill out table of criteria (Table 2).

Table 1. Criteria indicators

	Analysis rate, %	Analysis accuracy, %	Optimization rate, %	Optimization quality, %
A	100	40	30	50
B	30	60	100	60
C	70	100	55	45
D	65	60	40	100
E	70	70	70	70

The obtained $IP = 0.06$ characterizes the admissible coherence of judgments, since the necessary condition for this condition $IP < 0.1$. We turn to comparisons

Table 2. Definition of criteria weights

Criteria	Accuracy of analysis	Speed of the analysis	Optimization speed	Optimization quality	Own vector	Weight criterion w_i
Accuracy of the analysis	1	3	5	9	3.41	0.56
Speed of the analysis	1/3	1	3	7	1.63	0.27
Optimization speed	1/5	1/3	1	5	0.76	0.13
Optimization quality	1/9	1/7	1/5	1	0.24	0.04

of alternatives according to the criterion of "accuracy of analysis". To do this, we create a matrix of pairwise comparisons (all procedures are identical to similar procedures when determining the weight of the criteria) (Table 3).

Table 3. Definition of criteria weights

	A	B	C	D	E	Vector	Weight	V_X
A	1	3	1/3	3	5	1.72	0.26	0.21
B	1/3	1	1/5	1	3	0.72	0.11	0.17
C	3	5	1	3	7	3.16	0.48	0.29
D	1/3	1	1/5	1	3	0.72	0.11	0.15
E	1/5	1/3	1/7	1/3	1	0.32	0.05	0.18

Alternative C dominates this indicator, its weight is 48%. Alternative E has the smallest value for this criterion is 5%. In a similar way, we find the "weight" of all alternatives for each of the selected criteria. After that, when all the "weights" of each element of the hierarchy are found, we can go to the last stage - determining the best alternative. At this stage, the MAI needs to realize the synthesis of the "weights" found in the previous stages of the alternatives that are analyzed for the weighting of each criterion within the given hierarchy. The formula for determining the total score is as follows: $V_X = w_1 V_{X1} + w_2 V_{X2} + w_3 V_{X3} + w_4 V_{X4}$, where w_i is the weight of the i-th criterion, and V_{Xj} is the importance of the alternative of X for the i-th criterion. The priority of the alternative C is the highest (it is 29%), respectively, this is an alternative solution - best in the analytical hierarchy method that was applied to this task with these criteria. So, with the help of three different methods of analysis, namely the objectives tree, context diagrams and the method of the analytical hierarchy, this system was considered and analyzed. Target has helped identify the main and secondary tasks of system design, and prioritize them. Using context diagrams, data streams that connect processes in the system were analyzed, their structure and purpose are shown. A diagram of the hierarchy of tasks was constructed, which shows all the relationships and dependencies of the processes and sub processes of the system. The method of the analytical hierarchy helped to determine the hierarchy and priorities of the internal processes of the system, to analyze the importance of these processes, and to determine the optimal way of distributing software and hardware resources between them.

6   The Created Software Description

The software for analyzing the video stream data has the meaning of "V-EYE" and the full name "Video Eye". The following software is required for operation of the program: any operating system (Linux, Windows, Mac OS); Java Virtual Machine; Tomcat Server version is no lower than 5.0; Web browser (Chrome, Firefox, IE version not lower than 6.0). The product is developed in the IntelliJ IDEA environment using the following programming languages:

– Java is the logical part and the backend of the system;
– HTML and CSS is visualization of the web interface;
– JavaScript is front-end part of the system.

The system is designed to analyze video data. It solves the following classes of problems:

– object recognition on video;
– video color recognition;
– recognition of people;
– quantitative video analysis.

Functional constraints in the analysis are:

– limited performance and number of analysis objects (in the case of large-scale and loaded image objects);
– lack of sound stream analysis (small percentage of information compared to image);
– limited spectrum of final results.

It is obvious, the first versions of the program will contain some limitations that can eventually be addressed in subsequent updates. The logical structure of the software product is built on a modular system, in which each module contains a code of tasks, combined by a common task of execution. The structure is represented by the following modules:

1. face contains two interfaces and their implementation classes for face analysis;
2. model contains classes for each type of analysis:
   – ObjectCounter is quantitative analysis;
   – RecognisedFace is face detection;
   – RecognizedObject is object recognition;
   – RecognizedColour is color recognition;

3. activity is a module for working with connected devices and libraries, which
   contains the following classes:
   – FDCamera2Activity is working with a webcam;
   – FDOpenCVActivity is working with the OpenCV library;
   – MainActivity is the main module for the module.

A fragment of the MainActivity class code:

```
public class MainActivity extends Activity implements View.OnClickListener {
@Override
protected void onCreate(@Nullable Bundle savedInstanceState) {
super.onCreate(savedInstanceState);
setContentView(R.lawet.main_activity);
findViewById(R.id.opencv).setOnClickListener(this);
findViewById(R.id.camera).setOnClickListener(this);
}
@Override
public void onClick(View v) {
switch (v.getId()) {
case R.id.opencv:
startActivity(new Intent(this, FDOpenCVActivity.class));
break;
case R.id.camera:
startActivity(new Intent(this, FDCamera2Activity.class));
break; }    }
```

As we can see, using this fragment, the logical hierarchical structure is clearly
observed - depending on the need, the main class can cause one or another
class to perform the work required at the moment. The system consists of two
parts - the logical server part (backend) and the web interface (front-end), the
connection between them is done using the architectural style of REST, i.e.,
by POST, GET queries. Since the software is designed in Java, which is cross-
platform, that is, it can be launched on any platform, any hardware device
with the following hardware parameters will fit into its operation: the processor
frequency is 3.4 GHz, the amount of RAM is 2 GB, The storage capacity of the
media is 60 GB and with an Internet connection with a speed of at least 10 MB/s.
To run an application on a server, we need to have any of the modern OS, JVM,
JRE, Apache HTTP Server and Tomcat Server installed on it. After running the
program on the server, it can be accessed by its internal and external addresses.
A user needs to use a modern web browser only to use the program. The system's
input is a video file of any format, or a live streaming of video from webcam
(at the user's choice. Output - Video analysis results are displayed side by side,
and contain information that was obtained as a result of system operation. This
document provides an instruction for using the video data analysis information
system and the application of its capabilities for quantitative analysis, analysis
of people, colors and objects. The main part provides data on how to use the
product, its capabilities, technical data, and requirements for the platform to
run the program. In addition, information and examples of necessary inputs and
outputs, work program performance checks, and the entire product as a whole are

given. The developed software product has the commercial name "V-EYE", and the full name "Video EYE". With its development, the following languages and development environments were used: Java (logical part and backend part with Spring framework, development environment of Intellij Idea); HTML and CSS (markup and graphic design of the web front-end interface part, Sublime Text 3 development environment); Javascript (logical front-end part with AngularJS2 framework, Sublime Text 3 development environment). The program is designed to analyze the video stream. In addition, the analysis can be done as a finite video file, and streaming video from a webcam. It can be used to search and identify people, search for objects, analyze objects, graphically analyze videos, define the subject and context of the video, etc. The program allows we to analyze videos in four parameters: quantitative analysis, human recognition, colour recognition and object analysis. The software product can handle such tasks as: recognize the number of objects on the video; analysis of objects on video; recognizing people on video; color recognition for video. The following methods are used for analysis:

- Friedman chain algorithm - to determine the boundaries of individual objects, and to allocate them to the image.
- Kenny's algorithm - search for boundaries of objects using gradients.
- K-means algorithm is the clustering of objects and combining them into homogeneous groups.

When the system is used, the analysis time is reduced by several orders of magnitude, because the system is capable of working with streaming video, that is, processing up to 60 frames per second, while a person will need approximately 10 s to perform the analysis, in addition, it will take some time to record the results. The software product will work on the server around the clock, and users will be able to use it at any convenient time. For the system, provision is made for the creation of backups, which will allow to quickly restoring the system performance in case of unforeseen circumstances. Additionally, a separate session is created for each user, which allows one user to make the system work for others in case of mistakes. The developed program can be used in the fields of security, for monitoring the territory, in industry, in various artistic projects, in road control systems, and so on. Limitations in application are the dependence of the accuracy of the analysis on the quality of the image, the limited performance for large format video and video with high frame rates and the rejection of sound data in the analysis. The software is recommended for analyzing large volumes of video files, or for analyzing large numbers of them, for commercial and research purposes, and in cases where it is necessary to reduce the use of human resources, or to accelerate the performance of analysis tasks. For the normal functioning of the system, the following requirements are required for the hardware device on which it will be located: RAM 2 GB; the processor frequency is 3.4 GHz; memory capacity - 60 GB; Internet access speed is 10 MB/s. These are the minimum requirements for an adequate system operation, the improvement of which will lead to faster system operation. It is also recommended to use solid-state SSD drives instead of conventional hard drives.

There is no direct need for peripherals, but Webcam needs to be used for web video camera analysis feature. The following software is required to operate the system: JVM; Apache HTTP Server; Tomcat Server; Any of the modern web browsers (Chrome, Mozilla Firefox, Safari, Edge). The software can be run on any of the current operating systems - Windows 7 or higher, Linux Gnome and above, OS X 10.6 and above. To use the ability to analyze video from a webcam, its drivers are required. For full operation of the system, we need to connect to the Internet at a speed of at least 10 MB/s.

7   Results Analysis and Discussion

To verify the correct operation of the system, we will conduct an analysis of an example of the application of all functionalities of the system. Open the home page of the web interface of the system. Press the "Analyze" button, and then go to the analysis page. There are two ways to upload video to the system - from the data carrier and through the webcam. When choosing a video upload method, a file selection. After the file is selected, its analysis will begin and one of the four methods of output can be selected - color and object analysis (Fig. 9), quantitative analysis and people recognition (Fig. 10).

Fig. 9. Colour and objects analysis

Fig. 10. Quantitative analysis

All necessary functions work and produce the correct results, thus it can be concluded that the analysis of the control sample has been successfully performed and the system is functioning properly. Consequently, when performing the work, a software product was designed and created, namely an information system for analyzing the video stream data. It was designed and thought out its logical structure and interconnections between parts of the system. System modules were developed and tested, after which they were assembled into a holistic system, according to the hierarchical structure of the product and the interconnections between them were established. In the development and testing of the system, the hardware and software required for functioning was determined, the basic requirements for them were determined. After completion of the development, a control case analysis was carried out, which showed that the system is fully functional and performs all the tasks set before it.

8   Conclusions

The information system for analyzing video stream data was developed, which is intended for conducting various types of video stream analysis. According to the trends of technology development, the pace of video content generation, both for people and work, for which the video is the main way of obtaining information from external environment, independently, without human help, is constantly increasing. This usually leads to increase in demand for systems for working with video data. After analyzing literary and online sources, it was discovered that no references to analogue systems exist. That fact makes this system quite relevant, since in the absence of competitors, it will cover the entire analysis services area that it offers. The nearest analogues - systems for analyzing graphic images were analyzed, which allowed us to highlight their weaknesses and explore perspective solutions. This made it possible to implement the best of them in the system being created and prevent similar problems. In addition, a systematic analysis of the software product was conducted, which allowed to make detailed consideration and design of its structure, modules and their interconnection. In addition, a hierarchy of tasks was constructed, in accordance with the degrees of the importance of processes. Using the method of the analytical hierarchy, an optimal alternative was identified for allocating resources among the main processes in the operating system. The optimal tools for development the system were chosen, which allowed us to create a fast, reliable, optimized and user-friendly system with a comfortable web-based interface for users. After the development and testing of the system, the optimal hardware and software requirements for complete and uninterrupted operation of the product were identified. In addition, for the convenience of users, all the technical requirements and system parameters were described. Upon completion of the system, a control case analysis was performed, which showed that the system performs all the functions provided and works properly. Despite the fact that the system works correctly, and performs the task, there are still many ways to improve it. For example, adding the analysis of the audio channels of the video stream, despite the fact that their share in the total amount of information is up to 10%, analysis of these channels, can increase the accuracy of the analysis, as well as increase the amount of information received by the system as a result of the analysis.

In addition, it is possible to create a solution that will use several such products on different servers at once to analyze one video stream, which will allow we to perform a better analysis, faster and as a result - to get more data. In general, we can conclude that developed system functions properly, perform all the foreseen tasks, while working quickly and uninterruptedly. Since it does not have analogues that perform the same tasks, its development is undoubtedly relevant and necessary. However, in spite of the lack of competitors, similar systems can be designed and developed soon, requiring further improvement of the system and increase of number of tasks that it can solve.

References

1. Nascimento G, Ribeiro M, Cerf L, Cesário N, Kaytoue M, Raïssi C, Meira W (2014) Modeling and analyzing the video game live-streaming community. In: Latin American web congress, pp 1–9
2. Lypak H, Rzheuskyi A, Kunanets N, Pasichnyk V (2018) Formation of a consolidated information resource by means of cloud technologies. In: 2018 international scientific-practical conference on problems of infocommunications science and technology, PIC S and T
3. Rzheuskyi A, Kunanets N, Stakhiv M (2018) Recommendation system: virtual reference. In: 13th international scientific and technical conference on computer sciences and information technologies (CSIT), vol 1, pp 203–206
4. Kaminskyi R, Kunanets N, Rzheuskyi A (2018) Mathematical support for statistical research based on informational technologies. In: CEUR workshop proceedings, vol 2105, pp 449–452
5. Obermaier J, Hutle M (2016) Analyzing the security and privacy of cloud-based video surveillance systems. In: Proceedings of the 2nd ACM international workshop on IoT privacy, trust, and security, pp 22–28
6. Xu D, Wang R, Shi YQ (2014) Data hiding in encrypted H. 264/AVC video streams by codeword substitution. IEEE Trans Inf Forensics Secur 9(4):596–606
7. Saxena M, Sharan U, Fahmy S (2008) Analyzing video services in web 2.0: a global perspective. In: Proceedings of the 18th international workshop on network and operating systems support for digital audio and video, pp 39–44
8. Brône G, Oben B, Goedemé T (2011) Towards a more effective method for analyzing mobile eye-tracking data: integrating gaze data with object recognition algorithms. In: Proceedings of the 1st international workshop on pervasive eye tracking & mobile eye-based interaction, pp 53–56
9. Reibman AR, Sen S, Van der Merwe J (2005) Analyzing the spatial quality of internet streaming video. In: Proceedings of international workshop on video processing and quality metrics for consumer electronics (2005)
10. Perniss P (2015) Collecting and analyzing sign language data: video requirements and use of annotation software. In: Research methods in sign language studies, pp 56–73
11. Tran BQ (2014) U.S. Patent No. 8,849,659. U.S. Patent and Trademark Office, Washington, DC
12. Badawy W, Gomaa H (2015) U.S. Patent No. 9,014,429. U.S. Patent and Trademark Office, Washington, DC
13. Badawy W, Gomaa H (2014) U.S. Patent No. 8,630,497. U.S. Patent and Trademark Office, Washington, DC

14. Golan O, Dudovich B, Daliyot S, Horovitz I, Kiro S (2014) U.S. Patent No. 8,885,047. U.S. Patent and Trademark Office, Washington, DC
15. Chambers CA, Gagvani N, Robertson P, Shepro HE (2012) U.S. Patent No. 8,204,273. U.S. Patent and Trademark Office, Washington, DC
16. Maes SH (2011) U.S. Patent No. 7,917,612. U.S. Patent and Trademark Office, Washington, DC
17. Peleshko D, Ivanov Y, Sharov B, Izonin I, Borzov Y (2016) Design and implementation of visitors queue density analysis and registration method for retail video-surveillance purposes. In: Data stream mining & processing (DSMP), pp 159–162
18. Maksymiv O, Rak T, Peleshko D (2017) Video-based flame detection using LBP-based descriptor: influences of classifiers variety on detection efficiency. Int J Intell Syst Appl 9(2):42–48
19. Rusyn B, Lutsyk O, Lysak O, Lukeniuk A, Pohreliuk L (2016) Lossless image compression in the remote sensing applications. In: International conference on data stream mining & processing (DSMP), pp 195–198
20. Kravets P (2010) The control agent with fuzzy logic. In: Perspective technologies and methods in MEMS design, MEMSTECH 2010, pp 40–41
21. Babichev S, Taif MA, Lytvynenko V, Osypenko V (2017) Criterial analysis of gene expression sequences to create the objective clustering inductive technology. In: Proceedings of the 2017 IEEE 37th international conference on electronics and nanotechnology, ELNANO 2017, Article no. 7939756, pp 244–248
22. Nazarkevych M, Klyujnyk I, Nazarkevych H (2018) Investigation the Ateb-Gabor filter in biometric security systems. In: Data stream mining & processing, pp 580–583
23. Lytvyn V, Sharonova N, Hamon T, Vysotska V, Grabar N, Kowalska-Styczen A (2018) Computational linguistics and intelligent systems. In: CEUR workshop proceedings, vol 2136
24. Vysotska V, Fernandes VB, Emmerich M (2018) Web content support method in electronic business systems. In: CEUR workshop proceedings, vol 2136, pp 20–41
25. Lytvyn V, Peleshchak I, Vysotska V, Peleshchak R (2018) Satellite spectral information recognition based on the synthesis of modified dynamic neural networks and holographic data processing techniques. In: International scientific and technical conference on computer sciences and information technologies (CSIT), pp 330–334
26. Su J, Sachenko A, Lytvyn V, Vysotska V, Dosyn D (2018) Model of touristic information resources integration according to user needs. In: International scientific and technical conference on computer sciences and information technologies, pp 113–116
27. Vysotska V, Hasko R, Kuchkovskiy V (2015) Process analysis in electronic content commerce system. In: International scientific and technical conference computer sciences and information technologies (CSIT), pp 120–123

Methods and Means of Web Content Personalization for Commercial Information Products Distribution

Andriy Demchuk[1]🆔, Vasyl Lytvyn[1,2]🆔, Victoria Vysotska[1(✉)]🆔,
and Marianna Dilai[1]🆔

[1] Lviv Polytechnic National University, Lviv, Ukraine
andriydemchuk@gmail.com, {vasyl.v.lytvyn,victoria.a.vysotska}@lpnu.ua,
mariannadilai@gmail.com
[2] Silesian University of Technology, Gliwice, Poland

Abstract. In this paper we address the problem of designing an information system for methods and means of commercial distribution of information products using a personalized approach to visitors based on categories and tags for visitors that have interest to these information products. The designed system is the methods and means of reorganization in the online store, with the core of the automatic recommendation of tags (categories) in the form of a neural network with controlled learning that provides the intelligence of the system as a whole. The developed system has classes and subclasses to which real information products with logical links between them and intellectual analysis of the content. The system of commercial distribution of information products in the future will be able to bring real income to its owner, which will be in demand among users of the World Wide Web.

Keywords: Web resource · Web content · SEO technology ·
Web technology · Content monitoring · Content personalization ·
Content distribution · Neural network · Machine learning

1 Introduction

This topic is relevant for e-business, which has grown rapidly and has become an integral part of every business that targets a large number of consumers [1]. E-business is the conduct of business processes on the Internet [2]. These electronic business processes include the purchase and sale of goods, supplies and services, customer service, payment processing, production management, cooperation with business partners, information exchange, work automated services, recruiting and more [3]. An e-business can include a number of functions and services ranging from the development of the Internet services [4]. Today, most corporations constantly rethink their business from the point of view of the Internet, namely its accessibility, broad reach and ever-changing opportunities [5]. Almost everyone has the opportunity to create their own business on

V. Lytvynenko et al. (Eds.): ISDMCI 2019, AISC 1020, pp. 332–347, 2020.
https://doi.org/10.1007/978-3-030-26474-1_24

the Internet; you just need to select the appropriate field [6]. The most famous example is an online store, that is, a web resource that contains a catalogue of products with the ability to order and pay for those or other goods that would be ordered by a buyer who is an e-commerce product [7]. The world of e-commerce is interesting, and it is constantly changing and evolving to meet the needs of consumers [8]. Although it is great for consumers over the Internet, it can be a problem for business owners who often do not have time or resources to keep up with the latest improvements to system upgrades, upgrade platforms or software [9]. The e-commerce website, which provides a user-friendly experience, including the ability to quickly find the right stuff, is more in favors of a competitive advantage [10]. With the increased demand for these services, the need for improved quality offers is increasing, and this tendency will always be a part of any business, which emphasizes the relevance of this work [11]. The purpose of the intellectual system of the Internet commerce is to provide unique content based on the approach of personalization and the use of tags. As a rule, each user has different requirements for information for the request. But typical search engines return the same result for one query submitted by different users. Web-personalization is used to solve the problem of information overload and provide users with relevant information. Personalizing the Internet increases search engine accuracy, simplifies the search process, saves time and provides relevant information to users. Personalization creates a sense of individuality and uniqueness. Customers feel special and important as the company pays special attention to them. Moreover, by segmenting and targeting different buyers, personalization meets the different needs of each client, thereby optimizing client experience as well as the same average experience for everyone.

2 Literature Review

Over the past few years, the effectiveness of site search and navigation in the online retailer and e-commerce site has been increasing [1–3]. This is an important part of a successful trading strategy [4,5]. Companies tend to increase their personalization investment content on site as they recognize the benefits that this effective technology can bring to their business [6–9]. The keywords that consumers enter in the real-time search box can give the retailer a deeper understanding of the behaviour of their customers and provide the company with invaluable data that should be explored and used [10]. Used effectively, this information is invaluable as it can increase sales and improve customer retention [11]. People who use personalization on the site are more likely to buy compared to those who use standard navigation [12]. When people make purchases with intent, that is, they know what they want and if they can quickly and easily find it, they are more likely to make the purchase [13]. That is if the user immediately presents the novelties in the categories that interest him/her and they will be located in the blocks of content that can be seen on the main page, the chances of attracting attention will increase sharply [14]. Consumers are often at the last stage of the purchase because they are collecting information; they want to view the content of the product to meet the purchase criteria, such as price, availability, size, and shipping cost [15–20]. These consumers have entered the online store and are trying to find something specific in accordance with their category of product and range of interest. The search box comes with the expectation of providing relevant results and, accordingly, promotes search as a desirable choice. The expected result is a continuous path to the detailed product information page.

A personalized approach to the user's site leads to a higher sales rate. To put it simply, customers who cannot find the information they are looking for will ultimately leave the site unsatisfied and will never browse this site again, given the extremely competitive online market. A recent study revealed that up to 40% of visitors typically use a site search box, and each of these users shows the intention to purchase products by entering a name or code [4]. The site's as well as the search pages are the best places to insert blocks with recommendations. For these reasons, obtaining the desired search results is simply necessary for successful sales – this is where artificial intelligence helps. Sellers now turn to machine learning to improve the recommendations for consumers who use the site. Machine science can improve search results every time a user visits the site. It can also generate a search ranking that allows the site to sort results by relevance or by their estimated relevance. This estimate may take into account specific search terms, as well as features of a specific user profile (such as age range, previous purchases, and previous search terms). In a word, personalization algorithms allow you to associate users with a list of products that they are most likely to be interested in, and can also predict what customers might want to see, even if they do not yet know about it. In addition to the simple text entry of categories and tags, image-based and product descriptions become an increasingly viable option for adding automation and decision-making systems. Scientific advances in context recognition through deep neural networks now provide technology for automatically adding tags to product descriptions of e-commerce sites [22–33]. Furthermore, these methods can be used to classify facial expressions and recognize emotions [21].

A typical neural network has from a few dozens to hundreds, thousands, or even millions of artificial neurons, which are called units that are arranged in series of layers, each connecting to a layer on both sides [21] (Fig. 1).

In the above neural network, there are three neurons in the input layer and two neurons in the output layer. The number of neurons in the input and output levels does not change. The number of elements in the input and output templates for a particular neural network can never change. In order to use the neural network, you must express your problem in such a way that the array of numbers is floating point. Similarly, the problem should be an array with floating point numbers. This is, in fact, all that neural networks can do for you. They take one array and turn it into another. Neural networks do not call subprograms or perform any other tasks with traditional programming, neural networks recognize templates. Typing text in the neural network is particularly difficult. There are some problems that you have to deal with. Individual words have different

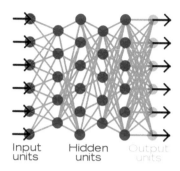

Fig. 1. View of the neural network

lengths. Neural networks require fixed input and output sizes. Should the letters "A"–"Z" be stored in one neuron, 26 neurons, or in another way together? The use of a recurrent neural network, such as the Elman neural network, solves part of this problem. In the last section, we do not need the size of the entrance window with the Elman neural network. We will create an Elman neural network, which has only sufficient input neurons to recognize Latin letters, and it will use the context layer to memorize the order. Text processing will use one long stream of letters. Each entry represents the number of one particular word. The entire input vector will contain one value for each unique word. Consider the following lines (Fig. 2).

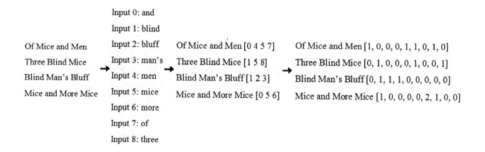

Fig. 2. Example of dictionary for neural network

We have certain unique words. This is our dictionary. Four lines above are encoded as follows. Of course, we must fill in the missing words with zero. Note that we now have a constant vector length of nine. Nine is the total number of words in our "dictionary". Each component number in the vector is an index in our dictionary of available words. Each vector component keeps counting the number of words for this vocabulary. Each line usually contains only a small subset of the dictionary. As a result, most vector values will be zero. As you can see, one of the most difficult aspects of machine learning programming translates your problem into an array of fixed numbers with floating point numbers. If you set the perfect result, you use so called "controlled learning". If you have not achieved the ideal results, you will use uncontrolled training. Supervisory training teaches the neural network to produce the ideal output. Uncontrolled training usually teaches a neural network to group incoming data into several groups defined by the initial number of neurons. Supervised and uncontrolled training is an iterative process. Concerning guided training, the each iteration of the workout calculates how close the actual result is to the ideal output. This proximity is expressed as a percentage of error. The each iteration modifies the internal matrices of the neural network weight to get the level of error at a rather low level.

3   Problem Formulation

Consumers visiting the site are not the first to be recognized by the system by the primary criteria that allow the system to provide a set of goods, information that is considered most relevant to that user. However, the introduction of tags and categories is expensive both in time and in efforts, as evidenced by the large flow of research. The cost of categorizing and adding tags not only affects how much the consumer is on the site before finding the right information, but also whether the successful marketing is ultimate. Long and unsuccessful searches can have a negative impact on e-commerce because consumers can leave a website without buying anything because they simply have not found it [1–5].

In general, there are two ways to improve personalization. The first is to provide a more relevant set of content. The second is to classify the content or to show it before in the process of rolling the pages so that consumers can avoid the selection of incorrect content or avoid spending time scrolling to the end to find the right content. Indeed, the best results from adding tags and categories are an important issue, as recent research has shown that they have a significant impact on consumer behavior and enterprise profits. At first glance, this seems to be a trivial problem – considering the totality of goods, simply placing them in descending order is of great importance so that consumers receive the corresponding goods at a minimal price. However, the difficulty is to determine the correct match for each document for each individual user. If custom preferences are similar, then we have a generally correct sequence of selected documents for this query and the problem of sorting is simpler. Instead, if custom preferences are different, that is, if two users use the same query to search for different data, the problem becomes much more complicated. A user searching for Java can search for coffee, find information about the Java programming language, or leave on the Java islands. The relevant values of the documents depend on a particular user. For example, a coffee shop is related to a person looking for coffee, but not for those who are looking for recreational tours. The optimal order must be individual. The potential solution to this problem is the personalization of the sequence of content. Indeed, all major search engines have experimented and in some cases have added their own customized content selection algorithms, even to users who first visit the site. Some researchers have even questioned the value of personalizing search results, not just in the context of search engines, but in other contexts. In fact, this discussion of personalization search results is part of a larger discussion of the results of personalization in the context of marketing mixed variables. Given the confidentiality, consumers and administrators should understand that storing long user histories helps to identify situations where personalization helps [1–9, 27–33].

In this paper, we are considering the personalization of content by sorting, that is, optimal individual sorting of the sequence of documents for obtaining the desired results. Our goal is to build a framework for displaying personalized content and personalization in general. In the process, we seek to determine the quantitative role of users and the level of requests for personalization. This is a complex problem for three reasons. Firstly, we need a structure with an extremely high accuracy of forecasting. The problem with personalized content in the system is one of the predictions – which result or document will be the most optimal for the user? Organize the list of selected results so that the top of the page shows the best options for each user. Traditional econometric tools used in marketing literature are not able to answer this question since they are designed for a causal conclusion for obtaining the best objective estimates, and not for prediction. Therefore, they perform weak forecasts. Objective assessments are not always the best indicators since they tend to have high dispersion, and vice versa. Thus, the problems of prediction and causation are fundamentally different and often contradictory. So, we need a methodological basis optimized for prediction accuracy. Second, we must include a large number of attributes and allow complex interactions between them. The standard approach to modeling in the marketing literature is to adopt a fixed functional form for the simulation of the output variable, include a small set of explanatory variables, allow multiple interactions between them, and then define the parameters associated with predefined functional forms. However, this approach has poor predictive power. Thirdly, our approach must be effective and scalable. Efficiency refers to the time of execution and deployment of a model that should be as short as possible with minimal loss of accuracy. Thus, an efficient model may be less accurate but it will bring enormous benefits during use. Scalability and efficiency are needed to run the model in real time. To solve these problems, we are turning to methods of machine learning, initiated by computer scientists. They are specially designed to provide extremely high precision, work with a large number of data attributes, and efficient scaling.

4   Materials and Methods

An internet client uses a website to make purchases online (Fig. 3). The most common use cases are a product review, shopping, and customer registration. The product review unit can be used by the client if the customer wants to find and see some products. This unit can also be used as part of the purchase order block. A customer registration unit allows a customer to register on a website, for example, to receive discounts or closed sales invitations. The settlement block is part of the purchase. In addition to the client's Internet, there are several other actors described hereunder with detailed use cases. The product review block includes adding product search tags and expanding with several additional blocks – the customer can search for products, browse the list of products, view the recommendations, add products to the basket or wish list. All these blocks extend the usability, as they provide some additional features that allow the customer to find the product. The search tag addition block is expanded by auto-generating the recommended tags and includes product creation.

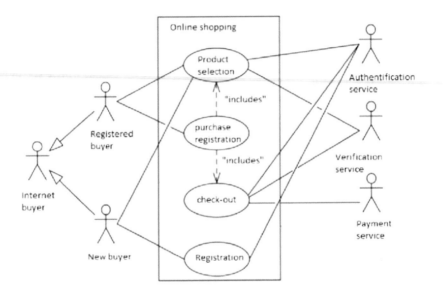

Fig. 3. UML Diagram for Online Store

The customer authentication unit includes a recommended product review block and added to desires one, as both require customer authentication. At the same time, the product can be added to the shopping cart without verifying the user's authenticity. To address the tasks described above, an e-commerce website needs to develop an intelligent decision-making system that provides an expanded product description, when adding content to the reader of each new product in the e-commerce application, and will provide the appropriate recommended categories and tags using synonymous variations. The synonym row will be determined by means of a neural network and a "list of algorithms". An example of application will be implemented based on Net CMS Sitecore, which has the personalization tools available to expand and add their own development based on the available core-functionality (Fig. 4).

The home page includes blocks of typed content: videos, courses, books, articles published by the personalization unit, which includes different versions of the above content, and thus includes the "altered content on the homepage" block. Moreover, the block of personalized content is expanded by the block of saved tags to the site, because on the basis of these tags there are the rules specified in the personalization. The personalization block is also expanded by pointing blocks of categories and subcategories that are also stored in the site CMS and therefore associated with the block "saved tags in the site" (Fig. 5).

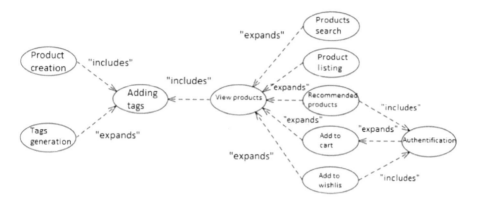

Fig. 4. Using an online store

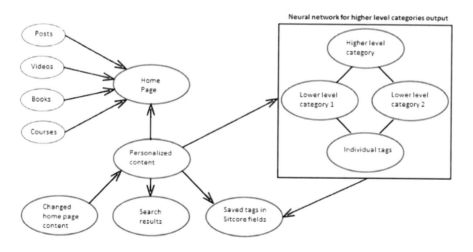

Fig. 5. Personalize content

In another diagram, we present a product add-on unit that expands the description of this product. In turn, the product description is expanded by the block of "auto-generated tags". The purpose of the entity-link diagram is to introduce some common data for the online store – the client, web user, account, cart, product, categories, tags, and the relationship between them. Each customer has a unique identifier and is associated with exactly one account. A customer can register as a web user to be able to buy goods on the Internet. The web user has a login name, which is also a unique identifier. A web user can be at several states – new, active, temporarily blocked or banned, and be linked to the shopping cart. The customer may not have orders or history of previous searches or page views. Customer orders are sorted and unique (Fig. 6).

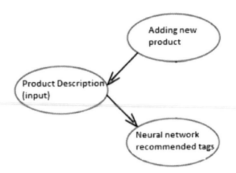

Fig. 6. Adding a product

One cart can contain many products. The products have tags that mark them, as well as the categories to which they relate. Each product can be classified into one category and has many tags (Fig. 7).

5   Experiment

The Tagger class reads the file vocabulary as a resource flow (if, for example, you put lexicon.txt in the same JAR file that compiles Tagger and Tokenizer classes) or as a local file. Each line in the lexicon.txt file passes through the utility method parseLine, which handles the input string, using the first token in the line as a hash key, and places the rest of the tokens in the array, which is a hash value. So, we will process the string "fair JJ NN RB" as a hash key "fair", and the hash value is an array of rows (currently only the first value is used, but we save other values for future use) (Fig. 8):

When the tag handles the list of word tokens, each token accesses the hash table and stores the first possible tag type for that word. In our example, the word "fair" will be assigned (possibly temporarily) the "JJ" tag. We now have a list of word tokens and a linked list of possible tag types. We will consider in detail the first rule: i is the cycle variable in the range [0, the number of tokens of the word is 1], and the word is the current word in index i. In English, this rule states that if the determinant (DT) in the word token of the index $i - 1$ is accompanied by the last time the verb (VBD) or the current verb (VBP),

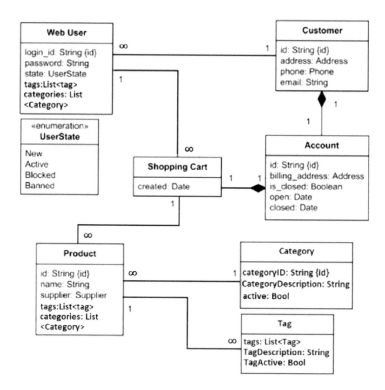

Fig. 7. ER Chart for Online Store

Tag	Description	Examples
NN	singular noun	dog
NNS	plural noun	dogs
NNP	singular proper noun	California
NNPS	plural proper noun	Watsons
CC	conjunction	and, but, or
CD	cardinal number	one, two
DT	determiner	the, some
IN	preposition	of, in, by
JJ	adjective	large, small, green
JJR	comparative adjective	bigger
JJS	superlative adjective	biggest
PP	proper pronoun	I, he, you
RB	adverb	slowly
RBR	comparative adverb	slowest
RP	particle	up, off
VB	verb	eat
VBN	past participle verb	eaten
VBG	gerund verb	eating
VBZ	present verb	eats
WP	wh* pronoun	who, what
WDT	wh* determiner	which, that

Fig. 8. The most commonly used tags

then the type of the tag changes to the index i for "NN". I list the remaining seven rules in the short syntax here rules:

- 2: convert the noun to the number (CD) if "." appears in the word;
- 3: convert the noun to a valid timestamp in the past, if words.get (i) ends with "ed";
- 4: convert any type into the adverb, if it ends with "ly";
- 5: convert the noun (NN or NNS) to the adjective, if it ends with "al";
- 6: convert the noun to the verb if the preceding word is "would";
- 7: if the word has been classified as a total, and ends with the symbol "s", the type is set to the plural common noun (NNS);
- 8: converting the noun into the active verb (for example, gerund).

6   The Results Analysis and Discussion

The maturity of Sitecore, as Content Management System allows us to do without creating a separate database with all the necessary links. And it is because Sitecore already contains 3 databases of its own: webmaster, the core, which stores all the necessary information for the functioning of web solutions (applications). In Sitecore each item and its field, which can be an administrator or content editor, is seen and kept in the above database as an abstract unit (item), which has its own ID and some general field. One of these fields is another abstract unit that will be a template for a new one. In essence, this is an imitation implementation. Sitecore supports a multi- inheritance example of working with interfaces in. Net. When creating a new template, you can specify which other templates this will impose. This is one of the elements needed to solicit or eliminate the need to create a separate database for web-based solutions. Another important element in the functioning of Sitecore is some types of data that allow you to realize relationships to one another and one to many. In any case, using Sitecore nothing restricts the web solution. For high-level tasks, we can use the available methods and tools at Sitecore. For tasks of a lower level – we can make an update to the bases in which Content Management system stores all data (Fig. 9).

```
SELECT TOP 1000 [Id]
      ,[ItemId]
      ,[Language]
      ,[Version]
      ,[FieldId]
      ,[Value]
      ,[Created]
      ,[Updated]
  FROM [Sitecore_pubs_master_v72].[dbo].[VersionedFields]
```

Fig. 9. SQL query to derive the relationship of fields and units in Sitecore

Thus, one of our diagrams of the relationship of entities will take the following form in realization. In the tree of abstract units, we will create an instance of the product.

And after a significant implementation of the server code for the homepage, which can be found in <Base.cs> and <Page.cs>, we get the following page with personalized blocks. The path analyzer displays a digital track (or sequence) of users, displaying it in an easy-to-use, visual way. These visual paths represent vital concepts for the purpose of identifying opportunities. If, for example, many visitors look at certain pages, which go for clicks, not conversion, it may mean, that something is not working smoothly. Analyzers to give you an idea about what you need to start create rules by yourself or Follow Sitecore offers. Profiles creation opportunities are theoretically unlimited. After using the Chrome browser as one unique client, and purchasing activity on the site, we will review the results of the user profile. Summarizing the details of the user profile:

- Surfing behavior: new visitors vs. returning visitors, websites visited, etc.
- Purchasing behavior: popular brands, price ranges, related products, etc.
- Geographic data
- Demographics: age and gender
- Sociological data: family situation, profession, interests, etc.

In general, you have decided to define these profiles; it depends heavily on the sector. You should always ask yourself the question, which relates to your customers, which is functionally possible and practically possible. In a practical example, the above is as follows: let us give you an example of a short article for learning the neural network (Fig. 10).

Meeting Modern Data Protection Requirements

How Business Suite Helps You Comply with the Latest Data Protection Regulations
August 24, 2017

As the volume of data collected by organizations continues to increase, so too do regulations designed to protect data from misuse, particularly when it comes to personal data. One of these is the European General Data Protection Regulation (GDPR), which goes into full effect on May 25th, 2018, and has global implications — it applies to any company that processes the personal data of people in the EU, whether or not that company is physically located within the EU. Learn how basic technical features and security safeguards included with SAP Business Suite applications help you comply with key areas of the GDPR data protection legislation and avoid the risk of steep fines due to violations.

Modern business systems are a treasure trove of highly sensitive information, such as the names, contact information, and various financial and health details for an organization's current and former employees and family members, as well as valuable information about business partners, shareholders, and customers. As the volume and types of data collected continue to increase through smart devices, social media, and other technologies, so too have laws and regulations designed to protect this data from misuse.

Fig. 10. An example of an article that is presented as input data for the study of the neural network AutoRecommendTags.exe

Let us launch the AutoRecommendTags console to teach the neural network to get the source data in the form of a table of proximity to the list specified in the tag system for content (Fig. 11).

```
C:\Windows\System32\cmd.exe
* =================== Tags recommendations report =
* model = [W52/WSJTP.TRAIN.W5]
* testset = [/home/me/WSJTP/WSJTP.TEST]
*
* =================== TAGGING SUMMARY ===================
#WORDS = 129654
#KNOWN = 126005 / 129654 --> (97.1856 %)
#UNKNOWN = 3649 / 129654 --> (2.8144 %)
#AMBIGUOUS = 45779 / 129654 --> (35.3086 %)
* =============== ACCURACY PER LEVEL OF AMBIGUITY =====
POS HITS TRIALS ACCURACY MFT-ACCURACY
*
# 15 / 15 = 100 % 100 %
$ 943 / 943 = 100 % 100 %
££ 1044 / 1045 = 99.9043 % 99.0431 %
( 186 / 186 = 100 % 100 %
) 187 / 187 = 100 % 100 %
, 6876 / 6876 = 100 % 100 %
. 5381 / 5381 = 100 % 100 %
: 752 / 752 = 100 % 100 %
CC 3237 / 3250 = 99.6 % 99.5692 %
CD 4789 / 4823 = 99.295 % 90.6075 %
DT 11117 / 11183 = 99.4098 % 98.453 %
EX 126 / 126 = 100 % 100 %
FW 7 / 30 = 23.3333 % 20 %
IN 13322 / 13492 = 98.74 % 98.3398 %
JJ 7617 / 8215 = 92.7206 % 85.2708 %
JJR 388 / 423 = 91.7258 % 95.9811 %
JJS 262 / 267 = 98.1273 % 95.5056 %
LS 10 / 15 = 66.6667 % 0 %
MD 1264 / 1267 = 99.7632 % 99.8421 %
NN 17257 / 17834 = 96.7646 % 91.9143 %
NNP 12717 / 13177 = 96.5091 % 85.0118 %
NNPS 98 / 170 = 57.6471 % 49.4118 %
* =============== OVERALL ACCURACY ===================
*
125966 / 129654 = 97.1555 % 91.3015 %
*
C:\Storage\Library\projects\Naz_mag\new>
```

Fig. 11. Intermediate Neural Network Learning Outcomes

Let us evaluate the source data (Fig. 12):

```
# single output {
  'governance': 0.00004324968926091062,
  'risk': 0.0077025285780033991,
  'compliance': 0.0002575132225946431,
  'risk management',
  'data management': 0.2071775132225946431,
  'big data': 0.008160047807935744,
  'administration': 0.00015069427192724994 }
```

Fig. 12. Output data of neural network learning according to input data

7   Conclusions

In this paper we have considered the problem of designing an information system for methods and means of commercial distribution of information products, using a personalized approach to visitors based on categories and tags interesting for visitors of information products. The designed system is the methods and means of re-organization in the online store, with the core of the automatic recommendation of tags (categories) in the form of a neural network with controlled learning that provides the intelligence of the system as a whole. The developed system has classes and subclasses to which real information products with logical links between them and intellectual analysis of the content. The systems of commercial distribution of information products in the future will be able to bring real income to its owners and will be in demand among the users of the World Wide Web.

Providing a user-friendly site is a key issue because online stores can help customers find the things they are looking for in a more versatile way. This allows visitors to manage their own buying experience, which helps to increase customers' loyalty and makes them more inclined to return to the site for more purchases, which, in turn, greatly facilitates trade. Artificial intelligence technologies will provide customers with better services and individual impressions. They also maximize the marketing efforts of the company, minimizing the need to spend money on ineffective advertising campaigns. It should also be noted that the topic of the Internet commerce in the context of e-business is more than ever relevant nowadays, in the time of rapid development of information technology. It is already very popular to order copyrighted information products. Therefore, those, who understand this trend in the market of commerce in general and are able to successfully fit in, will receive serious dividends.

References

1. Mobasher B (2007) Data mining for web personalization. In: The adaptive web. Springer, pp 90–135
2. Dinucă C, Ciobanu D (2012) Web content mining. Ann Univ Petroşani Econ 12:85–92
3. Xu G, Zhang Y, Li L (2011) Web content mining. Springer, pp 71–87
4. Khribi MK, Jemni M, Nasraoui O (2008) Automatic recommendations for e-learning personalization based on web usage mining techniques and information retrieval. In: International conference on advanced learning technologies, pp 241–245
5. Ferretti S, Mirri S, Prandi C, Salomoni P (2016) Automatic web content personalization through reinforcement learning. J Syst Softw 121:157–169
6. Lavie T, Sela M, Oppenheim I, Inbar O, Meyer J (2010) User attitudes towards news content personalization. Int. J Hum-Comput Stud 68(8):483–495
7. Fredrikson M, Livshits B (2011) Repriv: re-imagining content personalization and in-browser privacy. In: Symposium on security and privacy, pp 131–146
8. Chang CC, Chen PL, Chiu FR, Chen YK (2009) Application of neural networks and Kano's method to content recommendation in web personalization. Expert Syst Appl 36(3):5310–5316

9. Partovi H, Brathwaite R, Davis A, McCue M, Porter B, Giannandrea J, Li Z (2009) U.S. Patent No. 7,571,226. Washington, DC: U.S. Patent and Trademark Office
10. Kane FJ, Hicks C (2009) U.S. Patent Application No. 11/966,817
11. Mirri S, Prandi C, Salomoni P (2013) Experiential adaptation to provide user-centered web content personalization. In: Proceedings of IARIA conference on advances in human oriented and personalized mechanisms, technologies, and services (CENTRIC2013), pp 31–36
12. Fernandez-Luque L, Karlsen R, Bonander J (2011) Review of extracting information from the social web for health personalization. J Med Internet Res 13:e15
13. Hauser E (2011) U.S. Patent No. 8,019,777. Washington, DC: U.S. Patent and Trademark Office
14. Ho SY, Bodoff D, Tam KY (2011) Timing of adaptive web personalization and its effects on online consumer behavior. Inf Syst Res 22(3):660–679
15. Uchyigit G, Ma MY (2008) Personalization techniques and recommender systems, vol 70. World Scientific, London
16. Kothari N, Harder M, Howard R, Sanabria A, Schackow S (2006) U.S. Patent Application No. 10/857,724
17. Zhang H, Song Y, Song HT (2007) Construction of ontology-based user model for web personalization. In: International conference on user modeling. Springer, Berlin, pp 67–76
18. Chien H (2012) U.S. Patent No. 8,254,892. Washington, DC: U.S. Patent and Trademark Office
19. Linden GD, Smith BR, Zada NK (2011) U.S. Patent No. 7,970,664. Washington, DC: U.S. Patent and Trademark Office
20. Mehtaa P, Parekh B, Modi K, Solanki P (2012) Web personalization using web mining: concept and research issue. Int J Inf Educ Technol 2(5):510
21. Lytvyn V, Sharonova N, Hamon T, Vysotska V, Grabar N, Kowalska-Styczen A (2018) Computational linguistics and intelligent systems. In: CEUR workshop proceedings, vol 2136
22. Vysotska V, Fernandes VB, Emmerich M (2018) Web content support method in electronic business systems. In: CEUR workshop proceedings, vol 2136, pp 20–41
23. Kanishcheva O, Vysotska V, Chyrun L, Gozhyj A (2018) Method of integration and content management of the information resources network. In: Advances in intelligent systems and computing, vol 689. Springer, pp 204–216
24. Korobchinsky M, Vysotska V, Chyrun L, Chyrun L (2017) Peculiarities of content forming and analysis in internet newspaper covering music news. In: Proceedings of the international conference CSIT computer science and information technologies, pp 52–57
25. Naum O, Chyrun L, Kanishcheva O, Vysotska V (2017) Intellectual system design for content formation. In: Proceedings of the international conference CSIT computer science and information technologies, pp 131–138
26. Vysotska V, Lytvyn V, Burov Y, Gozhyj A, Makara S (2018) The consolidated information web-resource about pharmacy networks in city. In: CEUR workshop proceedings (computational linguistics and intelligent systems), vol 2255, pp 239–255
27. Vysotska V, Hasko R, Kuchkovskiy V (2015) Process analysis in electronic content commerce system. In: 2015 Xth international scientific and technical conference computer sciences and information technologies (CSIT), pp 120–123
28. Lytvyn V, Vysotska V (2015) Designing architecture of electronic content commerce system. In: Proceedings of the X-th international conference CSIT 2015 computer science and information technologies, pp 115–119

29. Zhezhnych P, Markiv O (2018) Linguistic comparison quality evaluation of web-site content with tourism documentation objects. In: Advances in intelligent systems and computing, vol 689, pp 656–667
30. Basyuk T (2015) The main reasons of attendance falling of internet resource. In: Proceedings of the X-th international conference computer science and information technologies, CSIT 2015, pp 91–93
31. Gozhyj A, Chyrun L, Kowalska-Styczen A, Lozynska O (2018) Uniform method of operative content management in web systems. In: CEUR workshop proceedings (computational linguistics and intelligent systems, vol 2136, pp 62–77
32. Kravets P (2010) The control agent with fuzzy logic. In: Perspective technologies and methods in MEMS design, MEMSTECH 2010, pp 40–41
33. Davydov M, Lozynska O (2017) Linguistic models of assistive computer technologies for cognition and communication. In: Proceedings of the international conferecne CSIT computer science and information technologies, pp 171–175

Automated Monitoring of Changes in Web Resources

Victoria Vysotska[1]([✉])[iD], Yevhen Burov[1][iD], Vasyl Lytvyn[1,2][iD], and Oleg Oleshek[1][iD]

[1] Lviv Polytechnic National University, Lviv, Ukraine
{victoria.a.vysotska,yevhen.v.burov,vasyl.v.lytvyn}@lpnu.ua,
oleshek95@gmail.com
[2] Silesian University of Technology, Gliwice, Poland

Abstract. The system providing the automated monitoring of changes in web resources was designed and developed. The entire process is divided into five parts. Each subsequent stage included some parts of the previous one. The practical value of this system resides in the ability to detect, in the shortest possible time, the changes in web-resource. In case, if unwanted changes were detected, the user will be able to perform the correction without the need for constant monitoring of content by developers. This facilitates the process of restoring the satisfactory state of web site after unauthorized changes were made. It also allows to track the competitors' web pages in order to timely improve their own products.

Keywords: Web-resource · Content · SEO-technology · Information technology · Web-technology · Changes monitoring

1 Introduction

Everyone knows that information technology is developing at a rapid pace [1]. However, not everyone is aware that the IT industry's development entails not only the dynamic development of technologies in related fields, but also all in other areas of human activity [2]. Wherever possible, it is advisable to introduce the automation, which increases productivity and improves the quality of manufactured products [3]. This in turn creates a demand for innovation, which further stimulates the development and implementation of intelligent technologies [4]. At the time of these dynamic changes, every conscious member of society should be aware of the latest developments and innovations in interesting or similar areas of life [5]. The need for consistently receiving the fresh information about the world requires a modern person to master and use the new methods of data collection [6]. It is not enough to read newly published textbooks or magazines today to keep pace with the innovations of the modern world, because printing these papers in place where they can get readers will require an incredible amount of effort and resources [7]. It is because of the effectiveness on achieving its main goal - the spread of knowledge- the Internet has become so popular [8].

Today, most people cannot imagine being without it [9]. It has rapidly burst into our lives and has long ceased to be merely entertainment, but has become a necessity from which people are reluctant to refuse [10]. Almost immediately after the opportunity to share files on the global web arose the problem of the aesthetic appearance of these files and the process of providing it - web design [11]. The want to be interesting for the maximum number of people hungry for knowledge and to promote themselves before the competitors is struggle, which governs the appearance, structure and content of Internet pages [12]. Making a dry text document "candy" is certainly good, but much effort is also needed to maintain the once achieved result [13]. You can come up with sophisticated data processing processes, with lots of checks to prevent the penetration of unwanted content to the site, but they cannot give you the one hundred percent protection of your Internet resource, and this protection works only temporarily. Therefore, if it is impossible to avoid harm from malware, you need to minimize it. In order to reduce such damage, the information system of automated monitoring of changes in web resources will be applied [14].

2   Literature Review

Working with a Subversion product is not very different from the working with the rest of the centralized version control systems. Users copy files from the repository, create working copies for themselves, make changes to these copies, and then synchronize their own files with reference in the repository, describing the changes made. System provides the possibility of simultaneous access of several users to the repository. Typically, the copying model - change - merge is used to provide this capability. In addition, it is possible to block a file by the user during its editing, which prevents simultaneous change in the code by different people. In modern versions of the preservation of changes in documents, the delta-compression mechanism is used: that is, there are differences between the old and the new versions of the system, and only these changes are written into the database, which provides a mechanism for the effective use of free disk space [6]. Among the main features of the system presented are [15–17]:

1. In the centralized repository, the history of changes to all files in the desktop is stored, even taking into account documents updates such as moving, changing attributes, and editing the owner's date.
2. There is a mechanism for transferring changes between duplicate files.
3. Emulation procedure:
   (a) New branches are created by copying the old ones and by adding fresh changes.
   (b) The combination of branches includes automatic finding of changes and their transfer.

4. All copies and changes in objects are saved as cheap (interconnected) documents that do not require large amounts of temporary and permanent resources of the system. Changes, no matter how large and global they are, would not be dropped in the repository in the atomic transactions presented.
5. To minimize the network load between the server and the client, only the differences between the work copy and the original are sent.
6. There is no significant difference in performance between text and binary files.
7. Ability to choose between two storage formats.
8. A set of regular files. International message support.
9. Possibility of mirroring the repository.

As mentioned above, the CVS system we used as the basis of the SVN. We just a not copied the core and the graphical shell, but extended the capabilities of the original product, getting rid of the defects of the predecessor and maximally optimizing useful functions [18]. Below are some basic innovations:

– Improved mechanism for working with binary and text files.
– Extends the revision control area by tracking changes not only in files but also in directories.
– Each file or directory has a set of properties that are also subject to version control.
– The process of data synchronization transfers only changes made by the user, which provides a reduced load on network traffic.
– Added ability to block files over which work goes.

The possibility of branching is an extremely useful feature of the version control system, because the usual methods of tracking changes already include the possibility of creating independent branches. Subversion supports a wide branching and merging mechanism see in Fig. 1 [1–5].

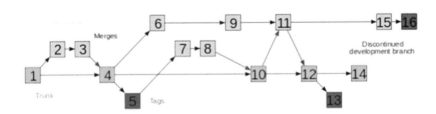

Fig. 1. Example branching of the project branches in Subversion

The structure of the project in the form of a branched tree with a characterization of all key stages of development is conditionally demonstrated in Fig. 1. The main branch of the project is highlighted in green. The secondary branches are yellow, and later merged with the main one. This process is called the merger or synchronization with the main branch - the red arrows and blues are marked with the labels, that is, the project branches that will not undergo changes in the future. The red arrows show the merging of changes [19–22]. The process of working with the version control system includes following steps:

1. Create a new or download a current working copy of the project.
2. Make the necessary changes to the working copy. If this is a change in the structure of directories or files (renaming, copying, transferring, adding new ones or deleting old ones) then they are executed using the tools built into the SVN itself.
3. If it is necessary, you can at any time cancel the changes or add new ones created by other users to your working copy.
4. Change the changes to the repository.

3   Problem Statement

All of the below listed products have been used for a long time in the industry. This allows to analyse in detail their weaknesses and strengths in order to form the main function of the monitoring system. The advantages and disadvantages of Subversion, SiteLock and Content Downloader are shown in Table 1.

Table 1. Advantages and disadvantages of the considered software products

Name	Advantages	Disadvantages
Subversion	Built in windows. Ability to view all changes that have undergone the document. Centralized data storage. Saving only changes, not files	Imperfect mechanism of simultaneous work. The impossibility of completely deleting data. To identify changes, you need to add new data to the repository
SiteLock	Maintaining a variety of web site scans. Self-response in the event of a threat. Ability to improve the basic characteristics of servers	Orientation to the CMS system. Not the inclusion of most services in the basic product kit. Restrictions on monitoring only the user's own web sites
Content downloader	Opportunity of different types of parsing with many custom parameters. Maintaining many formats for storing data. Ability to schedule regular parsing without user involvement	Limited system only for data parsing

By analyzing three independent products that provide a variety of end-user capabilities, we distinguish the features that are required for automated monitoring of changes in web resources:

– Subversion – a powerful mechanism for finding differences between files and presenting these differences to the user.
– Content Downloader – demonstrates all possible aspects of parsing.
– SiteLock – Web application with a cabinet for tracking changes, their presentation and changes to the monitoring process.

4   Materials and Methods

The first step in working with this information system is the user's login or registration procedure, if this is the first time a user has used the system (Figs. 2 and 3). In the second step, the user is asked which elements he wants to validate, including the URL of the resource, provide their own template for downloading the content or use several of standard ones, and the period of time for which the verification will take place. After successfully adding all the desired pages, the user is given the opportunity to view the newly added items and make corrections if necessary. This option will always be available to the user, which will ensure the proper ease of use and the flexibility of the monitoring process. If necessary, it will be possible to pause the check of a certain record until it is in a satisfactory state. The final stage that will occur between assigning user resources and editing them is the monitoring process. It is based on the work of the cron server, which allows you to periodically run scripts to run. With the passage of the verification time set in the first step, the system will compare the data stored by the user and those that are currently located at the given URL address. In the case of the differences, it will immediately alert the user. In case if no changes are found, it will only update the date of the last satisfactory validation. The central element in this diagram is the database, which is queried at every step.

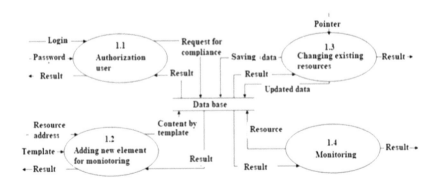

Fig. 2. Detailed data flow diagram

Figures 4, 5 and 6 describe in detail each step in the work of the information system. The details of the user authorization process do not contain much new

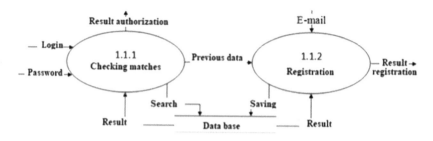

Fig. 3. Detailed data flow diagram for the "User Authorization Process"

Fig. 4. Detailed data flow diagram for the process "Adding new monitoring elements"

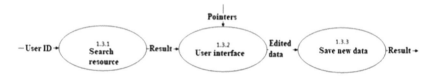

Fig. 5. Detailed data flow diagram for the "Changing Available Resources process"

Fig. 6. Detailed data flow diagram for the Monitoring process

information about this process, but it is an important step in the operation of any multi-user system.

When adding elements for monitoring, the user will be given the option to choose among the standard tracking elements: page headers, keywords, header, page body, all h1–h6 header tags, or self-specifying headers using a regular expression. The last function is focused on people who are familiar with the mechanism of the operation of regular expressions because in the case of setting the wrong pattern the system will not be able to function properly. The user interface is presented in the form of a table showing all the pages that the user can use to track changes, with the ability to perform the following commands:

– delete the resource - suppress the monitoring of this content and wipe it from the table;

- verify - the ability to instantly check the specified elements to match the initial data and in case of detected differences, immediately review the differences;
- upgrade - will be useful after the user has made changes to prevent the monitoring in the bundled data;
- status of the entry - checked or suspended- for the possibility of stopping the checking in the process of making the desired changes.

The monitoring procedure actively interacts with the database due to the need to receive various types of data, depending on the current state of work. To begin with, we check whether it is time to check this resource, then we get the templates on which you need to perform parsing. Next, the verification process follows, and the results of it are stored in the database. No matter what the result of the test is, we still have to put it back to the base. Consequently, we describe using the methodology of structural system analysis, the mechanism of interaction with external objects, elements, and the internal structure of the system and the work of each component. Based on the above, it became clear that the key element in the system's operation is the database. Thus, necessary to optimize the process of working with it and provide the ability to create backups in case of problems:

- The entity is represented in the form of a rounded rectangle around the edges.
- Each entity has a name. This name is placed in the middle of the entity, in its upper part, separated from the attributes by a solid line.
- The name of the entity is written in singular. All attributes are capitalized.
- An attribute should be placed next to the character representing its type.
- The key attribute cannot be empty. One connection can combine only two entities.
- At the initial stages of the design of the diagram, a many-to-many connection is allowed, but in final versions it needs to be replaced by many-to-one, possibly with the help of an additional intermediate entity.

For the automated monitoring system, which already contains all the necessary scenarios - the algorithms for processing input data, it is necessary for the end user to specify the URL of the address of the page and the regular expression that determines which elements are to be tracked. Each user can ask a set of pages and a set of regular expressions for them. Therefore, a situation may arise when different users will set one page for monitoring with different regular expressions. For the unambiguous identification of the user and the elements that he wants to control, the intermediate resource is specified. This entity contains information about a web page linked to it by regular expressions and the user who is assigned it. The system, based on the data provided, will conduct a compliance check and form a report based on it. Given the ability to process different data on one page, system need to generate different reports for each user. The choice of the final solution, taking into account the various criteria, is a difficult planning and decision making task.

Figure 7 shows the diagram of the analytical hierarchy analysis method for the automated monitoring system for changes in web resources. The main goal is to monitor the web resource. At the second level, the listed criteria are necessary for the selection of monitoring elements. Including:

- Presence describes whether the web resource has an element that the system will attempt to track.
- Validity is in the presence of all elements, which of them will give the most accurate results in the process of detecting changes or which contains the most important information and has the advantage of tracking for the end user.
- Staticity is an indicator that is responsible for the number of changes in the normal operation of the web resource in the content of the tracked element. If the content of such an item is constantly changing, then tracking its status is not possible.

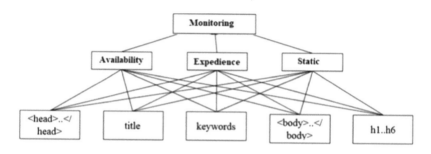

Fig. 7. Method of analysis of hierarchies

Alternatives are shown in Fig. 7. They represent the important elements of any internet page, which can be selected to check the page's compliance with the reference version. The head and body tags form the basis of the content of the resource, but they can contain dynamic data and therefore be unsuitable for monitoring changes, as the new results of the audit will be constantly issued. For this purpose, additional title, keywords and h1..h6 elements were created that are responsible for the page title (as the page is in the browser tab), the display of this resource to the corresponding user queries, and all headings contained in the content respectively. These three elements are intended to at least partially replace head and body in case of impossibility of their use.

5 Experiment

Periodic monitoring and verification by the system is carried out through the Cron server program. Cron is a utility in UNIX system systems that is used for running tasks at a specific time. Regular actions are described by the instructions placed in crontab files and in special directories. It determines whether the time has come for a particular resource to be checked and if it does, then the regular expression parses the content of this resource and compares it with the corresponding content in the database. If discrepancies are found then the system sends to the user who added the resource the message. The main interface of the program is a table with information about each resource that the user wanted to monitor and elements to change the state of this resource (Fig. 8). Each page the user added is displayed as its URL address on the Internet. If some pages have the same domain, they will be sorted by this domain. The information on the page is the date of its last check, the interval of verification and the items being checked. The buttons on the right side of the table allow to interact with the system. With their help, it is possible to permanently delete a certain resource from the monitoring system, to immediately compare the content of the page with the one stored in the database, update the content in the database and suspend the verification of the resource. In the case when the user manually wants to check the data on a specific resource and the changes were found, an additional link to the page will appear where all changes will be highlighted (Fig. 9). All changes that occurred within the part of the page that corresponds the regular expression the user selected are shown in Fig. 10. In the top, there is an explanation of how to understand the content of this page and the shortcut keys to move instantly between the elements. Below are data downloaded from the resource and data from the system database. The differences are highlighted in colour. After reviewing the changes, the user is able to immediately correct the changes. If these changes are not desirable, he can immediately restore the original state of the data without spending time finding a problem. We update the database by notifying that acceptance of the differences. The next time the system will compare online data with a newer version of database data. If the user had not reacted to the change message within a certain amount of time and new scheduling run is detected, and new changes are found, all of those changes are saved to a separate resource. This allows to track all changes to the page throughout the monitoring period (Fig. 11). The information system of automated monitoring of changes in web resources is currently hosted on the *freehost.com.ua* web hosting server and is freely accessible. Anyone can find it by domain *sitechecker.sc.ua* and test the functionality.

#	Адрес	Остання перевірка	Інтервал	Елементи моніторингу	
1	guide.karpaty.ua	30/5 13:41			
1.1	guide.karpaty.ua/uk/categories/marshruty#%D0%B2%D0%B6%D0%B1%D0%B8%D0%B1%D0%B8	30/5 13:41	Постійно	title	✕ 🔍 ⊘ ✓
1.2	guide.karpaty.ua/uk/categories/marshruty#%D0%B2%D0%B6%D0%B1%D0%B8%D0%B1%D0%B8	30/5 13:41	Постійно	body	✕ 🔍 ⊘ ✓
2	inneti.com.ua	30/5 13:41			
2.1	inneti.com.ua/	30/5 13:41	Постійно	body	✕ 🔍 ⊘ ✓
3	uk.wikipedia.org	30/5 14:33			
3.1	uk.wikipedia.org/wiki/%D0%9B%D1%8C%D0%B2%D1%96%D0%B2	30/5 14:33	Щогодини	head	✕ 🔍 ⊘ ✓
4	www.studentinfo.net	30/5 13:41			
4.1	www.studentinfo.net/	30/5 13:41	Постійно	body	✕ 🔍 ⊘ ✓

Fig. 8. Main table

Запис було змінено.

Для перегляду змін перейдіть за цим посиланням.

#	Адрес	Остання перевіка	Інтервал	Елементи моніторингу
1	guide.karpaty.ua	30/5 13:41		
2	inneti.com.ua	30/5 13:41		
3	uk.wikipedia.org	30/5 14:33		
4	www.studentinfo.net	30/5 13:41		

Fig. 9. Links to view differences

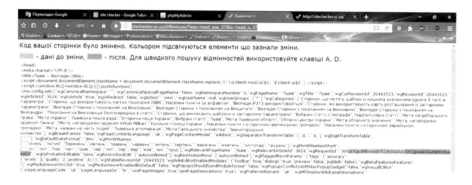

Fig. 10. View differences

		Головна Додати Мої записи **Журнал**
		За добу
#	Домен / Сторінка	Перевірено
1	guide.karpaty.ua	19/5 17:22
1.1	guide.karpaty.ua/uk/categories/marshruty#%D0%B2%D0%B8%D0%B1%D0%B8%D0%B1%D0%B8	19/5 17:22
1.2	guide.karpaty.ua/uk/categories/marshruty#%D0%B2%D0%B8%D0%B1%D0%B8%D0%B1%D0%B8	19/5 17:22
1.3	guide.karpaty.ua/uk/categories/marshruty#%D0%B2%D0%B8%D0%B1%D0%B8%D0%B1%D0%B8	19/5 16:01
2	inneti.com.ua	19/5 16:01
2.1	inneti.com.ua/	19/5 16:01
2.2	inneti.com.ua/	19/5 15:52
3	uk.wikipedia.org	30/5 13:33
4	www.studentinfo.net	30/5 13:33

Fig. 11. Resource change log

6   The Created Software Description

Anyone can visit the monitoring system home page and see the information posted there. The main page briefly describes what the system is, why it is needed and how to work with it. Also, for the convenience of the video instruction for working with the monitoring system. The page menu for work is in the upper right corner. The main page plays the role of a business card, the purpose of which is to interest people before starting to work with the system, so for its review it is not necessary to undergo a registration process different from the next stages of working with this resource. Having read the general information about the system and having expressed the desire to continue the work, the user undergoes a registration/authorization procedure. The next step is to specify the elements and parameters by which he want to track the changes. The *Add* page from the main menu corresponds to this. The address entry field specifies the template according to which the user must verify for correct link recognition. The best solution is to open the page user wants in another browser tab and copy its address. The same template is present in the field for a regular expression. Among the suggested tracking elements are standard templates such as head, title, keywords, body, and a set of all possible titles. Also, the user is given the opportunity to create its own regular expression on which elements of the monitoring will be selected. The last item is a drop-down menu is used to select the period of the verification (see Fig. 12).

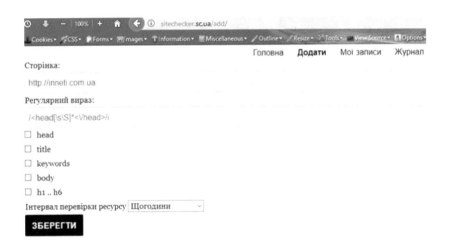

Fig. 12. The window for adding a new element for monitoring

After adding all the requested parameters and obtaining a satisfactory response from the system, the user can go to My Posts page to view all the saved pages. The data added by the user are displayed as a table. The main elements of the user's table are:

- # is the field with the sequence number of the page on which the previously selected items are tracked;
- the address is the URL resource link;
- the last check - the field indicating the date when it was last checked for the original content of the page;

– interval - displays the information for the time interval from last check and next;
– monitored elements - shows those elements that are monitored on a specific link, then the block with the buttons of manipulation is presented.

Among these functions is the page removal that is, the termination of its monitoring and erasure of information about it from the table, followed by a button to scan or verify the web document. This feature is needed to visually display to the user how the system works. If there is no match to the initial and current state of the page content, a link to the additional page will be displayed for detailed review of the differences.

The next function is to update the record - serves to rewrite the information in the database related to current record, will be useful in case of changes on the page, however, these changes are acceptable for further work of the resource. The last option is to pause checking. The icon for this process changes its status depending on the status of the resource (Fig. 13). As a result, the detected changes in the resource are presented to the user. He is given the opportunity to see the existing changes using a link indicating the changes (Fig. 14). The user may repeat this action if he receives the message by e-mail. However, he should be careful when receiving multiple emails. In this case, to view all changes that occurred on the page he must go to *Journal* page (Fig. 15) By clicking on this link the user will open a new window with a small instruction on the output data and the data corresponding to the content of the page on which the changes were found. This page displays the site code with colour-differentiated non-standard elements. After analyzing this data, the user will be able to quickly restore the original resource state or approve the changes by returning to the previous menu and updating the given entry. The *Journal* keeps a history of all user resource changes. To view changes in comparison with the reference option, the one that the user created on the *Add page*, or if the button is updated, in the field of this entry on the *My Posts* page.

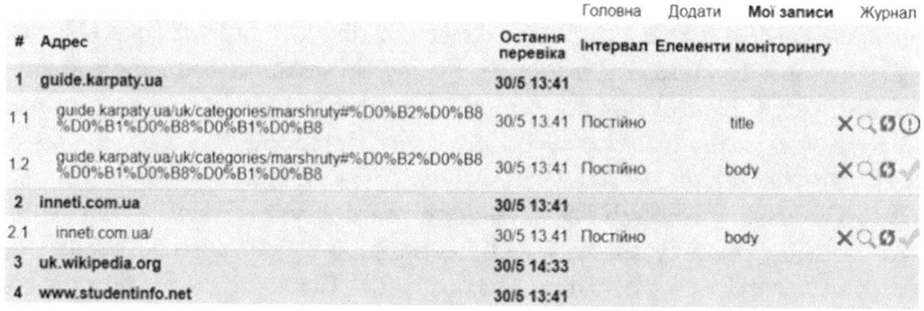

Fig. 13. Displays all traceable items

Fig. 14. Difference on the page

#	Домен / Сторінка	Перевірено	
1	**guide.karpaty.ua**	**19/5 17:22**	
1.1	guide.karpaty.ua/uk/categories/marshruty#%D0%B2%D0%B8%D0%B1%D0%B8%D0%B1%D0%B8	19/5 17:22	
1.2	guide.karpaty.ua/uk/categories/marshruty#%D0%B2%D0%B8%D0%B1%D0%B8%D0%B1%D0%B8	19/5 17:22	
1.3	guide.karpaty.ua/uk/categories/marshruty#%D0%B2%D0%B8%D0%B1%D0%B8%D0%B1%D0%B8	19/5 16:01	
2	**inneti.com.ua**	**19/5 16:01**	
3	**uk.wikipedia.org**	**30/5 13:33**	
3.1	uk.wikipedia.org/wiki/%D0%9B%D1%8C%D0%B2%D1%96%D0%B2	30/5 13:33	
4	**www.studentinfo.net**	**30/5 13:33**	

(Головна   Додати   Мої записи   **Журнал**   За 2 тижні)

Fig. 15. Page of the magazine

7   The Results Analysis and Discussion

To check the system's performance, we will give it a static resource for monitoring, located on the global web. This resource will use the page for the specification of tracked web resource (Fig. 16). After saving, we will go to the page with all the saved records and make sure that the template creation process was successful (Fig. 17). The date and time of creating a new record correspond to the system time on the user's computer. After clicking the button to check, we get a negative result to change the Fig. 18. So far, the system has shown itself perfectly, however, this trial has not ended. Next, we make changes to the content of the page www.studentinfo.net/ on my own. To complicate the task, we place the text "sitechecker.sc.ua" inside the <head> tag so that these changes are not visible to the user. Although these changes are not too critical, if the system is able to correctly identify them, it will be possible to say with certainty that it functions in accordance with the stated purpose. Result of re-checking is shown on Fig. 19. As it was described above, in the case of detection of changes, the system proposes to review in detail what was changed. Consequently, discrepancies were found in the online version and database. The next step is to check the correctness of the operation of the comparison algorithm see Fig. 20. From this figure, it is clear that the system has successfully tested the functioning of the main components. Although this was only a test of individual records, it was not

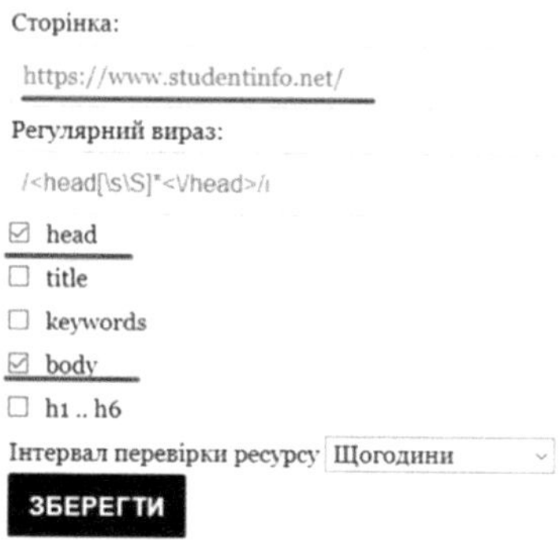

Сторінка:

https://www.studentinfo.net/

Регулярний вираз:

/<head[\s\S]*<\/head>/i

☑ head
☐ title
☐ keywords
☑ body
☐ h1 .. h6

Інтервал перевірки ресурсу Щогодини ⌄

ЗБЕРЕГТИ

Fig. 16. Create reference value

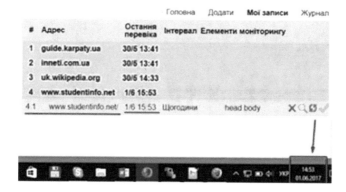

Fig. 17. Create template reference

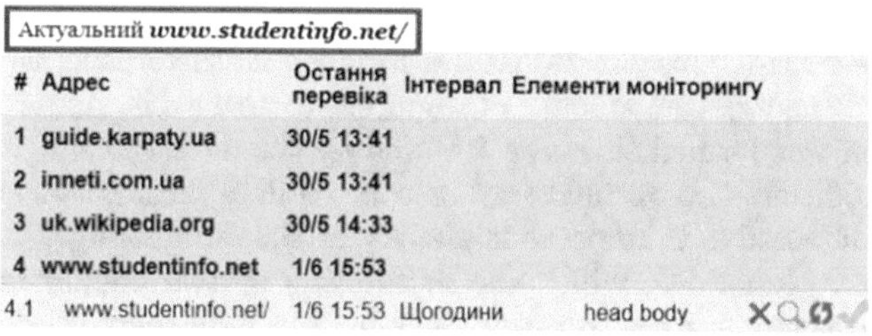

Актуальний *www.studentinfo.net/*				
# Адрес	Остання перевіка	Інтервал	Елементи моніторингу	
1 guide.karpaty.ua	30/5 13:41			
2 inneti.com.ua	30/5 13:41			
3 uk.wikipedia.org	30/5 14:33			
4 www.studentinfo.net	1/6 15:53			
4.1 www.studentinfo.net/	1/6 15:53	Щогодини	head body	✕ ⚲ ↻ ✓

Fig. 18. The check did not detect any changes

possible to say exactly how the system would behave in real conditions, with a heavy load on the part of users. The test results provide an optimistic estimate in this prediction.

Запис ***https://www.studentinfo.net/*** було змінено.

Для перегляду змін перейдіть за цим посиланням.

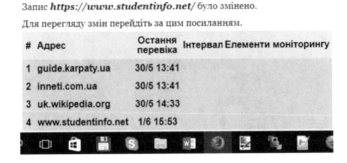

#	Адрес	Остання перевіка	Інтервал	Елементи моніторингу
1	guide.karpaty.ua	30/5 13:41		
2	inneti.com.ua	30/5 13:41		
3	uk.wikipedia.org	30/5 14:33		
4	www.studentinfo.net	1/6 15:53		

Fig. 19. The check found the changes

Fig. 20. Correct recognition of changes

8   Conclusions

In the paper, the main elements of the information system for monitoring the changes in web resources were designed and graphically represented using diagrams. Relying on the functions, which the final product must possess, phased steps of constructing this functional are formed. We determine the priority of each task and the required amount of resources to achieve it. As a result, a clear plan of action was formed that was followed by the development process of the system. In addition, the optimal way to use the final product was determined, which made it possible to improve it already at the design stage.

We describe functions, which the monitoring system should have and the way in which they interact, software tools for the implementation of the tasks and their technical characteristics. In the stage of practical implementation, we implemented a thought-out programmatic implementation of the monitoring system for changes in web resources. As a result, the software monitoring system was implemented. We provide the description of instructions for the average user.

References

1. Iturrioz J, Azpeitia I, Díaz O (2014) Generalizing the like button: empowering websites with monitoring capabilities. In: Proceedings of the 29th annual ACM symposium on applied computing, pp 743–750

2. Maggi F, Robertson W, Kruegel C, Vigna G (2009) Protecting a moving target: addressing web application concept drift. In: International workshop on recent advances in intrusion detection. Springer, Berlin, pp 21–40
3. Christensen JH (2009) Using RESTful web-services and cloud computing to create next generation mobile applications. In: The 24th ACM SIGPLAN conference companion on Object oriented programming systems languages and applications, pp 627–634
4. Neugschwandtner M, Neugschwandtner G, Kastner W (2007) Web services in building automation: mapping KNX to oBIX. In: International conference on industrial informatics, pp 87–92
5. Chen L, Hind JR, Li Y, Xiao L (2013) U.S. patent no. 8,613,039. U.S. Patent and Trademark Office, Washington, DC
6. Breiter G, Jall D, Mueller M, Neef A, Reitz M (2014) U.S. patent no. 8,812,424. U.S. Patent and Trademark Office, Washington, DC
7. Quintero AH, Fedor JS, Quan AG, Richardson K, Scott DW, Piper KA (2005) U.S. patent no. 6,910,071. U.S. Patent and Trademark Office, Washington, DC
8. Leshko I, Firstenberg Y, Kumar N (2013) U.S. patent application no. 13/528,873
9. Dan A et al (2004) Web services on demand: WSLA-driven automated management. IBM Syst J 43(1):136–158
10. Shelby Z (2010) Embedded web services. Wirel Commun 17(6):52–57
11. Garanina N, Sidorova E, Kononenko I, Gorlatch S (2017) Using multiple semantic measures for coreference resolution in ontology population. Int J Comput 16(3):166–176
12. Colton P, Sarid U (2014) U.S. patent no. 8,880,678. U.S. Patent and Trademark Office, Washington, DC
13. Colton P, Sarid U (2009) U.S. patent no. 7,596,620. U.S. Patent and Trademark Office, Washington, DC
14. Colton P, Sarid U (2012) U.S. patent no. 8,291,079. U.S. Patent and Trademark Office, Washington, DC
15. Colton P, Sarid U (2015) U.S. patent no. 8,954,553. U.S. Patent and Trademark Office, Washington, DC
16. Lytvyn V, Sharonova N, Hamon T, Vysotska V, Grabar N, Kowalska-Styczen A (2018) Computational linguistics and intelligent systems. In: CEUR workshop proceedings, vol 2136
17. Vysotska V, Fernandes VB, Emmerich M (2018) Web content support method in electronic business systems. In: CEUR workshop proceedings, vol 2136, pp 20–41
18. Kanishcheva O, Vysotska V, Chyrun L, Gozhyj A (2018) Method of integration and content management of the information resources network. In: Advances in intelligent systems and computing, vol 689. Springer, pp 204-216 (2018)
19. Lytvyn V, Vysotska V (2015) Designing architecture of electronic content commerce system. In: Proceedings of the X-th international scientific and technical conference computer science and information technologies, CSIT 2015, pp 115–119
20. Vysotska V, Hasko R, Kuchkovskiy V (2015) Process analysis in electronic content commerce system. In: 2015 X-th international scientific and technical conference computer sciences and information technologies (CSIT), pp 120–123
21. Kravets P (2010) The control agent with fuzzy logic. In: Perspective technologies and methods in MEMS design, MEMSTECH 2010, pp 40–41
22. Gozhyj A, Chyrun L, Kowalska-Styczen A, Lozynska O (2018) Uniform method of operative content management in web systems. In: CEUR workshop proceedings computational linguistics and intelligent systems, vol 2136, pp 62–77

Recognition of Human Primitive Motions for the Fitness Trackers

Iryna Perova$^{(\boxtimes)}$ (iD), Polina Zhernova (iD), Oleh Datsok (iD), Yevgeniy Bodyanskiy (iD), and Olha Velychko (iD)

Kharkiv National University of Radio Electronics, Kharkiv, Ukraine
rikywenok@gmail.com

Abstract. Principle elements of a healthy lifestyle and harmful risk factors caused to cardiovascular diseases are being described. Significance of diurnal monitoring of essential characteristics of trustworthy physical state assessment by a fitness tracker is being shown. It had been shown that assessment of registered physiological parameters has to be performed taking into account the personal data such as age, anthropometric data and level of recommended physical activity. The basic block diagram of the typical fitness tracker and its functions are being described. The overall factors such as sensor placement, type of physical activity and human primitive motion, impacting to the fitness trackers measurement accuracy are being analyzed. The two approaches of human primitive motion detection are being proposed. The first one is based on clusterization procedure of original multidimensional time series with 72,4% accuracy of true detection and the second one performs classification procedure of preprocessed multidimensional time series with 85,5% accuracy of true detection.

Keywords: Physical activity intensity · Fitness tracker ·
Three-axis accelerometer · Human primitive motions · Clusterization ·
Multidimensional time series

1 Introduction

Modern life requires a person to achieve their own goals large investments of their time, their work, and, most importantly, health. Such factors as sedentary working conditions, irregular and low-quality nutrition, environmental problems and other factors can affect physical and mental well-being.

Modern medicine and pharmacology are often impotent in cases where the human body is practically unable to combat an illness. The basis of a healthy lifestyle is the ability of a person to perform special actions focused on preserving his health and preventing diseases.

It is obvious that a healthy lifestyle may be provided by avoiding harmful habits such as smoking, addictive drugs and alcohol consumption. As shown in Fig. 1, moderate and balanced, correlated with individual physiological features nutrition, creating environmentally friendly living conditions and physical

© Springer Nature Switzerland AG 2020
V. Lytvynenko et al. (Eds.): ISDMCI 2019, AISC 1020, pp. 364–376, 2020.
https://doi.org/10.1007/978-3-030-26474-1_26

activity according to a person's physiology and age are the essentials of healthy style.

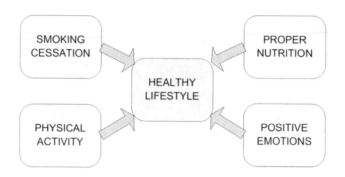

Fig. 1. The basis for a healthy lifestyle. Scheme

Insufficient physical activity is one of the high-risk factor and mortality owing to it arises at 6 % (Fig. 2). More significant reasons are high blood glucose levels, systematic smoking and high blood pressure (from 6% to 13%).

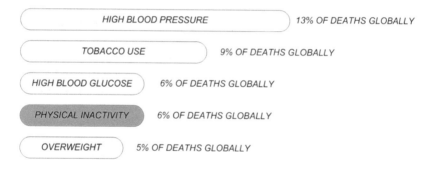

Fig. 2. Risk factors for death in the world [1]

Decreasing physical activity intensity is negatively affected the health of people, having cardiovascular and non-communicable diseases.

The intensity and type of physical activity are selected depending on the physical state and age of a person. The additional psychological criterion of choice is deciding what does the person likes that causes a minimum of discomfort or brings pleasure and that is more suitable to the particular person's temperament. The safety process of physical activity changing requires to control the cardiovascular system stress load by heart rate measuring. The optimal safety load is being rated at 60–85% of the maximum age-related value.

2 Review of the Literature

In the purpose of quantitive assessment of physical load, the metabolic equivalent (MET) may be applied. MET is the ratio between the person's metabolic rate during the physical activity and the person's metabolic rate at rest. One MET is the quantity of energy, expending for one-hour activity at rest and measuring at kcal/kg/hour. At this rate, for a moderately active person, the physical activity intensity is approximately equal to 3–6 MET and for highly active person MET exceeds 6 [2].

In accordance with published by the World Health Organization data, it is customary to distinguish three age groups of type and physical activity intensity [3].

The first group contains children and adolescents at 5 to 17 years old. The persons from this group are recommended the daily moderate or high physical activity, including the games, sport, competitions, etc. The persons at 18 to 64 years old have belonged into the second group and they have been recommended the age-related moderate physical activity such as health-improving training, walking or cycling. The third group consists of over 65-year-old persons, which are recommended the daily physical activity, depending on their health state and organism abilities.

Daily monitoring of personal physical loads strongly encouraged for all groups and must be performed for persons with cardiovascular diseases or high-risk liability to cardiovascular diseases. In addition, control of the physical activity intensity is the primary component during the various sport training programs, particularly in track and field.

A number of publications are dedicated to data processing for fitness trackers. Significance of a person's location for step counts, calorie burn, and heart rate was considered in [4]. Authors had proposed approach that opportunistically fills gaps in a user's location traces for minute-level biometric data calculation. Other researchers had developed the approach of precision improving, based on regression models [5]. Recognition of human motions using the Gaussian Mixture Modeling and Regression is described in [6]. The same principle is used in [7] for recognition of motion primitives and the creation of activity models.

3 Problem Statement

It is obvious, that algorithms of biometric data calculations are not perfect and needed in modification and improvement. One of the general sources of errors is originated from erroneous identification of activity caused by the human primitive motions. Clearly interpretation of human primitive motions will enlarge the precision of physiological parameter definition and improve technical fitness trackers characteristics.

4   Tracking of Physical Activity

Fitness trackers or fitness bracelets are designed for measuring and collecting related to physical activity data. This function may be realized by many portable devices, recording biometric data during physical activity. A typical block diagram of a fitness tracker, as shown in Fig. 3, consists of a sensor, a microcontroller, an OLED indicator and Bluetooth module for data transfer by the wireless.

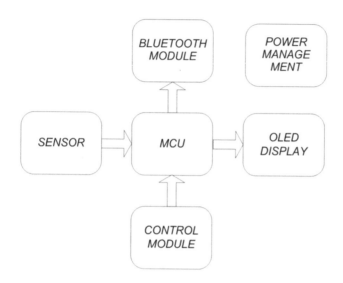

Fig. 3. Typical fitness bracelet

The basic functions of this device include step counter, calorie calculating rated to physical activity, heart rate monitoring during activity and at rest, named as a sleep monitor. Physical activity intensity is estimated by two transducers: accelerometer specifies the direction of motion, gyroscope determines space orientation. Results gotten from both transducers are transformed into a number of passed steps and burned calories.

The errors of measurement passed steps, burned energy and heart rate depending on the number of factors such as wrong recognition motion of a person in a car or bus as physical activity when the human body is immovable relatively the moving inertial system. It is specified by several factors. Firstly, each type of sensor has its own inner measurement error. The use of modern highly sensitive accelerometers with temperature error compensation allows us to minimize the influence of this factor. Secondary, result of measurements depends on sensor placement (pick-a-wrist, pick-a-back, pick-a-leg, pick-a-neck, etc.).

Despite on immovable mass centre of the human body motion activity during performance ordinary actions like toothbrushing, cooking, musical studies had been registered with intensive limb motions in pick-a-wrist sensor's location. The algorithm of movement and gesture recognition, including the human primitive motion, used by the manufacturing company, impact to the fitness tracker efficiency. Such developments are the intellectual property of the company and unable for changing. Some models of devices have self-learning abilities of gesture recognition with the following correction owing to the fact of generation certain patterns of typical movements. Therefore, any kind of registered movement may be wrongly recognized as step activity because of measured by sensors acceleration and velocity despite on biomechanical mismatches between step movement and gesture [8].

Anthropometric parameters, the habit of body, gait or having something at hands impact to measuring result in an arbitrary sensor position. In practice, measuring results are various during walking, at the race and athletic track when step length is shorter but step frequency is higher.

The fact that the accuracy of determining the intensity of physical exertion and human primitive motion is decisive in choosing a fitness bracelet is confirmed by the results of studies of their consumer properties. According to [9], the most significant functions of the fitness bracelet for the elderly are the ability to measure the number of steps and pulse rate, the measurement accuracy of these indicators and the possibility of wearing it on the arm. Less important are such features as sleep monitoring, the presence of a waterproof case and battery life. The least demanding respondents in this group are the design, price, and the possibility of wearing on a belt and in a pocket. For other age groups, priorities may be placed differently, but the requirements for high accuracy of the performance of the main measured parameters are preserved.

The significance of human primitive motion definition is the primary in the fitness tracker choice has been verified with an investigation of their consumer properties. According to investigations [9], the most significant fitness tracker functions for seniors are the ability to measure passed steps, heart rate, data accuracy and wrist wear facility. Sleep monitor, water protection, design, price, wearing in a pocket or a belt and storage energy time requirement are insignificant. Other ageing groups had managed another propitiates but high accuracy of measurement has been rated at the top.

The data for analysis had been obtained with the three-axis digital accelerometer (measuring range is $1.5\,g$ to $+\,1.5$ g, sensitivity is 6 bits per axis), built into measuring module, fixed on the wrist by such way that the X-axis coincides with the forearm direction, the Y-axis is oppositely directed to the thumb, and the Z-axis is perpendicular to the longitudinal plane of the forearm (Fig. 4).

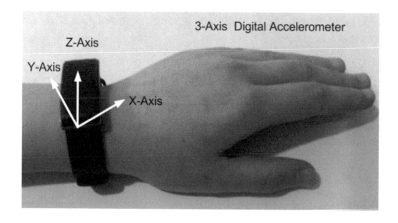

Fig. 4. Three-axis digital accelerometer placement

5   Multidimensional Time Series Analysis

5.1   Description of Time Series the Fitness

Time series data was gotten from the accelerometer and represent a three-dimensional array of human motions acceleration projected to x, y, z-axes [6,7]. As seen from Table 1, the dataset contains the numbers of samples which had been recognized as Human Motion Primitives (HMR) during the simple human activity.

Table 2 demonstrates x, y and z values the dataset section of climb stairs motion. The total number of observed patients is 277.

Entire parameters have been measured by an attached to the user's right wrist accelerometer according to the next scheme:

- the x-axis coincides with the forearm direction;
- the y-axis is directed to the left;
- the z-axis is perpendicular to the longitudinal plane of the forearm.

Illustration of accelerometer three axis values for different HMP are presented in Figs. 5, 6 and 7.

5.2   Experiment Results of Time Series Classification-Clusterization

Classification/clusterization of multidimensional time series may be performed by two approaches. The first one does not use the data preprocessing. The second one applies the multidimensional time series preprocessing by calculating the statistics of the time series rows.

Clusterization of Multidimensional Time Series. In the case, when time series were not being preprocessed the patient's data is presented with a multidimensional matrix of N patients, having n time series with a different number of time instants:

$$X(k) = \{x_{il}(k)\}, \qquad (1)$$

where $k = 1, ..., N$ is the number of patients in the matrix ($N = 277$ for current investigation), $i = 1, ..., n$ is the number of patient's time series (for x, y and z values), $l = 1, ..., q$ is the number of time instants personal for each patient ($q = 147...9318$). To make possible each other comparison of time series the datasets must have the equivalent number of time instants ($q = 147$).

Table 1. Human Motion Primitives (HMP)

Number of samples (patients)	Motion type	Description
12	Brush teeth	To brush one's teeth with a toothbrush (complete gesture)
25	Climb stairs	To climb a number of steps of a staircase
25	Comb hair	To comb one's hair with a brush (complete gesture)
25	Descend stairs	To descend a number of steps of a staircase
25	Drink glass	To pick a glass from a table, drink and put it back on the table
5	Eat meat	To eat something using a fork and knife (complete gesture)
3	Eat soup	To eat something using a spoon (complete gesture)
25	Getup bed	To get up from a lying position on a bed
26	Liedown bed	To lie down from a standing position on a bed
18	Pour water	To pick a bottle from a table, pour its content in a glass on the table and put it back on the table
25	Sitdown chair	To sit down on a chair
25	Standup chair	To stand up from a chair
13	Use telephone	To place a telephone call using a fixed telephone (complete gesture)
25	Walk	To take a number of steps

Table 2. Human Motion Primitives (HMP)

x	y	z
5	39	34
2	41	34
5	39	34
12	38	34
9	38	30
10	36	29
10	36	30
12	36	30

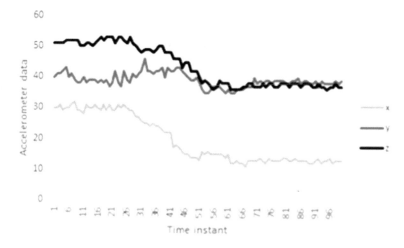

Fig. 5. x, y and z values of StandUp chair motion

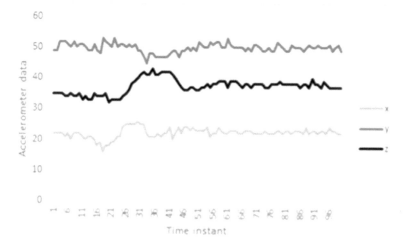

Fig. 6. x, y and z values of Brush Teeth motion

At the first step, the cluster centres, corresponding to fourteen HMP groups ($m = 14$) from Table 1, are being settled randomly.

$$C_m = \{c_{il}\}. \tag{2}$$

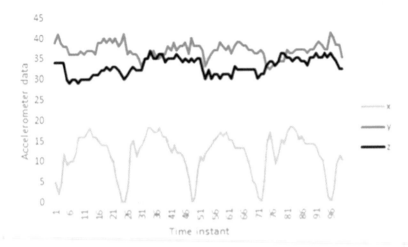

Fig. 7. x, y and z values of Climb Stairs motion

The next step calculates distances between multi-dimensional time series and certain cluster centers in the Frobenius metrics:

$$dist_m(k) = \sqrt{Trace((X(k) - C_m)(X(k) - C_m)^T)}, \tag{3}$$

that is used to obtain the degree of membership to each HMP-group [10]:

$$\mu_m(X(k)) = \frac{(dist_m(k))^{-2}}{\sum_{j=1}^{m}(dist_j(k))^{-2}} \tag{4}$$

Further, the positions of centers are recalculated with the formula [10–12]:

$$C_m(k) = C_m(k-1) + \eta(k)\mu_m^2 X(k)(X(k) - C_m(k-1)) \tag{5}$$

where $0 < \eta(k) \leq 1$ is the learning rate parameter.

Clusterization precision is equal to 72,4%. Results of obtained clusters visualization are displayed in Fig. 8.

Classification of Multidimensional Time Series. Multidimensional time series preprocessing had been implemented with calculation the time series rows statistics, using formulas (6)–(11).

The average with pre-robust filtering capacities is the following:

$$f_i^1 = \bar{x}_i = med\, x_{il}(k) \tag{6}$$

Sampling the second central moment is:

$$f_i^2 = \frac{1}{q}\sum_{l=1}^{q}(x_l(k) - \bar{x}_i)^2. \tag{7}$$

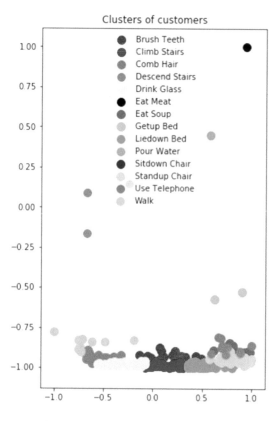

Fig. 8. Results of HMP recognition

Sweep is equal to:

$$f_i^3 = |x_i^{max}(k) - x_i^{min}(k)|. \tag{8}$$

Autocorrelation coefficients are defined with formulas:

$$f_i^4 = \frac{1}{q-1} \sum_{l=1}^{q-1} (x_l(k) - \bar{x}_i)(x_l(k+1) - \bar{x}_i) \tag{9}$$

$$f_i^5 = \frac{1}{q-2} \sum_{l=1}^{q-2} (x_l(k) - \bar{x}_i)(x_l(k+2) - \bar{x}_i) \tag{10}$$

$$f_i^{3+R} = \frac{1}{q-R} \sum_{l=1}^{q-R} (x_l(k) - \bar{x}_i)(x_l(k+R) - \bar{x}_i) \tag{11}$$

For each patient's dataset, a parameter $F_i(k) = \{(f_{il}^1, ..., f_{il}^{3+R})\}$ has been calculated during the preprocessing procedure.

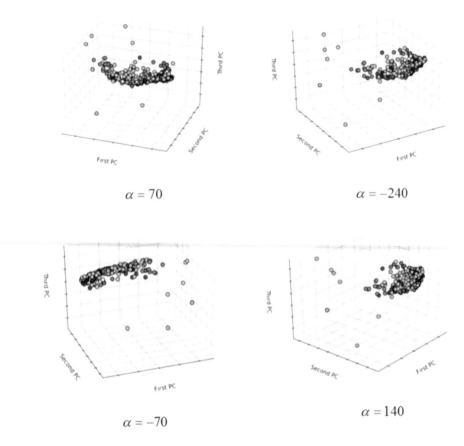

$\alpha = 70$

$\alpha = -240$

$\alpha = -70$

$\alpha = 140$

Fig. 9. Results of HRM classification in the three principal components space at different angles of view α

For time series classification initial data have been grouped into training and test datasets. Centres of m HMP-group in the training dataset were been calculated with the expression:

$$c_{il} = \frac{1}{N_m} \sum_{k=1}^{N_m} F_i(k) \tag{12}$$

where N_m is the number of patients in each of m HMP-group in the training dataset.

The matrix of centres coordinates is described as the multidimensional set $c_m = \{c_{il}\}$. Distances $F_i(k)$ between patients dataset from the test dataset and centres c_m have been calculated with formula (3) and membership degrees of patient's dataset to each HMP-group have been found with expression (4).

The membership degrees $\mu_m(F(k))$ match to the unity partition condition. Recognizing of HMP in the test dataset is performed in accordance with the highest membership degree value with precision at 85,5%. Visualization of the classification results in the three principal components space at different angles of view α is presented in Fig. 9.

6   Conclusion

The principal components of the healthy lifestyle and high-risk factors, leading to cardiovascular diseases had been described. The recommended levels of physical activity for age-groups had been considered and the significance of essential physical activity parameters monitoring had been expressed. Therefore, diurnal monitoring devices, controlling the level of physical activity are extremely important for certain ageing groups, and persons with the high-risk cardiovascular diseases, sportsmen and the ones who adhere to a healthy lifestyle. The block diagram of standard fitness tracker, its basic functions and primary source of errors during the measuring had been considered. It was demonstrated that the human primitive motions are recognized at 85,5% accuracy by the clusterization methods by using multidimensional time series preprocessing.

References

1. (2009) Global health risks: mortality and burden of disease attributable to selected major risks. World Health Organization, Geneva
2. What is Moderate-intensity and Vigorous-intensity Physical Activity? https://www.who.int/dietphysicalactivity/physical_activity_intensity/en/. Accessed 06 Oct 2014
3. Global recommendations on physical activity for health. www.who.int/dietphysicalactivity/factsheet_recommendations/. Accessed 14 May 2015
4. Vhaduri S, Poellabauer C (2018) Opportunistic discovery of personal places using smartphone and fitness tracker data. In: IEEE international conference on healthcare informatics, pp 103–114
5. Andalibi V, Honko H, Christophe F, Viik J (2015) Data correction for seven activity trackers based on regression models. In: 37th annual international conference of the IEEE engineering in medicine and biology society, pp 1592–1595
6. Bruno B, Mastrogiovanni F, Sgorbissa A, Vernazza T, Zaccaria R (2012) Human motion modelling and recognition: a computational approach. In: IEEE international conference on automation science and engineering (CASE), pp 156–161
7. Bruno B, Mastrogiovanni F, Sgorbissa A, Vernazza T, Zaccaria R (2013) Analysis of human behavior recognition algorithms based on acceleration data. In: IEEE international conference on robotics and automation (ICRA), pp 1602–1607
8. Schrack V, Zipunnikov C (2015) Crainiceanu electronic devices and applications to track physical activity. JAMA 313(20):2079–2080
9. Rasche P, Wille M, Theis S, Schäfer K, Schlick CM, Mertens A (2015) Activity tracker and elderly. In: 2015 IEEE international conference on computer and information technology, pp 1411–1416

10. Bezdek J, Keller J, Krisnapuram R, Pal N (1999) Fuzzy models and algorithms for pattern recognition and image processing, vol 4, Springer, 777 p
11. Shakhovska N, Medykovsky M, Stakhiv P (2013) Application of algorithms of classification for uncertainty reduction. Przeglad Elektrotechniczny 89(4):284–286
12. Perova I, Bodyanskiy Y (2015) Adaptive fuzzy clustering based on Manhattan metrics in medical and biological applications. Bulletin of the National University "Lvivska Polytechnika", vol 826, pp 8–12

Development of Information Technology for Virtualization of Geotourism Objects on the Example of Tarkhany Geological Section

Ainur Ualkhanova[1]([✉]) [ID], Natalya Denissova[1] [ID], and Iurii Krak[2] [ID]

[1] D. Serikbaev East Kazakhstan State Technical University,
Ust-Kamenogorsk, Kazakhstan
AUalkhanova@ektu.kz
[2] Taras Shevchenko National University of Kyiv, Kyiv, Ukraine
krak@univ.kiev.ua

Abstract. Currently, the popularization of geological science, the development of geotourism and the use of modern information technologies is an urgent and popular problem. At the same time, the East Kazakhstan region has a great potential for the development of geotourism. The application of information technologies can attract domestic and foreign tourists to the region and provide stakeholders with detailed information about the geological object with the follow-up using its virtual model.

This article considers the possibility of adaptation and application of photogrammetry technology for three-dimensional (3D) modeling of geological objects of East Kazakhstan, which are of interest for geotourism. We examine algorithms for photographing objects to obtain photos suitable for further processing in order to obtain a virtual model of the object. Features and problems of building of three-dimensional virtual models of geological objects are revealed, possibilities of the software product Agisoft Photoscan which allows to receive a point cloud, virtual three-dimensional models of objects in a semi-automatic mode based on their two-dimensional photos, including those received with the unmanned aerial vehicles (drones) are investigated.

Keywords: Photogrammetry · 3D model · 3D modeling ·
Point cloud · Geoportal · Unmanned aerial vehicle · Texturing ·
Orthomosaic

1 Introduction

East Kazakhstan region is unique in its tourism opportunities. A wide variety of landscape and climatic zones, the unique geological objects, monuments of history and culture determine the attractiveness of East Kazakhstan in terms of tourism. There are places of world importance: Belukha Mountain, Kokkolsky waterfall, one of the largest and most beautiful in the Altai, Lake Markakol,

V. Lytvynenko et al. (Eds.): ISDMCI 2019, AISC 1020, pp. 377–388, 2020.
https://doi.org/10.1007/978-3-030-26474-1_27

Lake Zaysan, the Irtysh River, Tarkhany geological section, Delbegetey mountain, Crown mountain, an array of Blazing hills, Kings' Valley, Kiin-Kerish. In this regard, the popularization of geology and the use of modern information technologies, including 3D-visualization of unique geological features will contribute to the development of geotourism in the region. According to the state program Digital Kazakhstan and Rouhani Ganguro, a great attention has been paid to digitization in Kazakhstan and, in particular, to the establishment of regional and industry-specific geoportals. Thus, the geoportals promote solving problems on integration of three-dimensional visualization of geological objects and geoinformation systems (GIS).

Therefore, a very urgent problem is to develop an information and analytical system based on GIS technologies taking into account the specifics of geological problems and features of the East Kazakhstan region. Integration of geographic information systems with 3D-modeling and 3D-panning technologies will allow visualizing geological objects with reference to geographical coordinates and combining data into a single information and analytical system.

2   Review of the Literature

The feasibility of using geographic systems in different fields of science is beyond doubt. Many works are devoted to the consideration of GIS application in various sectors of agriculture. For example, the researchers in the papers [1–3] note the importance of geoportals in tourism in terms of their economic efficiency and stages of implementation, in their works, moreover, it is said large value of such a resource for the development of tourism in the scale of individual regions. Papers [4,5] are engaged in the comparison of geoportals. The uniqueness, scientific and cultural significance of the natural paleontological object - the Tarkhany geological section is revealed in the works [6,7]. The possibilities of using unmanned aerial vehicles to create 3D models of geological objects and monuments of historical and cultural heritage were considered in the works [8–11]. Note that problems of the 3D model creation, reconstruction, localization, classification and recognition some special objects were solved in author's papers [12–15] and was significantly used for this research. The problems of creating three-dimensional geological sections based on photogrammetry were investigated in the works [16,17].

3   Problem Statement

The given work aims to develop information technology for obtaining a virtual model of a geological object for further placement in an integrated information system.

3D-modeling of the territories with unique geological objects located in East Kazakhstan region and their further integration into the geographic information system and information-analytical system will contribute to the development of the tourism sector of East Kazakhstan, as the three-dimensional view of the area on the screen is more visible for people inexperienced in cartography, clearer and more attractive.

4 Materials and Methods

The existing computer technologies allow us to see how this or that object or territory looks with the necessary level of detail, as well as to simulate its changing due to the course of some processes or realizing our decisions in practice. Methods of modeling and visualization of three-dimensional models implemented in modern geographic information systems allow to:

- create models of arbitrarily large areas;
- create models with a high degree of detail;
- provide images of objects with interactive reference information;
- simulate changes in terrain;
- assess the impact of the changes.

And the user of such systems is given the opportunity to:

- "have a look" at the study area or its individual objects from different distances and from different angles of view, at different times of the day;
- "walk" or "fly" over the territory of interest;
- perform various measurements and calculations;
- observe and assess the impact of changes;
- "see" the positional object orientation invisible under normal conditions (for example, underground structures);
- add or remove features to assess their impact on the terrain or its characteristics;
- and much more.

Three-dimensional maps created by means of GIS, provide high visibility and interpretability of data, enable the most complete transfer of information about changes in objects and the studied environment in process of time, as well as implementation of a number of applications which cannot be solved with traditional two-dimensional maps.

Theoretically, there are two approaches to obtain three-dimensional models of objects based on conventional two-dimensional images. The first allows to create the so-called pseudo-three-dimensional models with the illusion of volume, but in reality they are not three-dimensional. This approach in most cases is implemented by stitching photos with special programs-staplers in 360° panoramas and is actually used when creating a circular review, when the rotation of the object around its axis allows you to get a fairly complete representation of its appearance. Circular reviews are widely used in the Internet, where they give interactivity to web-pages, and are absolutely indispensable in interactive catalogs, as well as used in interactive presentations, encyclopedias, educational programs, etc. The second approach provides the creation of real three-dimensional models that can be exported to popular 3D-formats and continue their process-ing in three-dimensional modeling programs. If desired, such models can usually be exported to Java Applet for viewing in Internet browsers. In both cases, the basis of the models is a set of photos, and the process of their creation can be divided into two stages: directly photographing the object and generating the model.

These three-dimensional models are created in the following way: with the help of special built-in algorithms, applications convert the two-dimensional information contained in the photos into accurately calculated three-dimensional points, lines and planes, generate a polygonal model, and then carry out its tex-turing. In this case, not all operations are carried out by the programs in a fully automatic mode — they often require the user to participate in a partic-ular stage of modeling — in masking and removing the background, forming the frame of the model and/or its texturing. Therefore, this process cannot be called fast, although in general generating a 3D model takes much less time than a three-dimensional modeling.

We considered the technology of obtaining the so-called "real three-dimensional object model". This technology, which allows to obtain a three-dimensional model of a geological object based on its two-dimensional pho-tographs, in other words, photogrammetry, is described in [16,17]. Photogram-metry is becoming a highly effective alternative to laser scanning technol-ogy to create virtual models of geological objects, including outcrops. Three-dimensional digital model of a geological object can be created by using photos taken from different angles and following processing of these images using soft-ware. Three-dimensional image is provided by at least two images of the same scene taken from different positions. Knowing the position, orientation and focal length of the camera for each image allows you to calculate the position of any point in space from its two-dimensional coordinates in two images.

On the other hand, when the camera parameters (i.e. the position, orien-tation and focal length of the photo) are unknown, they can be obtained by two-dimensional coordinates of equivalent points in different photos, which in turn allow calculating the coordinates of points in space. From the above it is clear that determining the position of a set of points in different, but overlapping photos of the same scene is of great importance.

This question is under study from the beginning with Structure From Motion (SFM) algorithms designed to correlate points in images of the same scene taken from different positions and/or in different ones. SFM algorithms are currently implemented in various programs (e.g. Bundler, Microsoft Photosynth, Photo-modeller, Agisoft PhotoScan). For a given set of partially overlapping images, the SFM algorithms automatically detect a set of common points in each pair of images and then recognize the camera settings for each photo. This, in turn, allows you to define the 3D coordinates of each point, recognizable by at least two photos, and therefore create a point cloud representing the surfaces of objects within the target scene. Overlapping photos should be taken from several points of view. Therefore, you should use the same camera to minimize errors from using different lenses and camera sensors. It is also recommended to set a fixed focal length. Photos should also be taken sequentially or at least under the same conditions. These guidelines ensure that each part of the scene represented by a similar pixel pattern in different photos makes it easier for SFM algorithms to recognize points. These procedures provide and maximize point recognition, allowing you to create denser point clouds.

The algorithms of shooting various objects were worked out and adjusted while performing practical shooting and further processing of images, in order to obtain images suitable for further processing by special programs designed to create 3D models and orthophotos.

The most important steps of photography algorithms for a small object are as follows:

- choosing and ensuring the right lighting;
- camera setting, shooting with the same focal length of the lens;
- shooting around the perimeter of the object with a small offset, no more than 7°, the overlap of the photos should be about 80%;
- at the final stage, the circular shooting is carried out at an angle of 45°.

However, to obtain photographs of geological objects, as a rule, ground photography is a rather laborious and sometimes impossible process, since geological objects occupy vast territories of several hectares. In this case, photography with controlled unmanned aerial vehicles (drones) is a proper and modern solution. The possibilities of using remote-controlled pilotless vehicles to display geological structures are described in [11].

The drone can be controlled manually using the remote control and with software that allows you to set flight routes and remotely control the drone from your mobile device. Software control of the quad copter can be carried out with such applications as DJ IGo, Altizure, and 4dPixCapture. We consider the quad copter control using the DJI Go app (Fig. 1).

The DJI Go app is as much an integral part of the quad copter as any other part of it. Without it, the copter is able to perform only short-range flights, while the pilot remains "blind", because he does not have information about the current state of the drone and cannot control its camera. To use the application, you need to install it on your mobile device, and then connect to the Wi-Fi network created by the remote control. The application is a screen on which the video is broadcast from the quad copter, there are all sorts of information and buttons on the borders of the screen: spaced there are buttons for shooting and recording video, on the left – a window with a radar and a map with a marked copter on it, on top – flight modes, battery charge, and GPS signal quality, at the bottom – the current speed and altitude. This program allows you to set the trajectory of the geological object and survey parameters. Once the photos of the simulated object have been obtained, you can start processing them to create a 3D model of the object. It should be noted that obtaining a high-quality model will require about 50–100 photos of the object.

Currently, there are a number of software products for three-dimensional photogrammetry. In such programs the information on each photo is written to a

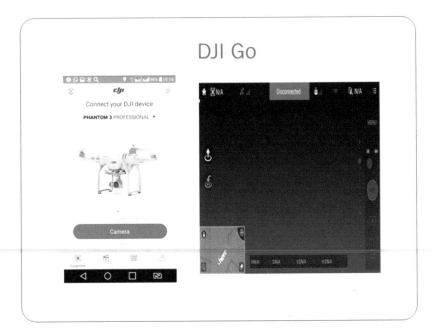

Fig. 1. Software for drone control

special file: the height, the rotation angle of the camera, the data of longitude and latitude. The program uses machine vision and photogrammetry technologies to find common points in many photos. As a result, each pixel in the photo is color matching in other photos. Each match becomes a key point. If a key point is found in three or more photos, the program builds this point in space. The more such points, the better the coordinates of the point in space are determined. Therefore, the more intersections between photos, the more accurate the model will be. The intersection of 60 to 80% is optimal. The spatial coordinates of each point are calculated by triangulation: the visuality line is automatically drawn from each survey point to the selected point, and their intersection gives the desired value. Moreover, algorithms are used in photogrammetry, the purpose of which is to minimize the sum of the squares of the set of errors. Generally, the Levenberg-Marquardt algorithm (or the ligament method) based on the solution of nonlinear equations by the least squares method is used for the solution. During photo processing, an extended point cloud (a collection of all 3D points) is created, which is used to generate a surface composed of polygons. Finally, it calculates the resolution and determines which pixels in the photo correspond to which polygon. To do this, the 3D model is deployed in a plane, and then the spatial position of the point is put in accordance with the original photo to set the color [18].

5   Experiment

Let us consider the stages of creating a 3D model and orthophoto on the example of modeling the territory of a unique geological object – Tarkhany geological section. In the world there is only one geological section with the similar properties, in the mountains of the Ardennes (Belgium), but Tarkhany geological section exceeds it several times in fossil fauna. The fossil organisms of this section are the reference to determine the boundaries of the geological age of rocks between Devon and Carbon. Tarkhany geological section is located on the Western edge of the upper end of the Tarkhanka village, located in the picturesque valley of the Ulba River 27 km upstream. Tarkhany geological section is a natural rock yield in the sculptural terrace of the right side of the valley of the Ulba River. The picturesque valley of the Ulba attracted the attention of its ancient inhabitants – the djungars, from who gave the name of Tarkhan (Prince, leader). Since the time of Ivan Grozny a great landowner was called the Tarkhan [7].

There are various photogrammetry packages that have a wide range of costs, easy to install and use, and can export results. Agisoft Photoscan was chosen for our investigation because of its user-friendly interface, academic licensing and tools for exporting results in a variety of formats. You should use only original images for PhotoScan as they appear on the digital camera. The use of photographs with geometric transformations or cropping is likely to result in negative or extremely inaccurate results. To restore a textured 3D object model, Photo Scan requires four processing steps. Pictures of the geological object are provided by the university scientists A. Okhotenko and A. Bubnyak. Shooting Tarkhany geological section was carried out with the help of an unmanned aerial vehicle Phantom 3 Standard (Fig. 2) in cloudy and windless weather.

Fig. 2. A Controlled unmanned aerial vehicle Phantom 3 Standard

Fig. 3. A set of Tarkhany geological section photos

After receiving a set of photos (Fig. 3) of the object under study, the construction of a three-dimensional virtual object model was carried out in the laboratory of the University using Agisoft Photoscan.

6   Results and Discussion

The first stage of building a three-dimensional model using Agisoft Photoscan program is to determine the positions and parameters of the external and internal orientation of the cameras. At this stage, PhotoScan finds common points of photos and determines all camera parameters: position, orientation, internal geometry. The stage results in a sparse cloud of common points in the 3D model space and data on position and orientation of the camera (Fig. 4).

PhotoScan does not use the point cloud for further processing, and it uses it only for visual assessment of the quality of photo alignment. However, the point cloud can be exported for further use in external applications. A set of data on the position and orientation of the cameras is used in the further stages of processing.

In the second stage, PhotoScan builds a dense point cloud based on the camera positions and photos calculated in the first stage of processing. Before

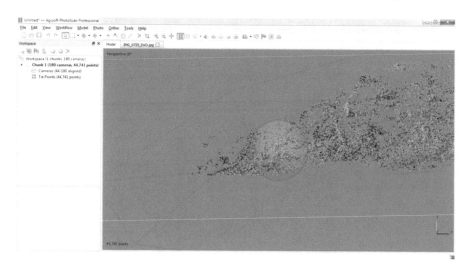

Fig. 4. Sparse point cloud

moving to the next stage of creating a 3D model or before exporting a model, a dense point cloud can be edited and classified.

In the third stage, PhotoScan builds a three-dimensional polygon model describing the shape of the object based on a dense point cloud. It is also possible to quickly build a model based on only a sparse point cloud. PhotoScan offers two basic algorithmic methods for constructing a polygon model: an elevation Map for flat surfaces (such as landscape or bas-relief) and a Random map for any type of surface. Photoscan provides some editing tools for the restored model, allowing you to optimize the model, remove isolated components of the model, fill holes, smooth, etc. Also, a polygon model export to make changes in an external editor and then a model import back into PhotoScan are provided.

The texturing and/or building orthophotomap are performed at the final stage. The created three-dimensional model can be exported to various formats, such as pdf. Figure 5 shows a textured 3D model of the territory of the geological object of Tarkhany Geological Section.

If in the process of photographing the binding to the terrain is set, for example, using the GPS, creating a 3D model can be automatically bound to the ground point. When shooting from a drone, it is easy to implement this, since unmanned aerial vehicles are equipped with built-in GPS. The Photoscan program automatically determines its geographical coordinates for each specific ground point. If there is a binding to the terrain, then you can create an orthophotomap, in other words, an ortho mosaic, and export it to the .kmz and further integrate it into GIS (Fig. 6). Building 3D models and orthomosaics of all necessary geological objects and integrating them with GIS can be made similarly, according to the technology described above.

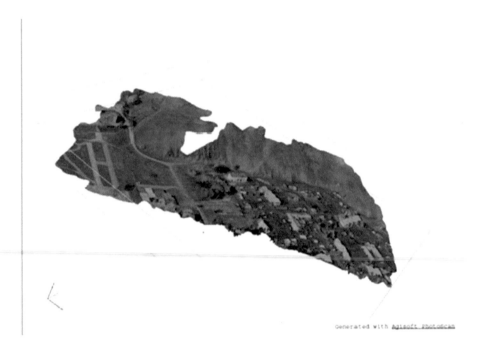

Fig. 5. A three-dimensional model of the Tarkhany geological section

Fig. 6. Orthomosaic in GoogleEarth

7   Conclusions

The described technology of construction of spatial models of geological objects based of photogrammetry, including the algorithm of photographing from ground cameras and cameras installed on the remote-controlled pilotless aircrafts, as well as the use of software that implements algorithms Structure from motion. They can be used both for creating 3D geological objects models occupying vast territories, and for creating small 3D models of objects located either on the route or on the object, for example, infrastructure facilities, monuments, and paleontological samples. Obtained by the described method, 3D models and orthomosaic can be exported to various formats, which allow you to integrate them with GIS, such as a popular Google Earth.

Based on the obtained data, the features of various GIS will be studied in the future, the most applicable GIS for the integration of 3D visualization elements will be made, the integration technology for visualization elements of geological objects and GIS will be developed.

References

1. Shaytura S (2015) Use of geoportals in e-Commerce. Vestnik MGOU 2:120–126 (in Russian)
2. Ushakova E (2013) Effective implementation of geographic information management systems for regional tourism development resources. Rossiyskoe predprinimatelstvo 21(243):76–85 (in Russian)
3. Robinson B, Ushakova E (2013) Issues of improving the management of regional resources for tourism development. Vestnik SGGA 3(24):63–71 (in Russian)
4. Koshkarev A (2008) Geoportal as a tool for managing spatial data and geoservices. Prostranstvennyye dannyye 2:6–14 (in Russian)
5. Safaryan A (2016) Geo-portal for tourism as a tool of research results visualization and promotion of destination. Serv Russia Abroad 4(65):56–70 (In Russian)
6. Mikhaylova N, Beysembayeva R (2016) Paleontological nature monument as an object of scien-tific tourism in the East Kazakhstan region. Nauka i mir 12(40):36–38 (in Russian)
7. Mikhailova N, Beisembaeva R, Salykbaeva G (2016) Features of the development of scientific tourism in the East Kazakhstan region. Vestnik VKGTU 1:19–22 (in Russian)
8. Kurkov V, Valdez P, de Jesus M, Blyakharsky D (2016) Creation of three-dimensional models of objects of monuments of historical and cultural heritage using unmanned aerial vehicles of aircraft and multirotor types. Izvestiya vysshikh uchebnykh zavedeniy. Geodeziya i aerofotos'yemka 2(60):94–99 (in Russian)
9. Shevchenko O, Borichevsky A (2015) Use of unmanned aerial vehicles for monitoring the use of territories. Ekonomika i ekologiya territorial'nykh obrazovaniy 3:150–152 (in Russian)
10. Saraev D (2017) Using of modern technologies for the construction of 3D-models of the area. Interexpo Geo-Siberia 10:126–128 (in Russian)
11. Yathunanthan V et al (2014) Semi-automatic mapping of geological structures using UAV-based photogrammetric data: an image analysis approach. Comput Geosci 69:22–32
12. Krak YV, Barmak AV, Baraban EM (2011) Usage of NURBS-approximation for construction of spatial model of human face. J Autom Inf Sci 43(2):71–81
13. Kryvonos IG, Krak IV (2011) Modeling human hand movements, facial expressions, and articulation to synthesize and visualize gesture information. Cybern Syst Anal 47(4):501–505

14. Tlebaldinova AS, Krak YV, Barmak AV, Denisova NF (2015) Localization and recognition of vehicle number plates by means of the method of support vector machine and histograms of oriented gradients. J Autom Inf Sci 47(10):24–31
15. Karymsakova IB, Krak IV, Denissova NF (2017) Criteria for implants classification for coating implants using plasma spraying by robotic complex. Eurasian J Math Comput Appl 5(3):44–52
16. Tavani S et al (2014) Building a virtual outcrop, extracting geological information from it, and sharing the results in Google Earth via OpenPlot and Photoscan: an example from the Khaviz Anticline (Iran). Comput Geosci 63:44–53
17. Bistacchi A et al (2015) Photogrammetric digital outcrop reconstruction, visualization with textured surfaces, and three-dimensional structural analysis and modeling: innovative methodologies applied to fault-related dolomitization (Vajont Limestone, Southern Alps, Italy). Geosphere 6(11):2031–2048
18. Ualkhanova A, Denisova N (2018) About three-dimensional visualization of geological objects based on photogrammetry. Vestnik VKGTU 3:240–251 (in Russian)

Information System for Connection to the Access Point with Encryption WPA2 Enterprise

Lyubomyr Chyrun[1(✉)] , Liliya Chyrun[2] , Yaroslav Kis[2] ,
and Lev Rybak[2]

[1] IT Step University, Lviv, Ukraine
chyrunlv@gmail.com
[2] Lviv Polytechnic National University, Lviv, Ukraine
lchirun21@gmail.com, yaroslav.p.kis@lpnu.ua, rybaklevko@ukr.net

Abstract. This work is to create an information system that would provide a connection to the access point with encryption WPA2 Enterprise. The practical value of the results obtained is to develop a more secure and perfect information system for working with a wireless access point. The introduction of this technology makes it possible to avoid data theft and speed up the Internet connection. Basically, it is aimed at the activity of private enterprises, but it can find its application and in scientific activity in the future. The developed information system is used by Point Dume Limited, which has already developed an analytical information system for mobile platforms. This information system is used in more than 100,000 access points in Japan, Hong Kong, the United Kingdom and the United States.

Keywords: Encryption WPA2-enterprise · Access point · Connection

1 Introduction

Any interaction between a client who wants to use a wireless connection and an access point is built on two principles: encryption and authentication. Authentication is a process of mutual representation and confirmation between a client and an access point that is, providing one another with information about the possibility of interaction. Encryption is the kind of algorithm used to transfer data used, how the encryption key is generated, and when it changes [1]. There are three kinds of authentication: open, with a general key and with expanding authentication protocol [2].

The untwisted authentication is the so-called open network in which all connected devices are authorized. This allows any device to try to establish a connection to an access point. However, this connection will only be available when the device keys Wired Equivalent Privacy (WEP) meets the WEP access point criterion. In turn, devices that do not support WEP will not even try to reach

© Springer Nature Switzerland AG 2020
V. Lytvynenko et al. (Eds.): ISDMCI 2019, AISC 1020, pp. 389–404, 2020.
https://doi.org/10.1007/978-3-030-26474-1_28

this access point. Also, open authentication does not require any external servers to connect [1–5].

A common key is a kind of key in which the correct connection of the device is checked by the password and the login to the access point. Authentication to a shared key is provided in accordance with the IEEE 802.11bstandard. With this type of authentication, an unencrypted call is sent to the device that tries to establish a connection to the access point. After this, the device asking for authentication handles text encryption for the call and sends it back to the access point. In the case of the correct passage of this process, the device that sent the request for access will be allowed to make authentication. Similarly, to open authentication in this case, there is no dependence on external servers. Connect with Extensible Protocol Om Authentication is a kind of connection in which the correctness is checked with the help of an external server. This type of authentication is best suited to provide the highest level of security for the wireless network. The client and server Radius performs inter authentication, resulting in the creation of a unique dynamic WEP key, with the use of an expanded authentication protocol. At the same time, the WEP key is sent to the access point from the server and RADIUS. To handle all single-digit data signals received or sent by the client to the server, the access point begins to use this key. In addition, WEP key encryption is implemented using the client's unicast key and is consistently sent to the client [2]. But except correct authentication, even when used open network is needed also a match for the encryption algorithm. There are such algorithms Encryption: NONE, WEP, CKIP, TKIP and AES. None is a complete absence of at least some encryption, all data is transmitted in the open form [6–8].

2   Literature Review

Wired Equivalent Privacy (WEP) is one of the classic protocols, which was one of the most popular and most reliable in the past [1–5]. But eventually, this protocol is much less secured than it was thought. For data encryption secret keys are used. Both sides of the connection should have information about these keys. We used two types of keys: 64-bit and 128-bit. The longer one is a bit more reliable, however, still does not provide sufficient level security. Although actually used 40 bit and 104 bit length, and others were taken for a variable that is called the vector of initialization IV (Vector Initialization (IV)). The first problem found was the shortness of length IV, because when using 24-bit keys, it is possible to get only 16.7 million variations, which is unacceptably small, because this figure can be reached in a few hours [9–15].

Another problem is reuse values initialization vector. Generally, the standard does not indicate that the value needs to be changed. For repeated use vector of initialization, it issues are the same keys protection data, and since he's up to that still short, these keys are repeated after a relatively short time in a busy network. One of the main cryptographic weaknesses of any security system is the reuse of keys. When sending a package it encrypted with help combinations vector of initialization and secret key. That it is different for each package, but the identification key. As a result, about to hold batch data acquire look like coincidence that causes problems with read out the entrance message for third-party Users who do not know the key. Except this, the initialization vector also does key flow vulnerable. Different wireless adapters can generate are the same vector sequences or use a constant meaning for them the designation, and in turn with the 802.11 standard, does not indicate in any way what is being done or changing the IP. As a result, traffic can be written and used for decryption encrypts text. Another problem is that encryption key RC 4 partially consists of the initialization vector. This leads to the fact that it is easy to successfully attack the network because the known 24-bit number of each key and the key RC4 are weak.

Next, the lack of such a protocol is a lack of creeps protection integrity because stream ciphers are combined with unencrypted data. It various improvements aimed at enhancing WEP system, but despite this, it remains extremely vulnerable. The systems relied on WAP others began to be updated and then WEP in 2004 Prot count has ceased to be used [5]. A new generation of protocols started with WPA. For WPA the length of the password has changed and may have a size from 8 to 63 bytes, which makes it difficult to select from WEP. More modern algorithms of encryption were used TKIP. This encryption algorithm began to apply a key system for each packet, which greatly increased security over a fixed key than WEP. Another positive change in WPA was the verification of message integrity, which allowed determining whether the packets that were transmitted between the access point and the client were changed or captured. The disadvantage of WPA was the lack of security of an additional system that used a secured wireless connection setting (WPA - Wi-Fi Protected Setup) to attach devices to modern access points [16–19].

Subsequently, WPA began to displace more modern protocol WPA 2. To change the encryption algorithm TKIP, came a more modern CCMP (message authentication coding protocol with blocking channels). This has led to greater reliability of this protocol, which can be attributed to its benefits. However, the main disadvantage of this was the ability to attack with the help of secured installation of a wireless connection (WPA - Wi-Fi Protected Setup), however, it should be noted that the time needed to use this vulnerability is already 2 to 14 h of continuous computer use effort [5].

The most up-to-date protocol is WPA2. Select two of its subspecies: Personal and Enterprise. The difference between these protocols lies in the places where the encryption keys are used, which are used in the mechanization of the AES algorithm. When using the Personal protocol, you use a static key and password that has a length greater than 8 characters and can be manually changed to the access point settings. It is the same for all users. The key to this problem is that when it is lost and consistently changed the password manually, this password must also be changed for all network users. This condition can easily be fulfilled with a small number of such users. For example, this network is used at home, but for corporate purposes, when several hundred devices are connected, this task is greatly complicated. Also, the problem is the low protection of such a network from third-party users who can simply pick up a static key. For corporate (Enterprise) the protocol uses a dynamic key that is individual for each client at a given time. Also, such a key is periodically updated for work without developing an Internet connection. This key is generated using the authorization server. This is usually a radius server. The special part of the operating system on the client side interacts with the authorizing part - the AAA server. If the system is properly organized, the role of the middleman to establish a connection between the client and the server assumes the access point. During this data with the sides the client are transmitted formed in the protocol 802.1h, and on the controller side they already have rotate in Radius - pack you need for the server [20–25].

If you use the EAP authorization mechanism, the access point requires authorization from the client in the infrastructure RADIUS server after it has passed the local authentication to the access point itself [26–32]. This access to a closed network using EAP is possible only by the theft of not only the login and password to this network but also the necessary connection certificate, which is possible only when the system is responsible for the protection of the network as a whole [33–39].

3 Problem Statement

The theme of security and reliability wireless Internet connection improvement is very relevant in a world as the impact of this kind of connection only growing and replacing or significantly reduces the proportion of other types. Wireless Internet connection is commonly used to transfer digital data streams on radio channels in areas where installing cable connection is impossible, impractical or simply uneconomical. Current implementation wireless and allow the network connection to move freely between access points covered by this network and receive data using mobile devices equipped with transceiver parts and accordingly access. Establishing a system with an improved connection is also appropriate for Ukraine because the number of wireless devices is increasing every day and the demand for improved their Internet connection is increasing accordingly. The developed system will solve the problem of data leakage, re-registration of users and overloading the network with unwanted devices. There are a number of phenomena that are problematic when working with wireless networking, such as security bypass, access to personal data and overload of the local area network. These problems are chosen for analysis and more detailed study in this paper. These problems are generated by the weaknesses in the protocols themselves, namely, the ways in which they are encrypted and authenticated. The following types of shielding are distinguished: NONE, WEP, CKIP, TKIP and AES. And the following types of authentication: with a public key, with a common key and with expanding authentication protocol. In order to protect against the maximum number of problems, the goal is to create a system for work on the most advanced level of encryption and authentication currently WPA2 Enterprise, which will use the AES encryption method and the method of authentication with the extended protocol and will guarantee the reliability, speed and security of data transmission.

The main results of the work are the following:

Made improvements to:

- Algorithms encryption and tools authentication are used by access points.
- Protocols that use a local network.
- Dissemination Information on installing a connection with an access point between its users.

Get further development:

- System for security and speed of data transmission in a local network
- Recommendations for using networks with more advanced security protocols.

4 Materials and Methods

To begin with, to construct a tree of goals (Fig. 1), one must formulate the general purpose. When working with our system it is: create an information system that would allow connecting the device to the access point with the level of protection WPA2 Enterprise. Its achievements are achieved through the implementation of the following objectives of the first level: to develop a user interface, install a component and required to connect the device and install the app access point. Tier 1 targets also include second-tier goals. For example, to install the components required to connect devices, you need to install a certificate and set up a network profile. For the purposes of the second level is also necessary and goals of the third level. Now, after constructing a goal tree, it is necessary to determine alternative ways to build the system as a whole, for this I will use the method of analytical hierarchy. This method is used when choosing a solution from a set of objects for multi-criterion and multi-criteria problems. When using this method, qualitative characteristics of alternatives are considered and decisions are made using pairwise comparisons. The greatest advantage of such a method is the absence of the dimension of the measured values for their pairwise comparison, thus the need to lead them to the same unit of measurement disappears. That is, the absence of physical or objective units of measurement is not a problem, only the mutual importance, probability or attractiveness of parameters is considered. The method of an analytical hierarchy is a procedure used to identify a problem in the form of a hierarchical presentation. Several types of such hierarchies are distinguished: dominant, horal and modular.

In general, the method consists in decomposing the problem into smaller components and analyzing them in the form of pairwise comparisons. For comparisons, as a rule, subjective sources of information are used - expert judgments that are presented in the form of a quantitative assessment on a scale from 1 to 9. These figures represent an objective digital representation of the data transmitted verbatim, in informal form. Subjective expert judgments are used, because static data is scarce, and the amount of experience is not sufficient. For people,

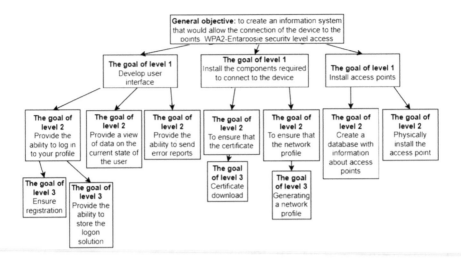

Fig. 1. Tree of goals

two features of analytical thinking are distinguished: the analysis and observation of the result obtained and the ability to link these observations with each other and synthesize them. The unit denotes the same importance in the decision, and the ninth is the obvious advantage of one of the parameters over another. Also, the method of an analytical hierarchy may include the process of synthesizing multiple judgments, finding compromise solutions, and obtaining the best criterion [12].

To begin with, you need to construct a graphical hierarchy for the method of the analytical hierarchy with three factors and three choices (Fig. 2). For the first criterion, choose the programming language, the second - the operating system, and the third - the development environment. Let's make such expert assessments: the programming language is significantly more important than the development environment; the operating system is equally important with the programming language; the development environment is less important than the operating system. Now it is necessary to construct a matrix of comparisons (Table 1). Mark A is the programming language, B is the operating system, and C is the development environment.

Table 1. Priority subcriteria and alternatives significance

Name	Priority		Name	Priority		Name	Priority	
	Internal	Global		Internal	Global		Internal	Global
C++	59.1	28.35	Windows	47.98	19.46	Qt Creator	42.89	4.91
Python	25.8	12.38	MacOS	40.57	16.46	Visual studio	42.89	4.91
C#	7.8	3.74	Linux	11.46	4.65	XCode	14.20	1.63
Java	7.1	3.41	–	–	–	–	–	–

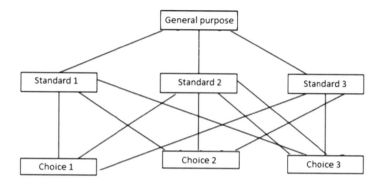

Fig. 2. Hierarchical structure

Table 2. Priority criterion

Categories	I		II		III		IV	
	Priority	Rank	Priority	Rank	Priority	Rank	Priority	Rank
A	47.9%	1	72.4%	1	42.9%	1	59.1%	1
B	40.05%	2	19.3%	2	42.9%	2	25.8%	2
C	11.45%	3	8.25%	3	14.2%	3	7.8%	3
D	–	–	–	–	–	–	7.1%	4

For each of the criterion, I will form a number of sub-criteria, on the basis of which the list of alternatives will be formulated. By using a pairwise comparison of these criterions, one can choose the best alternatives. For the programming language criterion, we select the following sub-criterion: C++, Python, Java, C#.

For operating systems are Windows, Linux, MacOS. For development environment are Visual Studio, Qt Creator, XCode. Based on the sub-criterion, it is now possible to form a plurality of alternatives and to calculate their priority. Suppose there are such alternatives: C++, Windows, Qt Creator; Python, MacOS, XCode; Java, Qt Creator, Linux. Let's make such expert assessments: Windows is significantly ahead of MacOS and is far ahead of Linux; Linux is less important than MacOS; Qt Creator is equal to Visual Studio and more important than XCode; Visual Studio is also more important than XCode; C++ is significantly overwhelming C#, Java and is more important than Python; C# and Java are equal in values; Python is significantly ahead of Java and is more important than C#. Next it is necessary to find similarity matrices of comparisons, normalized values of weight, proper and vectors, indexes of coherence, the ratio of coherence and criterion priorities for each criterion. On the basis of these data, using the analytical hierarchy method, one can choose the best alternative. Made necessary second criterion is data for criterion operating systems. Mark A is Windows, B is MacOS, C is Linux. First, we will form the matrix of comparisons (Table 1). By analogy, it has similar operations for two other criterions:

programming languages and development environment (Table 1). Mark A is Qt Creator, B is Visual Studio, C is XCode. Denote A is C++, B is Python, C is C#, D is Java and form the matrix of comparisons. Now we need to calculate the priority of sub criteria relative to global criterion. First, make it to the criterion programming language (Table 2). Global priority is 47,976 % required each local priority multiply him to get for each global criterion. Similarly, for the criterion: operating system and development environment (Table 2). Now for each alternative, you need to calculate the value for each of the alternatives selected and choose the best basis based on these results. We will mark them in order and will be reflected in the table (Table 3).

Table 3. The list of alternatives

Alternative	Value	List
Alternative 1	52.73%	C++, Windows, Qt Creator
Alternative 2	30.46%	Python, MacOS, Pycharm Editor
Alternative 3	12.96%	Java, Qt Creator, Linux

As can be seen from the results, the best alternative among the selected ones is the first, namely, C++, Windows, and Qt Creator. With a decent option, which was chosen to build the system, detail its structure with the help of a functional top-level functional diagram decomposed to the first level, and then specify each of its elements using Workflow - diagrams. For the context diagram, the central element is a functional block, which briefly describes the entire essence of the function execution in one sentence. This functional block contains 4 key streams that are left, right, top and bottom. To the left are the inputs, which are those inputs that are necessary to get started and which carry their cost completely. On the right there are outputs, that is, the result of a functional block execution. Below are the mechanisms that reflect who should perform what works and use what the means at the same time, they transfer their cost only partially. At the top, there are controls that show which controls are used (Fig. 3). Next, for each of the stages of decomposition hold (Figs. 4, 5, 6 and 7).

5   Experiment

The program for connecting to an access point with WPA2- Enterprise protection level has the following attributes:

- The name of the executable file is RoamVU.exe.
- The size of the executable file is 1,038,336 bytes.
- Executable file icon.
- File version - 1.0.0.
- Product Version 1.0.0.

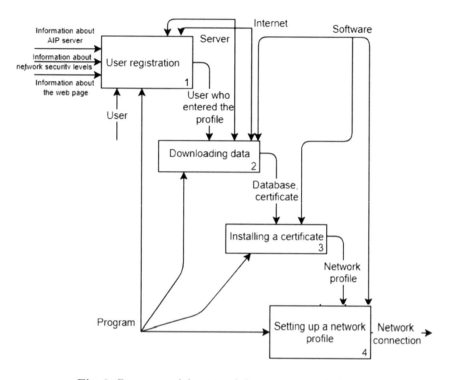

Fig. 3. Decomposed functional diagram of the first level

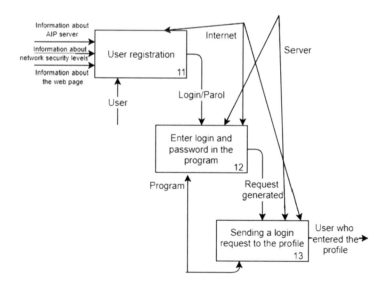

Fig. 4. Decomposed registration process

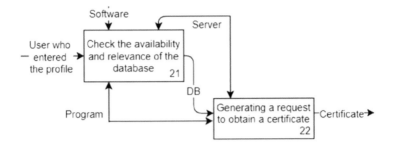

Fig. 5. Decompiled data download stage

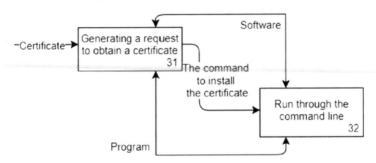

Fig. 6. Decompiled process of installing a certificate

- The internal name is RoamVU.
- The source file name is RoamVU.
- The source file name is RoamVU.exe.
- Product name – RoamVU.
- Description of the file version - 1.0.0.
- Manufacturer - Indeema Software Inc.
- Language - English (United States).

The software system tools to be used by the program are guaranteed to be presented in all versions of the operating system Windows 10 and in Windows 7 versions of professional and enterprise. In addition, an application requires several libraries and dependencies that are distributed and decompressed with the installation files for the program. List of components required for the RoamVU program:

- Media (qgenericbearer.dll, qnativewifibearer.dll).
- Icons (qsvgicon.dll, qgif. Dll, qsvg. Dll, qico. Dll).
- Platforms (qwindows.dll).
- SSL (ssleay32.dll, libeay32.dll).
- C++ libraries (libgcc_s_dw2-1.dll, libstdc ++ - 6.dll, libwinpthread-1.dll, d3dcompiler_47.dll).
- Qt libraries (Qt5Core.dll, Qt5Gui.dll, Qt5Widgets.dll, Qt5Xml.dll, Qt5S-ql.dll).

The programming language used to write the program is C++. The development environment is Qt Creator 4.5.0. Compiler - MinGW 5.8.0 32-bit.

The ultimate purpose of the program is to establish an association with an access point with the WPA2 Enterprise protection level. Intermediate stages necessary to implement this act website, login account, downloading and installing the certificate database and load generation network profile based on this database. Also, in order for the program to perform the necessary functions, it is necessary to set up an access point with the appropriate level of protection and make that point in the database.

The first functional constraint is the obligation to have the Internet at the first connection, as well as its availability at the registration site. The second one is the use of exclusively Windows-based operating systems. The third is the availability of 60 MB of RAM required for the program. The program has a graphical interface. The user enters his login and password and clicks the Sign In button to access the profile. The request is sent to the server, and if the registration was previously made successfully, the user is allowed to enter his profile. After that, the certificate is downloaded to install and install. A database is downloaded, data about access points is read from it and recorded in the network profile, which is installed in the system and there is an opportunity to connect to the access point. The program works in the background, but the windows About, Account Information on the system icons are unlocked and there is an opportunity to view user data. But the main property that appears is precisely the possibility of submissive to the access point. If this is nearby and is in the database, then such a connection will be made automatically by the operating system.

To create a graphical interface, the tools provided by the Qt Creator environment are used. When you install a certificate, you use the Windows command-line, WinApi, and the location for installing certmgr.msc. When establishing a network profile using database SQLite, team strips Windows, WinApi and net of known networks. To read data from the database, internal SQL Qt Creator libraries are used. For attachment, the means of the operating system are used systems Windows, provided there is an access point. To send a problem report, the default mail-messenger installed on the system will be used. The program is divided into three main blocks: for working with a graphical interface, for communication with the server and for general classes. For graphic interfaces, 9 classes have been implemented.

- CAboutWindow is a class for the graphic representation of the About window.
- CAccountInformationWindow is a class that graphically presents user information and its network connection.
- CExitWindow ia a class for graphical representation of the window to exit.
- CForgotPasswordMessage is a class for a graphical representation with fields for sending a request in the event of a lost password.
- CMessageBox is a class for representing windows with different text information.
- CSignInWindow is a class to represent the login window of the program. Also controls the password loss button.

- CTermOfUsageWindow is a class that displays a window with information about the rules of use.
- CWidgetBaseClass is a basic class for all graphic windows that control certain properties common to them.

To communicate with the server, use 3 classes.

- CSDK is the class that is responsible for sending requests to the server and processing responses.
- CAccountData is an auxiliary class, a decoder for profile information that allows you to write data that came from the server into the program field fields.
- CCertificateVersioning is an auxiliary class - a descriptor for information about the relevance of this certificate, which reads the response from the server into the fields of the program class and allows them to work in this way.

Common classes include 6 classes.

- CActivationTimer is a class that checks the status of the timer every 24 h.
- CCertificationInstaller is the class that is responsible for setting up a certificate.
- CDataBaseManager is a class for working with the database.
- CLogFileController is the class that is responsible for writing data to a log file.
- CMainProgramController is the base class controller for the rest of the classes.
- CTrayIcon is a class used to control and interact with the system icon.

To send reports are used in setting the default one email programs: OutLook, Windows Mail, and others. The necessary hardware is IBM PC with a processor up to or equal to 80386, operating memory of at least 128 MB, 70 MB of free disk space, availability of Windows operating system, availability of Internet connection. You can download the installer for the program from the Microsoft Store. Calling the program comes with .exe-file that is installed in the place selected during installation. An entrance id data is the username and password to log into the program account. The program does not provide any weekend data.

6   The Results Analysis and Discussion

Launch the program and when you first start the window displays the conditions of use (Fig. 7). Agree with them and go to the start menu, enter data. We are entering the profile successfully and see a message about successful activation. Now be downloaded a certificate and the operating system will display a warning about installing the certificate system. Now you can go to certmgr.msc and check the availability of the installed certificate. Also, in the folder/temp can be a database of Old network profile. You can also look at the presence of a network profile (Fig. 8). After that, the program runs in the background.

Fig. 7. Window Terms of Use and Start menu

Fig. 8. Network profile information and RoamVU in the background

7 Conclusions

During this work, the purpose was to create an information system that would provide a connection to the WPA2 Enterprise network. This system is relevant,

relevant and in-demand at the moment. The alternative variants of construction of the system as a whole, using the method of analytical are determined. The specification of the functioning of the system was performed by displaying it as a functional digraph.

A detailed description of the chosen C++ programming language has been made. It has been compared with other relevant languages, namely Java, Python, C#. As a result, it was substantiated the relevance. Chosen development environment Qt Creator with other commonly used Visual Studio. As a consequence, we can conclude that the right choice. Another described file extensions used in the program, namely p.12, Xml. The reasons for this choice are explained because of their essentially complete non-alternative. After that, the characteristics of the selected tools and their use were specifically described for the chosen system. Summing up the preliminary conclusions, one can declare the expediency of choosing the software for solving the task. A description of the created software was carried out in accordance with the standard GOST 19.402-78 "Program Description" as a result, it allows you to get acquainted with the developed software tool in more detail, namely: to understand the algorithms and methods of the product developed, to see the interrelationships between the classes of the program, you mean hardware and communication with other programs. This allows a deeper understanding of the essence of the developed information system. A detailed user manual has been described in detail in accordance with IEEE STD 1063-2001 Standard for Software User Documentation, which will make it easier for the user to understand how to use the program. In addition, a benchmark analysis was conducted to reflect the overall workflow of the program. An information system was developed, analyzed from the standpoint of system analysis, described in detail in the form of instructions for the user and in the form of description in accordance with the standard, demonstrating a control example of the program's performance. Taking all this into account, we can conclude that it was came now grounded and developed information system according to the requirements. The ultimate goal was achieved, namely, an information system was created that could provide a connection to the WPA2 Enterprise network.

References

1. Harris M et al (2012) Mobile and connected device security considerations: a dilemma for small and medium enterprise business mobility?
2. Trudeau P, Laroche S (2014) U.S. Patent No. 8,885,539. U.S. Patent and Trademark Office, Washington, DC
3. Sari A, Karay M (2015) Comparative analysis of wireless security protocols: WEP vs WPA. Inter J Commun Netw Syst Sci 8(12):483
4. Tsitroulis A, Lampoudis D, Tsekleves E (2014) Exposing WPA2 security protocol vulnerabilities. IJICS 6(1):93–107
5. Vanhoef M, Piessens F (2017) Key reinstallation attacks: forcing nonce reuse in WPA2. In: SIGSAC conference on computer and communications security, pp 1313–1328

6. Li J, Garuba M (2008) Encryption as an effective tool in reducing wireless LAN vulnerabilities. In: International conference on information technology: new generations, pp 557–562

7. Rumale AS, Chaudhari D (2011) IEEE 802.11 x, and WEP, EAP, WPA/WPA2. Tech Appl 2(6):1945–1950

8. Arana P (2006) Benefits and vulnerabilities of Wi-Fi protected access 2 (WPA2). INFS 612:1–6

9. Lashkari AH, Danesh MMS, Samadi, B (2009) A survey on wireless security protocols (WEP, WPA and WPA2/802.11 i). In: International conference on computer science and information technology, pp 48–52

10. Suresh L, Schulz-Zander J, Merz R, Feldmann A, Vazao T (2012) Towards programmable enterprise WLANS with Odin. In: Proceedings of the first workshop on hot topics in software defined networks, pp 115–120

11. Khasawneh M, Kajman I, Alkhudaidy R, Althubyani A (2014) A survey on Wi-Fi protocols: WPA and WPA2. In: International conference on security in computer networks and distributed systems. Springer, Heidelberg, pp 496–511

12. Robyns P, Bonné B, Quax P, Lamotte W (2014) Short paper: exploiting WPA2-enterprise vendor implementation weaknesses through challenge response oracles. In: ACM conference on security and privacy in wireless & mobile networks, pp 189–194

13. Vanhoef M, Piessens F (2017) Key reinstallation attacks: forcing nonce reuse in WPA2. In: ACM SIGSAC conference on computer and communications security, pp 1313–1328

14. Nussel L. (2010) The evil twin problem with WPA2-enterprise. SUSE Linux Products GmbH

15. Katz FH (2010) WPA vs. WPA2: is WPA2 really an improvement on WPA? In: Annual computer security conference, Coastal Carolina University, Myrtle Beach, SC

16. Harris AM, Patten PK (2014) Mobile device security considerations for small-and medium-sized enterprise business mobility. Inf Manag Comput Secur 22(1):97–114

17. Crainicu B (2008) Wireless LAN security mechanisms at the enterprise and home level. In: Novel algorithms and techniques in telecommunications, automation and industrial electronics. Springer, Dordrecht, pp 306–310

18. Abo-Soliman MA, Azer MA (2017) A study in WPA2 enterprise recent attacks. In: International computer engineering conference (ICENCO), pp 323–330

19. Radivilova T, Hassan HA (2017) Test for penetration in Wi-Fi network: attacks on WPA2-PSK and WPA2-enterprise. In: International conference on information and telecommunication technologies and radio electronics (UkrMiCo), pp 1–4

20. Lorente EN, Meijer C, Verdult R (2015) Scrutinizing WPA2 password generating algorithms in wireless routers. In: USENIX workshop on offensive technologies

21. Visan SA (2013) WPA/WPA2 password security testing using graphics processing units. J Mob Embed Distrib Syst 5(4):167–174

22. Sangani NK, Vijayakumar B (2012) Cyber security scenarios and control for small and medium enterprises. Informatica Economica 6(2):58

23. Bartoli A, Medvet E, Onesti F (2018) Evil twins and WPA2 Enterprise: a coming security disaster? Comput Secur 4:1–11

24. Gali TAB, Amin Babiker A, Mustafa N (2015) A comparative study between WEP, WPA and WPA2 security algorithms. Int J Sci Res. ISSN 2319-7064

25. Adnan AH, Abdirazak M, Sadi AS, Anam T, Khan SZ, Rahman MM, Omar MM (2015) A comparative study of WLAN security protocols: WPA, WPA2. In: International conference on advances in electrical engineering (ICAEE), pp 165–169

26. Nandi S (2019) Elliptic curve cryptography based mechanism for secure Wi-Fi connectivity. In: Distributed computing and internet technology: international conference, ICDCIT 2019, Bhubaneswar, India, vol 11319. Springer, p 422
27. Vysotska V, Fernandes VB, Emmerich M (2018) Web content support method in electronic business systems. In: CEUR workshop proceedings, vol 2136, pp 20–41
28. Vysotska V, Hasko R, Kuchkovskiy V (2015) Process analysis in electronic content commerce system. In: Proceedings of the international conference on computer sciences and information technologies, CSIT 2015, pp 120–123
29. Naum O, Chyrun L, Kanishcheva O, Vysotska V (2017) Intellectual system design for content formation. In: Proceedings of the international conference on CSIT computer science and information technologies, pp 131–138
30. Lytvyn V, Sharonova N, Hamon T, Vysotska V, Grabar N, Kowalska-Styczen A (2018) Computational linguistics and intelligent systems. In: CEUR workshop proceedings, vol 2136
31. Lytvyn V, Vysotska V (2015) Designing architecture of electronic content commerce system. In: Proceedings of the X-th international conference on computer science and information technologies, CSIT 2015, pp 115–119
32. Gozhyj A, Vysotska V, Yevseyeva I, Kalinina I, Gozhyj V (2019) Web resources management method based on intelligent technologies. Adv Intell Syst Comput 871:206–221
33. Gozhyj A, Kalinina I, Vysotska V, Gozhyj V (2018) The method of web-resources management under conditions of uncertainty based on fuzzy logic. In: 2018 IEEE 13th international Scientific and Technical Conference on Computer Sciences and Information Technologies, CSIT 2018 – Proceedings, vol 1, pp 343–346
34. Lytvyn V, Vysotska V, Dosyn D, Burov Y (2018) Method for ontology content and structure optimization, provided by a weighted conceptual graph. Webology 15:66–85
35. Kravets P (2010) The control agent with fuzzy logic. In: Perspective technologies and methods in MEMS design, MEMSTECH 2010, pp 40–41
36. Lytvyn V, Vysotska V, Kuchkovskiy V, Bobyk I, Malanchuk O, Ryshkovets Y, Pelekh I, Brodyak O, Bobrivetc V, Panasyuk V (2019) Development of the system to integrate and generate content considering the cryptocurrent needs of users. East-Eur J Enterp Technol 1(2–97):18–39
37. Lytvyn V, Kuchkovskiy V, Vysotska V, Markiv O, Pabyrivskyy V (2018) Architecture of system for content integration and formation based on cryptographic consumer needs. In: 2018 IEEE 13th international scientific and technical conference on computer sciences and information technologies, CSIT 2018 – Proceedings, vol 1, pp 391–395
38. Rusyn B, Lytvyn V, Vysotska V, Emmerich M, Pohreliuk L (2019) The virtual library system design and development. Adv Intell Syst Comput 871:328–349
39. Rusyn B, Vysotska V, Pohreliuk L (2018) Model and architecture for virtual library information system. In: International scientific and technical conference on computer sciences and information technologies, CSIT 2018 – Proceedings, vol 1, pp 37–41

Development of System for Managers Relationship Management with Customers

Yaroslav Kis[1]⬤, Liliya Chyrun[1(✉)]⬤, Tanya Tsymbaliak[1]⬤,
and Lyubomyr Chyrun[2]⬤

[1] Lviv Polytechnic National University, Lviv, Ukraine
yaroslav.p.kis@lpnu.ua, lchirun21@gmail.com
[2] IT Step University, Lviv, Ukraine
tanya.tsymbaliak@gmail.com, chyrunlv@gmail.com

Abstract. In the given work the questions of automation of work of managers in interaction with clients were investigated. Existing solutions in the CRM market were also explored and a target audience was identified whose issues were not solved by existing solutions. In this way, a software management system for customer relationship management was developed, as well as integration with Nova Poshta. After the introduction of the CRM system, the company identified a number of key points. Negative conversion from customers has grown. This is largely due to automation of the business process, which allowed managers to concentrate more on customer interaction. The company's total turnover has increased. The training cycle for new staff has decreased. Also, the time taken to collect and send the parcel was considerably reduced by integrating with the Nova Poshta. Thus, we can say that after the introduction of the CRM-system, the company achieved significant results.

Keywords: CRM system · Content management

1 Introduction

Customer Relationship Management (CRM) is a concept that embraces concepts used by companies to manage their customer relationships, including collecting, storing and analyzing information about consumers, suppliers, partners, and information about their relationship with them [1]. The CRM system is primarily aimed at automating the relationship between managers and clients. At the same time, after the introduction of the system, the number of sales increases by 20%. This is achieved by automating the business process. In other words, the system helps the manager to lead the client from the first contact to the final sale and in the future to re-sale [2]. CRM provides a model of interaction in which the centres of the entire business philosophy is the client, and the main areas of activity are measures to support effective marketing and warehouse work, sales and customer service. Support for these business goals includes collecting, storing

© Springer Nature Switzerland AG 2020
V. Lytvynenko et al. (Eds.): ISDMCI 2019, AISC 1020, pp. 405–421, 2020.
https://doi.org/10.1007/978-3-030-26474-1_29

and analyzing information about consumers, suppliers, partners, as well as the company's internal processes. Functional capabilities to support these business goals include sales management, marketing management, and customer service management, warehouse and call centres [3]. CRM needed by companies operating in a highly competitive market. Accordingly, increasing the level of customer loyalty offers advantages, as it allows you to increase the number of potential customers, the number of re-purchases, and reduce the cost of attracting new customers. This leads to an increase in company revenue [4].

2   Literature Review

CRM system allows you to keep in touch with a large number of customers, not to forget about them. Such a system gets a special significance in a large company, where it is simply impossible to track the demands of each client and to process tasks in a timely manner [5]. The CRM system allows you to organize the client's process and distinguish the different stages of sales [6]. Of particular importance is the acquisition of long sales cycles or large companies that have many warehouses with different products [7]. An example is a large enterprise that has several warehouses with different products and deals with wholesale sales [8]. For example, in the case when a customer makes a large order of products from different warehouses, the manager must contact each warehouse to check the availability of each unit of product, inform the customer about the availability of goods and place an order [9]. After that collect goods from all warehouses in one parcel and send it to the client. This procedure is quite lengthy, the company does not effectively use the time managers [10]. Under this scheme, the manager spends a lot of time on one order, specifying the information about the availability of the product with his employees. It should be understood that sales do not always involve a customer who has to buy something from the company [11]. This may be hiring new employees, attracting new partners, buying agro-products, etc. Accordingly, in different cases, sales will have different stages. The main goal of most companies is to build long-term mutually beneficial relationships with customers. For this, the CRM system provides the following capabilities [12].

- **Create a single information field.** This will not duplicate client information and each manager will be able to add additional data [13–15].
- **Ensure coherence of the interaction channels.** This way, several managers will be able to contact one customer. In this case, contacts can take place in different ways: telephone call, mail, personal meeting, any messaging program, etc. Communication with the client is critical to eliminating incomplete, outdated or conflicting information [16–19].
- **Increase the speed of decision making.** This can be achieved through the delegation of authority from the top down [20]. The CRM strategy implies that when you interact with a client on any channel, your employee is provided with complete information about all the relationships with him, based on

which he can make decisions. Data on this, in turn, is also stored and available at all subsequent contacts.

- **Increase the efficiency of attracting new orders.** By using direct marketing methods based on existing customer data. And also at the expense of systematic step-by-step work with potential clients in several directions [21–23].
- **Increase employee efficiency.** On the one hand, the functional CRM-system allows you to simplify and automate the execution of routine operations. On the other hand, the use of CRM-solutions has a positive effect in companies that work with a large number of customers at the expense of rational distribution of efforts among specialists [24, 25].
- **Minimize the human factor.** Thanks to the use of CRM-technologies, the company disappears rigid "tie" of the client to a specific employee and minimizes the role of the personal (human) factor (in the negative sense of the concept). This allows the company to maintain a history of customer relationships in case of personnel changes [26, 27].
- **Optimize the work of the company.** With the help of CRM- system, the manager will not need to call the warehouse at any time to clarify the availability of certain goods, as well as staff members, can see detailed information about the order and immediately create Nova Poshta bill in the system to send the order to the client [28].

Also, due to the use of automated centralized data processing, it is possible due to the processing speed - to make early detection of risks and potential opportunities [29].

There are three CRM approaches, each of which can be implemented separately from others [30, 31]:

- The operative approach is the automation of consumer business processes, which provides operative access to information during contact with the client in the process of sale and service. Includes marketing, sales, and service [32];
- The collaborative approach is a joint analysis of data characterizing the activities of the client and the firm. Getting new knowledge, conclusions, recommendations, etc. Uses complex mathematical models for searching statistical laws and choosing the most effective marketing, sales, customer service strategy [33];
- The analytical approach ensures the direct involvement of the client in the company's activities and the ability to influence the processes of product development, its production, service [34].

3   Problem Statement

At present, the main goal of each company is to effectively interact with customers. Successful cooperation leads to further cooperation and recommendations for the company's goods or services. Special attention deserves companies with direct and long-term contact with the end customer. These are companies that provide consulting and audit services or long-term sales.

The development of the company is connected with the expansion of the client base, as a consequence, an increase in the importance of processes of interaction with customers. However, such companies often encounter a large number of barriers associated with a weak technical implementation of business processes for customer interaction and the lack of optimized work warehouses of the enterprise. That leads to various losses and a significant reduction in the company's work.

One of the most obvious drawbacks is the lack of centralized control over the processes of interaction with customers. This leads to the fact that with the participation of several persons in the process, one of them in the performance of their duties controls the activities of another person. Another problem is the lack of client management tools. Often in companies that do not pay close attention to customer interaction, the data is stored in Excel tables and the Access database, for use with the warehouse, often used 1C. Such storage methods have many disadvantages. Excel spreadsheets have limitations on the maximum size of the letter, which imposes restrictions on the creation of large tables. Similarly, there are problems monitoring the incorrect input of data, which leads to the presentation of one client as several different. In addition, Excel does not contain tools for constructing complex reports and analyzes. Excel and Access tools do not offer a convenient solution for storing information about negotiations with customers and personal information about them. This leads to the fact that employees responsible for negotiating with clients often store important data in their personal records. As a result, new employees face the difficulty of obtaining important customer data that they will have to work with, and in the event of a company employee's release, the data will be irretrievably lost. To work with the warehouse at the moment the most convenient is the 1C system. It is convenient to keep a list of goods that are in stock, to make purchases and sales of goods, to perform redundancy of goods. But this system also has its disadvantages, namely, it is not oriented to the automated work of the warehouse, that is, it is convenient to record only the goods balances and accounting. Employees of the warehouse cannot provide custom roles, draw up an appropriate work plan for each employee. It takes a lot of time to process the parcel, because there is no integration of postal companies. All these disadvantages make it impossible to optimize the work of the warehouse. Identified problems can be divided into several groups according to the causes of their occurrence.

- Lack of centralized control of business processes of interaction with customers leads to a temporary cost of finding an executor at this stage of the course of the business process and ascertaining the status of the current process.
- Lack of storage of customer data and negotiations leads to duplication of data, temporary costs of finding the necessary information and data loss.
- The lack of automation of the workflow associated with business processes of customer interaction leads to a loss in time for document creation and the search for information to fill them out.

So, today, the development and implementation of an information system for automation of customer interaction is an urgent task that solves important problems in a company that deals with long-term sales. To solve these problems a system that combines the elements of automation of business processes of customer interaction and process processes is required. Many companies are now interested in optimizing their employees. In recent years, many studies have been conducted on enterprise performance analysis, the implementation of CRM systems in organizations and the analysis of implementation results. The main purpose of creating an information system is to analyze the results of research in this area, to formulate conclusions about similar existing systems, to develop a conceptual model based on own research and to create a software product.

4   Materials and Methods

Tree goals with quantitative indicators used as a means of making decisions, and is called the "decision tree". The main advantage of "decision trees" over other methods is the ability to link the matching of goals with actions to be implemented in the present. Each company every year tries to increase its income. You can do this in two ways. The first is that an entrepreneur can hire more sales managers, spend more on advertising for goods, which in our time is not at all cheap. That is, an entrepreneur needs to spend a lot of money on increasing staff and on the road and not always effective advertising. The second option is to optimize the enterprise by implementing CRM system. It will help to optimize the work of managers with clients, in general work units of the enterprise, to coordinate communication channels, to provide a single data warehouse. In my opinion, this is the most effective, which will ensure an increase in sales in the company and, accordingly, increase profits. This is all pictured on the tree for the purposes of Fig. 1.

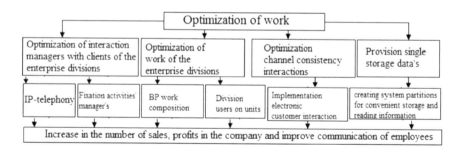

Fig. 1. Tree of enterprise goals

Thus, the main goal of the implementation of this CRM system is the single source of information for all employees of the company/enterprise, optimization of the relationship between employees and ensuring the efficient work of staff members, speeding up the processing of each client from the first application to the receipt of the order. We defined the main goals of the development of the information system, then we need to determine what needs to be paid the most attention during development. To do this, we use the decision-making method when evaluating alternatives by a single criterion. The task is to evaluate the four options that are most needed for long-term sales companies. Option A is for fixing all manager activity, to monitor every step of the job. Option B is for optimization of the work of the warehouse of the enterprise, the creation of the BC for the warehouse, integration with Nova Poshta. Option C is for introduction of IP telephony to record and save conversations with customers. Option D is for introduction of electronic interaction with the client, that is, communication via email. Consequently, option A is not much more important for an enterprise than variant B and less important than variant C (not much), option A and D are equally important. The multiplicative matrix of pairwise comparisons provided that the expert gives estimates on the fundamental scale is shown in Table 1.

Table 1. Multiplicative matrix of pairwise comparisons with key and vectors of priorities

	A	B	C	D	V	P
A	1	3	0.33	1	1	0.22
B	0.33	4	5	1	1.602	0.36
C	3	0.2	1	0.2	0.589	0.14
D	1	1	5	0.33	1.275	0.28
The sum	5.33	8.2	11.33	2.53	4.466	1

In order to get l_{max} you need to find the value of the main vector and the vector of priorities.

- $l_{max} = 5.33 * 0.22 + 8.2 * 0.36 + 11.33 * 0.14 + 2.53 * 0.28 = 5.405;$
- Coherence Index: $CI = \frac{5.405 - 4}{3} = 0.468;$
- Concordance ratio: $MRCI = 0.89; CR = \frac{0.468}{0.89} = 0.526.$

So, after finding the main vector and the priority vector, it turned out that these values are the largest for option B, that is, it is most important for entrepreneurs. Therefore, during the development of the system, we will focus on developing an information system to optimize the work of the warehouse.

The conceptual model is a systematic, meaningful description of a simulated system (or problem situation) in an informal language. An unformalized description is developed by the simulation model and includes the definition of the main elements of the simulated system, their characteristics and the interaction between the elements. It can use tables, charts, charts, etc. An unformalized description of the model is required both by the developers themselves (when verifying the adequacy of the model, its modifications, etc.) and for mutual understanding with specialists of other profiles. The conceptual model contains the source information for the system analyst who performs the formalization of the system and uses it for a certain methodology and technology that is, on the basis of an informal description, the development of a more rigorous and detailed formalized description is carried out.

The conceptual model of this information system is a database, which is presented in the form of tables for the easy perception of information. To construct a conceptual model of a database, entities, attributes, relationships between entities, primary and external keys are determined. Essences and relevant attributes in this database are:

- Contacts (unique identifier, name, surname, patronymic, contact type, counterparty, gender, age);
- Counterparties (unique identifier, name, type of counterparty, contact, description);
- Sales (unique identifier, headline, sales number, counterpart, contact, start date, end date, status, stage, product);
- Tasks (unique identifier, title, start date, end date, task type, status, contact, counterpart, sale, account);
- Products (unique identifier, name, price, stock);
- Parcels (unique identifier, sale, contact, type of payment, type of delivery, payer type, delivery technology, date of departure, volume, number of places, sender, sender city, sender's telephone, recipient, recipient city, recipient's phone, weight, form payment, description).

In the context diagram for our information system, we will display custom roles, namely the role of the administrator, manager, employee of the warehouse (Fig. 2). Also, we will depict all possible ways of interaction of users with the system, information flows from the user to the system and vice versa. In Fig. 3 depicts DFD information system. The following processes are presented: receipt of an order from the site, processing the order by the manager, collecting the parcel by the warehouse employee, forming a Nova Poshta consignment. The information system has three user roles: administrator, manager, employee warehouse.

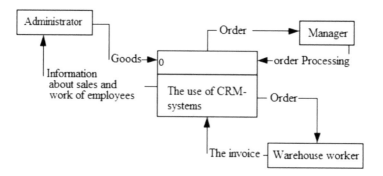

Fig. 2. Contextual DFD diagram

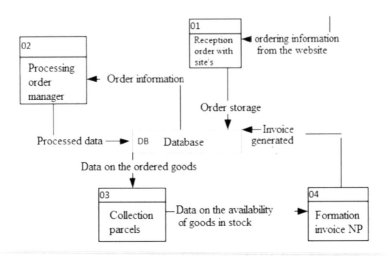

Fig. 3. DFD of the first level

In the CRM system from the side of the site gets an order, all information about this order is stored in the database, appointed responsible manager who has to process this order and pass it to the staff of the warehouse to collect the parcel and form a Nova Poshta bill. This information system allows business employees to view contacts who made purchases, or when they came to the company, see applications from customers, create sales, and staff members (according to the composition they work on) are able to see the packages they need to make, as well as can form Express Mail of Nova Poshta (Fig. 4).

The main goal of CRM implementation is to increase sales and profits. The CRM system allows you to build a sales management system, increase customer loyalty and increase sales. Loyalty, customer loyalty to the company is an important component of CRM. A loyal customer is a regular customer; he likes the products and services provided by the company. The most loyal customer brings constant income to the company, maintains a positive image of the company, and attracts new clients. Loyalty is possible due to the fact that the company provides not only high-quality products and services to the client, but also comes to meet its personal interests (personalization of the client). CRM-systems are needed for companies who want to build an effective sales management system and in which the client is the only source of income for the company. New CRM technologies in the sales, marketing, and service department are the key to the company's welfare. Also, CRM is needed for companies operating in a highly competitive market, because increasing customer loyalty is an additional leverage in the competition. Increasing loyalty allows you to increase the number of regular customers, the number of re-purchases, and reduce the cost of attracting new customers. It helps keeping contacts together with many customers, not to forget about them, so these systems are especially relevant in companies with a large number of customers. Finally, CRM systems will be useful in organizations that have a long-term

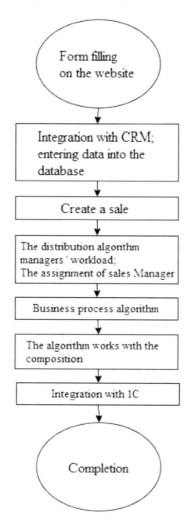

Fig. 4. The general algorithm of work of CRM-system at integration with an online store/site

sales process and include several steps, such as customer search and engagement, product presentation, preparation and negotiation of the transaction, contract, payment and delivery options, warranty service, and more.

The basic algorithm of work of the CRM-system when integrating with an online store involves the passage of several stages. These stages can be seen in Fig. 4. In this case, there are several smaller algorithms, such as integration with the online store, business process, and distribution of sales for the manager.

5   Experiment

CRM-system is implemented on the basis of client-server architecture. As a client, it can de considered a web application and mobile applications. To operate the CRM- system, the following main sections were implemented: Tasks; Contacts; Counterparties; Sales; Products; Users; Administration; Parcels. Each section provides access to the database. Tasks - a section where the manager can view the existing tasks for execution. It is important that each task can have several states: "New", "In Progress", "Undone", "Completed". Contacts - A section where the manager can access information about existing clients of the company or potential. It should provide search by name, surname or means of communication. From each entry in this section, you can go to the detailed view of the contact.

Counterparties - A section where the company manager can access comprehensive company information. At the same time, each company has its own type that can be customized in the directory directories. For example, the type may be: "Client", "Competitor", "Partner". There is a connection between the contact and the counterpart. The company may have many contacts, but one main one.

Sales are a section where the manager can see and keep all available deals. Each sale has its own status, which can be set automatically due to the functional business process. Products are a section that contains all existing products of a company or service. Ability adds a new product through manual data entry and automatic filling due to integration with a website or online store. It should also include the possibility of tracking stockpiles. This can be realized through integration with 1C.

Users are a section where you can view all existing managers of the company.

The administration is a section where you can configure managers' permissions.

Parcels are a section for staff warehouse, it is convenient to keep the inventory of products in store, store information about the order and create overhead Nova Poshta.

The CRM system implements integration with the customer's site. When integrating the system it is assumed that on the side of the site there will be a part of the script, which is responsible for sending data from the form to the CRM-system. That is, all the products that are on the site will automatically be transferred to the CRM system. When a customer makes a product order on a website, the ordering information, customer data and address of the order recipient are received in the CRM. According to this business process, the first task involves the top manager to appoint a responsible sales manager. In this case, you can apply different options for distributing managers for sales. It can be assumed that sales will automatically be allocated depending on the location or the load. In accordance for a manager a new sale is created, a task, a reminder that it is necessary to process the order, call the customer to clarify the details of the order, agree on the date of sending the order. As you progress through the business process, the tasks will be created automatically. Each subsequent task is created depending on the result of performing the previous task. Also as the promotion of the business process will change and the stage of sales. When the manager fully processes the order, he transfers the sale to the stage "Ready to pack". After that, the warehouse staff will check for the existing product. If a certain quantity of goods that the customer needs will not be in stock, but within a few days the product will appear, then such order will not be cancelled, and the required product will be reserved on the system. If the goods ordered by the client are in different warehouses of the company, then the employees of the relevant warehouses have the task of drawing up the parcel for the client. After the warehouse staff has collected the necessary goods in one parcel, they mark the parcel with the "Picked up" check and appoint a person who will check the collected parcel. If all products are in accordance with the customer's order, the checker puts in the checked-in checked checkbox. When the package is completed, managers place an appropriate entry in the system in the form of the next state of the sale "Send". Managers create in the system the waybill Nova Poshta, all product data, namely weight, number of units, price automatically tightened from the section Products, the recipient information is taken from the sale. The manager checks the correctness of the parcel data and prints a Nova Poshta bills. Immediately chickens the sender sends the parcel to the Nova Poshta, with the client will be automatically sent information about the fact of sending the ordered product.

6 Analysis of the Results and Discussion

The directory section contains all types of contacts, products, sales, counterparties. Actually here managers can add new types and in future use them in the corresponding section. Now let's go to the system under another user, who acts as a warehouse employee. Only sections that contain parcel information are available to this user. When an order comes from a site, the manager processes it and it falls into the section "Not collected parcels" for the employee of the corresponding warehouse. In Fig. 5 shows a detailed representation of an unverified parcel. An employee sees the list of goods ordered by the customer, his task is to check whether there are goods in stock if there is something to assemble them in a parcel. In order to make it easier for the employee to look for the goods in the warehouse, he can print the list of products by clicking on the "Print" button. After the worker collects the parcel, he will mark the "Completed" check mark on each editing card in front of each product and assign the person responsible for the check. After that, the parcel falls into the section "Unverified" in Fig. 6. When the responsible person has checked the collected parcel, it falls into the section "Checked" in Fig. 7. If all the goods are in the parcel, it is checked responsible and packed, then such parcel falls into the section "Collected" in Fig. 8. In Fig. 9 shows the section "Parcels", the records of which are formed on the basis of assembled parcels. This section is designed to create and save Nova Poshta bills. The overhead can be generated as an HTML page and a PDF file.

The User's Guide to this section provided a checklist of one of the possible variants of the business process in the company's warehouse. In this example, the work of the warehouse was demonstrated, namely, receipt of an order for a warehouse employee, processing an order, collecting a parcel, checking the parcel by the responsible person, generating an express Nova Poshta bill.

Fig. 5. Editing card "Not collected parcels"

Fig. 6. Unscheduled parcels section

As a result of the test, a new "Nova Poshta" bill was created, which means the system is working properly, the business process is working out to the end. This greatly facilitates the work of the staff members, because the system has everything you need to send the parcel. The employee does not need to go to the personal office of Nova Poshta and fill in all the data there. The system automatically transfers the buyer information from the "Contacts" section to

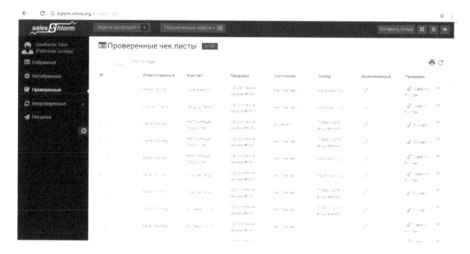

Fig. 7. Checked Parcels section

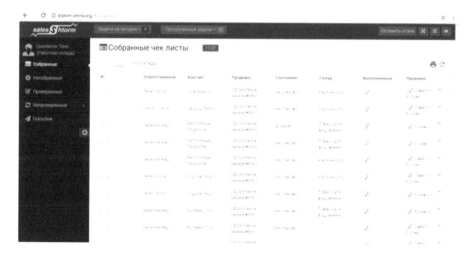

Fig. 8. The "Collected Parcels" section

the page that edits the express bills, as well as the product information from the "Products" section; the warehouse staff will only check the correctness of the data and generate the bills. This saves time considerably.

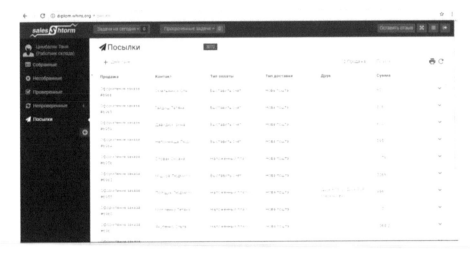

Fig. 9. Parcel section

7   Conclusions

In the given work the questions of automation of work of managers in interaction with clients were investigated. Existing solutions in the CRM market were also explored and a target audience was identified whose issues were not solved by existing solutions. In this way, a software management system for customer relationship management was developed, as well as integration with Nova Poshta. After the introduction of the CRM-system, the company identified a number of key points. First, ice conversion from customers has grown. This is largely due to the automation of the business process, which allowed managers to concentrate more on customer interaction. Secondly, the company's total turnover has increased. Thirdly, the training cycle for new staff has decreased. Also, the time taken to collect and send the parcel was considerably reduced by integrating with the Nova Poshta. Thus, we can say that after the introduction of the CRM-system, the company achieved significant results.

The main goal of CRM- system implementation for the enterprise is to increase the effectiveness of managers, the implementation of business processes, and improve the work of the warehouse. Many executives do not think about implementing the CRM system, but rather buy new equipment, hire a larger number of employees, but the company's profits do not increase, because there is no sale, there are no buyers for the relevant product. In our opinion, the CRM system will help increase sales without increasing the number of employees. After implementing CRM -systems at a certain enterprise, the waiting time from the customer's request to receive the goods will be significantly reduced, this will ensure the optimization of the managers through business processes, that is, the manager will constantly be reminded of every step of the business process and he will not be able to forget about the order. Integration with Nova Poshta is also greatly facilitated by staff members. Now, employees of the warehouse do not need to go to the Nova Poshta office with a prepayment and wait in line, and they will be able to form a consignment note to each parcel on their own. It saves the time of the employees, and also reduces the waiting time of the client on his order. Accordingly, the service is improving and the number of clients is increasing.

References

1. Khodakarami F, Chan YE (2014) Exploring the role of customer relationship management (CRM) systems in customer knowledge creation. Inf Manag 51(1):27–42
2. Thakur R, Workman L (2016) Customer portfolio management (CPM) for improved customer relationship management (CRM): are your customers platinum, gold, silver, or bronze? J Bus Res 69(10):4095–4102
3. Trainor KJ, Andzulis JM, Rapp A, Agnihotri R (2014) Social media technology usage and customer relationship performance: a capabilities-based examination of social CRM. J Bus Res 67(6):1201–1208
4. Stein AD, Smith MF, Lancioni RA (2013) The development and diffusion of customer relationship management (CRM) intelligence in business-to-business environments. Ind Mark Manag 42(6):855–861
5. Nguyen B, Mutum DS (2012) A review of customer relationship management: successes, advances, pitfalls and futures. Bus Process Manag J 18(3):400–419
6. Richards KA, Jones E (2008) Customer relationship management: finding value drivers. Ind Mark Manag 37(2):120–130
7. Fournier S, Avery J (2011) Putting the relationship back into CRM. MIT Sloan Manag Rev 52(3):63
8. Kim HW, Pan SL (2006) Towards a process model of information systems implementation: the case of customer relationship management (CRM). ACM SIGMIS Database: DATABASE Adv Inf Syst 37(1):59–76
9. Campbell AJ (2003) Creating customer knowledge competence: managing customer relationship management programs strategically. Ind Mark Manag 32(5):375–383
10. Chen Y, Li L (2006) Deriving information from CRM for knowledge management - a note on a commercial bank. Syst Res Behav Sci: Off J Int Fed Syst Res 23(2):141–146
11. Daghfous A, Barkhi R (2009) The strategic management of information technology in UAE hotels: an exploratory study of TQM, SCM, and CRM implementations. Technovation 29(9):588–595
12. Zeng YE, Wen HJ, Yen DC (2003) Customer relationship management (CRM) in business-to-business (B2B) e-commerce. Inf Manag Comput Secur 11(1):39–44
13. Rigby DK, Ledingham D (2004) CRM done right. Harv Bus Rev 82(11):118–130
14. Hung SY, Hung WH, Tsai CA, Jiang SC (2010) Critical factors of hospital adoption on CRM system: organizational and information system perspectives. Decis Support Syst 48(4):592–603
15. Chen IJ, Popovich K (2003) Understanding customer relationship management (CRM) People, process and technology. Bus Process Manag J 9(5):672–688

16. Ryals L (2005) Making customer relationship management work: the measurement and profitable management of customer relationships. J Mark 69(4):252–261
17. Kumar V (2010) Customer relationship management. In: Wiley international encyclopedia of marketing
18. Richard JE, Thirkell PC, Huff SL (2007) An examination of customer relationship management (CRM) technology adoption and its impact on business-to-business customer relationships. Total Qual Manag Bus Excel 18(8):927–945
19. Garrido-Moreno A, Padilla-Meléndez A (2011) Analyzing the impact of knowledge management on CRM success: the mediating effects of organizational factors. Int J Inf Manag 31(5):437–444
20. Lo AS, Stalcup LD, Lee A (2010) Customer relationship management for hotels in Hong Kong. Int J Contemp Hosp Manag 22(2):139–159
21. Ang L (2011) Community relationship management and social media. J Database Mark Cust Strat Manag 18(1):31–38
22. Ernst H, Hoyer WD, Krafft M, Krieger K (2011) Customer relationship management and company performance - the mediating role of new product performance. J Acad Mark Sci 39(2):290–306
23. Malthouse EC, Haenlein M, Skiera B, Wege E, Zhang M (2013) Managing customer relationships in the social media era: Introducing the social CRM house. J Interact Mark 27(4):270–280
24. Nguyen TUH, Waring TS (2013) The adoption of customer relationship management (CRM) technology in SMEs: an empirical study. J Small Bus Enterp Dev 20(4):824–848
25. Vysotska V, Fernandes VB, Emmerich M (2018) Web content support method in electronic business systems. In: CEUR workshop proceedings, vol 2136, pp 20–41
26. Vysotska V, Hasko R, Kuchkovskiy V (2015) Process analysis in electronic content commerce system. Proceedings of the international conference on computer sciences and information technologies, CSIT 2015, pp 120–123
27. Naum O, Chyrun L, Kanishcheva O, Vysotska V (2017) Intellectual system design for content formation. In: Proceedings of the international conference on computer science and information technologies CSIT, pp 131–138
28. Korobchinsky M, Vysotska V, Chyrun L, Chyrun L (2017) Peculiarities of content forming and analysis in internet newspaper covering music news In: Proceedings of the international conference on computer science and information technologies, CSIT, pp 52–57
29. Kanishcheva O, Vysotska V, Chyrun L, Gozhyj A (2018) Method of integration and content management of the information resources network. In: Advances in intelligent systems and computing, vol 689. Springer, pp 204–216
30. Su J, Sachenko A, Lytvyn V, Vysotska V, Dosyn D (2018) Model of touristic information resources integration according to user needs. 2018 IEEE 13th international scientific and technical conference on computer sciences and information technologies, CSIT 2018 – Proceedings, vol 2, pp 113–116
31. Lytvyn V, Sharonova N, Hamon T, Vysotska V, Grabar N, Kowalska-Styczen A (2018) Computational linguistics and intelligent systems. In: CEUR workshop proceedings, vol 2136
32. Lytvyn V, Vysotska V (2015) Designing architecture of electronic content commerce system. In: Proceedings of the X-th international conference on computer science and information technologies, CSIT 2015, pp 115–119

33. Gozhyj A, Vysotska V, Yevseyeva I, Kalinina I, Gozhyj V (2019) Web resources management method based on intelligent technologies. Adv Intell Syst Comput 871:206–221
34. Gozhyj A, Kalinina I, Vysotska V, Gozhyj V (2018) The method of web-resources management under conditions of uncertainty based on fuzzy logic. In: 2018 IEEE 13th international scientific and technical conference on computer sciences and information technologies, CSIT 2018 – Proceedings vol 1, pp 343–346

Probabilistic Inference Based on LS-Method Modifications in Decision Making Problems

Peter Bidyuk[1](✉) , Aleksandr Gozhyj[2] , and Irina Kalinina[2]

[1] National Technical University of Ukraine
"Igor Sikorsky Kyiv Polytechnic Institute", Kyiv, Ukraine
pbidyuke_00@ukr.net
[2] Petro Mohyla Black Sea National University, Nikolaev, Ukraine
alex.gozhyj@gmail.com, irina.kalinina1612@gmail.com

Abstract. The article provides information on the modification of the Lauritzen-Spiegelhalter (LS) method for constructing a probabilistic inference in the Bayesian network. Modification of the method consists in a new way of filling tables of conditional probabilities. The method consists of two stages: the first stage – the construction of a combined tree, the second stage – the construction of the distribution algorithm. At the first stage, the construction of a combined tree is performed by clicking on the primary structure of the network and filling the vertices of this tree with tables of conditional probabilities of the network. The second stage of the LS-method is the refinement of the distribution algorithm. The modified method allows us to more accurately calculate a posteriori probabilities and build a probabilistic inference. The example of evaluation of decision-making options for the construction of solar and wind power plants is considered. The features of the application of the method are analyzed.

Keywords: Bayesian networks · Probabilistic inference · Method of Lauritzen – Spiegelhalter

1 Introduction

One of the main tools for data analysis today is Bayesian Networks (BN). Most of the intelligence analysis tools are based on two technologies: machine learning and visualization (visual representation of information). Bayesian Networks is an approach that combines these two technologies. The main feature of Bayesian Networks in the tasks of intellectual data analysis is the ability to detect non-obvious, non-trivial, and unknown relationships between factors.

The main idea of Bayesian Networks is to present a description of the system being studied in the form of a graph edges of which reflect causal relationships that connect the main elements of the system. The universal nature of Bayesian Networks, as a data processing instrument, provides a possibility of solving the

© Springer Nature Switzerland AG 2020
V. Lytvynenko et al. (Eds.): ISDMCI 2019, AISC 1020, pp. 422–433, 2020.
https://doi.org/10.1007/978-3-030-26474-1_30

main tasks of the intellectual analysis of data. The versatility of using Bayesian networks to analyze processes of different nature and the functioning of technical systems allows taking into consideration and using the variety of data – expert assessments and statistical information. The variables used in Bayesian networks can be both discrete and continuous, and the nature of their receipt in the analysis and decision-making can be in real-time and in the form of statistical arrays of information and databases. Due to the presentation of the interaction between process factors in the form of causal relationships in the network achieved the highest level of visualization and, as a consequence, a clear understanding of the essence of the interaction of the process factors between themselves. It is the feature that distinguishes Bayesian networks from other methods of intellectual data analysis. There are such advantages of Bayesian networks as the possibility of taking into account the uncertainties of statistical, structural and parametric nature [1], the construction of models in the presence of hidden vertices and incomplete observations, as well as the formation of an inference using various methods is approximate and accurate.

The basic idea of constructing a graphical model in the form of a network is the concept of modularity that means the expansion of a complex system into simple elements. In order to combine individual elements into the system, we use the results of the theory of probabilities that provide the model for practical capability in general, and also enable to combine graphic models with databases. Such a graph-theoretical approach for model constructing gives a researcher the ability to construct process models on a set of highly interacting variables, as well as create data structures for the subsequent development of effective algorithms for their processing and decision-making. In general, we can say that Bayesian network is a high-resource method of probabilistic modeling of arbitrary nature processes with uncertainties of various types, which provides a sufficiently precise description of their functioning, estimates of forecasts, and building management systems.

There are many examples of Bayesian networks in various fields such as energy generating, defense, robotics, medicine, banking, financial analysis [3–11]. But regardless of their name, these networks do not necessarily mean a close connection with Bayesian methods. The name is associated, first of all, with the Bayesian rule for the formation of a probabilistic inference [2]. In the literature, Bayesian networks are sometimes referred to as Bayesian belief networks (BBNs), causal networks, or probabilistic networks. Bayesian Networks are a convenient tool for describing the operation of complex processes and events with uncertainties. They were particularly useful in developing and analyzing machine learning algorithms.

The formalism of constructing generalized graphic models combines many methods of statistical simulation, such as factor analysis, distribution analysis, distribution mix models, hidden Markov models, Kalman filters, Eizing models, and some others. All these models can be considered as part of the graphic models of the Bayesian type as separate examples of general formalism. The advantage of this approach is that the methods of research processes and data processing developed in one area can be successfully transferred into others.

Despite of the fact that Bayesian networks are given a lot of attention in the literature, many issues related to the principles of their construction, their training and use, as well as the organization of probabilistic inferences remain unexplored.

2   Problem Statement

The model construction of investigated process in the form of a Bayesian network allows considering comprehensively the process and predicting development of the situation, which in the end facilitates the search for a rational or optimal solution. Bayesian network analysis and modeling enable not only to identify existing interconnections and relationships in the system, but also to predict and presage phenomena and situations based on different models of probabilistic inference.

The task of this work is to develop and investigate a modified probabilistic inference algorithm based on the LS-method (Lauritzen-Spiegelhalter) and to determine the special features of its application.

3   Bayesian Approach to the Probabilistic Inference Formulation

The Bayesian approach to constructing models and formation of a probabilistic inference implies the pursuance of certain stages (Fig. 1). In Bayesian statistics, it is assumed that the information comes from two sources: a priori information from the researcher in relation to the problem under study and statistical (experimental) data obtained as a result of experiments.

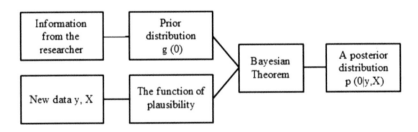

Fig. 1. Information flows in the system of modeling and decision making based on the Bayes theorem

A priori information from the researcher is the results of previous studies; theoretical foundations of the studied phenomena, processes and objects; additional informal data obtained from different sources. Overall – this is information that should not be based on experimental data related to a particular task.

In the center of the Bayesian methodology is the Bayesian theorem (BT). We write it on the basis of signs that are widely used in econometric analysis of data:

$$p(\theta|y, \mathbf{X}) = \frac{g(\theta)\, p(y|\theta, \mathbf{X})}{p(y)} \propto g(\theta)p(y|\theta, \mathbf{X})$$

where y is the vector of measurements of the main (dependent) variable of the investigated process; $\mathbf{X}$ is a matrix of measurements of independent variables (regressors) that determines behavior of the main variable; θ is the vector of random parameters that (together with $\mathbf{X}$) determines the function of the density of distribution of probabilities of a variable (in the classical regression model θ is the vector of parameters of the regression model); $p(\cdot)$ is function of probability distribution density (PDD); $p(\theta)$ is a priori probability distribution density for the values of a random parameter θ based on a priori knowledge of the researcher before the use of the data $(y, \mathbf{X})$; $p(y|\theta, \mathbf{X})$ is the conditional data density y at specific values θ and x, in other words, this is a function of probability for data y; $p(y)$ is marginal plausibility for (since the effect of independent variables and parameters is removed).

Density $p(\theta|y, \mathbf{X})$ is the posterior distribution for θ, which is based on updating the a priori distribution due to the data $(y, \mathbf{X})$. The BT shows how to integrate information coming from two sources (a priori distribution and data) in order to clarify the a priori distribution. In fact, the BT not only provides the opportunity to combine information from two sources, but other combination violates the logical (and mathematical) nature of the rules of operation with probability of distributions.

In the expression (1), the first version of the theorem is written as equality, as it contains a denominator $p(y)$ (unconditional density for y), which plays the role of a normalizing constant, and ensures that the a posteriori conditional density for θ is appropriate and integrates to 1 over the area of parameter definition. The second version of the BT is presented with the accuracy of the constants of proportionality, without the standardization, which is often done to simplify the representation of expressions. While solving practical problems the numerator is calculated initially and then, if it is necessary, normalizing constant. In many instances there is a need for computing ratios results and thus normalizing constant is reduced.

Methods for forming probabilistic inference in the Bayesian networks can be divided into two groups: methods of exact conclusion and approximate approaches [11]. There are the following algorithms for the exact inference: Pearl distribution algorithm for single-connected networks, clustering of a tree click, cutest conditioning algorithm, variable elimination algorithm, symbolic probabilistic inference, differential method. Approximation inference algorithms include the following methods and algorithms: exact inference partially, variational algorithms, methods, stochastic sampling, search-based methods, based on heuristic search algorithms that are used in the transition from the problem of calculating probability inference to an optimization problem.

4 LS-Method and Its Modification

The idea of the LS-method (Lauritzen - Spiegelhalter) for forming probabilistic inferences is the basic idea of clustering methods, their usage for implementation of probabilistic inference must first bring Bayesian network structure to form joint tree, and then use the algorithm to propagate messages in a tree "top-down" to successively list the tables of conditional probabilities of the vertices of a tree [9,10].

In general terms, the LS-method involves two stages. At the first stage, the construction of a combined tree is performed by clicking on the primary structure of the network and filling the vertices of this tree with tables of conditional probabilities of the network. At the second stage, the values of the probabilities of the vertex states are calculated on the basis of the algorithms for propagation of the probability values for the combined tree.

Based on the LS-method, more than one method for calculating probabilistic inference was created. The following algorithm of the modified method also uses the basic ideas of this clustering method. The proposed modification is based on the new principle of filling in the conditional probability tables describing the clicks of the combined tree [12–15].

4.1 Modified Algorithm of the LS–Method

The method consists of two stages: the first stage is the construction of a combined tree, the second stage is the construction of the distribution algorithm.

The first stage of implementing the LS-method is the construction of a combined tree.

At the first stage, the construction of a combined tree is performed by clicking on the primary structure of the network and filling the vertices of this tree with tables of conditional probabilities of the network. This stage is implemented in six steps:

1. *Moralization of the graph.*
2. *Adjusting the graph to the unplugged form.*
3. *Graph triangulation.*
4. *Identification of a click.*
5. *Building a combined tree.*
6. *Fill in the combined tree tables.*

Let's consider each of the steps in details.

1. *Moralization of the graph.*
   At this step, all the peaks of the Bayesian network, which have parents, are sequentially removed. If the parent vertices are not interconnected, then a link is introduced between them ("the neighbor").
2. *Adjusting the graph to the unplugged form.*
   For all tops, the "father-to-child relationship" is replaced by the "neighbor" link.

3. *Graph triangulation.*
   3.1 All the vertices of the BN are sequentially moved until all vertices are considered, following these steps:
   - Check whether the neighbors of the analyzed node are adjacent to each other. If so, then such a vertex is *sympissional* that forms a click together with its neighbors. Such a vertex is excluded from consideration along with its arcs;
   - If after the inspection the set of all network peaks remaining for consideration is non-empty, then proceed to step 2;
   - If the vertices are not left, then the graph is triangulated.
   3.2 In the network that is left for consideration, the vertices are sequentially sorted as follows:
   - Performing the search for the vertex with the largest number of neighbors; this vertex becomes simplified by inserting additional edges between its non-neighboring vertices (now it forms a clique with its neighbors), and then this vertex is excluded from consideration.
   - After considering the entire network to the initial moralistic unidirectional graph add additional edges and such a graph will be triangulated [16].
4. *Identification of a click.*
   All clicks received in the triangulation stage are sequentially removed. For the current click, do the following:
   4.1 Check if it is a subset of other irreversible clicks. If so, it is destroyed.
   4.2 If there is at least one vertex in it and the other irreplaceable clicks, then an edge containing the separator is inserted between the corresponding clicks – the intersection of the sets of vertices of those clicks.
5. *Building a combined tree.*
   All clicks are interconnected in series by selecting the arcs with the largest separators, and the other edges are removed. A tree that binds all clicks with the maximum separator powers will be a combined tree. For convenience, the root of the combined tree is selected by clicking with the largest number of vertices. If there are several vertices, then the most clicks with the most ribs are preferred.
6. *Fill in the combined tree tables.*
   Filling starts from the tree letters and all clicks are sequentially moved. Examples of "unmarked" vertices in unprocessed cliques:
   6.1 If there are some unmarked vertices that are not found in other raw clicks, then the table should be like – in this case, this vertex is "marked" and the clique is considered as to be processed.
   6.2 If there is no such table in the Bayes primary network, or such a vertex is not one, then the aggregate probability distribution table is used – in this case, all the clicks are "highlighted" and the clique is processed.
   6.3 If there are already "marked" vertices in the cache and only one "unmarked" vertex remains, then a lookup table from the Bayesian primary network is used – in this case, this vertex is "marked" and the clique is considered to be processed.

6.4 If clique is already "marked" top, and there are several "unchecked" peaks, then use the table of cumulative probability distribution of "unchecked" peaks and then the summit is "marked" and click is considered processed.

6.5 If in the raw clique all vertices are already marked, then the clique is filled in with a table of view and is considered as to be processed.

The Second Stage of the LS-Method is the Distribution Algorithm. For each observation of a variable, one table is selected that contains this variable. We set all occurrences that contradict the observation to zero:

1. Ascending upward.

Each tree leaf sends messages to its parent variable. Messages are the result of marginalization that is the summation of the table variables that are not contained in the separator. After sending the message, the sender divides its current table of conditional probabilities into this mode. When the recipient receives a message, it multiplies it into its probabilistic table and thus a new table is obtained. When it receives a message from all its descendants, it also sends messages to its parent and divides its table on the sent message. The process continues until the root of the coherent tree receives a message from all its descendants [16, 22].

2. Climbing downward.

The root sends messages to each of its descendants. It divides its probabilistic table into a message received from the descendant, marginalizes the table by separator and sends the result. When a descendant receives a message from its father, it multiplies it into current conditional probability table and thus forms its new table. Then it marginalizes it (by separator) and sends it to his descendant. The process continues until all the leaves receive the message.

Thus, the result will be re-listed tables for each variable, subject to the presence of observations, which then needs to be normalized.

3. Calculation of probabilities of vertices states.

For each vertex, a click is found in which it is contained. If there is a click that differs from a leaf, then this click will be selected. Otherwise any of them will be selected.

5   Example of the Modified LS-Method

As an example of constructing a probabilistic inference we consider the task of evaluating the options for the constructing of renewable energy systems. The following attributes are used as process variables:

SE - conditions for solar power plants;
WE - conditions for wind power plants;
OE - availability of other power plants;
PC - the cost of building a power plant;
NP - the need for the population;
NI - the need of industry;
TT - relief of the terrain;
RS - result for solar power plants;
RW - result for wind power plants.

To construct the network topology, a heuristic method for constructing a Bayesian network is used [16,22]. Figure 2 shows the resulted structure of the Bayesian network. The values of the probabilities of the vertices of this Bayesian network are presented in Table 1.

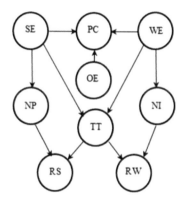

Fig. 2. Bayesian network structure for evaluating options for building renewable energy systems

To triangulate the graph, it is necessary to eliminate the vertex OE and divide the graph into two sub-graphs (Figs. 3 and 4). Each of the sub-graphs consists of five vertices and is complete. Each sub-graph represents a causal network for making decisions about the constructing of a solar or wind power plant.

Now calculate the probabilities for each state of the vertices of the causal graphs of Fig. 4. The result of modeling the Bayesian network will be the resulting probability values for solar and wind power plants. They are given in Table 2.

The calculations are performed using the B-net program. Further improvement of the LS method is possible by splitting the Bayesian network and more accurately calculating the probabilities.

Table 1. The values of the probabilities of the vertices in Bayesian network

Variable	Description	State	Probability
SE	Conditions for solar power plants	S1	0.764
		S2	0.412
WE	Conditions for wind power plants	S1	0.786
		S2	0.395
OE	Availability of other power plants	S1	1
		S2	0
PC	The cost of building a power plant	S1	0.863
		S2	0.489
		S3	0.243
NP	The need for the population	S1	1
		S2	0
NI	The need of industry	S1	1
		S2	0
TT	Relief of the terrain	S1	0.786
		S2	0.279
RS	Result for solar power plants	S1	-
		S2	-
RW	Result for wind power plants	S1	-
		S2	-

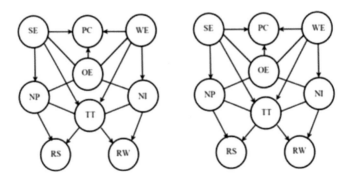

Fig. 3. Bayesian network for both moralization and non-directional graphs

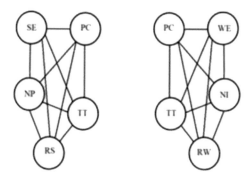

Fig. 4. Bayesian network in the case of both the causal network for evaluating options for building a solar power plant and the causal network for assessing options for building a wind power plant

Table 2. The values of the probabilities of the vertices in Bayesian network

Variable	Description	State	Probability
SE	Conditions for solar power plants	S1	0.764
		S2	0.412
WE	Conditions for wind power plants	S1	0.786
		S2	0.395
PC	The cost of building a power plant	S1	0.863
		S2	0.489
		S3	0.243
NP	The need for the population	S1	0.685
		S2	0.432
NI	The need of industry	S1	0.795
		S2	0.341
TT	Relief of the terrain	S1	0.786
		S2	0.279
RS	Result for solar power plants	S1	0.716
		S2	-
RW	Result for wind power plants	S1	0.607
		S2	-

6 Conclusions

The idea of the LS-method for generating probabilistic inference in the Bayes networks has been widely used in modeling processes in the form of causal relation-ships. On the basis of the general idea – the propagating of messages on the combined tree up and down were proposed by many similar methods and computational algorithms. In the article the modified LS-method of probabilistic inference is considered in detail, its step-by-step algorithm is presented and an example of its application in the problem of evaluating options for making decisions about building solar and wind power plants. The results of application of the method in this example for the various combinations of observations for the vertices of the corresponding network, that testify the effective use of the proposed algorithm are also given.

References

1. Cheng J, Greiner R, Kelly J, Bell DA, Liu W (2002) Learning Bayesian networks from data: an information-theory based approach. Artif Intell J (AIJ) 137:43–90
2. Jouffe L, Munteanu P (2001) New search strategies for learning Bayesian networks. In: Tenth international symposium on applied stochastic models and data analysis (ASMDA 2001), Compiegne, France, vol 2, pp 591–596
3. Cooper G, Herskovits E (1992) A Bayesian method for the induction of probabilistic networks from data. Mach Learn 9:309–347
4. Kim JH, Pearl J (1983) CONVINCE: a conversational inference consolidation engine. IEEE Trans Syst Man Cybern 17(2):120–132
5. Friedman N, Linia M, Nachman I (2000) Using Bayesian networks to analyze gene expression data. J Comput Biol 7:601–620
6. Falzon L (2006) Using Bayesian network analysis to support center of gravity analysis in military planning. Eur J Oper Res 170(2):629–643
7. Pourret O, Naim P, Marcot B (2008) Bayesian networks: a practical guide to applications. Wiley, Chichester, p 448
8. Sebastiani P, Ramoni M (2008) Bayesian inference with missing data using bound and collapse. J Comput Graph Stat 9(4):779–800
9. Lauritzen SL, Spiegelhalter DJ (1988) Local computations with probabilities on graphical structures and their application to expert systems. J R Stat Soc 50(2):157–194
10. Spiegelhalter DJ, Cowell RG (1992) Learning in probabilistic expert systems. Bayesian Stat 4:447–465
11. Olesen KG, Lauritzen SL, Jensen FV (1992) HUGIN: a system for creating adaptive causal probabilistic networks. In: Uncertainty in Artifcial Intelligence: The Eighth Conference, pp 223–229. Morgan Kaufmann, CA
12. Spiegelhalter D, Dawid P, Lauritzen S, Cowell R (1993) Bayesian analysis in expert systems. Stat Sci 8:219–282
13. Spiegelhalter DJ, Lauritzen SL (1990) Sequential updating of conditional probabilities on directed graphical structures. Networks 20:579–605
14. Lauritzen SL, Spiegelhalter DJ (1988) Local computations with probabilities on graphical structures and their application to expert systems. J R Stat Soc Ser B-Stat Methodol 50(2):157–194
15. Kozlov AV, Singh JP (1994) A parallel Lauritzen-Spiegelhalter algorithm for probabilistic inference. In: Supercomputing 1994, Washington, DC, pp 320–329
16. Terent'yev AN, Bidyuk PI, Korshevnyuk LA (2007) Bayesian network as an instrument of intelligent data analysis. J Autom Inf Sci 39:28–38
17. Babichev S, Taif MA, Lytvynenko V, Osypenko V (2017) Criteria analysis of gene expression sequences to create the objective clustering inductive technology. In: IEEE 37th international conference on electronics and nanotechnology, ELNANO 2017, pp 244–248, Article no 7939756 (2017)

18. Babichev S, Lytvynenko V, Korobchynskyi M, Taiff MA (2017) Objective cluster-
    ing inductive technology of gene expression sequences features. In: Communications
    in computer and information science, vol 716, pp 359–372
19. Gilks WR, Richardson S, Spiegelhalter DJ (2000) Markov Chain Monte Carlo in
    practice. CRC Press LLC, New York
20. Bidyuk P, Gozhyj A, Kalinina I, Gozhyj V (2017) Method for processing uncertain-
    ties in solving dynamic planning problems. In: XII-th international conference on
    computer science and information technologies, CSIT 2017, Lviv, LP, pp 151–156
21. Bidyuk P, Gozhyj A, Kalinina I, Gozhyj V (2017) Analysis of uncertainty types
    for model building and forecasting dynamic processes. In: Advances in intelligent
    systems and computing, vol 689, pp 66–82. Springer, Heidelberg
22. Zgurowskii MZ, Bidyuk PI, Terentyev OM (2008) Method of constructing Bayesian
    networks based on scoring functions. Cybern Syst Anal 44(2):219–224

Information Technologies for Environmental Monitoring of Plankton Algae Distribution Based on Satellite Image Data

Oleg Mashkov[1], Victoria Kosenko[2], Nataliia Savina[3], Yuriy Rozov[4], Svitlana Radetska[4], and Mariia Voronenko[4(✉)]

[1] State Ecology Academy of Postgraduate Education and Management, Kyiv, Ukraine
mashkov_oleg_52@ukr.net
[2] State University of Telecommunications, Kyiv, Ukraine
koseno4ka.4ever@gmail.com
[3] National University of Water and Environmental Engineering, Rivne, Ukraine
n.b.savina@nuwm.edu.ua
[4] Kherson National Technical University, Kherson, Ukraine
rozov.yuriy@kntu.net.ua, rad_svete@rambler.ru, mary_voronenko@i.ua

Abstract. The paper discusses the features and specifics of using the methods of remote sensing of the Earth when monitoring the ecological and technical state of water management process systems. A technology for monitoring surface waters according to remote sensing of the Earth is proposed. A method of satellite monitoring of algal bloom intensity (monitoring of planktonic algae clusters) has been proposed and substantiated. An approach to flood risk assessment using satellite observations is proposed. The technique of quantitative assessment of water quality according to the space monitoring of surface waters is substantiated. As a result of the research, it was found that when assessing the complex effects of pollutants on the ecological state of aquatic ecosystems using aerospace technologies, it is advisable to take into account changes in biological indicators (indicators of biomass and species composition of phytoplankton and higher aquatic plants). A method has been developed for predicting the long-term risks of emergency situations of a hydrological and hydrometeorological nature based on physical and mathematical modeling and the use of satellite observations and spatially distributed data. According to the results of further studies based on the results obtained in this work, it is planned to create flood risks distribution maps, quantitative predictions of surface water quality deterioration, as well as an assessment of the risks of air and soil pollution.

Keywords: Aerospace technologies · Water system · Remote sensing of the Earth · Ecological state of ecosystems · Environmental risks · Surface water · Satellite monitoring · Algal blooms

© Springer Nature Switzerland AG 2020
V. Lytvynenko et al. (Eds.): ISDMCI 2019, AISC 1020, pp. 434–446, 2020.
https://doi.org/10.1007/978-3-030-26474-1_31

1   Introduction

World experience has proved that it is necessary to apply modern innovative means and technologies to improve the quality, efficiency, and effectiveness of the environmental monitoring system [1,2]. These include: automated and automatic measuring systems; aerospace technology using satellites [3], aircraft and unmanned aerial vehicles [4], automated remote sensing data processing systems [4], geoinformational analytical systems for information processing [5], with the consideration of the laws of its change both in time and in space; integrated multi-level environmental monitoring and control systems; methods and technologies for analyzing environmental monitoring data and determining the level of technological and environmental safety, etc.

The development of the scientific foundations for the creation and implementation of such systems, methods, and technologies complies with European and global approaches to environmental management, and also complies with the requirements and directives of the Association Agreement between Ukraine and the EU. The results of this study will significantly expand the possibilities of international cooperation of Ukraine in the field of environmental protection and will help to bring the state of the environment in line with European and international requirements.

Currently, the problem of intense water blooming is characteristic of a wide variety of water areas. It is known that intensive blooming is characteristic, first of all, of reservoirs with weak currents in particular, such as the cascade of the Dnieper reservoirs. Recently, this factor is characteristic of other aqua technosystems. The water blooming is a consequence of the massive development of microscopic algae (usually blue-green) and it is accompanied by a significant deterioration in water quality.

The cause of this phenomenon is a whole complex of factors, such as climate change, the flow of large amounts of various mineral and organic substances into the water. Therefore, control of using of remote sensing systems (Earth remote sensing) [4] is relevant in two aspects. Firstly, it is control of the hydrothermal mode, the development of proposals to improve the ability to cool the circulating water. Secondly, it is control of the ecological state, namely the level of blooming and overgrowing of water by aquatic plants. Thermal control is also important at hydropower facilities.

2   Review of the Literature

The [6] is devoted to formation of the scientific foundations of modern environmental monitoring, which developed the basic principles for the establishment of the environmental monitoring system, as well as the international aspects of the global monitoring system are partly reflected.

The development of multichannel complexes of remote sensing (RS), which are stimulated by the needs of the national economy, including the problems of environmental monitoring, are described in [7].

In [8], the use of multichannel complexes RS for the detection of subsurface water bodies of natural and anthropogenic origin is described.

The concept of creating a space monitoring system for the territory of Ukraine, based on the use of the complex systems theory, was proposed in [9], where technical issues of building subsystems and their functioning are considered. The paper proposed the concept of creating a space monitoring system for a country's territory based on the use of the complex systems theory [9], the main features of which are: (1) the separability into a finite number of subsystems, each of which consists of interconnected elements; (2) the functioning of the subsystem in interaction under the control of one of the central elements; (3) combining them with a communication system that circulates information in a multi-level consumer system. The monitoring system is proposed to be built on a multilevel hierarchical principle with various spacecraft operating in various orbits with a wide class of remote equipment and a network of ground-based data receiving and processing centers in the interests of various consumers.

Questions of the application of multi-spectral methods of the Earth remote sensing in the nature management problems are considered in [3, 4, 10–13].

The [14–17] is devoted to the study of aerospace observations from using the earth's surface to control of the ecological and technical condition of aquatic technetium systems. The power model of geotechnical systems is systematically resolved into work [18].

The works [19, 20] suggest the use of computerized regional systems of state monitoring of surface water in the form of models, algorithms and software. Information technology for the design of data processing systems for data monitoring of water quality and the integration of mathematical models in geoinformation systems for surface water monitoring is proposed in [21, 22].

In [23], the authors evaluated the effectiveness of the use of space information in water and water management, substantiated the use of remote sensing of the Earth in solving water management and water conservation problems.

However, in the described works there is no description of the methodology for assessing the complex effects of pollutants on the ecological state of aquatic ecosystems using aerospace technologies. It is advisable to take into account changes in biological indicators and methods for predicting the long-term risks of emergency situations of a hydrological and hydrometeorological nature based on physical and mathematical modeling and the use of satellite observations and spatially distributed data. This work is dedicated to solving these problems.

3   Problem Statement

The study of the problems of aquatic ecosystems status assessment causes difficulties due to the difficulty of conducting of relevant research in terrestrial conditions. Traditional methods of studying of aquatic ecological systems state do not provide an opportunity to obtain a spatial picture of the phenomena and processes that occur in aquatic ecological systems.

Recently, aerospace technology and geo-information systems have been used to solve problems of monitoring of aquatic ecological systems state. Therefore, there is a problem of assessing the features of using the Earth remote processing methods to control the ecological and technical state of aquatic ecosystems.

4   Materials and Methods

The existing EU Water Framework Directive, when assessing the ecological status and monitoring the environment, provides an assessment of the hydromorphological characteristics of water bodies and streams [24]. Therefore, remote sensing methods are essential for monitoring changes in the coastline.

In addition, phytoplankton has a direct impact on the quality of drinking water: suspension, color, toxicity; it causes fish freeze because of significant development of biomass, disruption of sewage treatment plants work, pollution of the coasts and beaches. Indicators of the quantitative development of phytoplankton are widely used in determining the trophic status of water bodies and for making decisions on the ecological rehabilitation of water bodies. However, obtaining data on the status of phytoplankton is a quite time-consuming and costly process. All this requires the use of modern technologies for obtaining relevant monitoring information.

The proposed method uses the data of the Earth remote sensing in the wavelength range of 8–14 µm (long-wave infrared radiation), which allows obtaining information about the thermophysical properties of objects on the Earth's surface, including water bodies. It should be noted that the raster data of thermal radiation with an average spatial resolution of the TIRS sensor installed on the Landsat-8 satellite are presented in two spectral ranges (10.3–11.3 µm and 11.5–12.5 µm).

To convert the raster data of thermal radiation into a temperature distribution, it is proposed to use the Planck inverse equation for thermal radiation of a gray body. This conversion makes it possible, taking into account the coefficient of thermal radiation, to evaluate the ability of different surfaces to radiate heat. The resulting heat maps allow us to study the heterogeneity of the temperature fields of the techno-systems and show the effectiveness of new structures for regulating the thermal regime in cooling ponds.

A fragment of the Landsat 8 satellite image to study the spatial distribution of blue-green algae blooming in the Kiev reservoir in the area of Kozarovichi-Lyutezh settlements [13] is presented in Fig. 1.

It can be seen that changes in watercolor (blooming) appear due to the mass reproduction of microscopic algae. Processing satellite images in order to identify water blooming areas consist in analyzing changes in the diffuse reflection of light from the surface and under the surface layers of water with increasing phytoplankton concentration in them.

The analysis shows that, the red spectral region of the visible range 600 ... 700 nm and the near infrared range are used for observations. Vegetative Index The Normalized Difference Vegetation Index (NDVI) is well suited for identifying flowering areas. However, to assess the state of the reservoir, it is recommended to use other indicators, such as turbidity – normalized differential turbidity index

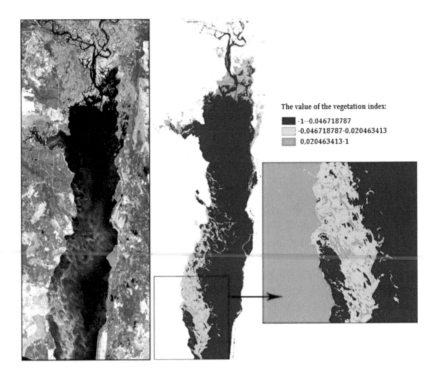

The value of the vegetation index:

■ -1--0,046718787
☐ -0,046718787-0,020463413
▨ 0,020463413-1

Fig. 1. Study of the spatial distribution of blue-green algae blooming areas on the territory of the Kiev reservoir (a is fragment of the Landsat satellite image 8 August 13, 2013; b is distribution of the values of the vegetation index in the blooming area in the area of Kozarovichi - Lyutezh settlements)

(NDTI), algoindex – normalized differential algal index (NDAI), etc. Moreover, the additional decoding sign for water identification "blooming" can be texture. Specific threadlike textures are most often characteristic of intense blooming areas.

On satellite images, it can be seen that areas of intense blooming are elongated along the currents and participate in vortex movements. The wind has a significant effect on the algae transfer. During periods of prolonged warm, sunny and windless weather, cyanobacteria are combined into aggregates that float to the surface, forming surface or subsurface accumulations. However, it should be borne in mind that the data obtained using satellite observations should be confirmed by the results of field ground surveys.

5 Experiment, Results and Discussion

5.1 Flood and Flooding Risk Assessment Using Satellite Observation Data

The ability to identify the risks of flooding is determined by the relationship between changes in the spectral characteristics of the surface reflection and the response of ecosystems to external factors (Q_{stress}). To describe the spectral reflection of a specific type of surface N (where N is a class, in accordance with the preliminary classification of land covers), an integrated indicator - the spectral reflection index SRI was introduced. This index is represented as a fixed combination of spectral characteristics in separate bands of the spectrum r_λ. In general, its appearance can be represented as: $SRI_\tau = f(r_\lambda)_\tau$, where τ is the moment of shooting.

In our case, any of the NDVI (or ARVI) indices, EVI, allows us to determine the trends in the studied ecosystems. Also, the PRI (Photochemical Reflectance Index) can be useful, as well as the "Stressful" indexes: the SIPI (Structure Intensive Pigment Index), the NDWI water index and the Plant Stress Index. Depending on the controlled parameter, any of the existing spectral indices can be used as an integrated indicator (PRI, SIPI or NDNI indices - Normalized Difference Nitrogen Index). Considering the possibility of multiple surveys obtaining, we introduce an index reflecting the changes in the studied spectral parameters over the observation period – the normalized spectral reflection index [25]:

$$SRI*_\tau = \frac{\max\{SRI_\tau\} - SRI_\tau}{\max\{SRI_\tau\} - \min\{SRI_\tau\}} \tag{1}$$

Then you can consider the informative attribute $\Delta SRI*$, which is the difference between the average value $SRI*$ for the observation period and the value obtained as a result of the experiment $SRI*_\tau$. The equation that determines the probability of stress based on the totality of the spectral characteristics of the earth's surface is derived from the use of Bayes' rule:

$$P(\Delta SRI * (x,y)|Q_{stress}) = \frac{P_s(x,y) \cdot \prod_N P_N(\Delta SRI * |Q_{stress})}{\int_{x,y} P_N(\Delta SRI * |Q)dP_s(x,y)} \tag{2}$$

$$= \frac{P_S(x,y) \cdot P_N(\Delta SRI * |Q_{stress})}{P_N(\Delta SRI * |Q_{stress})P_S(x,y) + P_N(\Delta SRI * |Q_0)P_0(x,y)}$$

In this equation, the Q_{stress} index refers to areas under the influence of stress factors, and the Q_0 index denotes the class of pixels in which the action of such factors is not present. The probability $P_S(x,y)$ is determined on the basis of the distribution of observational data, that is, semi-empirical. The ratio of the probabilities $P_S(x,y)$ and $P_0(x,y)$ is determined as:

$$\lim_{x,y,\tau} (P_S(x,y)_\tau + P_0(x,y)_\tau) = 1 \tag{3}$$

To determine the probability $P_S(x, y)$, one can use the rule based on the use of the Gauss function:

$$P_S(x, y) = P_{\min} + (P_{\max} - P_{\min}) \cdot e^{d_s^2/2\sigma_p^2} \tag{4}$$

Here $P_S(x, y)$ is the probability of a threat occurring; P_{max} is the maximum possible probability of occurrence of a threat in the studied place, which depends on the sensor type, physical and geographical features of the region and surface type (P_{max} for Landsat TM and ETM sensors is 0.25–0.3); P_{min} is minimum probability (P_{min} is close 0.01); $d_s(x, y)$ is distance from the nearest place that is under the threat; σ_p is an empirical indicator, determined on the basis of field studies, based on the characteristics of the vegetation cover of the research area and the type of sensor (for example, for Landsat TM and ETM in the region of research, the indicator σ_p is 1.1–1.5 km).

Thus, for the region of research and sensors, Landsat TM and ETM $P_S(x, y)$ can be determined using the formula:

$$P_S(x, y) = 0,01 + 0,26 \cdot e^{d_s^2/1,69} \tag{5}$$

The task of identifying areas within the N classes with x, y coordinates (under the influence of stress caused by the influence of the factors Q_{stress}) can be reduced to the problem classification of images within the selected periods i, spectral bands r_λ, types of sensors and the region of research. These risks are evaluated as the probability of negative consequences.

Thus, a risk value of 0.5 means that in conditions of reliable exceeding of the average level of seasonal fluctuations of the amount of precipitation, a flood event will be recorded. Based on local and regional risks, flood, and flood risks were assessed. The calculated results are presented in Figs. 2 and 3.

The approach to the assessment of risks of flooding and the method of calculating the spatial distribution of regional indicators of risks of flooding using satellite observations in the optical range are proposed.

5.2   Proposed Method for Quantitative Assessment of Water Quality by Observations and Measurements

The idea of the method under consideration is to develop a formal approach to the use of heterogeneous data sets for assessing water quality indicators in terms of risk. The probability of deterioration in the presence of elevated concentrations of certain pollutants is assessed.

At the same time, we evaluate the quality of water by classes (first through fifth) and categories (first through seventh). Water quality is determined by the water class or water category. Analytically, water quality is determined by sets of indices, which are grouped into three arrays: mineral-salt indices (three indicators), sanitary and hygienic criteria (another name is trophosaprobiotic indices, 20 of them) and specific indices of toxic pollutants (another name are indices of specific substances of toxic and radiation exposure, this group includes 15 indicators).

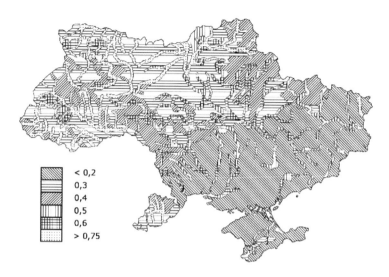

Fig. 2. Scenario calculation of indicators of flood risk for the period up to 2030

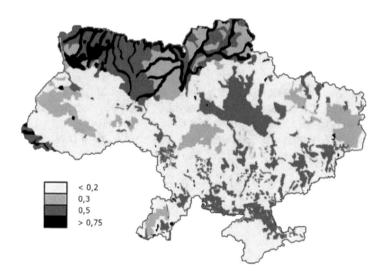

Fig. 3. Scenario calculation of risk indicators for flood processes for the period up to 2030

In real situations, we have the ability to measure limited sets of indicators that are indirectly related to these indices. The task of reliable assessment of water quality becomes very difficult. However, we can estimate the probability of water quality change in accordance with the class (or category) with changes in the observed indicators [26].

Thus, the problem under consideration can be reduced to a formal algorithm for obtaining dimensionless interval estimates using rank criteria sets based on the theory of fuzzy sets.

The risk assessment algorithm is divided into several stages and can be represented in a relatively simple form. The set of indices for risk assessment is defined as follows [27]:

$$M = (x_1, x_2, ..., x_n) = \{x_i\}, i = 1, 2, ..., n \tag{6}$$

where n is the number of selected estimated parameters, x_i is a parameter from the set of $x_{\min}$ risk/pollutant parameters (in most real cases they operate with a set of several known pollutants, for example, we will take into account those that we can see by means of remote sensing: transparency, suspended matter, phytoplankton biomass, trophicity, surface-active organic substances, synthetic surfactants, $n = 6$).

Based on the water quality criteria introduced by most of the constituent documents, for example, the European Water Directive, many criteria for risk assessment should be defined as follows [18]:

$$D = (d_1, d_2, ..., d_m) = \{d_i\}, j = 1, 2, ..., m \tag{7}$$

where m is the number of classes or quality categories (the risk of assignment to which we will determine) d_j, the corresponding quantities of pollutants for which the assessment is carried out, usually $m = 5$ (which corresponds to the number of classes).

Further, the risks associated with water pollution are distributed by a fixed number of degrees (risk assessment intervals): low risk, acceptable risk, unacceptable risk, high risk and catastrophic risk.

The matrix Z, which will link the risk assessment indices (pollutants) M and the water quality criteria D, will look like this:

$$Z = \begin{bmatrix} z_{11} \cdots W \cdots z_{1m} \\ A \cdots C \cdots T \\ z_{n1} \cdots E \cdots Z_{nm} \end{bmatrix} \tag{8}$$

where z_{ij} is the assessment of the partial risk i^{th} by the criterion of the risk of individual pollution j^{th} from the total set of indices used to assess quality (38 parameters).

Here, to determine the weight of the indices V_i, the next coefficients of variation will be used:

$$V_i = \frac{\sqrt{\frac{1}{n} \sum_{i=1}^{n} (x_i - \frac{1}{n} \sum_{i=1}^{n} x_i)}}{\frac{1}{n} \sum_{i=1}^{n} x_i} \tag{9}$$

where $0 \leq V_i \leq 1$.

Then the risk assessment matrix F can be represented as follows:

$$F = V \bullet Z = (f_1 \ldots f_2 \cdots, \ldots f_m) \tag{10}$$

In this case, we can calculate the quantitative indicators of the risk of contamination (in accordance with the definition entered) RI by a simple algorithm:

$$RI = \frac{\sum_{j=1}^{n} f_i \times j}{\sum_{j=1}^{n} f_i} \tag{11}$$

5.3 Discussion

Figure 4 represents the results of modeling of risk of the terrestrial water resources quality degradation according to satellite observations using the above algorithm.

Thus, this algorithm can be proposed for assessing the risk of pollution of water bodies in accordance with the criteria for assessing water quality and sets of indicators, which are obtained from measurements and observations. The proposed method requires testing using field-based spectrometric measurements. In the future, this technique can be improved by using data from field spectrometric measurements and applying spatial modeling.

Methods for remote assessment of the ecological and technical state of water technology systems are proposed. A comprehensive approach for predicting the risks of natural disasters based on physical, mathematical, and geospatial modeling using Earth remote sensing data is proposed. A method for determining the density of planktonic algae based on satellite imagery data has been developed.

Using remote sensing methods, maps of the distribution of turbidity, algoindex, lakes and thermal heterogeneity maps in the surface layer of reservoirs can be obtained. Requirements for the protection of aquatic ecosystems using aerospace technology are formed. The development of scientific foundations of multispectral methods and technical means of monitoring the ecological status of aquatic ecosystems is a prerequisite and the basis for the effective management of their ecological safety.

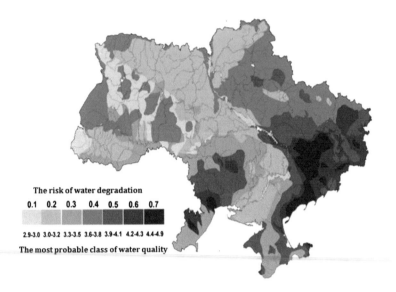

Fig. 4. The risk of deterioration of the surface water resources quality according to satellite observations MODIS, MISR and AIRS 2002–2014 (model grid 50 × 50 km)

6   Conclusions

As a result of research, it was found that when assessing the complex impact of pollutants on the ecological state of aquatic ecosystems using aerospace technologies, it is advisable to take into account changes in biological indicators (biomass indicators and species composition of phytoplankton and aquatic higher plants).

In this paper, the authors developed a method for predicting the long-term risks of emergency situations of a hydrological and hydrometeorological nature based on mathematical modeling using satellite observations and spatially-distributed data.

In accordance with the Water Framework Directive 2000/60/EC, the monitoring of integral indicators of water pollution should be based on their ecotoxicity, which is determined using biotesting and allows for the synergistic interaction of pollutants to be taken into account.

The result of the proposed information technology application are maps of planktonic algae distribution and a method for quantitative assessment of their density based on satellite imagery. RS-methods can provide distributions of turmoilty, algo-index, lake-index and maps of thermal inhomogeneity in the surface layer of reservoirs and methods of remote estimation of the ecological and technical condition of aquatic ecosystems. According to the results of the research, requirements for the technology of protection of water ecosystems using aerospace technologies have been formed, namely, the development of scientific foundations of multispectral methods and technical means for controlling the ecological status of water ecosystems, which is a prerequisite and basis for the effective management of their environmental safety. As a result of the conducted researches, it was found out that in assessing the complex influence of pollutants on the ecological status of aquatic ecosystems, using aerospace technologies, it is expedient to take into account changes in biological parameters.

References

1. Bilyavsky GO, Furduy RS, Kostikov IYu (2005) Fundamentals of ecology. K.: Lybid, p 408
2. Bondar OI, Korin'ko IV, Tkach VM (2005) Environmental monitoring. DEI-GTI, K.-H., p 126
3. Kostyuchenko YV, Kopachevsky IM, Solovyov DM (2011) The use of satellite observations for the assessment of regional hydro-hydrogeological risks. Space Sci Technol 17(6):19–29
4. Toom LM, Tomiltsev AI, Tomchenko OV (2016) Assessment of the status of water protected areas using the methods of remote sensing of the earth. Hydropower of Ukraine, no 3–4, pp 51–56
5. Pichugin MF, Mashkov OA, Saschuk IM (2006) Processing of geophysical signals in modern automated complexes, Zhytomyr, ZVIERE-2006, p 176
6. Klymenko MO (2002) Environmental Monitoring. UDUVGP, Rivne, p 232
7. Krasovsky GY (1989) Aerospace monitoring of surface waters (practical aspects). Scientific Council on Space Research for the National Economy of the ISS, p 231
8. Delb S, Gamba P, Roccato D (2000) A fuzzy approach to recognize hyperbolic signatures in subsurface radar images. IEEE Trans Geosci Remote Sens 38(3):1447–1451
9. Bondur VG (1995) Principles of building a cosmic earth monitoring system for environmental and natural resource purposes. Izv. Universities, Geod. and aerial photography, no 2, pp 14–38
10. Lyalko VI, Fedorovsky OD, Popov MO (2006) Multi-spectral methods of remote sensing of the earth in the problems of nature use. K.: Scientific thought, p 357
11. Koshan SS (2009) Remote sensing of the earth: theoretical basis. K.: High school, p 511
12. Lyalko VI (2002) Application of multi-zonal space images for assessing the ecological state of forests. In: Information technologies for environmental safety management, resources and emergency measures, proceedings. Krym, Sibirskaya Rybache, pp 75–77
13. Stankevich SA, Kozlova AO (2006) Features of the calculation of the index of species diversity by the results of the statistical classification of aerospace images 19(58):144–150
14. Matsnev AI, Protsenko SB, Sabli LA (2017) Monitoring and engineering methods of environmental protection: teaching. manual. OJSC "Rivne Printing House", p 504
15. Mashkov OA, Vasiliev VE, Frolov VF (2014) Methods and technical means of ecological monitoring. K.: DEA "Ecological sciences", no 1/(5), pp 57–67
16. Romanenko VD, Shcherbak VI, Yakushin VM (2012) The threats of anthropogenic impact on the landscape and biological diversity of the lakes of Shatsk National Nature Park. RVB "Tower", Lutsk, no 9, pp 319–324
17. Tomchenko OV, Silayeva AA, Protasov OO (2017) Use of space observation data from the earth's surface to assess the transformation of the lithologic zone of the cooling water reservoir provided the water level is reduced. In: Clean Water: fundamental, applied and industrial aspects, Kyiv, Ukraine, 26–27 October, pp 207–209

O. Mashkov et al.

18. Pivnyak G, Busygin B, Garkusha I (2010) Geographic information technology monitoring and mapping of coal fires in Ukraine, according to the space survey. In: 12th international symposium on environmental issues and waste (SWEMP), Praga, pp 416–422
19. Mokin VB (1998) New models of processes in a river. In: 20th international scientific symposium of students and young research workers, Zielona Gora, Poland, vol 5, Part II, pp 67–71
20. Mokin VB (2000) The algorithms of river water quality control. In: XXI International scientific symposium AQUA. Politechnika Warszawska, Plock, pp 20–27
21. Mokin VB, Mokin BI (2000) River water control of sewage disposal detection. In: XVI World congress IMEKO-2000. Abteilung Austauschbau and Messtechnik Karlsplatz, Vienna, Hofburg, Austria, pp 297–301
22. Mokin VB, Mokin BI (2003) Control over volume and quality of sewage water in the river waterway. In: XVII IMEKO-2003. HMD Croatian Metrology Society, Dubrovnik, Croatia, vol 19, pp 2090–2093
23. Lyalko VI, Artemenko IG, Zholobak GM, Kostyuchenko YV, Levchik OI, Sakhatsky OI (2009) Evaluating vegetation cover change contribution into greenhouse effect by remotely sensed data: case study for Ukraine. Regional Aspects of climate-terrestrial-hydrologic interactions in non-boreal Eastern Europe. In: Springer with NATO Public diplomacy division, pp 157–164
24. (2017) Method of remote estimation of ecological status and quality of water of inland water bodies. Issue "Ecology and Environmental Protection". K.: Academiperiodica, p 32
25. Shapar AG, Eemets NA, Bugor AN (2013) Analytical component (knowledge base) of the environmental monitoring system. IPPI of the National Academy of Sciences of Ukraine, no 17, pp 181–187
26. Mashkov OA, Kachalin IG, Sinitsky RN (2005) Design and development of an automated system for the collection and processing of geophysical information, Kyiv, vol 29, pp 57–64
27. Isaenko BM, Lisichenko GV (2016) Monitoring and methods for measuring environmental parameters: training manual - K.: Vt. Nats. aviation Un "NaU-Print", p 312

Ontology-Based Intelligent Agent for Determination of Sufficiency of Metric Information in the Software Requirements

Tetiana Hovorushchenko$^{(\boxtimes)}$ ⓘ, Olga Pavlova ⓘ, and Dmytro Medzatyi ⓘ

Khmelnytskyi National University, Institutska str., 11, Khmelnytskyi 29016, Ukraine
{tat_yana,medza}@ukr.net, olya1607pavlova@gmail.com

Abstract. This research is devoted to implementation of the ontology-based intelligent agent (OBIA) for the determination of the sufficiency of metric information in the software requirements. The method of actions of such an OBIA is developed. The intelligent agent, who works on the basis of the developed method, determines the sufficiency of metric information in the software requirements, performs a numerical assessment of the sufficiency level of metric information, and offers a visual list of missing indicators necessary for the calculation of metrics. Functioning of the realized agent allows increasing the sufficiency level of metric information in the software requirements. The developed intelligent agent allows to partially eliminate the person from the processes of processing the information, to avoid the losses of important information and to increase the amount of metric information at the phase of requirements gathering, which in the complex provides increasing the software quality. During the experiments, the intelligent agent investigated the requirements for software of the system of improving the safety of computer systems' software, which resulted in the sufficiency of metric information in requirements increased by 44%.

Keywords: Software requirements · Metric information · Sufficiency · Ontology-based intelligent agent (OBIA)

1 Introduction

Statistics show [1–3], that today there are problems in the software quality assurance - large projects are still executed with a lag behind the schedule or exceeding the cost estimates, the software is often lacking the necessary functionality, its performance is low, and its quality doesn't suit customers (Fig. 1 [1]). A significant amount of bugs is made in the software at the phase of requirements gathering. The statistics show that the inaccurate gathering of requirements is one of the primary causes of software failures (Fig. 2) [1,4]. Software projects, the requirements of which contain insufficient, inaccurate, incomplete and contradictory information, cannot be successful [5,6]. Ideally, all problems encountered

© Springer Nature Switzerland AG 2020
V. Lytvynenko et al. (Eds.): ISDMCI 2019, AISC 1020, pp. 447–460, 2020.
https://doi.org/10.1007/978-3-030-26474-1_32

	CHAMPIONS	UNDER-PERFORMERS
Average percentage of projects completed on time	88%	24%
Average percentage of projects completed within budget	90%	25%
Average percentage of projects that meet original goals/business intent	92%	33%
Average percentage of projects experiencing scope creep	28%	68%
Average percentage of projects deemed failures	6%	24%
Average percentage of budget lost when a project fails	14%	46%

Fig. 1. Project performance averages of champions (organizations with 80% or more of projects being completed on time and on budget, and meeting original goals and business intent, and having high benefits realization maturity) versus Underperformers (organizations with 60% or fewer of projects being completed on time and on budget, and meeting original goals and business intent, and having low benefits realization maturity) [1]

during the requirements phase should be resolved before the design starts. The late discovery of requirements errors is the most expensive to correct [7].

Therefore, it's necessary to investigate the requirements and to assess the degree of sufficiency of information about future software in the requirements for ensuring the required functionality and quality of the software. Particular attention is required to the requirements, which characterize the quality of software, because high quality of software (as the ability of software to satisfy the claimed and predicted needs of customers [8]) is one of the main requirements of users for modern software.

Today, there are a number of models, which can evaluate and predict software quality. Most of these models are based on the use of different software metrics (software metric is a measure that gives the numerical value of any software property [9]).

Estimating the software quality based on the use of metrics is as follows: metrics are calculated based on the indicators (as a weighted arithmetic mean of the values of indicators of each metric with taking into account their weighting factors); comprehensive evaluation of software quality is calculated based on these metrics. The analysis of known methods and tools of metrics calculation, which was conducted in [10], has shown, that now the determination of indicators for the software metrics occurs only at the quality assurance phase for the finished source code. But all the necessary indicators should already be filled in the software requirements, i.e. it's possible to determine the sufficiency of information for further evaluation and assurance of the software quality already on the basis of requirements.

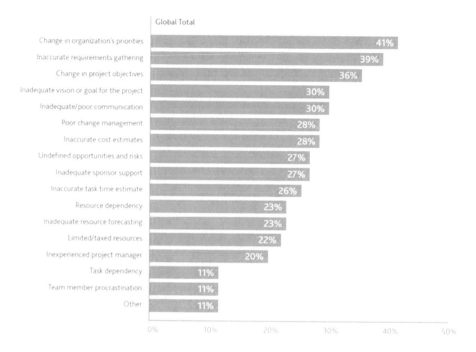

Fig. 2. The primary causes of software failures [1]

The *sufficiency of metric information in the software requirements* is the availability of all the information elements (indicators), which are required for determining the software metrics and quality.

If there are no some indicators in the requirements, then the information in requirements is insufficient for determination of software metrics and quality. It's necessary to generate inquiries to developers regarding improving the requirements for eliminating the insufficiency of information in the software requirements. On the basis of such the inquires, requirements' developers must make the necessary metric information in the software requirements.

2   Review of the Literature

Today, in the era of the development of the semantic web, taking into account the need of minimization of the human factor in the processing of requirements, developing the knowledge-based systems and the ontologies to manage the requirements and model problem domains is the requirement of time. Ontologies provide the automatic understanding and analysis of information, elimination of a man and minimization of information losses in the process of information processing and knowledge gaining [11]. The use of ontologies for software engineering avoids the re-conceptualization of the problem domain, reduces the use of resources in the early phases of the software development life cycle, reduces the number of errors at the requirements gathering phase [12]. Ontologies could improve some of the activities involved in the requirements phase such as: effectively support automatic requirements analysis, detecting incompleteness, insufficiency and inconsistency in requirements, and evaluate the quality of requirements [7].

Ontologies provide access, understanding and analysis information by intelligent agents (IA). During the actions, IA uses information from its environment, analyzes it (based comparison of it with the known facts) and decides on further actions [13]. The intelligence of agent is in the formation of new knowledge, in particular, certain conclusions and recommendations.

Many studies are devoted to the use of ontologies and the development of OBIAs for software engineering domain - Fig. 3.

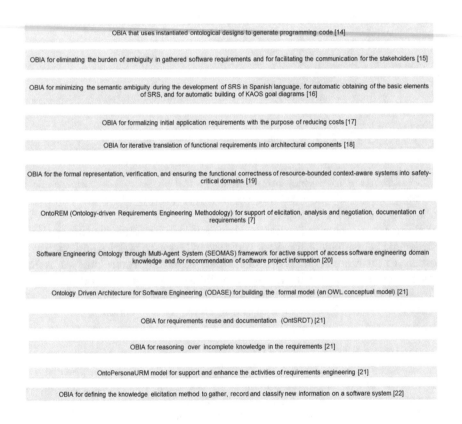

Fig. 3. OBIA for requirements and software engineering

All considered OBIAs are designed either for the automated development of source code by the requirements, or for the formalization of available requirements, or, maximally, for elimination of the uncertainty in the requirements.

They don't perform verification and validation of requirements to the needs of customers; don't check whether all the customers' needs have been reflected in such requirements and don't show the needs of customers, which were not reflected in the requirements; don't identify informational losses during the requirements gathering; don't assess the sufficiency of the available information in the requirements (including metric information).

One of the solutions for assessing the sufficiency of information in the software requirements is paper [23], in which the ontological approach to the assessment of information sufficiency for software quality determination is proposed, but automation of such an assessment isn't realized in this paper. This approach will be developed during further solving the task of determination of the sufficiency of metric information in the software requirements in the part of its automation.

3 Problem Statement

The need of the software quality assurance, of the possibility of evaluating the software quality already on the basis of requirements, the need of increasing the sufficiency of information in the software requirements is the *actual scientific and applied problem*. One of the ways of solutions to this problem is developing the ontology-based intelligent agent for the determination of the sufficiency of metric information in the software requirements. This is the *purpose of this study*.

For achieving this goal, the following *tasks* must be solved:

1. Developing the method of actions and realization of OBIA for determination of sufficiency of metric information in the software requirements.
2. Conducting the experiments using the developed OBIA for determination of the sufficiency of metric information in the software requirements.

4 Ontology-Based Intelligent Agent for Determination of Sufficiency of Metric Information in the Software Requirements

The proposed OBIA uses the base (ideal) ontology for quality and complexity of software project and software as the known fact (this ontology is in the agent's knowledge base) during actions. Exactly ontology will detect the missing indicators in the requirements and will determine the software metrics, which cannot be determined without such indicators. For developing such an ideal ontology, 24 metrics of software quality and complexity with exact or predicted values at the design phase have been selected in [10]. These metrics depend on 72 indicators, including 42 different indicators. A fragment of the base ontology for software complexity, realized in Protégé 4.2, is presented on Fig. 4.

The agent extracts metric information (indicators for metrics) from the requirements for real software, forms a real ontology, which is compared with the ideal ontology. Based on this comparison of ontologies in Protégé 4.2, the

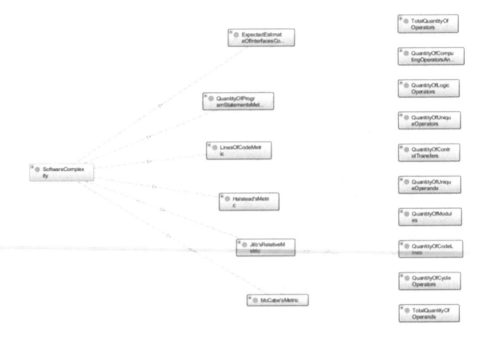

Fig. 4. Fragment of the base ontology for software complexity

OBIA forms a list of missing indicators. The OBIA analyzes the list of missing indicators, calculates the number of missing indicators, and generates the numerical assessment of the sufficiency level of metric information in the software requirements and the recommendations (in a visual form) about improving the sufficiency level.

In [24] we have developed method of activity of ontology-based intelligent agent for evaluating the initial stages of the software lifecycle, which is used for assessing the sufficiency of information regarding non-functional requirements. In this paper, this method has gained further development in the part of automation parsing the software requirements, and in the part of determining the sufficiency of exactly metric information in the software requirements. The improved *method of actions of OBIA for determination of the sufficiency of metric information in the software requirements* is schematically shown on Fig. 5.

5   Experiments and Results

On the basis of the presented method, OBIA for determination of the sufficiency of metric information in the software requirements was implemented in PHP with using Protégé 4.2 (for work with ontologies).

For the experiment, the automatic analysis (parsing) of the requirements for the software of the system of improving the safety of computer systems' software

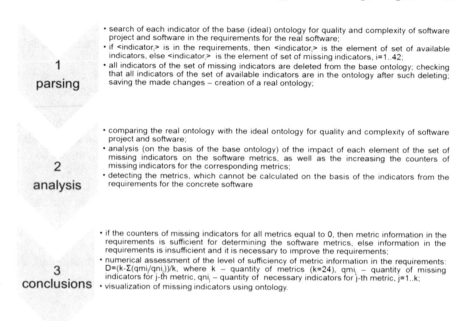

Fig. 5. Method of actions of OBIA for determination of sufficiency of metric information in the software requirements

was performed. This parsing made it possible to construct the real ontology for quality and complexity software project and software.

In accordance with the developed method of actions, implemented OBIA performs the comparison of real ontology with the ideal ontology. As a result of this comparison, OBIA provided the set of missing indicators in the requirements for the software of the system of improving the safety of computer systems' software:

1. For software project complexity metrics: QuantityOfModules, QuantityOf ProceduresToReadFromDataStructure.
2. For software project quality metrics: QuantityOfModules, TypeOfModuleInputData, TypeOfModuleOutputData, QuantityOfCodeLines, QuantityOfPotentialAccessToGlobalVariables, ProjectDuration.
3. For software complexity metrics: QuantityOfModules, QuantityOfCodeLines, TotalQuantityOfOperators, TotalQuantityOfOperands.
4. For software quality metrics: QuantityOfCodeLines, ProjectDuration, CostOfOneLine.

OBIA detects the metrics, which cannot be calculated on the basis of the indicators of the requirements for the software of the system of improving the safety of computer systems' software:

1. Software project complexity metrics: Chepin's metric, Jilb's absolute metric, McClure's metric, Kafur's metric, i.e. all 4 software project complexity metrics cannot be calculated.
2. Software project quality metrics: Coupling metric, Global variables calling metric, Time of models modification, Quantity of found bugs in models, i.e. 4 from 5 software project quality metrics cannot be calculated.
3. Software complexity metrics: Lines of code, Halstead's metric, McCabe's metric, Jilb's relative metric, Quantity of program statements, Expected estimate of interfaces complexity, i.e. all 6 software complexity metrics cannot be calculated.
4. Software quality metrics: Software design total time, Design stage time, Design cost, Quality audit cost, Realization productivity, Code realization cost, Effort applied by Boehm, Development time by Boehm, i.e. 8 from 9 software quality metrics cannot be calculated.

Numerical assessment of the sufficiency level of metric information in the requirements is the following:

$$D = (24 - (1/5 + 1/2 + 1/5 + 2/4 + 0/2 + 2/5 + 1/2 + 2/3 + 1/2 + 1/1$$
$$+ 3/5 + 1/3 + 1/3 + 2/2 + 2/3 + 2/2 + 2/3 + 2/2 + 2/4 + 2/2 + 2/3 + 0/5$$
$$+ 1/2 + 1/2))/24 = 0,4486 = 44,86\%$$

Thus, the *realized OBIA provides the following conclusion*: "The available indicators in the analyzed requirements is insufficient for determining 22 metrics (from 24). The sufficiency level of metric information in the analyzed software requirements is 44.86%. There is a need of improving the requirements by supplementing indicators for metrics".

OBIA also provides the *visualization of missing indicators* (Figs. 6, 7, 8 and 9), where missing indicators are indicated as a strikeout in the base ontology, and the metrics, which cannot be calculated on the basis of the available indicators, are outlined by the circle in the base ontology.

This visualization displays the list of missing indicators; displays influence of indicators on one or another metric; shows the priority of supplementing the indicators in the requirements with the purpose of increasing the sufficiency level of metric information. Used ontologies show the correlation of metrics by the indicators, i.e. instruct the developer what requirements should be first of all supplemented and by what indicators - for a more rapidly increasing the sufficiency level of metric information.

The decision on the need for improving the requirements is taken by the customer on the basis of the OBIA's conclusion about the sufficiency level of metric information. The customer of the system of improving the safety of computer systems' software decided that, given the low sufficiency level of metric information, requirements need to be improved (by supplementing the indicators).

After improving the requirements for software of the system of improving the safety of computer systems' software, the re-analysis (re-parsing) of requirements

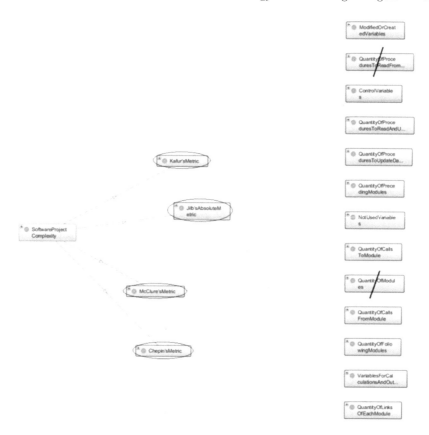

Fig. 6. Visualization of missing indicators for software project complexity metrics

was performed. This parsing made it possible to construct the real ontology for quality and complexity software project and software (version 2).

In accordance with the developed method of actions, implemented OBIA performs the comparison of real ontology (version 2) with the ideal ontology. As a result of this comparison, OBIA provided the set of missing indicators in the improved requirements for the software of the system of improving the safety of computer systems' software:

1. For software project complexity metrics: QuantityOfProceduresToReadFrom-DataStructure.
2. For software project quality metrics: TypeOfModuleInputData, QuantityOf-PotentialAccessToGlobalVariables.
3. For software complexity metrics: TotalQuantityOfOperands.
4. For software quality metrics: CostOfOneLine.

OBIA detects the metrics, which cannot be calculated on the basis of the indicators from the improved requirements for the software of the system of improving the safety of computer systems' software:

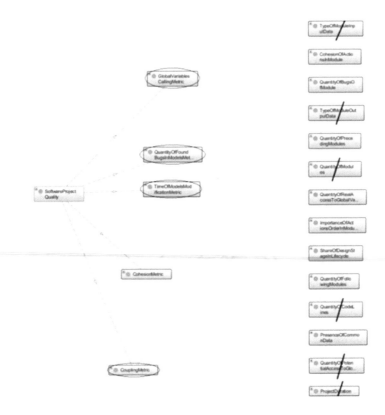

Fig. 7. Visualization of missing indicators for software project quality metrics

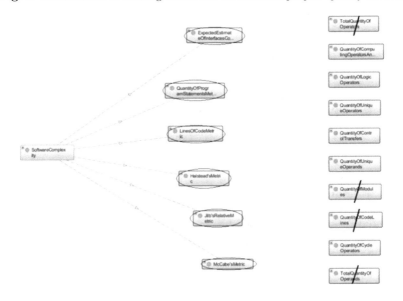

Fig. 8. Visualization of missing indicators for software complexity metrics

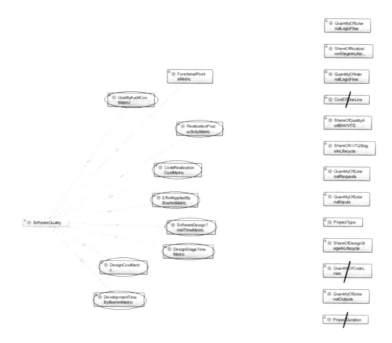

Fig. 9. Visualization of missing indicators for software quality metrics

1. Software project complexity metrics: Kafur's metric, i.e. 1 from 4 software project complexity metrics cannot be calculated after improving the requirements.
2. Software project quality metrics: Coupling metric, Global variables calling metric, i.e. 2 from 5 software project quality metrics cannot be calculated after improving the requirements.
3. Software complexity metrics: Halstead's metric, Expected estimate of interfaces complexity, i.e. 2 from 6 software complexity metrics cannot be calculated after improving the requirements.
4. Software quality metrics: Design cost, Quality audit cost, Code realization cost, i.e. 3 from 9 software quality metrics cannot be calculated after improving the requirements.

Numerical assessment of the sufficiency level of metric information in the requirements (after improvement) is the following:

$$D = (24 - (0/5 + 0/2 + 0/5 + 1/4 + 0/2 + 1/5 + 1/2 + 0/3 + 0/2 + 0/1$$
$$+ 1/5 + 0/3 + 0/3 + 0/2 + 1/3 + 0/2 + 0/3 + 1/2 + 1/4 + 0/2 + 1/3$$
$$+ 0/5 + 0/2 + 0/2))/24 = 0,8931 = 89,31\%$$

Thus, the OBIA provides the following conclusion: "The available indicators in the analyzed requirements is insufficient for determining 8 metrics (from 24). The sufficiency level of metric information in the analyzed software requirements is 89.31%. There is a need of improving the requirements by supplementing indicators".

OBIA also provides the visualization of missing indicators, similar to visualization for the first experiment. The customer of the system of improving the safety of computer systems' software decided that, given the sufficiency level of metric information more than 85%, requirements don't require further improvement.

6   Discussion

The developed method of actions and implemented OBIA provided the automated analysis of software requirements for the sufficiency of their information through the use of ontologies. Exactly ontologies provided the reveal of duplications and gaps in knowledge, and the visualization of missing logical connections. The implemented OBIA is aimed at automating the determination of the sufficiency of metric information in the software requirements, in contrast to the developed in [7,14–22] intelligent agents. Thus, the developed OBIA is the evolution of the concept of assessing the sufficiency of information, which was presented in [23]. The intelligence of the realized agent is in the conclusion about the sufficiency or insufficiency of metric information, the about sufficiency level of metric information in the requirements, and in the visual recommendations for the improvement of software requirements.

The conducted in Sect. 5 experiment has shown, that the supplementing the requirements for software of the system of improving the safety of computer systems' software by 5 indicators increased the sufficiency level of metric information by 44.45% - from 44.86% to 89.31%. Consequently, the developed OBIA provides improving the sufficiency level of metric information in the software requirements, in contrast to the described in [7,14–22] known IA. The developed OBIA can be used for any software.

7   Conclusions

In this paper, the method of actions of OBIA for determination of sufficiency of metric information in the software requirements has been further developed in terms of automation of parsing the software requirements, and in terms of determination of the sufficiency of metric information in the software requirements. On the basis of this method, the OBIA was implemented, which provides:

1. Automated analysis (parsing) of software requirements.
2. Conclusion about the sufficiency or insufficiency of metric information in the software requirements.
3. Numerical assessments of the sufficiency level of metric information.
4. The list of attributes that should be supplemented in the software requirements with the purpose of improving the sufficiency level of their information.

5. Visual hints on the priority of supplementing the indicators in the requirements.
6. Minimization of the human factor during information processing and knowledge gaining.

During the experiment, the automatic analysis (parsing) of requirements for software of the system of improving the safety of computer systems' software was performed using the developed OBIA. The performed experiment showed that all the results of the functioning of realized OBIA in the complex provide: the avoidance of the losses of important information, the minimization the occurrence of errors at the early phases of the software development lifecycle; increasing the sufficiency level of metric information in the software requirements; improving the software quality.

References

1. (2017) PMI's pulse of the profession 9th global project management survey. https://www.pmi.org/-/media/pmi/documents/public/pdf/learning/thought-leadership/pulse/pulse-of-the-profession-2017.pdf. Accessed 12 Apr 2019
2. Lynch J (2018) Project Resolution Benchmark Report. https://www.standishgroup.com/sample_research_files/DemoPRBR.pdf. Accessed 12 Apr 2019
3. Johnson J (2018) CHAOS report: decision latency theory: it is all about the interval. The Standish Group
4. Sarif S, Ramly S, Yusof R, Fadzillah NAA, Sulaiman NY (2018) Investigation of success and failure factors in IT project management. Adv Intell Syst Comput 739:671–682
5. Levenson NG (2012) Engineering a safer world: systems thinking applied to safety, Cambridge, USA
6. Rosato M (2018) Go small for project success. PM World J 7(5):1–10
7. Meziane F, Vadera S (2010) Artificial intelligence applications for improved software engineering development: new prospects. Advances in intelligent information technologies, pp 278–299
8. ISO/IEC TR 19759:2015 (2015) Software Engineering. Guide to the software engineering body of knowledge (SWEBOK)
9. ISO/IEC/IEEE 24765:2010 (2010) Systems and software engineering. Vocabulary
10. Pomorova O, Hovorushchenko T (2011) Research of artificial neural network's component of software quality evaluation and prediction method. In: The 2011 IEEE 6-th international conference on intelligent data acquisition and advanced computing systems: technology and applications proceedings, Prague, vol 2, pp 959–962
11. Burov E (2014) Complex ontology management using task models. Int J Knowl Based Intell Eng Syst 18(2):111–120
12. Burov E, Pasitchnyk V, Gritsyk V (2014) Modeling software testing processes with task ontologies. Br J Educ Sci 2(6):256–263
13. Wooldridge M, Jennings NR (1995) Intelligent agents - theory and practice. Knowl Eng Rev 10(2):115–152
14. Freitas A, Bordini RH, Vieira R (2017) Model-driven engineering of multi-agent systems based on ontologies. Appl Ontol 12(2):157–188

15. Ossowska K, Szewc L, Weichbroth P, Garnik I, Sikorski M (2017) Exploring an ontological approach for user requirements elicitation in the design of online virtual agents. In: Information systems: development, research, applications, education, vol 264, pp 40–55

16. Lezcano-Rodriguez LA, Guzman-Luna JA (2016) Ontological characterization of basics of KAOS chart from natural language. Revista Iteckne 13(2):157–168

17. Garcia-Magarino I, Gomez-Sanz JJ (2013) An ontological and agent-oriented modeling approach for the specification of intelligent ambient assisted living systems for Parkinson patients. In: Hybrid artificial intelligent systems, vol 8073, pp 11–20

18. Wilk S, Michalowski W, O'Sullivan D, Farion K, Sayyad-Shirabad J, Kuziemsky C, Kukawka B (2013) A task-based support architecture for developing point-of-care clinical decision support systems for the emergency department. Methods Inf Med 52(1):18–32

19. Rakib A, Faruqui RU (2013) A formal approach to modelling and verifying resource-bounded context-aware agents. Lecture notes of the institute for computer sciences social informatics and telecommunications engineering, vol 109, pp 86–96

20. Pakdeetrakulwong U, Wongthongtham P, Siricharoen WV, Khan N (2016) An ontology-based multi-agent system for active software engineering ontology. Mobile Netw Appl 21(1):65–88

21. Strmečki D, Magdalenić I, Kermek D (2016) An overview on the use of ontologies in software engineering. J Comput Sci 12(12):597–610

22. Anquetil N, Oliveira KM, Dias MGB (2006) Software maintenance ontology. In: Ontologies for software engineering and software technology, pp 153–173

23. Hovorushchenko T, Pomorova O (2016) Ontological approach to the assessment of information sufficiency for software quality determination. CEUR-WS 1614, pp 332–348

24. Hovorushchenko T, Pavlova O (2019) Method of activity of ontology-based intelligent agent for evaluating the initial stages of the software lifecycle. Advances in intelligent systems and computing, vol 836, pp 169–178

Information Technology of Control and Support for Functional Sustainability of Distributed Man-Machine Systems of Critical Application

Viktor Perederyi[1]([✉])(iD), Eugene Borchik[2](iD), and Oksana Ohnieva[1]([✉])(iD)

[1] Kherson National Technical University, Kherson, Ukraine
viperkms1@gmail.com, Oksana_Ognieva@meta.ua
[2] Maritime Institute of Postgraduate Education named after F.F. Ushakov,
Kherson, Ukraine

Abstract. The paper considers information technology for control and support of the functional sustainability (FS) of man-machine systems of critical application, which complements the theory and methods for solving the tasks of ensuring system's fault tolerance and liveness, based on the interaction of a set of operation security indicators, as well as human factor indicators, in managing and making decisions on its each hierarchical level. A Bayesian Trust Network was built to evaluate the entire system's FS, with the help of which, on the basis of expert knowledge, an assessment of the state probability of both individual components of the structural organization and a comprehensive FS assessment of the entire distributed system was performed. For practical substantiation of the obtained results, an experiment was carried out, the results of which confirmed the practical importance of the proposed information technology, which can be used for the control and support of the FS of man-machine distributed systems of critical application.

Keywords: Information technology · Man-machine systems ·
Complex organizational and technical objects · Decision making ·
Functional sustainability · Human factor ·
Fault tolerance of the system · System's liveness

1 Introduction

Each distributed man-machine system of critical application has its own characteristics, which are stipulated by the scope of its application. The importance and challenging nature of the tasks solved by means of such systems in real time have brought about high requirements for the reliability of these systems, where it is often impossible to perform maintenance during the operation and a fault of the entire distributed system or its individual components can lead to negative consequences.

© Springer Nature Switzerland AG 2020
V. Lytvynenko et al. (Eds.): ISDMCI 2019, AISC 1020, pp. 461–477, 2020.
https://doi.org/10.1007/978-3-030-26474-1_33

The high cost and potential insecurity of complex distributed man-machine systems of critical application, operating in extreme conditions require an adequate level of reliability and security of use. Herewith, traditional methods based on multiple reservations, the introduction of integrated control and elements with increased reliability to the systems, worsened the technical and economic characteristics of projected systems, without leading to a necessary reduction in the likelihood of occurrence of hazardous situations. The need to introduce additional hardware redundancy to ensure the reliability of the system has become a fundamental limitation of this approach.

Modern man-machine systems of critical application, in particular the control systems in the energy, transport, petrochemical industry, have a branched hierarchical structure, and are characterized by human involvement in the processes of control and decision-making in conditions of uncertainty, under the influence of various external and internal factors in it in real time mode.

Thus, functional sustainability is, first of all, aimed at improving the characteristics of fault tolerance and liveness, but not necessarily of reliability indicators of separate components, but at the complex of safety indicators of the system operation, taking into account the human factor in the management and decision-making at each hierarchical system level.

At this, the main difficulties are connected not only with the improvement of technical and software tools, but also with underdeveloped theory of functional sustainability with the methods of accounting of the human factor, so the formation of the basic indicators of functional sustainability is in the development stage and is an important area of scientific research in designing and functioning of distributed man-machine systems of critical application.

2   Literature Review

A topical issue for today is the development of a complex of analytical models and methods for monitoring the safe and stable process of functioning of distributed man-machine systems and the interaction of agents in assessing their stable functional state.

Analysis of works of functional sustainability problems of information systems was studied in works by both domestic and foreign scientists, and it allows to reveal the tendencies of their development based on the principles of intellectualization and adaptability.

The realization of the functional sustainability, which is achieved by use of complex technical systems various and already existing ones types of redundancy (structural, temporal, informational, functional, load, etc.) through the redistribution of resources for the purpose of reflecting the consequences of extraordinary situations, is considered in [1,2].

In [3] the close connection of the concept of "functional sustainability" with the concepts of "reliability", "liveness" and "stability" is shown. The principal difference between them is described: where the methods of providing functional sustainability are directed not to reduce the number of faults and violations (as traditional methods for increasing the reliability, liveness and stability of technical systems), but to ensure performing the most important functions, when these violations have already occurred.

Indicators, which directly characterize the quality of functioning without taking into account possible extraordinary situations, such as the speed of information transmission, the quality of the information transmitted, the security of information, the availability of information to the user and which is often called tactical and technical characteristics of the system, are determined in works [4].

Analysis of ways to ensure the reliability of distributed information systems proves that this problem is solved comprehensively by using highly reliable elements and due to the implementation of the principle of fault tolerance [5–7].

The problems of the emergence and impact of information security threats, in the form of the possibility of cybernetic attacks in distributed information systems, for the management and efficiency of decision-making are explored in [8,9].

Analysis of different concepts of sustainability, and methods for their determination showed that the classical stability theory operates mainly with dynamic systems, which are described by the system of differential equations in different modifications: linear, nonlinear, digital, stochastic, adaptive, optimal and other systems.

However, the problem of determining the functional sustainability of complex organizational and technical systems, the class of which includes distributed man-machine systems of critical application, currently remains unsolved.

3   Problem Statement

The research of modern scientifically substantiated approaches to increasing the efficiency of the operation of complex organizational and technical systems, to which full-scale distributed man-machine systems of critical application can be attributed, have allowed to conclude that in recent years a new priority approach has been formed which is connected with providing the feature of functional sustainability in these systems.

However, for the development of the theory of functional sustainability of distributed man-machine systems and the analysis of literary sources carried out above, the present paper proposes an information technology for controlling and maintaining the functional sustainability of distributed man-machine systems of critical application, aimed at improving the features of fault-tolerance and liveness, on the basis of the complex indicators of work safety, taking into account the human factor in the management and decision-making at each hierarchical system level.

To do this, the following partial tasks can be formulated:

- to determine the set of indicators of sustainability and safety of the system operation characterizing its fault tolerance and liveness,

- to determine the complex of indicators of the human factor in management and decision-making, at each hierarchical system level, which directly affect the stability of functional sustainability in general,
- to carry out an expert assessment, according to the degree of importance, of a set of indicators characterizing the fault tolerance and liveness of the system by experts, the corresponding hierarchical system level,
- to carry out an expert assessment by the experts of the corresponding hierarchical system level, of a set of indicators of the human factor according to the degree of importance,
- to develop a block diagram of the information technology of interaction and distribution of individual modules of the set of the main indicators of functional sustainability in the general hierarchical structure of a complex organizational and technical object,
- on the basis of the scheme of the information technology, in order to assess the functional sustainability of the system, to apply the Bayesian Trust Network (BTN), in designing of which conditional probability tables are used based on the knowledge of experts,
- using the proposed BTM to assess the probability of the state of functional sustainability as separate components of the structural organization as well as a comprehensive assessment of the FS of the entire distributed system,
- to carry out an experiment using the proposed information technology for assessing the FS of the entire distributed system.

4   Materials and Methods

In order to maintain fault tolerance, liveness and information security of a distributed system of critical application, it is necessary to monitor, evaluate and maintain the core performance of its FS in real time. For this system of the appropriate level of the hierarchical structure, it is necessary to monitor and evaluate data integrity; data availability; data confidentiality; data consistency; data reference integrity; external influence; hardware failure; software failure; actions of employees; actions of administrators; analyze data for security; verify the assignment of access rights.

For comprehensive control and support of the main indicators of the entire distributed system's FS, it is advisable to apply additional safety and information security methods: monitoring a user's login with administrator rights; analyzing the system for viruses; performing preliminary data substitution in the query and checking for SQL-injections; monitoring incorrect user actions; hardware failures; the availability of data and resources; checking data integrity (using hash); checking the information for the occurrence of data substitution.

To solve these tasks, an information technology for monitoring and maintaining the functional sustainability of distributed man-machine systems of critical application is presented in Fig. 1.

In accordance with Fig. 1, the main components of the structural organization of the system are COTO and ADDSS.

COTO represents a group of resources of a complex organizational and technical object taking into account psycho-functional characteristics of users when making decisions in the management of the technological process. The ADDSS group includes the main components of the adaptive distributed decision support system, which comprise adaptive user interface (DM); distributed database, which is intended to store alternatives for decision making and automation of their updating; system of WEB search - a search engine that searches for and filters decisions and knowledge that is not in the database; dynamic database - a special database, in which solutions from the WEB are temporarily stored.

Real-time monitoring and control modules scan the FS state of the COTO and DSS sectors assigned to them.

At the end of scanning and control of all vertices of the system, the assessment of the probabilities of possible states of the monitoring and control modules of the system FS during its operation is performed. If, according to the results of the control, no threats are detected, the module evaluates the state of the system sectors as completely protected.

In case of presence of failures or threats, the corresponding module calculates an estimate of the probability of the degree of protection of the system sectors, after which certain measures are taken to eliminate them and restore the full system FS.

Eventually, the results of assessments of the probability of the degree of security of all modules are sent to the central COMPLEX INFORMATION SECURITY module, where control, evaluation and support of the degree of integrated security of the man-machine-system of critical application are carried out.

To date, Bayesian trust networks are one of the most appropriate models for dealing with incomplete, inaccurate, and contradictory information. The mathematical apparatus of Bayesian networks is based on a probabilistic approach and is able to maximize the use of information coming from dedicated sources to achieve maximum effect [10].

In view of this, in this work, on the basis of the information technology scheme (Fig. 1), the Bayesian Trust Network (BTN) was used to assess the functional sustainability of the system (Fig. 2). When building its structure and tables of conditional probabilities, the knowledge of experts was used.

The vertices $A, B, C, D, E, G, F, H, K2, K3, K1, K6, K8, K11, K10, K9, K7, K12, K5, K4$ are of "Decision" type, and correspond to homonymic modules of FS monitoring and control and can take three states: "high security", "medium security", "low security". Below, we will denote these states as hs, ms, ls respectively, and the vertices (variables) $A, B, C, D, E, G, F, H, K2, K3, K1, K6, K8, K11, K10, K9, K7, K12, K5, K4$ are denoted as M_i $(i = \overline{1, 20})$ respectively.

These vertices (modules) are a set of basic indicators of the system's FS and are presented by monitoring modules: module A is a module of preliminary substitution to data queries and SQL - injection checks; module B - module for analyzing the system for viruses; module C - module for monitoring login with administrator rights; module D - hardware failure monitoring module; module E - module for

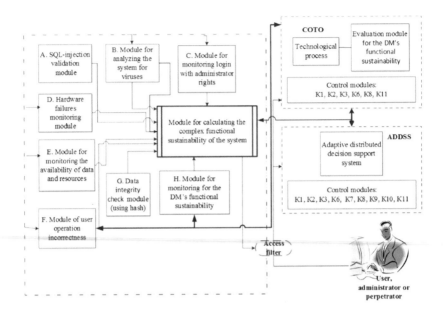

Fig. 1. Information technology for control and support of functional sustainability of distributed man-machine systems of critical application

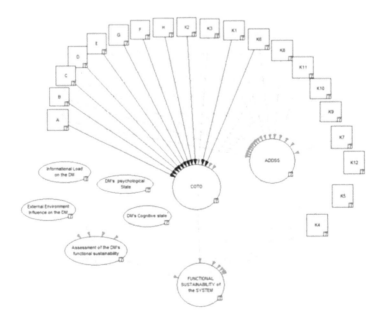

Fig. 2. Bayesian trust network for assessing the functional sustainability of the system

monitoring the availability of data and resources; module F - module of user operation incorrectness; module G - data integrity check module (using hash); module H - module for monitoring for the DM's functional sustainability; control modules: module $K1$ - data integrity control; module $K2$ - data accessibility control; module $K3$ - data confidentiality control; module $K4$ - data consistency control; module $K5$ - reference value control; module $K6$ - External influence control; module $K7$ - hardware failure control; module $K8$ - software failure control; module $K9$ - control of the influence of user actions; module $K10$ - control of the influence of administrator actions; module $K11$ - data protection control; module $K12$ - power failure control.

Modules scan sectors assigned to them. Herewith, the probability P_i of the module M_i security is estimated by the following formula:

$$P_i = \frac{L_i}{N_i} \tag{1}$$

where L_i - number of normally functioning sectors; N_i - total number of sectors checked by the module M_i.

Based on calculated probabilities P_i it is possible to determine the states of the vertices A, B, C, D, E, G, F, H, $K2$, $K3$, $K1$, $K6$, $K8$, $K11$, $K10$, $K9$, $K7$, $K12$, $K5$, $K4$ of BTN using the table of conditional probabilities of possible states of the FS monitoring and control modules. For this, experts working in distributed organizational and technical systems of critical application of the higher level were proposed to conduct a comprehensive assessment by the degree of importance of possible probabilities of the states of these modules. Expert assessment results are presented in Tables 1 and 2.

Table 1. Table of conditional probabilities of possible states of the FS monitoring modules

Vertex	A	B	C	D	E	F	G	H
State	Security probability							
High	0,95-1	0,9-1	0,85-1	0,95-1	0,9-1	0,8-1	0,9-1	0,95-1
Medium	0,4-0,95	0,4-0,9	0,4-0,85	0,45-0,95	0,4-0,9	0,4-0,8	0,4-0,9	0,4-0,95
Low	0-0,4	0-0,4	0-0,4	0-0,45	0-0,4	0-0,4	0-0,4	0-0,4

We assume that the network vertices "COTO", "ADDSS", "Assessment of the DM's functional sustainability" and "FUNCTIONAL SUSTAINABILITY of the SYSTEM" are binary, i.e. have two states. Variables "COTO" and "ADDSS" may take the values: "Protected" and "Unprotected", and the variables "Assessment of the DM's functional sustainability" and "FUNCTIONAL SUSTAINABILITY of the SYSTEM" may take the values "stable" and "unstable".

Table 2. Table of conditional probabilities of possible states of the FS monitoring modules

Vertex	K1	K2	K3	K4	K5	K6	K7	K8	K9	K10	K11	K12
State	Security probability											
High	0,85-1	0,9-1	0,85-1	0,8-1	0,95-1	0,9-1	0,9-1	0,9-1	0,7-1	0,85-1	0,9-1	0,9-1
Medium	0,15 -0,85	0,1 -0,9	0,15 -0,8	0,2 -0,8	0,1 -0,9	0,1 -0,9	0,1 -0,9	0,1 -0,9	0,3 -0,7	0,1 -0,8	0,1 -0,9	0,1 -0,9
Low	0-0,15	0-0,1	0-0,15	0-0,2	0-0,05	0-0,1	0-0,1	0-0,1	0-0,3	0-0,15	0-0,1	0-0,1

Since the vertices considered have a large number of Parent links, filling the tables of conditional probabilities would have required a huge number of table values. To reduce the complexity of filling the conditional probability tables, it is convenient to assign Noisy MAX type to these vertices. Using Noisy MAX type vertices is also convenient because it is much easier for experts to assess the impact of only one factor on an expected event than to estimate the combined effect of several factors on it [11].

To describe the noisy vertices "COTO", "ADDSS" and "FUNCTIONAL SUSTAINABILITY of the SYSTEM" experts of the distributed system of the lower level were suggested to assess conditional probabilities of possible states of the specified vertices by the degree of importance. The results are presented in Tables 3, 4 and 5.

Table 3. COTO vertex description

Parent	A			B			C			D			E		
State	ls	ms	hs	ls	ms	hs	ls	ms	hs	ls	ms	hs	ls	ms	hs
Unprotected	0.7	0.15	0	0.65	0.1	0	0.8	0.1	0	0.6	0.05	0	0.7	0.05	0
Protected	0.3	0.85	1	0.35	0.9	1	0.2	0.9	1	0.4	0.9	1	0.3	0.95	1
Parent	F			G			H			K1			K2		
State	ls	ms	hs	ls	ms	hs	ls	ms	hs	ls	ms	hs	ls	ms	hs
Unprotected	0.75	0.15	0	0.85	0.1	0	0.8	0.2	0	0.6	0.1	0	0.75	0.1	0
Protected	0.25	0.85	1	0.15	0.9	1	0.2	0.8	1	0.4	0.95	1	0.25	0.9	1
Parent	K3			K6			K8			K11					
State	ls	ms	hs	ls	ms	hs	ls	ms	hs	ls	ms	hs			
Unprotected	0.8	0.1	0	0.55	0.05	0	0.6	0.1	0	0.7	0.15	0			
Protected	0.2	0.9	1	0.45	0.95	1	0.4	0.9	1	0.3	0.85	1			

Since vertices "COTO" and "ADDSS" were assigned Noisy MAX type, it was assumed that the state of M_j vertex influences the probabilities of the states of the vertices "COTO" and "ADDSS" regardless the states of the vertices M_i,

Table 4. ADDSS vertex description

Parent	A			B			C			D			E		
State	ls	ms	hs	ls	ms	hs	ls	ms	hs	ls	ms	hs	ls	ms	hs
Unprotected	0.7	0.05	0	0.55	0.1	0	0.8	0.2	0	0.65	0.1	0	0.7	0.1	0
Protected	0.3	0.95	1	0.45	0.9	1	0.2	0.8	1	0.35	0.9	1	0.3	0.9	1
Parent	F			G			H			K1			K2		
State	ls	ms	hs	ls	ms	hs	ls	ms	hs	ls	ms	hs	ls	ms	hs
Unprotected	0.6	0.05	0	0.55	0.1	0	0.7	0.15	0	0.4	0.05	0	0.5	0.1	0
Protected	0.4	0.95	1	0.45	0.9	1	0.3	0.85	1	0.6	0.95	1	0.5	0.9	1
Parent	K3			K6			K8			K11			K10		
State	ls	ms	hs	ls	ms	hs	ls	ms	hs	ls	ms	hs	ls	ms	hs
Unprotected	0.7	0.2	0	0.6	0.1	0	0.55	0.1	0	0.6	0.1	0	0.45	0.15	0
Protected	0.3	0.8	1	0.4	0.9	1	0.45	0.9	1	0.4	0.9	1	0.55	0.85	1
Parent	K9			K7											
State	ls	ms	hs	ls	ms	hs									
Unprotected	0.7	0.15	0	0.65	0.05	0									
Protected	0.3	0.85	1	0.35	0.95	1									

Table 5. FUNCTIONAL SUSTAINABILITY of the SYSTEM vertex description

Parent	K7			K12		
State	ls	ms	hs	ls	ms	hs
Unprotected	0,25	0,07	0	0,2	0,05	0
Protected	0,75	0,93	1	0,8	0,95	1
Parent	K5			K4		
State	ls	ms	hs	ls	ms	hs
Unprotected	0,25	0,05	0	0,3	0,1	0
Protected	0,75	0,95	1	0,7	0,9	1
Parent	COTO			ADDSS		
State	Unprotected	Protected		Unprotected	Protected	
Unstable	0,15	0		0,13	0	
Stable	0,85	1		0,87	1	
Parent	Assessment of the DM's functional sustainability					
State	Unstable			Stable		
Unstable	0,2			0		
Stable	0,8			1		

$i = \left(\overline{1,20}, i \neq j\right)$. That is why the probabilities P_{COTO} and P_{ADDSS} of the vertices "COTO" and "ADDSS" being in "Protected" state can be calculated by the following formulas:

$$P_{\text{COTO}} = \prod_{i=1}^{14} P\left(\text{COTO} \mid M_i\right) \tag{2}$$

$$P_{\text{ADDSS}} = \prod_{i=1}^{17} P\left(\text{ADDSS} \mid M_i\right) \tag{3}$$

where $P\left(\text{COTO} \mid M_i\right)$ and $P\left(\text{ADDSS} \mid M_i\right)$ are the probabilities of the vertices "COTO" and "ADDSS" being in "Protected" state under the condition, that the node M_i is in one of the following states: "high security", "medium security", "low security".

In man-machine systems of critical application, the most unstable and unpredictable link in the system FS is the decision maker at each level of control of an organizational and technical object, which requires a separate consideration of the assessment of the DM's FS when the system functions in real time.

The evaluation module of the DM's FS is represented by the vertex of the network "Assessment of the DM's functional sustainability". The functional sustainability of the DM is mainly influenced by 4 factors (Fig. 1): the external environment, the information load on the DM, psychological and cognitive states. This is reflected in the Bayesian network by the fact that the vertex "Assessment of the DM's functional sustainability" has 4 parent vertices: "External Environment Influence on the DM"; "Informational Load on the DM", "DM's psychological State", "DM's Cognitive state". In Fig. 2, these vertices are designated as EE, IL, PS, CS respectively. We assume that the network variables EE, IL, PS, CS are binary and may take values: "positive" and "negative".

The experts of the distributed system of the lower level carried out an evaluation, according to the degree of importance, of the conditional probabilities of the possible states of a Noisy MAX type node "Assessment of the DM's functional sustainability". The results are presented in Table 6.

We apply the nodes of the type "Chance - Genera" for Parent nodes EE, IL, PS, CS.

The method for evaluating the probabilities of the states of the vertices EE and IL is described detailed in [12]. The following is a summary of this approach.

Let us consider the node EE. The following factors influence the random variable "External Environment Influence on the DM" [13]: noise intensity IN; intensity of vibration IV; workplace illumination E; temperature T; humidity H; atmospheric pressure fluctuations $\Delta(P)/\Delta t$. Normative values of these factors are given in [14], the impact of environmental factors in this case is considered to be positive (suitable) for DM.

The more external factors differ from the normative ones, the more likely that their impact on the DM will be negative. The question arises: what is the probability P_2 that the random variable "External Environment Influence on the DM" becomes negative?

Table 6. Description of the node "Assessment of the DM's functional sustainability"

Parent	Informational load on the DM		External environment influence on the DM		Other causes
State	Negative	Positive	Negative	Positive	
Unstable	0,67	0,0	0,85	0,0	0,01
Stable	0,33	1,0	0,15	1,0	0,99
Parent	DM's cognitive		DM's psychological state		Other causes
State	Negative	Positive	Negative	Positive	
Unstable	0,62	0,0	0,73	0,0	0,01
Stable	0,38	1,0	0,27	1,0	0,99

To answer this question, we will create a system for forecasting probability values, based on the fuzzy logical conclusion on the Mamdani algorithm on the fuzzy knowledge base [15], in which the values of the input and output variables are given by fuzzy sets. Taking into account that according to [12] the most noticeable influence on the DM is exerted by the factors of noise intensity, workplace illumination and intensity of vibration, the following fuzzy knowledge base is proposed by the experts:

RULE 1: IF u_1 is "not a norm" THEN ν is a "high"

RULE 2: IF u_2 is "not a norm" And u_1 is "norm" And u_3 is "norm" THEN ν is "above average"

RULE 3: IF u_3 is "not a norm" And u_1 is "norm" And u_2 is "norm" THEN ν is "above average"

RULE 4: IF u_2 is "not a norm" And u_3 is "not a norm" THEN ν is "high"

RULE 5: IF u_4 is "not a norm" And u_1 is "norm" And u_2 is "norm" And u_3 is "norm" THEN ν is "average"

RULE 6: IF u_5 is "not a norm" And u_4 is "norm" And u_1 is "norm" And u_2 is "norm" And u_3 is "norm" THEN ν is "below average"

RULE 7: IF u_6 is "not a norm" And u_4 is "norm" And u_1 is "norm" And u_2 is "norm" And u_3 is "norm" THEN ν is "below average"

RULE 8: IF u_1 is "norm" And u_2 is "norm" And u_3 is "norm" And u_4 is "norm" And u_5 is "norm" And u_6 is "norm" THEN ν is "low"

Here, u_i $\left(i = \overline{1,6}\right)$ denotes the linguistic variables "noise intensity", "intensity of vibration", "workplace illumination", "temperature", "humidity", "atmospheric pressure fluctuations" respectively, and through ν - is the linguistic variable "the probability that the random variable" "External Environment Influence on the DM" takes the value "negative".

To describe the linguistic variables u_i $\left(i = \overline{1,6}\right)$ we will use the term set "norm", "not a norm", and for the variable ν the term set "low," "below average," "average", "above average", "high".

The membership functions u_i $\left(i = \overline{1,6}\right)$ of the linguistic variables will be set in the form of the bilateral Gauss distribution, but of the term of linguistic variable ν it will be set in the form of the symmetric Gauss distribution.

To adjust the fuzzy F model, we will solve the problem of mathematical programming of minimizing the value of the standard error, which could result from differences between the experimental results and results those calculated with Mamdani algorithm. Taking into account the experimental data and the experts' knowledge of the influence of environmental factors on the DM we solve the problem using Fuzzy Logic Toolbox and Optimization Toolbox packages.

A similar approach based on a fuzzy logic conclusion on the Mamdani algorithm is used to estimate the probability $\mathbf{P}_1$ that the random value of "Informational Load on the DM" takes the value "negative".

The psychological and cognitive state of the DM is determined with the tests. Let us denote $\mathbf{P}_3$ the probability that the random value "DM's Psychological State" takes the value "negative", and denote $\mathbf{P}_4$ the probability that the random value "DM's Cognitive state" takes the value "negative". In estimating these probabilities, we will assume that $\mathbf{P}_3 = \dfrac{q_3}{Q_3}$, $\mathbf{P}_4 = \dfrac{q_4}{Q_4}$, where q_3 is the number of tests that characterize the DM's psychological state negatively; Q_3 is a total number of tests of the DM's psychological state; q_4 is a number of tests that characterize the DM's cognitive state negatively; Q_4 is a total number of tests for estimating DM's cognitive state.

5   Experiment

To test the performance of the proposed information technology, an experiment was conducted, where at a certain time moment the result of scanning the sectors of the FS control and monitoring modules showed that the BTN vertices $A, B, C, E, G, F, H, K2, K3, K1, K6, K8, K11, K10, K9, K7$ are in the "high security" state, and the vertex B (module for analyzing the system for viruses) is in the "medium security" state. According to the formulas (2), (3) with the help of Tables 3, 4 and Fig. 2, the probability that the "COTO" and "ADDSS" vertices are in the "protected" state was calculated:

$$P_{\text{COTO}} = \prod_{i=1}^{3} 1 \cdot 0,9 \cdot \prod_{i=5}^{14} 1 = 0,9$$

$$P_{\text{ADDSS}} = \prod_{i=1}^{3} 1 \cdot 0,9 \cdot \prod_{i=5}^{17} 1 = 0,9$$

Let the DM's FS at this time point is evaluated by the probability of 90%. We will evaluate the system FS in cases where $K4$, $K5$, $K12$ nodes are in the states: "low security", "medium security", "high security". The calculation results are shown in Fig. 3, 4 and 5.

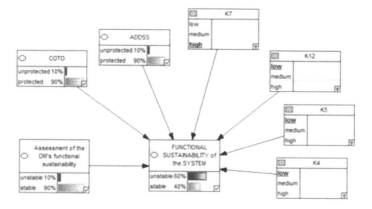

Fig. 3. System FS probability at low security of the nodes K4, K5, K12

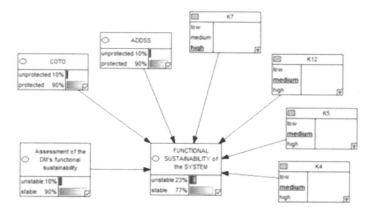

Fig. 4. System FS probability at medium security of the nodes K4, K5, K12

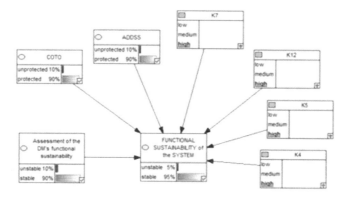

Fig. 5. System FS probability at high security of the nodes K4, K5, K12

If the $K4$, $K5$, $K12$ nodes are in the "high security" state, then the probability that the system is in the "stable" state is 95% (Fig. 5). According to the results of studies of literature sources for many systems, such a value of the FS probability is acceptable. Therefore, the system can be considered functionally stable.

If nodes $K4$, $K5$, $K12$ are in the state of "medium security" and "low security", then the probability that the system is in the state "stable" is 40% and 72%, respectively (Figs. 3 and 4). Since these values are less than 95%, the system FS cannot be considered. In these cases, it is necessary to eliminate the danger indicated by the results of scanning the sectors of the FS control and monitoring modules: system infection with viruses; inconsistency of data; lack of reference value; power failures. It is also necessary to normalize the values of the factors influencing the information load and the influence of the external environment on the DM.

6   Results and Discussion

On the basis of the information technology scheme for evaluating the FS system, a Bayesian Trust Network, taking into account the integrated assessment of FS indicators was proposed by the experts in the form of BTN conditional probability tables.

It is proposed to use the Noisy MAX type for BTN vertices with a large number of parent vertices. This made it possible to reduce the laboriousness of filling in the tables of conditional probabilities, since in this case; it is enough for the expert to evaluate the effect of the value of the parent vertex on the node being examined, regardless of what other values are taken by parent vertices of this node. At the same time, the tables of conditional probabilities of Noisy MAX vertices are considerably simplified.

For the BTN parent nodes "External Environment Influence on the DM" and "Informational Load on the DM", the state of which depends on many factors, a method was proposed for estimating the probability of states, based on a fuzzy logical conclusion on the Mamdani algorithm using a fuzzy knowledge base, in which input and output variable values are given by fuzzy sets. This made it possible, when a number of factors deviating from the normative values, to assess the probability of their negative influence on DM.

At one of the engineering and technical objects, on a regional scale, an experiment was conducted to test the developed information technology of control and support of the power system FS. In the first case, during the operation of the power system, according to the results of scanning the sectors of the FS control and monitoring modules, it was determined that all modules are in a "high security" state with the exception of the module B (module for analyzing the system for viruses), which was in the "medium security" state. The calculation by formulas (2), (3) shows that the security of the "COTO" and "ADDSS" nodes in this case was 90%. The information load on the DM turned out to be so intense that its functional stability was at the level of 90%. The calculation on the BTN showed (Fig. 5) that the probability of the system FS in the considered case turned out to be 95%. For the power system, such a value is quite acceptable in order to consider it functionally sustainable.

At another time point (rush hour), scanning showed that, as previously, all the control and monitoring modules of the power system are in a "high security" state, and module B is in a "medium security" state. Therefore, the security of the "COTO" and "ADDSS" nodes as in the previous case is 90%. The information load and the degree of influence of external factors on the DM did not change, so that its functional stability remained at the level of 90%. But at the same time, the modules $K12$ - power failures, $K5$ - reference value, and $K4$ - data consistency were in a state of "low security". This is explained by the fact that during the rush hour the hardware of the distributed system suffers from a large uneven load, which leads to voltage instability; and the overload of the network and the channel bandwidth capability leads to the deterioration of the modules $K4$ and $K5$. The calculation on the BTN showed (Fig. 3) that in this case the probability of the system FS was equal to 40%. For a power system, such a value is absolutely unacceptable in order to consider it functionally sustainable.

To restore and maintain the initial state of modules $K4$ and $K5$ and the complex system FS as a whole, the consistency and value of the data are achieved using the method of synchronous (real time) replication. In this case, the data are allocated between the various nodes of the distributed database, which are connected by a stable communication channel, and synchronized immediately after changing the state of the system FS.

As it can be seen from the experiment, the results of calculations for the proposed BTN adequately describe the degree of FS depending on the state of the DM's FS and the results of scanning of control and monitoring modules of the entire power system.

Based on the proposed BTN, an algorithm was developed for estimating the probability of system FS.

1. Based on the results of scanning the sectors of the FS control and monitoring modules, the probabilities of the safe operation of these modules are estimated.
2. At the first stage, the values of factors influencing the information load on the DM are measured, factors of the external environment influence on the DM; tests are conducted to assess the DM's psychological and cognitive states.
3. Based on the data obtained at the first stage, using the formulas (2), (3), the probabilities are calculated that the "COTO" and "ADDSS" vertices are in "protected" state.
4. Based on the data obtained at the second stage, using the Mamdani fuzzy inference, the probabilities of the states of the vertices of the Bayesian network "External Environment Influence on the DM" and "Informational Load on the DM" are estimated.
5. According to the results of testing the DM, the probabilities of the states of the vertices "DM's psychological state" and "DM's Cognitive state" are calculated.

6. Using the proposed Bayesian network, the total probability of the system FS is calculated.

If this probability is greater than or equal to a certain critical value (the value of this variable depends on the degree of importance of the organizational and technical object), then the system can be considered functionally stable, otherwise it cannot. If the system turned out to be functionally unstable, then the corresponding threats are eliminated, which are indicated by the results of scanning the sectors of FS control and monitoring modules, and the normal values of the factors influencing the information load and environmental factors on the DM are normalized. After that, again, the estimation of the probability of the system FS according to the described algorithm is repeated.

7 Conclusions

The paper considers information technology for FS control and support of man-machine systems of critical application, which complements the theory and methods for solving the issues of ensuring system's fault tolerance and liveness, based on the interaction of a set of operation safety indicators, and human factor indicators, in managing and making decisions on its each hierarchical level.

To achieve these results a set of indicators of sustainability and safety of the system operation, characterizing its fault tolerance and liveness has been determined; a set of indicators of the human factor in management and decision making directly affecting the sustainability of the FS as a whole are defined at each hierarchical level of the system; an expert assessment was carried by the experts of the corresponding hierarchical system level, of a set of indicators of the human factor according to the degree of importance; the information technology was developed for the distribution and interaction of individual control modules for a set of basic FS indicators in a general hierarchical structure of a complex organizational and technical object, on the basis of which a Bayesian Trust Network (BTN) was built to evaluate the entire system's FS.

With the help of the proposed BTN, on the basis of expert knowledge, an assessment was made of the state probability of both individual components of the structural organization and a comprehensive FS assessment of the entire distributed system. For practical substantiation of the obtained results, an experiment was carried out, the results of which confirmed the efficiency of the proposed information technology, which can be used for the control and support of the FS of man-machine systems of critical application.

References

1. Dotson W, Norwood F, Taylor C (1993) Reliability polynomial for a ring network. IEEE Trans Commun 41(6):825–827
2. Koval DO, Xinlie Z, Propst J (2002) Spreadsheet reliability model applied to gold book standard network. In: 2002 IEEE industrial and commercial power systems technical conference, pp 66–72

3. Mashkov OA, Barabash OV (2003) Topological criteria and indicators of functional stability of complex hierarchical systems. Collection of scientific works of the National Academy of Sciences of Ukraine, IPEM - "modeling and information technologies", vol 25, pp 29–35 (in Ukrainian)
4. Floyd S (2001) Difficulties in simulating the internet. IEEE/ACM Trans Netw 9(4):392–403
5. Chae YL, Hee KC (2001) Multicast routing considering reliability and network load in wireless ad-hoc network. In: 2001 IEEE VTS 53rd vehicular technology conference, VTC 2001 spring, vol 3, pp 2203–2207
6. Liu B, Iwamura K (2000) Topological optimization models for communication network with multiple reliability goals. Comput Math Appl 39(7–8):59–69
7. Shao F-M, Zhao L-C (1997) Optimal design improving a communication network reliability. Microelectron Reliabil 37(4):591–595
8. Michele R (2004) Accurate analysis of TCP on channels with memory and finite round-trip delay. IEEE Trans Wirel Commun 3(2):627–640
9. Lukova-Chuiko NV (2018) Methodological bases for ensuring the functional stability of distributed information systems to cyber threats (PhD Thesis in specialty 05.13.06). State University of Telecommunications, Kyiv (in Ukrainian)
10. Murphy K (2001) A brief introduction to graphical models and bayesian networks. Technical report 2001-5-10, Department of Computer Science, University of British Columbia, Canada
11. Fefelov AO (2008) Models and methods for solving the problems of technical diagnostics on the basis of artificial immune systems and bayesian networks (PhD Thesis). National Technical University of Ukraine, Kyiv Polytechnic Institute, Kyiv (2008) (in Ukrainian)
12. Perederyi V, Borchik E (2019) Information technology for determination, assessment and correction of functional sustainability of the human-operator for the relevant decision-making in human-machine critical application systems. In: Theoretical and practical aspects of the development of modern science: the experience of countries of Europe and prospects for Ukraine: monograph. Baltija Publishing, Latvia, Riga, pp 490–509
13. Perederyi VI, Babichev SA, Lytvynenko VI (2012) The use of the Bayesian network to assess the degree of significance of influencing factors on the DM in automated systems for making relevant decisions. Bull Natl Univ "Lviv Polytechnic" (733):120–128. Computer Science and Information Technologies, Lviv (in Ukrainian)
14. Lomov BF (1982) Handbook of Engineering Psychology. Machine Building, Moscow (in Russian)
15. Shtovba SD, Introduction to the theory of fuzzy sets and fuzzy logic (in Russian). http://matlab.exponenta.ru/fuzzylogic/book1

Football Predictions Based on Time Series with Granular Event Segmentation

Hanna Rakytyanska$^{(\boxtimes)}$ and Mykola Demchuk

Vinnitsa National Technical University, Vinnitsa, Ukraine
h_rakit@ukr.net, demchuk_mykola@gmail.com

Abstract. The method of data quality management in mining fuzzy relations from time series of sports data is proposed. The fuzzy relation-based model of time series with granular event segmentation is developed. Correction of the available expert and experimental information is carried out at the stage of structural tuning of the fuzzy knowledge base. To improve the data set, it is advisable to extract events or specific areas of interest from the time series, rather than analyze the time series as a whole. Experts identify significant events in the form of trends and patterns at the stage of time series segmentation. At the stage of time series granulation, the events in the form of collection of time windows are transformed into fuzzy terms. Expert relationships "fuzzy events – match outcome" are subject to further correction in order to improve the predictive accuracy at the cost of removal of different types of biases in expert judgments. Biases in expert judgements (favorite/outsider bias, home team bias) can be revealed in paired comparative assessments chosen by an expert according to the 9-mark Saaty's scale. The genetic algorithm is used for off-line finding the area of unbiased assessments with further adaptive correction of paired comparisons matrices embedded into the neural network. As a result, predictive ability of the method outperforms bookmakers' predictions and, hence, provides the profitability against market odds.

Keywords: Football prediction · Sports data mining ·
Data quality management · Time series segmentation and granulation ·
Fuzzy relational model

1 Introduction

Prediction of football matches results is of interest when demonstrating the abilities of different data mining techniques compared to competitive sports betting market. The problem is what data should be collected and what is the best way to process them by discovering the unknown regularities and finding the appropriate explanations in the form of applied knowledge [1]. Since making sense of sports data is a challenging task, it is necessary to recognize hidden trends and patterns in teams' performances to see the underlying relationships within the

© Springer Nature Switzerland AG 2020
V. Lytvynenko et al. (Eds.): ISDMCI 2019, AISC 1020, pp. 478–497, 2020.
https://doi.org/10.1007/978-3-030-26474-1_34

highly stochastic sports data [1, 2]. Numerous data mining techniques have been proposed in attempt to generate profit against published market odds [3]. Due to the randomness inherent to sports data, obvious benefits of pure machine learning techniques in automated knowledge extraction are achieved at the cost of accuracy and interpretability [4]. At the same time, experts and fans have gained considerable experience, supported by extensive statistical analysis. Following [4], instead of relying on even large datasets, a smart data approach to data and knowledge engineering in sports domain is required. Development of such approach implies data quality management focused on what expert and experimental data are really required, rather than what data are ready available [4].

2 Review of the Literature

Bookmakers' predictions are built on the grounds of statistical methods and expert judgments affected by objective and subjective factors. Given the average number of goals scored by a team per game, the probability of the expected number of goals for the upcoming game is calculated using the Poisson distribution based models and converted into the bookmakers' odds [5–7].

Pure data-driven models heavily rely on data preprocessing which includes feature selection and missing data correction [3]. To provide the best predictive accuracy, the general machine learning framework for sports result prediction includes the option of testing different feature selection techniques combined with candidate classification models [8, 9]. The neural network models are constructed after correlation based feature subset selection [8]. To optimize the network structure, the artificial neural networks are reinforced with the automatic weighted feature selection process based on sequential forward selection or relief attribute evaluation [8, 10]. To incorporate expert opinions in the neural network, a hybrid model with a feature set of betting odds and public data features is proposed in [8]. Finally, the network design can be enhanced by learning the statistical parameters of the model [10].

Some other machine learning techniques consider a data-driven Bayesian network [11, 12], a decision tree learner [11, 13] and SVM [13]. These models do not take into account subjective factors that remain unobserved but may influence the match outcome. Comparison of the current studies with previous results is difficult because of the difference in testing datasets collected from sports websites [3]. Nevertheless, studies show that pure data-driven models are not able to outperform the bookmakers' predictions. For pure data-driven models, the predictions remain considerably less accurate compared to the market predictions observed in bookmakers' odds [3].

Compared to the pure data-driven machine learning techniques, more accurate predictions can be achieved by integrating data with expert knowledge [4, 14]. The extended Bayesian network with integrated expert-driven nodes is proposed in [11, 14] to improve data set by adding expert knowledge for unobserved data (stress and fatigue, coach support, transferred players adaptation etc.). A compound framework using Bayesian inference and rule-based reasoning is proposed in [15]. In highly stochastic environment of sports matches, the strategy of the team game can be described by logic rules [15]. In developing fuzzy rule based systems, the relationships between factors affecting the sports game and the result are verified through the experimental design techniques for randomized experiments where levels of variables are described by fuzzy terms [16]. Cooperation of causal knowledge and real world data improved the model accuracy such that the profitability of generated forecasts could compete with bookmakers' performance [4, 14–16].

Furthermore, mining information from the betting market led to development of the pure expert driven systems aimed to extract predictions directly from the bookmakers' odds [17, 18]. The market odds inevitably contain the bookmakers' built-in profit margin leading to the difference between the expected and actual winning probabilities [4, 19]. The next widely accepted type of bias observed in bookmakers' odds is the home-team bias. Profitable opportunities of a betting market are extended to favorite/outsider bias, where bets on popular teams tend to have higher expectations than bets on less famous teams [4]. In this case, top teams are overestimated; teams placing lower positions in tournament tables are underestimated. As a result, the different types of biases in expert judgments inevitably leads to the outcome bias [19].

The next group of methods is developed for mining information from time series data. In this case, matches outcomes are already influenced by the factors identified as uncontrollable variables [16]. Sports data can be collected in the form of the ratings of abilities assigned to players/teams and updated between games on the grounds of past performance [20, 21]. Rating systems are commonly used to detect statistical relationships while developing logit regression models [21].

Statistical methods are widespread for discovering trends and patterns in temporal data. Data grouping methods and serial correlation are used to detect relationships between patterns observed in the past statistics [22, 23]. Team performance is defined by the clustered and switched play types, which alternate more frequently than an independent stochastic process [22]. In modeling trends in time series data, polynomial classifiers are used to identify the winning team [23]. In this case, the investigated groups represent different types of win-draw, win-defeat and draw-defeat combinations observed in the course of the championship. A decisiveness measure based on the entropy of probability distribution is proposed in [24, 25] to quantify the impact of a particular game on the match outcome. The teams rating in the tournament table at the time of the game determines the measure of match importance. Given the previous matches results, the degree of influence of a particular game on the current tournament position is computed using Monte Carlo simulation [24]. Finally, logistic regression is used to evaluate the odds of the possible match outcomes.

Despite obtaining promising results, relying purely on datasets can be inappropriate due to the randomness in time series data. Regardless the size of the dataset, pure data-driven techniques may fail to discover the underlying relationships, which govern a process generating time series [4]. At the same time, experts can often understand and identify the unknown regularities that can be revealed in available experimental data.

3   Problem Statement

In [26], we suggest to predict the football matches results by modeling a time series using fuzzy IF-THEN rules. The rule set generated by experts of betting market is tuned using the genetic and neuro approach. Parametric tuning of the expert rules in the presence of different types of biases cannot guarantee accurate predictions. Errors in expert judgments are manifested in the offset assessments while establishing relationships in the patterns described by fuzzy rules. The stage of data preprocessing is omitted since the forecasts are generated based on a comparison of the observed time series with the fuzzy time series formulated by experts. Despite the fact that generated forecasts agree well with the market odds, the rule-based model cannot outperform the bookmakers' predictions.

This paper presents a further development of [26]. Compared to [26], the method of data quality management is proposed at the level of adjustment of the structure of a fuzzy relational model [27]. Data quality management implies expert information correction and experimental data set improvement. Time series modeling is realized by two-stage tuning of fuzzy knowledge bases that corresponds to the stages of structural and parametric identification [28]. Domain experts are involved in developing fuzzy time series models at the stage of structural tuning. It is supposed that the expert rule set is subject to further correction to improve the predictive accuracy at the cost of removal of the different types of biases in expert judgments [29,30]. In fuzzy time series modeling, rule set refinement represents the compound framework of granular and relational models, using which the semantically sound description of time series is ensured [31,32].

Time series modeling implies the stages of segmentation and granulation [33,34]. Time series segmentation is carried out in order to identify events or specific areas of interest in terms of the impact on the match outcome [35,36]. To improve the data set, it is advisable to extract certain patterns from the time series, rather than analyze the time series as a whole [35]. Experts, as a rule, identify significant events in the form of trends and patterns at the stage of time series segmentation. At the stage of granulation, the events in the form of collection of time windows are transformed into fuzzy terms [33,34]. To describe the behavior of a time series, fuzzy terms represent the estimates of the amplitude trend of the time window [33,34].

Following [37], a method of partitioning the universe of discourse of time series based on relational partition with granular representation of fuzzy terms is proposed. To improve the predictive accuracy in the framework of fuzzy relational model, the granular representation of fuzzy terms consists of generating a set of fuzzy granules with the same linguistic interpretation.

Expert data correction is carried out at the stage of mining fuzzy relations between granulated events and match outcome. The method of Saaty's paired comparisons is used for construction of "causes (events) – match outcome" fuzzy relations [38–40]. The matrix of pairwise comparisons reflects the degree of advantage of individual areas of the time series in terms of the possible match outcome. In this case, the biases in expert judgements can be revealed in paired comparative assessments chosen by an expert according to the 9-mark

Saaty's scale [41]. Thus, the essence of tuning consists of finding such parameters of granular representation of fuzzy terms and such paired comparative assessments of fuzzy terms (events), which minimize the difference between model and experimental data. The genetic algorithm is used for hitting the area of unbiased assessments in off-line mode with further adaptive correction of paired comparisons matrices embedded into the neural network.

The aim of the work is to develop a method of data quality management in mining fuzzy relations from time series of sports data. To provide the reliable forecasts, correction of the available expert and experimental information is carried out at the stage of structural tuning of the fuzzy knowledge base. Predictive ability of the method should outperform the bookmakers' predictions and, hence, should provide the profitability against market odds. To achieve this aim, it is necessary to develop the fuzzy relation-based model of time series with granular event segmentation; to develop a genetic and neuro algorithm for tuning the matrix of fuzzy relations based on fuzzy events paired comparisons.

4 Materials and Methods

4.1 Structure of the Prediction Model

The aim is to calculate the result of match between teams T_1 and T_2, which is characterized as the difference of scored and conceded goals y, $y \in [\underline{y}, \bar{y}] = [-5, 5]$. The value of y is determined on the following five levels [26]: d_1 is a big loss (bL), $y = -5, -4, -3$; d_2 is a small loss (sL), $y = -2, -1$; d_3 is a draw (D), $y = 0$; d_4 is a small win (sW), $y = 1, 2$; d_5 is a big win (bW), $y = 3, 4, 5$.

It is supposed that the football game result (y) is influenced by the patterns and trends observed in the series of results of five previous matches of team T_1 ($s = 1$), team T_2 ($s = 2$), personal meetings between teams T_1 and T_2 ($s = 3$).

Segmentation of a time series corresponds to the description of the following events.

The number of the following outcomes and the average difference of the scored and conceded goals in the won and lost matches define the shape space of a time series:

- x_1^s, x_2^s are the number of losses and the average amplitude of the lower segment of a time series: $x_1^s \in [0, 5]$, $x_2^s \in [-5, 0]$;
- x_3^s is the number of draws: $x_3^s \in [0, 5]$;
- x_4^s, x_5^s are the number of wins and the average amplitude of the upper segment of a time series: $x_4^s, x_5^s \in [0, 5]$.

The length and the average amplitude of the following series describe the segments of a time series that correspond to the stable clustered play type:

- x_6^s, x_7^s are the parameters of the series of losses: $x_6^s \in [0, 5]$, $x_7^s \in [-5, 0]$;
- x_8^s, x_9^s are the parameters of the series without wins: $x_8^s \in [0, 5]$, $x_9^s \in [-4, 0]$;
- x_{10}^s is the length of the series of draws: $x_{10}^s \in [0, 5]$;

- x^s_{11}, x^s_{12} are the parameters of the series without losses: $x^s_{11} \in [0,5]$, $x^s_{12} \in [0,4]$;
- x^s_{13}, x^s_{14} are the parameters of the series of wins: $x^s_{13}, x^s_{14} \in [0,5]$.

The amplitude of the following transitions describes the segments of a time series that correspond to the unstable results with the switched play types:

- x^s_{15}, x^s_{16} and x^s_{17} are the amplitudes of the loss-win, loss-draw and draw-win transitions in positive dynamics: $x^s_{15} \in [0,10]$, $x^s_{16}, x^s_{17} \in [0,5]$;
- x^s_{18}, x^s_{19} and x^s_{20} are the amplitudes of the win-loss, win-draw and draw-loss transitions in negative dynamics: $x^s_{18} \in [-10,0]$, $x^s_{19}, x^s_{20} \in [-5,0]$.

The hierarchical interconnection between input variables x^s_1–x^s_{20} and output variable y corresponds to the following tree of logical inference:

$$z_s = f_s(x^s_1, ..., x^s_{20}), \quad s = 1,2,3; \tag{1}$$

$$y = F(z_1, ..., z_3), \tag{2}$$

where $z_1, ..., z_3$ are the intermediate match predictions based on the previous results of games of team T_1, team T_2, personal meetings of teams T_1 and T_2.

Intermediate predictions $z_1, ..., z_3$ are defined on the grounds of the time series proximity measure, which should also consider the relative location of time series.

Prediction $z_1(z_2)$ is defined as an average difference between points of the time series corresponding to the results of the team $T_1(T_2)$ and points of the time series corresponding to the results of the personal meetings between teams T_1 and T_2. Prediction $z_1(z_2)$ reflects the degree of advantage of team $T_1(T_2)$ over the history of personal meetings of teams T_1 and T_2 depending on the results of previous matches $x^1_1 - x^1_{20}(x^2_1 - x^2_{20})$ of the team $T_1(T_2)$.

Prediction z_3 is defined as an average difference between points of the time series corresponding to the previous results of both the teams T_1 and T_2. Prediction z_3 reflects the recent ratio of strengths between the teams T_1 and T_2 depending on the results $x^3_1 - x^3_{20}$ of the retrospective personal meetings of the teams T_1 and T_2.

For the intermediate predictions $z_1, ..., z_3$, the change range is $[-5,5]$. The value of z_s is determined on the following five levels. If $z_s = 0$, then the team T_1 has no advantage over the team T_2, i.e. the teams have equal chances. If $z_s > 0$, then the team T_1 has an advantage over the team T_2. If $z_s < 0$, then the team T_2 has an advantage over the team T_1. Depending on the distance between time series, the change range is $[0,2]$ or $[-2,0]$ for weak advantage; $[2,5]$ or $[-5,-2]$ for strong advantage. The degree of advantage of the team over the rival can be directly transformed into the possible match outcome which is evaluated using above mentioned fuzzy terms bL, sL, D, sW, bW.

The variables x^s_1–x^s_{20} are evaluated using the following fuzzy terms: *very small* (vS), *small* (S), *average* (A), *big* (B), *very big* (vB).

4.2   Fuzzy Relations Construction

We shall denote: $\{c_{11}, ..., c_{1h_1}, ..., c_{n1}, ..., c_{nh_n}\} = \{C_1, ..., C_N\}$, $N = h_1 + ... + h_n$ is a set of input terms for evaluation of variables x_i, $i = \overline{1, n}$; $\{e_1, ..., e_q\}$ is a set of output terms for evaluation of variable y; h_i and q are the numbers of input and output terms.

We shall describe the dependency $y = f(x_1, ..., x_n)$ between inputs $\mathbf{X} = (x_1, ..., x_n)$ and output y with the help of the multi-dimensional matrix of fuzzy relations:

$$\mathbf{R} \subseteq C_I \times e_p = [\mathbf{r}_p, \ p = \overline{1, q}] = [r_I^p, \ I = \overline{1, N}, \ p = \overline{1, q}]. \tag{3}$$

Here $\mathbf{r}_p = (r_1^p, ..., r_N^p)^T$ is the vector-column of the relational matrix $\mathbf{R}$.

Following [38], to obtain matrix $\mathbf{R}$, it is necessary to form the matrix of paired comparisons for each effect e_p, which reflects the influence of terms (events) $C_1, ..., C_N$ while forming the vector-column $\mathbf{r}_p$.

We shall denote: $\{\tilde{C}_1^p, ..., \tilde{C}_N^p\} = \{\tilde{c}_{11}^p, ..., \tilde{c}_{1h_1}^p, ..., \tilde{c}_{n1}^p, ..., \tilde{c}_{nh_n}^p\}$ is the set of terms (events) selected for the effect e_p.

For the effect e_p, the interconnection between the candidate terms C_I and the selected terms $\tilde{C}_K^p$ is described using the fuzzy matrix of paired comparisons:

$$\mathbf{A}_p \subseteq C_I \times \tilde{C}_K^p = [\alpha_{IK}^p = a_{IK}^p/9, \ I, K = \overline{1, N}, \ p = \overline{1, q}], \tag{4}$$

where $\alpha_{IK}^p \in [0, 1]$ is the fuzzy measure of advantage of the term (event) $\tilde{C}_K^p$ over the term (event) C_I for the class e_p; $a_{IK}^p \in [1, 9]$ is the paired comparative assessment chosen by an expert according to the 9-mark Saaty's scale [41].

To obtain the elements α_{IK}^p of the fuzzy matrix (4), the degrees of advantage a_{IK}^p are to be normalized by way of dividing into the highest comparative assessment:

$1/9$ – if term (event) $\tilde{C}_K^p$ has no *advantage* over term (event) C_I;
$3/9, 5/9, 7/9, 9/9$ – if $\tilde{C}_K^p$ has a *weak, an essential, an obvious, an absolute* advantage over C_I.

Values of $2/9, 4/9, 6/9, 8/9$ correspond to intermediate comparative assessments.

In accordance with [38], the properties of symmetry, transitivity and diagonality for the fuzzy matrix (4) can be formulated as follows:

$$\frac{a_{IK}^p}{9} = \frac{1}{9a_{KI}^p}; \ \frac{a_{IL}^p a_{LK}^p}{9} = \frac{a_{IK}^p}{9}; \ a_{II}^p = \frac{1}{9}, \ I, K, L = \overline{1, N}.$$

Following [38], these properties allow defining all elements of fuzzy matrix (4) by using elements of only a single resolution row. If the L-th row is known, i.e., the normalized elements $\dfrac{a_{LK}^p}{9}$, $K = \overline{1, N}$, then an arbitrary fuzzy assessment $\dfrac{a_{IK}^p}{9}$ is defined as follows:

$$\frac{a_{IK}^p}{9} = \frac{a_{LK}^p}{9a_{LI}^p}; \ I, K, L = \overline{1, N}, \ p = \overline{1, q}.$$

After defining matrix (4), the correlation (3) can be rewritten as follows:

$$\mathbf{R} \subseteq C_I \times e_p = [\mathbf{r}_p, \; p = \overline{1,q}] = [\mathbf{A}_p \circ \mathbf{w}_p, \; p = \overline{1,q}], \tag{5}$$

where $\mathbf{w}_p = (w_1^p, ..., w_N^p)^T$ is the vector-column of weights of the terms selected in the class e_p, i.e. $w_K^p = 1$; $\circ$ is the operation of max-min composition [27].

As a result of the composition of the matrix $\mathbf{A}_p$ and the vector-column $\mathbf{w}_p$, the elements of the vector-column $\mathbf{r}_p$ are calculated by the formula:

$$r_I^p = \max_{K=\overline{1,N}} (\alpha_{IK}^p), \; I = \overline{1,N},$$

which is derived from the relation (5).

4.3 Fuzzy Relational Model with Granular Parameters

We shall denote: $\{C_{I1}, ..., C_{Ig_I}\}$ is the set of fuzzy granules interpreted as the fuzzy term C_I; g_I is the number of fuzzy granules.

Given matrices $\mathbf{A}_p$, $p = \overline{1,q}$, the extended compositional rule of inference [27] can be interpreted in the form of the following system of fuzzy relational equations:

$$\mu^{C_I}(x_i) = \max_{J=\overline{1,g_I}} (\mu^{C_{IJ}}(x_i)), \; I = \overline{1,N}; \tag{6}$$

$$\mu_p^{\tilde{C}_K}(x_i) = \max_{I=\overline{1,N}} [min(\mu^{C_I}(x_i), \alpha_{IK}^p)], \; K = \overline{1,N}; \tag{7}$$

$$\mu^{e_p}(y) = \min_{i=\overline{1,n}} \{ \max_{\substack{K=\overline{\Delta+1,\Delta+h_i} \\ if\, i=1:\Delta=0 \\ if\, i=2:\Delta=h_1 \\ if\, i>2:\Delta=h_1+...+h_{i-1}}} [min(\mu_p^{\tilde{C}_K}(x_i), w_K^p)]\}, \; p = \overline{1,q}; \tag{8}$$

where $\mu^{C_{IJ}}(x_i)$ is a membership function of a J-th granule of the fuzzy term C_I granular representation, $J = \overline{1,g_I}$; $\mu^{C_I}(x_i)$ is a membership function of a variable x_i to the fuzzy term C_I, $I = \overline{1,N}$; $\mu_p^{\tilde{C}_K}(x_i)$ is a membership function of a selected fuzzy term $\tilde{C}_K^p$ in the class e_p; $\mu^{e_p}(y)$ is a membership function of a variable y to the class e_p, $p = \overline{1,q}$.

We use a bell-shaped model of the membership function for a fuzzy granule C_{IJ}:

$$\mu^{C_{IJ}}(x_i) = \frac{1}{1 + ((x_i - \beta_{IJ})/\sigma_{IJ})^2}, \tag{9}$$

where β_{IJ} is a coordinate of function maximum; σ_{IJ} is a parameter of concentration.

By analogy with the fuzzy rule-based model [26], decision classes e_p are formed by digitizing the range $[\underline{y}, \bar{y}]$ into q levels $e_p \in [\underline{y}_p, \bar{y}_p]$, $p = \overline{1,q}$.

The operation of defuzzification is defined as follows [26]:

$$y = \frac{\sum_{p=1}^q \underline{y}_p \cdot \mu^{e_p}(y)}{\sum_{p=1}^q \mu^{e_p}(y)}. \tag{10}$$

Correlations (6)–(10) define the fuzzy relational model of the object as follows:

$$y = f(\mathbf{X}, \mathbf{A}^*, \mathbf{B}, \mathbf{\Omega}), \tag{11}$$

where $\mathbf{A}^* = (a^1_{L_1 1}, ..., a^1_{L_1 N}, ..., a^q_{L_q 1}, ..., a^q_{L_q N})$ is the vector of elements of resolution rows of paired comparisons matrices (4); L_p is the number of the resolution row in the matrix $\mathbf{A}_p$; $\mathbf{B} = (\beta_{11}, ..., \beta_{1 g_1}, ..., \beta_{N 1}, ..., \beta_{N g_N})$ and $\mathbf{\Omega} = (\sigma_{11}, ..., \sigma_{1 g_1}, ..., \sigma_{N 1}, ..., \sigma_{N g_N})$ are the vectors of parameters of granular representation of fuzzy terms $C_1, ..., C_N$; f is the operator of "inputs–output" connection corresponding to formulas (6)–(10).

4.4   Problem of Fuzzy Prediction Model Tuning

Using the fuzzy relational model (11) and the tree of logic inference (1), (2), the prediction model can be represented in the following form:

$$y = F_y(x^s_1, ..., x^s_{20}, \mathbf{A}^*_s, \mathbf{B}_s, \mathbf{\Omega}_s, \mathbf{R}_s), \quad s = 1, 2, 3, \tag{12}$$

where F_y is the operator of inputs-output connection corresponding to correlations (1), (2); $\mathbf{A}^*_s = (a^{s,1}_{L_1 1}, ..., a^{s,1}_{L_1 N}, ..., a^{s,q}_{L_q 1}, ..., a^{s,q}_{L_q N})$ are the vectors of comparative assessments for the intermediate predictions in the correlation (1); $\mathbf{B}_s = (\beta^s_{11}, ..., \beta^s_{1 g_1}, ..., \beta^s_{N 1}, ..., \beta^s_{N g_N})$ and $\mathbf{\Omega}_s = (\sigma^s_{11}, ..., \sigma^s_{1 g_1}, ..., \sigma^s_{N 1}, ..., \sigma^s_{N g_N})$ are the vectors of parameters of the fuzzy terms granular representations for the intermediate predictions in the correlation (1); $\mathbf{R}_s = (r^s_{11}, ..., r^s_{q1}, ..., r^s_{1m}, ..., r^s_{qm})$ are the matrices of fuzzy relations for the final prediction in the correlation (2).

The fuzzy model (12) is tuned using the tournament tables' data [26]. Training data can be obtained in the form of M pairs of experimental data $\langle \hat{\mathbf{X}}_l, \hat{y}_l \rangle$, $l = \overline{1, M}$, where $\hat{\mathbf{X}}_l = (\hat{x}^{s,l}_1, ..., \hat{x}^{s,l}_{20})$ are the parameters of series of previous matches for teams T_1 and T_2 in the experiment number 1, $\hat{y}_l$ is the game result in the experiment number l.

The essence of tuning the prediction model (12) is as follows. It is necessary to find the vectors of parameters of granular representation of fuzzy terms $\mathbf{B}_s$, $\mathbf{\Omega}_s$, the vectors of comparative assessments $\mathbf{A}^*_s$, and the relational matrices $\mathbf{R}_s$, which provide the minimum distance between theoretical and experimental results:

$$\sum_{l=1}^{M} (F_y(\hat{x}^{s,l}_1, ..., \hat{x}^{s,l}_{20}, \mathbf{A}^*_s, \mathbf{B}_s, \mathbf{\Omega}_s, \mathbf{R}_s) - \hat{y}_l)^2 = \min_{\mathbf{A}, \mathbf{B}, \mathbf{\Omega}, \mathbf{R}}, \quad s = 1, 2, 3. \tag{13}$$

The genetic and neuro approach is used for tuning the fuzzy relational model [26].

4.5   Genetic and Neuro Tuning of the Prediction Model

The chromosome is defined as a vector-string of binary codes of the parameters of granular representation of the fuzzy sets β_{IJ}, σ_{IJ}, $I = \overline{1, N}$, $J = \overline{1, g_I}$, comparative assessments of the resolution rows of paired comparisons matrices

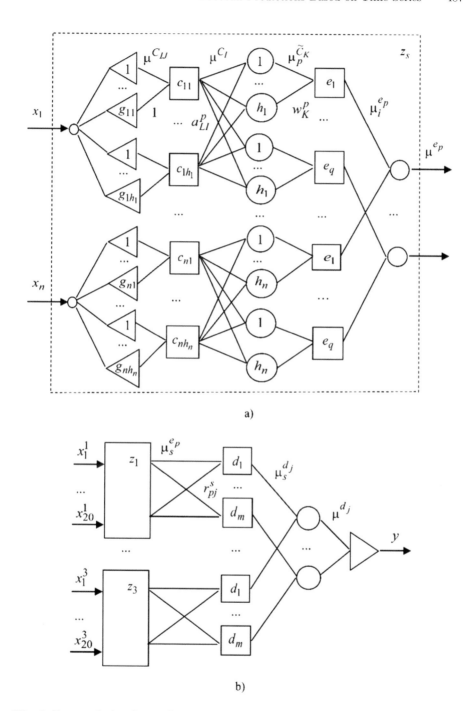

Fig. 1. Fuzzy relational neural network with granular parameters for the intermediate predictions (a); for the final prediction (b)

a^p_{IK}, $p = \overline{1,q}$, $I, K = \overline{1,N}$, and fuzzy relations r_{pj}, $j = \overline{1,m}$. Since the resolution row is formed for the worst element over which other elements have an advantage regarding effect e_p, the comparative assessments a^p_{IK} are coded by integer numbers within the interval $[1, 9]$. The crossover operation is carried out by exchanging the parts of the chromosomes in the vectors of parameters $\mathbf{B}$, $\boldsymbol{\Omega}$, the vectors of assessments $\mathbf{A}^*$ and the vectors-columns $\mathbf{r}_p$ of the relational matrix $\mathbf{R}$. The cyclic crossing over operation, which provides both the positive and negative offset, is used to remove the biases in the expert paired assessments. Fitness function is built on the basis of criterion (13).

Fuzzy relational neural network with granular parameters is presented in Fig. 1. For intermediate predictions, the neuro-fuzzy model (Fig. 1a) is obtained by embedding the matrix of paired comparisons into the neural network so that the weights of arcs subject to tuning are comparative assessments and parameters of granular representation of fuzzy terms. For final predictions, the neuro-fuzzy model (Fig. 1b) aggregates the intermediate networks by embedding the matrices of fuzzy relations.

For tuning the parameters of fuzzy relations, the recurrent relations are used:

$$a^{s,p}_{LI}(t+1) = a^{s,p}_{LI}(t) - \eta \frac{\partial \varepsilon_t}{\partial a^{s,p}_{LI}(t)}; \quad r^s_{pj}(t+1) = r^s_{pj}(t) - \eta \frac{\partial \varepsilon_t}{\partial r^s_{pj}(t)};$$

$$\beta^s_{IJ}(t+1) = \beta^s_{IJ}(t) - \eta \frac{\partial \varepsilon_t}{\partial \beta^s_{IJ}(t)}; \quad \sigma^s_{IJ}(t+1) = \sigma^s_{IJ}(t) - \eta \frac{\partial \varepsilon_t}{\partial \sigma^s_{IJ}(t)}, \tag{14}$$

which minimize the criterion

$$\varepsilon_t = \frac{1}{2}(\hat{y}_t - y_t)^2,$$

where $\hat{y}_t$, (y_t) is the experimental (theoretical) difference of the scored and conceded goals at the t-th step of training; $a^{s,p}_{LI}(t)$ are the comparative assessments for the intermediate predictions at the t-th step of training; $r^s_{pj}(t)$ are the fuzzy relations for the final prediction at the t-th step of training; $\beta^s_{IJ}(t)$, $\sigma^s_{IJ}(t)$ are the parameters of granular representation of fuzzy terms at the t-th step of training.

The partial derivatives included in (14) are calculated according to [26].

5   Experiment

The model was used to predict the outcomes of the Finland Veikkaus League games, which are characterized by a relatively small number of sensations [42]. 12 teams take part in the championship, playing 33 tours of 6 games. To train the prediction model, we used the results of 925 matches from 2011 to 2015 seasons.

The results of time series segmentation for team T_1, team T_2, teams T_1 and T_2 are presented in Fig. 2 in the form of the partial dependencies (x^s_i, z_s) which reflect the impact of the factors (events) x^s_i on the intermediate predictions z_s. The partial dependencies (z_s, y) which reflect the impact of the intermediate predictions z_s on match outcome y are presented in Fig. 3.

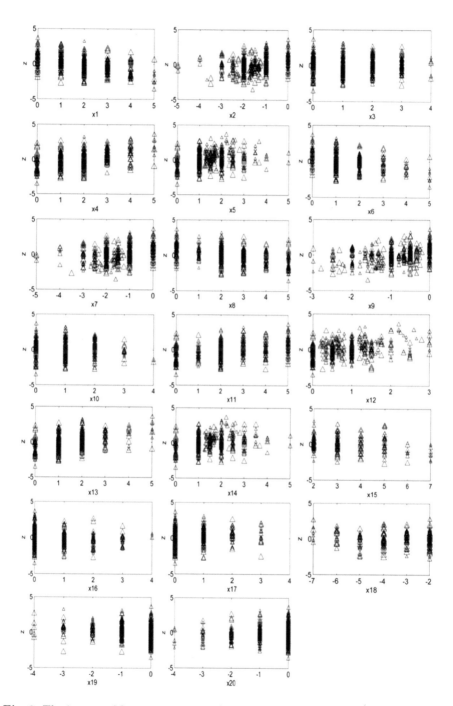

Fig. 2. The impact of factors x_i on partial match prediction z for teams T_1; T_2; T_1, T_2

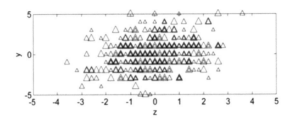

Fig. 3. The impact of intermediate predictions z for teams T_1; T_2; T_1, T_2 on match outcome y

Once the segmentation has been completed, the granulation and selection of the events influencing match outcome are carried out. For each factor x_i, fuzzy events are granulated within segments defined by the time window and the amplitude change range which correspond to the fuzzy term C_I. Time window granulation can be interpreted as in-game time series segmentation driven by in-game events such as series of defense actions and attacks. In this case, amplitude granulation may reflect a widespread situation when the score does not correlate with in-game events.

The results of tuning both the comparative assessments, which reflect the degree of advantage of terms (events) for intermediate predictions and the fuzzy relations between intermediate predictions and possible match outcome are presented in the tables which are shown in Figs. 4 and 5 respectively.

To provide the unbiased assessments, fuzzy relations "terms (events) – intermediate predictions – match outcome" are defined as the relative frequency of occurrence of an outcome d_j in case of a prediction z_j influenced by an event C_I [43]. In this case, behavior of a time series is described using assessments of the 9-mark Saaty's scale which can be interpreted as *never* (1), *seldom* (3), *unspecific* (5), *often* (7), *always* (9) [43].

6   Results and Discussion

To test the prediction model we used the results of 541 matches from 2016 to 2018 seasons [42]. The fragment of testing data and prediction results are shown in Fig. 6, where: $\tilde{d}$ are the predictions extracted from bookmakers' odds; $\hat{y}$, $\hat{d}$ are the historical results; y, d are the prediction results generated by the proposed fuzzy model.

Figure 6 reflects the following types of errors in prediction results marked using symbols (!, +, −). The model is on par with bookmakers in two cases. For expected outcomes, the bookmakers and the model demonstrate the correct predictions, i.e., $\tilde{d} = \hat{d} = d$. For unexpected outcomes or sensations, the bookmakers and the model generate the erroneous predictions, i.e., $\tilde{d} \neq \hat{d}$, $d \neq \hat{d}$ (marked as !).

Input	Fuzzy events vS, S, A, B, vB	Intermediate prediction z				
		bL	sL	D	sW	bW
x_1	0–1, 1–2, 2–3, 3–4, 4–5	1 3 3 2 1	5 8 7 6 4	7 9 9 5 3	6 7 8 4 1	5 4 3 1 1
x_2	−5–4, −4–3, −3–2, −2–1, −1–0	1 3 4 5 3	3 5 7 8 6	2 3 5 9 6	1 1 4 7 5	1 1 1 3 3
x_3	0–1, 1–2, 2–3, 3–4	4 3 3 1	8 7 6 3	9 7 5 3	9 8 6 4	5 4 3 2
x_4	0–1, 1–2, 2–3, 3–4, 4–5	4 5 3 2 1	9 8 7 3 1	8 9 9 6 5	8 7 6 4 2	3 5 4 3 1
x_5	0–1, 1–2, 2–3, 3–4, 4–5	6 4 2 1 1	9 7 6 3 1	8 9 8 4 3	6 8 7 3 1	3 5 4 2 1
x_6	0–1, 1–2, 2–3, 3–4, 4–5	5 6 5 3 2	8 9 7 4 3	7 8 6 2 1	6 7 5 3 1	5 4 2 1 1
x_7	−5–4, −4–3, −3–2, −2–1, −1–0	1 2 4 5 3	2 5 6 7 8	3 4 7 8 9	1 2 5 6 7	1 1 2 5 4
x_8	0–1, 1–2, 2–3, 3–4, 4–5	1 3 3 2 1	5 6 7 5 4	8 9 8 7 5	4 5 7 6 4	1 4 5 3 1
x_9	−3–2.4, −2.4–1.8, −1.8–1.2, −1.2–0.6, −0.6–0	1 1 2 4 5	2 4 6 7 8	3 3 7 8 9	2 3 5 6 7	1 1 3 5 6
x_{10}	0–1, 1–2, 2–3, 3–4	5 4 3 1	7 9 6 3	9 8 4 2	8 7 5 3	5 4 3 1
x_{11}	0–1, 1–2, 2–3, 3–4, 4–5	2 3 2 1 1	6 7 6 4 3	8 9 8 6 5	5 6 7 5 4	3 4 5 3 1
x_{12}	0–0.6, 0.6–1.2, 1.2–1.8, 1.8–2.4, 2.4–3	4 3 3 1 1	7 6 5 2 1	8 7 6 3 1	9 8 7 4 2	6 5 3 1 1
x_{13}	0–1, 1–2, 2–3, 3–4, 4–5	3 3 2 1 1	8 7 6 3 1	9 8 5 4 2	8 9 7 3 1	5 6 4 2 1
x_{14}	0–1, 1–2, 2–3, 3–4, 4–5	4 3 3 1 1	7 8 7 3 1	9 7 6 5 3	8 9 7 4 1	5 6 3 1 1
x_{15}	2–3, 3–4, 4–5, 5–6, 6–7	3 3 1 1 1	7 6 4 1 1	9 8 6 5 3	8 7 5 4 2	4 3 1 1 1
x_{16}	0–1, 1–2, 2–3, 3–4	5 3 2 1	8 7 6 4	9 8 7 3	7 6 4 1	6 5 3 1
x_{17}	0–1, 1–2, 2–3, 3–4	4 2 1 1	8 7 5 1	9 8 6 1	8 6 4 2	5 4 2 1
x_{18}	−7–6, −6–5, −5–4, −4–3, −3–2	1 1 1 3 1	2 4 5 6 7	3 5 7 8 9	1 3 4 5 6	1 1 3 5 3
x_{19}	−4–3, −3–2, −2–1, −1–0	1 1 3 5	1 4 6 8	3 5 7 9	3 4 5 8	1 1 3 4
x_{20}	−4–3, −3–2, −2–1, −1–0	1 1 2 3	3 5 6 7	4 7 7 8	2 4 5 6	2 3 3 4

Fig. 4. Comparative assessments of fuzzy events for intermediate predictions

Input	Intermediate predictions	Decision classes for match outcome y				
		bL	sL	D	sW	bW
	bW	never	very seldom	seldom	often	unspecific
	sW	seldom	unspecific	very often	very often	unspecific
z	D	seldom	often	nearly always	very often	often
	sL	often	very often	nearly always	often	unspecific
	bL	unspecific	unspecific	seldom	very seldom	never

Fig. 5. Occurrence frequency of match outcomes in case of intermediate predictions

Two other cases correspond to the no coincidences of the expert and model results. The model outperforms bookies if the model prediction is correct, and the bookies prediction is erroneous, i.e., $\tilde{d} \neq \hat{d}$, $\hat{d} = d$ (marked as +). The bookies outperform the model if the bookmakers' prediction is correct, and the model prediction is erroneous, i.e., $\tilde{d} = \hat{d}$, $\hat{d} \neq d$ (marked as −).

To assess the performance of the relation-based model, the results were compared with some other time-series based models, which select trends and patterns from the past statistics. The models were validated in terms of how accurately they can predict football matches results in comparison with bookmakers' predictions. We have developed a regression-based model, which uses the same set

Game	Previous games results															$\tilde{d}$	$\hat{y}$	$\hat{d}$	y	d
	Team T1					Team T2					T1 and T2									
	1	2	3	4	5	1	2	3	4	5	1	2	3	4	5					
Tour 32, 33, Season 2018																				
Lahti–Kemi	1	-3	-3	-2	0	2	-2	-4	-2	-1	0	0	-1	1	0	sW	-1	sL	1	sW !
Mariehn–KuPS	-2	3	0	0	1	1	2	3	0	0	-1	-3	-1	1	1	sL	-1	sL	-1	sL
RoPS–Ilves	1	1	-1	-3	0	-1	-2	4	-1	1	2	-1	-1	1	2	sW	1	sW	1	sW
SJK–Int Turku	-1	-1	1	0	-1	-2	2	0	2	-1	-2	-1	5	1	-3	sW	0	D	0	D +
TPS–HJK	-1	-1	3	0	1	1	1	0	1	3	-1	-1	-3	-5	1	bL	-4	bL	-3	bL
Honka–VPS	2	1	0	0	1	-1	-1	-3	2	-1	-1	0	-1	-2	-3	sW	2	sW	2	sW
VPS–RoPS	-1	-3	2	-1	0	1	-1	-3	0	2	-1	-1	-3	1	2	sL	-1	sL	-1	sL
Ilves–Lahti	-2	4	-1	1	1	-3	-3	-2	0	-2	0	-2	0	1	0	sW	-1	sL	1	sW !
Turku–Honka	2	0	2	-1	-1	1	0	0	1	1	-1	0	0	-1	-2	sL	-2	sL	-1	sL
KuPS–TPS	2	3	0	0	1	-1	3	0	1	-1	1	0	1	0	3	sW	1	sW	2	sW
Kemi–Mariehn	-2	-4	-2	-1	-2	3	0	0	1	-2	-1	-1	-1	3	-3	sL	2	sW	-2	sL !
IIJK–SJK	1	0	1	3	2	-1	1	0	-1	1	4	1	2	1	6	sW	1	sW	2	sW
Tour 32, 33, Season 2017																				
JJK–VPS	-1	1	0	-2	-2	1	-3	0	-1	0	0	-3	-2	-1	2	sL	2	sW	-1	sL !
Lahti–IFK	-1	-3	-3	-1	-1	0	-3	4	1	-1	3	1	0	0	0	sW	1	sW	1	sW
Mariehn–Ilves	3	-1	1	1	-1	1	1	0	1	0	0	-3	1	2	1	sW	-1	sL	-1	sL +
Kemi–KuPS	-1	0	-4	-4	0	1	-1	0	1	1	-3	-3	-1	-2	-2	sL	-2	sL	-3	bL
SJK–Turku	0	0	-2	-1	0	-2	3	0	-1	2	1	-3	1	1	2	sW	5	bW	1	sW
HJK–RoPS	2	3	2	4	1	-3	3	0	2	0	-1	3	1	0	0	bW	-1	sL	2	sW !
VPS–Lahti	-3	0	-1	0	-2	-3	-1	-1	0	5	-2	0	-1	1	4	sL	1	sW	-1	sL !
Ilves–Kemi	1	0	1	0	-3	0	-4	-4	0	-1	0	-2	2	0	-1	sW	1	sW	1	sW
Turku–IIJK	3	0	-1	2	-5	3	2	4	1	3	0	0	1	0	-2	sL	-2	sL	-2	sL
KuPS–JJK	-1	0	1	1	2	1	0	-2	-2	0	1	0	4	1	0	bW	1	sW	2	sW
RoPS–Mariehn	3	0	2	0	-3	-1	1	1	-1	1	3	-1	0	0	-1	sL	-3	bL	-1	sL
IFK–SJK	-3	4	1	-1	0	0	-2	-1	0	3	-2	0	2	0	-1	D	0	D	0	D
Tour 32, 33, Season 2016																				
VPS–Lahti	3	1	1	0	-1	-2	0	0	0	2	1	-4	2	0	2	sW	-1	sL	1	sW !
KuPS–Kemi	2	-1	-2	-1	-1	-3	0	-2	-2	-2	2	2	-2			sW	1	sW	1	sW
Mariehn–Ilves	2	0	2	0	0	-2	0	0	2	1	2	1	2	0	2	sW	1	sW	1	sW
RoPS–IFK	0	1	-1	0	-1	-2	0	1	0	2	1	0	0	0	2	D	0	D	0	D
HJK–SJK	2	0	-1	1	-1	5	0	-2	1	1	1	1	-3	2	1	sW	0	D	1	sW !
Ilves–HJK	0	0	2	1	0	0	-1	1	-1	-3	1	-4	1	0	-1	sL	-2	sL	-1	sL
Turku–RoPS	0	2	0	0	1	1	-1	0	-1	1	-1	-1	2	0	4	D	0	D	1	sW–
Lahti–Mariehn	0	0	0	2	0	0	2	0	0	-1	0	-3	0	2	-1	sL	-2	sL	-1	sL
Kemi–VPS	0	-2	-2	-2	-2	1	1	0	-1	0	0	1	-2	-2		sL	-3	bL	-2	sL
SJK–PK35W	0	2	1	1	1	-1	-2	-2	-1	-2	1	2	0	4	3	bW	5	bW	3	bW
IFK–KuPS	0	1	0	2	2	-1	-2	-1	-1	0	-1	1	-2	-2	0	sW	-2	sL	1	sW !
Ilves–IFK	0	2	1	0	2	1	0	2	2	0	2	1	-2	0	0	sW	0	D	0	D +

Fig. 6. Testing data fragment: Finland, Veikkaus League, 2016–2018

Model	Match outcome	The model is on par with bookmakers		Book-makers outperform the model	The model outperforms book-makers	Probability of correct prediction	
		expected outcomes	sensations			bookmakers	model
Regression-based model	big loss	26/39 0.667	8/39 0.205	4/39 0.102	1/39 0.026	(26+4)/39 0.769	(26+1)/39 0.692
	small loss	91/147 0.619	40/147 0.272	11/147 0.075	5/147 0.034	(91+11)/147 0.694	(91+5)/147 0.653
	draw	78/133 0.586	43/133 0.323	9/133 0.068	3/133 0.023	(78+9)/133 0.654	(78+3)/133 0.609
	small win	107/174 0.615	42/174 0.241	19/174 0.109	6/174 0.035	(107+19)/174 0.724	(107+6)/174 0.649
	big win	32/48 0.667	10/48 0.208	5/48 0.104	1/48 0.021	(32+5)/48 0.771	(32+1)/48 0.687
Fuzzy rule-based model	big loss	27/39 0.692	7/39 0.179	3/39 0.077	2/39 0.051	(27+3)/39 0.769	(27+2)/39 0.744
	small loss	92/147 0.626	37/147 0.252	10/147 0.068	8/147 0.054	(92+10)/147 0.694	(92+8)/147 0.680
	draw	79/133 0.594	41/133 0.308	8/133 0.060	5/133 0.038	(79+8)/133 0.654	(79+5)/133 0.632
	small win	112/174 0.644	38/174 0.218	14/174 0.080	10/174 0.057	(112+14)/174 0.724	(112+10)/174 0.701
	big win	34/48 0.708	9/48 0.187	3/48 0.063	2/48 0.042	(34+3)/48 0.771	(34+2)/48 0.750
Fuzzy relation-based model	big loss	28/39 0.718	4/39 0.103	2/39 0.051	5/39 0.128	(28+2)/39 0.769	(28+5)/39 0.846
	small loss	95/147 0.646	25/147 0.170	7/147 0.048	20/147 0.136	(95+7)/147 0.694	(95+20)/147 0.782
	draw	81/133 0.609	30/133 0.226	6/133 0.045	16/133 0.120	(81+6)/133 0.654	(81+16)/133 0.729
	small win	117/174 0.672	27/174 0.155	9/174 0.052	21/174 0.121	(117+9)/174 0.724	(117+21)/174 0.793
	big win	35/48 0.729	6/48 0.125	2/48 0.042	5/48 0.104	(35+2)/48 0.771	(35+5)/48 0.833

Fig. 7. Comparison of predictive accuracy of the time series based models

of features (events) extracted from the time series by data grouping methods [22, 23]. We have considered the predictive ability of the rule-based model [26], which integrates the expert knowledge in the form of linguistic description of time series with machine learning techniques.

A comparison of the predictive accuracy of the time series based models for the testing data is shown in Fig. 7. It is shown, that the regression model [22, 23] is less accurate than bookmakers' predictions; the performance of the fuzzy rule-based model [26] is on par with bookmakers' predictions; the fuzzy relation-based model outperforms bookmakers' predictions. For the testing data, bookmakers demonstrated the average predictive accuracy at the level of 0.722. The regression model achieved the average accuracy at the level of 0.658. The fuzzy models achieved the average accuracy at the level of 0.701 and 0.797 for rules and relations, respectively.

Thus, the fuzzy relation-based model demonstrates 7.5% higher profitability against published market odds. At the same time, the profitability of the rule-based model [26] is on par with bookmakers' predictions. The best prediction results can be obtained for the loss and win with big score d_1 and d_5, and the worst results of prediction can be obtained for the draw result d_3. Following the trends of betting market, the model demonstrates higher profitability at the start of the season, and the worst results are expected at the end of the season.

7 Conclusions

The fuzzy relational model of time series with granular event segmentation is proposed. Time series modeling implies the stages of segmentation and granulation. At the stage of segmentation, experts identify data features in the form of certain regions of interest or events. To improve data set, it is more expedient to extract certain patterns from time series rather than analyze time series as a whole. At the stage of granulation, the events in the form of collection of time windows are transformed into fuzzy terms, which include both the amplitude and trend description. The underlying relationships, which govern a process generating time series, connect the fuzzy events with the possible match outcome. The fuzzy relational model allows refining the expert information and the experimental data set at the stage of structural tuning.

The method of data quality management in mining fuzzy relations from time series of sports data is proposed. To improve the predictive accuracy, the expert rule set is subject to further correction by removal the different types of biases in expert judgments. In mining fuzzy relations, the biases in expert judgements can be revealed in paired comparative assessments chosen by an expert according to the 9-mark Saaty's scale. To provide the unbiased predictions, the stage of tuning consists of finding such paired comparative assessments, which minimize the difference between model and real matches results. Predictive accuracy of the method outperforms the bookmakers' performance and demonstrates profitability against market odds.

The proposed approach can find application in mining information from time series driven by event segmentation in medicine, economics, geosciences etc. Further research is required to provide the automatic time series segmentation since it is not always possible for the experts to identify an area of a time series as the event with high degree of certainty [35]. To improve predictive accuracy, the automatic segmentation should provide the optimal time series segmentation with further linguistic interpretation of the extracted patterns.

References

1. Schumaker R, Solieman O, Chen H (2010) Sports data mining, vol 26. Integrated series in information systems. Springer, Heidelberg
2. Severini TA (2014) Analytic methods in sports: using mathematics and statistics to understand data from Baseball, Football, Basketball, and other sports. CRC Press, New York
3. Haghighat M, Hamid R, Nourafza N (2013) A review of data mining techniques for result prediction in sports. Adv Comput Sci Int J 2(5):7–12
4. Constantinou A, Fenton N (2017) Towards smart-data: improving predictive accuracy in long-term football team performance. Knowl-Based Syst 124:93–104. https://doi.org/10.1016/j.knosys.2017.03.005
5. Rue H, Salvesen O (2000) Prediction and retrospective analysis of soccer matches in a league. Statistician 3:339–418. https://doi.org/10.1111/1467-9884.00243
6. Karlis D, Ntzoufras I (2003) Analysis of sports data by using bivariate Poisson models. Statistician 52(3):381–393. https://doi.org/10.1111/1467-9884.00366
7. Dixon M, Pope P (2004) The value of statistical forecasts in the UK association football betting market. Int J Forecast 20:697–711. https://doi.org/10.1016/j.ijforecast.2003.12.007
8. Tax N, Joustra YP (2015) Predicting the Dutch football competition using public data: a machine learning approach. Trans Knowl Data Eng 10(10):1–13. https://doi.org/10.13140/RG.2.1.1383.4729
9. Bunker RP, Thabtah F (2019) A machine learning framework for sport result prediction. Appl Comput Inform 15(1):27–33. https://doi.org/10.1016/j.aci.2017.09.005
10. Ivankovic Z, Rackovic M, Markoski B, Radosav D, Ivkovic M (2010) Analysis of basketball games using neural networks. In: 11th IEEE international symposium on computational intelligence and informatics. IEEE, Budapest, pp 251–256 https://doi.org/10.1109/CINTI.2010.5672237
11. Joseph A, Fenton NE, Neil M (2006) Predicting football results using Bayesian nets and other machine learning techniques. Knowl-Based Syst 19(7):544–553. https://doi.org/10.1016/j.knosys.2006.04.011
12. Baio G, Blangiardo M (2010) Bayesian hierarchical model for the prediction of football results. J Appl Stat 37(2):253–264. https://doi.org/10.1080/02664760802684177
13. Baboota R, Kaur H (2019) Predictive analysis and modelling football results using machine learning approach for English Premier League. Int J Forecast 35(2):741–755. https://doi.org/10.1016/j.ijforecast.2018.01.003
14. Constantinou A, Fenton N, Neil M (2016) Integrating expert knowledge with data in Bayesian networks: preserving data-driven expectations when the expert variables remain unobserved. Expert Syst Appl 56:197–208. https://doi.org/10.1016/j.eswa.2016.02.050
15. Min B, Kim J, Choe Ch, Eom H, (Bob) McKay RI (2008) A compound framework for sports results prediction: a football case study. Knowl-Based Syst 21(7):551–562. https://doi.org/10.1016/j.knosys.2008.03.016
16. Liu F, Shi Y, Najjar L (2017) Application of design of experiment method for sports results prediction. Procedia Comput Sci 122:720–726. https://doi.org/10.1016/j.procs.2017.11.429
17. Forrest D, Goddard J, Simmons R (2005) Odds-setters as forecasters: the case of English football. Int J Forecast 21:551–564. https://doi.org/10.1016/j.ijforecast.2005.03.003

18. Graham I, Stott H (2008) Predicting bookmaker odds and efficiency for UK football. Appl Econ 40:99–109. https://doi.org/10.1080/00036840701728799
19. Constantinou AC, Fenton NE, Neil M (2013) Profiting from an inefficient association football gambling market: prediction, risk and uncertainty using Bayesian networks. Knowl-Based Syst 50:60–86. https://doi.org/10.1016/j.knosys.2013.05.008
20. Hvattum L, Arntzen H (2010) Using ELO ratings for match result prediction in association football. Int J Forecast 26(3):460–470. https://doi.org/10.1016/j.ijforecast.2009.10.002
21. Leitner C, Zeileis A, Hornik K (2010) Forecasting sports tournaments by ratings of (prob)abilities: a comparison for the EURO 2008. Int J Forecast 26(3):471–481. https://doi.org/10.1016/j.ijforecast.2009.10.001
22. Emara N, Owens D, Smith J, Wilmer L (2017) Serial correlation in National Football League play calling and its effects on outcomes. J Behav Exp Econ 69:125–132. https://doi.org/10.1016/j.socec.2017.01.007
23. Martins R, Martins A, Neves L, Lima L, Flores E, do Nascimento M (2017) Exploring polynomial classifier to predict match results in football championships. Expert Syst Appl 83:79–93. https://doi.org/10.1016/j.eswa.2017.04.040
24. Scarf Ph, Shi X (2008) The importance of a match in a tournament. Comput Oper Res 35(7):2406–2418. https://doi.org/10.1016/j.cor.2006.11.005
25. Geenens G (2014) On the decisiveness of a game in a tournament. Eur J Oper Res 232(1):156–168. https://doi.org/10.1016/j.ejor.2013.06.025
26. Rotshtein A, Posner M, Rakityanskaya A (2005) Football predictions based on a fuzzy model with genetic and neural tuning. Cybern Syst Anal 41(4):619–630. https://doi.org/10.1007/s10559-005-0098-4
27. Yager R, Filev D (1994) Essentials of fuzzy modeling and control. Wiley, New York
28. Rotshtein A, Rakytyanska H (2012) Fuzzy evidence in identification, forecasting and diagnosis, vol 275. Studies in fuzziness and soft computing. Springer, Heidelberg. https://doi.org/10.1007/978-3-642-25786-5
29. Rotshtein A, Rakytyanska H (2013) Expert rules refinement by solving fuzzy relational equations. In: 6th IEEE conference on human system interaction proceedings. IEEE, Sopot, pp 257–264. https://doi.org/10.1109/HSI.2013.6577833
30. Rotshtein A, Rakytyanska H (2014) Optimal design of rule-based systems by solving fuzzy relational equations. In: Hippe Z, Kulikowski L, Mroczek T, Wtorek J (eds) Issues and challenges in artificial intelligence, vol 559. Studies in computational intelligence. Springer, Cham, pp 167–178. https://doi.org/10.1007/978-3-319-06883-1_14
31. Rotshtein A, Rakytyanska H (2012) Fuzzy genetic object identification: multiple-inputs multiple-outputs case. In: Hippe Z, Kulikowski J, Mroczek T (eds) Human-computer systems interaction - Part II, vol 99. Advances in intelligent and soft computing. Springer, Heidelberg, pp 375–394. https://doi.org/10.1007/978-3-642-23172-8_25
32. Rakytyanska H (2017) Optimization of fuzzy classification knowledge bases using improving transformations. East-Eur J Enterp Technol 5(2):33–41. https://doi.org/10.15587/1729-4061.2017.110261
33. Duan L, Yu F, Pedrycz W, Wang X, Yang X (2018) Time-series clustering based on linear fuzzy information granules. Appl Soft Comput 73:1053–1067. https://doi.org/10.1016/j.asoc.2018.09.032
34. Lu W, Pedrycz W, Liu X, Yang J, Li P (2014) The modeling of time series based on fuzzy information granules. Expert Syst Appl 41(8):3799–3808. https://doi.org/10.1016/j.eswa.2013.12.005

35. Ares J, Lara JA, Lizcano D, Suárez S (2016) A soft computing framework for classifying time series based on fuzzy sets of events. Inf Sci 330:125–144. https://doi.org/10.1016/j.ins.2015.10.014

36. Fuchs E, Gruber Th, Pree H, Sick B (2010) Temporal data mining using shape space representations of time series. Neurocomputing 74(1–3):379–393. https://doi.org/10.1016/j.neucom.2010.03.022

37. Lu W, Chen X, Pedrycz W, Liu X, Yang J (2015) Using interval information granules to improve forecasting in fuzzy time series. Int J Approx Reason 57:1–18. https://doi.org/10.1016/j.ijar.2014.11.002

38. Rotshtein A, Rakytyanska H (2008) Diagnosis problem solving using fuzzy relations. IEEE Trans Fuzzy Syst 16(3):664–675. https://doi.org/10.1109/TFUZZ.2007.905908

39. Rakityanskaya A, Rotshtein A (2007) Fuzzy relation-based diagnosis. Autom Remote Control 68(12):2198–2213. https://doi.org/10.1134/S0005117907120089

40. Rotshtein A, Rakytyanska H (2009) Adaptive diagnostic system based on fuzzy relations. Cybern Syst Anal 45(4):623–637. https://doi.org/10.1007/s10559-009-9130-4

41. Saaty T (1968) Mathematical models of arms control and disarmament. Wiley, New York

42. Football Results Online. https://www.myscore.com.ua/football/finland/veikkausliiga. Accessed 14 Apr 2019

43. Mordeson J, Malik D, Cheng S-C (2000) Fuzzy mathematics in medicine. Phisica-Verlag, Heidelberg

Computational Intelligence and
Inductive Modeling

Recognition of Visual Objects Based on Statistical Distributions for Blocks of Structural Description of Image

Volodymyr Gorokhovatskyi[1] [ID], Svitlana Gadetska[2]([✉]) [ID],
and Roman Ponomarenko[1] [ID]

[1] National University of Radio Electronics, 14,
Nauky Avenue, Kharkiv 61166, Ukraine
gorohovatsky.vl@gmail.com, roman.ponomarenko@nure.ua
[2] Kharkiv Educational and Research Institute of Banking University,
55, Peremogy Avenue, Kharkiv 61174, Ukraine
svgadetska@ukr.net

Abstract. In this paper, we investigate a problem of recognizing visual objects in computer vision systems with the use of a feature space as a set of key image points. The set of key point descriptors is determined by binary detectors ORB, BRISK, AKAZE. In order to extract from the existing description the more detailed information of distinguishing objects and reducing the volume of computing compared with the procedure of voting descriptors, it is proposed to form a new system of attributes for the structural description of the object on the basis of the preliminary formation of statistical distributions for the system of fragments description of the image. On the basis of cross-correlation distribution processing for a case of a one-bit size of blocks of binary descriptors, a new system of integrated features was constructed for efficiently calculating the value of the relevance of descriptions in the process of object recognition. By means of software modeling, an experimental evaluation of the time of calculating the relevance value in the built feature space is performed in comparison with the use of acceptable distribution cases and the tradition-based approach of voting for a set of descriptors. Experiments have shown that the implemented block system creates a representative feature space, in which images with the close structure actually demonstrate similarity. The effectiveness of the proposed statistical system of attributes in terms of a significant increase in the rate of performance with an adequate level of recognition quality is confirmed.

Keywords: Computer vision · Structural image recognition ·
Key point descriptors · Binary descriptor ·
Statistical distribution of fragment value · Block representation ·
Spatial analysis · Cross-correlation processing · Common descriptor ·
Relevance of descriptions · Time of relevance calculation

© Springer Nature Switzerland AG 2020
V. Lytvynenko et al. (Eds.): ISDMCI 2019, AISC 1020, pp. 501–512, 2020.
https://doi.org/10.1007/978-3-030-26474-1_35

1   Introduction

Methods for recognizing visual objects by description in the form of a set of key points (KP) have been applied in modern systems of computer vision due to properties of invariance of values of KP descriptors to geometric transformations of objects, the possibility of confident recognition in conditions of partial visibility, resistance to various types of noise [1–6]. The structural features, apart from the values of the KP descriptors and their coordinates, can be also attributes of projections, responses of specialized spatial filters, etc. [1]. Current task remains to extract from the existing description the deeper information according the properties of distinguishing objects. According to the results of the analysis or the purposeful transformation of the description, a new system of attributes with the best properties concerning the recognition efficiency or processing speed can be formed (in comparison with the traditional approaches). The main requirement for applying modifications of the analysis methods or processing of structural data is to ensure the invariance of the newly created features regarding the arbitrary order of the descriptors in the description [7]. The bit-nature of the KP descriptors in the binary vector space B^n (n is the degree of the number 2) allows the descriptor to be considered and handled as a tuple of successive elements (for example, bytes) whose range of values is known. This allows us to consider the structural description (that is, the set of binary vectors) in the aspect of the newly created system of features and carry out its spatial and statistical analysis based on the attributes of the internal data structure.

2   Review of the Literature

The traditional method for calculating the relevance of two descriptions O_1, O_2, called voting [1,3], is to count the number r of "equivalent" elements of one of the descriptions, considered as an etalon, in the structure of another description. Mathematically, this is a task of determining the similarity degree of two sets of vectors, which can also be solved on the basis of a metric or measure of similarity [2,4,8–10]. Both of these approaches from a computational point of view lead to the comparison of elements of sets on the principle of "everyone with each". It requires, with a large number of descriptors in descriptions that reaches several hundreds of elements, a significant amount of computation. One of the more effective ways of processing is the detection and application of cluster systems in the feature space. To do this, the previous clustering of the set of descriptors for the base of etalon images is carried out [7]. The data of the analyzed description are related to one of the specified clusters by calculating the relevance in the space cluster - etalon. Such processing reduces the computing costs by tens or even hundreds of times [7,8]. Further improvement of recognition indicators can be achieved by using the tools of statistical data distribution, their intellectual analysis and cross-correlation processing. These approaches help to acquire new data properties for image recognition, as well as generalize information, significantly reducing its volume [9–11].

3   Problem Statement

When comparing descriptions as sets of descriptors, it is expedient to use the model of the spatial arrangement of the data components of the description in the same way as the scheme of analysis of the values of spatial fragments of the signal [1,3]. Considering the spatial attributes of data allows to decrease their range of values, that greatly simplifies the analysis of their statistical characteristics and the calculation of the relevance of descriptions. The particular advantages of spatial analysis are to reduce the dimension of the processed signals in the same way as the formation of one-dimensional projections for two-dimensional objects. Due to the introduction of spatial analysis we obtain a new data structure with its properties and values ranges. The synthesized structure can be processed by several independent channels, deciding on their subset or preferring the most likely of them [3,9].

The description Z of a recognizable visual object will be defined as a finite set $Z = \{z_v\}_{v=1}^{s}$, $z_v \in B^n$, $Z \subset B^n$ of s descriptors KP (e.g., generated by ORB, BRISK, AKAZE detectors [9,12]) with binary components. Classification involves determining the belonging of the description to one of the classes of the etalon set $\Theta = \{Z^1, Z^2, ..., Z^J\}$ and is carried out by calculating the relevance of the object description with the etalons.

The purpose of the article is to develop the method of structural recognition of images [3] on the basis of the implementation of spatial-statistical analysis for the tuple of the description blocks values in the form of a set of descriptors. The tasks of the research are the creation of productive models for the formation of structural descriptions, as well as the efficient calculation of their relevance, the study of their properties and the experimental evaluation of the effectiveness of the proposed approaches for the results of image processing.

4   Materials and Methods

For a fixed n we will consider a description of an object as a binary data matrix $D = \{\{d_{i,j}\}_{i=1}^{s}\}_{j=1}^{n}$ formed by a sequence of s descriptors KP, which is input for analysis in arbitrary order. For simplicity, we consider the values to be the same for all etalons within the base for classification. It is easy to achieve by selecting a fixed number of descriptors from any etalons.

Each of the rows of the matrix D, which is a descriptor of KP, will be presented as a tuple of non-intersecting blocks that are sequentially placed one after the other. We obtain a new data space, the advantage of which is the ability to analyze the internal structure and statistical properties of the blocks description for its classification as one of the etalon classes from Θ. Fragments descriptors represent the newest system of features that provide greater flexibility and new possibilities for data analysis [3,9].

After introducing $\forall z_v \in Z$ in the form of m blocks, we have a fixed structure $z_v = z_v^1 \& z_v^2 \& ... \& z_v^m$ of the same type elements, where z_v^k is the block with number k. The value of each z_v^k in a sequence can be written by the number of the specified range. As a result, the description Z acquires the form of a matrix of s rows (number of descriptors) by m elements (number of blocks) in a row (Table 1).

Table 1. Block representation of the data structure D

Descriptor numbers	Block numbers				
	1	...	k	...	m
1	z_1^1	...	z_1^k	...	z_1^m
...	...	...	...	...	...
s	z_s^1	...	z_s^k	...	z_s^m

Considering that the set of descriptors reflects the properties of the analyzed object, we will build the distributions of data blocks based on the analysis of the matrix D [9]. We map a set of binary vectors to a set of integer power vectors containing statistical distributions of the values of their fragments. We define a mapping $\Omega : Z \longrightarrow Q$, $Z \subset B^n$ from a set of binary vectors to a set Q of integer vectors of power $w \leq n$ containing statistical distributions of their fragments values. The mapping Ω is intended to provide a distinction between visual objects according to the probabilistic distributions of data blocks of their descriptions.

Define the distribution $q \in Q$ as a vector of integers $q = \{q_1, ..., q_w\}$, where q_i is the number of values i, $i = \overline{1, w}$, for k-fragment among all fragments with number k of the set Z:

$$q_i = card\{z_v^k \in z_v, z_v \in Z | z_v^k = i\} \tag{1}$$

In this case, the condition $\sum_i q_i = s$ holds, since the sum of the values of distribution components q is equal to the number s of the description Z.

The value w is determined by the range of data values for the fragment and depends on m. For example, for the BRISK detector when splitting into bytes for the case $n = 512$ we have $m = 64$, $w = 256$.

For each of m blocks, based on the analysis Z, we get the distribution $q = \{q_1, ..., q_w\}$. As a result of the transformation Ω the set Z will be described by a matrix $Q = \{\{q_{i,v}\}_{i=1}^w\}_{v=1}^m$. The matrix Q reproduces the statistical properties of the description Z in the form of a tuple of distributions of its components. Statistical data distributions realize the fundamental concept of machine learning, generalizing knowledge about objects [11, 13].

Notice that the size of the data of the matrix Q for small fragment sizes m is much smaller compared to the matrix D. This contributes to a significant reduction in the volume of computing in determining the relevance degree of descriptions. However, with a rising m the situation changes slightly, if $m = 8$ then data volumes approximately coincide, and for $m > 8$ the volume of data of distributions begins to exceed the volume of the initial description.

We will now conduct cross-correlation processing of matrix D and build on the basis of one-bit distributions an integrated system of attributes. For the description matrix D we calculate the value of the sum of the columns (1-bit fragment) and obtain the vector $t = (t_1, ..., t_j, ..., t_n)$, where $t_j = \sum_{i=1}^{s} d_{i,j}$, $j = \overline{1, n}$ (in fact, t_j is the sum of units in the column). They are the features of the lowest level that preserve the invariant properties of the set Z.

On the basis of features t_j, we calculate the attributes u_k for blocks which are the sets of columns (features of a higher level):

$$u_k = \sum_{j=k}^{k+b-1} t_j, \quad t_j = \sum_{i=1}^{s} d_{i,j} \tag{2}$$

where $b = n/m$ is a size, $k = 1, b+1, 2b+1, ..., n-b+1$ is a number of a fragment.

The features (2) are the result of cross-correlation processing [10] with a rectangular mask of size $s \times b$. Expression (2) implements the convolution of data inside a block of size b, which gives us a generalized vector of features. As a result of the calculation (2) we obtain an integer vector u_k of dimension m. The coordinates of the vector $u = (u_1, ..., u_k, ..., u_m)$ can be used as independent statistical features, which are calculated by the columns of the matrix D. Processing (2) is determined by the parameter b and implements spatial processing on the set of descriptors. Consider a fairly simple computation model u, since the components u_k are determined solely by the addition of integers for an arbitrary fragment size.

On the basis of representation (2) a hierarchical method of recognition can be applied. It uses a system of features u_k with different degrees of data integration for comparison with the etalons. Model (2) implements the decreasing of informational redundancy of the spatial signal due to the allowable decrease (in terms of quality of distinction) of the degree of distribution of the characteristic representation of the description. The range of values for the features u_k can be directly determined by the size of the fragment as $u_k \in \{0, ..., sb\}$. The vector u can be normalized by the number of descriptors or the fragment size.

One way to simplify the analysis of block data is to use the values of individual parameters of distribution Q. Defining of a common descriptor (CD) for constructing a concentrated description can be based, for example, on the use of the most commonly used templates that are most commonly found in the description fragments system [9].

To do this, determine the maximum value of the fragment distribution in each column of the matrix Q and form a chain of d CD as the result of the clutch of the most commonly used values:

$$\begin{cases} d_v = \arg \max_{i=1,...,w} q_{i,v}, v = \overline{1, m}, \\ d = d_1 \& d_2 \& ... \& d_m \end{cases} \tag{3}$$

The value d_v as the maximum argument in (3) is obtained for each of the fragments. Thus, the KP for the description Z has here the form of a vector containing the values that are most commonly found in the corresponding block of each descriptor.

In model (3) we should consider only sufficiently significant values with limit significance $q^* = \max\limits_{i=1,\ldots,w} q_{i,v}$, for example, provided $q^* \geq 0{,}5$. It is possible at the previous stage to form a list of fragments for a concrete etalon which have a rather high level of support q^* within the description. Restriction on the value of support provides an adaptation in terms of more accurate reproduction of the contents of the description. Products of the form (3) correspond to the principle of the maximum a posteriori probability. For essentially different descriptions, such an analysis brings the necessary result [9].

Determine relevance as a measure of difference between matrices $Q = \{q_{i,j}\}$, $i = \overline{1,w}$, $j = \overline{1,m}$, for descriptions of comparable objects. We compute the relevance r for the descriptions a and b as the Manhattan distance between the matrices $Q(a)$, $Q(b)$:

$$r[Q(a), Q(b)] = \sum_{i=1}^{w} \sum_{j=1}^{m} |q_{i,j}(a) - q_{i,j}(b)| \tag{4}$$

To estimate the value (4) within its defined range, we also introduce the normalized distance:

$$r^* = r/r_{max}, \tag{5}$$

where r_{max} is the maximum possible distance for a fixed number s of description points with size of the descriptor equals n and the number of bits in the distribution equals b.

As an example, for a one-bit distribution at $s = 700$, $n = 512$ we have $r_{max} = 1400 \times 512$, for two-bit distribution $r_{max} = 1400 \times 256$, for 8-bit distribution $r_{max} = 1400 \times 64$.

5   Experiment

The proposed method of recognition by data of distributions is applied on the example of images of icons ("Kazan godmother", "Sistine madonna", etc.) in the size of 400×540 in the environment of Visual Studio 2017 using the library Open CV [14], an illustration of the image with the coordinates of KP is shown on Fig. 1. BRISK descriptors with parameter $n = 512$ are applied. In Fig. 2 an example of a two-bit distribution for descriptor blocks describing two icons (blue and yellow) for the first pair of bits is given. As we can see from Fig. 2, the distributions are quite similar. It can be explained by the same type of image. Our other experiments showed that for very different images (the icon and the emblem of Kharkiv city) experimental divisions in two bits differ significantly.

Experimental Table 2 contains the values of the normalized Manhattan distance for the descriptions of the icons, depending on the number of descriptors (16 ... 700) and the number of bits (1 ... 8) in the distribution blocks. The measure of similarity by the voting method (counting the number of similar descriptors with an equivalence threshold 20% of the maximum value of the Heming distance) for these two icons is 0,12.

Fig. 1. Image icons with coordinates KP

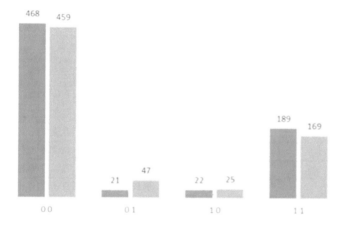

Fig. 2. An example of distributions by values of two bits

As a result of comparison of the values of the Table 2 with the corresponding values obtained for essentially different image contents (icon and city emblem), we can see that the values in the column for 1 bits are approximately 50% higher, and in the column for 8 bits - 20% higher. This fact confirms the sufficient possibilities for distinguishing objects of the applied distributions model even in the case of a small number of bits of a fragment. We now analyze the available computing costs for the implementation of the considered methods in comparison with the traditional approaches. Calculation time in seconds using software models is fixed in the Table 3. Figures 3 and 4 in a comparative way contain the values of spatial features of the images $u_1, ..., u_5$ of the icon (yellow) and the emblem of Kharkiv (blue) for 1-bit (Fig. 3) and 8-bit forms (Fig. 4). Table 3 contains the value of Manhattan distance (5) for full description of the icon and emblem within the proposed system of attributes u, normalized to the total number sb of analyzed bits.

Table 2. Distance value (5) depending on the number of distribution bits

Number of descriptors	The number of distribution bits			
	1	2	4	8
700	0,059	0,084	0,143	0,319
500	0,061	0,086	0,149	0,326
100	0,071	0,110	0,204	0,450
50	0,106	0,165	0,290	0,570
30	0,108	0,175	0,320	0,600
16	0,127	0,210	0,402	0,699

Table 3. Estimates of time (in seconds) in computer modelling

Number of descriptors	Measure (5)	The number of distribution bits			
		1	2	4	8
700	10,700	0,008	0,011	0,028	11
500	6,400	0,006	0,010	0,017	6,2
100	0,480	0,002	0,002	0,007	1
50	0,110	0,001	0,002	0,004	0,915
30	0,080	0,001	0,001	0,003	0,527
16	0,045	<0,001	<0,001	0,002	0,315

6   Results and Discussion

From the Table 1, we can see that the measure on the basis of distributions has sufficient possibilities for distinguishing similar images. As the number of distribution bits increases from 1 to 8, the distance increases, that is, the recognition efficiency improves. At the same time, with a decrease in the number of KP of description, the distance also increases, but it can be said about the possible decrease in the recognition accuracy, since two completely different images may have, on a small sample of features, both similarity and difference. Values of the Table 1 show that 50–100 of KP are enough to distinguish the image of the icons. For other types of images or other number of etalons, these values may differ (Table 4).

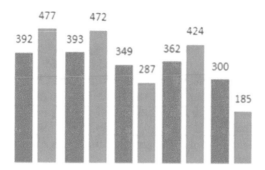

Fig. 3. The values of features $u_1, ..., u_5$ for a 1-bit representation of the icon and emblem

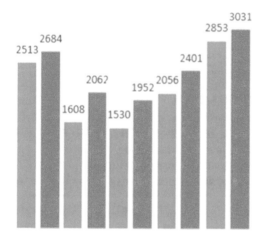

Fig. 4. Comparative values of features $u_1, ..., u_5$ for an 8-bit representation

Table 4. Value of the distance (5) for the images of the icon and the emblem with features (2)

Number of descriptors	Total number of bits for a feature u			
	1	2	4	8
700	0,096	0,085	0,071	0,063
500	0,070	0,061	0,051	0,047
100	0,015	0,013	0,011	0,009
50	0,008	0,007	0,006	0,005

As we see from the Table 2, the proposed division-based approach has significant computational advantages over voting, since the time to determine its relevance is less than 1300 times for a 1-bit representation and 400 times for a 4-bit one. The time for the 8-bit version is almost equal to the cost of the traditional method; however, the main costs are directed to the formation of distributions, but not to the relevance calculation. It is prospective to use the models for relevance determination without generating distributions for a recognizable image, based solely on etalon distributions calculated at the pre-processing stage [9]. It is clear, however, that the time spent on calculating relevance for all models is proportional to the number of descriptors in the descriptions and depends on the program execution mode.

According to the Table 3, we note that, for the same type of images (two different icons), the calculated distance is approximately less than 50%. It emphasizes the sufficient sensitivity of the proposed features (2) to the content of the image. Despite the results of small absolute values of distance obtained in the Table 2, experiments showed the faultless result of classification within the set of 10 icons and emblems of Ukrainian cities as well as the mixed set of 20 images.

The time to calculate the distance for the analyzed processing cases (Table 3) is practically the same, and even for a quantity of 700 points is within 1 ms. It emphasizes computational simplicity and, as a result, the high speed of the proposed methods. A more detailed analysis of the computational cost for determining the distance at showed that for 8-bit features the time is approximately 5 times smaller than for 1-bit ones. It indicates a reserve for further improvement of performance speed characteristics. Taking into account that in comparison with the traditional calculation using the voting model, the time of relevance calculation in the space of integrated features is generally decreases thousands of times, it is possible to draw conclusions about the perspective application of spatial models (2).

As you can see from the data of the Table 3, the value of the relevance measure of the descriptions in the created feature space is predicted to decrease with decreasing in the number of points. It indicates a reduction in the distribution capacity of the recognition system and the possible reduction of the level of error-free recognition.

Our experiments have shown that for such information-rich images as icons and emblems, a sufficient number of points for error-free classification is the number exceeding 100.

From the Table 3, we also see that with increasing degree of integration, which is associated with an increase in the number of bits from 1 to 8, the distributive properties of the features are slightly reduced, but insignificantly. For example, the distance for 8-bit features at $s = 700$ was 0.063 compared to 0.096 for a 1-bit ones. From the obtained experimental results, we can conclude that it is expedient to use the features u with the number of bits 1, 2, 4, 8. Compared to the distribution features Q, the proposed system of spatial features u is much easier to calculate and has almost the same classification performance indicator, so it can be successfully applied in practice to accelerate recognition.

7 Conclusions

A generalized statistical analysis of the fragment system for a set of descriptors constituting a structural description of an image is a promising direction of reducing the computational cost of recognition. Probabilistic data representation contributes not only to significantly accelerating (more than a thousand times) processing procedures, but also provides a sufficient level of image diversity. Fragments of key point descriptors can be considered as a system of spatial attributes, which simplifies the processing and retains the sensitivity of differentiating data.

Spatial methods for handling descriptors by processing or analyzing substructures of data reveal statistical patterns and are stable when recognizing the characteristics of objects. The implementation of the block data structure makes it possible to flexibly manage the process of determining the relevance of descriptions. The bit form of descriptors facilitates the use of simple tools for computational processing costs with appropriate implementation in applied tasks.

The experiments showed that the implemented block system creates a rather representative space of features, since semantically related images in it demonstrate the same similarity.

The prospects of the further investigation are related to the synthesis of intelligent classification procedures based on the decisions of the processing data block system to provide sufficient recognition performance in arbitrary image collections. Another direction can be the introduction of intellectual analysis, which reveals hidden patterns or knowledge in the existing structural descriptions of objects.

References

1. Putiatin EP, Averin SI (1990) Obrabotka izobrazhenij v robototekhnike [Image processing in robotics]. Engineering Publications, Moscow
2. Duda RO, Hart PE, Stork DG (2008) Pattern classification. Wiley, Hoboken
3. Gorokhovatskyi VA (2014) Strukturnyj analiz i intellektual'naya obrabotka dannyh v komp'yuternom zrenii [Structural analysis and intellectual data processing in computer vision]. SMIT, Kharkiv
4. Peters JF (2017) Foundations of computer vision: computational geometry, visual image structures and object shape detection. Springer, Cham
5. Lowe DG (2004) Distinctive image features from scale-invariant keypoints. Int J Comput Vis 60(2):91–110
6. Muja M, Lowe DG (2012) Fast matching of binary features. In: Proceedings of the 9th conference on computer and robot vision, pp 404–410
7. Gorokhovatskyi O, Gorokhovatskyi V, Peredrii O (2018) Analysis of application of cluster descriptions in space of characteristic image features. Data 3(4):52. https://doi.org/10.3390/data3040052 https://www.mdpi.com/2306-5729/3/4/52
8. Grycuk R (2016) Novel visual object descriptor using surf and clustering algorithms. J Appl Math Comput Mech 15(3):37–46

9. Gorokhovatskyi V, Gadetska S, Ponomarenko R (2018) Statisticheskie rasprede-leniya i cepnoe predstavleniya dannyh pri opredelenii relevantnosti strukturnyh opisanij vizual'nyh ob"ektov [Statistical distributions and chain representation of data in determining the relevance of structural descriptions of visual objects]. Control Navig Commun Syst Acad J 6(52):87–92. https://doi.org/10.26906/SUNZ.2018.6.087

10. Shapiro L, Stockman J (2006) Komp'yuternoe zrenie [Computer vision]. Translated by BINOM Publishers, Moscow

11. Bishop C (2006) Pattern recognition and machine learning, 1st edn. Springer, New York

12. Leutenegger S, Chli M, Siegwart RY (2011) BRISK: binary robust invariant scalable keypoints. In: IEEE international conference on computer vision (ICCV), pp 2548–2555

13. Han J, Kamber M (2006) Data mining: concepts and techniques. Morgan Kaufmann Publishers, Amsterdam

14. Open source computer vision library (2018). https://docs.opencv.org/master/index.html. Accessed 3 Sept 2018

Hybrid Methods of GMDH-Neural Networks Synthesis and Training for Solving Problems of Time Series Forecasting

Volodymyr Lytvynenko[1] , Waldemar Wojcik[2] , Andrey Fefelov[1] ,
Iryna Lurie[1] , Nataliia Savina[3] , Mariia Voronenko[1(✉)] , Oleg Boskin[1] ,
and Saule Smailova[4]

[1] Kherson National Technical University, Kherson, Ukraine
immun56@gmail.com, fao1976@ukr.net, lurieira@gmail.com,
mary_voronenko@i.ua, aandre.lenoge@gmail.com
[2] Lublin University of Technology, Lublin, Poland
waldemar.wojcik@pollub.pl
[3] National University of Water and Environmental Engineering, Rivne, Ukraine
n.b.savina@nuwm.edu.ua
[4] D. Serikbayev East Kazakhstan State Technical University,
Ust-Kamenogorsk, Republic of Kazakhstan
saule_smailova@mail.ru

Abstract. In this paper, for solving the problem of forecasting non-stationary time series, hybrid learning methods for GMDH-neural networks are proposed. Training methods combine artificial immune systems with members of the evolutionary algorithm family, in particular, gene expression programming systems. The following hybrid computational methods for the synthesis and training of GMDH-neural networks have been developed: a method in which candidate models are represented as gene expression, and training is performed by clonal selection; the method of two-phase structural-parametric synthesis, in which the structural component is formed by programming the expression of genes, and the parameterization is performed by clonal selection; a method based on cooperative-competitive processes of interaction of the elements of the immune system, in which the structure and parameters of the GMDH-neural network are represented by the entire population element-wise. Comparative experimental studies of the quality of the proposed computational methods for solving forecasting problems were carried out.

Keywords: Group method of data handling ·
Clonal selection method · Gene expression programming

1 Introduction

The idea of combining various computational methods and the formation of hybrids is based on the assumption that the resulting new computational method

© Springer Nature Switzerland AG 2020
V. Lytvynenko et al. (Eds.): ISDMCI 2019, AISC 1020, pp. 513–531, 2020.
https://doi.org/10.1007/978-3-030-26474-1_36

should have a higher performance than the components it contains. Being embodied in life, this idea led to the creation of a hybridization technology that allows you to synthesize entire classes of algorithms that are capable of solving complex problems at a qualitatively new level. According to this technology, the strengths of the methods involved in hybridization form a combined result that characterizes the main advantages of the hybrid approach, namely: obtaining better solutions, obtaining solutions in less time, solving large-scale problems.

A remarkable feature of the hybridization technology is its versatility, which is manifested in the ability to combine computational methods used to solve problems of different classes, for example, approximation and combinatorial optimization. In this case, an artificial neural network, for example, can serve as an approximation method designed to solve the task of forecasting a time series. Then, to guarantee an optimal forecast, it is necessary to carry out a complete enumeration of all configuration options for this network. However, this task, like many similar ones, belongs to the NP-complete class, whose search space is too large to be able to find its exact solution in a reasonable time. In this case, it is possible to involve a heuristic procedure for learning the network, which will make it possible to find a solution, although inaccurate, but close to optimal and in less time. In addition, heuristics make it possible to increase substantially the dimension of the problem, i.e. consider a greater number of network parameters and synthesize more complex architectures. The hybridization technology allows you to combine these methods with other computational paradigms, increasing efficiency and expanding their field of application.

In this paper, to solve the problem of non-stationary time series predicting, a type of neural networks is chosen - GMDH-neural networks. A hybrid version of an artificial immune system and one of the representatives of the family of evolutionary algorithms—the gene expression programming system—was used as a heuristic method for teaching the GMDH neural network. It is expected that the developed new hybrid technology will increase the efficiency of the predictive model and maintain sufficient stability of the algorithm when working with various types of input information. A number of comparative experiments have been carried out involving synthetic and real time series.

2   GMDH Networks

One of the approaches to modeling of complex dynamic systems, where, of course, include solving problems of forecasting, is the group method of data handling (GMDH) [1,2]. This method is based on the selection procedure, which implements the process of sequential testing of models from a variety of candidates in accordance with the selected criterion. In general, the relation between input and output variables is represented by a Volterra functional series, the discrete analogue of which is the Kolmogorov–Gabor polynomial:

$$
\begin{aligned}
y = a_0 + \sum_{i=1}^{M} a_i x_i &+ \sum_{i=1}^{M} \sum_{j=1}^{M} a_{ij} x_i x_j \\
&+ \sum_{i=1}^{M} \sum_{j=1}^{M} \sum_{k=1}^{M} a_{ijk} x_i x_j x_k + ...
\end{aligned}
\tag{1}
$$

where $\{x_i, x_j, x_k, ...\}$ is the vector of input variables, and $\{a_0, a_i, a_{ij}, a_{ijk}, ...\}$ is the coefficient vector (weights). Components of the input vector can be not only independent variables, but also functional or finite-difference forms of expressions. Formula (1) involves the calculation of all possible combinations of elements of the input vector, and, therefore, is a complete mathematical description of the simulated system. Selection of partial descriptions is based on the so-called external criteria, which is a unique feature of GMDH. In this case, the training data sample is divided into training, test and test. The training sample is used to calculate the values of coefficients, the test sample is used to select candidate models, and the test sample is used to select the final model.

GMDH-neural networks is the function approximation method, based on the GMDH algorithm, presented in the form of a neural network, where the partial descriptions are the key elements (neurons). GMDH-neural networks have an architecture similar to the direct propagation neural network architecture, in which the neurons of the hidden layer instead of the classical activation functions use polynomials of various degrees (mainly linear, quadratic or cubic).

3   Problem Statement

The synthesis process of the predictive model considered earlier has a number of drawbacks.

1. The possibility of hitting the local minimum when searching for the optimal model. While since its emergence the GMDH algorithm has undergone a number of modifications that increase its accuracy, in many cases, in particular with long-term forecasting, the use of GMDH yields low results. In studies [3,4] there are several reasons for this situation, including: insufficient functional diversity of candidate models, the possibility of losing significant variable in the selection process, inaccuracies associated with the calculation of parameters by the least squares method.
2. The absence of a stochastic component in the work and the presence of a fixed structure. The deterministic nature of learning of the GMDH neural network training, among other things, causes the problem mentioned in the previous paragraph, i.e. "Stuck" in the local optima of the solution space.
3. Lack of clear criteria for choosing the form of a polynomial, a partial description for each network node. Currently, as a result of research [5–7], many variants of partial descriptions have been proposed, the use of which as activation functions of neurons shows a certain degree of increase in the performance of the algorithm.
4. The tendency to build a complex network structure, even for relatively simple tasks. The reason for this is the constant increase in the structural complexity of the network with each new iteration through the mechanism of adding layers and the absence of a mechanism for structural reduction or compression. In [8,9], several ways to reduce the complexity of the model are considered, including a change in approach to the selection of candidate models and the

use of polynomials of different order in the formation of the first and subsequent layers of the network. However, a systematic approach to reducing the structural complexity of the model was not proposed.

In order to reduce the number of shortcomings of the standard approach to learning GMDH neural networks, the idea of creating a hybrid technology with other heuristic methods, in particular, with evolutionary algorithms, seems quite logical. In [10–13], a description is given of existing examples of the implementation of such a technology, as a result of which it was possible to achieve a significant increase in the performance of the GMDH algorithm.

4   Materials and Methods

The use of evolutionary computing (EC) in solving problems is a relatively new direction in science, which is based on the mechanisms of the natural evolution of living organisms. Characteristic features of EC make them a very convenient tool that can work in dynamic conditions, manipulate data containing a significant amount of noise, solve multi-purpose optimization problems, and, also, carry out a global search for an optimal solution. The development of the direction of EC led to the emergence of a number of computational methods, among which the most widely used were: genetic algorithms (GA), evolutionary strategies (ES), evolutionary programming (EP), genetic programming (GP), programming of gene expression (PGE), etc. More recently, a new branch of computational intelligence has emerged - artificial immune systems (AIS). Despite the fact that AIS don't formally refer to EC, they also use an evolutionary approach to find their solutions. All computational methods based on EC, based on three characteristic components that distinguish them from other search methods. First, they work with a population of solutions, thereby conducting a collective search. Like natural populations of animals or cells, the machine population of individuals in such algorithms tends to adapt to annoying external influences. Secondly, the decision population is exposed to stochastic operators, such as: selection, recombination, and mutation. Some of the methods may include special operators. For example, in PGE transpositions are used, and in AIS - suppression. Thirdly, on the basis of the values of the objective function, a measure of its fitness in relation to an external stimulus, which determines the likelihood of an individual's transition to the next generation, is put in accordance with each individual. It should be noted that the algorithms implementing the EC methods are universal, i.e. suitable for solving any problem that can be converted to the kind of problem of searching for an extremum of a certain function. The only thing that requires mandatory clarification in a specific implementation of the algorithm is the method of encoding the decisions (individuals of the population) and the type of the objective function.

Synthesis of GMDH neural network from the point of view of EC can be represented as an optimization problem, where the search for the structure of a polynomial is performed along with the definition of its parameters. Such a task is best solved by a complex of methods, applying each of them to the subtask with which it, due to its peculiarities, can handle the most quality. Thus, by adapting each of the methods to solve the problem of learning the GMDH neural network, it is possible to construct a hybridization technology with higher quality indicators than its individual elements.

4.1   Programming of Gene Expression

One of the relatively new methods of EC, well-proven in solving problems of structural optimization, is the PGE algorithm. PGE is the development of an earlier method of GP with multiple modifications, which made it possible to get rid of a number of drawbacks inherent in GP [14]. The pseudocode of the PGE algorithm is shown in the Fig. 1.

```
generation = 0
Create initial population
cycle while stop condition = false
      generation = generation + 1
      Expand linear individuals to the level of expressions.
      Define Expression Values
      Calculate the fitness of individuals
      Select parent pairs from population.
      Recombine the population (with probability p_cross)
      Conduct a population mutation (with probability p_mut)
      Perform is / RIS population transposition
end of cycle while
```

Fig. 1. Pseudocode of algorithm PGE

The character of PGE focused on the synthesis of structures suggests that replacing simple mathematical functions in a functional subset with more complex polynomials will create the necessary basis for further integration of PGE into the method of teaching GMDH neural networks.

4.2   Artificial Immune Systems

Modern research in the field of immunology has allowed to form a special look at the natural immune system of animals and humans and to consider it as a powerful computational structure. The further development of the immune computational paradigm led to the creation of a number of methods under the general name - artificial immune systems [15, 16]. The specific properties of the immune system as a distributed, decentralized and heterogeneous structure formed the basis of various models describing the processes of searching, detecting, recognizing and protecting the body from foreign agents - viruses and bacteria. One of such models is clonal selection [17], which function is consistent adaptation of the population of antibodies—variants of the solution of the problem to an incoming antigen—the objective function. The pseudocode of the clonal selection algorithm is shown in Fig. 2.

```
generation = 0
Create initial population
cycle while stop condition = false
    generation = generation + 1
    Calculate the avidity of antibodies to the antigen (f)
    Choose antibodies with the greatest affinity
    Create clones of selected antibodies (in the amount of n~f)
    Conduct a mutation of clones (pmut !~f)
    Calculate avinnost clones to antigen (fc)
    Choose the clones with the greatest affinity
    Transfer clones to the main population
    Replace antibodies with the least affinity
end of cycle while
```

Fig. 2. Pseudocode of clonal selection algorithm

In general, formally, the clonal selection algorithm (CSA) can be represented as follows:

$$CLONALG = (Ab^0, Ag, L, N, n, \partial, d, e) \tag{2}$$

where Ab^0 is the initial antibody population; Ag is the antigen population; N is the amount of antibodies in a population; L is the antibody receptor length; n is the number of antibodies selected for cloning (with the highest affinity); ∂ is the multiplying factor regulating the number of clones of selected antibodies; d the number of antibodies to be replaced with new ones (with the lowest affinity); e is the break criterion.

If this method is used to solve optimization problems, but not recognition, then the population of antigens is replaced by the function of affinity ($f \Rightarrow$ max or $f \Rightarrow$ min). CSA is quite convenient for solving parametric optimization problems, it is an effective method of global search and can be used when training a GMDH neural network as an alternative to the least squares method.

4.3 Structural-Parametric Synthesis of GMDH Neural Networks Using Evolutionary Computational Methods

This section proposes a hybridization technology that includes three options for combining the hereinbefore approaches. According to the first variant, a simultaneous synthesis of the structure and adjustment of the parameters of the GMDH neural network is performed using CSA with individuals represented as PGE chromosomes. This hybrid is called an algorithm for clonal programming of gene expression (CPGE). The second option involves the sequential synthesis of the network structure using the PGE algorithm and the parameterization of the resulting structure by clonal selection. This variant is called the clonal-genetic hybrid learning algorithm of the GMDH neural network. To implement the third option, the cooperative immune algorithm described in [18] was used. In this case, the cooperative algorithm is involved both in the synthesis of the network structure and in the configuration of its parameters. EC is universal and for their application to the solution of a specific problem it is necessary to clarify the details regarding the method of coding solutions and the type of the objective function. In this case, you must additionally add features of the implementation of new algorithms in the context of the proposed hybridization technology.

4.3.1 Algorithm of Clonal Programming of Gene Expression

The coding of individuals for this hybrid is carried out according to the PGE algorithm scheme, taking into account the design features of GMDH neural networks. The functional alphabet is represented by two polynomials: linear and quadratic (2), $F = \{L, Q\}$, and the terminal alphabet is represented by variables and constants $F = \{L, Q\}$. The number of variable symbols in the terminal alphabet is equal to the number of independent variables of the model. In forecasting problems, the group of observations immediately preceding the current one, which, in turn, is considered to be the dependent variable, is often chosen as the arguments of the time series model. Constants are set by real numbers in the interval, the value of which is set in the parameters of the algorithm. Each individual contains a head and a tail, but in contrast to the PGE algorithm, in the construction under consideration added a section that includes the coefficients of partial descriptions ai. A generalized diagram of the structure of the individual is shown in Table 1. Here $(f/t)_i$ – are the elements, respectively, of a functional or terminal set that can be used in the head of an individual; t_i – are the elements of only the terminal set, which limited the tail composition of the individual. The size of the partition coefficients is determined as follows:

Table 1. The structure of the individual KPGE

Head	Tail	Coefficients
$(f/t)_1, (f/t)_2, ..., (f/t)_h$	$t_1, t_2, ..., t_t$	$a_1, a_2, ..., a_k$

$k = h * A$, where A – is maximum number of coefficients in polynomials of the functional alphabet. In this case, $A = 6$ is the number of coefficients of a quadratic polynomial. Moreover, if the construction of a particular solution uses partial descriptions with a smaller number of coefficients, then the corresponding section of the individual line will contain insignificant elements, which is quite y consistent with the PGE concept.

The time series value y_i $(\ldots, y_{t-n}, \ldots, y_{t-1}, y_t, \ldots)$ there is some function of n previous values of the same series. I.e:

$$y_t = f(y_{t-1}, y_{t-2}, ..., y_{t-n}) = f(x_1, x_2, ..., x_n) \tag{3}$$

This function provides the unambiguous prediction of the next value of a series by its n previous values. Next, the training sample is ordered by (3), which allows to obtain all the observation points necessary to calculate the mean square error of the model.

Despite the fact that the KPGE algorithm works according to the clonal selection method, the specific design of antibodies allows importing transposition operators characteristic of PGE. IS- and RIS-transpositions further enhance the effect of hypermutation in terms of supporting the diversity of the population. The main cycle of the CPGE algorithm in the form of a flowchart is shown in Fig. 3.

4.3.2   Clonal Genetic Hybrid Algorithm

Features of the implementation allow us to divide this hybrid into structural and parametric components, each of which is a separate optimization method. Each method uses its own type of individuals, and, therefore, the coding of the complete solution is distributed between the components of the algorithm.

The coding of the structural component of the solution is performed according to the PGE scheme. Functional and terminal alphabets are identical to those considered earlier in the CPGE algorithm. However, in this case, unlike KPGE, the antibody string does not contain a section of partial description coefficients, since Parameterization of the solution occurs in a separate cycle with a fixed structure of the individual.

Similar to coding, the evaluation of individuals is divided into the evaluation of structures and the evaluation of parameterized (essentially complete) decisions. In the first case, the linear correlation coefficient, which shows the statistical relationship between the observed value and the model, is taken as the basis for calculating the suitability of individuals. Such a choice is due to the assumption that the construction of a point forecast, as well as its estimation using the gc criterion, for a non-parametrized model cannot provide a sufficient quality of selection, since single "emissions" can significantly worsen the value of the assessment.

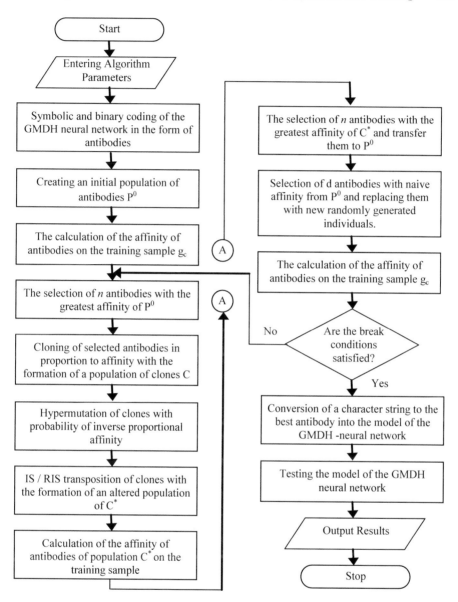

Fig. 3. Block diagram of the KPGE algorithm

5 Experiment

For the experimental studies, three time series were selected, one of which was created on the basis of solving the system of Lorentz differential system, and the other two are observations of real processes. The Lorentz system, which is a fairly popular test case for developing one or another prediction method, describes the

522   V. Lytvynenko et al.

change in three variables: $o\,(t)$, $q\,(t)$, $p\,(t)$. In this work, to construct the Lorentz series, difference equations of the form:

$$
\begin{aligned}
o\,(t+1) &= o\,(t) + \Delta t \cdot \sigma \cdot (p\,(t) - o\,(t)) \\
p\,(t+1) &= p\,(t) - \Delta t\,(o\,(t) \cdot q\,(t) - r \cdot o\,(t) + p\,(t)) \\
q\,(t+1) &= q\,(t) + \Delta t\,(o\,(t) \cdot p\,(t) - b \cdot q\,(t))
\end{aligned}
\tag{4}
$$

where $\Delta t = 0.012$, $\sigma = 16$, $r = 45.92$, $b = 4$, and the variable is selected as the predicted value $o\,(t)$. The series contains 500 observations, its graph is presented in Fig. 4a. For convenience of calculation, the series values are normalized in the interval $[0,1]$.

The first 350 observations (70%) of the series are used as a training set; the remaining 150 observations (30%) are selected for testing. The size of the required time window (immersion depth n) is calculated on the basis of the partial autocorrelation function (PACF) presented in Fig. 4b (here the maximum significant lag value located above the level indicated by the dotted line is 7, i.e. $n = 7$).

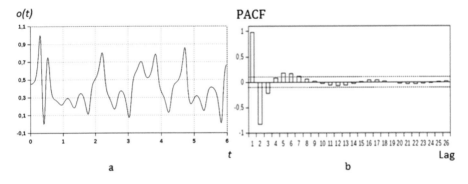

Fig. 4. The time series of changes of the variable $o(t)$ in the Lorentz system (a), PACF of the series of changes of the variable $o(t)$ in the Lorentz system (b)

The source of the second and third rows is the Neural Forecasting Competition [19]. The data are real observations of the load dynamics of various types of highways and infrastructure. This includes information on traffic congestion, tunnels, subways, aviation and rail systems, etc. Time series are grouped by range and frequency, and have different lengths.

The graph of the hourly load line of the Paris Metro line (NNGC1-1.F008) is shown in Fig. 5a. Observations were made for several days from 02.01.2005 to 29.03.2005 from 5.00 to 24.00 h. The series contains 902 values. As in the previous case, the series is normalized in the interval $[0,1]$. The first 630 values (70%) constitute the training sample, the last 272 values - the test sample.

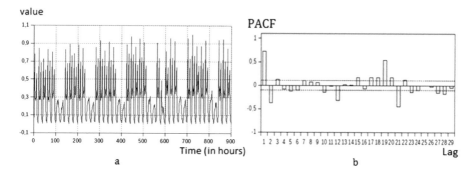

Fig. 5. The time series of passenger traffic of the metro (NNGC1-1.F008) (a), PACF of the passenger traffic of the metro (NNGC1-1.F008) (b)

The PACF of the NNGC1-1.F008 series is shown in Fig. 5b. The PACF chart shows that the series contains repetitive effects. This character of the process does not allow one to unambiguously choose the time lag, therefore, to begin with, we take $n = 12$. In further experiments, the selected lag value will be automatically corrected using reinforcement training.

The third row (NNGC1-1.E009) contains daily data of car traffic through the tunnel in both directions (Fig. 6a). The observations were conducted from 01.11.2003 to 16.11.2005 and include 747 values. Of these, the first 520 observations constitute the training sample, the remaining 227 are left for testing. PACF series NNGC1-1.E009 is shown in Fig. 6b. Here, as in the previous case, the PACF fluctuation is observed, therefore, we take $n = 8$.

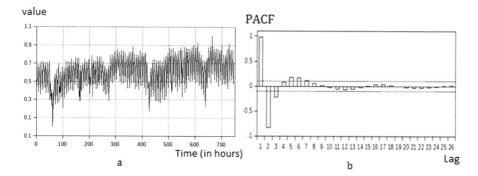

Fig. 6. Tunnel load time series (NNGC1-1.E009) (a), PACF tunnel loading series (NNGC1-1.E009) (b)

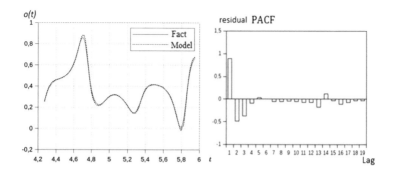

Fig. 7. The graph of one-step prediction by the KPGE algorithm of the series of changes in the variable $o(t)$ in the Lorentz system and the PACF forecast residuals

6   Results and Discussion

To evaluate the performance of the hybrid algorithms developed, experimental studies have been conducted on the prediction of the three time series presented above. To compare the quality of the results obtained, experiments on the prediction of the same series using the classical GMDH, neural networks of the radial basis, wavelet-neural networks, and the PGE method were additionally conducted. The results of the experiments are shown below in the graphs and in the summary tables.

The parameters of the algorithms for each of the three time series are presented in Table 2. For convenience, we denote the time series of the change in the variable $o(t)$ in the Lorentz system as "Row No. 1"; a number of metro passenger traffic (NNGC1-1.F008) as "Row No. 2"; The tunnel loading row (NNGC1-1.E009) as "Row No.3".

Figures 7, 8, 9, 10, 11, 12, 13, 14 and 15 show the results of each of the three proposed hybrid algorithms on a test data sample, as well as the PACF residuals of the obtained prediction. PACF residuals are used here as one of the tests for the prediction of adequacy, allowing you to identify the autocorrelation unaccounted for by the prediction algorithm when building the model.

In Figs. 7, 10 and 13, it can be seen that a number of residues contain strong autocorrelation.

In [20], this behavior is described, which is characteristic of chaotic processes, which include the Lorentz system. According to [20], the remnants of chaotic series also exhibit chaotic properties and can be predicted as a normal time series.

In Figs. 8, 11, and 14, it can be seen that the PACF residuals, when shifted by 20 points, show a small but fairly clear autocorrelation. Moreover, a similar result is shown not only by the developed hybrid algorithms, but also by those prediction algorithms with which comparative experiments are carried out in this research.

Table 2. Parameters of hybrid MSUA-neural algorithms for the task of predicting test time series

Parameter name	Parameter value		
	Range no. 1	Range no. 2	Range no. 3
Algorithm KPGE			
The size of the main population of antibodies	50	100	100
Clone population size	30	500	300
Maximum number of generations	300	300	300
Coefficient of selection of the best antibodies	0,7	0,7	0,7
Antibody replacement rate	0,3	0,3	0,3
The size of the tournament in the selection	20	8	10
Probability of mutation	0,8	0,8	0,8
Probability of IS/RIS transpositions	0,1	0,1	0,1
The size of the time lag	7	12	8
Individual head length	8	8	8
The accuracy of the representation of constants	8 bit	12 bit	12 bit
Constant change interval	$[-2;2]$	$[-2;2]$	$[-2;2]$
Functional alphabet	$\{L,Q\}$	$\{L,Q\}$	$\{L,Q\}$
Clonal genetic hybrid algorithm			
Structural synthesis phase			
The size of the main chromosome population	100	100	100
Maximum number of generations	200	300	300
Selection rate of the best chromosomes	0,9	0,9	0,8
Chromosome replacement rate	0,1	0,2	0,2
The size of the tournament in the selection	10	8	10
Crossover probability	0,8	0,75	0,7
Probability of mutation	0,05	0,01	0,01
Probability of IS/RIS transpositions	0,15	0,15	0,15
The size of the time lag	7	12	8
Individual head length	8	8	8
Functional alphabet	$\{L,Q\}$	$\{L,Q\}$	$\{L,Q\}$
Parametric synthesis phase			
The size of the main population of antibodies	50	100	100
Clone population size	300	300	300
Maximum number of generations	100	300	300
Coefficient of selection of the best antibodies	0,7	0,7	0,7
Antibody replacement rate	0,3	0,3	0,3
The size of the tournament in the selection	20	8	8
Probability of mutation	0,8	0,85	0,8
The accuracy of the representation of constants	12 bit	12 bit	12 bit
Constant change interval	$[-2;2]$	$[-1;1]$	$[-1;1]$
Cooperative immune algorithm			
Number of generations	9000	10000	10000
Probability of mutation	0,5	0,2	0,2
Compression period (generations)	5	5	5
The size of the time lag	7	12	8
The accuracy of the representation of constants	12 bit	16 bit	16 bit
Constant change interval	$[-2;2]$	$[-2;2]$	$[-2;2]$
Functional alphabet	$\{L,Q\}$	$\{L,Q\}$	$\{L,Q\}$

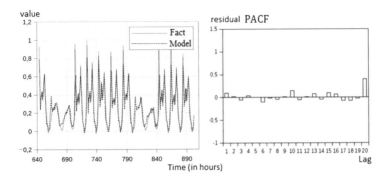

Fig. 8. Graph of one-step forecast by the KPGE algorithm of a number of metro passenger traffic (NNGC1-1.F008) and PACF forecast balances

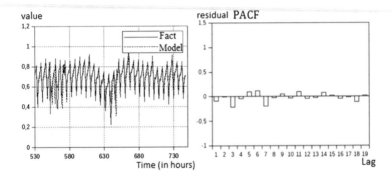

Fig. 9. Graph of one-step forecast by the KPGE algorithm of the tunnel loading series (NNGC1-1.E009) and the PACF forecast residuals

Studies have shown that the reason for this behavior of PACF in this case was the incorrectly chosen initial time lag ($n = 12$). It does not allow the algorithm to form an adequate model of the series, since some important arguments are not taken into account in the short lag.

The following are some numerical estimates of the quality of prediction by the developed hybrid algorithms, as well as their comparison with other methods (Table 3). In addition to the root-mean-square error (RMSE), the following additional criteria were used [21]: mean absolute error (MAE); average relative percentage error (MAPE); maximum absolute error (WAE); Maximum Absolute Percentage Error (WAPE); Theil's coefficient (U).

In order to exclude random deviations of the results, all experiments were conducted 50 times with subsequent averaging of the indicators. Experiments show that, in general, the hybridization technology makes it possible to improve the quality of forecasting according to most criteria. According to the results of the comparison, it is also possible to conclude that for all time series the model

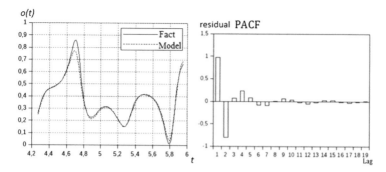

Fig. 10. Graph of one-step forecast by clonal-genetic hybrid algorithm for a series of changes of variable $o(t)$ in the Lorenz system and PACF forecast residuals

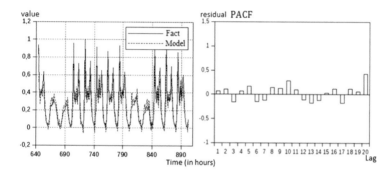

Fig. 11. Graph of one-step forecast by the clonal-genetic hybrid algorithm of a number of passenger traffic of the metro (NNGC1-1.F008) and PACF forecast residues

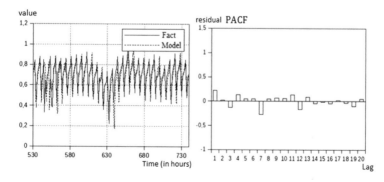

Fig. 12. Graph of one-step forecast by clonal-genetic hybrid tunnel loading series algorithm (NNGC1-1.E009) and PACF forecast residuals

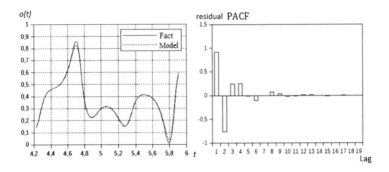

Fig. 13. The one-step forecast graph by the cooperative immune algorithm of the series of changes in the variable $o(t)$ in the Lorentz system and the PACF forecast

Table 3. Comparative assessment of the quality of predicting obtained by various computational methods

Computational method	Indicators					
	RMSE	MAE	MAPE, %	WAE	WAPE, %	U
Row no. 1						
Algorithm KPGE	0,013	0,008	5,72	0,05	260	0,016
Clonal genetic algorithm	0,025	0,015	8,00	0,10	268	0,031
Cooperative immune algorithm	0,012	0,008	6,70	0,04	235	0,015
Algorithm GMDH	0,032	0,016	13,5	0,14	524	0,040
RBF neural network	0,040	0,023	15,0	0,15	324	0,050
Wavelet neural network	0,026	0,016	8,68	0,09	286	0,032
PGE algorithm	0,035	0,025	14,4	0,13	339	0,043
Row no. 2						
Algorithm KPGE	0,086	0,057	51,9	0,47	1193	0,12
Clonal genetic algorithm	0,084	0,061	53,00	0,33	2232	0,12
Cooperative immune algorithm	0,062	0,047	63,70	0,18	3411	0,09
Algorithm GMDH	0,134	0,080	70,4	0,65	4211	0,19
RBF neural network	0,170	0,110	64,6	0,65	1351	0,25
Wavelet neural network	0,133	0,083	59,4	0,77	1193	0,19
PGE algorithm	0,116	0,076	69,4	0,58	2177	0,17
Row no. 3						
Algorithm KPGE	0,078	0,047	8,19	0,52	156	0,056
Clonal genetic algorithm	0,087	0,052	8,63	0,49	145	0,062
Cooperative immune algorithm	0,072	0,043	7,81	0,52	162	0,051
Algorithm GMDH	0,100	0,068	11,4	0,36	149	0,071
RBF neural network	0,160	0,130	21,6	0,52	129	0,110
Wavelet neural network	0,107	0,073	12,3	0,53	162	0,076
PGE algorithm	0,116	0,080	13,7	0,41	124	0,080

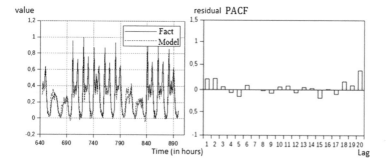

Fig. 14. Graph of one-step forecast by the cooperative immune algorithm of a number of passenger metro traffic (NNGC1-1.F008) and PACF forecast residues

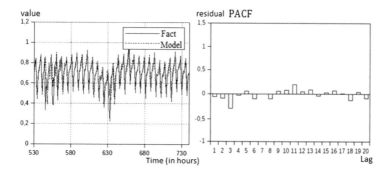

Fig. 15. Graph of one-step forecast by a cooperative immune algorithm of a series of tunnel loading (NNGC1-1.E009) and PACF forecast residues

obtained using the cooperative immune algorithm is the most qualitative. From Table 3 it can be seen that the forecast of the number of passenger traffic of the metro (NNGC1-1.F008), designated as number 2, turned out to be the worst of the three, which is confirmed by all computational methods involved in the experiments.

7   Conclusions

The analysis of the problem of synthesizing the structure and tuning the parameter setting of GMDH neural networks has been carried out. The main disadvantages of the classic GMDH in the context of solving the problem of forecasting time series are revealed. To eliminate these drawbacks, the proposed technology hybridization GMDH neural network with evolutionary algorithms and immune systems. Based on the hybridization technology, new methods of adaptive structural-parametric synthesis of computational models for solving forecasting problems have been developed.

A hybrid computational method for the synthesis and training of GMDH neural networks has been developed, in which candidate models are represented as gene expression, and training is performed by clonal selection. This hybrid is called clonal programming of gene expression (CPGE). A hybrid computational method for the two-phase structural-parametric synthesis of GMDH neural networks has been developed, in which the structural component is formed using programming of gene expression, and the parameterization is performed by clonal selection. The hybrid is called clonal-genetic hybrid algorithm.

A hybrid computational method, based on cooperative-competitive processes of the interaction of the elements of the immune system has been developed, in which the structure and parameters of the GMDH neural network are represented by the entire population element-wise. Comparative experimental studies of the performance of the proposed hybrid computational methods for solving forecasting problems have been carried out.

References

1. Ivakhnenko AG (1971) Heuristic self-organising systems in cybernetics. Technique, Kiev, 392 p (in Russian)
2. Ivakhnenko AG, Ivakhnenko GA, Muller JA (1994) Selforganisation of neuronets with active neurons. Pattern Recogn Image Anal 4:177–188 (in Russian)
3. Anastasakis L, Mort N (2001) The development of self-organization technique in modelling: a review of the group method of data handling (GMDH). Research report no 813, Department of Automatic Control and Systems Engineering, The University of Sheffield, Sheffield
4. Taušer J, Buryan P (2011) Exchange rate predictions in international financial management by enhanced GMDH algorithm. Prague Econ Pap Univ Econ Prague 2011(3):232–249
5. Park HS, Oh SK, Ahn TC, Pedrycz WC (1999) A study on multi-layer fuzzy polynomial inference system based on extended GMDH algorithm. In: Proceedings of the 1999 IEEE international conference on fuzzy systems, FUZZ-IEEE 1999, vol 1, pp 354–359
6. Ivakhnenko AG, Zholnarskiy AA (1992) Estimating the coefficients of polynomials in parametric GMDH algorithms by the improved instrumental variables method. J Autom Inf Sci c/c Avtomatika 25(3):25–32 (in Russian)
7. Sarychev AP (1984) Stable estimation of the coefficients in multilayer GMDH algorithms. Sov Autom Control c/c Avtomatika 17(5):1–5 (in Russian)
8. Parker RG, Tummala MJ (1992) Identification of volterra systems with a polynomial neural network. In: Proceedings of the 1992 IEEE international conference on acoustics – speech and signal processing, ICASSP 1992, vol 4, pp 561–564
9. Dolenko SA, Orlov YV, Persiantsev IG (1996) Practical implementation and use of group method of data handling (GMDH): prospects and problems. In: Proceedings of the 2nd international conference on adaptive computing in engineering design and control - ACEDC 1996. PEDC, University of Plymouth, UK, pp 291–293
10. Onwubolu GC (2009) Hybrid self-organizing modeling systems. Springer, Berlin, pp 233–280
11. Moroz OV, Stepashko VS (2015) An overview of hybrid structures of GMDH-like neural networks and genetic algorithms. Inductive modeling of complex systems. K.: MNNTS IT•S NAN ta MON Ukrayiny, Vyp 7, pp 173–191 (in Ukrainian)

12. Sharma A, Onwubolu G (2009) Hybrid particle swarm optimization and GMDH system. In: Onwubolu GC (ed) Hybrid self-organizing modeling systems, vol 211. Studies in computational intelligence. Springer, Heidelberg, pp 193–231
13. Onwubolu GC (2007) Design of hybrid differential evolution and group method in data handling for modeling. In: International workshop on inductive modeling, IWIM 2007, Prague, Czech, 23–26 September, pp 87–95
14. Ferreira C (2001) Gene expression programming: a new adaptive algorithm for solving problems. Complex Syst 13(2):87–129
15. Dasgupta D (1999) Artificial immune systems and their applications. Springer, Heidelberg, 306 p
16. De Castro LN, Timmis JC (2002) Artificial immune systems: a new computational intelligence approach. Springer, Heidelberg, 357 p
17. De Castro LN, Von Zuben FJ (2001) The clonal selection algorithm with engineering applications. In: Proceedings of GECCO 2000, pp 36–37
18. Fefelov AO, Lytvynenko VI, Bidyuk PI (2006) Cooperative algorithm for solving the problem of signal approximation, processing of signals, images and recognition of images: VIII Allukrainian mizhnar conference, 11–15 October 2006, pp 41–44 (in Ukrainian)
19. Artificial neural network and computational intelligence forecasting competition. http://www.neural-forecasting-competition.com/
20. Ardalani-Farsa M, Zolfaghari S (2010) Chaotic time series prediction with residual analysis method using hybrid Elman-NARX neural networks. Neurocomputing 2010(73):2540–2553
21. Shcherbakov MV, Brebels AC, Shcherbakova NL, Tyukov AP, Janovsky TA, Kamaev VA (2013) A survey of forecast error measures. World Appl Sci J 24:171–176

An Evaluation of the Objective Clustering Inductive Technology Effectiveness Implemented Using Density-Based and Agglomerative Hierarchical Clustering Algorithms

Sergii Babichev$^{1,2(\boxtimes)}$ (ID), Bohdan Durnyak^2(ID), Iryna Pikh^2(ID),
and Vsevolod Senkivskyy^2(ID)

^{1} Jan Evangelista Purkyne University in Usti nad Labem,
Usti nad Labem, Czech Republic
sergii.babichev@ujep.cz
^{2} Ukrainian Academy of Printing, Lviv, Ukraine
durnyak@uad.lviv.ua, pikhirena@gmail.com, senk.vm@gmail.com

Abstract. The paper presents the results of the research concerning comparison analysis of the efectiveness of OPTICS and DBSCAN density-based and agglonarative hierarchical clustering algorithms within the framework of the objective clustering inductive technology. Implementation of this technology allows us to determine the optimal parameters of appropriate clustering algorithm in terms of the maximum values of the complex balance criterion which contains as the components both the internal and the external clustering quality criteria. The data from the Computing School of East Finland University database were used as the experimental one during the simulation process. The results of the simulation have shown high effectiveness of the proposed technique. The investigated objects were divided into clusters correctly in all cases. Moreover, the results of the simulation have shown also higher effectiveness of the density-based clustering algorithms in comparison with agglomerative hierarchical algorithm use due to more level of the detail during the objects clustering.

Keywords: OPTICS and DBSCAN clustering algorithm ·
Agglomerative hierarchical clustering · Internal ·
External and balance clustering quality criteria · Objective clustering ·
Inductive technology

1 Introduction

One of the current directions of modern Data Science is data clustering [1–4]. Implementation of this technique allows us to divide the objects or features into groups taking into account the level of their mutual similarity. There are a lot

V. Lytvynenko et al. (Eds.): ISDMCI 2019, AISC 1020, pp. 532–553, 2020.
https://doi.org/10.1007/978-3-030-26474-1_37

of clustering algorithms nowadays. Choice of the appropriate algorithm is determined by type and particularities of the investigated data. So, in the papers [1, 2, 4] the authors presented the results of the research concerning application of both k-means and fuzzy c-means clustering algorithms to cluster analysis of complex data. The paper [5] is devoted to implementation of the DBSCAN clustering algorithm to detect communities in social networks. The tasks concerning the implementation of both the self-organizing SOTA and hierarchical clustering algorithms are solved in [3, 6, 7]. The results of the research concerning practical implementation of self-organizing neural networks (Kohonen Map) are presented in [8, 9]. However, it should be noted that result of appropriate clustering algorithm operation in the most cases depends on its initial parameters. Setup of these parameters is not easy task and this step is usually implemented empirically during the simulation process taking into account the aim of the solved task [10, 11].

Determination of the clustering quality is another task which has not unambiguous solution nowadays. There are a lot of internal clustering quality criteria which allows us to estimate the character of both the objects distribution within the clusters and the clusters distribution in the features space. These criteria are implemented as the functions in the package clusterCrit [12] of R software [13]. However, the results of the simulation, which were carried out by the authors in the [14] have shown inconsistence of these criteria in the case of the use of similar datasets. As the simulation results the authors in this paper have proposed the complex internal clustering quality criterion which was calculated as the multiplicative combination of the Calinski-Harabasz [15] and WB-index [16].

However, the internal criteria do not always allow us to divide the objects into clusters objectively. One of the current problems of the existing clustering algorithms is the reproducibility error, in other words, successful clustering results obtained on one dataset do not repeat while using another similar dataset. Reduction of this error can be achieved by careful verification of the obtained model using "fresh information", which was not used during the model making. A higher degree of coincidence between the clustering results on the similar data corresponds to a higher degree of the obtained model objectivity. This idea is the basis of the objective clustering inductive technology, the main conception of which was presented in [17, 18] and further developed in [19–21]. Implementation of this technology involves determination of the optimal clustering based on the extremum value of the complex balance criterion which contains as the components both the internal and external clustering quality criteria. The practical implementation of the objective clustering inductive technology based on the k-means, agglomerative hierarchical, self-organizing SOTA and DBSCAN clustering algorithms were presented in [22–24]. In [25–27] the authors present the result of the research concerning implementation of this technology within the framework of the information technology of gene expression profiles processing for purpose of gene regulatory networks reconstruction.

In this work we continue the research concerning implementation of the objective clustering inductive technology based on existing clustering algorithms.

The aim of the paper is carrying out the investigations concerning the practical implementation of OPTICS density-based clustering algorithm within the framework of the objective clustering inductive technology and comparison analysis of OPTICS, DBSCAN and agglomerative hierarchical clustering algorithms which are implemented using the objective clustering inductive technology.

2   Materials and Methods

2.1   Objective Clustering Inductive Technology

Three fundamental principles are the basis of the objective clustering inductive technology [17,21]:

- The principle of sequential enumeration, i.e., sequential enumeration of clustering within the admissible range for purpose of selection the best clustering in terms of the used criteria.
- The principle of external edition, i.e., a necessity of the use of two "equal power" subsets which contains the same quantity of pairwise similar objects.
- The principle of inconclusive of solution, i.e., generation a set of intermediate clustering in order to select from them the optimal variant in terms of the goal of the current task.

The internal clustering quality criterion was calculated as multiplicative combination of Calinski-Harabasz criterion [15] and WB-index [16] as follows:

$$QC_{int} = \frac{K(K-1)QCW^2}{(N-K)QCB^2} \tag{1}$$

where K is the number of clusters, N is the number of the investigated objects. The components QCW and QCB of the criterion (1) are calculated as an average distance from the objects to the mass centers of the clusters, where these objects are:

$$QCW = \frac{1}{N}\sum_{s=1}^{K}\sum_{i=1}^{N_s} d(x_i^s, C_s) \tag{2}$$

and as an average distance between mass centers of the clusters in current clustering:

$$QCB = \frac{2}{K(K-1N)}\sum_{i=1}^{K-1}\sum_{j=i+1}^{K} d(C_i, C_j) \tag{3}$$

where x_i^s is the i-th object in the cluster S; N_s is the number of the objects in the cluster S; C_i, C_j and C s are the mass centers of the clusters i, j and S respectively; $d(\cdot)$ is the metric used to estimate the proximity level between the investigated objects.

The external clustering quality criterion within the framework of the proposed technology is determined as the normalized difference of the internal criteria which are calculated using the equal power subsets A and B:

$$QC_{ext}(A, B) = \frac{|QC_{int}(A) - QC_{int}(B)|}{QC_{int}(A) + QC_{int}(B)} \tag{4}$$

The complex balance criterion is calculated based on the Harrington desirability function [28], the plot of which is shown in Fig. 1.

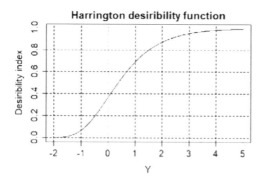

Fig. 1. Harrington desirability function

Calculation of this criterion assumes the following steps:

1. Transformation of scales of the internal and the external clustering quality criteria into reaction scale Y in the following way:

$$Y = a - b \cdot QC$$

where the parameters a and b are determined empirically taking into account the boundary values of the appropriate clustering quality criterion:

$$\begin{cases} Y_{max} = a - b \cdot QC_{min} \\ Y_{min} = a - b \cdot QC_{max} \end{cases}$$

2. Calculation of the Y_i nondimensional parameter for each of the used criteria:

$$Y_i = a - b \cdot QC_i$$

3. Calculation of the private desirabilities for each of the criteria:

$$d_i = exp(-ext(-Y_i))$$

4. Calculation of the complex balance clustering quality criterion as the geometric average of all private desirabilities:

$$QC_{bal} = \sqrt[r]{\prod_{i=1}^{r} d_i} \qquad (5)$$

where r is the number of both the internal and external clustering quality criteria.

Implementation of the objective clustering inductive technology involves the following steps [21]:

1. Preparing, analyzing and preprocessing the investigated data. The data is formed as a matrix, where number of rows is a number of the objects and number of columns is a number of the features which characterized the objects. The preprocessing stage involves the following: missing values processing, normalization, filtering, et al.
2. Choice of the similarity metric taking into account both the type and particularities of the investigated vectors (Euclidean, Manhattan, correlation et al.).
3. Division of the initial dataset into two equal power subsets (contains the same quantity of the pairwise similar objects) using the algorithm, presented in [21].
4. Choice of the clustering algorithm. Setup of its initial parameters and range of these parameters change.
5. Implementation of the clustering algorithm on the equal power subsets within a given range of the algorithm parameters change. Fixation of the clustering at each step of this procedure implementation. Calculation of both the internal and the external clustering quality criteria by the formulas (1) and (4) in the cases if the quantity of the clusters in the different clustering are the same ones.
6. Calculation of the complex balance clustering quality criterion by the formula (5).
7. Results analysis. Fixation of intermediate solutions which correspond to the maxima values of the complex balance criterion.
8. Determination of the algorithm parameters which correspond to the optimal clustering in terms of the aim of the current task.
9. Data clustering with the use of the current clustering algorithm using determined before parameters. Final results formation.

2.2   DBSCAN Density-Based Clustering Algorithm

DBSCAN (Density-Based Spatial Clustering of Applications with Noise) clustering algorithm was proposed in 1996 as a solution of the problem to divide the data into clusters of arbitrary shapes [30,31]. The following definitions are the basis of this algorithm operation [30]:

Definition 1. The *Eps-neighborhood* of a point p is defined as follows:

$$Eps(p) = \{q \in D | d(p,q) \leq EPS\}$$

Definition 2. A point q is directly density-reachable from a point p if the following conditions are performed:

$$\begin{cases} q \in Eps(p) \\ N_{EPS}(p) \geq MinPts \end{cases}$$

where $N_{EPS}(p)$ and $MinPts$ are the number of points and the minimum number of points within Eps-neighborhood of a point p respectively.

Definition 3. A point q is density-reachable from a point p if there is a chain of points $q_1, ..., q_n$, $q_1 = p$, $q_n = q$ such that q_{i+1} is directly density-reachable from q_i.

Definition 4. A point q is density-connected with a point p if there is a point k such that both the points q and p are density-reachable from the point k.

Definition 5. A cluster C is a non-empty subset of a set of points D if the following conditions are performed:

1. $\forall p, q : if\ p \in C$ and q is density-reachable from p, then $q \in C$;
2. $\forall p, q :$ if q is density-connected with p, then $p, q \in C$

Definition 6. Let C_i, $i = \overline{1, k}$ is a set of the allocated clusters. The noise is the set of points of the database D, which not belonging to any cluster C_i:

$$noise = \{p \in D | \forall i : p \notin C_i,\ i = 1, k\}$$

Result of DBSCAN clustering algorithm operation depends on the two parameters: EPS and $MinPts$. To determine the optimal EPS value for appropriate $MinPts$ the technique based on sorted k-dist graph was proposed in [30]. However, it should be noted, that implementation of this technology does not allow us to determine the EPS value exactly. This fact influences the quality of the algorithm operation. The implementation of the proposed technique allows us to determine only the range of the EPS values change for appropriate $MinPts$ value. To solve this problem, we propose the technique of the DBSCAN clustering algorithm optimal parameters determination based on the objective clustering inductive technology. The algorithm of this technique implementation assumes the following steps:

1. Formation of the initial data as a matrix, where number of rows is the number of the studied objects and number of columns is a number of the features, which characterized the objects.
2. Determination of the affinity functions in dependence on type of the studied data. Division of the initial data into two equal power subsets.
3. Formation of the internal, external and complex balance clustering quality criteria.
4. Setup of DBSCAN clustering algorithm. Determination of the range of the $MinPts$ value change. Creation of the sorted k-dist graph within this range. Determination of both the range and step of the EPS value change.
5. Setup of the initial value of the $MinPts$ algorithms parameter ($k = min(MinPts)$).
6. Setup of the initial value of the EPS algorithms parameter ($e = min(EPS)$).
7. Data clustering on the two equal power subsets concurrently. Clusters formation.

8. Calculation of both the internal and external clustering quality criteria by formulas (1)–(4).
9. If the condition $e \leq max(EPS)$ is true increasing the EPS value $(e = e+de)$ and repetition of the steps 7 and 8 of this procedure. Otherwise, calculation of the complex balance criterion by the formula (5).
10. If the $MinPts$ value is less than maximum $(k \leq max(MinPts))$ increasing the $MinPts$ value $(k = k+1)$ and transition to the step 6 of this algorithm. Otherwise, creation of the charts of the complex balance criterion versus the EPS for each of the $MinPts$ values.
11. Analysis of the obtained results. Fixation of the optimal clustering.

2.3   OPTICS Density-Based Clustering Algorithm

The OPTICS (Ordering Points To Identify the Clustering Structure) clustering algorithm was proposed in [29] as the logical continue of the DBSCAN algorithm. Similarly to the DBSCAN algorithm, the OPTICS clustering algorithm also needs two main parameters: Eps-neighborhood (EPS) and $MinPts$. The difference of the OPTICS algorithm from the DBSCAN one is the following: its implementation does not need the assign of cluster memberships [29]. Instead, the algorithm implementation assumes the store of both the order in which the objects are processed and the information which would be used to assign cluster memberships. This information consists of only two values for each object: a core-distance and a reachability-distance, which are introduced in the following definitions [29]:

Definition 1: (core-distance of the p object)
    Let p is one of the objects from database D, ε is a distance value, $N_\varepsilon(p)$ is a number of objects within ε-neighborhood of the p object, $MinPts$ is a minimum number of the objects within the ε-neighborhood of the p object. Then, the core-distance of p is defined as follows:

$$core\_dist_{\varepsilon,MinPts}(p) = \begin{cases} undefined, \ if \ N_\varepsilon(p) < MinPts \\ MinPts\_dist(p), \ otherwise \end{cases}$$

where $MinPts\_dist(p)$ is the distance from p to its $MinPts$ neighbor.

Definition 2: (reachability-distance of the p and o objects)
    Let p and o are the objects from database D. Then, the reachability distance of p with respect to o is determined in the following way:

$$reachability\_dist_{\varepsilon,MinPts}(p) = \begin{cases} undefined, \ if \ N_\varepsilon(o) < MinPts \\ max(core\_dist(p), \ dist(o,p)), \ otherwise \end{cases}$$

    It is obvious, the reachability-distance of an object p with respect to another object o is the smallest distance such that p is directly density-reachable from o. In this case o is a core object and the reachability-distance cannot be smaller than the core-distance of o because for smaller distances no objects which are directly density-reachable from o. Otherwise, if o is not a core object even at the generating distance ε, the reachability-distance of p with respect to o is undefined.

Thus, the augmented cluster-ordering which consists of the ordering of the points taking into account both the reachability-values and the core-values is the result of the OPTICS algorithm operation. The detail description of the OPTICS clustering algorithm is presented in [29]. The results of the algorithm operation can be shown as the reachability plot, which presents itself distribution of the reachability-distance versus the cluster-order of the investigated objects. Illustration of the reachability distance plot in the case of the use if three-cluster 2-d synthetic data is presented in Fig. 2 [29].

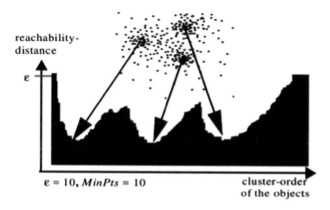

Fig. 2. Illustration of the reachability distance chart versus the cluster-order of the objects in three-cluster structure

As it can be seen, the reachability distance chart presents itself the hierarchical structure of the clusters. Since the points belonging to the single cluster have a small reachable distance to the nearest neighbour, the clusters look like valleys on the reachability graph. The deeper valley corresponds to the higher density of the objects distribution within the cluster. The analysis of the Fig. 2 allows us to conclude that as in the case of the DBSCAN clustering algorithm use, the result of the OPTICS clustering algorithm operation also depends on two parameters: ε-neighborhood radius and $MinPts$ value. However, as opposed to the DBSCAN algorithm in the case of the use of OPTICS algorithm the ε-neighborhood value can be taken as the maximum distance between investigated objects. This value does not influence to the reachability distance plot significantly. So, the use of the reachability distance plot allows us to estimate the range of the ε-neighborhood value variation in the case of the use of different $MinPts$ values. Then, we can change the ε-neighborhood value within this range with little step in order to form the cluster structure for each of the $MinPts$ values using appropriate clustering quality criteria. Within the framework of this work this procedure is implemented based on the objective clustering inductive technology (OCIT). Algorithm of this procedure implementation assumes the following steps:

1. Formation of the initial dataset as a matrix, where rows are the investigated objects and columns are the features which these objects characterize.
2. Determination of the affinity (similarity) function in dependence on type of the investigated data. Division of the initial dataset into two equal power subsets (contains the same quantity of pairwise similar objects). Formation of the internal, the external and the complex balance clustering quality criteria.
3. Setup of the OPTICS clustering algorithm. Calculation of the dissimilarity matrix between the investigated objects. Setup of the ε-neighborhood value (Eps) as the maximum value of the dissimilarity matrix. Setup the range of the $MinPts$ value variation. Creation of the reachability distance charts for the $MinPts$ boundary values. Fixation of both the range and step of the Eps value variation.
4. Data clustering within the range of the Eps value variation for each of the $MinPts$ values. Calculation of both the internal and external clustering quality criteria at each step of this procedure implementation.
5. Calculation of the complex balance criterion for each of the $MinPts$ values.
6. Results analysis. Fixation of the intermediate solutions (new range of the Eps value variation for each of the $MinPts$ values) for the following analysis. These solutions correspond to the maxima values of the complex balance criterion.
7. Fixation of the final solution by comparison analysis of the intermediate solutions taking into account the goal of the current task.

2.4   Agglomerative Hierarchical Clustering Algorithm

Implementation of the agglomerative hierarchical clustering algorithm involves step-by-step grouping of the mutually nearest objects. As a result of this technique application, we have the dendrogram. The cutting oh this dendrogram allows us to form the cluster structure. There are different methods to implement the agglomerative hierarchical algorithm. We used within the framework of our research the tools of the "cluster" package of R software [31]. We test the following methods of the cluster structure formation: unweighted pair-group (average); single linkage (single); complete linkage (complete); ward's method (ward); weighted average linkage (weighted) and flexible method which uses a constant version of the Lance-Williams formula [33]. The optimal clustering was determined in the case of each of the used methods in terms of the complex balance criterion maximum value within the framework of the objective clustering inductive technology. The procedure of this process implementation assumes the following steps:

1. Formation of the initial dataset and data pre-processing (if it is necessary).
2. Determination of the affinity (similarity) function in dependence on type of the investigated data. Division of the initial dataset into two equal power subsets. Formation of the internal, the external and the complex balance clustering quality criteria.

3. Formation of vector of the methods of agglomerative hierarchical algorithm implementation.
4. Data clustering within the range of the used methods change. Calculation of both the internal and external clustering quality criteria at each step of this procedure implementation.
5. Calculation of the complex balance criterion for each of the used clustering method.
6. Results analysis. Fixation of the intermediate solutions for the following analysis.
7. Fixation of the final solution by comparison analysis of the intermediate solutions taking into account the goal of the current task.

3 Experiments and Results

The simulation process was performed based on R software [13] using tools of both dbscan and cluster packages [31,32]. The data Aggregation [34], Compound [35], Multishapes [36] and Jain [37] of the school of computing of the Eastern Finland University were used during the simulation process. These data are presented in the two-dimensional space and they include the clusters of different shapes. The character of the objects distribution in the appropriate datasets are presented in Fig. 3.

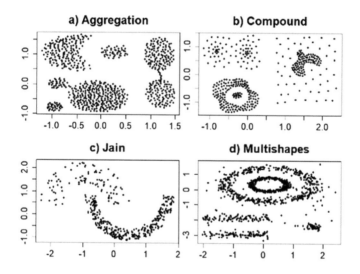

Fig. 3. Two-dimensional synthetic datasets

3.1 The Results of the DBSCAN Clustering Algorithm Operation

Figure 4 presents the sorted k-dist graphs for Aggregation data in the cases of MinPts = 3 and MinPts = 8. The simulation results have shown that increase of

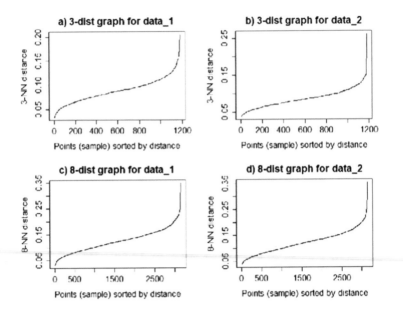

Fig. 4. Sorted k-dist graph for "Aggregation" data

the MinPts value does not reasonable. The similar graphs were obtained in the case of other data use.

The analysis of the obtained results allows us to determine both the ranges and steps of the EPS value changes for each type of the investigated data. These parameters are presented in the Table 1.

Table 1. The range and step of the EPS value change

EPS value	Data			
	Aggregation	Compound	Multishapes	Jain
EPS_{min}	0.1	0.1	0.1	0.15
EPS_{max}	0.25	0.6	0.4	0.5
Step	0.005	0.01	0.005	0.005

Charts of the complex balance criterion versus the EPS for different MinPts value in the case of the "Aggregation" data use are presented in Fig. 5. The similar charts were obtained for the other investigated data. The analysis of the obtained charts allows us to select the subset of the intermediate solutions (new less ranges and steps of the EPS value change), which correspond to the maximum values of the complex balance criterion.

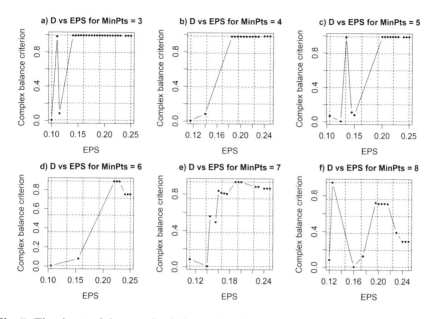

Fig. 5. The charts of the complex balance criterion versus the EPS for different MinPts values in the case of "Aggregation" data use

Then, the detail analysis of the selected solutions is performed in order to determine the optimal clustering in terms of the goal of the current task. The optimal parameters of DBSCAN clustering algorithm operation, which were determined within the framework of the proposed technology for the investigated data are presented in Table 2. Figure 6 present the results of the DBSCAN clustering algorithm operation.

Table 2. The optimal parameters of DBSCAN clustering algorithm operation

Parameters	Data			
	Aggregation	Compound	Multishapes	Jain
EPS	0.136	0.157	0.237	0.305
MinPts	5	5	4	3

3.2   The Results of the OPTICS Clustering Algorithm Operation

The reachability distance charts for equal power subset 1 of the Aggregation data in the case of OPTICS density based clustering algorithm using are presented in Fig. 7. The MinPts value was changed from 3 to 6 during the simulation process.

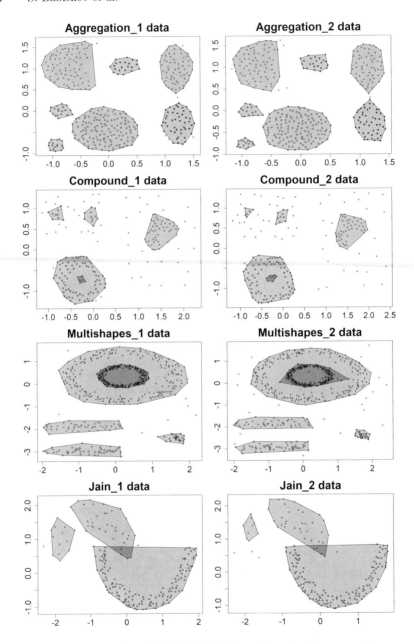

Fig. 6. The results of the DBSCAN clustering algorithm operation

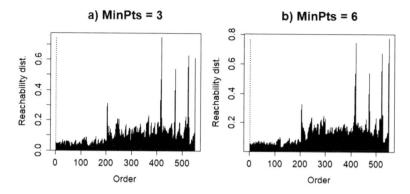

Fig. 7. Reachability distance charts for Aggregation data

The same charts were created for another of the investigated data. The range of the eps value variation was determined as the result of the obtained charts analysis: $eps_{min} = 0.05$; $eps_{max} = 0.15$. The step of the *eps* value change was determined in the following way:

$$d_{eps} = \frac{eps_{max} - eps_{min}}{n}$$

where n is the number of the investigated objects. Figure 8 presents the charts of the complex balance criterion versus the *eps* values. This criterion was calculated as the result of hereinbefore described hybrid model implementation.

The analysis of the obtained charts allows us to select the intermediate solutions (*eps* values for each of the $MinPts$ values) which correspond to the maxima of the complex balance criterion. The final step of this procedure implementation is the analysis of the selected solutions in order to select from them the best variants taking into account the aim of the current task. Table 3 presents the final parameters of the OPTICS clustering algorithm which were determined during the simulation process using investigated datasets. The results of the algorithm operation are presented in Fig. 9.

Table 3. The optimal parameters of OPTICS clustering algorithm operation

Parameters	Data			
	Aggregation	Compound	Multishapes	Jain
EPS	0.134	0.17	0.264	0.305
MinPts	5	4	4	3

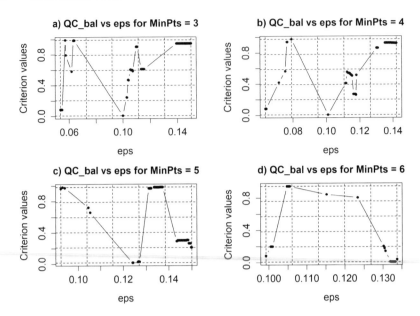

Fig. 8. Charts of the complex balance criterion vs the *eps* for different *MinPts* values in the case of the Aggregation data use

3.3   The Results of the Agglomerative Hierarchical Clustering Algorithm Operation

Figure 10 shows the charts of the complex balance criterion versus the method of the objects grouping in the case of agglomerative hierarchical clustering algorithm use for Aggregation data. The same charts were obtained for other datasets. The level of the obtained dendrogram cutting was changed from 2 to 10. In other words, we formed the number of clusters from 2 to 10 for each of the applied methods. Then, we calculated the complex balance criterion at each level of the cluster structure formation for each of the used methods. The intermediate solutions are the result of this stage implementation. Finally, we have determine the optimal method and the clusters quantity for each of the used data (Table 4). The results of the agglomerative hierarchical algorithm operation within the framework of the objective clustering inductive technology are presented in Fig. 11.

4   Discussion

The analysis of the obtained results allows us to conclude about high effectiveness of the proposed technique, since the data were divided into clusters correctly in all cases. However, the number of clusters are differ in the cases

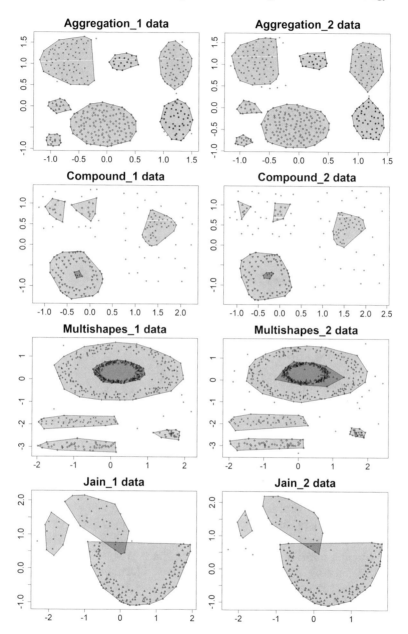

Fig. 9. The results of the OPTICS clustering algorithm operation

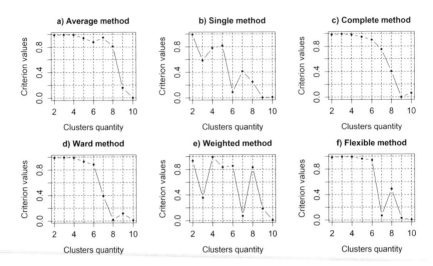

Fig. 10. Charts of the complex balance criterion vs the method of the similar objects grouping for Aggregation data

Table 4. The used methods and the clusters quantity in the case of agglomerative hierarchical clustering algorithm use

Parameters	Data			
	Aggregation	Compound	Multishapes	Jain
Method	Average	Complete	Average	Ward
Number of clusters	7	3	3	3

of density based and agglomerative hierarchical clustering algorithms use. The DBSCAN and OPTICS clustering algorithms allows us to obtain more higher detail level during the objects grouping. Some of the objects were identified as the noise. This fact can be explained in the following way. The density of the noise objects distribution in the feature space is significantly less to compare with density of other objects distribution. Thus, the algorithms identify these objects as the noise. In the case of *Aggregation* data use we have as the result seven clusters in all cases. In the case of *Compound* data processing the OPTICS clustering algorithm has shown more effectiveness since the number of objects which are identified as noise is less in comparison with result obtained by DBSCAN clustering algorithm use. The number of the clusters was 4 in the cases of density based algorithms use and 3 in the case of agglomerative hierarchical algorithm application. However, the objects are distributed into clusters correctly too. The same results are observed in the cases of *Multishapes* and *Jain* data use. *Multishapes* data contains clusters different shapes and sizes. As it can be seen, the studied objects are distributed into clusters correctly in

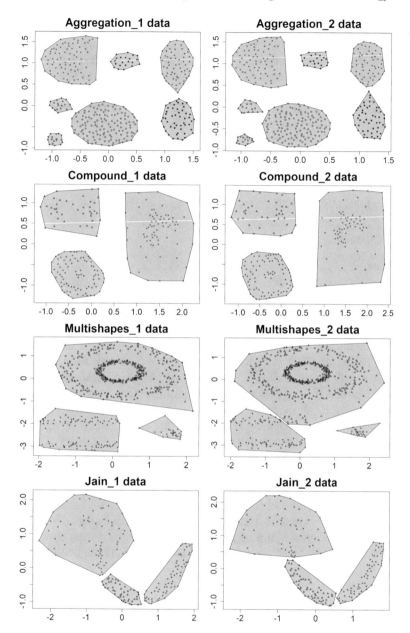

Fig. 11. The results of the agglomerative hierarchical clustering algorithm operation

the cases of OPTICS and DBSCAN clustering algorithms use. The apply of the agglomerative hierarchical algorithm has not allowed us to obtain the necessary detail level during the object grouping. But, we have obtained the same results using two equal power subsets. This fact indicates decreases of the reproducibility error. Five clusters were allocated in this case. The objects of *Jain* data were distributed into three clusters. However, the character of the object grouping is differ in the cases of density based and agglomerative hierarchical clustering algorithms use. The OPTICS and DBSCAN clustering algorithms are more effective in this case. It should be noted also that the OPTICS algorithm is more preferable in comparison with DBSCAN one, since it has less sensitivity level to the initial parameters and setuping process of this algorithm is easer in comparison with implementation of this process in the case of DBSCAN algorithm use.

5   Conclusions

Comparison analysis of effectiveness of OPTICS and DBSCAN density-based and agglomerative hierarchical clustering algorithms within the framework of the objective clustering inductive technology has been performed during the simulation process. The data of the school of computing of the Eastern Finland University have been used as the experimental ones. The optimal parameters of the clustering algorithm have been determined for each of the investigated data. The value of these parameters corresponded to the maximum value of the complex balance criterion. The results of the simulation have shown the high effectiveness of the proposed technique, since the data were divided into clusters correctly in all cases. However, the simulation results have shown also higher effectiveness of the density based clustering algorithms in comparison with agglomerative hierarchical one due the higher detail level of the object grouping. The density based clustering algorithms allow us also to allocate the objects which are identified as noise in terms of density of their distribution. It is not possible in the case of agglomerative clustering algorithm use.

The results of the simulation have shown also that OPTICS algorithm is more preferable in comparison with DBSCAN one, since it has less sensitivity level to the initial parameters and setuping process of this algorithm is easer in comparison with implementation of this process in the case of DBSCAN algorithm use.

References

1. Li C, Liu L, Sun X, Zhao J, Yin J (2019) Image segmentation based on fuzzy clustering with cellular automata and features weighting. EURASIP J Image Video Process 2019(1), article no 37. https://doi.org/10.1186/s13640-019-0436-5
2. Bi Y, Wang P, Guo X, Wang Z, Cheng S (2019) K-means clustering optimizing deep stacked sparse autoencoder. Sens Imaging 20(1), article no 6. https://doi.org/10.1007/s11220-019-0227-1
3. Chang Y-S, Yoon SH, Kim JR, Baek S-Y, Cho YS, Hong SH, Kim S, Moon IJ (2019) Standard audiograms for koreans derived through hierarchical clustering using data from the Korean national health and nutrition examination survey 2009–2012. Sci Rep 9(1), article no 3675. https://doi.org/10.1038/s41598-019-40300-7

4. Wan R, Xiong N, Hu Q, Wang H, Shang J (2019) Similarity-aware data aggregation using fuzzy c-means approach for wireless sensor networks. EURASIP J Wirel Commun Netw 2019(1), article no 59. https://doi.org/10.1186/s13638-019-1374-8
5. Khatoon M, Banu WA (2019) An efficient method to detect communities in social networks using DBSCAN algorithm. Soc Netw Anal Min 9(1), article no 9. https://doi.org/10.1007/s13278-019-0554-1
6. Gómez SLS, Rodríguez JDS, Rodríguez FJI, Juez FJC (2017) Analysis of the temporal structure evolution of physical systems with the self-organising tree algorithm (SOTA): application for validating neural network systems on adaptive optics data before on-sky implementation. Entropy 19(3), article no 103. https://doi.org/10.3390/e19030103
7. Ros F, Guillaume S (2019) A hierarchical clustering algorithm and an improvement of the single linkage criterion to deal with noise. Expert Syst Appl 128:96–108. https://doi.org/10.1016/j.eswa.2019.03.031
8. Frid A, Manevitz LM, Mosafi O (2019) Kohonen-based topological clustering as an amplifier for multi-class classification for Parkinson's disease. In: 2018 IEEE international conference on the science of electrical engineering in Israel, ICSEE 2018, article no 8646026. https://doi.org/10.1109/ICSEE.2018.8646026
9. Silva EDS, da Silva EGP, Silva DDS, Novaes CG, Amorim FAC, dos Santos MJS, Bezerra MA (2019) Evaluation of macro and micronutrient elements content from soft drinks using principal component analysis and Kohonen self-organizing maps. Food Chem 273:9–14. https://doi.org/10.1016/j.foodchem.2018.06.021
10. Tkachenko R, Izonin I (2019) Model and principles for the implementation of neural-like structures based on geometric data transformations. Adv Intell Syst Comput 754:578–587. https://doi.org/10.1007/978-3-319-91008-6_58
11. Vitynskyi P, Tkachenko R, Izonin I, Kutucu H (2018) Hybridization of the SGTM neural-like structure through inputs polynomial extension. In: Proceedings of the 2018 IEEE 2nd international conference on data stream mining and processing, DSMP 2018, article no 8478456, pp 386–391. https://doi.org/10.1109/DSMP.2018.8478456
12. Compute clustering validation indices. https://cran.r-project.org/web/packages/clusterCrit/clusterCrit.pdf
13. Ihaka R, Gentleman R (1996) R: a language for data analysis and graphics. J Comput Graph Stat 5(3):299–314
14. Babichev S, Taif MA, Lytvynenko V, Osypenko V (2017) Criterial analysis of gene expression sequences to create the objective clustering inductive technology. In: Proceedings of 2017 IEEE 37th international conference on electronics and nanotechnology, ELNANO 2017, article no. 7939756, pp 244–248. https://doi.org/10.1109/ELNANO.2017.7939756
15. Calinski T, Harabasz J (1974) A dendrite method for cluster analysis. Commun Stat 3:1–27
16. Zhao Q, Xu M, Fränti P (2009) Sum-of-squares based cluster validity index and significance analysis. Lecture notes in computer science (including subseries lecture notes in artificial intelligence and lecture notes in bioinformatics), vol 5495, pp 313–322. https://doi.org/10.1007/978-3-642-04921-7_32
17. Madala HR, Ivakhnenko AG (1994) Inductive learning algorithms for complex systems modeling. CRC Press, Boca Raton
18. Ivakhnenko AG (1987) Objective clustering based on the theory of self-organization models. Automatics 5:6–15 (in Russian)

19. Stepashko V (2017) Inductive modeling from historical perspective. In: Proceedings of the 2017 12th international scientific and technical conference on computer sciences and information technologies, CSIT 2017, vol 1, article no 8098845, pp 537–542. https://doi.org/10.1109/STC-CSIT.2017.8098845

20. Yefimenko S, Stepashko V (2015) Intelligent recurrent-and-parallel computing for solving inductive modeling problems. In: Proceedings - 2015 16th international conference on computational problems of electrical engineering, CPEE 2015, article no 7333385, pp 236–238. https://doi.org/10.1109/CPEE.2015.7333385

21. Babichev S, Lytvynenko V, Korobchynskyi M, Taiff MA (2017) Objective clustering inductive technology of gene expression sequences features. Commun Comput Inf Sci 716:359–372. https://doi.org/10.1007/978-3-319-58274-0_29

22. Babichev S, Taif MA, Lytvynenko V (2016) Estimation of the inductive model of objects clustering stability based on the k-means algorithm for different levels of data noise. Radio Electron Comput Sci Control 4(4):54–60

23. Babichev S, Taif MA, Lytvynenko V (2016) Inductive model of data clustering based on the agglomerative hierarchical algorithm. In: Proceedings of the 2016 IEEE 1st international conference on data stream mining and processing, DSMP 2016, article no 7583499, pp 19–22. https://doi.org/10.1109/DSMP.2016.7583499

24. Babichev S, Lytvynenko V, Skvor J, Fiser J (2018) Model of the objective clustering inductive technology of gene expression profiles based on SOTA and DBSCAN clustering algorithms. Adv Intellt Syst Comput 689:21–39. https://doi.org/10.1007/978-3-319-70581-1_2

25. Babichev S, Lytvynenko V, Skvor J, Korobchynskyi M, Voronenko M (2018) Information technology of gene expression profiles processing for purpose of gene regulatory networks reconstruction. In: Proceedings of the 2018 IEEE 2nd international conference on data stream mining and processing, DSMP 2018, article no 8478452, pp 336–341. https://doi.org/10.1109/DSMP.2018.8478452

26. Babichev S, Korobchynskyi M, Mieshkov S, Korchomnyi O (2018) An effectiveness evaluation of information technology of gene expression profiles processing for gene networks reconstruction. Int J Intell Syst Appl 10(7):1–10. https://doi.org/10.5815/ijisa.2018.07.01

27. Fefelov AO, Lytvynenko VI, Taif MA, Savina NB, Voronenko MA, Lurie IA, Boskin OO (2019) Hybrid immune algorithms in the gene regulatory networks reconstruction. In: CEUR Workshop Proceedings, vol 2353, pp 193–210

28. Harrington J (1965) The desirability function. Ind Qual Control 21(10):494–498

29. Ankerst M, Breunig MM, Kriegel H-P, Sander J (1999) OPTICS: ordering points to identify the clustering structure. In: ACM special interest group on management of data record SIGMOD, vol 28, no 2, pp 49–60. https://doi.org/10.1145/304181.304187

30. Ester M, Kriegel HP, Sander J, Xu X (1996) A density-based algorithm for discovering clusters in large spatial datasets with noise. In: Proceedings of the second international conference on knowledge discovery and data mining, Portland, Oregon, pp 226–231

31. Density Based Clustering of Applications with Noise (DBSCAN) and Related Algorithms. https://cran.r-project.org/web/packages/dbscan/dbscan.pdf

32. Maechler M et al: "Finding groups in data": cluster analysis extended Rousseeuw et al. https://cran.r-project.org/web/packages/cluster/index.html

33. Nguyen T-D, Schmidt B, Kwoh C-K (2014) SparseHC: a memory-efficient online hierarchical clustering algorithm. Procedia Comput Sci 29:8–19. https://doi.org/10.1016/j.procs.2014.05.001

34. Gionis A, Mannila H, Tsaparas P (2007) Clustering aggregation. ACM Trans Knowl Disc Data 1(1), article no 1217303. https://doi.org/10.1145/1217299.1217303

35. Zahn CT (1971) Graph-theoretical methods for detecting and describing gestalt clusters. IEEE Trans Comput C–20(1):68–86. https://doi.org/10.1109/T-C.1971.223083

36. Factoextra : Extract and Visualize the Results of Multivariate Data Analyses. https://rpkgs.datanovia.com/factoextra/index.html

37. Jain AK, Law MHC (2005) Data clustering: a user's dilemma. Lecture notes in computer science (including subseries lecture notes in artificial intelligence and lecture notes in bioinformatics), vol 3776, pp 1–10

Research of Efficiency of Information Technology for Creation of Semantic Structure of Educational Materials

Olexander Barmak[1], Iurii Krak[2(✉)], Olexander Mazurets[1],
Sergey Pavlov[3], Andrzej Smolarz[4], and Waldemar Wojcik[4]

[1] National University of Khmelnytsky,
11, Institutes Street, Khmelnytskyi 29016, Ukraine
barmakov@khnu.km.ua
[2] Taras Shevchenko National University of Kyiv,
64/13 Volodymyrska Street, Kyiv 01601, Ukraine
krak@univ.kiev.ua
[3] Vinnytsia National Technical University,
95 Khmelnytske shose, Vinnytsia 21021, Ukraine
psv@vntu.vinnica.ua
[4] Lublin University of Technology, 38D Nadbystrzycka, 20618 Lublin, Poland
{a.smolarz,waldemar.wojcik}@pollub.pl

Abstract. The information model of semantic structure of informational education material and the corresponding information technology for creation of semantic structure of educational materials are considered the article. Research of efficiency of information technology for creation of semantic structure of informational educational materials with the help of corresponding software is the ultimate goal of this article. The information model of semantic structure of informational education material is the formal representation of semantic structure of informational education material of the course of discipline, and in the given form it allows to use it as the model for implementation of the corresponding information technology. The information model includes the set of headings, the set of terms, the set of words and the set of relations. The established effectiveness of the proposed technology allows use it to solution a number of urgent tasks, such as semantic assistance in creating tests, determination the conformity of sets of test tasks to educational materials, determination the conformity of educational materials to content requirements, etc.

Keywords: Educational materials · Dispersion evaluation ·
Semantic structure · Key terms

1 Introduction

At the present stage, the wide introduction of information technologies into communication systems [1–3], including voice and sign communications [4,5], is taking place. The central link of information exchange in such systems is digital

© Springer Nature Switzerland AG 2020
V. Lytvynenko et al. (Eds.): ISDMCI 2019, AISC 1020, pp. 554–569, 2020.
https://doi.org/10.1007/978-3-030-26474-1_38

text [6,7]. To automate the analysis of such digital texts, an important aspect is the automated definition of its semantic structure [8]. Particular importance is the automation of the semantic structure definition when working with digital texts of a large volume, such as educational materials [9].

The specialized educational virtual environments such as Moodle [10] are used to develop and use courses in modern higher education. When used, the potential quality of educational services received directly depends on the quality of the educational materials. In the conditions of narrow specialization of courses of educational disciplines, their numbers and intensive updating, the only way to assess the quality of courses and their elements is to automate the solution of a range of problems in the field of modern higher education. These tasks include: determination the conformity of educational materials to content requirements, determination the conformity of sets of test tasks to educational materials, semantic assistance in creating tests, automation of the creation of abstracts and annotations to the elements of educational materials, etc.

2   Problem Statement

The accepted approach is to apply educational materials in the form of digital documents of a defined structure as an educational tool. However, in all the above-mentioned tasks, for the achievement of the relevant results, not the actual digital document or its content, but its semantic model is used. Formalization of the creation of such a semantic model is ensured through the application of some model (for example, ontology [11,12]) as the method of formal description of knowledge contained in educational materials [13]. This model of the educational material may consist of keywords, key terms, educational material structure, keyword and key terms attributes, which define their properties and provide relation to the elements of the learning material structure [14]. The use of such a model of educational material is a means both for revealing the meaning of the educational material and for solving a number of practical problems.

The main stages of creation the model of educational material are the search for key semantic terms [15] in the content of the educational material and the construction of its logical structure. The input data is an electronic document of the educational material, so the automation of the implementation of the above steps requires software processing of the corresponding digital files (usually .docx format). The problem of automating the creation of the logical structure of the educational material (for example: Discipline/Chapter/Topic) is proposed to be solved by defining the hierarchy of content blocks in a digital document in the style of a text editor (respectively: Heading 1/Heading 2/Heading 3), thus forming the upper level structure in models of educational materials of the corresponding academic discipline. The automated search of key terms in the content of the educational material provides the creation of the lower level of the model of educational materials [16].

The characteristic feature of the elements of educational materials used for analysis in the search for key terms is a rather small extent of content. The small extent of content and the narrow semantic orientation of the analysis elements reduces the effectiveness of using commonly used text analysis methods, such as frequency estimation TF, TFIDF estimation and dispersion estimation DE [17]. This necessitates the development of a specialized method for the automated definition of key terms in the content of educational materials.

Therefore, the purpose of research is to develop an information technology for the automated creation of semantic structure of information educational materials and to investigate its effectiveness with the help of corresponding software.

3   Information Model

Information technology for creation the semantic structure of information educational materials is based on the use of a corresponding information model.

Information educational material (IEM) is in most cases formed as a textbook, manual or lecture notes, and is the main bearer of information in an educational course, which designed to acquire knowledge and part of the skills of the subject studying the course.

To achieve the goal set in the research, the semantic structure of IEM can be presented in the form (Fig. 1):

$$\left(M_{Heading} \bigcup M_{Term} \bigcup M_{Word} \bigcup M_{Rel} \right) \subset IEM \tag{1}$$

where IEM – informational education material; $M_{Heading}$ – set of headings; M_{Term} – set of terms; M_{Word} – set of words; M_{Rel} – set of relations.

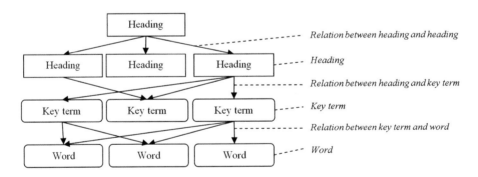

Fig. 1. Example of using elements of the set of entities of the IEM to represent the semantic structure of the IEM

The following are considered separate components of the considered sets: the set of headings $M_{Heading}$, the set of terms M_{Term}, the set of words M_{Word} and the set of relations M_{Rel}.

Taking into account the existing generally accepted requirements for the structure of the course's information educational material (for example: Title of Discipline/Chapter/Topic), the correspondence of the headings of the content fragments of the IEM to the hierarchical model is determined. Thus, each element of the set of headings $M_{Heading}$ is the cortege of the following form:

$$M_{Heading} = (ID, Name, Grade) \qquad (2)$$

where the attribute ID – a unique identifier of the element ($M_{ID} \in Z$); $Name$ – the name of the header; $Grade$ – the header level in the hierarchical structure.

Key terms are semantically meaningful names of concepts, the understanding of which is mandatory for the effective learning of content of a certain fragment of the IEM, the acquisition of relevant knowledge and skills. In contradistinction to words, the terms are semantically integral expressions. Each element of the set of terms M_{Term} is a cortege of the following form:

$$M_{Term} = (TermName, TermNorm, TermNum, TermLem) \qquad (3)$$

where $TermName$ – the symbolic name of the term; $TermNorm$ – the symbolic name of the term in the normalized form; $TermNum$ – the number of words in the term ($M_{TerNum} \in Z$); $TermLem$ – the boolean parameter of the lemation.

The plural of words M_{Word} is formed by including all the elements corresponding to the unique words in the text. Each element of the set of words M_{Word} is a cortege of the following form:

$$M_{Word} = (WordName, WordNorm, WordPart, WordLem) \qquad (4)$$

where $WordName$ – the symbolic name of the word; $WordNorm$ – the symbolic name of the word in a normalized form; $WordPart$ – word's part of the language; $WordLem$ – the Boolean parameter of the lemation ($M_{WordLem} = \{0,1\}$).

The set of relations M_{Rel} includes elements of the semantic structure of the IEM that determine the presence and character of binary unidirectional relations between the two elements of the $M_{Heading}$; M_{Word} and M_{Term} sets. The set relations M_{Rel} includes elements that define:

$$M_{Rel} = (TypeRel, Obj1, Obj2, Feature) \qquad (5)$$

where $TypeRel$ – the integer ($TypeRel \in Z$); indicating the type of relation; $Obj1$ – the first entity of the relation; $Obj2$ – the second entity of the relation; $Feature$ – the attribute that indicates the relation property.

Depending on the attributes of the $Obj1$ and $Obj2$ belonging to separate sets from the $MHeading$, $MWord$ and $MTerm$ list, the $TypeRel$ attribute accepts the values given in Table 1.

According to the types of elements that are related with the elements of set M_{Rel}, its structure can be represented as:

$$M_{Rel} = M_{Rel:H-H} \bigcup M_{Rel:H-T} \bigcup M_{Rel:T-W} \qquad (6)$$

where $M_{Rel:H-H}$ – the set of relations between headings and headings; $M_{Rel:H-T}$ – the set of relations between headings and key terms; $M_{Rel:H-T}$ – the set of relations between key terms and words.

Table 1. List of values of the *TypeRel* attribute for the elements of set of relations M_{Rel}

TypeRel value	Affiliation *Obj1*	Affiliation *Obj2*	*Feature* value
1	$M_{Heading}$	$M_{Heading}$	Non-available (null)
2	$M_{Heading}$	M_{Term}	Numerical indicator of the importance of the key term
3	M_{Term}	M_{Word}	Serial number of the word in the term

Thus, the formal representation of the semantic structure of the IEM of course of the academic discipline in the given form allows it to be used as a model for the implementation of the corresponding information technology. The formal representation of the semantic structure of such IEM consists of the set of headings, the set of terms, the set of words, and the set of relations. The set of relations includes elements of the semantic structure of the informational education materials, which determine the presence and character of relations between other elements of the semantic structure of the informational education materials. The set of relations include relations between headings, relations between headings and key terms, relations between key terms and words.

4 Scheme of Information Technology

In accordance with the general scheme of information technology for the automated creation of the semantic structure of IEM (Fig. 2), it's possible to identify two sequential stages of information transformation. The result of these actions is the filling of all sets of the information model of the semantic structure of the IEM, which makes it possible to use it to solve the described application problems.

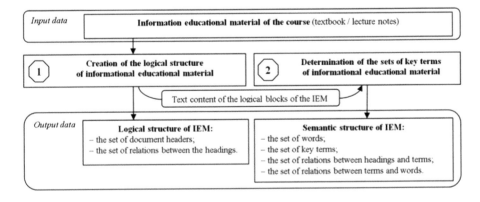

Fig. 2. The general scheme of information technology

The *input data* of the information technology for the automated creation of the semantic structure of IEM is the file of the electronic document (for example, .docx format) with poorly structured text content of the IEM containing the structure of the document in the form of headings, as well as the corresponding text theoretical information.

In the process of execution Block 1 (creation of the logical structure of informational educational material) of information technology, creation of the levels of the logical structure of the IEM is carried out. These levels correspond to the author's specified hierarchical structure of the IEM in the form of a system of headings of different levels used in electronic documents with poorly structured text content, and is implemented by determining the set of existing documents headings $M_{Heading}$ and determining their relations by generating the set of unidirectional relations between the headings $M_{Rel:H-H}$. Also, in this step, the corresponding texts of the headings are determined automatically for further analysis.

When execution Block 2, the automated determination of the sets of key terms M_{Term} of the IEM for the logically separate fragments of the text content of the IEM, determined in the previous stage, is carried out. As a result of processing, the set of key terms sorted by the semantic value for each of the fragments are determined. The determination of sets of key terms proceed using the dispersion estimation method, filtering according to portraits of key terms, evaluating the semantic importance of key terms, and automatically limiting the sets of key terms. Consequently, the lower, semantic level of the structure of the IEM corresponds to the set of key terms M_{Term}, which is matched to elements of the set of headings by means of the corresponding relations $M_{Rel:H-T}$, which form the set of relations between the headings and the key terms. Each term is an ordered collection of words, defined by the set of the words M_{Word} and the set of relations between the words and the terms $M_{Rel:T-W}$.

Accordingly, the *output data* of the information technology are defined elements of all sets of the information model of the semantic structure of the IEM.

Thus, the information technology developed allows to automate the creation of the logical structure of the informational educational material and to fill all the sets of the information model of the semantic structure of the IEM. This makes it possible to use it to solve engineering problems.

5   Steps of Automated Determination of Semantic Terms Sets

In the automated determination of the set of semantic terms in the content of educational materials, the input data is the content of the educational material or its defined part in the form of .docx file of any hierarchy of elements; the output data is the set of semantic terms of the educational material M_{Term} and the set of relations between headings and key terms $M_{Rel:H-T}$; the process of automated determination of the set of semantic terms consists of several stages of the transformation of information.

Portrait of Key Terms. According to the results of the analysis of more than 1300 elements of educational materials with experts (compilers) representative sets of key terms, it has been established that all elements of the set of sets M_T correspond to the following regularities:

- The number of words in the term $n = 1..6$.
- If the term is a word ($n = 1$), then it is included in the set of the nouns M_N.
- If the term is a phrase ($n > 1$), then it consists of elements of the set M_M. The set M_M consists of sets of semantically meaningful elements (set of nouns M_N and set of adjectives M_A) and semantically binding elements (set of conjunctions M_S, set of numeral M_{Num} and set of prepositions M_P).
- If $n > 1$, then the phrase contains at least one element from the set of nouns M_N.
- If $n > 1$, then the first ($k = 1$) and the last ($k = n$) words are elements of the set of semantically meaningful elements.
- If $n > 1$, there are no punctuation marks between the elements of the phrase (except for the hyphen inside complex nouns, which is part of the word).
- All elements (symbols, words) of the same term in the text have the identical stylistic properties, so they belong to the same container *TextRange* in the structure of the digital document.

As a result of the use of this method, the aim is to create sets of terms M_T, which correspond to the above laws [16].

Data Processing Sequence. Figure 3 provides the sequence of steps for the automated creation of sets of semantic terms in the content of educational materials, which shows the sequence of stages of data transformation to achieve the ultimate goal.

Segmentation by paragraphs and selection of paragraph for analysis (Step 1) consists in analyzing the structure of a digital document. It is based on the natural correspondence of the hierarchical system of headings of educational materials as electronic documents to the upper levels of the logical structure of the educational material of the discipline. For example, the titles of the disciplines correspond to elements of the standard "Heading 1" style, titles of chapters – "Heading 2", topic titles – "Heading 3", etc. (Table 2). Consequently, the structure of educational materials as digital documents is regulated in the languages of marking of digital documents and is implemented through the system of headings. Output data of Step 1 are defined fragments of content of the digital document of the IEM, which will be further processed individually.

Segmentation by phrases (Step 2) is used to split a fragment of content in a digitized document, which is processed, into smaller fragments – phrases, or containers. The phrase is a semantically integral node, distinguished by stylistic text formatting or punctuation, and localizes the location of separate terms. According to the document's object model, MS Office uses *Sections* object to localized parts of the document that have different formatting. The Section objects are contained in the *Document* object (Fig. 4) in the Collection *Selections*. *Sections* contain smaller elements of the structure – *Paragraphs*. And *TextRange* is the

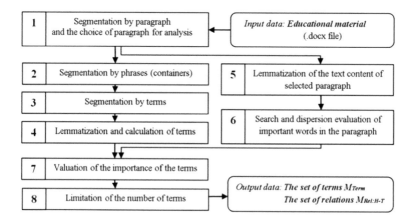

Fig. 3. Data processing sequence for automated definition of semantic terms in IEM

Table 2. Example of correspondence of the upper levels of the structure of the IEM for standard digital document styles.

Level in the hierarchy *Grade*	Level of logical structure IEM	The name of document style
1	Educational discipline	Heading 1
2	Chapter	Heading 2
3	Topic	Heading 3

lowest level of document structure that defines a text fragment of the identical style within *Paragraph*.

Consequently, the set of phrases include continuous ordered sequences of words that do not extend beyond the limits of the *TextRange* digital document containers and are not interrupted by punctuation marks. Getting as a result of the Step 2 sets of phrases allows to further process each of the phrases separately for the search terms.

Segmentation by terms (Step 3) intended for create the set of all possible terms that are present in the content being analysed.

At first, the set of terms of the educational material M_T includes all possible continuous sequences of words that are not beyond the limits of the phrases and are consistent with the condition:

$$M_T = \{\langle x_1, x_2, x_3, x_4, x_5, x_6 \rangle \,|\, x_1, x_2, x_3, x_4, x_5, x_6 \in M_M, \\ \langle x_1, x_2, x_3, x_4, x_5, x_6 \rangle \cap M_N \neq \oslash\} \tag{7}$$

where M_M – the set of semantically meaningful elements (nouns M_N and adjectives M_A) and semantically binding elements (conjunctions M_S, numeral M_{Num} and prepositions M_P); $M_M = M_N \cup M_A \cup M_S \cup M_{Num} \cup M_P \cup \oslash$; M_N – set of nouns.

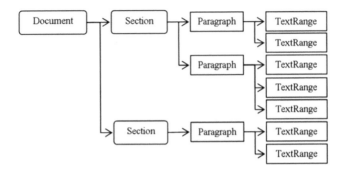

Fig. 4. General structure of the object model of MS Office document

Segmentation by terms is executed using the database of the corpus of words of actual language and as output data forms the set of terms M_T contained in the processed fragment of the digital document of the educational material.

Lemmatization and calculation of terms (Step 4) allows, on the basis of the set of terms M_T, to form the set of lematic-independent terms M_{T1} and compare each of them the number of occurrences in the text. To do this, at first makes the lemmatization of each word in each phrase in the set M_T. Under the lemmatization, means the translating of words to the infinitive state – for example, the nouns are translated into the singular form. After that, the resulting set is processed and compacted in such the way that all identical duplications of terms is deleted, and for each term, the value K_n is summed up. This value K_n reflects the found number of occurrence of each term n in the input set M_T.

Since at the stage of forming the set of terms M_T, all possible variants of the terms from the phrases were added to it, without absorption by smaller words, in this step an analysis of the need for such an absorption is made as follows. If there is a term n_1 in the set M_{T1} (K_{n1} – the amount of occurrence of the term n_1 in the set M_{T1}), which is an ordered set of x_1 words, and the term n_2 (K_{n2} – the number of occurrence of the term n_2 in the set M_{T1}), which is an ordered set of x_2 words, and n_1 is a subset of n_2 and $x_1 < x_2$, then, with the correctness of the equation $2x_1 > x_2$, the term is deleted from the resulting set. In order to facilitate further processing from the resulting set M_{T1}, it is also advisable to delete all terms in which $K_n = 1$, because the once use of the term excludes the fact of considering the concept of term in this fragment of the IEM.

The resulting set of lematic-independent terms M_{T1} contains terms that at the same time form the set M_{Term}. These are the terms used in the educational material with a quantitative usage indicator, but the importance of these terms has not yet been determined.

Lemmatization of the text content of selected paragraph (Step 5) transforms the text of a defined content fragment of the digital document with educational material being analyzed to the corresponding sequence of words in the infinitive state, which is the input data of this step. They allow the further evaluation of the word dispersion.

Search and dispersion evaluation of important words in the paragraph (Step 6) is intended to evaluate the importance of each word in the text fragment using the dispersion evaluation method.

This method is an estimate of the discriminant weight of words. The method of dispersion evaluation allows to exclude from the general set of widely used words in the text of word, which are arranged evenly. The method of dispersion evaluation has shown its high efficiency in previous researches [17].

According to the existing mathematical model [18], if some word A in a text consisting of N words is indicated by A_k^n, where the index k – the number of the occurrence of the given word in the text, and n – the position of the given word in the text, then the interval between successive occurrences words in such notation will be the value $\Delta A_k^m = A_{k+1}^m - A_k^n = m - n$, where the word "$A$", which occurrences $k + 1$ and k times, is located on the m-th and n-th positions in the text. Thus, the dispersion evaluation is calculated by the equation $\sigma = \sqrt{\left(\Delta A^2 - (\Delta A)^2\right)/(\Delta A)}$, where (ΔA) – the average value of the sequence ΔA_1, ΔA_2, ΔA_k; (ΔA^2) – sequences A_1^2, A_2^2, A_k^2; K – the number of occurrence of the words A in the text.

The input data of this step is the lemmatized text content of an investigated fragment of the digital document of the educational material; the output data is an ordered set of words, each of which compares the estimation of its dispersion, which is positioned as an estimate of the importance of the given word in the investigated fragment of the IEM.

Valuation of the importance of the terms (Step 7) as the input data has a set of lematic-independent terms M_{T1} with the matching of each of them the number of occurrences in the investigated text, and an ordered set of words with a matching of each of them with an estimate of its importance (dispersion) in the investigated text.

The importance ν_n for each term n from the set M_{T1} is calculated by the formula:

$$\nu_n = \sum_{i=1}^{x_n} \frac{K_n \sigma_n}{k_n} \tag{8}$$

where K_n – the number of occurrences of term n in the set M_{T1}; k_n – the number of occurrences of the i-th word of the term n in the lemmatized text fragment; σ_n – dispersion evaluation of the i-th word of the term n; x_n – the number of words in the term n.

The output data of this step is the set of lematic-independent terms M_{T1} with the matching for each of them the number of occurrences in the investigated text and the value of the importance evaluation, sorted by decreasing the nominal value of the importance evaluation.

Limitation of the number of terms (Step 8) is intended to the creation of the set of key terms by input data – the set of lematic-independent terms M_{T1}. The set of key terms is created on the basis of lematic-independent terms from the set M_{T1} with the highest values of importance evaluation. The number of key terms is recognized through a well-known semantic text processing knowledges, keyword density. The keyword density D is the ratio of the number of words in the key terms in the text to the total number of words in the text:

$$D = \sum_{i=1}^{n} \frac{K_n x_n}{X_{txt}} \tag{9}$$

where K_n – the number of occurrences of term n in the set M_{T1}; x_n – the number of words in the term n; X_{txt} – the total number of words in the text; n – current number of terms in the set M_{TK}.

The algorithm is based on the fact that the terms from the set M_{T1} with the highest values of importance evaluation are added to the empty resulting set of key terms M_{TK} until the equality is satisfied (for example, for $D = 7\%$): $D \leq 0.07$.

The output data of the step and respectively of the method of keyword density are: a set $M_{TK} = M_{Term}$ of the key terms corresponding to the investigated content fragment of the digital document of the educational material; the set of relations between the headings and the key terms $M_{Rel:H\text{-}T}$, for each element of which the attribute $Feature = \nu$ is defined as the numerical indicator of the importance of this term in this investigated fragment of the IEM.

Thus, the proposed approach allows, on the basis of a digital document of the IEM, to automate the creation of the corresponding set of key terms M_{Term} and the set of relations between the headings and key terms $M_{Rel:H\text{-}T}$.

6   Efficiency of Information Technology

Efficiency of practical applying of the considered information technology can be estimated by using values precision and recall (Fig. 5).

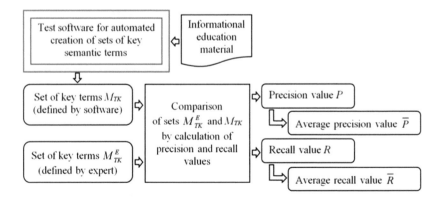

Fig. 5. Schema of research of efficiency of information technology

In order to check the effectiveness of the developed information technology for creation of semantic structure of educational materials, was conducted the comparison of the automated creation of sets of key semantic terms with sets of experts (authors) for a test samples of digital documents of educational materials.

6.1 Software Realization of Information Technology

In accordance with the proposed information technology, test software was developed that implements the processing of digital content of educational materials as outlined above.

Digital files with educational materials .docx format are organized using an open XML format that stores documents as collections of separate files and folders in a compressed package. In order to implement software processing of digital documents, it is expedient to use specialized software packages that provide object-oriented tools for programmatic work with content of relevant files, such as Spire.Doc.dll [19], Microsoft.Office.Interap.Word.dll, Document-Format.OpenXml.dll [20], etc. Within the developed test software, the extension Spire.Doc.dll was used to provide both an analysis of the levels of the document structure Heading and access to content elements, in particular TextRange, which is the lowest level of the document structure that defines fragments of text of the identical style. Moving the functions of automatically matching text block styles to their properties from the level of the functional code of the application code to the functional level of the software library has made it possible to simplify operations of the system with the digital document and the programming process.

The developed test software product is processing on the input data as an educational material file, automatically generates the structure of the digital document for the selection of the element for analysis, after which the segmentation is executed by phrases and terms, the terms are lemmatized and their set is compactified. On the basis of automatically lemmatized text the search and dispersion evaluation of the importance of words in the selected fragment are calculated, after which the importance of the terms is evaluated, and their number is limited in accordance with the above mathematical models. In particular, processing of the topic "Neural Networks Cognitron and Neocognitron" from the educational discipline "Methods and systems of artificial intelligence" by the developed test software product was analyzed. The final result of the test software product working is the set of key terms and an evaluation of the importance of each term in the investigated fragment of the IEM. In this occurrence the following terms were included at the key words density of 7% in the set of key terms: cognitron, neuron, neocognitron, pattern, complex node, input image, learning, simple node.

6.2 Research of Efficiency of Information Technology

Efficiency of practical applying of the considered information technology can be estimated by using values precision and recall [21].

The precision P (the ratio of the number of relevant key terms found automatically to the total number of key terms found in the text under investigation) and the recall R (the ratio of the number of relevant key terms found automatically to the total number of relevant key terms in the text under investigation) are calculated as follows:

$$P = \frac{\left| M_{TK}^E \cap M_{TK} \right|}{\left| M_{TK} \right|}, R = \frac{\left| M_{TK}^E \cap M_{TK} \right|}{\left| M_{TK}^E \right|} \tag{10}$$

where M_{TK}^E – the set of relevant key terms generated by the expert; M_{TK} – the set of automatically found key terms.

Accordingly, $\overline{P}$ the average search precision $\overline{R}$ and the average search recall are calculated using the following formulas:

$$\overline{P} = \frac{\sum_{i=1}^k P_k}{k}, \overline{R} = \frac{\sum_{i=1}^k R_k}{k} \tag{11}$$

where k – the number of educational materials in the test sample.

6.3   Results of Research of Information Technology Efficiency

In order to determine the effectiveness of the practical applying of the considered information technology, a test sample of 50 files from various IEM was processed by the test software.

Table 3. Comparative analysis table of sets of terms.

Num	Key term	Defined by expert	Defined by software
1	Cognitron	+	+
2	Neocognitron	+	+
3	Neuron	+	+
4	Exciting neuron	+	−
5	Braking neuron	+	−
6	Complex node	+	+
7	Simple node	+	+
8	Pattern	−	+
9	Input image	−	+
10	Learning	−	+

For example, as the result of the testing of the before-mentioned educational material "Neural Networks Cognitron and Neocognitron", the set of key terms were obtained and compared with the set of experts' key terms. The comparison results are given in Table 3. In this case, the search precision was 0.625, and the search recall was 0.714.

In general, result of investigation was next. The average precision was 0.732, the average recall was 0.697. The minimum precision was 0.512, the minimum recall was 0.581; the maximum precision was 0.929, the maximum recall was 1.000 (Fig. 6).

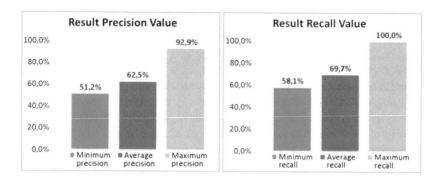

Fig. 6. Final results of investigation, values precision and recall

Analysis of the results revealed that the absence of the terms found by the program in the author's set does not always characterize the disadvantage of this information technology. Some semantically important terms are ignored by the experts subjectively. Another category is the terms on which the experts accent excessive attention, but they are semantically secondary in the educational material.

7   Conclusion

The information model of semantic structure of IEM and the corresponding information technology for creation of semantic structure of educational materials was considered. This information model is the formal representation of semantic structure of informational education material of the course of discipline, and in the given form it allows to use it as the model for implementation of the corresponding information technology. The formal representation of the semantic structure of such informational education materials includes the set of headings, the set of terms, the set of words and the set of relations.

The developed information technology allows to automatically finding the sets of key semantic terms in information educational materials with high efficiency. Developed software in accordance with this information technology as a result of processing the input data in the form of a digital document with information educational material allows obtaining the output data in the form of the logical structure of information educational material and filling all sets of information model of semantic structure of the IEM.

The test software that allows automating the determination of sets of key semantic terms using this information technology is considered. Conducted investigations confirmed possibility of effectively forming the set of key semantic terms of educational materials, average evaluated search precision metrics 73.2% and average search recall 69.7%.

The potential information technology efficiency is higher than the calculated values as some semantically important terms are ignored by the experts subjectively, in addition, some terms on which the experts accent excessive attention, but they are semantically secondary in IEM. The established effectiveness of the proposed technology allows use it to solution a number of urgent tasks, such as semantic assistance in creating tests, automation of the creation of abstracts and annotations to the elements of educational materials, determination the conformity of educational materials to content requirements, determination the conformity of sets of test tasks to educational materials, etc.

Next research is aimed at analyzing the impact on the effectiveness indicators of the proposed information technology of relationships between the number of key semantic terms in the resulting sets and the values of the density of keywords, and at the perfect of the considered information technology in order to improve the results.

References

1. Krak IuV, Barmak OV, Romanyshyn SO (2014) The method of generalized grammar structures for text to gestures computer-aided translation. Cybern Syst Anal 50(1):116–123
2. Mallat S, Hkiri E, Zrigui M, Maraoui M (2015) Semantic network formalism for knowledge representation: towards consideration of contextual information. Int J Semant Web Inf Syst 11(4):64–85
3. McCrae JP, Moran S, Hellmann S, Brummer M (2015) Multilingual linked data. Semant Web 6(4):315–317
4. Krak IV, Kryvonos IG, Barmak OV, Ternov AS (2016) An approach to the determination of efficient features and synthesis of an optimal band-separating classifier of dactyl elements of sign language. Cybern Syst Anal 52(2):173–180
5. Kryvonos IG, Krak IV, Barmak OV, Shkilniuk DV (2013) Construction and identification of elements of sign communication. Cybern Syst Anal 49(2):163–172
6. Semantic analyzer "Istio". https://istio.com/. Accessed 25 Feb 2019
7. System for semantic analyze "Serpstat". https://serpstat.com/. Accessed 25 Feb 2019
8. Kastrati Z, Imran AS, Yildirim-Yayilgan S (2016) A semantic and contextual objective metric for enriching domain ontology concepts. Int J Semant Web Inf Syst 12(2):1–24
9. Snytyuk VE, Yurchenko KN (2013) Intelligent management of knowledge assessment. Cherkassy
10. Moodle – Open-source learning platform. https://moodle.org/. Accessed 25 Feb 2019
11. Guo H, Gao S, Krogstie J, Tretteberg H, Wang AI (2015) An evaluation of ontology based domain analysis for model driven development. Int J Semant Web Inf Syst 11(4):41–63

12. Oberle D (2014) How ontologies benefit enterprise applications. Semant Web 5(6):473–491
13. Angrosh MA, Cranefield S, Stanger N (2014) Contextual information retrieval in research articles: semantic publishing tools for the research community. Semant Web 5(4):261–293
14. Krak I, Barmak O, Mazurets O (2016) The practice investigation of the information technology efficiency for automated definition of terms in the semantic content of educational materials. In: CEUR Workshop Proceedings, vol 1631, pp 237–245
15. Beliga S, Mestrovic A, Martincic-Ipsic S (2016) Selectivity-based keyword extraction method. Int J Semant Web Inf Syst 12(3):1–26
16. Krak Y, Barmak O, Mazurets O (2018) The practice implementation of the information technology for automated definition of semantic terms sets in the content of educational materials. In: CEUR Workshop Proceedings, vol 2139, pp 245–254
17. Ventura J, Silva J (2007) New techniques for relevant word, ranking and extraction. In: Proceedings of 13th Portuguese conference on artificial intelligence, Guimarães, Portugal, 3–7 December 2007, pp 691–702
18. Ortuño M, Carpena P, Bernaola P, Muñoz E, Somoza AM (2002) Keyword detection in natural languages and DNA. Europhy Lett 57(5):759–764
19. Create .NET apps with NuGet. Spire.Doc for .NET. https://www.nuget.org/packages/Spire.Doc/. Accessed 23 Feb 2019
20. Document Format Open Xml. https://www.nuget.org/packages/DocumentFormat.OpenXml/. Accessed 23 Feb 2019
21. Manning C, Raghavan P, Schutze H (2008) Introduction to information retrieval. Cambridge University Press, Cambridge

Simulation of a Combined Robust System with a P-Fuzzy Controller

Bohdan Durnyak[1]([⊠]) [iD], Mikola Lutskiv[1]([⊠]), Petro Shepita[1]([⊠]),
and Vitalii Nechepurenko[2]([⊠]) [iD]

[1] Ukrainian Academy of Printing, Lviv, Ukraine
`durnyak@uad.lviv.ua`, `lutolen@i.ua`, `pshepita@gmail.com`
[2] Military-Diplomatic Academy named after Eugene Bereznyak, Kyiv, Ukraine
`nechepurenko_v@ukr.net`

Abstract. A combined robust control system with a fuzzy P-controller
has been presented in the paper. A bias signal which creates a forward-
looking ratio proportional to the input task has been introduced to the
control action of a fuzzy P-controller. The results of the simulation modelling have been performed during the simulation process. It has been
shown that a fourfold increase of the object transfer coefficient causes
the change of the overregulation from 5 to 20%. The results of the simulation have shown too that fivefold increase/decrease of the time constant
influences a little to the quality of regulation process. This fact is a main
advantages of the proposed fuzzy controller technique.

Keywords: Fuzzy logic · Controllers · Combined · Object · Model ·
Parameters · Variable · Robust · Simulation · Scheme

1   Introduction

Nowadays the current requirements concerning high quality of product and
reducing the cost for their manufacture require the statement and solving new
tasks for the design of control systems of technological processes and objects
with incomplete information about the object, changes of its parameters and
the effects of various factors. The main disadvantages of the existing control
systems with a traditional controller are the sensitivity to changing the object
parameters. So, significant fluctuations appear in the system when occur the disturbances. This fact impairs the quality of control and finished products. Managing such objects based on the principles and methods of adaptive control system
and identification and intellectual control are very complex and costly, what limits significantly their application for the simple objects. Therefore, the task of
constructing and analyzing a combined fuzzy system, determining and adjusting
the parameters of its modeling and studying its robustness when changing the
parameters of an object is very important.

The basics of modeling and fuzzy control have been presented in the monographs [7,10–14]. In these works the theory of both the fuzzy models and systems was presented. In [1,3,4,8] various versions of fuzzy controllers, rules base,

© Springer Nature Switzerland AG 2020
V. Lytvynenko et al. (Eds.): ISDMCI 2019, AISC 1020, pp. 570–580, 2020.
https://doi.org/10.1007/978-3-030-26474-1_39

structural schemes of different controllers and their analysis were presented. Linguistic description, structural schemes of digital fuzzy controllers and systems, and the results of simulation in the form of transient graphs were described in the monographs [2,4,9,10,15–17].

For purpose the simulation and analysis of fuzzy systems Simulink model [9,12,18,20,21] of the MATLAB software was used. However, in the library in the section Fuzzy logic controller there is no direct access to its specific parts, in particular to the blocks of normalization and de-normalization of the input and output signals. This fact greatly limits the ability of both the simulating and studying of the effect of changing the parameters of the object and the fuzzy system. There are various methods of normalization and de-normalization in fuzzy models [8,22], but they are presented in the general sight. In [5,6,23] were shown that the choice of parameters of normalization and de-normalization greatly affect to the quality and properties of fuzzy systems. So, this stage is very important one to construct and adjust the object.

2 Materials and Methods

Taking into account the novelty and practical part of the task of constructing a combined robust control system with P - fuzzy controller let's firstly to consider the traditional systems. Controller (law of management) usually is being chosen taking into account the properties of the object of regulation and the given quality indicators, which depend on the parameters of the object, the nature and size of the perturbation, as well as the type of controller and parameters of its debugging. Calculation of the parameters of the controller debugging involves the definition of their values, when which the transient processes and static accuracy in a closed system satisfy the accepted quality indicators. This operation is performed based on the dynamic properties and parameters of the regulation object. The available simplified calculation methods make it possible to determine the numerical values of the adjustment parameters of regulators using formulas or graphs that connect these values with the object parameters [1,2,4]. Simple proportional controllers are used for the objects with medium capacity, small order, a little delay and with smooth load changes. The proportional controller forms the regulatory action (control) on the object [1,2]:

$$V(t) = k_p e(t),\tag{1}$$

where $e(t)$ is the control error, k_p is the coefficient of the controller transmitting.

Deviation of the adjustable value from the setting value (error of the signal):

$$e(t) = y_0(t) - y(t),\tag{2}$$

where $y_0(t)$ is a given value of the regulated quantity, $y(t)$ is the output of the system (the regulated quantity).

The zero value of the signal $e(t) = 0$ is required for a zero deviation of regulation. When zero value of the deviated signal according to expression (1) the control action of the controller is $V(t) = 0$. However, when zero deviation value the regulator should form a steady value of control. Consequently, the known main disadvantage of P-controllers is low static precision which limits their application. The main advantage of P-controllers is a relatively quick correction of deviations, i.e., a good operation in a dynamic mode [1,2]. In order to increase the accuracy of the system in the state of equilibrium, i.e., after the completion of the transition process, it is entered the I-component control [1,2,8]. The controlling action of the P-controller is instantaneously changed due to changes in the adjustable value from the set value. Whereas, in the I controller the control gradually increases, i.e., the regulatory action "lags" from the variation of the deviation, which can also influence to the real systems and often leads to the appearance of weakly damped oscillations of the regulated values relative to its predetermined value. This fact is a significant disadvantage of the system with the PI controller [1,2]. To improve the dynamic properties of the system it is introduced a differential component of the law of control. Such controllers are called PID controllers, which are more complex than the previous ones.

The second approach to the creation of controllers has been proposed. Implementation of this approach involves that in addition to the proportional controller the initial value (shift) of the control signal, which sets the average level of the output signal of the controller, is additionally introduced. In this case the regulatory effect on the object is determined as follows:

$$V(t) = k_p e(t) + U_0, \tag{3}$$

where U_0 is the initial value of the controller control signal.

Consider the operation of the control system with the proposed controller for a steady-state equilibrium regime, in which the initial regulated value is calculated in the following way:

$$y = k_0 V, \tag{4}$$

where k_0 is the transfer coefficient of the object.

After adjusting the control action (3), the adjusted value can be calculated as follows:

$$y = k_p k_o e + k_o U_0. \tag{5}$$

The initial value of the given control signal of the controller depends on the value of the adjustable value is calculated by the formula:

$$U_0 = k_c y_0. \tag{6}$$

where k_c is the desired transmission ratio.

After substituting the initial value of the signal we have:

$$y = k_p k_o e + k_o k_c y_0. \tag{7}$$

On the basis of hereinbefore, the dependence of the regulated quantity on its predetermined value in a closed system is written as follows:

$$y = \frac{k_p k_o + k_o k_c}{1 + k_p k_c e y_0}. \tag{8}$$

Similarly, we define the static error of a closed system in the following way:

$$e = \frac{1 + k_o k_c}{1 + k_p k_c e}.$$ (9)

For an adjustable value to be equal to its given value in expression (8), equality must be performed:

$$k_p k_o + k_c k_o = 1 + k_p k_o.$$ (10)

Then, we define the desired value of the transfer coefficient as follows

$$k_c = \frac{1}{k_o}.$$ (11)

If the coefficient of transmission k_c is defined, then the static error (9) in the constructed in such a way system equals zero. Consequently, it is theoretically possible to obtain a zero static error in the d system without entering an I-component control algorithm. Since the transfer coefficient k_c is in the numerator of the system Eq. (8), the management of the channel of the organization of the initial value of the control signal of the regulator will not affect the stability of the closed system, which is a rather positive property of the constructed system. To improve the dynamic properties of the constructed system and provide robustness with respect to the variation of the parameters of the control object, apply the fuzzy controllers which form a controlling action on basis of fuzzy logic [2,8,11]. In general, the synthesis of a fuzzy controller consists in choosing the belonging function of the numeral - sets of linguistic variables, fuzzy output algorithms and optimizing the main parameters of the controller by minimizing the selected quality criterion for a closed control system. Consider the problem of constructing the proposed combined fuzzy P-controller for the control object, which is described by the second-order transfer function:

$$W_o(s) = \frac{Y(s)}{V(s)} = \frac{k_o}{(T_1(s) + 1)(T_2(s) + 1)},$$ (12)

where k_o is the transfer coefficient of the object, T_1, T_2 are the time constants. The rule of regulating action to the object is the following:

$$V(s) = k_p e(s) + U_0(s),$$ (13)

For synthesis of a fuzzy controller it is assumed that the number of thermosets which is used to estimate the linguistic variable is equal two. The fuzzy version of controller P is based on knowledge of the state of the control process with the use of linguistic variables [2,5,8]. If the error is negative/positive, then the control is described by a fuzzy rule base as the follows:

$$\begin{aligned} R1 : \ & if \ (E = N) \ Then \ (U_n = A) \\ R2 : \ & if \ (E = P) \ Then \ (U_n = B) \end{aligned}$$ (14)

where E is the normalized input of the control error, U_n is the normalized control, N, P are the fuzzy linguistic sets which qualitatively evaluate the control error (N is a negative error, P is positive).

Fuzzy linguistic models and sets A and B which are described by relativity functions given by the (L-left, P-right) external relativity functions [2,5,8], are used to construct the fuzzy controller model. The main parameter of the membership function is the width of the window which is assumed to be equal to 1. Since the membership functions of the fuzzy sets are normalized and their values are in the range [0, 1], then the error signal must be normalized (scaled). There are various methods of data normalization [2,8], for which it is necessary to know the range of changes of both the input and output signals of the controller (the minimum and maximum values of the regulation error). We propose to estimate the maximum range of error correction $E_n = y_0$ on the basis of which a normalized error for an optional task is determined:

$$E = \frac{1}{y_0} e, \tag{15}$$

where y_0 is the setting value of the regulated value at the input of the system, which is known and which can vary widely depending on the mode of operation of the system.

Formulated in accordance with the fuzzy algorithm (14) normalized control can be changed in the range $[-1, 1]$. Therefore, to create a physical regulating action on the object it is necessary to carry out de-normalization (scaling) and to take into account the initial value of the control signal:

$$V = MU_n \frac{1}{k_o} y_0, \tag{16}$$

where M is the scaling factor which is the main parameter for setting up a fuzzy controller.

3   Experiment

Based on the rules (14), the selected membership functions, fuzzification and de-fuzzification operations, expressions of both the normalization (15) and de-normalization (16), a structured scheme of the Fuzzy Control System model with a fuzzy P-controller using the Simulink package of MATLAB software is constructed. This model is shown in Fig. 1. The fuzzy controller consists two blocks: a normalized algorithm masked in the Subsystem, Fuzzification, and De-fuzzification blocks in the Enabled Subsystem sub-block. On the right side of the Subsystem is a control object model which is provided by Transfer function blocks. The normalization of the error is carried out by the expression (15) and is implemented by means of Simulink and disguised in the subsystem block. The main blocks of the fuzzy controller are the fuzzification and the de-fuzzification blocks masked in the Enabled Subsystem (Fig. 2). On the left side there is a fuzzification block which performs the fuzzification of the normalized error signal

E. This block consists two subblocks of Triangular MF (membership functions) of type N and P, which are activated by the X signal from the Ramp unit. Defazzification of the fuzzy signal is carried out by the Mamdani method [2,8] by cropping the membership function to the U_p level which defines the input signal E using the Saturation Dynamic blocks. Modified in this way the functions of dependence and the set $A*$, $B*$ are being given to the input of the operator MAX. At the input of this block a normalized U_n control is formed. This signal is transferred to the first input of the $DotProduct$ multiplication unit for the de-normalization process (multiplication by the scaling factor M). A physical regulatory effect on the object which consists of a de-normalized control U and an initial set-point of the control signal is created on the input of the simulator.

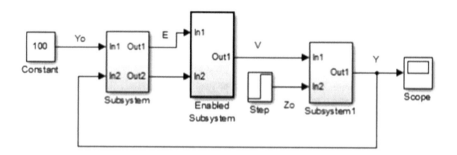

Fig. 1. Fuzzy control systems with P-fuzzy controller

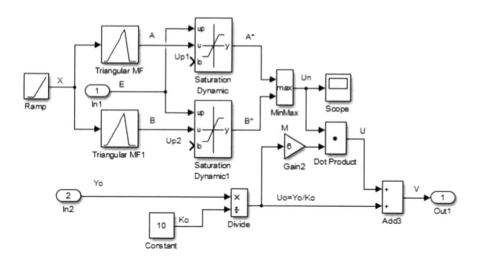

Fig. 2. Block chart of fuzzification and de-fuzzification model

4   Results and Discussion

The quality of the regulatory process of the proposed system with a fuzzy P-controller was investigated with the use of the method of imitation modeling. The object parameters $k_o = 10$, $T_1 = 5c$, $T_2 = 3c$ were specified for example. The value of the adjustable magnitude $y_0 = 100$ was accepted. The main parameter for a fuzzy P-controller tooling is the scaling multiplier M. The results of the analysis of the plots, which are shown in Fig. 3 allows us to conclude that the combined fuzzy P-controller in the control system with an inertial object of the second order provides a given overriding and quite small static errors 0.00065% and 0.0062%. The time of rise of the adjustable magnitude in this case is 4.2 s and 7.2 s respectively, while the regulation time is 17 s. In the case of the action of a rather large disturbance onto the object $z_0 = 30$, the overregulation is 8% and static error is 4,8%. If the system quality (overregulation and accuracy) with two membership functions of fuzzy P-controller satisfies the customer, then the use of the developed combined fuzzy controller can be recommended as the simplest one. The work of a fuzzy controller for non-stationary objects of regulation where the parameters varies in the wide ranges was investigated. The object gain for the preliminary tooling of the controller $M = 6$ and $M = 4$ was increased in four times ($k_o = 40$). The results of the imitation modeling are shown in Fig. 4. When increasing the object gain the quality of the regulatory process depends on the parameters of the regulator tooling. In this case, the overregulation increases from 19% to 25%, but the static error is quite small. The increasing effect of the object gain on the regulation quality from the initial value of the control signal $U_0 = y_0/40$, provided that it is a constant, was investigated. The results of the imitation modeling are shown in Fig. 5. When increasing of the object gain is from 10 to 40 the regulatory process quality is significantly disapproving. In this case the overregulation is 65%, a static error is greatly increased. It is 11.9% and 17.65% respectively. However, the decrement of the transition process is large. Therefore, the initial value of the control signal significantly affects the regulation quality. This fact should be taken into account when designing a fuzzy controller and its tooling. The influence of the object time constant on the regulation quality was investigated in a fivefold increase and reduced constant time: $T_1 = 25\,s$, $T_2 = 15\,s$, $k_p = 10$, $M = 4$. The results of the imitation modeling when the constant time variations are shown in Fig. 6. The analysis of the obtained results allows us to conclude that the constant time changing of the regulation object has little effect on the regulation quality. The overregulation is 6,5% and 4% respectively. Instead, the time of regulation varies widely from 4 to 40 s, which is a logical physical phenomenon due to the change in the inertial properties of the object. The conclusion about the proposed fuzzy P-controller which provides stable operation of the system and regulation quality has robust properties and is easier than traditional adaptive systems, was based on modeling results and the built transient characteristics of the system when changing the objects parameters of regulation in wide limits.

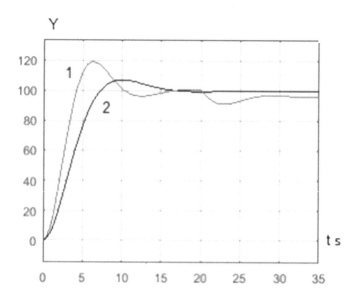

Fig. 3. Transient characteristics of the system: 1 - for 20% overregulation 2 - for 5% overriding

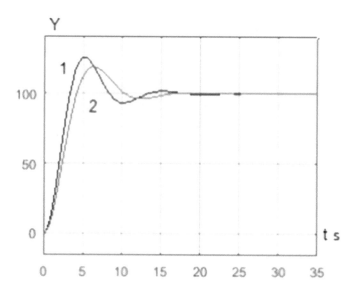

Fig. 4. Transient characteristics of the system at four times multiplication of the object gain when tolling is $1 - M = 6$, $2 - M = 4$.

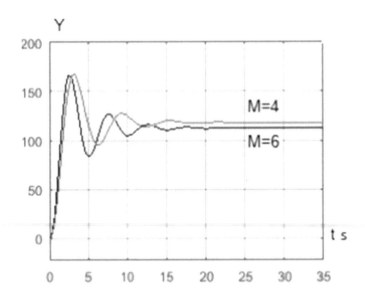

Fig. 5. Transient characteristics of the system when increasing of the object gain is from 10 to 40 for different tooling of the controller M = 4, M = 6.

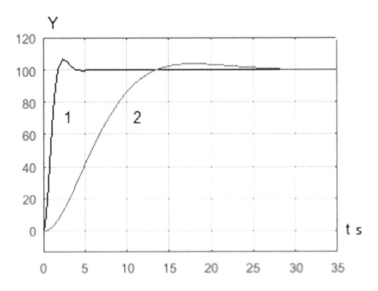

Fig. 6. Transient characteristics of the system for different constant time of the object.

5   Conclusions

The combined robust system with a fuzzy P-controller in which an additional bias signal was introduced into the regulatory action of the object has been constructed. Implementation of this technique creates an advance proportional to the input task in which it is theoretically possible to obtain a zero static error in the system without introducing into the control algorithm of the I-component, which is "lagging" from the error of regulation. Structural scheme of fuzzy control system model with fuzzy controller based on the Simulink package of the Matlab software has been constructed. Denominational (zooming) control has been made taking into account the additional bias signal. The results of simulation for object of the second order at variation of object parameters have been presented. A fourfold increase in the transfer factor causes the change of overregulation from 5% to 20%. A fivefold increase/decrease in time has little effect on the quality of regulation. Based on the simulation modeling, transient characteristics of the system, it has been established that the proposed fuzzy P-controller with a biasing signal creates and ensures stable operation of the system and good quality of regulation when changing the parameters of the object within wide limits and has robust properties. The advantages of the combined P-controller enhance its use for nonstationary objects control.

References

1. Brzozka J (2004) Regulatory i układy automatyki. Mikom, Warszawa
2. Ivakhi, O, Nakonechny M (2017) Fundamentals of construction of control systems with fuzzy logic. National University Lviv Polytechnic University, Lviv: Raster-7, 129 p. ISBN 978-617-7497-13-3
3. Gostev VI (2008) Fuzzy controllers in automatic control systems. Monograph, Kyiv, Radio Amateur, 972 p. ISBN 978-966-96178-2-0
4. Kwaśniewski J, Sterowniki P (2008) PLC w praktyce inżynierskiej. Wydawnictwo BTC Legionowo, 344 p. ISBN 978-83-60233-35-1
5. Lutskiv M (2017) Systems of automatic control with a simplified version of the P-fuzzy controller. Computer technologies of printing: sb. sciences works, Lviv, vol 1, no 37, pp 25–31
6. Lutskiv M (2017) De-normalization of management in systems with fuzzy controller. Computer technologies of printing: sb. sciences Works, Lviv, vol. 2, no 38, pp 16–24
7. Rutkowski L (2012) Metody i techniki sztucznej inteligencji. Wydawnictwo Naukowe PWN, Warszawa
8. Piegat A (1999) Modelowanie i sterowanie rozmyte Akademicka Oficyna. Wydawnicza EXIT, Warszawa
9. Asai K, Watada D, Terano EdT, Sukeno M (1993) Applied fuzzy systems. Mir, Kyiv, 368 p
10. Sivanandam SN, Sumathi S, Deepa S (2006) Introduction to fuzzy logic using MATLAB. Springer, Heidelberg ISBN 3540357807
11. Chen G, Trung TP (2000) Introduction to fuzzy sets, fuzzy logic, and fuzzy control systems. Mathematics. CRC Press, Boca Raton, 328 p

12. Shtovba SD (2007) Designing fuzzy systems by MATLAB. Hotline, Telecom, Moscow, 288 p
13. Asiain MJ, Bustince H, Mesiar R, Kolesarova A, Takac Z (2018) Negations with respect to admissible orders in the interval-valued fuzzy set theory. IEEE Trans Fuzzy Syst https://doi.org/10.1109/TFUZZ.2017.2686372
14. Valdez F, Melin P (2008) Comparative study of PSO and GA for complex mathematical functions. J Autom Mob Robot Intell Syst 2(1):43–51
15. Martinez R, Castillo O, Aguilar L (2009) Optimization of type-2 fuzzy logic controllers for a perturbed autonomous wheeled mobile robot using genetic algorithms. Inf Sci 179–192:2158–2174
16. Tanaka K, Wang HO (2001) Fuzzy control systems design and analysis: a linear matrix inequality approach. Wiley, New York
17. Seraya OV, Demin DA (2012) Linear regression analysis of a small sample of fuzzy input data. J Autom Inf Sci 44(7):34–48. https://doi.org/10.1615/jautomatinfscien.v44.i7.40
18. Roldan-López-de-Hierro A-F, Karapınar E, Manro S (2014) Some new fixed point theorems in fuzzy metric spaces. J Intell Fuzzy Syst 27(5):2257–2264
19. Soualhi A, Medjaher K, Zerhouni N (2015) Bearing health monitoring based on Hilbert-Huang transform, support vector machine, and regression. IEEE Trans Instrum Measur 64(1):52–62. https://doi.org/10.1109/TIM.2014.2330494 Article no 6847199
20. Mendel JM (2001) Uncertain rule-based fuzzy logic systems: introduction and new directions. Prentice Hall PTR, Los Angeles
21. Sharma A, Barve A (2012) Controlling of quad-rotor UAV using PID controller and fuzzy logic controller. Int J Electr Electron Comput Eng 1(2):38–41
22. Kolbari H, Sadeghnejad S, Parizi AT, Rashidi S, Baltes JH (2016) Extended fuzzy logic controller for uncertain teleoperation system. In: Proceedings of the 4th international conference on robotics and mechatronics (ICROM 2016), Tehran, Iran, October 2016, pp 78–83
23. Mahmoud MS, Saif A-WA (2012) Robust quantized approach to fuzzy networked control systems. IEEE J Emerg Sel Topics in Circuits Syst 2(1):71–81

Investigation of Random Neighborhood Features for Interpretation of MLP Classification Results

Oleksii Gorokhovatskyi$^{(\boxtimes)}$ ⓘ, Olena Peredrii ⓘ, Volodymyr Zatkhei ⓘ,
and Oleh Teslenko ⓘ

Simon Kuznets Kharkiv National University of Economics, Kharkiv, Ukraine
`oleksii.gorokhovatskyi@gmail.com`, {`elena_peredriy,zathey_va`}`@ukr.net`,
`tov1967@meta.ua`

Abstract. In this paper, we propose an investigation of the random neighborhood features and their classification results in order to get explanations for a particular decision and increase the interpretability of the black box model. It is shown (by examples from the known iris and Wisconsin breast cancer datasets), that even if outputs of neural network model seem to be confident, it doesn't necessarily mean that decision is stable enough. Investigation of statistic properties of decisions near the input feature vector shows the importance of some features over others. After that, we build random forest (as a method with a high level of interpretability) on a set of generated neighbor input vectors. Investigation of separate nodes of solution routes in each decision tree independently shows that it is possible to find counterfactuals of feature vectors, classified by a model as a different class. As a result, the human gets not only the explanation of current classification but also examples of feature vectors representing other classes, that differs from initial input feature vector just with particular features keeping these changes as low as possible.

Keywords: Classification explanation · Black box model ·
Multilayered perceptron · Random neighborhood · Decision tree ·
Interpretability · Counterfactual

1 Introduction

The question if a human can trust the decisions made by automatic artificial intelligent (AI) systems is one of the most interesting last years with the ever-increasing amount of AI tools, embedded in daily life. Majority of the modern models, especially those based on neural networks, represent a "black box" model, which can produce a high-quality solution of a problem (typically classification or regression), but not able to introduce any explanation of result provided. The property of the algorithm/model to explain its decision in a human-friendly form is called an interpretation, which forms the basis of the explainable (transparent) artificial intelligence (XAI) idea [1,2]. Neural networks can provide the best accuracy amongst all

ⓒ Springer Nature Switzerland AG 2020
V. Lytvynenko et al. (Eds.): ISDMCI 2019, AISC 1020, pp. 581–596, 2020.
https://doi.org/10.1007/978-3-030-26474-1_40

existing machine learning methods (which explains their popularity), but at the same time, they are least explainable. The decision tree, on the other hand, is one of the most explainable models but its performance is lowest. Random forest, statistical models, Bayesian nets, Markov models, etc. are in between high accuracy and high explainability in this hierarchy [1].

Usually, the problem of explanation arises when black box models are used in sensitive environments like identification, medical diagnostic and health care, etc. and this kind of problem remains very important for humanity. Cases, when a black box model provides the result that is better than provided by a human on the same initial data are of particular interest. They might be helpful for humans to learn some new principles or dependencies only if black box model is interpretable. Additionally, the fairness problem arises at the usage of the black box model. Does this model take into account only features that it really should? Doesn't it take into account some accidental features, which should be ignored or were unknown previously? If the model is interpretable, answers to these questions may be found.

2   Review of the Literature

The quantity of scientific researches about interpretability of black box models increases over recent years.

One of the most famous methods of searching interpretable explanations nowadays is LIME - Local Interpretable Model-Agnostic Explanations [3,4] which could theoretically be applied to any model. The goal of the method's implementation is to search for the relation between the input vector features, which is human-friendly (the examples with input data in the form of text and images are shown in the paper [3]). The idea of LIME is to construct a separate interpretable model, which uses artificial data, located in the neighborhood of the input feature vector as input data. The disadvantage of the LIME method is the need to select only some of the features of the input vector for analysis. Their number should be small enough to guarantee user-friendly interpretation, but sufficient to fit the original model. Considering all available features is possible only if their number is small. Issues with the criteria for choosing neighbor inputs around the input data, as well as the instability of explanations that can vary significantly even for two very close inputs are often referred to be other drawbacks of LIME [5].

Among other methods which are known to construct interpretations, it is possible to point out SHAP (SHapley Additive exPlanations) [5–7]. The core of it is the application of the game theory to features analysis, which represents each feature as a player for which you want to calculate the contribution (Shapley value) in the overall solution. The main drawback of this approach is a significant computational complexity since it is necessary to review all possible combinations of all features. Unlike LIME, this method has a complete theoretical background.

"What-If" tool [8], presented by Google, allows visualizing work of model on a dataset. One of its functions is searching for counterfactual for input feature vector - the closest vector from the existing dataset that requires minimal changes and represents other class.

Numerous methods of visualization have also been actively used, the values of weight coefficients in the structure of the neural network may reveal the "perception" by the network of certain (usually the most important) features of the input vector [9–15].

3   Problem Statement

A goal of the paper is the investigation of the possibility to search for local explanations of particular classification result provided by a black box model. Importance of the separate input features, as well as the overall confidence of decision made, is based on the neighborhood of the input features numeric vector. The contribution of the paper is not only in obtaining the explanation but also in search for the closest counterfactual cases when minimal (numerical) changes in initial feature vector lead to the classification result change. Set of vectors to search counterfactuals inside is based on real cases of black box model result, which shows its real work, but not on items in existing limited (train or test) dataset. The significance of such feature changes is a subject for further human analysis, because in most cases only the expert having information about feature values acquisition, can say whether such changes are meaning.

We will use multilayer perceptron (MLP) neural network as a black box model to find explanations for, statistic methods to investigate neighborhood of input features vector, and decision trees as good transparent models.

4   Materials and Methods

4.1   Structure of Network and Training Policy

Multilayered perceptron with a single hidden layer is one the simplest known models of neural networks, taking into account its possibilities for solving classification problems with classes that are not linearly separable. The first layer of the perceptron accepts an input feature vector, which is typically normalized before. In our case, we divide each feature value on its mean value, calculated across the whole training set. These mean values are used later to scale any input vector before classification. We will use Andrews curves [16] for visualization of multidimensional data, which transform a multidimensional point into a Fourier series function. Figure 1 shows the Andrews curves for the famous Iris dataset [17] and the corresponding curves for the Wisconsin Breast Cancer Diagnostic dataset [18] are presented in Fig. 2.

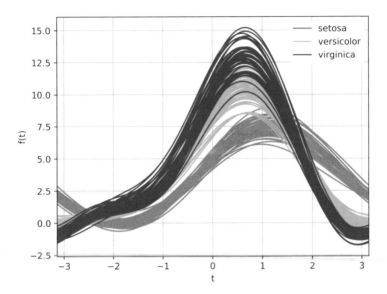

Fig. 1. Andrews curve visualizations for iris dataset entries

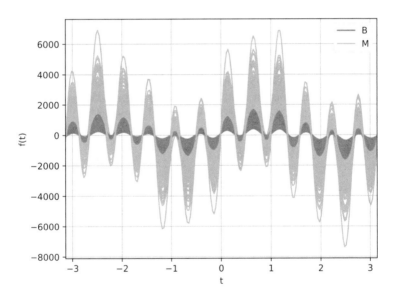

Fig. 2. Andrews curve visualizations for Breast Cancer Wisconsin Diagnostic dataset entries

The classic version of the sigmoid activation function is used for each neuron:

$$y(x) = \frac{1}{1 + e^{-z}}, \; z = wx + b$$

where x is the input value, w is the weight, b is the bias.

The output value of the j-th H_j neuron of hidden layer can be written as $H_j = f\left(\sum_i (w_{ij}x_i)\right)$, where x_i is the input value (or output from corresponding neuron of a previous layer), correspondingly, output of k-th neuron Y_k of last layer can be written as $Y_k = f(\sum_j(w_{jk}H_j))$.

Output vector o is formed at the last layer, its length is equal to the available amount of classes K, e.g. vector for first class is $o_1 = (1, 0, ..., 0)$, the second one – $o_2 = (0, 1, ..., 0)$ etc.

The number of hidden network layers and the number of neurons in each such layer are selected as a result of some modeling experience. In our investigations, we used a single hidden layer with two neurons. The network is trained with the teacher using the error backpropagation algorithm based on gradient descent and includes traditional three stages:

- forward propagation of the input signal and getting output values for each neuron;
- error backpropagation;
- weights and biases correction.

Error for k-th neuron of output layer, which relates to i-th neuron of previous layer is calculated as follows:

$$\delta_k = (y_{desired\_k} - y_k)y_k(1 - y_k) \tag{1}$$

where $y_{desired\_k}$ is the desired output value for neuron k.

Error for j-th neuron of hidden layer, which relates to i-th neuron of previous layer is calculated as follows:

$$\delta_j = h_j(1 - h_j)\sum_k w_{jk}\delta_k \tag{2}$$

Weights correction is done according to the following:

$$w = w + \Delta w, \; \Delta w = \eta\delta x + m\Delta w, \; b = b + \eta\delta$$

where η is the learning rate ($\eta = 0.2$ value was used in our implementations), δ – error values according to (1) or (2), x – the output value of previous neuron, m – momentum value, which helps to avoid stuck at local minimum, (our value during modeling was $m = 0.8$).

The network learning stops if the maximum number of iterations I exceeds some limit or the number of errors in the classification of the training set is less than the given value E. Comparison of the current outputs of the last layer with the desirable ones was made using the Manhattan distance, the total absolute error should be less, than the 0.1. However, if the current number of iterations exceeds I_c and training is not finished, the process is accelerated by changing the comparison distance, which takes into account only the position of the maximum value in the vector obtained from the results of the output of the network:

$$d = \sum_{k=1}^{K} |\hat{o}_{desired\_k} - \hat{o}_k|, \; \hat{o}_i = \begin{cases} 1, \; y_i = \max_i y_i \\ 0, \end{cases} \tag{3}$$

where $y = (y_1, y_2, ..., y_k)$ is the output vector.

4.2   Neighborhood Exploration

Let's consider the vector $y = (y_1, y_2, ..., y_k)$, formed at the output of the multilayer perceptron, whose value allows us to make the final decision regarding the class to which the input vector $x = (x_1, x_2, ..., x_F)$ belongs to, F is the amount of features. How can you understand the level of decision "confidence" of the network when forming this vector? Often this task is facilitated by using the softmax activation function on the last layer of the network, which provides representations of output values as probabilities with a sum equal to 1. But as practice shows, the high probability level doesn't always correspond to the high level of decision confidence.

Instead of making the decision based on the input vector $x = (x_1, x_2, ..., x_F)$ only, let's expand the behavior of the model onto N values that are uniformly distributed on a surface of the F-dimensional ellipsoid around x vector in a multidimensional space. In order to get a vector $\tilde{x} = (\tilde{x}_1, \tilde{x}_2, ..., \tilde{x}_F)$, that is close to x, we will use the expression [19]:

$$\tilde{x} = (\tilde{x}_1, \tilde{x}_2, ..., \tilde{x}_F) = \begin{pmatrix} x_1 \\ x_2 \\ ... \\ x_F \end{pmatrix} + \frac{1}{\sqrt{u_1^2 + u_2^2 + ... + u_F^2}} \begin{pmatrix} u_1 r_1 \\ u_2 r_1 \\ ... \\ u_F r_F \end{pmatrix}$$

where $u = (u_1, u_2, ..., u_F)$ is the unit vector of random variables, distributed normally, $r_1, r_2, ..., r_F$ – radiuses for each dimension, $t_i = \alpha s_i, \overline{i = i, F}$, where s_i is the reach (the difference between maximum and minimum values) of normalized feature i from training set, α is reach scale factor.

Collecting the results of the classification of N separate experiments allows us to estimate the closeness of the other existing classes to the class-decision. The result of the classification can be considered confident if all the experiments yielded the same result.

If the classification results of all N experiments indicate not a single class, one can evaluate the importance of each input feature by constructing S^k vectors, which are compiled from the weighted sum of the deviations of each feature for each of the classes received separately:

$$S^k = \left(s_1^k, s_2^k, ..., s_F^k \right)$$

$$= \left(\frac{1}{N_k} \sum_{N_k} \left(x_1 - \tilde{x}_1^k \right), \frac{1}{N_k} \sum_{N_k} \left(x_2 - \tilde{x}_2^k \right), ..., \frac{1}{N_k} \sum_{N_k} \left(x_F - \tilde{x}_F^k \right) \right), \quad (4)$$

$$\forall k, k = \overline{1, K'}, K' \le K$$

where K' is the quantity of different classes, obtained in N results of classifications, $N_k < N$ – amount of experiments, showing that k class is the result, $\sum_{k=1}^{K'} N_k = N$.

4.3   Decision Trees

Decision trees and random forests [20,21] are better interpreted compared to neural networks [1,2] and may provide a basis for an understanding reasoning for making a decision.

The disadvantage of random forests is their tendency to overfit, which in our case is not so important. We will use the decisions provided by separate forest trees to explain the classification results of a more general model with further verification of this decision by the model itself. So there is no concept of a test set, and we will build decision trees on experiments in the neighborhood of the input vector $x = (x_1, x_2, ..., x_F)$, whose solution we initially need to explain.

The assessment of the quality of the partition during the construction of the decision tree was done out with the help of the Gini coefficient. The number of trees and their depth are experimentally chosen. Keeping depth as less as possible seems to be a good idea because this is directly about human-friendly explanation: perception of a lot of features in the deep tree might be complex for a human.

Let's consider one of decision tree d that was built on N experiments in neighborhood of the input vector x. Let k_i be the current class, provided as solution by tree d. Let us denote as $(r_1, ..., r_g)$ – set of rules for corresponding nodes $g \leq G$, where G is the depth of the tree. These rules explain the decision made by tree in favor of class k_i.

In order to get another route in the same tree d that points to some another class k_j, $i \neq j$ we traverse the route of the tree from a bottom node by node. The rule in each node compares some input feature with value, that was assigned earlier as a result of decision tree building. We iterate the decision route from the last node to the first, change value of the feature in a node g to make rule r true for another branch and check immediately if the class is changed. If this happened, we have successfully found an explainable route with minimal changes in initial input features.

Decision trees allow us to get an explanation and easily find such changes in specific features of the initial input vector, that makes sense about the importance of features and confidence of decision made. Random forest performs the accumulation of different solutions from all decision trees.

5   Experiment

5.1   Datasets

We will test our suggestions on two popular tabular machine learning datasets: Iris [17] and a Wisconsin Breast Cancer Diagnostic datasets [18]. Both are classification tasks.

Iris dataset contains 150 measurements, each is characterized by four features (sepal length, sepal width, petal length, petal width) and belongs to one of the three classes (setosa, virginica, versicolor).

The use of multilayer perceptron is widespread enough for the classification of this dataset, the experimental strategies and the structure of the neural network are different from paper to paper. For example, the investigation of the structure of a perceptron with a single hidden layer and 3 and 4 neurons is described in [22], 120 feature vectors of the dataset were used for training, 30 for testing. Paper [23] describes the process of training a neural network with a single hidden layer, containing 3 neurons and a single output value, 75 measurements were used for training and testing. The complete overview of possible combinations of network structures with one layer or two is presented in [24], this information is used to find the most efficient structure in terms of accuracy, 120 measurements were selected as test ones. Investigation [25] contains the discussion of the genetic algorithms used to obtain the optimal network structure, resulting in a perceptron with three hidden layers, each of which contains 4 neurons. Results of correct classification rates in different documented investigations are in the range 90–100%.

Wisconsin Breast Cancer Diagnostic dataset contains 569 feature vectors of 30 features derived from a digital image. 212 measurements correspond to malignant formations, 357 – to benign ones. The usage of a multilayered perceptron is not typical for this dataset.

The investigation of the classification process with the use of k-means and SVM with various partitions between training and test sets (from 50/50% to 80/20%) is presented in [26] with the classification accuracy over 99%. Nice overview of different machine learning approaches for solving the problem of breast cancer classification with an accuracy of over 97% is shown in [27,28].

5.2   Quality of Trained Networks

We will train a lot of multilayered perceptron neural networks on [17] and [18] datasets and make sure that the quality of these networks is good enough compared to previously published results. We also need to find such initial input vectors in both datasets, which cause problems during training and/or classification.

Table 1 contains information about values of parameters I, E and I_c that were used during training. That is important to note that expression (3) was used in all cases related to testing set items classification. 10 000 neural networks with 4 neurons in the input layer, 2 hidden layer neurons, and 3 outputs have been trained on iris dataset. 120 training and 30 test measurements were randomly

Table 1. Training parameters

Parameter	Iris dataset	Wisconsin breast cancer dataset
Maximum amount of iterations I	100000	300000
Maximum amount of errors E	2	5
Amount of iterations I_c	10000	20000

selected for each of these models. The average classification accuracy for the training set was 0.97, for test set – 0.95. Input vectors in [17] have no unique identifier, so we enumerate them from #1 (values 5.1, 3.5, 1.4, 0.2, setosa) to #150 (5.9, 3.0, 5.1, 1.8, virginica).

During training procedures we had classification problems with such input vectors: #84: 6.0, 2.7, 5.1, 1.6, versicolor (7844 trained networks after learning were unable to classify this item correctly); #71: 5.9, 3.2, 4.8, 1.8, versicolor (5770 networks); #73: 6.3, 2.5, 4.9, 1.5, versicolor (3883 networks).

Such input vectors from testing sets caused classification errors: #84: 6.0, 2.7, 5.1, 1.6, versicolor (2082 networks were unable to classify this item correctly); #71: 5.9, 3.2, 4.8, 1.8, versicolor (2042 networks); #134: 6.3, 2.8, 5.1, 1.5, virginica (1839 networks).

10 independent experiments of 10-fold cross-validation were performed for Wisconsin breast cancer dataset [18]. All networks had 30 neurons in the input layer, 2 hidden layers of neurons and 2 outputs. The average classification accuracy of training samples was 0.9925, for test set – 0.9669.

These input items caused classification problems during training: #892189: (90 networks after training were unable to classify this item correctly); #859983 (83 networks); #868202 (64 networks), problematic cases during processing of the test part of dataset are: #892189, #855133, #868202, #859983 and #855167 (for all of them all 10 networks were unable to classify these items correctly).

As a result of this stage, we have a confirmation that selected architectures of multilayer perceptron and their training policies for both problems are good enough to use these models for searching for explanations.

6   Results and Discussion

6.1   Neighborhood Analysis

Let's investigate the decision making for input measurement #84 of iris dataset, that was found to be problematic both during training and classification. The correct class label for this vector is "versicolor". The vector obtained as output from one of the trained networks is (0; 0.01; 0.99) and indicates the class "virginica". Applying investigation of model's behavior in the neighborhood of this input with reach scale factor $\alpha = 0.1$ we obtained 180 additional neighbor results with the label "versicolor" and 820 labeled "virginica" cases for $N = 1000$ experiments. Andrews curves visualization of the initial values of the classified input feature vector (labeled "virginica_result") and for the two closest results for the virginica and versicolor classes are shown in Fig. 3.

As it can be seen, the curves are very close, which makes a confident decision only by the outputs from the last neural network layer impossible. Bar chart of S^k (4) features differences for two classes, found in the neighborhood of input vector is presented in Fig. 4. As one can see, the major deviations between input vectors in different classes are about two last features (labeled by 2 and 3). Both values of these features are typically greater for virginica class and less for versicolor in comparison to the initial input vector.

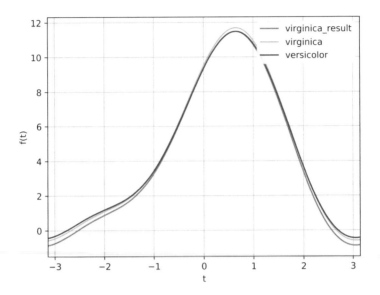

Fig. 3. Visualization of Andrews curves of input vector and two closest neighbors belonging to different classes

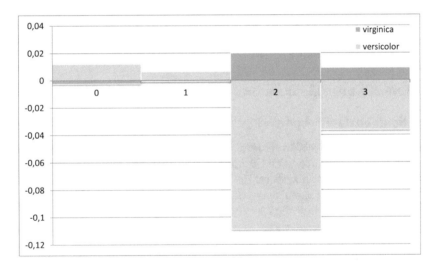

Fig. 4. Visualization of bar chart of S^k features deviation (iris dataset)

Analysis of the classification result for input vector #868202 from Wisconsin breast cancer dataset follows. The correct class label for this vector is "malignant" (M). The vector obtained as output from the trained network is (0.05; 0.95), indicating the "benign" (B) class. Inspection of $N = 1000$ classification results around the input vector at scale factor $\alpha = 0.1$ allowed to get 132 results labeled with "M" and 868 with label "B".

Figure 5 contains the visualization of Andrews curves for the initial values of the classified initial feature vector (B_result) and for the two closest B and M results. As can be seen graphically, the curves are close enough to understand, that the result of the classification in favor of B is not confident enough. Additionally, visualization of S^k components (Fig. 6) shows, that features #5–7, 13, 16–19, 22, 25, 26 seem to be the most influential at decision making.

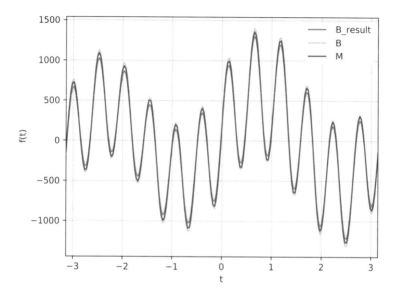

Fig. 5. Visualization of Andrews curves of input neighbor vectors belonging to different classes

Looking at examples, which caused no problems during training of MLP, e.g. #32 or #871201 for iris/breast cancer datasets correspondingly, we saw that all N random neighbor cases were associated with a single (correct) class.

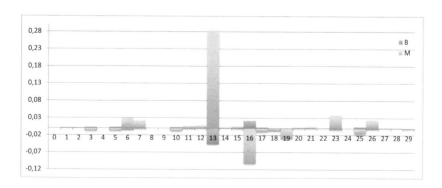

Fig. 6. Visualization of bar chart of S^k features deviation (Wisconsin breast cancer dataset)

6.2 Decision Tree Analysis

Let's consider searching for an explanation of the solution for the examples given above with the help of decision trees. Graphic rendering of a tree below is constructed in the space of normalized features (as decision trees were built on them), the explanation of solutions is denormalized (to make it better for human perception).

The initial values of the input vector #84 (6.0; 2.7; 5.1; 1.6) in the normalized form are: (1.02; 0.88; 1.36; 1.33). 3 decision trees out of 50 constructed were helpful to find not only an explanation of solution but also modify the input normalized vector in a way to obtain the opposite classification result. Example of such decision tree is presented in Fig. 7.

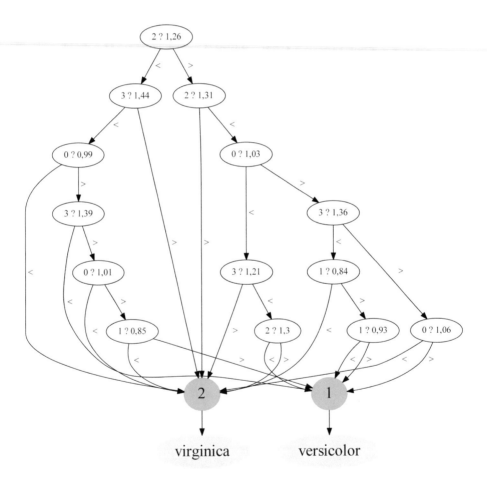

Fig. 7. One of decision trees, constructed for explanations and searching of the counterfactual (iris dataset)

Example of solution explanation output for this case is provided below. As one can see from lines #9–11 current "virginica" result was obtained because the second feature is greater than 4.72 (corresponding normalized value is 1.26, top node in Fig. 7), then the same feature is compared to 4.91 value (corresponding normalized one is 1.31, line #11). Counterfactual with changed features is provided in lines #12–17. As you can note from line #17, if you decrease feature #2 on a 0.3822 value, the result of classification by the neural network (8.3747E-06; 0.7668; 0.4222) points to another class. At this stage human can make a decision himself: whether provided value 0.3822 for this second feature is significant or not.

As one can see from the tree (Fig. 7), features #2 and #3 are the top nodes, that commonly matches Fig. 4, showing, that they are more valuable than other ones.

#1 *Start of the tree*
#2 *Normalized inputs*
#3 1.0248 0.8812 1.3613 1.3333
#4 *Denormalized inputs* :
#5 6 2.7 5.1 1.6
#6 *NN outputs* :
#7 $8.9507E - 06$, 0.01250, 0.9895
#8 *Result of tree = virginica*
#9 *Current resolution explanation*
#10 $2 : 5.1 > 4.72(1.26)$
#11 $2 : 5.1 > 4.91(1.31)$
#12 *– Another classcase*
#13 *Updated features denormalized* :
#14 6, 2.7, 4.71776278991597, 1.6
#15 *Updated outputs of NN* :
#16 $8.3747E - 06$, 0.7668, 0.4222
#17 *Denormalized differences* $= 0, 0, -0.3822, 0$
#18 *End of the tree*

The result of solution explanation for item #868202 of Wisconsin breast cancer diagnostic dataset is presented in the Appendix. In this case, 42 trees out of 50 were able to provide counterfactual with other class as a classification result.

7 Conclusions

Searching for human-friendly explanations of results, provided by black box models, is a promising technology, which might make embedding of artificial intelligence tools into sensitive data processing fields much easier.

The proposed approach is based on the investigation of the neighborhood (points on a surface of the multidimensional ellipsoid) of the input feature vector in order to understand whether a solution, suggested by model, is confident enough. It is shown, that even if outputs of a model vote in favor of some class surely, it doesn't necessarily mean that decision is stable. Investigation of classification results in feature vectors near input one may help to see if the decision is made on a border between different classes. There is no special artificial selection of significant features. A simple gathering of statistic properties according to (4) shows the early information about features, which are more important than others.

Additionally, we tried to answer the question "What should be changed in the input feature vector to get it classified as another class?". We tested the random forest usage in order to get direct explanations of a solution. Random forest is built in the neighborhood of the input vector, but the path solution of each decision tree is analyzed separately. Changing feature by feature in this path sometimes makes possible to find a similar route, pointing to the different solution. After this, a human can look at these different input feature vectors, explanations for them and make a final decision if this difference is significant.

The difficulties of the proposed approach relate to the necessity of parameters tuning: N, α, quantity of trees in the forest. Performance of the model is a potential issue too because we need computational time to perform classification of neighbor vectors.

In the case when all N experiments in some reach α are classified as representatives of a single class, the decision is considered to be confident. Searching for explanations based on the proposed approach is not applicable here, because the algorithm requires a counterfactual to find differences between features. Reach factor α might be increased in this case to find such examples.

Further development of the proposed method may be related to applications for image classification result explanations.

A   Appendix

Classification explanation output for item #868202 of Wisconsin breast cancer diagnostic dataset.

Start of the tree
Normalized inputs:
0.9044 1.1627 0.8887 0.7730 0.9391 0.5513 0.5283 0.5508 0.8755 0.9661 0.5842
1.1276 0.5074 0.4891 1.0612 0.4725 0.7466 0.7644 0.7998 0.7049 0.8905 1.2938
0.8575 0.7405 1.0698 0.5960 0.7964 0.8144 0.9739 0.9622
Denormalized inputs:
12.77 22.47 81.72 506.3 0.09055 0.05761 0.04711 0.02704 0.1585 0.06065 0.2367
1.38 1.457 19.87 0.007499 0.01202 0.02332 0.00892 0.01647 0.002629 14.49 33.37
92.04 653.6 0.1419 0.1523 0.2177 0.09331 0.2829 0.08067
NN outputs:
0.05463, 0.9453
Result of tree = B (1)
Current resolution explanation
13: 19,87 < 28,05 (0.69)

16: 0,02332 > 0,02 (0.64)
6: 0,04711 < 0,068 (0.77)
17: 0,00892 > 0,0069 (0.6)
7: 0,02704 < 0,034 (0.71)
28: 0,2829 > 0,26 (0.91)
-Another class case
Updated features denormalized:
12.77, 22.47, 81.72, 506.3, 0.0905, 0.0576, 0.0683, 0.0347, 0.1585, 0.0606, 0.2367, 1.38, 1.457, 28.0929, 0.0074, 0.0120, 0.0200, 0.0069, 0.0164, 0.0026, 14.49, 33.37, 92.04, 653.6, 0.1419, 0.1523, 0.2177, 0.0933, 0.2643, 0.0806
Updated outputs of NN:
0.9993, 0.0006
Denormalized differences = 0, 0, 0, 0, 0, 0, 0.0212, 0.0076, 0, 0, 0, 0, 0, 8.2229, 0, 0, −0.0032, −0.0019, 0, 0, 0, 0, 0, 0, 0, 0, 0, 0, −0.0185, 0
End of the tree

References

1. Gunning D: Explainable Artificial Intelligence (XAI). https://www.cc.gatech.edu/~alanwags/DLAI2016/(Gunning)%20IJCAI-16%20DLAI%20WS.pdf. Accessed 17 Feb 2019
2. Gunning D: Explainable Artificial Intelligence (XAI). https://www.darpa.mil/attachments/XAIProgramUpdate.pdf. Accessed 17 Feb 2019
3. Ribeiro MT, Singh S, Guestrin C: "Why should i trust you?" explaining the predictions of any classifier. https://arxiv.org/pdf/1602.04938.pdf. Accessed 17 Feb 2019
4. Ribeiro MT: LIME - local interpretable model-agnostic explanations. https://homes.cs.washington.edu/~marcotcr/blog/lime/. Accessed 17 Feb 2019
5. Molnar C: Interpretable machine learning. a guide for making black box models explainable. https://christophm.github.io/interpretable-ml-book. Accessed 17 Feb 2019
6. SHAP. https://github.com/slundberg/shap. Accessed 17 Feb 2019
7. Lundberg SM, Lee S-l (2017) A unified approach to interpreting model predictions. In: Guyon I, Luxburg UV, Bengio S, Wallach H, Fergus R, Vishwanathan S, Garnett R (eds) Advances in neural information processing systems 2017. Curran Associates, Inc, Red Hook, pp 4765–4774
8. The what-if tool: code-free probing of machine learning models. https://ai.googleblog.com/2018/09/the-what-if-tool-code-free-probing-of.html. Accessed 10 Mar 2019
9. Saliency methods. https://github.com/PAIR-code/saliency. Accessed 17 Feb 2019
10. Simonyan K, Vedaldi A, Zisserman A: Deep inside convolutional networks: visualising image classification models and saliency maps. https://arxiv.org/pdf/1312.6034.pdf. Accessed 17 Feb 2019
11. Sundararajan M, Taly A, Yan Q (2017) Axiomatic attribution for deep networks. In: Precup D, Teh YW (eds) International conference on machine learning. PMLR, Sydney, pp 3319–3328
12. Erhan D, Bengio Y, Courville A, Vincent P: Visualizing higher-layer features of a deep network. https://pdfs.semanticscholar.org/65d9/94fb778a8d9e0f632659fb33a082949a50d3.pdf. Accessed 17 Feb 2019

13. Perazzi F, Krähenbühl P, Pritch Y, Hornung A (2012) Saliency filters: contrast based filtering for salient region detection. In: 2012 IEEE conference on computer vision and pattern recognition. IEEE, New York, pp 733 – 740
14. Kindermans P, Schutt KT, Alber M, Muller K, Erhan D, Kim B, Dahne S: Learning how to explain neural networks: PatternNet and PatternAttribution. https://arxiv.org/pdf/1705.05598.pdf. Accessed 17 Feb 2019
15. Exploring Neural Networks with Activation Atlases. https://distill.pub/2019/activation-atlas. Accessed 09 Mar 2019
16. Andrews DF (1972) Plots of high dimensional data. Biometrics 28:125–136
17. Iris Data Set. http://archive.ics.uci.edu/ml/datasets/Iris. Accessed 17 Feb 2019
18. Breast Cancer Wisconsin (Diagnostic) Data Set. https://archive.ics.uci.edu/ml/datasets/Breast+Cancer+Wisconsin+(Diagnostic). Accessed 17 Feb 2019
19. Hypersphere Point Picking. http://mathworld.wolfram.com/HyperspherePointPicking.html. Accessed 17 Feb 2019
20. Amit Y, Geman D (1997) Shape quantization and recognition with randomized trees. Neural Comput 9:1545–1588
21. Breiman L (2001) Random forests. Mach Learn 45:5–32
22. Abdulkadir RA, Imam KA, Jibril MB (2017) Simulation of back propagation neural network for iris flower classification. Am J Eng Res (AJER) 6(1):200–205
23. Swain M, Dash SK, Dash S, Mohapatra A (2012) An approach for iris plant classification using neural network. Int J Soft Comput (IJSC) 3(1):79–89. https://doi.org/10.5121/ijsc.2012.3107
24. Sharma L, Sharma U (2014) Neural network based classifier (pattern recognition) for classification of iris data set. Int J Recent Dev Eng Technol 3(2):64–66
25. Ramchoun H, Idrissi MAJ, Ghanou Y, Ettaouil M (2017) New modeling of multilayer perceptron architecture optimization with regularization: an application to pattern classification. Int J Comput Sci 44(3):261–269
26. Sridevi T, Murugan A (2014) An intelligent classifier for breast cancer diagnosis based on k-means clustering and rough set. Int J Comput Appl 85(11):38–42
27. Hazra A, Mandal SK, Gupta A (2016) Study and analysis of breast cancer cell detection using Naive Bayes, SVM and ensemble algorithms. Int J Comput Appl 145(2):39–45
28. Salama GI, Abdelhalim MB, Zeid MA (2012) Breast cancer diagnosis on three different datasets using multi-classifiers. Int J Comput Inf Technol 1(1):36–43

Optimizing the Computational Modeling of Modern Electronic Optical Systems

Lesia Mochurad$^{(\boxtimes)}$ ⓘ and Albota Solomiia$^{(\boxtimes)}$ ⓘ

Lviv Polytechnic National University, Lviv, Ukraine
{Lesia.I.Mochurad,Solomiia.M.Albota}@lpnu.ua

Abstract. The paper focuses on one of the classes of problems concerning mass calculations – the problems of calculating the electrostatic fields of modern electronic optical systems. In a mathematical modeling process of potential fields, it is necessary to solve systems of linear algebraic equations of large dimensions. The tasks of calculating electrostatic fields can be greatly simplified by maximizing the geometric symmetry within the configuration of the electrode surfaces. The usage of the group theory technique allows to ensure the stability of the calculations, create all the prerequisites for parallelizing the procedures for solving complex three-dimensional problems concerning electrostatics in general. The application of the mathematical modeling methods of electrostatic fields along with the modern trends within the computer system development has been considered in order to reduce the calculation time. The optimisation of the computational process in solving electronic optics tasks due to the rapid development of modern nanotechnology and new requirements for the speed of calculations, has been implemented using the OpenMP parallel programming technology. The classes of systems with symmetries of the eighth and sixteenth orders have been considered.

Keywords: Cmass calculations · Multi-core system · Mathematical modeling · System of linear algebraic equations · OpenMP technology

1 Introduction

Commonly, when solving practical problems of science and technology, there is an issue of effective organization of calculations. Some tasks involve the analysis of large data and decision-making in real time [1]. Increasing size of the input data and the mass of computations leads to an increase in the complexity of problem solving. Therefore, it is necessary to improve the existing approaches to the organization and performance of computations, and to develop new ones, using high-performance computing systems.

The problems of calculating the potential fields of electronic optics systems belong to the important class of mass calculations. In the process of mathematical modeling of modern systems of electronic optics, there is a problem of determining the electrostatic field created by the system of charged electrodes.

ⓒ Springer Nature Switzerland AG 2020
V. Lytvynenko et al. (Eds.): ISDMCI 2019, AISC 1020, pp. 597–608, 2020.
https://doi.org/10.1007/978-3-030-26474-1_41

The analytical calculation of the field is very complicated or impossible task. Therefore, numerical methods for calculating electrostatic fields are widely used in the simulation of electronic optics systems.

The methods of mathematical modeling allow to study the electrostatic fields of electrodes. Their size, shape and placement corresponds to the electrode configuration of the real investigational device. The simulation of most systems of electronic optics requires a numerical analysis of the parameters of the electrostatic field generated by these systems.

2   Review of the Literature

One of the ways of optimizing the computational process is its parallelization for the purpose of further implementation concerning the systems of parallel architecture [2]. Depending on the algorithm of the problem solving, it is necessary to solve the problem of choosing a computer system for its implementation. So, for the implementation of parallel algorithms, it is possible to use several computers in parallel mode or to refer to the so-called multithreading. The first way is helpful when, choosing different processors, achieving either the maximum efficiency of their download or increasing the speed of computing [3,4]. The second way enables the usage of modern multi-core processor architecture [5–7].

At present, the possibility of improving the efficiency of the computer system due to the increased clock speed of the processor is almost irrelevant or economically unprofitable. The idea of increasing productivity as a result of increasing the number of processors is promising. Building multiprocessor systems is the first and the most important direction in multithreading, but for such development a lot of money is required. The leading manufacturers of processors talk about the prospect of computer development, which consists of the construction of computer systems based on the multi-core processors. Therefore, the usage of multi-core systems is the best opportunity to increase the power of processors. Multi-core computing systems are increasingly being used in various subject areas and are gradually replacing mononuclear ones.

The electronic optics systems are the main components of modern research complexes by which the complex physical processes are studied [8,9]. The development of such complexes in Ukraine is directly fostered by OJSC "Selmi", Sumy. The rapid progress of nanotechnology imposes new requirements for electronic optics systems. Such system design is based on the simulation of the field created by a set of the charged electrodes.

3   Problem Statement

3.1   Physical Model Description of the Studied Phenomenon

Let the electrostatic field of the electronic optics system is determined by the system of N infinitely thin, ideally conductive electrodes $\{S_i\}$, which in their totality form a multiply connected surface $S := \bigcup\limits_{i=1}^{N} S_i$, where $S \cap S = 0$ at

$i \neq j$. A known potential, which is a constant value, is applied to each electrode $S_i \in \{S_i\}$. The distribution of potential is needed to be calculated.

Due to the geometry of electrodes and their potentials, it is always possible to calculate the distribution of potentials in space, and then represent these calculations graphically and obtain the desired image.

The possibility of preliminary calculation of the electrostatic field is an urgent task when designing the electronic optics systems. Its complexity is in the interaction between conductors and the redistribution of charge on their surface. The definition of this redistribution is a key step in the calculation of the resulting electrostatic field and determines the complexity of the problem: the system of equations, which describes the distribution of charge within the system of the charged conductors, has to take into account all conductors without possibility to reduce their number. An adequate summary of references is required to describe the current state-of-the-art or a summary of the results.

If charged electrodes in the process of designing of electronic optic systems are simulated by infinitely long cylindrical surfaces, the products of which are infinitely thin, uniformly charged along the length of the thread, parallel to one of the coordinate axes, then in a section with an arbitrary plane perpendicular to this axis, a certain set of open arcs is formed. In this case, the value of the potential of the field at an arbitrary point of space does not depend on one coordinate, but the necessary calculations are sufficient to conduct only in $\mathbf{R}^2$ [10]. Such approach is quite widespread in the practice of designing the electronic optics systems.

3.2 Description of the Mathematical Model of the Studied Phenomenon

In the process of mathematical modeling of the electronic optics system, it is necessary to find a function $u(P) \in H^1 \left(\Omega_S^-, \triangle \right)$ that meets the conditions

$$\triangle u = 0 \text{ on } \Omega_S^- := \mathbf{R}^3 \setminus \overline{S}; \tag{1}$$

$$\delta^{\pm} u = f\left(P\right) \text{ on } S; \tag{2}$$

$$\lim u\left(P\right) = 0, P \in \Omega, \tag{3}$$

where $\delta^{\pm} : H^1 \left(\Omega_S^- \right) \rightarrow H^{1/2}\left(S\right)$ – operators of the trace, $f\left(P\right)\left(P \in S\right)$ – a given limit value of the potential, which is constant on each of the electrodes, and

$$H^1 \left(\Omega_S^-, \triangle \right) := \left\{ u | u \in H^1 \left(\Omega_S^- \right), \triangle u \in L_2 \right\} \left(\Omega_S^- \right).$$

In general, these constant values do not have symmetry or antisymmetry.

It is known [11] that the problem (1)–(3) is equivalent to such an integral equation

$$(A\rho)(P) = \iint\limits_{S} K(P, M)\,\rho(M)\,\mathrm{d}S_M = f(P), \tag{4}$$

where $P \in S$, $K(P, M) := 1/\mathrm{dist}(P, M)$ – a fundamental solution of the Laplace equation in $\mathbf{R}^3$, and $\rho(M)$ – the S charge distribution density is required.

In the theory of integral equations, relatively to S, there is a statement, which expresses the solvability of the problem (4). If S – the Lipschitz surface is not closed in $\mathbf{R}^3$, then $A : H_{00}^{-1/2}(S) \to H^{1/2}(S)$ – isomorphism, and

$$m_1 \cdot \|\rho\|_{H_{00}^{-1/2}(S)} \leq \|A\rho\|_{H^{1/2}(S)} \leq m_2 \cdot \|\rho\|_{H_{00}^{-1/2}(S)}, \ 0 < m_1 \leq m_2.$$

The mathematical modeling of electrostatic fields is based on the determination of the field of charged electrodes by applying numerical methods to differential (1)–(3) or integral (4) forms of the equation that describe the main task of electrostatics. Numerical methods for solving an integral equation for a smooth boundary surface of a simple structure are well known. However, in the analysis of modern electronic optics systems there is a significant amount of charged electrodes of complex configuration. Using even the most economical method of collocation under the conditions of piecewise constant approximation of the required density of charge distribution requires a numerical solution of systems of linear algebraic equations (SLAE) of large orders with densely filled matrices. It, in turn, leads to numerical instability. Since most electron optic systems have geometric symmetry, this research was treated as a problem with an Abelian finite order symmetry group, which made it possible, with the same accuracy of approximation, to reduce the order of matrix equations, that approximate the corresponding integral, thus, reducing the volume of computations and expanding the class of systems of the specified form, while admitting numerical simulation within the boundary integral equation method.

4   Materials and Methods

Finding solutions of large SLAE is one of the main subtasks that arise in the process of mathematical modeling of the electrostatic fields by applying numerical methods to a differential (1)–(3) or integral (4) forms of an equation that describe the main task of electrostatics. Currently, "large" implies SLAE, which contains 1000 or more unknown ones. Each numerical method of mathematical physics leads to solving SLAE along with the matrices of a certain structure.

The most effective method for calculating the electrostatic field of electronic optical systems is the application of the method of integral equations [12]. It allows to reduce the dimension of the problem per unit and to find a solution within the unlimited domains. When analyzing modern systems of electronic optics, a considerable number of complex configuration electrodes was considered. Therefore, the usage of this method leads to the solving of SLAE of large dimensions with densely filled matrices. Because of Gauss' method for numerical solution of these systems, there is an instability of calculations.

It is noticed that the boundary surfaces of most systems of electronic optics have geometric symmetry. From the electrostatics point of view, quadrupole lenses, parallel capacitor, etc. can be included to such systems. In Fig. 1 there is one of the parallel capacitor, and in Fig. 2 – one of the possible configurations of the quadrupole system.

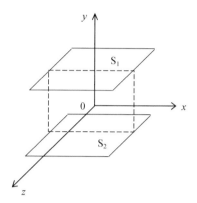

Fig. 1. Parallel capacitor

The existing symmetry, using the apparatus of the theory of groups [13], was treated as of great importance. So, for the classes of systems of electronic optics, the boundary surfaces of electrodes of which possess Abelian groups of symmetries of finite orders, it is possible to reduce the orders of matrix equations that approximate the corresponding integral [14]. And thereby the stability of the calculations is ensured.

Taking into account the symmetry in the geometry of the boundary surfaces of the electrodes by using the theory group technique allows to proceed from the system of integral equations [15–17], where integration is carried out along the entire boundary surface to an equivalent sequence of independent integral equations, with an integration only on some congruent component of the aggregate surface. This creates all the prerequisites for parallelization of the problem solving in general.

It is known [18] that the Gauss method requires $^2\backslash_3\, n + O(n)$ arithmetic operations, where n is the SLAE dimension. To achieve high accuracy of calculations in computing the electrostatic field of some class of systems of electronic optics, it is necessary to increase the number of points of collocation, respectively, the dimension of SLAE will be increased. An application of the group theory allows to reduce the computational complexity in the development of SLAE in N, where N – the order of the symmetry group, which has the configuration of the surfaces of the electrodes of the corresponding system of electronic optics.

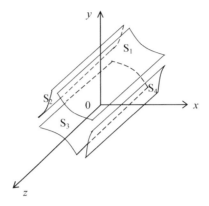

Fig. 2. Quadrupole lens

For numerical experiments, the classes of systems with symmetries of the eighth ($N = 8$) and sixteenth ($N = 16$) order are considered. In the first case, using the apparatus of the theory of groups, it was possible to proceed to the sequential solution of eight independent integral equations, and in the second – to the sixteen ones.

Based on the character table [13] the Fourier transformation matrix has been found. It is used in the analysis of electrostatic field systems, the boundary electrode surfaces of which have an Abelian symmetry group of the eighth order. There is one of the possible configurations in Fig. 1. This matrix can be presented as follows:

$$F_1 = \begin{pmatrix} 1 & 1 & 1 & 1 & 1 & 1 & 1 & 1 \\ 1 & 1 & -i & -i & -1 & -1 & i & i \\ 1 & 1 & -1 & -1 & 1 & 1 & -1 & -1 \\ 1 & 1 & i & i & -1 & -1 & -i & -i \\ 1 & -1 & 1 & -1 & 1 & -1 & 1 & -1 \\ 1 & -1 & -i & i & -1 & 1 & i & -i \\ 1 & -1 & -1 & 1 & 1 & -1 & -1 & 1 \\ 1 & -1 & i & -i & -1 & 1 & -i & i \end{pmatrix}.$$

The Fourier transformation matrix for the systems with a symmetry of the sixteenth order, one of the possible configurations is presented in Fig. 2.

$$F_2 = \begin{pmatrix}
1 & 1 & 1 & 1 & 1 & 1 & 1 & 1 & 1 & 1 & 1 & 1 & 1 & 1 & 1 & 1 \\
1 & -1 & 1 & -1 & 1 & -1 & 1 & -1 & 1 & -1 & 1 & -1 & 1 & -1 & 1 & -1 \\
1 & 1 & -1 & -1 & 1 & 1 & -1 & -1 & 1 & 1 & -1 & -1 & 1 & 1 & -1 & -1 \\
1 & -1 & -1 & 1 & 1 & -1 & -1 & 1 & 1 & -1 & -1 & 1 & 1 & -1 & -1 & 1 \\
1 & 1 & 1 & 1 & -1 & -1 & -1 & -1 & 1 & 1 & 1 & 1 & -1 & -1 & -1 & -1 \\
1 & -1 & 1 & -1 & -1 & 1 & -1 & 1 & 1 & -1 & 1 & -1 & -1 & 1 & -1 & 1 \\
1 & 1 & -1 & -1 & -1 & -1 & 1 & 1 & 1 & 1 & -1 & -1 & -1 & -1 & 1 & 1 \\
1 & -1 & -1 & 1 & -1 & 1 & 1 & -1 & 1 & -1 & -1 & 1 & -1 & 1 & 1 & -1 \\
1 & 1 & 1 & 1 & 1 & 1 & 1 & 1 & -1 & -1 & -1 & -1 & -1 & -1 & -1 & -1 \\
1 & -1 & 1 & -1 & 1 & -1 & 1 & -1 & -1 & 1 & -1 & 1 & -1 & 1 & -1 & 1 \\
1 & 1 & -1 & -1 & 1 & 1 & -1 & -1 & -1 & -1 & 1 & 1 & -1 & -1 & 1 & 1 \\
1 & -1 & -1 & 1 & 1 & -1 & -1 & 1 & -1 & 1 & 1 & -1 & -1 & 1 & 1 & -1 \\
1 & 1 & 1 & 1 & -1 & -1 & -1 & -1 & -1 & -1 & -1 & -1 & 1 & 1 & 1 & 1 \\
1 & -1 & 1 & -1 & -1 & 1 & -1 & 1 & -1 & 1 & -1 & 1 & 1 & -1 & 1 & -1 \\
1 & 1 & -1 & -1 & -1 & -1 & 1 & 1 & -1 & -1 & 1 & 1 & 1 & 1 & -1 & -1 \\
1 & -1 & -1 & 1 & -1 & 1 & 1 & -1 & -1 & 1 & 1 & -1 & 1 & -1 & -1 & 1
\end{pmatrix}.$$

For further parallelization of the computational process of solving non-dependent integral equations, it is efficient to use the program of parallel programming with the OpenMP specification [19,20]. The parallelism of the calculation of the electrostatic field for the class of the system of electron optics whose boundary surfaces of electrodes have an Abelian symmetry group of some finite order involves solving independent equations, which are allocated in separate threads and are in parallel. This problem is performed by pseudo comments of the OpenMP standard. Theoretically, it is enough to transfer the sequential solving of N integral equations in multithreading, if the equations have no dependencies, as it is in our case. Pseudo commentaries divide the sequence of integral equations into parallel blocks automatically and transfer each of them to the corresponding core of the processor. It is not necessary to describe the exchange of information.

For example,

```
#pragma omp parallel num_threads(n)
{
// This code will be executed by n threads
   # pragma omp for
// The parts of the cycle will be divided between n threads of the main group
of threads
...
}
```

To get the values of computing time the following snippet can be used:

```
...
// We remember the starting time
```

```
    DWORD dwStart = :: GetTickCount();
// We perform parallel computing
    Calculate ();
// We calculate the execution time
    m_dwRunTime = :: GetTickCount () – dwStart.
```

So, for the implementation of a parallel solution of the sequence of independent integral equations, one can refer to the so-called multithreading, which allows to use modern multi-core processor architectures. In this case, the application of methods for calculating electrostatic fields with modern trends in the development of computer technology is treated as of great importance.

5   Experiments

The software implementation of the proposed algorithm for calculating the electrostatic fields of the modern electronic optical systems with the available geometric symmetry of the surfaces of electrodes was developed in order to conduct experiments.

To prove the reasonability and estimation of the technique efficiency, a few numerical experiments were carried out. The configuration of the surfaces of electrodes is presented in Figs. 1 and 2. The boundary values of potential are the arbitrary known values. The approximation solution of integral equation of the sought density by piecewise-constant base functions.

For the evident presentation of the electrostatic field the equipotential surfaces are used. Usually, in this way, the potential fields are experimentally studied, and lines of tension are constructed as orthogonal lines to the equipotential surfaces.

6   Results and Discussion

In the Fig. 3 the calculated distribution of equal potential of electronic optic systems was shown (see case a, b – in Fig. 1 and case c, d – in Fig. 2). For the evident presentation of the electrostatic field the equipotential surfaces were used. The boundary values of potential are the arbitrary known values. The obtained results correspond to the physics of the investigated phenomenon.

In the Figs. 4 and 5 the dependence of the computation time on the number of threads when realizing parallel computations concerning solution of the sequence of eight and sixteen independent integral equations on the eight-core and sixteen-core processor, respectively, is introduced. The SLAE dimension – $n = 500$, accuracy of calculations $\varepsilon = 0,01$.

The proposed algorithm for calculating the electrostatic field of electronic optics systems allowed, with the same accuracy of calculations, to solve SLAE of the order of 500×500 instead of $- 4000 \times 4000$ (in the first case) and 8000×8000 (in the second case). As a result, the calculations have been simplified.

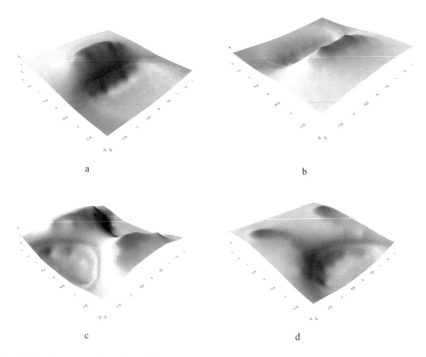

Fig. 3. Equipotential surface: (a) a case that corresponds to the anti-symmetric boundary value of the potential on the electrodes $f_1 = 1$, $f_2 = -1$; (b) $f_1 = 100$, $f_2 = 1000$; (c) $f_1 = f_3 = 1$, $f_2 = f_4 = -1$; (d) $f_1 = 10$, $f_2 = 20$, $f_3 = -100$, $f_4 = 1$

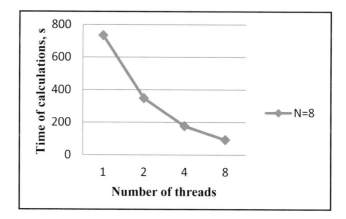

Fig. 4. Dependence of computing time on the number of threads on the eight-core processor

In contrast to existing algorithms, it allowed to avoid instability when solving a complex three-dimensional problem. The results of numerical experiments obtained here have a place for a whole class of systems of electronic optics, whose geometric form includes symmetric components. In this case, the limit values of the potential can be arbitrary, the antisymmetric ones were chosen in order to obtain asymptote expressively expressed in the cross section, where the potential is equal to zero. The disadvantage of the proposed algorithm is the inability to extend it to the case when the geometric form contains only asymmetric components.

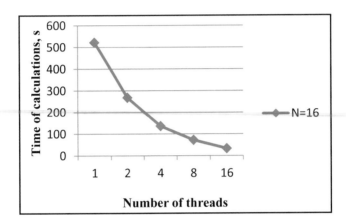

Fig. 5. Dependence of computing time on the number of threads on the sixteen-core processor

All in all, using the modern architecture of multi-core processors, it was possible to reduce the computational time when solving SLAE, approximating the relevant sequences of independent integral equations. For the classes of systems with symmetries of the eighth and sixteenth orders – approximately in eight and sixteen times, respectively. The obtained results refer to the possibility of further optimization of the software by varying the number of parallel threads and processor cores. The ratio of the sequential number to the number of parallel operations led to an increase in capacity of eight and sixteen times, respectively.

7   Conclusion

To optimize the computational process of calculating the electrostatic field of modern systems of electronic optics, it was taken into account that the application of the group theory technique to the systems of classes whose surfaces of electrodes have the Abelian symmetry groups of one or another finite order creates all the prerequisites for parallelizing the calculation of electrostatic fields. Due to the rapid development of multi-core architecture, the application of methods of mathematical modeling of fields of electronic optics systems with modern trends in the development of computer technology was considered [21].

The ratio of the sequential number to the number of parallel operations was discovered. It led to an increase in programming capacity in N – times, where N – the order of the symmetry group, which refers to the configuration of the surfaces of the electrodes of the corresponding system of electronic optics. As a result, the time of calculations was reduced when solving SLAE, which approximates the corresponding integral equations in about N times, provided that the number of equations coincides with the number of processor cores with different variations of parallel threads.

References

1. Shakhovska N (2017) The method of big data processing. In: 2017 12th international scientific and technical conference on computer sciences and information technologies (CSIT), September 2017, vol 1. IEEE, pp 122–126
2. Nemnyugin S, Stesyk O (2002) Parallel programming for multiprocessor computing systems. St. Petersburg (in Russian)
3. Ilyin V (1985) Numerical methods for solving problems in electrophysics. The main edition of physics and mathematics literature. Nauka, Moscow (in Russian)
4. Kumar V, Grama A, Gupta A, Karypis G (1994) Introduction to parallel computing. The Benjamin/Cummings Publishing Company, San Francisco
5. Voss M (2003) OpenMP share memory parallel programming. Toronto, Canada
6. Chapman B, Jost G, van der Pas R (2008) Using OpenMP: portable shared memory parallel programming (scientific and engineering computation). The MIT Press, Cambridge
7. Chandra R, Menon R, Dagum L, Kohr D, Maydan D, McDonald J (2000) Parallel programming in OpenMP. Morgan Kaufmann Publishers, Burlington
8. Rashmi S (2018) An affine arithmetic approach to model and estimate the safety parameters of AC transmission lines. Int J Image Graph Sig Proc 10(1)
9. Goel V, Raj K (2018) Removal of image blurring and mix noises using Gaussian mixture and variation models. Int J Image Graph Sig Proc 10(1)
10. Vladimirov V (1981) Equation of mathematical physics. Science (in Russian)
11. Sybil Y (1997) Three dimensional elliptic boundary value problems for an open Lipschitz surface. Math Studios 8(1):79–96
12. Mochurad L, Pukach P (2017) An efficient approach to calculating the electrostatic field of a quadrupole. Edition of Kherson National Technical University, Kherson, vol 1, pp 155–165 (in Ukrainian)
13. Serre J-P (2003) Linear representations of the finite groups. Moscow (in Russian)
14. Mochurad L, Pukach P (2017) A posteriori method for estimating error and parallelization of calculations for a class of problems of electronic optics of a lens. In: Scientific Edition of UNFU, a collection of scientific and technical works, Lviv, vol 27, pp 155–159 (in Ukrainian)
15. Zakharov Y, Safronov S, Tarasov P (1992) The abelian groups of finite order in numeral analysis of linear boundary tasks of the theory of potential. J Comput Math Math Phys 32:40–58 (in Russian)
16. Mochurad L, Ostudin B (2009) The research of the approximate solutions of a boundary task of theory of potential with the Abelian group of symmetry of the sixteenth order. In: Works of the international symposium of "The problems of optimization of calculations", 24–29 September 2009. Katsyveli, Institute for cybernetics of V.M. Hluskov NASU of NASU, Kyiv, pp 117–122 (in Ukrainian)

17. Mochurad L, Harasym Y, Ostudin B (2009) Maximal using of specifics of some boundary problems in potential theory after their numerical analysis. Int J Comput 8:149–156
18. Balandin M, Shurina E (2000) The methods for solving high-dimensional SLAE. NSTU, Novosibirsk (in Russian)
19. Aksak N, Volontsevich V, Osotov I, Filin Y (2007) The Research of capacity increase in programming based on multi-core cluster system. Edition of Lviv Polytechnic National University, computer systems and networks, Lviv, pp 8–11 (in Ukrainian)
20. Antonov A (2009) Parallel programming using the OpenMP technology, tutorial. Moscow State University, Moscow (in Russian)
21. Shakhovska N, Vovk O, Kryvenchuk Y (2018) Uncertainty reduction in big data catalogue for information product quality evaluation. East Eur J Enterp Technol 1(2–91):12–20

A Model of Logical Inference and Membership Functions of Factors for the Printing Process Quality Formation

Vsevolod Senkivskyy[1], Iryna Pikh[1], Svitlana Havenko[1],
and Sergii Babichev[1,2(⊠)]

[1] Ukrainian Academy of Printing, Lviv, Ukraine
senk.vm@gmail.com, pikhirena@gmail.com, havenko1559@gmail.com
[2] Jan Evangelista Purkyně University in Ústí nad Labem,
Ústí nad Labem, Czech Republic
sergii.babichev@ujep.cz

Abstract. Technological stages or separate procedures of the publishing and printing process involve relevant databases and a set of factors (criteria, parameters, requirements) to ensure their quality implementation. However, they are different in nature, purpose, setting type, peculiarities of impact and application methods, that complicates the generalizations and single formal reproduction of information. This fact causes the necessity for both bringing them to an universal metric database and using of the methods that would provide the research of the publishing and printing process from the standpoint of a single methodological basis, determining components of which are the methods of system and matrix analysis, the theory of hierarchical multilevel systems, the modelling theory, the operations research and the fuzzy set theory.

The paper presents the model of logic inference, which determines the predictive quality indicators of the process of flatbed offset printing technique on the basis of fuzzy sets. An universal term-set of values and its corresponding linguistic terms of linguistic variables has been formed. The membership functions of linguistic variables have been designed and calculated which are the basis for the structure of the knowledge matrices and fuzzy logic equations as a prerequisite condition of formation and definition of the integral quality indicator of the printing process.

Keywords: Printing process · Quality · Factor ·
Model of logical inference · Fuzzy logic · Linguistic variable ·
Membership function · Fuzzy set · Prediction

1 Introduction

We often talk about the quality of books without thinking about the following: what lies behind this general and multifaceted concept? How and at what stage of the book production is it achieved? Is it possible to provide its proper level a priori

© Springer Nature Switzerland AG 2020
V. Lytvynenko et al. (Eds.): ISDMCI 2019, AISC 1020, pp. 609–621, 2020.
https://doi.org/10.1007/978-3-030-26474-1_42

before the book will appear in the world? Is there a single criterion that can be used to predict the quality of printed products for different purposes? What factors are decisive on the stages of the production cycle of the publication preparation and release? Is it possible to express the publication quality as a single metric characteristic, since in addition to numerical parameters, the book is an artistic reflection of the text, graphic, symbolic, and art information available in it? Such a diversity of the printed matter confirms its versatility once again and, consequently, the multicriteria of quality assessments of all aspects of book publication.

Progress in improving the quality of the publishing and printing processes that become the basis of the proper level of the printed matter becomes of paramount importance in the modern conditions of production computerization, the active introduction of information technologies, and the limitation of energy resources. The improvement of the technical means and the technological component does not always lead to solving the problem and obtaining the expected result. In addition, quite often the control of finished products is carried out a posteriori and proceeds in the role of the necessary, but insufficient condition in manufacturing a good-quality product. The current solution of the quality component of the printing process through the introduction of new technical means equipped with microprocessor-based programmable modules and computerized sections, modern flow lines does not guarantee obtaining a qualitative result. To our mind, it is necessary to have a powerful and theoretically balanced information supplement to existing systems and tools using a new direction confirmed by the practice of foreign companies (non-printing profile) – the fuzzy control theory, the founder of which is the creator of the formal tools of fuzzy logic, American scientist Lotfi Zadeh.

The justification for applying this approach to publishing and printing technologies is consisted in the following: significant number of characteristics or requirements relating to the processes of preparation and production of the printed matter are mainly descriptive and therefore ideally suited for their use as the source factors (linguistic variables) of fuzzy control systems. The statement is strengthened by the absence of a theoretically substantiated mechanism which would cover and implement a predictive assessment of the effectiveness of the stages realization or individual procedures of the publishing and printing process, which prevents a priori the proper level of performance of production operations, and, as consequence of the quality of printed matter. The solution of this problem is possible by creation of informational means for forming and predictive evaluation of the printing processes quality on the basis of the fuzzy set theory and the methods of the system and technical oriented direction, which will facilitate obtaining the products of the appropriate level.

2   Problem Statement

The most important ingredient of the fuzzy set theory is the linguistic variables (LV), the defining component of which are the membership functions (MF), which are constructed with the use of term-set of values and linguistic terms (LT) of factors. In the works [1,2] the conception of a universal set D is introduced for the entire problem area. Then, the fuzzy subset M of the set D is determined through an universal set or through the scale D and the membership function $\mu_M(d)$ as follows:

$$M = \{(\mu_M(d), d), d \in D\} \qquad (1)$$

where $0 \le \mu_M(d) \le 1$.

The membership function defines the degree of affiliation of each element of the fuzzy set in the universal set $M \in D$. With the discreteness and finiteness of the basic scale (i.e. it is divided into quanta or parts or spaces), the fuzzy set M can be presented in the following way:

$$M = \{\mu_M(d_1)/d_1, \mu_M(d_2)/d_2, ..., \mu_M(d_n)/d_n\} = \sum_{i=1}^{n} \mu_M(d_i)/d_i \qquad (2)$$

The symbol "/" in the expression (2) conditionally "attaches" MF $\mu_M(d_i)$ to the element d_i. The symbol "Σ" symbolically means a set of pairs $\mu_M(d)$ and d_i. Finally, the MFs act as the identifiers of the input values of linguistic variables in the fuzzy format, i.e. the set of values of the variable d corresponds to the membership functions $\mu(d)$.

3   Review of the Literature

The analysis of works concerning the quality evaluation of the printing products testifies to the fact that this problem can not be solved without a complex system approach and taking into account all factors of production [3–11]. A complete assessment of the quality indicator of any print edition or technical process can be given separately or in combination with metrological, statistical, psychophysiological, economic and other methods [12–17]. The methods, which are considered in these works, refer to the qualimetric methods for assessing the quality of printed products [9,10,17] and they have specialized means and technical tools of quality control of publications. They are the least formalized, since they are based on subjective assessments. This category also includes psychophysiological and, to some extent, statistical methods for quality control [12,18,19]. The metrological group of methods for determining the image quality is designed on the basis of technical means of measurements and calculations. The most common among this group are the measurement methods that use densitometers and colorimeters to measure the optical characteristics of the image; the registration methods, which with the help of resolvometers determine the microscopic characteristics of the image; the most common class of calculation methods, based on the analysis of empirical and theoretical indicators [12,20,21]. The effective use of control methods of printing products is possible on the condition of their classification by methods of determining the quality indicator [6,17]. The division of methods is carried out on the prevailing presence in them of visual, electronic and statistical components. The electronic methods of measuring and converting numerical quality indicators using some statistical control results are used in the recent times [4,12,13]. Interesting are the methods that suggest generalising or combining partial quality indicators into one generalized indicator [4,9], which is formed on the basis of the weight of its components. In recent years, the information technology has got rapid development, in which it is suggested to use an integral parameter to determine the quality of the book [4,11], which takes into account the information components. Evaluating the quality of book publications in the suggested ways relates mainly to the construction of the book. In general, a methodology for solving the problems of evaluating the quality of publishing and printing processes based on the use of information technologies has been formed. The latent quality indicator of book editions has been suggested [22–25]. The models of factors of prognostication of the printing process quality [26] and the control system of the book production quality have been developed. It is recommended to determine the integral prognostication of the book quality through the logical accumulation of indicators for the implementation of the technological stages of the printed matter production.

The analysis of literature sources have shown that there are no informational approaches to solve the problem of the book quality editions, the basis of which is the printing process. The essence of the new methodology is the complex use of methods and techniques of the theory of operations research, system analysis, fuzzy sets, the theory of modelling, expert evaluation of publishing and printing processes, which enables the predictive evaluation of the printed matter quality.

The aim of the research are: development of the logical inference model which determines the quality of flatbed offset printing technique on the basis of the formed universal term-set of values; design and calculation of the values of the membership functions of linguistic variables which will serve as the basis for constructing knowledge matrices and fuzzy logical equations as a prerequisite for the formation and predictive evaluation of the quality of the printing process.

4   Materials and Methods

For purpose of logical structuring of the investigation we will refer the factors influencing the quality of the printing process to the categories that are relevant to the printing technical factors; materials; technological (functional) procedures. The most important element of such an approaches is the preservation and the further development of the principle of multilevel hierarchical structuring of the process and the related source data to it according to the rule "top down" and the formation of qualitative indicators in the opposite direction. Consequently, the dependency of the printing process quality can be presented by the function:

$$V = F_V(T, R, L) \tag{3}$$

where T, R, L are the sets of factors of the singled categories. Since the total set of factors used by us to study their influence on the printing process is divided into functional groups, we make their additional identification according to the categories in expression (3). In this case, each of the factors can be given a formalized record in the form of a mathematical variable and its linguistic decryption. This fact will provide their interpretation by means of fuzzy logic.

Finally, we present a subset of factors in the following way. We shall refer to the category of technical factors the linguistic variables of the set:

$$T = \begin{cases} t_1 - a\ type\ of\ a\ printing\ press; \\ t_2 - a\ type\ of\ a\ printing\ plate; \\ t_3 - characteristics\ of\ the\ bruzer\ tympan \end{cases} \tag{4}$$

the total influence of which on the printing quality will be defined by the function:

$$T = F_T(t_1, t_2, t_3) \tag{5}$$

Similarly, we describe the other groups. The set of linguistic variables referred to the category of "materials" is presented as follows:

$$R = \begin{cases} t_1 - characteristics\ of\ paper; \\ t_2 - characteristics\ of\ ink; \\ t_3 - moistrurizing\ solution \end{cases} \tag{6}$$

The corresponding function of the arguments for the printing quality formation related to paper, ink and moisturizing solution will be looked as follows:

$$R = F_R(r_1, r_2, r_3) \tag{7}$$

The linguistic variables of the technological procedures determine the functional components related to the printing speed and the deformation of a bruzer tympan:

$$L = \begin{cases} l_1 - deformation\ of\ a\ bruzer\ tympan\ in\ the\ contact\ with\ a\ printing\ plate \\ l_2 - deformation\ of\ a\ bruzer\ tympan\ in\ the\ printing\ contact\ zone; \\ l_3 - printing\ speed \end{cases}$$

Taking into account the arguments of the last expression, the function of the total influence of the technological procedures on the process of acquiring a part of quality will look as follows:

$$L = F_L(l_1, l_2, l_3) \tag{8}$$

The model of logical inference that will interpret the progress of evaluating the publication quality for the printing process taking into account the hereinbefore categories of the factors is presented in Fig. 1. In the given model the logic of forming the printing process quality reflects the generation of quality level in accordance with the selected groups of linguistic variables given by the expressions (4–8). The model provides a hierarchy of "bottom-up" results – from individual linguistic variables, each of which contributes to the formation of the integral quality of the process, which is additively transmitted through appropriate categories to the higher level of this process implementation.

It is well known that the transition from the linquistic interpretation of LV to the numerical expression of the influence degree of technological factors on the formation and evaluation of the quality level process which is expressed by a

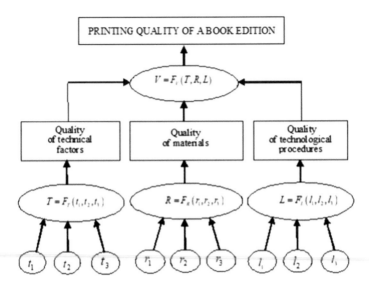

Fig. 1. The model of logical inference: the formation of the quality indicator of the publication printing process

numerical indicator, is carried out with the use of membership functions. These functions act as the determining characteristics of the research. The initial step for the MF obtaining is the setup of term-set of linguistic variables values within the technological ranges of their use, or in conditional units in the case of their absence. It should be noted that for the printing process the characteristics of equipment, materials and technological procedures in the overwhelming majority have numerical expressions in different scales. Thus, the universal set for the analysed linguistic variables can be presented as the real expression using numeric variables. Concurrently, we define the appropriate qualitative linguistic terms given by a fuzzy scale. The given initial data are be presented in Table 1.

5   Experiment

Based on the Table 1, we divide the universal term-set D into parts (quants) for each linguistic variable, at the points of division of which we determine the ranks $r_v(d_i)$ LV, that are identified with LT. The ranks of the factors are determined by the expert method or they calculated using the ranking method or the hierarchical analysis method [24]. Then, the numerical values of membership functions can be obtained from the following relationships:

$$\left.\begin{array}{l} \mu_1 = (1 + \dfrac{r_2}{r_1} + \dfrac{r_3}{r_1} + \dots + \dfrac{r_n}{r_1})^{-1}; \\[2mm] \mu_2 = (1 + \dfrac{r_1}{r_2} + \dfrac{r_3}{r_2} + \dots + \dfrac{r_n}{r_2})^{-1}; \\[2mm] \dotfill \\[1mm] \mu_n = (1 + \dfrac{r_1}{r_n} + \dfrac{r_2}{r_n} + \dots + \dfrac{r_{n-1}}{r_n})^{-1} \end{array}\right\} \qquad (9)$$

Table 1. Term-sets of the values of linguistic variables of the printing process

Variable	Linguistic essence of a variable	Universal set of values (the set D)	Linguistic terms (the set X)
t_1	A type of a printing press	(1–5) c.u.	Sheet-fed, web-fed
t_2	Characteristics of a bruzer tympan (Shore hardness)	(30–90) A.u.	Soft, medium, hard
t_3	A type of a printing plate (circulation stability)	(100–500) thousand prints	Low, medium, hard
r_1	Characteristics of paper (thickness)	(0,03–0,25) mm	Small, medium, large
r_2	Characteristics of ink (viscosity by Lorey)	(15–20) Pa s	Low, medium, high
r_3	Moisturizing lotion (acidity)	(1–7) pH	High, low, neutral
l_1	Deformation of a bruzer tympan in the contact zone with a printing plate	(1–4)%	Small, medium, big
l_2	Deformation of a bruzer tympan in the printing contact zone (total compression)	(1–4)%	Small, medium, big
l_3	Printing speed	(2000–18000) rpm	Low, medium, high

As the result, the linguistic term "quality of the publication printing process" V is expressed by the membership functions in the points $d_1, d_2, ..., d_n$ by the following fuzzy set:

$$V = \{\frac{\mu_v(d_1)}{d_1}, \frac{\mu_v(d_2)}{d_2}, ..., \frac{\mu_v(d_n)}{d_n}\} \qquad (10)$$

where $V \subset D$ and $\mu_v(d_i)$ is a degree of affiliation of the element $d_i \in D$ to the set V. For the level that determines the quality of the printing process, LT of which is given by the set (10), it is necessary to calculate the membership functions based on the Table 1, which will become numerical parameters of the fuzzy logic equations.

6   Results and Discussion

Dividing the intervals of the technological parameters values, which are presented in the Table 1, into three additional points for four quants, we receive five checkpoints with the extreme bounds in which the values of the membership functions will be calculated. Consequently, the initial database will be the information in Table 1, the universal term-set of values $D = \{d_1, d_2, ..., d_n\}$ and ranks of linguistic variables $r_v(d_i)$. Additionally, for some LVs, we will construct the combined graphs of the membership functions related to each of the qualitative terms, which provides a visual comparison of the MF values in the control and, which is essential, in random points of the intervals of the term-set.

Linguistic variable t_1 "a type of a printing press" is defined on the universal set $D(t_1) = [1; 2; 3; 4; 5]$ of conditional units by linguistic terms $X(t_1) = <sheet\text{-}fed, web\text{-}fed>$. The fuzzy set with the normalized values of the membership functions will look like this:

$$sheet\text{-}fed\,press = \{\frac{1}{1}; \frac{0,85}{2}; \frac{0,60}{3}; \frac{0,22}{4}; \frac{0,10}{5}; \} \, c.u$$

$$web\text{-}fed\,press = \{\frac{0,06}{1}; \frac{0,20}{2}; \frac{0,78}{3}; \frac{0,90}{4}; \frac{1}{5}; \} \, c.u$$

The graphs of the membership functions of the hereinbefore described linquistic variable is shown in Fig. 2a.

Linguistic variable t_2 "characteristics of a bruzer tympan" is defined on the universal set $D(t_2) = [30; 45; 60; 75; 90]$ units of Shore hardness. It is identified by the linguistic terms $X(t_2) = <soft, medium, hard>$. The formalized representation of LT "soft", "medium", "hard" are given as follows:

$$a\,soft\,bruzer\,tympan = \{\frac{1}{30}; \frac{0,60}{45}; \frac{0,45}{60}; \frac{0,20}{75}; \frac{0,10}{90}; \} \, A.u$$

$$a\,medium\,bruzer\,tympan = \{\frac{0,11}{30}; \frac{0,70}{45}; \frac{1}{60}; \frac{0,45}{75}; \frac{0,11}{90}; \} \, A.u$$

$$a\,hard\,bruzer\,tympan = \{\frac{0,11}{30}; \frac{0,22}{45}; \frac{0,44}{60}; \frac{0,78}{75}; \frac{1}{90}; \} \, A.u$$

The charts of this linquistic variable membership functions are presented in Fig. 2b.

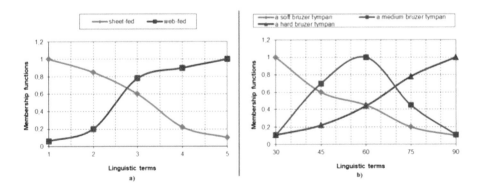

Fig. 2. The membership functions of linguistic variables: (a) "a type of a printing press"; (b) "characteristics of a bruzer tympan"

Linguistic variable t_3 "a type of a printing press" is defined by the parameter "circulation stability" on the universal set $D(t_3) = [100; 200; 300; 400; 500]$ thousand prints. It is described by the linguistic terms $X(t_3) = <low, medium, high>$ with the following values of the membership functions:

$$low\ circulation\ stability = \{\frac{0,11}{100}; \frac{0,30}{200}; \frac{0,60}{300}; \frac{0,85}{400}; \frac{1}{500};\}\ thousand\ prints$$

$$medium\ circulation\ stability = \{\frac{0,25}{100}; \frac{0,65}{200}; \frac{1}{300}; \frac{0,50}{400}; \frac{0,22}{500};\}\ thousand\ prints$$

$$high\ circulation\ stability = \{\frac{1}{100}; \frac{0,44}{200}; \frac{0,20}{300}; \frac{0,11}{400}; \frac{0,06}{500};\}\ thousand\ prints$$

The linguistic variable r_1 "characteristics of paper" is specified by one of the main technological parameters - "thickness", which is defined on the universal set $D(r_1) = [0, 03; 0, 09; 0, 15; 0, 20; 0, 25]$ mm. Linguistic terms, which characterize the paper thickness, are $X(r_1) = <small, medium, large>$. Fuzzy sets of LV r_1 are presented in the following way:

$$small\ thickness = \{\frac{1}{0,03}; \frac{0,89}{0,09}; \frac{0,67}{0,15}; \frac{0,33}{0,20}; \frac{0,11}{0,25};\}\ mm$$

$$medium\ thickness = \{\frac{0,22}{0,03}; \frac{0,60}{0,09}; \frac{1}{0,15}; \frac{0,80}{0,20}; \frac{0,22}{0,25};\}\ mm$$

$$large\ thickness = \{\frac{0,11}{0,03}; \frac{0,33}{0,09}; \frac{67}{0,15}; \frac{0,89}{0,20}; \frac{1}{0,25};\}\ mm$$

The charts of the appropriate memberships functions are presented in Fig. 3.

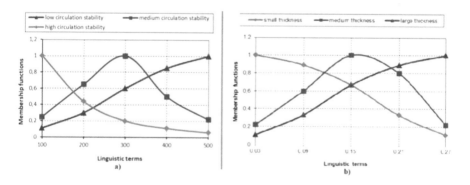

Fig. 3. The membership functions of the linguistic variables: (a) "a type of a printing plate"; (b) "characteristics of paper"

In the following, the graphical visualization of the membership functions will be omitted. Linguistic variable r_2 "characteristics of ink" LV is identified with viscosity, which is considered as one of the most important characteristics of ink, which determines its suitability for the selected printing technique, its stability while storing and its interaction with other inks. It is defined on the universal set $D(r_2) = [15; 16; 17; 5; 19; 20]$ Pa s by the linguistic terms $X(r_2) = <low, medium, high>$. The membership functions which were obtained are defined by the following equations:

$$low\ viscosity = \{\frac{1}{15}; \frac{0,45}{16}; \frac{0,2}{17,5}; \frac{0,11}{19}; \frac{0,05}{20};\}\ \text{Pa s}$$

$$medium\ viscosity = \{\frac{0,35}{15}; \frac{0,60}{16}; \frac{1}{17,5}; \frac{0,33}{19}; \frac{0,20}{20};\}\ \text{Pa s}$$

$$high\ viscosity = \{\frac{0,11}{15}; \frac{0,20}{16}; \frac{0,50}{17,5}; \frac{0,80}{19}; \frac{1}{20};\}\ \text{Pa s}$$

Linguistic variable r_3 "moisturizing solution" we have added as the additional characteristic – "acidity", which is defined by the set $D(r_3) = [1; 3; 5; 6; 7]$ pH. This variable is implemented with the use of the linguistic terms $X(r_3) = <high, low, neutral>$ and the membership functions in the following way:

$$high\ acidity = \{\frac{1}{1}; \frac{0,75}{3}; \frac{0,20}{5}; \frac{0,10}{6}; \frac{0,05}{7};\}\ \text{pH}$$

$$low\ acidity = \{\frac{0,06}{1}; \frac{0,35}{3}; \frac{1}{5}; \frac{0,65}{6}; \frac{0,05}{7};\}\ \text{pH}$$

$$neutral\ acidity = \{\frac{0,04}{1}; \frac{0,08}{3}; \frac{0,12}{5}; \frac{0,17}{6}; \frac{1}{7};\}\ \text{pH}$$

Linguistic variable l_1 "deformation of a bruzer tympan in the contact zone with a printing plate" is defined by the set in percents as follows: $D(l_1) = [1; 2; 2, 5; 3; 4]\%$. The linguistic terms of the deformation are expressed by the set $X(l_1) = <small, medium, big>$. The fuzzy sets of the membership functions are presented by the following equations:

$$small\ deformation = \{\frac{0,05}{1}; \frac{0,22}{2}; \frac{0,60}{2,5}; \frac{0,92}{3}; \frac{1}{4};\}\%$$

$$medium\ deformation = \{\frac{0,4}{1}; \frac{0,8}{2}; \frac{1}{2,5}; \frac{0,45}{3}; \frac{0,2}{4};\}\%$$

$$big\ deformation = \{\frac{1}{1}; \frac{0,45}{2}; \frac{0,2}{2,5}; \frac{0,10}{3}; \frac{0,05}{4};\}\%$$

Linguistic variable l_2 "deformation of a bruzer tympan in the printing contact zone" is identified by the deformation, which reflects the total compression of a bruzer tympan. This variable is defined in percent by the division points of the set $D(l_2) = [5; 6; 7; 9; 10]\%$. The linguistic terms, related to the bruzer tympan compression, are defined by the set $X(l_2) = <small, medium, big>$.

The fuzzy sets of the membership functions of the appropriate terms are formalized in the following way:

$$small\ deformation = \{\frac{0,10}{5}; \frac{0,28}{6}; \frac{0,65}{7}; \frac{0,90}{9}; \frac{1}{10};\}\%$$

$$medium\ deformation = \{\frac{0,30}{5}; \frac{0,75}{6}; \frac{1}{7}; \frac{0,50}{9}; \frac{0,20}{10};\}\%$$

$$big\ deformation = \{\frac{1}{5}; \frac{0,55}{6}; \frac{0,28}{7}; \frac{0,15}{9}; \frac{0,10}{10};\}\%$$

Linguistic variable l_3 "printing speed" is defined on the universal set $D(l_3) = [2000; 6000; 10000; 14000; 18000]$ rpm. It is described by the linguistic terms $X(l_3) = <low, medium, high>$ with values of the membership functions for each of the terms which are expressed by the fuzzy sets:

$$low\ speed = \{\frac{0,11}{2000}; \frac{0,30}{6000}; \frac{0,60}{10000}; \frac{0,85}{14000}; \frac{1}{18000};\}\ rpm$$

$$medium\ speed = \{\frac{0,25}{2000}; \frac{0,90}{6000}; \frac{1}{10000}; \frac{0,50}{14000}; \frac{0,22}{18000};\}\ rpm$$

$$high\ speed = \{\frac{1}{2000}; \frac{0,44}{6000}; \frac{0,20}{10000}; \frac{0,11}{14000}; \frac{0,06}{18000};\}\ rpm$$

7   Conclusions

The analysis of the of literary sources in this subject area has shown that nowadays there is no informational approach to the problem of the formation of the book quality based on the printing process. The paper presents one of the stages the new methodology, the essence of which consists in the complex use of methods and means of the theory of operations research, system analysis, the fuzzy set theory, the modelling theory and the expert evaluation of publishing and printing processes. Implementation of the proposed technique allows us to design the information technologies for the formation and predictive evaluation of the quality of printed products.

A multilevel model of logical inference has been developed during the investigation process. This model determines the formation of a quality indicator for the implementation of the printing process of publications, presented in the form of a fuzzy set. A universal term-set of values and its corresponding linguistic terms of linguistic variables has been formed. These values identify the factors influencing the quality of the book printing process. The obtained results contain the mathematical record of a linguistic variables, their technological essence, numerical indicators and types of equipment and materials which were obtained on the basis of real data of the printing process. A set of linguistic terms defined by the expert method has been presented. These terms reproduce the qualitative characteristics of the variables in the points of the universal set division.

The obtained membership functions of the linguistic variables have been presented for the set of linguistic terms with respect to the quants of division of the values intervals given by the universal term-set. A graphical visualization of the acquired values of membership functions has been performed. This procedure provides the comparability of their values obtained for certain terms at a particular point of division of the task interval. The obtained results can be serve as the basis for constructing knowledge matrices and fuzzy logical equations as a prerequisite for determining the predictive integral indicator of the printing process quality.

References

1. Zadeh L (1976) The concept of a linguistic variable and its application to the adoption of approximate solutions. Mir, Moscow, 165 p
2. Zadeh L (2001) The role of soft computing and fuzzy logic in the understanding, design and development of information intelligence systems. Artif Intell News 2(3):7–11
3. Durnyak B, Sabat V, Nazarenko O (2011) Ways of using information components for calligraphic models. Sci Pap 4(37):236–243. UAP, Lviv
4. Lutskiv MM (2012) Digital printing technology: monograph. Ukrainian Academy of Printing, Lviv, 488 p
5. Likhachev VV (1999) Fundamentals of quality management of printed products: textbook. MSUPA, Moscow, 88 p
6. Melnykov OV (2007) Technology of flatbed offset printing technique: textbook, 2nd edn (ed: Lazarenko ET). Ukrainian Academy of Printing, Lviv, 388 p
7. Yakutsevych S, Lazarenko E, Nazar I (2005) Quality of printed matter: evaluation indicators. Book Qualilogy (8):5–13
8. Kipphan, H (2003) Handbook of printed media. Technology and production techniques. Translated from German. MSUPA, Moscow, 1253 p
9. Nazarkevych LI, Lebed GG, Ershov EA (1981) Qualimetric assessment of the technical level of the gravure printing process. In: Proceedings of RSRI of printing, Moscow, vol 31, no 3, pp 35–46
10. Nazarkevych LI, Lebed GG, Moseev NP (1982) Qualimetric methods for assessing the quality of the technological processes and printing products. Printing industry. Background information, no. 2. SRC "Informpechat", Moscow, 44 p
11. Romeikov IV (1989) Generalized criteria for assessing the quality of the printed image. Technology of printing processes. Kniga, Moscow, pp 246–284
12. Sava VI (2000) Fundamentals of the technique of the book creation: textbook. Kameniar, Lviv, 136 p
13. Liberman NI (1977) Statistical methods of quality control of printed products: according to indicators of ink discrepancy and folding accuracy. Kniga, Moscow, 119 p
14. Petriashvili GG, Durnyak BV (2006) Parameters characterizing consumer features of book editions. J Sci Works (32):225–228. IMEE of NAS of Ukraine, Kyiv
15. Raikhman EP, Azgaldov GG (1973) About qualimetry. Izdatelstvo standartov, Moscow, 173 p
16. Shipkov KV (1970) Quality control of imprints. CBSRI of Printing, Moscow, 86 p
17. Durnyak BV, Nazarenko OM (2010) Peculiarities of using of discrete mathematical methods in the problems of evaluation of book editions. Model Inf Technol J Sci Works (57):264–271. IMEE, NAS of Ukraine, Kyiv

18. Yakutsevych S (2003) Imprint quality of web-fed offset printing technique. Sci Pap (6):57–59. UAP, Lviv
19. Yakutsevych S (2000) Imprint quality: the study of offset matte paper. Book Qualilogy (3):121–125. UAP, Lviv
20. Zaitsev SA, Gribanov DD, Tolstov AN, Merkulov RV (2002) Control and measuring devices and instruments: textbook. Academy, Moscow; ProfObrIzdat, 200, 464 p
21. Shapovalenko OH, Bondar VM (2002) Fundamentals of electric measurements: textbook. Lybid, Kyiv, 320 p
22. Durnyak BV, Senkivskyy VM, Pikh IV (2014) Information technology of prognostication and providing the quality of publishing and printing processes (methodology of problem solution). Technol. Complexes 1(9):21–24. RVV Lutsk STU, Lutsk
23. Pikh IV, Senkivskyy VM (2013) Information technology of modelling of publishing processes: textbook. Ukrainian Academy of Printing, Lviv, 220 p
24. Pikh IV, Durnyak BV, Senkivskyy VM, Holubnyk TS (2017) Information technology of formation of book edition quality. Monograph, Ukrainian Academy of Printing, Lviv, 308 p
25. Petriashvili GG (2006) Relationship of the consumer parameter of the book operation time with the parameters of the technological processes of the book production. Model Inf Technol J Sci Works (39):225–228. IMEE of NAS of Ukraine, Kyiv
26. Senkivska NYe (2011) Synthesis of a model of factors for prognostication of the quality of printing process (on the example of flatbed offset printing technique). Book Qualilogy 1(19):46–52. Ukrainian Academy of Printing, Lviv

Calculation the Measure of Expert Opinions Consistency Based on Social Profile Using Inductive Algorithms

Viacheslav Zosimov[ID] and Oleksandra Bulgakova[(✉)][ID]

V.O. Sukhomlynsky Mykolaiv National University,
Nikolska Street, 24, Mykolaiv 54030, Ukraine
zosimovvvv@gmail.com, sashabulgakova2@gmail.com

Abstract. The measure of expert opinions consistency is an essential component for building ranking models based on user ratings. The article describes a model building method for calculating the measure of opinions consistency for potential experts relative to the current user in the case when it is impossible to apply methods based on their past activities analysis. In the case when the current user does not have common evaluations with potential experts for some set of web resources, the measure of their opinions consistency can be calculated on the basis of social and personal factors. The article analyzes the results of constructing such models using various inductive modeling algorithms and neural network.

Keywords: Information search · Expert evaluations · Search engine ·
Search relevance · Inductive modeling · Ranking · GMDH ·
Neural network

1 Introduction

Search engine ranking algorithms takes into account a large number of factors, but the main weight has the page rank, which is calculated based on the analysis of both the quantity and quality of external links to the page. Such valuation methods are objective, but they can be easily falsified in the presence of a certain advertising budget due to the purchase of the required quantity of quality external links from the authority web-resources. From this it follows that they are aimed at satisfying the needs of advertisers, not users.

This problem was dealt with Brits R.A. His paper [1] presents the theoretical model of ranking, based on the statistics of visits to web resources and the time of viewing documents. However, it does not take into account the experts evaluations, as well as the level of consistency of opinion between the current user and experts. Also, ranking methods based on user ratings are widely used when ranking goods in online stores. However, they also take as their basis the number of assessments and its meanings.

Information retrieval - the process of finding unstructured documentary information that meets the user's information needs [2]. Given the fact that the user's

V. Lytvynenko et al. (Eds.): ISDMCI 2019, AISC 1020, pp. 622–636, 2020.
https://doi.org/10.1007/978-3-030-26474-1_43

information needs are individual and subjective concepts, such a definition indicates that search results ranking algorithms should be based on subjective for each user factors.

Considering all the above, the development of search results personalizing methods is one of the actual tasks for the development of the world wide web.

That is why a new research area Big data is created. This paradigm is already reflected in academic programs. The information, for example, structured and unstructured data, media or random processes that is practically impossible to be processed in traditional way refers to the Big data branch of science. The traditional monolithic systems are being replaced with the new asynchronous and parallel solutions facilitating working with Big data.

Big Data information technology is the set of methods and means of processing different types of structured and unstructured dynamic large amounts of data for their analysis and usage for decision support. This technology is an alternative to traditional database management systems and Business Intelligence solutions class. In addition, all of these allow us parallel data processing [2,3]. The system consists of several independent blocks that efficiently process information under conditions of its continuous growth and its distribution throughout the multiple cluster nodes. In such systems, the volumes of information increase exponentially, where unstructured data constitute a substantial part of it. Therefore, the issues of proper interpretation of data flow in systems of such type become more and more urgent [1].

The subject of research is the methods and tools for building, editing and adapting the information flow in distributed information systems.

The purpose of the paper is to study perspective directions and technologies to analyze the data structures in distributed information systems.

2   Levels of Experts Due to the Current User

For ranking data based on expert evaluations, such methods as the Kemeny median, the Kendal rank correlation coefficient, the Borda count method, etc. [13] are used. The use of such methods involves the presence of a predefined group of experts. However, in the real task of search results ranking, inputs use evaluations from users for whom their status is not defined as an expert. Obviously, it would not be right to accept the opinion of all users who rated the web resource as expert's. It is also evident that the evaluation value and the personal data specified in the registration form is not enough to identify the user as an objective domain expert. However, these data are sufficient to identify subjective expert groups for each user based on the consistency of user's evaluations with the evaluations of each expert.

A new technique for the creation of unique expert groups for each user is proposed, which involves three approaches depending on the availability of common evaluations for a certain set of web resources between the current user and potential experts:

1. In the presence of common evaluations between potential experts and the current user, the experts of the first level are determined.
2. In the absence of common evaluations with the current user, but in the presence of common evaluations between potential experts and members of the expert group, experts of the second level are determined.
3. In the absence of any common evaluations, the expert group is formed according to a model constructed on the basis of a user's social profile using inductive algorithms - third level experts.

The division of experts into levels is only a logical one. The weight of experts of all levels is equivalent and is taken into account for ranking without additional coefficients.

To construct methods for expert groups forming, a series of experiments was conducted on four data samples. In each sample of data, 20 users rated the quality and convenience of using for 20 web resources. Table 1 shows a fragment of the data sample 1. User U_0 is the current user for this set.

Table 1. The users rating for web resources data set

Users	Web resource												
	S_1	S_2	S_3	S_4	S_5	S_6	S_7	S_8	$\cdots$	S_{17}	S_{18}	S_{19}	S_{20}
U_0	10	10	8	5	7	4	6	8	$\ldots$	8	8	3	5
U_1	10	10	9	6	8	6	6	7	$\ldots$	8	8	4	4
U_2	8	8	9	5	6	4	5	5	$\ldots$	7	7	3	4
U_3	9	9	9	5	7	5	4	9	$\ldots$	7	6	3	3
U_4	10	10	9	7	7	6	6	8	$\ldots$	9	9	4	3
U_5	2	7	1	6	7	10	3	2	$\ldots$	7	2	7	9
U_6	9	10	7	6	8	3	5	8	$\ldots$	7	6	4	4
U_7	7	7	3	9	10	9	10	4	$\ldots$	5	2	9	9
U_8	8	8	6	4	7	4	5	8	$\ldots$	9	7	3	4
U_9	7	9	8	5	6	5	6	6	$\ldots$	9	7	4	3
U_{10}	7	9	7	4	8	5	6	6	$\ldots$	9	8	3	4
U_{11}	9	10	8	5	7	6	5	8	$\ldots$	7	9	4	4
U_{12}	8	8	9	6	7	4	4	7	$\ldots$	8	8	3	4
U_{13}	8	10	7	6	8	3	4	3	$\ldots$	3	4	9	10
U_{14}	7	7	5	4	8	7	7	7	$\ldots$	8	9	2	4
U_{15}	9	3	9	1	7	9	3	3	$\ldots$	5	3	7	10
U_{16}	4	2	10	3	1	9	8	4	$\ldots$	4	6	9	7
U_{17}	5	2	9	3	2	8	8	4	$\ldots$	4	4	9	10
U_{18}	7	7	10	6	4	5	9	8	$\ldots$	5	8	6	7
U_{19}	7	9	9	8	5	6	6	9	$\ldots$	6	7	7	9
U_{20}	8	7	9	7	4	7	4	8	$\ldots$	5	8	8	4

3   Calculation the First and Second Level Experts Weight

The weight is calculated by the method of average difference of evaluations, which includes:

- the calculation the average value of the evaluations difference modulus h, for each pair "the current user - a potential expert";
- the use of normalizing function (1) to bring data into a scale from 0 to 0.99:

$$G(h) = 1 - (1.1 \cdot \frac{h}{10}) \qquad (1)$$

- selection of users from Table 2 who have a strong connection with the current user at the Chaddock scale ($value > 0.7$).

Table 2. Potential experts weight in relation to U_0

User	User weight	Selected to the expert group
U_1	0.9285	+
U_2	0.8735	+
U_3	0.9010	+
U_4	0.9120	+
U_5	0.6370	−
U_6	0.9120	+
U_7	0.5765	−
U_8	0.8955	+
U_9	0.8955	+
U_{10}	0.9010	+
U_{11}	0.9285	+
U_{12}	0.9175	+
U_{13}	0.6975	−
U_{14}	0.8570	+
U_{15}	0.6315	−
U_{16}	0.5600	−
U_{17}	0.5270	−
U_{18}	0.7580	+
U_{19}	0.7195	+
U_{20}	0.7525	+

The results of the additional researches have shown that for the effective solving of the research tasks, it is advisable to use the method of the evaluations average difference, which gives significantly better results than the Kendal rank correlation coefficient and the like.

The presence of common evaluations in the first two approaches allows us to calculate the experts weight not using the intelligent data processing methods. However, much more interesting is the task of calculating the weight of experts in the absence of common assessments.

This method can be used for a low number or a lack of common evaluations. Based on the potential experts weight calculating model built on the basis of social personal profiles. Calculation the potential experts weight of on the basis of the previous activity analysis always gives the most accurate results. However, at the stage of putting into operation of the ranking system, and in the period from the beginning of operation to the accumulation of a certain assessments base, there will necessarily be situations in which there is not enough data for its application. The accumulation of a certain estimates base means that the database has common estimates in an amount sufficient to form expert groups for most users of the system. Therefore, to ensure the correct operation of the system at an early stage of operation, it was decided to develop a method for determining the users weight, which generally do not have common assessments.

The user's social profile is formed on the basis of the information he indicates when registering on the system. Taking into account the world and domestic experience in psychological research conducting [4], a number of socio-personal factors that can influence on the user's opinion formation has been selected. According to the research results, the factors that turned out to be informative during building a model for potential experts weight determining will be included in the registration form in the system as mandatory fields.

4   Calculation the Third Level Experts Weight

4.1   Gathering Input Data

As input data, random 20 social user profiles $(U_0 - U_{60})$ that participated in previous experiments described in [5] were used. Since their weight relative to the U_0 user has already been calculated on the data samples, the same group of experts as presented in [5] were used to construct the model.

For definition in terms of the users who do not have common evaluations either with the current user, or with the members of the expert group $U_{0,exp}$, we will call the potential expert of the third level.

It is necessary to build a model of the influence of social and personal factors of Internet users on the level of consensus between the potential third-level expert and current users.

The sample contains $n = 20$ points of observation and is divided into two parts: 2/3 points – training sample A, 1/3 points - test sample B: $n_A = 14$, $n_B = 6$.

The accuracy of the obtained models was based on the determination coefficient.

To simulate the degree of expert opinions consistency, 11 algorithms for constructing models were investigated (10 Group Method of Data Handling algorithms and 1 - neural network):

1. Combinatorial algorithm (Combi).
2. Iterative algorithms for inductive modeling.
   - with a linear partial description of the model:
     - Multilayered Iterative Algorithm (MIA);
     - Relaxation Iterative Algorithm (RIA);
     - Combined Iterative Algorithm (CIA);
   - with a quadratic partial description of the model:
     - MIA;
     - RIA;
     - CIA;
     - Multilayered Iterative-Combinational (MICA);
     - Relaxation Iterative-Combinational (RICA);
     - Generalized Iterative Algorithm (GIA);
3. Neural network (NN).

To construct a model for calculating the weight of experts, a number of subjective x_m attributes were selected that can directly or indirectly influence the visitor's assessment:

x_1 - expert activity expressed in the total number of expert assessments;

x_2 - the difference in age, which, in contrast to the absolute value of full years, is a more subjective indicator relative to the user.

x_3 - revenue. The difference is in thousands;

x_4 - answer to questions $3 + 3 \times 3 = ?$

x_5 - sex;

x_6 - the presence of pets;

x_7 - education. The difference between the education of the user and the expert is expressed in numerical value. The grading scale is 1 to 6, where 1 is secondary education, and 6 is a degree.

x_8 - education direction: humanitarian, technical;

x_9 - marital status;

x_{10} - the presence of a vehicle;

x_{11} - playing sports.

4.2   Algorithms

Group Method of Data Handling algorithms. The world-wide known Group Method of Data Handling (GMDH) [6] is one of the most successful methods of inductive modeling. It is used in various tasks of data analysis and knowledge discovery, forecasting and systems modeling [7–9], classification and pattern recognition [10–12].

As of today, a great variety of GMDH algorithms of the sorting-out and iteration types was developed and explored [13]. The sorting-out algorithms are effective as the tool for structural identification but only for limited number of arguments (etc. Combi). Iterative algorithms are capable of working with a lot of arguments but the specific of their architecture do not guarantee constructing the true model structure. Until recently, these two classes of algorithms were developed without combining their strengths.

GIA GMDH Algorithm Description
Formally, in general case, a layer of the GIA GMDH may be defined as follows [14]:

1. the input matrix is $X_{r+1} = (y_1^r, ..., y_F^r, x_1, ..., x_m)$ for a layer $r + 1$, where $x_1, ..., x_m$ are the initial arguments and $y_1^r, ..., y_F^r$ are the intermediate ones of the layer r;
2. the operators of the kind

$$
\begin{aligned}
y_l^{r+1} &= f(y_i^r, y_j^r), \quad l = 1, 2, \ldots, C_F^2, \quad i, j = \overline{1, F}, \\
y_l^{r+1} &= f(y_i^r, x_j), \quad l = 1, 2, \ldots, Fm, \quad i = \overline{1, F}, \ j = \overline{1, m}
\end{aligned}
\tag{2}
$$

may be applied on the layer r + 1 to construct linear, bilinear and quadratic partial descriptions:

$$
\begin{aligned}
z = f(u, v) &= a_0 + a_1 u + a_2 v; \\
z = f(u, v) &= a_0 + a_1 u + a_2 v + a_3 uv \\
z = f(u, v) &= a_0 + a_1 u + a_2 v + a_3 uv + a_4 u^2 + a_5 v^2.
\end{aligned}
\tag{3}
$$

3. for any description, the optimal structure is searched by combinatorial algorithm; e.g., for the linear partial description the expression holds:

$$
f(u, v) = a_0 d_1 + a_1 d_2 u + a_2 d_3 v
\tag{4}
$$

where d_k, $k = 1, 2, 3$, $d_k = \{0, 1\}$ are elements of the binary structural vector $d = (d_1\, d_2\, d_3)$ where values 1 or 0 mean inclusion or not a relevant argument. Then the best model will be described as $f(u, v, d_{opt})$, where

$$
d_{opt} = \arg \min_{l=\overline{1,q}} CR_l, \quad q = 2^p - 1\, f_{opt}(u, v) = f(u, v, d_{opt})
\tag{5}
$$

4. the algorithm stops when the condition $CR^r > CR^{r-1}$ is checked, where CR^r, CR^{r-1} are criterion values for the best models of $(r - 1)$-th and r-th layers respectively. If the condition holds, then stop, otherwise jump to the next layer.

The GIA structure is schematically represented in Fig. 1.

Main Particular Cases of the GIA GMDH Architecture [14]
The following six typical algorithms can be generated as particular cases of the generalized architecture:

Iterative GMDH algorithms:

– Multilayered Iterative Algorithm MIA in which pairs are only between the intermediate arguments;
– Relaxation Iterative Algorithm RIA in which pairs are between the intermediate and initial arguments;
– Combined Iterative Algorithm CIA in which pairs are possible between both the intermediate and intermediate and initial arguments.

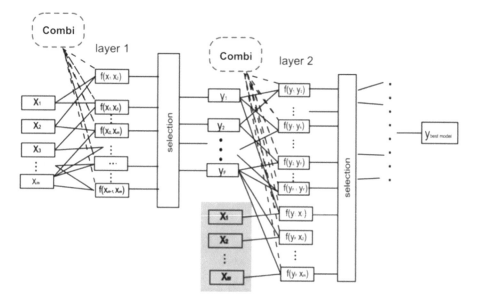

Fig. 1. The generalized architecture of GIA GMDH

Iterative-combinatioral GMDH algorithms with combinatioral optimization of partial descriptions:

- Multilayered Iterative-Combinational MICA;
- Relaxation Iterative-Combinational RICA;
- Combined Iterative-Combinational CICA (being practically the same as GIA).

Any of these variants may be formed and used with the help of the specialized modeling software complex [7,14] based on the GIA GMDH with implementation of the following features: automatic and interactive options for organization of user interface; administration through the web interface; ensuring multi-access. Best constructed models are presented by system for the graphic and semantic analysis as well as selection of the most informative arguments.

Neural Network Approach. A multi-layered perceptron with three hidden layers and a hyperbolic tangent as a function of activation was used to construct the model. In the first layer - 5 neurons, the second - 5, the third - 3 neurons. At the output of the network was one neuron. This network structure was selected during numerical experiments [15,16] as optimal in terms of the number of neurons, layers, and the meaning of the error.

4.3   Results of the Experiments

Below are presented the results of the experiments. The Tables 3, 4 and 5 shows the main results of the simulation of the compared algorithms.

Combinatorial (Combi GMDH)
The constructed model has the form:

$$R^2 = 59.42\% : \hat{y} = 0,579 + 0,0052x_1 + 0,0322x_3 + 0,307x_4 - 0,061x_5 - 0,0434x_7$$
$$-0,1757x_{10} - 0,0257x_{11}$$

Table 3. Construction model of the iterative GMDH algorithms

Algorithm	R^2, %	Model
Iterative algorithms with linear partial description of model		
MIA	42,2	$\hat{y} = 0,691 + 0,0005x_1 - 0,003x_2 + 0,005x_3 + 0,141x_5 + 0,0007x_6 -$ $- 0,011x_7 + 0,085x_8 + 0,063x_9$
RIA	17,7	$\hat{y} = 0,8 + 0,031x_3$
CIA	55,1	$\hat{y} = 0,652 + 0,005x_1 - 0,0046x_2 + 0,008x_3 + 0,105x_5 + 0,00017x_6 -$ $- 0,026x_7 + 0,113x_8 + 0,093x_9 - 0,0738x_{10}$
Iterative algorithms with quadratic partial description of the model		
MIA	56,3	$\hat{y} = 0,8407 - 0,0438x_3 + 0,1938x_5 + 0,052x_6 + 0,017x_7 - 0,0807x_9 -$ $- 0,005x_6x_8 - 0,0559x_5x_9 - 0,0094x_6x_7 + 0,003x_3x_5 - 0,007x_5x_8 -$ $- 0,0717x_6x_9 + 0,0186x_5x_6 + 0,0178x_8x_9 + 0,0059x_3x_7 + 0,004x_3x_8 +$ $+ 0,0109x_6^2 + 0,031x_7x_9 + 0,12x_9^2 - 0,00009x_5x_7 + 0,0076x_7x_8 +$ $+ 0,005x_7^2 + 0,0027x_3^2 + 0,0012x_8^2$
RIA	54,3	$\hat{y} = 0,969 + 0,1584x_5 - 0,0174x_7 - 0,1077x_9 - 0,0612x_5x_7 + 0,278x_5^2 -$ $- 0,343x_5x_9 + 0,0377x_7x_9 + 0,003x_7^2 + 1,077x_9^2$
CIA	77,2	$\hat{y} = 0,419 - 0,099x_3 + 6,184x_5 - 0,085x_7 + 0,0834x_3x_5 - 0,0007\,x_3^2 +$ $+ 0,01x_3x_7 - 6,153x_5^2 + 0,0024x_7^2$
RICA	54,3	$\hat{y} = 0,969 + 0,1584x_5 - 0,0174x_7 - 0,1077x_9 - 0,0612\,x_5x_7 + 0,278\,x_5^2 -$ $- 0,343x_5x_9 + 0,0377\,x_7x_9 + 0,003x_7^2 + 1,077x_9^2$
GIA	80,3	$\hat{y} = 0,89 + 0,106x_5 - 0,1448x_7 + 0,0637x_8 + 0,000159x_1^2 - 0,0007x_2^2 -$ $- 0,005x_8^2 + 0,0003x_1x_8 + 0,0053x_2x_7$
MICA	71,8	$\hat{y} = 0,904 + 0,003x_1 - 0,0289x_3 + 0,271x_5 - 0,002x_6 - 0,118x_7 +$ $+ 0,089x_8 + 0,182x_{10} + 0,0024x_1x_7 - 0,119x_{10}^2 - 0,00016x_3x_6 -$ $- 0,228x_{10}x_5 + 0,0238x_{10}x_3 - 0,00001x_1x_8 + 0,0228x_3x_5 - 0,076x_5x_8 -$ $- 0,014x_8^2 + 0,0085x_3x_8 - 0,001x_3^2 - 0,085x_{10}x_8 - 0,00018x_6x_8 +$ $+ 0,009x_3x_7 - 0,0001x_1x_3 - 0,1013x_5^2 + 0,00001x_1x_6 + 0,011x_7x_8 +$ $+ 0,002x_6x_7 - 0,0092x_7^2 + 0,000018x_1^2$

Table 4. The modeling results of the neural network

№	Input data	y (real data)	$\hat{y}$ (model data)	ε, error $(y-\hat{y})$
1		1	0,905	0,095
2		0,9285	0,894	0,035
3		0,8735	0,752	0,122
4		0,901	0,883	0,018
5		0,912	0,81	0,102
6		0,637	0,628	0,009
7	A	0,912	0,865	0,047
8		0,5765	0,558	0,019
9		0,8955	0,817	0,079
10		0,8955	0,961	-0,066
11		0,901	0,911	-0,010
12		0,9285	0,886	0,043
13		0,9175	0,928	-0,011
14		0,6975	0,737	-0,040
15		0,857	0,856	0,001
16		0,6315	0,63	0,001
17	B	0,56	0,68	-0,120
18		0,527	0,574	-0,047
19		0,758	0,788	-0,030
20		0,7195	0,814	-0,094
21	C	0,7525	0,75	0,002
	R^2	81,96 %		

Graphic representation of the results are shown in Figs. 2, 3.

Construction Model of the Iterative GMDH Algorithms with Combinatioral Optimization.

Table 3 and Figs. 4, 5 and 6 show the modeling results of the iterative algorithms.

Table 3 shows that the generalization (algorithm GIA) allows the most accurately find model than other modifications of iterative algorithms.

Figure 4 shows the results and the value of the selection criteria, AR by layers (the generalized algorithm reaches the minimum in the 7-th layer).

Graphic representation of the results for GIA are shown in Figs. 4, 5 and 6.

Neural Network (NN)

Neural network training was carried out using the method of backpropagation error. The input data was randomly divided into three parts in the following proportions: 70% - test, 25% - check, 5% - control. Table 4 show the modeling results of the Neural network. Table 5 show the summary of simulation results for all algorithms and availability of input parameters in the model. The analysis of the obtained results allows us to conclude that the best simulation results were obtained using GIA and the neural network. But the neural network identifies the hidden dependencies between the input and output (black box system) and after the training is able to predict the next value. However, a mathematical model that will display the connection between input and output is difficult

Table 5. The summary of simulation results for all algorithms

Algorithm	R^2, %	Availability of input parameters in the model											Number of extra parameters
		x_1	x_2	x_3	x_4	x_5	x_6	x_7	x_8	x_9	x_{10}	x_{11}	
Combi	59,4	+		+	+	+		+			+	+	4
Iterative algorithms with linear partial description of model													
MIA	42,2	+	+	+		+	+	+	+	+			3
RIA	17,7			+									10
CIA	55,1	+	+	+		+	+	+	+	+	+		2
Iterative algorithms with quadratic partial description of the model													
MIA	56,23			+		+	+	+	+	+			5
RIA	54,3					+		+		+			8
CIA	77,2			+		+		+					8
MICA	71,8	+		+		+	+	+	+		+		4
RICA	54,3					+		+		+			8
GIA	80,3	+	+			+		+	+				6
Neural network													
NN	81,9	NA											NA

to obtain, since for each network connection on each layer, a large number of weights is determined. Therefore, the mathematical model in expanded form will be immensible for perception and use. In GIA we can use the mathematical model, which is more convenient for this task, because, in the context of this work, the resulting model is an intermediate step for solving a given problem in a mode of limited time, unlike the neural network. In addition, as can be seen from the obtained dependence (GIA model), among the 11 indicators are significant only 6:

x_1 - expert activity expressed in the total number of expert assessments;

x_2 - the difference in age, which, in contrast to the absolute value of full years, is a more subjective indicator relative to the user.

x_5 - sex;

x_7 - education. The difference between the education of the user and the expert is expressed in numerical value. The grading scale is 1 to 6, where 1 is secondary education, and 6 is a degree;

x_8 - education direction: humanitarian, technical.

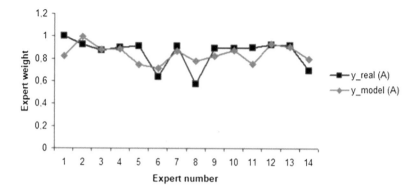

Fig. 2. The value of the true and model results on the sample A (Combi GMDH)

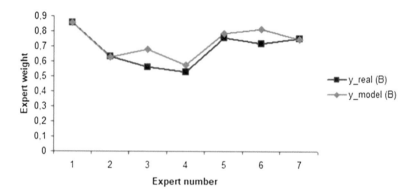

Fig. 3. The value of the true and model results on the sample B (Combi GMDH)

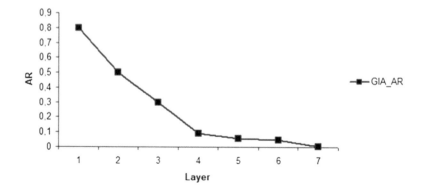

Fig. 4. The value of the evaluation criterion (AR criterion) in layer for GIA

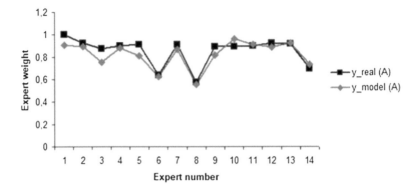

Fig. 5. The value of the true and model results on the sample A

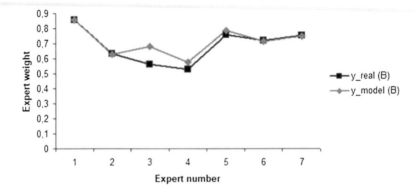

Fig. 6. The value of the true and model results on the sample B

5   Conclusions

A comparative analysis of the methods of inductive modeling and neural network method showed that the neural network gives the most accurate results. However, the task was to construct a model depending on the consistency degree of thoughts on the socio-personal factors of users, and the neural network method does not allow for such a model. The best model was found by a generalized iterative algorithm that contains all the previous iterative structures.

Personalizing search results is an effective method of dealing with search spam and artificial promotion of web resources. Modern search engines such as Google uses a single ranking algorithm for all user search queries. Such algorithms take into account a large number of objective factors characterizing the web resource rank, but do not take into account the user's subjective information needs.

Artificial promotion and search spam are based on the task of the affecting the main ranking factors values, close to ideal. In the personalized ranking method based on expert assessments there are no objective factors manipulation of which can lead to an increase of a web resource rank in search results.

The developed method of search results ranking gives to each one user the list of web resources that has its own unique item order. Such an effect is achieved through the use of expert group members evaluations that is unique to each user. And also due to that each evaluation is included in the web resources final rank calculating model with its unique weight, which is a measure of consistency of thoughts of the current user and expert, calculated on the basis of analysis of their previous activity in the system.

References

1. Pogodaev AK, Gluhov AI, Pogodaev AK (2005) Formalnyiy podhod k samoot-senke deyatelnosti promyishlennyih predpriyatiy. Sovremennyie slozhnyie sistemyi upravleniya In sb. tr. nauch.-praktich. konferentsii. VGASU, T2, Voronezh, pp 133–138. (in Russian)
2. Arrow KJ (1951) Social choice and individual values. Wiley, New York, 124 p
3. Gibbard A (1973) Manipulation of voting schemes: a general result. In: Economet-rica, vol 41, 587 p
4. Melnik SN (2004) Psihologiya lichnosti uchebnoe posobie. TIDOT DVGU, Vladi-vostok (in Russian)
5. Zosimov V, Bulgakova O (2018) Usage of inductive algorithms for building a search results ranking model based on visitor rating evaluations. In: Proceedings of the 13th international scientific and technical conference on computer sciences and information technologies CSIT. IEEE, Lviv, pp 466–470
6. Ivakhnenko AG (1968) Group method of data handling as a rival of the stochastic approximation method. Sov Autom Control 3:58–72
7. Zosimov V, Khrystodorov O, Bulgakova O (2018) Dynamically changing user inter-faces: software solutions based on automatically collected user information. Pro-gram Comput Softw 44:492–498
8. Amarger R, Biegler JLT, Grossmann IE (1990) An intelligent modelling interface for design optimization. Carnegie Mellon University, Pittsburgh
9. Glushkov SV, Levchenko NG (2014) Intelligent modeling as a tool to improve the management of the transport and logistics process. In: Proceedings of the international scient.-tekhn. conference of the eurasian scientific association, pp 1–5. (in Russian)
10. Huang W, Oh SK, Pedrycz W (2002) Fuzzy polynomial neural networks: hybrid architectures of fuzzy modeling. IEEE Trans Fuzzy Syst 10(5):607–621
11. Kondo T (1998) GMDH neural network algorithm using the heuristic self-organization method and its application to the pattern identification. In: Pro-ceedings of the 37th SICE annual conference SICE 1998. IEEE, Piscataway, pp 1143–1148
12. Onwubolu GC (2007) Design of hybrid differential evolution and group method of data handling for inductive modeling. In: Proceedings of the II international workshop on inductive modelling IWIM-2007. CTU in Prague, Czech Republic, pp 87–95

13. Stepashko V (2013) Ideas of academician O.H. Ivakhnenko in the inductive modelling field from historical perspective. In: Proceedings of the 4th international conference on inductive modelling ICIM-2013. IRTC ITS NASU, Kyiv, pp 30–37
14. Stepashko V, Bulgakova O, Zosimov V (2018) Construction and research of the generalized iterative GMDH algorithm with active neurons. In: Shakhovska N, Stepashko V (eds) Advances in intelligent systems and computing, vol 689. Springer, Cham, pp 492–510
15. Bulgakova O, Samoylenko O (2007) Comparing NN and GMDH methods for prediction of socio-economic processes. In: Proceedings of the II international workshop on inductive modelling IWIM-2007, 19–23 September 2007, Prague, Czech Republic. Czech Technical University, Prague, pp 217–220
16. Schmidhuber J (2015) Deep learning in neural networks: an overview. Neural Netw 61:85–117

Online Robust Fuzzy Clustering of Data with Omissions Using Similarity Measure of Special Type

Yevgeniy Bodyanskiy$^{(\boxtimes)}$ (ID), Alina Shafronenko$^{(\boxtimes)}$ (ID), and Sergii Mashtalir$^{(\boxtimes)}$ (ID)

Kharkiv National University of Radio Electronics, Kharkiv, Ukraine
{yevgeniy.bodyanskiy,alina.shafronenko,sergii.mashtalir}@nure.ua

Abstract. The task of clustering is important and the most difficult part of the overall problem of Data Mining because it is based on the self-learning paradigm, i.e. implies the absence of pre-tagged training sample. In real conditions, this task is complicated by the fact that in having data arrays some of the observations can be corrupted by anomalous outliers and some - contain missing data, that is, the "object-property" table has "empty" cells. In addition, data can be arrive in online mode on processing, especially for tasks related with Data Stream Mining and Big Data. In the paper the problem of fuzzy adaptive online clustering of data distorted by outliers that are sequentially fed to the processing when the original sample volume and the number of distorted observations are apriori unknown is considered. The probabilistic and possibilistic adaptive online clustering algorithms for such data, that are based on the similarity measure of a special type that weaken or overwhelming outliers are proposed. The computational experiment confirms the effectiveness of approach under consideration.

Keywords: Computational and artificial intelligence ·
Fuzzy neural networks · Machine learning ·
Self-organizing neural network

1 Introduction

The problem of clustering data sets are often occurs in many practical problems, and its solution has been successfully used mathematical apparatus of computational intelligence [1] and first of all, artificial neural networks and soft computing methods (in the case of overlapping classes) is usually assumed that original array is specified a priori and processing is made in batch mode. Here as one of the most effective approach based on using FCM [2,3,6–8], which is reduced as a result to minimize the goal function with constraints of special form.

Real data often contains abnormal outliers of different nature, for example, measurement errors or distributions with "heavy tails". In this situation classic FCM is not effective because the goal function based on the Euclidean metric, only reinforces the impact of outliers. In such conditions it is advisable to

© Springer Nature Switzerland AG 2020
V. Lytvynenko et al. (Eds.): ISDMCI 2019, AISC 1020, pp. 637–646, 2020.
https://doi.org/10.1007/978-3-030-26474-1_44

use robust goal functions of special form [9–16], the overwhelming influence of outliers. For information processing in a sequential mode, in [17,18] adaptive procedures on-line fuzzy clustering have been proposed, which are in fact on-line modifications of FCM, where instead of the Euclidean metric robust goal functions, weaken the influence of outliers where used. The task is substantially more complicated, if the raw data in the table "object-property" includes missing observations, i.e. the corresponding cell is empty. And although for today there exist methods for solving the problem of data clustering with missing observations when the data simultaneously contain both outliers and missing observations [19–22] the online algorithms are unknown. In this regard, consideration of the problem of fuzzy clustering of data with outliers and missing observations in real time is interesting and important today.

2   Problem Statement

The problem of robust on-line clustering based on objective functions is proposed using similarity measure of special form, which allows to synthesize efficient and numerically simple algorithms is proposed.

Baseline information for solving the task of clustering in a batch mode is the sample of observations, formed from N n-dimensional feature vectors $X = \{x_1, x_2, ..., x_N\} \subset R^n$, $x_k \in X$, $k = 1, 2, ..., N$. The result of clustering is the partition of original data set into m classes $(1 < m < N)$ with some level of membership $U_q(k)$ of k-th feature vector to the q-th cluster $1 \leq q \leq m$. Incoming data are previously centered and standardized by all features, so that all observations belong to the hypercube $[-1, 1]^n$. Therefore, the data for clustering form array $\tilde{X} = \{\tilde{x}_1, ..., \tilde{x}_k, ..., \tilde{x}_N\} \subset R^n$, $\tilde{x}_k = (\tilde{x}_{k1}, ..., \tilde{x}_{ki}, ..., \tilde{x}_{kn})^T$, $-1 \leq \tilde{x}_{ki} \leq 1$, $1 < m < N$, $1 \leq q \leq m$, $1 \leq i \leq n$, $1 \leq k \leq N$ that is, all observations $\tilde{x}_{ki}$ are available for processing.

We have developed numerically simple on-line procedure for partitioning sequentially fed to the data processing $\tilde{x}_k$ on m perhaps overlapping classes, while it is not known in advance whether $\tilde{x}_k$ undistorted or contains missing observations and outliers. Furthermore, it is assumed that the amount of information under processing is not known in advance and is increased with time.

3   Adaptive Fuzzy Robust Data Clustering Based on Similarity Measure

As already mentioned, to solve the problem of fuzzy clustering of data containing outliers the special goal functions of the form [6,9,23,24] can be used, by some means these anomalies overwhelming, and the problem itself is associated with the minimization of these functions. From a practical point of view it is more convenient to use instead of the objective functions, based on the metrics, the so-called measures of similarity (SM) [2], which are subject to more lenient conditions than metrics:

$$\begin{cases} S(\tilde{x}_k, \tilde{x}_p) \geq 0; \\ S(\tilde{x}_k, \tilde{x}_p) = S(\tilde{x}_p, \tilde{x}_k); \\ S(\tilde{x}_k, \tilde{x}_p) = 1 \geq S(\tilde{x}_p, \tilde{x}_k) \end{cases}$$

(no triangle inequality), and clustering problem can be "tied" to maximize these measures.

If the data are transformed so that $-1 \leq \tilde{x}_{ki} \leq 1$ the measure of similarity can be structured so as to suppress unwanted data lying at the edges of interval $[-1, 1]$.

Figure 1 illustrates the use of similarity measure as the Cauchy function with different parameters width $\sigma^2 < 1$.

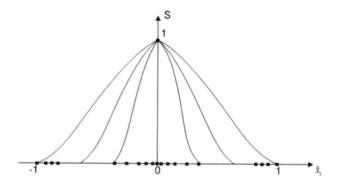

Fig. 1. Similarity measure based on the Cauchy

Picking up the width parameter σ^2 functions

$$S(\tilde{x}_k, w_q) = \cfrac{1}{1 + \cfrac{\|\tilde{x}_k - w_q\|^2}{\sigma^2}} = \frac{\sigma^2}{\sigma^2 + \|\tilde{x}_k - w_q\|^2} = \frac{\sigma^2}{\sigma^2 + D^2(\tilde{x}_k, w_q)} \quad (1)$$

possible exclude the effect outliers, which in principle can not be done using the Euclidean metric

$$D^2(\tilde{x}_k, w_q) = \|\tilde{x}_k - w_q\|^2. \quad (2)$$

Further, by introducing the goal function based on similarity measure (1)

$$E_S(U_q(k), w_q) = \sum_{k=1}^{N} \sum_{q=1}^{m} U_q^\beta(k) S(\tilde{x}_k, w_q) = \sum_{k=1}^{N} \sum_{q=1}^{m} \frac{U_q^\beta(k)\sigma^2}{\sigma^2 + \|\tilde{x}_k - w_q\|^2},$$

probabilistic constraints

$$\sum_{q=1}^{m} U_q(k) = 1.$$

Lagrange function

$$L_S(U_q(k), w_q, \lambda(k)) = \sum_{k=1}^{N}\sum_{q=1}^{m} \frac{U_q^{\beta}(k)\sigma^2}{\sigma^2 + \|\tilde{x}_k - w_q\|^2}$$

$$+ \sum_{k=1}^{N}\lambda(k)(\sum_{q=1}^{m}U_q(k) - 1) \qquad (3)$$

(here $\lambda(k)$ indefinite Lagrange multipliers) and solving the system of Karush-Kuhn-Tucker equations, we arrive at the solution

$$\begin{cases} U_q(k) = \dfrac{(S(\tilde{x}_k, w_q))^{\frac{1}{1-\beta}}}{\sum_{l=1}^{m}(S(\tilde{x}_k, w_l))^{\frac{1}{1-\beta}}}; \\[2em] \lambda(k) = -(\sum_{l=1}^{m}(\beta S(\tilde{x}_k, w_l))^{\frac{1}{1-\beta}})^{1-\beta}; \\[1.5em] \nabla_{w_q}L_S(U_q(k), w_q, \lambda(k)) = \sum_{k=1}^{N}U_q^{\beta}(k) \\[1em] \times \dfrac{\tilde{x}_k - w_q}{(\sigma^2 + \|\tilde{x}_k - w_q)^2} = \vec{0}. \end{cases} \qquad (4)$$

The last Eq. (4) has no analytic solution, so to find a saddle point of the Lagrangian (3) we can use the procedure of Arrow-Hurwitz-Uzawa, as a result of which we obtain the algorithm

$$\begin{cases} U_q(k+1) = \dfrac{(S(\tilde{x}_{k+1}, w_q))^{\frac{1}{1-\beta}}}{\sum_{l=1}^{m}(S(\tilde{x}_{k+1}, w_l))^{\frac{1}{1-\beta}}}; \\[2em] w_q(k+1) = w_q(k) + \eta(k+1)U_q^{\beta}(k+1) \\[1em] \times \dfrac{\tilde{x}_{k+1} - w_q}{(\sigma^2 + \|\tilde{x}_{k+1} - w_q\|^2)^2} \\[1em] = w_q(k) + \eta(k+1)\varphi_q(k+1)(\tilde{x}_{k+1} - w_q). \end{cases} \qquad (5)$$

where

$$\varphi_q(k+1) = \frac{\tilde{x}_{k+1} - w_q}{(\sigma^2 + \|\tilde{x}_{k+1} - w_q\|^2)^2}$$

is the neighbourhood robust functions of WTM-self-learning rule. Assuming the value fuzzifier as $\beta = 2$, we arrive at a robust variant of FCM:

$$\begin{cases} U_q(k+1) = \dfrac{S(\tilde{x}_{k+1}, w_q)}{\sum_{l=1}^{m}S(\tilde{x}_{k+1}, w_l)}; \\[1.5em] w_q(k+1) = w_q(k) + \eta(k+1)\dfrac{U_q^2(k+1)}{(\sigma^2 + \|\tilde{x}_{k+1} - w_q\|^2)^2} \end{cases}$$

Further, using the concept of accelerated time, it's possible to introduce robust adaptive probabilistic fuzzy clustering procedure in the form

$$
\begin{cases}
U_q^{(\tau+1)}(k) = \dfrac{(S(\tilde{x}_k, w_q^{(\tau)}))^{\frac{1}{1-\beta}}}{\sum_{l=1}^{m}(S(\tilde{x}_k, w_l^{(\tau)}))^{\frac{1}{1-\beta}}}; \\[4mm]
w_q^{(Q)}(k) = w_q^{(Q)}(k+1); \\[2mm]
w_q^{(\tau+1)}(k+1) = w_q^{(\tau)}(k+1) + \eta(k+1)\dfrac{(U_q^Q(k))^{\beta}}{(\sigma^2 + \|\tilde{x}_{k+1} - w_q^{(\tau)}(k+1)\|^2)^2} \\[2mm]
\times(\tilde{x}_{k+1} - w_q^{(\tau)}(k+1)).
\end{cases}
\tag{6}
$$

Similarly, it's possible to synthesize a robust adaptive algorithm for possibilistic [10,11] fuzzy clustering using criterion

$$
E_S(U_q(k), w_q, \mu_q) = \sum_{k=1}^{N}\sum_{q=1}^{m} U_q^{\beta}(k) S(\tilde{x}_k, w_q) + \sum_{q=1}^{m}\mu_q \sum_{k=1}^{N}(1 - U_q(k))^{\beta}
$$

Solving the problem of optimization, we obtain the solution:

$$
\begin{cases}
U_q(k+1) = (1 + (\dfrac{S(\tilde{x}_{k+1}, w_q(k))}{\mu_q(k)}))^{-1}; \\[3mm]
w_q(k+1) = w_q(k) + \eta(k+1)U_q^{\beta}(k+1) \\[2mm]
\times\dfrac{\tilde{x}_{k+1} - w_q(k)}{(\sigma^2 + \|\tilde{x}_{k+1} - w_q(k)\|^2)^2}; \\[3mm]
\mu_q(k+1) = \dfrac{\sum_{p=1}^{k+1} U_q^2(p) S(\tilde{x}_p, w_q(k+1))}{\sum_{p=1}^{k+1} U_q^2(p)}
\end{cases}
\tag{7}
$$

And, finally, introducing the accelerated time we obtain the procedure

$$
\begin{cases}
U_q^{(\tau+1)}(k) = \dfrac{1}{1 + (\dfrac{S(\tilde{x}_k, w_q^{(\tau)}(k))}{\mu_q^{(\tau)}(k)})^{\frac{1}{\beta-1}}} \\[4mm]
w_q^{(Q)}(k) = w_q^{(Q)}(k+1); \\[2mm]
w_q^{(\tau+1)}(k+1) = w_q^{(\tau)}(k+1) + \eta(k+1)\dfrac{(U_q^Q(k))^{\beta}}{(\sigma^2 + \|\tilde{x}_{k+1} - w_q^{(\tau)}(k+1)\|^2)^2} \\[2mm]
\times(\tilde{x}_{k+1} - w_q^{(\tau)}(k+1)) \\[2mm]
\mu_q^{(\tau+1)}(k) = \dfrac{\sum_{p=1}^{k} U_q^{(\tau+1)}(p) S(\tilde{x}_p, w_q(k))}{\sum_{p=1}^{k} U_q^{(\tau+1)}(p)}.
\end{cases}
\tag{8}
$$

4   Experiments

Experimental research conducted on samples of data such as Iris UCI repository.

Iris. This is perhaps the best known database to be found in the pattern recognition literature. Fisher's paper is a classic in the field and is referenced frequently to this day. The data set contains 3 classes of 50 instances each, where each class refers to a type of iris plant. One class is linearly separable from the other 2; the latter are NOT linearly separable from each other. Predicted attribute: class of iris plant. This is an exceedingly simple domain. Number of Instances: 150 (50 in each of three classes). Number of Attributes: 4 numeric, predictive attributes (sepal length in cm, sepal width in cm, petal length in cm, petal width in cm) and the 3 classes: Iris Setosa, Versicolour, Virginica.

Wine. These data set contain results of a chemical analysis of wines grown in the same region in Italy and derived from three different cultivars. The analysis determined the quantities of 13 constituents found in each of the three types of wines. Attributes that donated by Riccardo Leardi same as alcohol, malic acid, ash, alkalinity of ash, magnesium etc. Number of Instances: 178. Number of Attributes: 13.

To estimate the quality of the algorithm we used quality criteria partitioning into clusters such as: Partition Coefficient (PC), Classification Entropy (CE), Partition Index (SC), Separation Index (S), Xie and Beni's Index (XB), Dunn's Index (DI).

> **Partition Coefficient (PC):** measures the amount of "overlapping" between clusters.
> **Classification Entropy (CE):** it measures the fuzzyness of the cluster partition only, which is similar to the Partition Coefficient.
> **Partition Index (SC):** is the ratio of the sum of compactness and separation of the clusters. It is a sum of individual cluster validity measures normalized through division by the fuzzy cardinality of each cluster. SC is useful when comparing different partitions having equal number of clusters. A lower value of SC indicates a better partition.
> **Separation Index (S):** on the contrary of partition index (SC), the separation index uses a minimum-distance separation for partition validity.
> **Xie and Beni's Index (XB):** it aims to quantify the ratio of the total variation within clusters and the separation of clusters. The optimal number of clusters should minimize the value of the index.
> **Dunn's Index (DI):** this index is originally proposed to use at the identification of "compact and well separated clusters". So the result of the clustering has to be recalculated as it was a hard partition algorithm.

The experimental results are presented in the Tables 1 and 2. We also compared the results of our proposed algorithm with other more well-known such as Fuzzy C-means (FCM) clustering algorithm.

Figure 2 shows the work of data clustering algorithm that is not corrupted by abnormal outliers and missing observations, Fig. 3 shows the work of data

Table 1. Results of experiments

Algorithms	Iris UCI repository					
	PC	CE	SC	S	XB	DI
Adaptive fuzzy robust clustering data based on similarity measure	0.0199	0.0122	−0.2439	0.0022	0.0015	1
FCM	0.8011	0.3410	0.2567	0.0030	7.1965	0.0080

Table 2. Results of experiments

Algorithms	Iris UCI repository					
	PC	CE	SC	S	XB	DI
Adaptive fuzzy robust clustering data based on similarity measure	0.0230	0.0219	−0.3007	0.0032	0.0065	0.9999
FCM	0.7411	0.2389	0.3112	0.0040	6.9945	0.0078

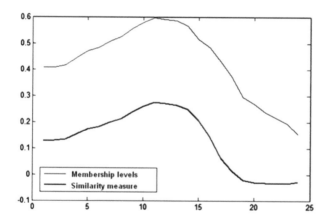

Fig. 2. Data clustering that is not corrupted by abnormal outliers, where solid line is membership level; bold line is function of similarity measure

clustering algorithm that corrupted by missing observations and Fig. 4 shows the work of data clustering algorithm that corrupted by abnormal outliers and missing observations.

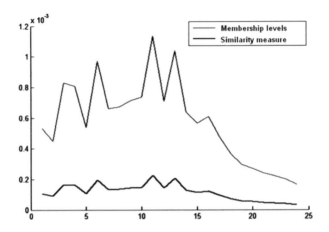

Fig. 3. The results of the data clustering algorithm that corrupted by missing observations

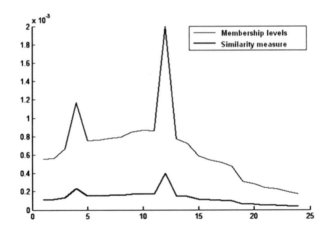

Fig. 4. The results of the data clustering algorithm that corrupted by abnormal outliers and missing observations

5   Conclusions

In the paper the task of fuzzy clustering of a stream of observations corrupted by abnormal outliers and missing observations is considered. The proposed approach is based on hybridization of probabilistic and possibilistic procedures of fuzzy clustering, robust estimation criteria, similarity measures of a special type and T. Kohonen self-learning according the "Winner Takes More" principle for optimizing the goal function of special type. Optimization of goal function under constraints using the method of Lagrange multipliers permitted to obtain analytical expressions for calculating membership levels and recurrent relations for tuning of centroids for formed clusters. The proposed algorithms are essentially gradient optimization procedures and are fast and easy in numerical implementation. It is shown that the proposed procedures are a generalization of the well-known algorithms of fuzzy clustering and coincide with them if the initial data are "clean" (without outliers and missing observations). The performed simulation experiments confirm the effectiveness of the developed approach. The subject of further research is the problem of fuzzy online clustering in conditions when not only the initial data are "corrupted" by outliers and missing observations, but also the number of clusters in the processed data is unknown. It is intended in future to develop Kohonen's neuro-fuzzy map with various numbers of neurons in the network.

References

1. Rutkowski L (2008) Computational intelligence: methods and techniques. Springer, Heidelberg
2. Sepkovski JJ (1974) Quantified coefficients of association and measurement of similarity. J Int Assoc Math 6(2):135–152
3. Bezdek JC (1981) Pattern recognition with fuzzy objective function algorithms. Plenum, New York
4. Yang YK, Shieh HL, Lee CN (2004) Constructing a fuzzy clustering model based on its data distribution. In: International conference on computational intelligence for modeling, control and automation (CIMCA 2004), Gold Coast, Australia
5. Nikolova M (2004) A variational approach to remove outliers and impulse noise. J Math Imaging Vis 20(1–2):99–120
6. Davé RN, Sen S (1997) Noise clustering algorithm revisited. In: NAFIPS 1997, 21–24 September 1997, pp 199–204
7. Gabrys B, Bargiela A (2000) General fuzzy min-max neural network for clustering and classification. IEEE Trans Neural Netw 11(3):769–783
8. Hathaway RJ, Bezdek JC, Hu Y (2000) Generalized fuzzy c-means clustering strategiesusing Lp norm distances. IEEE Trans Fuzzy Syst 8(5):576–582
9. Davé RN, Sen S (2002) Robust fuzzy clustering of relational data. IEEE Trans Fuzzy Syst 10(6):713–727
10. Comaniciu D, Meer P (2002) Mean shift: a robust approach toward features pace analysis. IEEE Trans Pattern Anal Mach Intell 24(5):603–619
11. Dave RN, Krishnapuram R (1997) Robust cluatering methods: a unified view. IEEE Trans Fuzzy Syst 5(2):270–293
12. Zhang J, Leung Y (2003) Robust clustering by pruning outliers. IEEE Trans Syst Man Cybern 33(6):983–999
13. Ren LX, Irwin G (2003) Robust fuzzy Gustafson-Kessel clustering for nonlinear system identification. Int J Syst Sci 34(14–15):787–803
14. Leski J (2003) Towards a robust fuzzy clustering. Fuzzy Sets Syst 137(2):215–233
15. Honda LK, Sugiura N, Ichihashi H (2003) Robust local principal component analyzer with fuzzy clustering. In: Proceedings of the international joint conference on neural networks, vol l, pp 732–737
16. Yang MS, Wu KL (2004) A similarity-based robust clustering method. IEEE Trans Pattern Anal Mach Intell 26(4):434–448
17. Keller L, Krishnapuram R, Pal NR (2005) Fuzzy models and algorithms for pattern recognition and image processing. Springer, New York

18. Shafronenko A, Dolotov A, Bodyanskiy Y, Setlak G (2018) Fuzzy clustering of distorted observations based on optimal expansion using partial distances. In: 2018 IEEE second international conference on data stream mining and processing (DSMP), pp 327–330
19. Pal NR, Pal K, Keller JM et al (2005) A possibilistic fuzzy c-means clustering algorithm. IEEE Trans Fuzzy Syst 13(4):517–530
20. Kokshenev I, Bodyanskiy Y, Gorshkov Y, Kolodyazhniy V (2006) Outlier resistant recursive fuzzy clustering algorithm. In: Reusch B (ed) Computational intelligence: theory and application. Advances in soft computing, vol 38. Springer, Heidelberg, pp 647–652
21. Cai W, Chen S, Zhang D (2007) Fast and robust fuzzy c-means clustering algorithms incorporating local information for image segmentation. Pattern Recogn 40(3):825–838
22. Banerjee (2009) Robust fuzzy clustering as a multi-objective optimization procedure. In: Proceeding soft the 28th annual meeting of the North American fuzzy information processing society (NAFIPS 2009), June 2009
23. Bezdek JC (2013) Pattern recognition with fuzzy objective function algorithms. Springer, New York
24. Ji Z, Liu J, Cao G, Sun Q, Chen Q (2014) Robust spatially constrained fuzzy c-means algorithm for brain MR image segmentation. Pattern Recogn 47:2454–2466
25. Krinidisand S, Chatzis V (2010) A robust fuzzy local information c-means clustering algorithm. IEEE Trans Image Process 19(5):1328–1337

Neural Network Approach for Semantic Coding of Words

Vladimir Golovko[1,2] (ID), Alexander Kroshchanka[1] (ID), Myroslav Komar[3(✉)] (ID),
and Anatoliy Sachenko[3,4] (ID)

[1] Brest State Technical University, Brest, Belarus
vladimir.golovko@gmail.com
[2] Państwowa Szkoła Wyższa im. Papieża Jana Pawła II, Biala Podlaska, Poland
[3] Ternopil National Economic University, Ternopil, Ukraine
mko@tneu.edu.ua, sachenkoa@yahoo.com
[4] Kazimierz Pulaski University of Technology and Humanities, Radom, Poland

Abstract. This article examines and analyzes the use of the word2vec method for solving semantic coding problems. The task of semantic coding has acquired particular importance with the development of search system. The relevance of such technologies is associated primarily with the ability to search in large-volume databases. Based on the obtained practical results, a prototype of a search system based on the use of selected semantic information for the implementation of relevant search in the database of documents has been developed. Proposed two main scenarios for the implementation of such a search. The training set has been prepared on the basis of documents in the English version of Wikipedia, which includes more than 100,000 original articles. The resulting set was used in the experimental part of the work to test the effectiveness of the developed prototype search system.

Keywords: Neural network · Semantic coding · Search system ·
Training set · Word2vec method

1 Introduction

The task of semantic coding has gained special importance with the development of search engines. The urgency of such technologies is primarily due to the possibility of searching in large-scale databases. In this case, it is not so much the finding of identical words as the realization of the search for relatives in a certain semantic metric of words. As a metric, the position of the target word is often chosen in the sentence relative to the other word.

It is intuitive to understand that words close to the meaning of a sentence must appear in the same or similar contexts. Contextually, in this case, words that are located in close proximity to the target or, in other words, the target word are understood. It is this idea that is based on the methods of semantic coding (for example, [1–3]). These methods allow a dictionary D of a fixed size, whose words are represented in a certain code, to translate it into a code of a

© Springer Nature Switzerland AG 2020
V. Lytvynenko et al. (Eds.): ISDMCI 2019, AISC 1020, pp. 647–658, 2020.
https://doi.org/10.1007/978-3-030-26474-1_45

smaller (reduced) dimension (Fig. 1). In parallel with this, due to the specifics of the implementation of such methods, the allocation of semantically meaningful information that can be used to perform search functions occurs.

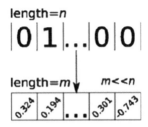

Fig. 1. Coding words with dimensional reduction

Due to the fact that words in the vocabulary of a certain language are almost always different in length, the realization of a task of comparing words is significantly complicated. Bringing each vocabulary word to a vector of a given dimension, the same in length for all words, allows you to compare the searched and verified words directly by computing any (for example, the Euclidean metric). Such technology allows not only to simplify search tasks, but also to make such search more intelligent.

To solve the search problem, you can use deep neural networks. Today, machine learning methods [4–7], in particular deep learning [8–10], along with advances in computing power, play an important role in intelligent data processing. Deep neural networks [11–13] have a greater efficiency of non-linear transformation and presentation of data compared to traditional neural networks. Such a network performs deep hierarchical transformations of the input image space. In addition to Hinton, deep neural networks were studied LeCun et al. [14], Bengio et al. [15–17], Erhan et al. [18], Deng and Yu [19], Schmidhuber [20], Fischer and Igel [21], Golovko [22,23].

Due to the multi-layered architecture, deep neural networks allow processing and analyzing a large amount of information [24–26], as well as simulating cognitive processes in various fields, in particular speech recognition [24,27], computer vision [28,29], Detecting Computer Attacks [30–32], Image Recognition [33–35], etc. [36].

Deep neural networks are also successfully used in industrial products that take advantage of the sheer volume of digital data. Companies such as Apple [37], Google [38], Microsoft [39], IBM [40] Facebook and others, who collect and analyze large amounts of data on a daily basis, are aggressively promoting projects related to deep learning. According to the scientists of the Massachusetts Institute of Technology, deep neural networks are included in the list of the 10 most promising high technologies that can in the near future largely transform the daily lives of most people on our planet. Deep learning has become one of the most sought after areas of information technology.

2   Problem Statement

One of the methods of semantic coding, widely used in practice, is word2vec. This approach was proposed by Nikolov in 2013 [1].

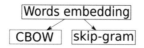

Fig. 2. Variants of the word2vec method [1]

Word2vec allows you to perform semantic analysis of the text with the selection of the closest words. There are two variants of the method word2vec (Fig. 2), differing in the context of the participation policy. Under the context, in this case, is the collection of words (left and right) surrounding the target word taken within a certain window. The first option, called skip-grams, is based on the training of the neural network model, which creates a context based on a single target word given to the model input (Fig. 3). The second variant, called the Continuous Bag of Words (CBOW), uses a neural network to generate a target word based on the supplied context (Fig. 4). Following Nikolov [1], the skip-gram option is more often used in the case of a small dimension of the learning set and allows for the representation of rare words and phrases. CBOW is faster than skip-grams, and shows better accuracy for high-frequency words.

At the heart of both options lies the use of a simple in its structure of the surface neural network (in fact, two-layer).

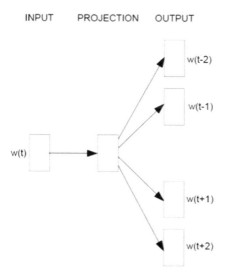

Fig. 3. The basic word2vec neural network algorithm (variant skip-grams) [1]

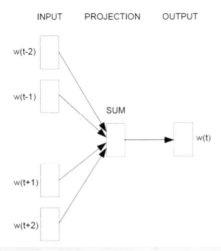

Fig. 4. The basic word2vec neural network algorithm (CBOW version) [1]

As the result of applying word2vec, an artificial neural network is trained to display a word written in the form of a unitary code (and, consequently, a dictionary) into a space of smaller dimension, which is then used to evaluate the semantic proximity of words. The resulting immersion can be used to form a list of semantically close words, as well as predictions of semantic relations. For example, the king for the queen too is the father for ?. In this case, the model will be able to select the correct answer.

Before direct training of the neural network, pre-processing of text data and generation of a training set in the form of pairs (target-word, contextual word) is performed. This view allows you to accelerate the training of the neural network.

The process of preprocessing text data can be conditionally divided into the following steps:

1. Tokenization. It is locked into parsing of the text with its word splitting and deletion of punctuation marks and special characters (the register is being changed), the input is text in the format txt, the output is the list of words of the text.
2. Optional deletes stop words (i.e., the most common and high-frequency words). This category can include articles, prepositions, particles, etc. Input – a list of words in the text, an output – a list with deleted noises. After this stage, the dictionary is actually formed.
3. Removing low frequency words. Often, such words refer to rarely used words or words written in a language other than the language of the text. Similar words are replaced by a special kind of token.
4. Pre-coding words. It is produced, for example, with unitary coding (one-hot-coding). Unitary encoding is carried out by generating for each word a vector of the dictionary's dimension, in which the value of 1 is set for the ordinal position of the word in the dictionary, and 0 for all other elements).

Input – dictionary. Output – the representation of words in the form of one-hot-vectors.

Let's consider the algorithms of applying each variant of the Word2vec method.

3   Algorithms for Using Skip-Grams and CBOW of Word2vec Method

3.1   The Skip-Gram Variant of the Word2vec Method

Training is carried out according to the following algorithm:

0 Clear the dictionary D.
1 For each document from the training set, the following is performed:
    1.1 if the format is non-textual, recognize the text of the document, otherwise – go to step 1.2. Input – document, output – document in txt format.
    1.2 Break the received text into words and delete punctuation marks and special characters. Input – text in txt format, output – a list of words.
    1.3 Remove stop words from the list of words. Input – a list of words in the document, the output is a list with deleted noises.
    1.4 Add new words to D dictionary.
    1.5 Encode document words using a one-hot vector. Input – dictionary. The output is the matrix W, composed of one-hot vectors.
    1.6 Positive Sampling. For the matrix W form a list of pairs of vectors representing semantically close words. Add a list to the tutorial set L.
    1.7 Negative sampling. Add a pair of semantically distant words to the learning set L.
2 For training set L and vocabulary D, learn the perceptron for the skip-gram model.
3 Save the weight ratios obtained.

3.2   The CBOW Variant of the Word2vec Method

Training is carried out according to the following algorithm:

0 Clear the dictionary D.
1 For each document from the training set, the following is performed:
    1.1 if the format is non-textual, recognize the text of the document, otherwise - go to step 1.2. Input – document, output – document in txt format.
    1.2 Break the received text into words and delete punctuation marks and special characters. Login - text in txt format, output - a list of words.
    1.3 Remove stop words from the list of words. Input – a list of words in the document, the output is a list with deleted noises.
    1.4 Add new words to D dictionary.
    1.5 Encode document words using a one-hot vector. Input – dictionary. The output is the matrix W, composed of one-hot vectors.

1.6 Run the passage through the text of the document with the window. For each window, enter the central word and contextual words in the learning set L as one-hot vectors from the matrix W.
2 For the learning set L and vocabulary D, learn the perceptron for the CBOW model.
3 Save the weight ratios obtained.

4 Solution to the Problem of Semantic Coding. Search of the System Development

The search system prototype, based on the use of the word2vec method, is presented in Fig. 5 [41].

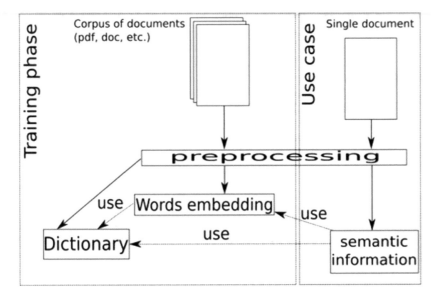

Fig. 5. Scheme of a prototype search system built on the basis of word2vec [41]

As it can be seen, the stage of functioning of the system is closely connected with the stage of adjustment of its parameters during the training of the corresponding neural network model. In our case, the training phase involves preprocessing and training the network used on a large set of documents. These documents can be presented in various formats (including in the form of images). Therefore, the preprocessing step may include recognizing individual words and letters of the source text presented in a binary (non-symbolic) format. For this purpose, traditional means (such as OCR) and neural network models (for example, deep convolutional neural networks) can be used. After performing preprocessing, the training of the corresponding neural network model is carried out. In the course of this process, an appeal to the generated dictionary may be periodically carried out. After training, the system is ready for use.

In the use case (Use case), the trained model is used to identify semantic information (namely, the proximity of various words of the text to the keyword and the search by pattern). Thus, there are two possible options for using the proposed system:

1. Search by keywords:
   – Submit the keywords to the system input from dictionary D.
   – To realize their conversion using saved weight coefficients. On the output are converted vectors.
   – Organize a metric comparison for the pool stored in the database documents. Use the Euclidean norm as a metric. The output is the results of calculating the metric for the words from each database document.
   – Select the documents that are most relevant to the request.
2. Search for similar documents:
   – Submit the document to the system input.
   – To carry out its pre-treatment (as at the stage of training). At the output, the matrix W is composed of one-hot-vectors.
   – Perform the multiplication of the matrix W by the weight coefficients stored at the learning stage (direct propagation). At the output - vectors of words of reduced dimension.
   – Calculate the signature of the document using the data from item 3. The dimension of the final signature vector coincides with the dimension of the vectors from item 3. At the output – the signature of the document.
   – Search the database of documents with saved signatures.
   – Select documents with similar signatures.

To illustrate the work of the word2vec method, we give an example of a two-dimensional rendering of reduced word codes obtained by us for a set of 100,000 English-language Wikipedia documents and a total vocabulary of 50,000 words. In this experiment, the simplified skip-grams architecture, presented in Figs. 3 and 4 and including 50,000 input neurons corresponding to the target word, 300 hidden and 50,000 output neurons corresponding to the contextual word. Thus, the weight coefficients of the first layer of the trained neural network are a dip matrix for generating a reduced code of the source words.

During training, pairs of words (target word, contextual word) were generated, which were fed to the neural network of mini-bats of 128 pairs in each. After learning to reduce the word codes, the t-SNE algorithm [42] was used to reduce the dimensionality of the data. A fragment obtained by a two-dimensional map of semantic similarity is depicted in Fig. 6 [41].

The illustration illustrates the fact that with the help of parameters of a trained neural network one can search for semantically similar words in the semantic plane, for example, the words **lake, river, sea** and **water** fall into one group, and the words **album, record, song** – to another.

Fig. 6. Map of semantic resemblance

Enter url and keywords for analysis

https://en.wikipedia.org/wiki/Sherlock_Holmes   fiction, book   Submit

book: ['books', 'novel', 'story', 'essay', 'article', 'poem', 'volume', 'novels']

fiction: ['novels', 'fantasy', 'stories', 'horror', 'story', 'literature', 'genre', 'poetry']

Sherlock Holmes (/ˈʃɜːrlɒk ˈhoʊmz/) is a fictional private detective created by British
author Sir Arthur Conan Doyle. Referring to himself as a "consulting detective" in the
stories, Holmes is known for his proficiency with observation, forensic science, and
logical reasoning that borders on the fantastic, which he employs when investigating
cases for a wide variety of clients, including Scotland Yard. First appearing in print in
1887 (in A Study in Scarlet), the character's popularity became widespread with the
first series of short **stories** in The Strand Magazine, beginning with "A Scandal in
Bohemia" in 1891; additional tales appeared from then until 1927, eventually totalling
four **novels** and 56 short **stories**. All but one are set in the Victorian or Edwardian eras,
between about 1880 and 1914. Most are narrated by the character of Holmes's friend
and biographer Dr. Watson, who usually accompanies Holmes during his investigations
and often shares quarters with him at the address of 221B Baker Street, London, where
many of the **stories** begin. Though not the first fictional detective, Sherlock Holmes is
arguably the best known, with Guinness World Records listing him as the "most
portrayed movie character" in **history**. Holmes's popularity and fame are such that
many have believed him to be not a fictional character but a real individual; numerous
literary and fan societies have been founded that pretend to operate on this principle.
Widely considered a British cultural icon, the character and **stories** have had a
profound and lasting effect on mystery writing and popular culture as a whole, with the
original tales as well as thousands written by authors other than Conan Doyle being
adapted into stage and radio plays, television, films, video games, and other media for

Fig. 7. Search system (displaying results)

Based on the obtained results we can implement a system that searches for related words and their illumination in the text under study, the result of which for the given URL-address and the list of searched words is presented in Fig. 7.

In addition, the word2vec method can be used to form a knowledge base in a specific subject area (for example, [43]) and extract semantic relationships in general [44].

5 Conclusions

This article discusses and analyzes the use of the word2vec method for solving semantic coding problems.

An analysis of the applicability of the word2vec method for performing semantic word coding, based on which a prototype of a search system was developed, based on the use of the obtained semantic information to carry out a relevant search in the basis of documents in two main scenarios. Based on the obtained practical results, a prototype of a search engine, based on the use of dedicated semantic information for the implementation of a relevant search in the document base, has been developed. Two major scenarios for implementing such a search are proposed. A training set has been prepared on the basis of the English-language version of Wikipedia documents, which includes more than 100,000 original articles. The set was used in the experimental part of the work to test the effectiveness of the developed prototype of the search engine.

Since the applied neural network structure for implementing word2vec coding is essentially superficial, the question of the comparative characterization of word2vec and semantic coding methods implemented using deep neural networks and convolutional neural networks remains open to research.

References

1. Mikolov T, Chen K, Corrado G, Dean J (2019) Efficient estimation of word representations in vector space. https://arxiv.org/abs/1301.3781. Accessed 14 Apr 2019
2. Mikolov T, Sutskever I, Chen K, Corrado GS, Dean J (2019) Distributed representations of words and phrases and their compositionality. https://arxiv.org/abs/1310.4546v1. Accessed 14 Apr 2019
3. Pennington J, Socher R, Manning C (2014) Glove: global vectors for word representation. In: Proceedings of the 2014 conference on empirical methods in natural language processing (EMNLP), pp 1532–1543
4. Lin J, Kolcz A (2012) Large-scale machine learning at twitter. In: Proceedings of the ACM SIGMOD international conference on management of data, Scottsdale, Arizona, USA, pp 793–804
5. Smola A, Narayanamurthy S (2010) An architecture for parallel topic models. In: Proceedings of the 36th international conference on very large data bases, Singapore, pp 703–710
6. Ng A et al (2006) Map-reduce for machine learning on multicore. In: Proceedings of advances in neural information processing systems, Vancouver, Canada, pp 281–288

7. Panda B, Herbach J, Basu S, Bayardo R (2012) MapReduce and its application to massively parallel learning of decision tree ensembles, in scaling up machine learning: parallel and distributed approaches. Cambridge University Press, Cambridge

8. Crego E, Munoz G, Islam F (2019) Big data and deep learning: big deals or big delusions? http://www.hufngtonpost.com/george-munoz-frank-islamanded-crego/big-data-and-deep-learnin_b_3325352.html. Accessed 14 Apr 2019

9. Bengio Y, Bengio S (2000) Modeling high-dimensional discrete data with multilayer neural networks. In: Proceedings of advances in neural information processing systems, Vancouver, Canada, vol 12, pp 400–406

10. Ranzato MA, Boureau YL, LeCun Y (2007) Sparse feature learning for deep belief networks. In: Proceedings of advances in neural information processing systems, Vancouver, Canada, vol 20, pp 1185–1192

11. Hinton GE, Osindero ES, Teh Y (2006) A fast learning algorithm for deep belief nets. Neural Comput 18:1527–1554

12. Hinton G, Salakhutdinov R (2006) Reducing the dimensionality of data with neural networks. Science 313(5786):504–507

13. Hinton GE (2010) A practical guide to training restricted Boltzmann machines. Machine Learning Group, University of Toronto, Technical report 2010-000

14. LeCun Y, Bengio Y, Hinton G (2015) Deep learning. Nature 521(7553):436–444

15. Bengio Y, Lamblin P, Popovici D, Larochelle H (2007) Greedy layer-wise training of deep networks. In: Scholkopf B, Platt JC, Hoffman T (eds) Advances in neural information processing systems, vol 11. MIT Press, Cambridge, pp 153–160

16. Bengio Y (2009) Learning deep architectures for AI. Found Trends Mach Learn 2(1):1–127

17. Bengio Y et al (2013) Representation learning: a review and new perspectives. IEEE Trans Pattern Anal Mach Intell 35(8):1798–1828

18. Erhan D, Bengio Y, Courville A, Manzagol P-A, Vincent P, Bengio S (2010) Why does unsupervised pre-training help deep learning? J Mach Learn Res 11:625–660

19. Deng L, Yu D (2014) Deep learning: methods and applications. Found Trends Signal Process 7(3–4):197–387

20. Schmidhuber J (2015) Deep learning in neural networks: an overview. Neural Netw 61(1):85–117

21. Fischer A, Igel C (2014) Training restricted Boltzmann machines: an introduction. Pattern Recogn 47(1):25–39

22. Golovko V, Kroschanka A (2016) The nature of unsupervised learning in deep neural networks: a new understanding and novel approach. Opt Mem Neural Netw 3:127–141

23. Golovko V (2017) Deep learning: an overview and main paradigms. Opt Mem Neural Netw 1:1–17

24. Hinton G et al (2012) Deep neural network for acoustic modeling in speech recognition. IEEE Signal Process Mag 29:82–97

25. Golovko V, Egor M, Brich A, Sachenko A (2017) A shallow convolutional neural network for accurate handwritten digits classification. In: Krasnoproshin V, Ablameyko S (eds) Pattern recognition and information processing, communications in computer and information science, vol 673. Springer, Cham, pp 77–85

26. Krizhevsky A et al (2012) Imagenet classification with deep convolutional neural networks. Proc Adv Neural Inf Process Syst 25:1090–1098

27. Dahl GE, Yu D, Deng L, Acero A (2012) Context-dependent pre-trained deep neural networks for large-vocabulary speech recognition. IEEE Trans Audio Speech Lang Process 20(1):30–41

28. Cirean D, Meler U, Cambardella L, Schmidhuber J (2010) Deep, big, simple neural nets for handwritten digit recognition. Neural Comput 22(12):3207–3220
29. Zeiler M, Taylor G, Fergus R (2011) Adaptive deconvolutional networks for mid and high level feature learning. Proc IEEE Int Conf Comput Vis 1(2):2018–2025
30. Dorosh V, Komar M, Sachenko A, Golovko V (2018) Parallel deep neural network for detecting computer attacks in information telecommunication systems. In: Proceedings of the 38th IEEE international conference on electronics and nanotechnology. TUU "Kyiv Polytechnic Institute", Kyiv, pp 675–679
31. Komar M, Sachenko A, Golovko V, Dorosh V (2018) Compression of network traffic parameters for detecting cyber attacks based on deep learning. In: Proceedings of the 9th IEEE international conference on dependable systems, services and technologies, Kyiv, Ukraine, pp 44–48
32. Komar M, Dorosh V, Sachenko A, Hladiy G (2018) Deep neural network for detection of cyber attacks. In: Proceedings of the IEEE first international conference on system analysis and intelligent computing, Kyiv, Ukraine, pp 186–189
33. Komar M, Golovko V, Sachenko A, Dorosh V, Yakobchuk P (2018) Deep neural network for image recognition based on the caffe framework. In: Proceedings of the IEEE second international conference on data stream mining and processing, Lviv, Ukraine, pp 102–106
34. Golovko V, Bezobrazov S, Kroshchanka A, Sachenko A, Komar M, Karachka A (2017) Convolutional neural network based solar photovoltaic panel detection in satellite photos. In: Proceedings of the 9th IEEE international conference on intelligent data acquisition and advanced computing systems: technology and applications, Bucharest, Romania, pp 14–19
35. Golovko V, Kroshchanka A, Bezobrazov S, Komar M, Sachenko A, Novosad O (2018) Development of solar panels detector. In: Proceedings of the IEEE international scientific-practical conference "problems of infocommunications, science and technology", Kharkiv, Ukraine, pp 761–764
36. Salakhutdinov R, Mnih A, Hinton G (2007) Restricted Boltzmann machines for collaborative filtering. In: Proceedings 24th international conference on machine learning, Corvalis, USA, pp 791–798
37. Efrati A (2019) How 'deep learning' works at Apple, beyond. https://www.theinformation.com/How-Deep-Learning-Works-at-Apple-Beyond. Accessed 14 Apr 2019
38. Jones N (2014) Computer science: the learning machines. Nature 505(7482):146–148
39. Wang Y, Yu D, Ju Y, Acero A (2011) Voice search. In: Tur G, De Mori R (eds) Language understanding: systems for extracting semantic information from speech, chap 5. Wiley, New York
40. Kirk J (2019) Universities, IBM join forces to build a brain-like computer. http://www.pcworld.com/article/2051501/universities-join-ibm-in-cognitive-computing-researchproject.html. Accessed 14 Apr 2019
41. Kroshchenko A, Golovko V, Bezobrazov S, Mikhno E, Rubanov V, Krivulets I (2017) The organization of semantic coding of words and search engine on the basis of neural networks. Vesnyk Brest State Tech Univ 5(107):9–12 (in Russian)
42. Van der Maaten L, Hinton GE (2008) Visualizing high-dimensional data using t-SNE. J Mach Learn Res 9:2579–2605

43. Pelevina M, Arefyev N, Biemann C, Panchenko A (2019) Making sense of word embeddings. https://arxiv.org/pdf/1708.03390.pdf. Accessed 14 Apr 2019
44. Xiong S, Wang X, Duan P, Yu Z, Dahou A (2017) Deep knowledge representation based on compositional semantics for Chinese geography. In: Proceedings of the 9th international conference on agents and artificial intelligence, pp 17–23

Complex Approach of High-Resolution Multispectral Data Engineering for Deep Neural Network Processing

Volodymyr Hnatushenko$^{(\boxtimes)}$ ⓘ and Vadym Zhernovyi ⓘ

Department of Computer Science and Information Technologies,
Oles Gonchar Dnipro National University, Dnipro, Ukraine
vvgnat@ukr.net, vadim.zhernovoy@gmail.com

Abstract. A lot of terabytes of complex geospatial data are acquired every day, and it is used in almost every field of science and solves such problems as vegetation health monitoring, disaster management, surveillance, etc. In order to solve mentioned problems this data usually requires multiple steps of pre-processing before inferencing via machine learning algorithms. These steps may include such families of algorithms as image tiling or data augmentation. However, various studies focused on the basic concepts and research on techniques for remote sensing very high-resolution data pre-processing is in scarce.

The current article proposes an approach for data engineering to improve results of processing via the deep learning techniques. The algorithm and dataset are developed, they combine image-tiling techniques and satellite imagery properties. A suggested solution is tested on featured deep convolutional neural networks, such as FuseNet and region-based Mask R-CNN. Described approach for data engineering demonstrates segmentation quality increase for 6%, which is a notable improvement, considering a number of objects of interest in modern high-resolution satellite imagery.

Keywords: Remote sensing · Deep learning · Image processing · Datasets · Region proposals

1 Introduction

High-resolution multispectral data is digital representation of Remote Sensing Imagery acquired by satellites. Satellite imagery are images of the surface of Earth or other planets obtained by government and commercial imaging satellite vehicles. Featured satellites acquire multiple terabytes of data every day. Total volume of Earth observation data will reach 1 Pbyte in 2020 [1].

Nowadays satellite imagery is applied to solve thousands of problems in many areas of knowledge such as civil building, vegetation health monitoring, military decision making, etc. Improvements in multispectral imagery classification may

V. Lytvynenko et al. (Eds.): ISDMCI 2019, AISC 1020, pp. 659–672, 2020.
https://doi.org/10.1007/978-3-030-26474-1_46

cause a positive influence on the other fields of Earth observations such as Meteorology, Climate change, Surveillance, etc.

Deep Learning is one of the most popular modern approaches for solving the following tasks: image classification, object detection, 3D reconstruction, etc. Deep learning is a narrowed part of machine learning. Many solutions are being developed every month to solve the mentioned problems in any field of science and engineering. The reason for that is the undeniable success of machine learning in object detection tasks which even surpasses human performance. It is achieved by the development of deep learning concept and the development of software and hardware solutions focusing on machine learning.

The approach is required to develop and apply algorithms for classification improvement's results on very high-resolution multispectral spatial imagery, using the full potential of spatial resolution and power of featured deep neural network frameworks. Firstly, such a solution is necessary due to the increasing rate of humanitarian crises. Natural disasters negatively affect people all over the world. The frequency and scale of disasters have a strong impact on the map: maps loose accuracy in the places where some disaster has happened. In the perspective of satellite imagery, it is not clear enough how to classify certain objects on the ground. The value of fast-responding mapping system is hard to overestimate in the time before or after the occurrence of the disaster. Having the sequence of classified images, it would be possible to estimate the damage. The new maps are currently needed, especially for regions that are affected by natural disasters very often. Maps for these regions are drawn by hand by volunteers who participate in mapping challenges. They contribute to OpenStreetMap by drawing roads and building on satellite images.

2   Review of Literature

Deep learning approach is based on a concept of artificial neural network (ANN). ANN is a mathematical model based on principles of biological neural system functionality. The software of hardware implementations of artificial neural networks represents a system of connected artificial neurons and solves different tasks in Pattern recognition particularly.

Neural networks can be applied in many fields of science and engineering. In deep learning and machine learning, neural networks are a special case of methods for pattern recognition. Deep neural networks are applied to solve Earth observation problems as well [2]. Pattern recognition is a field of science, which develops methods and algorithms for processing different kind of data including signals, objects, etc. Certain data is characterized by a set of features. Image recognition can be applied in remote sensing. In this case, satellite imagery data is treated as a 2-dimensional signal, which is represented by values of spectral components, and metadata is treated as a composite object. Using image recognition algorithms, these signals and objects can be determined as some class which is characterized by a set of features.

Semantic mapping [3–6] is an example of the task in Pattern Recognition for Remote Sensing. The main goal of Semantic mapping is automatically assigning areas of the populated area to some classes that usually corresponds to some land or water object on aerial or satellite imagery such as airports, parks, institutes, etc.

Artificial neural network (ANN) is a mathematical model based on principles of biological neural system functionality. Software or hardware implementations of artificial neural networks represent a system of connected artificial neurons and solve various tasks in Pattern recognition particularly.

Backpropagation method underlies ANN learning process. This method is used to calculate a gradient for modification of weight for ANN. Backpropagation is also known as the backward error propagation. An error is computed as an output of certain neuron and distributed backwards throughout the layers of a neural network. This technique is commonly used in training of deep neural networks particularly. Due to the development of gradient backpropagation and Graphics Processing Unit (GPU) accelerated computing, Deep Learning became the most popular and promising approach for semantic mapping and other tasks in Remote Sensing [7,8].

Dissimilar from classic neural networks, deep neural networks (DNN) are adjusted with multiple layers between input and output layers. DNN can generate complex non-linear relationships between layers and can be a composition of many different neural network models. Each DNN architecture usually is designed for some specific data. This is one of the main reasons why featured deep neural network solutions demonstrate impressive results on recognizing hundreds of objects on 'state-of-the-art' datasets but also may not work for remote sensing data at all. We can't rely on a ready-made Deep Learning solution without certain modifications while designing DNN for satellite imagery. The dataset or DNN architecture either must be modified. The most common issues that appear for DNN training are overfitting and inference time. Overfitting is a common problem for neural net models that are not appropriate for certain dataset. However, inference time or computation speed have the biggest impact while processing satellite imagery products due to the amount of data in every single product set, as well as the amount of data acquired per day by satellites and aircraft. If overfitting is not a problem, other problems called accuracy and speed trade-off arise. DNN must configure a lot of parameters such as layers' size, initial weights, the learning rate in order to meet the mentioned problem.

It was hardly possible to use deep learning algorithms effectively before 2012. The best neural network models reached the lowest error rate of 25%. However, in 2012 the neural network called AlexNet [9] achieved the error rate of 16%. AlexNet solution is based on deep convolutional neural network. Few years later, the error rate fell to several percent. Such image classification results were admitted as even more crucial than human eye capabilities.

AlexNet solution was designed to provide better performance with extended hardware support as it was originally developed with NVIDIA CUDA technology. It was the starting point for the huge amount of improvements. Nowadays, more powerful frameworks like Mask R-CNN [10] developed by Facebook FAIR group which is more flexible, are developed to meet a wide range of tasks. This featured solution provides high-quality masks, which is essential for tasks such as mapping. Furthermore, it demonstrates better performance and provides a solid baseline for further research in the direction of image recognition and classification. Mask R-CNN is a short name of a region based convolutional neural network with masks. Mask R-CNN implements segmentation mask generation layers for its predecessor - Faster RCNN. Region based convolutional neural network (RCNN) is a neural network architecture designed to solve object detection problem using region proposals and classification algorithms. At first RCNN, selective search algorithm was used to generate 2000 region proposals, which probably contained an object. These proposals are preprocessed for further processing by AlexNet in order to obtain feature vectors for each proposal. These feature vectors are used as an input for Support Vector Machines (SVM) which are trained to classify objects by these features. Figure 1 illustrates all essential parts of Mask R-CNN solution.

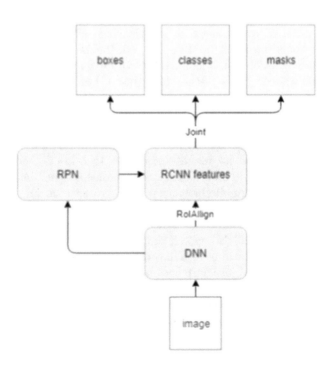

Fig. 1. Architecture of mask RCNN

However, the results demonstrated by all popular deep learning solutions, such as AlexNet, VGGNet, GoogleNet [11] or even solutions based on Mask R-CNN, have one common issue – the size of input image is limited to certain value. For example, for the input layer of VGGNet the size is set to $224 \times 224 \times 3$. Such image resolution is chosen considering the mechanism of convolution and performance. Such size of the input layer is suitable for successful processing of small RGB images such as map fragments acquired using Google Drive API ($640 \times 640 \times 3$). The majority of implementations of featured deep neural networks implement a module to resize the image to a format that is suitable for the input layer of the net. Resizing of satellite imagery to mentioned resolution results in significant information loss since many objects of interests may occupy several dozens of pixels. Even for the original satellite imagery size, it is hard to develop feature map due to mentioned sizes of objects. With resized satellite imagery to degrees of VGGNet, generating a feature map for classification may be impossible but such approach inevitably leads to significant information loss. In context, information loss means impossibility to extract valuable features from the satellite imagery for solving object detection tasks.

Another feature of Mask RCNN is Region Proposal Network (RPN) after the last convolution layer. RPN is able to produce proposals from convolution results, which are processed by Fast RCNN.

In the context of satellite imagery where the resolution may achieve more than 8192 pixels per dimension, such size of the input layer is not suitable for a wide range of processing tasks. Simple resizing of satellite imagery will result in significant loss of information. The problem becomes more remarkable considering the spatial resolution parameter. The better the technology of image acquisition – the smaller spatial resolution, the smaller spatial resolution – the fewer objects of the same type to be found on the image. Featured satellites provide image data with a spatial resolution of 30 cm per pixel. Considering the size of the input layer of the previously mentioned VGGNet framework, the spatial resolution of $224 \times 224 \times 3$ image may be $224 \times 0.3\,\mathrm{m} = 67.2\,\mathrm{m}$, which virtually may not fit many cases for regions of interests (e.g. military base, airport, many big city objects, etc.).

3   Problem Statement

The goal of current work is to research and develop a complex approach for application of the featured deep learning solutions to very high-resolution satellite imagery of WorldView-3. The approach must consider the unordinary nature of remote sensing imagery data structure and relevant data sparsity for deep neural network training. Contribution of the paper lies in the complexity of the suggested approach and covering all aspects of the object detection problem including dataset design techniques as well as research and setup of neural network architecture and its adaptation for inference on specific Remote Sensing imagery data.

In order to approach the dataset development problem from WorldView-3 satellite images, the structures of the satellite panchromatic and multispectral imagery are investigated, essential and important properties which are using metadata are highlighted and used for solving a problem. Pan-sharpening, image tiling and augmentation algorithms and techniques are investigated and used for sufficient training dataset construction.

Due to the lack of datasets that can solve segmentation and object detection problems on very high-resolution satellite images specifically, architectures of the featured deep neural network solutions are investigated and reconsidered in the context of developed dataset and specific satellite vehicle. Joint learning technique is approached and applied for the latest solutions, for instance segmentation in order to set up a specific deep neural network architecture to solve manmade stationary object detection on WorldView-3 imagery.

4   Materials and Methods

4.1   Structure of Satellite Imagery

As a rule, satellite imagery is a compilation of digital products gathered by satellite vehicles – not just images of Earth. Satellite imagery product usually consists of images, scene previews and supplementary materials that are usually called metadata.

Satellite images are files that contain raw data acquired by satellite vehicle sensors. In some cases, image data may be preprocessed for certain purposes by distributors. Images registered by satellite vehicles are often multispectral. Besides regular images that usually consist of three channels – red, green and blue, satellite images may contain additional visible channels (bands) – for example, purple or yellow, as well as non-visible infrared (IR) bands. Since infrared bands represent intensity values for solar radiation reflected from the surface of the planet, they are used to solve a lot of tasks in many fields of knowledge. Using information provided by IR bands, it is possible to monitor health parameters of vegetation, monitor Earth resources, monitor natural disasters and its consequences, etc. That, in combination with high spatial resolution nature of such images, contributes a lot in file size.

Since loading of satellite image is not an easy process and sometimes visual characteristics are the only thing which is interested, satellite imagery often contains image preview files. These files can be open virtually by any image reading software. Using previews whole scene can be displayed in a few seconds. Regions of interest (ROIs) can be estimated using image preview.

Metadata is sometimes referenced as Image Support Data (ISD). Metadata is a file that contains different characteristics of the image. All metadata information may conveniently be split into such categories:

- Satellite imagery product basic information. The image metadata provides basic information regarding the conditions of the image acquisition such as date, time, hardware registration parameters, etc. These parameters are more valuable for image pre-processing (scene tailoring, orthorectification, etc.).
- Digital image information. Metadata contains also digital characteristic of images: number of bands, number of rows/columns, color bit depth, etc. ISD may even contain algorithms, which were used for the creation of certain parts of whole satellite imagery product. Its structure varies for different satellite vehicles.
- Image acquisition information. Image acquisition information is usually physical parameters that influence image capturing by registers: mean position uncertainty, pixels direction, sun angles, track view angles, nadir angles, etc. These parameters are used mostly in satellite imagery product tailoring. However, they may influence inference of deep neural network system too.

4.2 Image Tiling

Image tiling is a set of algorithms, which is targeted to split image, that is too big for certain task. It doesn't matter if it is just a display of image or loading an image into memory, etc. In satellite imagery where the single file can be over 10 gigabytes in size, tile-based image processing is applied in many different cases.

Commercial satellites routinely provide panchromatic images with sub-meter resolution. The resulting overabundance of data brings with it the challenge of timely analysis. Fully automated processing still appears infeasible, but an intermediate step might involve a computer-assisted search for interesting objects. This would reduce the amount of data for an analyst to examine, but remains a challenge in terms of processing speed and working memory [12].

In current research tiling without paddings (TWOP) technique were developed. Suggested image tiling algorithm is designed in such way so each tile contains only image data. In order to cover all image area, algorithm is applied four times starting from each corner of the image (Fig. 2). Such approach is used in conjunction with overall solution to improve results of feature map building using deep neural network. In addition, algorithm saves coordinates of each image tile on source multispectral image, which are used later as anchors for bilinear interpolation.

4.3 Augmentation

Another approach for remote sensing data preprocessing is augmentation. Augmentation of satellite imagery may be achieved by application of different segmentation algorithms as well as affine transformations to source images in order to extend initial dataset.

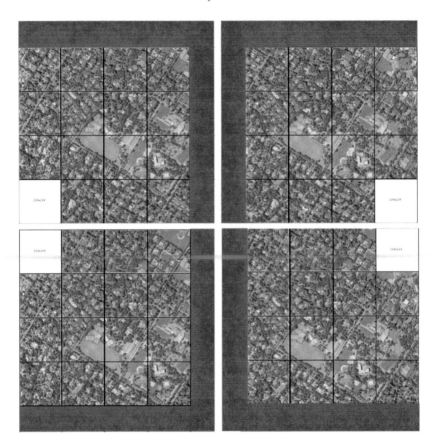

Fig. 2. Visual representation of suggested image tiling algorithm

In suggested solution, properties of WorldView-3 multispectral image data are used. WorldView-3 sensors acquire wide range of different wavelengths: panchromatic, multispectral, short-wavelength infrared (SWIR), CAVIS bands specifically designed for retrieval of atmospheric conditions related information. Pan-sharpening algorithm is used first in order to enhance resolution of initial 8-band multispectral image [13].

Four different band combinations of WorldView-3 image are used as a starting point of current research (Fig. 3, from left to right): True Color (RGB), False Color Infrared (FCI), Enhanced False Color Infrared (EFCI) and Bathymetry.

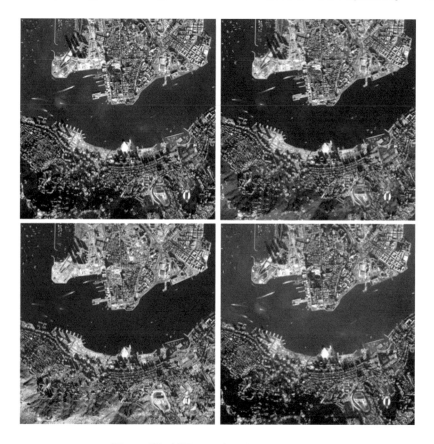

Fig. 3. WorldView-3 band combinations

5 Experiment

There are many satellite imagery datasets available on web [14]. However, a series of experiments in current work showed that in order to detect objects on WorldView-3 very high-resolution imagery data, a deep neural network must be trained on batches of images made by WorldView-3 registers particularly. So development of an approach for a new dataset engineering was a necessity.

For testing of suggested approach, an initial set of four WorldView-3 Earth observation images was prepared. It is augmented using multiple band combinations mentioned on Fig. 3 so the original number of images of the initial size is 300. Mentioned bands are successfully used for calculation of normalized difference metrics such as a normalized difference vegetation index (NDVI) or a normalized difference water index (NDWI) [15,16] and many other custom indices [17]. Considering scarcity of source images for engineering a dataset and also usefulness of these bands for extracting additional information in a form of normalized difference indices, they are chosen as an initial step for augmentation of the dataset. Then, the dataset is extended further using developed image tiling algorithm so the final size of the dataset is calculated using the next algorithm (Fig. 4):

```
def number_of_images(image_number,X,Y,x,y):
    tile_num = 0 #number of tiles for one image
    i = x
    while(i < X):
        i += x
        j = y
        while(j < Y):
            j += y
            tile_num += 1
    return image_number * tile_num * 4
```

Fig. 4. Dataset size calculation

In Fig. 4 N is a number of satellite images. X, Y correspond to satellite image size in pixels and x, y correspond to a size of an image tiling window. In current work the developed approach allowed to obtain a dataset of more than 10000 images from few original satellite images.

Deep neural network architecture was considered to use advantages of augmented versions of same source image. Featured architectures of deep neural networks for remote sensing data processing include: dual stream networks, stochastic ensembles, joint learning [18].

For current research Joint learning approach was chosen. This kind of learning allows to perform unsupervised learning of one deep neural network model to learn pseudo classes for very high-resolution satellite imagery and use them for updating another model. Through such updates pseudo classes of one network are close enough to sharply defined classed of another network which makes the real classes and features of whole architecture more discriminative [19].

The final deep neural net architecture involves Mask RCNN classifier which is initialized with ResNet-152 [20] feature extractor backbone pre-trained on ImageNet dataset [21]. One of the main feature of this convolutional neural network is stucked convolutional layers with shortcut connections which allow to work with general features of the image as well as regional features. Mask RCNN classifier was modified to use the frozen output layer of FuseNet [22] which supplies the classifier with decoded segmented images to use advantages of joint learning approach. FuseNet is based on auto-encoder neural network design and allows to scale the solution for different size of the input data and perform segmentation based on features built up through the unpooling operation. The decoder part of FuseNet, along with upsampling operation, performs restoration of the segmented image to its original spatial resolution. FuseNet was trained on brand new Microsoft dataset called Computer generated building footprints for

Canada [23]. There are 3 million images in the dataset and mostly include urban areas of Canada. For better determination of features the dataset also includes areas such as forests, mountains, coasts, etc.

Furthermore, mentioned architecture is modified to use image tiles from developed dataset rather than random patches extracted from high-resolution images. Other types of augmentation (e.g. with random rotations and flipping) are not used in current approach for measurements accuracy. The common feature of mentioned architecture is an ability to fuse multiple sources on different stages of deep neural network inference (Fig. 5).

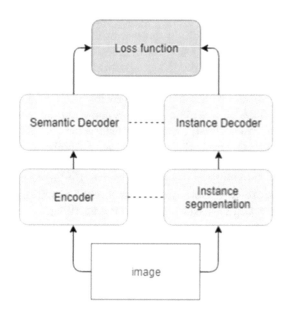

Fig. 5. Joint learning applied to suggested solution

6   Results and Discussion

Table 1 demonstrates results on the image from described dataset in comparison to traditional tiling with paddings (TWP) algorithm. Dataset column represents combinations of augmented data samples which are selected as an input for each stream of a network. TP, FP, FN are true positive (object of interest is present and detected), false positive (object of interest is not present, but detected) and false negative (object is present, but not detected) detections correspondingly. Completeness is calculated as a percentage of true positive detections to a total number of objects. Correctness is calculated as a percentage of true positive detections to all detections. Quality is calculated as a percentage of true positive detections to a sum of all positive detection and false negative occurrences.

Table 1. Object detection results

Dataset	TP	FP	FN	Completeness	Correctness	Quality
TWP+Binarization	138	12	19	0.89	0.92	0.76
TWOP+RGB	143	9	14	0.91	0.94	0.82
TWOP+RGB+FCI	141	10	16	0.90	0.94	0.80
TWOP+RGB+FCI+EFCI+ Bathymetry	137	7	20	0.87	0.96	0.75

Certain design of a dataset may influence performance of a computer vision system. For comparison of results a dataset of images tiled with existing algorithms and augmented with binarization samples were chosen. TWOP algorithm applied to RGB data without any additional augmentation shows a good improvement in a number of positive detections and even little decrease in a false positive detections. Additional augmentation with false color samples of a dataset tiled with TWOP technique does not show any improvements in detection using mentioned neural net. Further augmentation with all mentioned methods described in current article does not improve a number of positive detection but improves correctness of these results up to 96%. Considering completeness and correctness of detections equally important a TWOP-tiled dataset without any augmentation provides more quality results in comparison to TWOP-tiled augmented datasets and datasets tile with existing tiling.

7   Conclusions

In current work, a complex approach is suggested to work with very high-resolution multispectral data, which help to solve many problems in nature disasters monitoring, vegetation monitoring, civil building, surveillance, etc. Developed data processing methods provide capabilities to automate satellite geospatial imagery pre-processing to improve overall speed of processing.

For current research multispectral WorldView-3 imagery are used. Data from mentioned satellite vehicle provide a capability to investigate, extract and process a lot of Earth observation information to meet requirements for solving mentioned world-wide problems. Developed solution utilizes multiple algorithms, which are based on image tiling approaches, properties of satellite imagery high-resolution multispectral data, image processing approaches using region-based deep convolutional neural network processing for instance segmentation on images.

Image tiling techniques provide fast and flexible way to prepare datasets for image processing via deep learning networks with pre-trained feature extractors. In this article, a new image tiling algorithm is developed in order to consider a structure of satellite multispectral imagery. Existing solutions for image tiling with paddings make it hard to adopt a whole source image for building a good deep neural network feature map. Final feature map of a trained network influences an accuracy rate a lot so research and implementation of new image tiling algorithm targeting remote sensing high-resolution multispectral data was a necessity.

The neural network architecture used in current research is considered and chosen from the perspective of development of a quick and effective satellite imagery multispectral data pre-processing algorithm. Region-based deep convolutional networks, Mask R-CNN particularly, showed good results in segmentation tasks for building precise maps of certain locations.

Described approach for data engineering demonstrates 6% improvement in quality of segmentation, which is a noticeable improvement considering a number of objects of interest on modern high-resolution satellite imagery.

Suggested tile-merging algorithm has a room for further improvements. Due to modularity of suggested method, it can be modified in many different ways. It is possible to modify tile-merging algorithm in order to use benefits of different segmentation algorithm. On the next stage of research, segmentation approach will be used for remote sensing data augmentation instead of multispectral band combinations. Other segmentation algorithms may provide even better results depending on neighbor tile merging algorithm [24]. Mentioned possibilities is a course of future work – to find the best combinations of data augmentation approaches (including segmentation) and the state-of-the-art classification/detection algorithm in Remote Sensing.

References

1. Audebert N, Le Saux B, Lefèvre S (2018) Deep learning for remote sensing - an introduction. IRISA - Université Bretagne Sud, Atelier DLT Sageo, ONERA
2. Hordiiuk DM, Hnatushenko VV (2017) Neural network and local laplace filter methods applied to very high-resolution remote sensing imagery in urban damage detection. In: 2017 IEEE international young scientists forum on applied physics and engineering (YSF), Lviv, pp 363–366. https://doi.org/10.1109/YSF.2017.8126648
3. Zhang J (2010) Multi-source remote sensing data fusion: status and trends. Int J Image Data Fusion 1(1):5–24
4. Henderson FM, Lewis AJ (1998) Principles and applications of imaging radar. Manual of remote sensing, vol 2, 3rd edn. American Geophysical Union, Washington, DC
5. Câmara G et al (1996) SPRING: integrating remote sensing and GIS by object-oriented data modelling. Comput Graph 20(3):395–403
6. Xin H, Zhang L (2013) An SVM ensemble approach combining spectral, structural, and semantic features for the classification of high-resolution remotely sensed imagery. IEEE Trans Geosci Remote Sens 51(1):257–272
7. Jia X, Richards JA (1999) Segmented principal components transformation for efficient hyperspectral remote-sensing image display and classification. IEEE Trans Geosci Remote Sens 37(1):538–542
8. Alcantarilla PF, Nuevo J, Bartoli A (2011) Fast explicit diffusion for accelerated features in nonlinear scale spaces. IEEE Trans Patt Anal Mach Intell 34(7):1281–1298

9. Krizhevsky A, Sutskever I, Hinton GE (2012) Imagenet classification with deep convolutional neural networks. In: Advances in neural information processing systems
10. Kaiming H et al (2017) Mask R-CNN. In: 2017 IEEE international conference on computer vision (ICCV). IEEE
11. Szegedy C et al (2015) Going deeper with convolutions. In: Proceedings of the IEEE conference on computer vision and pattern recognition
12. Wassenberg J (2011) Efficient algorithms for large-scale image analysis. KIT Scientific Publishing, Karlsruhe Schriftenreihe automatische Sichtprufung und Bildverarbeitung
13. Hnatushenko VV, Vasyliev VV (2016) Remote sensing image fusion using ICA and optimized wavelet transform. In: International archives of the photogrammetry, remote sensing and spatial information sciences, vol XLI-B7, XXIII ISPRS Congress, Prague, Czech Republic, pp 653–659
14. Awesome satellite imagery datasets. GitHub. https://github.com/chrieke/awesome-satellite-imagery-datasets. Accessed 10 Mar 2019
15. Szabó S, Gacsi Z, Boglárka B (2016) Specific features of NDVI, NDWI and MNDWI as reflected in land cover categories. Acta Geographica Debrecina Landscape Environ 10(3–4):194–202
16. Longbotham N et al (2014) Prelaunch assessment of worldview-3 information content. In: 2014 6th workshop on hyperspectral image and signal processing: evolution in remote sensing (WHISPERS). IEEE
17. Samsudin HS, Shafri HZM, Hamedianfar A (2016) Development of spectral indices for roofing material condition status detection using field spectroscopy and WorldView-3 data. J Appl Remote Sens 10(2):025021
18. Han J, Zhang D, Cheng G, Guo L, Ren J (2015) Object detection in optical remote sensing images based on weakly supervised learning and high-level feature learning. IEEE Trans Geosci Remote Sens 53(6):3325–3337
19. Lu X, Yuan Y, Zheng X (2017) Joint dictionary learning for multispectral change detection. IEEE Trans Cybern 47(4):884–897
20. He K et al (2016) Deep residual learning for image recognition. In: Proceedings of the IEEE conference on computer vision and pattern recognition
21. Deng J et al (2009) Imagenet: a large-scale hierarchical image database. In: 2009 IEEE conference on computer vision and pattern recognition. IEEE
22. Hazirbas C et al (2016) Fusenet: incorporating depth into semantic segmentation via fusion-based CNN architecture. In: Asian conference on computer vision. Springer, Cham
23. Microsoft computer generated building footprints for Canada. https://github.com/Microsoft/CanadianBuildingFootprints. Accessed 8 Mar 2019
24. Hnatushenko VV, Kashtan VJ, Shedlovska YI (2017) Processing technology of multispectral remote sensing images. In: IEEE international young scientists forum on applied physics and engineering YSF-2017, Lviv, Ukraine, pp 355–358

Protein Tertiary Structure Prediction with Hybrid Clonal Selection and Differential Evolution Algorithms

Iryna Fefelova[1], Andrey Fefelov[1], Volodymyr Lytvynenko[1],
Róża Dzierżak[2], Iryna Lurie[1], Nataliia Savina[3], Mariia Voronenko[1(✉)],
and Svitlana Vyshemyrska[1]

[1] Kherson National Technical University, Kherson, Ukraine
fao1976@ukr.net, immun56@gmail.com, lurieira@gmail.com,
mary_voronenko@i.ua, printvvs@gmail.com
[2] Lublin University of Technology, Lublin, Poland
rozadzierzak@gmail.com
[3] National University of Water and Environmental Engineering, Rivne, Ukraine
n.b.savina@nuwm.edu.ua

Abstract. The paper deals with the problem of protein tertiary structure prediction based on its primary sequence. From the point of view of the optimization problem, the problem of protein folding is reduced to the search for confirmation with minimal energy. To solve this problem, a hybrid artificial immune system has been proposed in the form of a combination of clonal selection and differential evolution algorithms. The developed hybrid algorithm uses special methods of encoding and decoding individuals, as well as an affinity function, which allows reducing the number of incorrect conformations (solutions with self-intersections). To test the algorithm, Dill's hydrophobic-polar model on a two-dimensional square lattice was chosen. Experimental studies were conducted on test sequences, which showed the advantages of the developed algorithm over other existing methods.

Keywords: Protein folding · Hydrophobic-polar model ·
Clonal selection · Differential evolution · Artificial immune systems

1 Introduction

Among the open problems of computational biology, the most difficult is to determine the tertiary structure of a protein [1,2]. Understanding the mechanisms of protein self-organization processes, in addition to its fundamental significance, is of paramount importance for solving many practical tasks, such as developing drugs or creating new proteins with predetermined properties. Theoretical modeling of the folding of a polypeptide chain into its unique native conformation and the prediction of the result of such a process is an important tool in solving this problem.

© Springer Nature Switzerland AG 2020
V. Lytvynenko et al. (Eds.): ISDMCI 2019, AISC 1020, pp. 673–688, 2020.
https://doi.org/10.1007/978-3-030-26474-1_47

A direct experimental study of the function of the protein, unfortunately, remains too laborious and slow, and in the foreseeable future will not be able to meet the needs created by decoding genomes for such an analysis. An alternative to experimental study is the theoretical prediction of protein function. The close relationship between function and structure is well known. Therefore, the successful development of methods for predicting the spatial structure of proteins can provide a tool for successful solving of the problem of functional genome mapping. Following the fundamental correlation of the function of a protein, the problem is to clarify the detailed mechanism for implementing this function, already at the physicochemical level, and theoretical modeling of these processes can undoubtedly make a significant contribution to its solution. The problem is that science with all its computing power and experimental data has not learned to build models that describe the process of coagulation of a protein molecule and predict the tertiary structure of a protein based on its primary structure. However, it is wrong to assume that nothing happens in this area of science. Known patterns of folding (folding) of the protein developed methods for its modeling. There are two groups of methods to predict the tertiary structure of a protein. The first group includes de novo modeling methods when models are built only on the basis of the primary structure [3,4]. The significance of these methods consist in understanding the physicochemical principles of protein folding. Disadvantages - low accuracy and complexity of calculations. The second group includes methods of comparative modeling [5,6]. They are based on the phenomenon of homology. The "predictor" has the ability to compare the sequence of a protein, the structure of which must be modeled with a protein's structure that is known and which is supposedly a homolog. Based on their similarity, a model is built with subsequent adjustments. The popularity of these methods consist in their ability to solve specific practical problems, and the disadvantage is the inability to solve the problem of protein folding itself. Since the processes of protein coagulation have not been fully studied, researchers have proposed a number of simplified models based on the physical properties of molecules and leading to combinatorial optimization problems. The use of computer models of proteins leads to the need for the discretization of space, that is, the use of a limited set of singular points of space, each of which acts as a representative of some surrounding area, that is, in fact, a set of nodes of a certain spatial lattice. One example of such models is discrete lattice models [1]. Such lattices can be either regular (for example, cubic) or irregular (given by the structures of protein globules). These models have fundamental theoretical value, despite the fact that they cannot be directly applied to real proteins. When using lattices, the prediction of a tertiary structure is to find a bundle with minimum energy, which is calculated according to a certain rule.

2   Problem Statement

The structure of the amino acid residues in the chain. The hydrophobic-polar (HP) protein folding is the following: water-loving (hydrophilic) amino acids (polar monomers) and H - water-repellent (hydrophobic) amino acids (non-polar monomers). The amino acid sequence can be considered as a vector $S = (s_1, s_2, \ldots, s_n)$, $s_i \in \{H, P\}$, $i = \overline{1, n}$. Each item s_i (amino acid residue) is located in the site of a lattice in a way that the adjacent elements in the sequence correspond to adjacent lattice sites, forming a path. The structure of the protein is set using a path that does not contain self-intersections and is called convolution. Then in which node this element will appear depends on the tendency of the hydrophobic residues to attract each other in such a way that to avoid water as much as possible. H-H bonds occur between the contacting hydrophobic residues. The energy of each bond is taken as -1, and the energy of the structure is the sum of the energies of all the bonds in it. Formally, the energy of the sequence can be represented as [7]:

$$E\left(S\right) = - \sum_{1 \le j \le j-2 \le n-2} I\left(N_i, N_j\right) h\left(s_i, s_j\right)$$

$$I\left(N_i, N_j\right) = \begin{cases} 1, if\ node\ N_i\ adjacent\ to\ N_j \\ 0,\ else \end{cases}, \tag{1}$$

$$h\left(s_i, s_j\right) = \begin{cases} 1, if\ s_i\ and\ s_j\ hydrophobic \\ 0,\ else \end{cases}$$

where $N_i = (x_i, y_i)$ – coordinates of the site of a two-dimensional square lattice in which the element is located s_i.

The task of determining the structure of the protein is reduced to the search for the convolution with the lowest energy. It is proved that in this formulation the problem is NP-complete [8,9]. Thus, the purpose of this study is to develop a new, fast and accurate search method that takes advantage of the population-based approach and hybridization technology.

3   Literature Review

The hydrophobic-polar (HP) model [10] is an abstraction *protein structure prediction* problem (PSP). This model reflects the fact that hydrophobicity is one of the main driving forces in protein folding. Prediction of protein structures using the HP model is a complex combinatorial optimization problem, which has been demonstrated as NP-complete [8,9]. This means that, in general, exact methods are not able to cope with such a task in a reasonable time.

Therefore, to solve the protein folding problem, a number of metaheuristic algorithms and approaches have been proposed: ant colony algorithms [11], artificial immune systems [12], etc. These methods relate to the direction of "natural computing", i.e. simulate certain biological processes.

Various metaheuristic approaches have been applied to this problem, including genetic algorithms [13–15], hybrid and memetic algorithms [16,17], ant colony optimization [2], particle swarm optimization [18], differential evolution [19], tabu search [20], Monte Carlo methods [21], simulated annealing [22–24], the Firefly Algorithm (FA) [25], quantum optimization [26,27], immune algorithms [12]. In addition, simplified energy models are used to approximate tertiary protein structures, the most popular is the model of the hydrophobic-polar (H-P) energy [1,28–30], Miyazawa–Jernigan (M-J) energy model [31,32]. In the H-P model, proteins are represented by chains (inside a given lattice), the vertices of which are marked either as H (hydrophobic) or P (hydrophilic); It is believed that H-nodes attract each other, and P-nodes are neutral.

An analysis of the current state of research in the field of these problems indicates the presence of deficiencies related to the forecast accuracy and the time that it takes to obtain an optimal solution. Consequently, the development of new computational methods devoid of these drawbacks is relevant. In this paper, the authors focused on the two-dimensional square lattice model, which is a special case of the well-known hydrophobic-polar (HP-model) Dill [10]. To search the optimal protein conformation on the selected model, a hybrid algorithm of clonal selection and differential evolution is proposed.

4   Materials and Methods

4.1   Spatial Model

In accordance with the previously considered HP model, the sequence confirmation is represented as a path in a discrete lattice. In order to be able to use conformations as individuals of a population algorithm, the path must be encoded into a string that uniquely determines the shape of the molecule. For convolution coding, the internal coordinate method is often used [33], where the path is defined by a sequence of movements, and the position of each amino acid residue depends on the positions of its predecessors in the chain. This representation depends on the lattice shape and the selected direction of coding scheme (absolute or relative directions). For example, when using absolute directions on a two-dimensional square lattice, there are four possible movements along the way: north (N), south (S), west (W) and east (E). Displacements are specified in some fixed frame of reference, for example, associated with the lattice itself (Fig. 1a). When using relative directions, the reference system is not fixed, and the next direction depends on the previous movement (Fig. 1b). In this case, on the same grid, there are only three possible displacements: forward (F), left (L) and right (R).

Thus, in each of the variants, the individuals of the population algorithm are encoded with character strings using alphabets $A_{abs} = \{N, S, W, E\}$ $A_{rel}\{F, L, R\}$ and respectively. Moreover, the length of an individual is one less than the length of the sequence itself, since the position of the first amino acid residue in the chain is considered fixed.

Figure 2 shows an example of the optimal conformation of one of the test sequences used in the work. The sequence is described by the string HPHP-PHHPHPPHPHHPPHPH, consisting of twenty amino acid residues. While coding according to the scheme with absolute directions, the individual is written

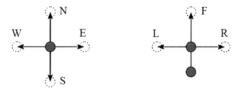

Fig. 1. Directional coding schemes in the method of internal coordinates: a - absolute directions; b - relative directions

in the form of the string SEENWNENNWSWSWNWSS. For a circuit with relative directions, the individual string will assume the form FLFLLRRLFLLRL-RRLLFL. In this paper, the authors chose a scheme with relative directions, since according to studies depicted in [33], this scheme has advantages on square lattices.

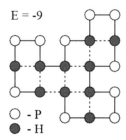

Fig. 2. An example of an optimal protein folding with a sequence of residues HPHPPHHPHPPHPHHPPHPH

It is obvious that the probabilistic nature of population algorithms will lead to the fact that most of the generated solutions will be unacceptable, i.e. contain self-intersections. As in many similar cases, such a situation is usually handled with fines, allowing the "wrong" individuals to participate in the processes of selection, mutation, and recombination. Along with fines, special modifications of operators are being developed that do not produce unacceptable solutions. In this study, a similar effect is achieved using the developed procedure for decoding individuals.

4.2   Clonal Selection Algorithm

Artificial Immune Systems (AIS) are related to biologically inspired computing. In them, the principles and concepts of modern immunology serve as a basis for creating computational methods and applications used in science and technology. It has been established that the immune system of animals and humans has special features of pattern recognition systems. Due to these properties, biological organisms are able to distinguish pathogens (viruses and bacteria) even if they are encountered for the first time. The process of recognition of antigens is reduced to the sequential adaptation of a population of antibodies, their affinity to the antigen increases from generation to generation.

The adaptive properties of the immune system can be used to solve optimization problems. AIS is essentially a simplified models of the immune system. They implement algorithms that describe basic biological processes, such as positive and negative selection, clonal selection, and immune networks. In [34], the functioning of the immune system is considered from the point of view of clonal selection. On the basis of this theory, the algorithm of optimization CLONALG was proposed in [35], which became widely used as one of the varieties of AIS. In the clonal algorithm, affinity values express the measure of an individual's proximity to the optimal solution and are calculated based on the objective function, i.e. minimization problem is considered:

$$f\left(\overline{x}\right) \rightarrow \min, \quad \overline{x} = (x_1, \ldots, x_l) \tag{2}$$

where $\overline{x}$ – the vector of parameters of the problem, on the basis of which the decision population is built $\overline{x}_i^G$, $i = 1, \ldots, P$, P – decision population size; G – present generation.

In CLONALG, depending on the type of problem, you can use different ways of representing solutions. Binary and real encoding are the most commonly used. Also, the conditions and goals of the task are decisive when choosing how to represent the immune operators, the type of affinity function, values of the parameters of the algorithm. The main CLONALG cycle consists of the sequential application of operations of selection, cloning and mutation over a series of generations (iterations). In the process of assessing a population of antibodies, the conditions are created for the selection of those individuals that are closest to the goal, i.e. to minimize the objective function. The selected antibodies are activated, which results in an increase in their numbers due to cloning. Clones mutate, exploring new areas of the search space, where theoretically there can be a global optimum. The mutation is represented by a simple scheme, according to which the value of each element of an antibody string varies randomly (with a given probability p_m) by the following formula:

$$x_{ij}^{G+1} = \begin{cases} randInit(), \; if \; randEvent\,(p_m) = 1 \\ x_{ij}^G, \; else \end{cases} , j = 1, 2, \ldots, l, \tag{3}$$

where $x_{i,j}$ – j-th element of the i-th individual in the population; l – individual line length; p_m – probability of changing the line element, called here as the mutation intensity; $rndInit()$ – random element initialization function; $rndEvent()$ – binary function of generating a random event with a given probability.

Problems arise in the cases where the surface under study, along with the global one, contains a large number of local optima. The problem of protein folding in the form of an HP model considered in this paper has these disadvantages, which makes the use of CLONALG in its pure form ineffective.

4.3   Differential Evolution Algorithm

The differential evolution algorithm (DE) belongs to the family of evolutionary algorithms [36]. Possessing high efficiency, DE has found application in many subject areas, as a method of global optimization. There are several versions of the DE, differing in the implementation details of the evolutionary operators. In this paper, the variant presented in [37] is used.

The main difference between the DE algorithm and other evolutionary algorithms is the implementation of the mutation operator. DE mutation is as follows:

$$\overline{\nu}_i^{G+1} = \overline{x}_{r_3}^G + F\left(\overline{x}_{r_1}^G - \overline{x}_{r_2}^G\right) \tag{4}$$

where $\overline{\nu}_i^{G+1}$, $i = 1, \ldots, P$ – mutational individual; $r_1, r_2, r_3 \in \{1, \ldots, P\}$ – indices of individuals who are randomly selected from a population of solutions in the current generation are such that $r_1 \neq r_2 \neq r_3 \neq i$; F – scale factor $(F \geq 0)$.

Not all components of the vector $\overline{\nu}_i^{G+1}$ are moving into a new generation. Formation of candidates is made by the following expression:

$$x_{ij}^{G+1} = \begin{cases} \nu_{ij}^{G+1}, \ if \ randEvent\,(p_{DE}) = 1 \vee j = k \\ x_{ij}^G, \ else \end{cases} \quad , j = 1, 2, \ldots, l, \tag{5}$$

where $k \in \{1, \ldots, l\}$ – random parameter index, selected once for each individual, the meaning of which is to ensure the transition of at least one component of the vector $\overline{\nu}_i^{G+1}$ to vector $\overline{x}_i^{G+1}$; p_{DE} – transition probability of the j-th vector component $\overline{\nu}_i^{G+1}$ to vector $\overline{x}_i^{G+1}$.

As in [36] expression (5) is associated with the crossover operator in evolutionary algorithms, p_{DE} which is called the crossing over probability. It is not difficult to notice that expression (5) is very similar to expression (3), therefore, in this paper, p_{DE} it is called the intensity of the DE mutation.

4.4   Hybrid Algorithm CLONALG+DE

In the classic case, the character encoding of individuals is used to represent the path in the lattice. However, the mutation in the DE algorithm works with real numbers. Thus, in this paper, it was necessary to create a real coding scheme. Each element of the solution vector x_{ij} is encoded with a real number in the range from 0 to 1, i.e. $x_{ij} \in [0.0, 1.0]$. In the process of decoding an individual, this gap is divided into as many parts as there are valid directions from the current lattice node. N_i A direction that is not self-intersecting is considered valid, i.e. when, as a result of moving in one of the directions $\{F, L, R\}$, the new node N_j will be not occupied by the sequence element S. Each part of the gap is assigned to one of the permissible directions (Fig. 3).

Elements $x_{i,j}$ are converted to characters in accordance with the display scheme presented in Fig. 3, i.e. when a value $x_{i,j}$ falls into one of the parts of the gap, the direction, in which the gap is fixed, is assigned to it. Individual decoding is performed sequentially from the beginning of the line. For each subsequent

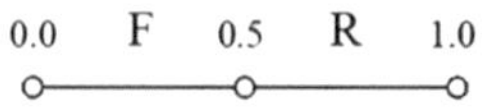

Fig. 3. An example of a scheme for mapping real numbers into direction symbols (if there are valid directions $\{F, R\}$)

movement, the allowable directions are calculated and the display scheme is applied.

Such an approach to decoding excludes self-intersections and does not require the development of special immune operators. However, in rare cases, it is possible to get into a dead end when it turns out that all directions are unacceptable before the next movement. To account for this situation, penalties are included in the affinity function:

$$f = \frac{1}{|E(S)| + l^*},\tag{6}$$

where l^* – length of the decoded part of the individual line to a dead end.

The pseudo-code of the clonal selection and differential evolution hybrid algorithm is presented in Fig. 4. Following the principles of functioning of the immune systems, the proposed algorithm works with objects of two types: antigens and antibodies. The antigens in an implicit form are the amino acid sequences of S. Antibodies are strings consisting of directions encoded with real numbers from 0 to 1. In each generation of G, there is the main population of antibodies $Ab^{(G)}$ of size P. The size of the main population is kept constant. The initial population $Ab^{(0)}$ at $G = 0$ is randomly generated by $initPop()$.

The $evaluate()$ function calculates the affinity values of each antibody $\overline{x}_i^G \in Ab^{(G)}$ based on the affinity function (6). Before calculating the affinity, antibodies are decoded into character strings of directions along which the protein conformations on the lattice are constructed and the energy is calculated. $TerminationCondition()$ stops under one of the conditions: the minimum energy value was found (E^{min}) (it is known for test tasks), the maximum number of generations reached (G^{max}). In $cloning()$ cloning, antibodies with high affinity for the antigen (i.e. low values of the f function) are involved. Since affinity is represented by a scalar value, selection in a population is not difficult. In the present work, the selection is made using a tournament, when individuals k_{Ab} are randomly selected from a population (k_{Ab} - size of a tournament) and a further comparison of their affinities determines the winner, who proceeds to the next stage. In this case, the tournaments themselves are repeated q times. The result is an intermediate population of clones $Ab^{(C)}$ of size P_C.

Mutation operators affect antibodies in a clone population. The first phase is represented by a simple $mutation()$ mutation, performed by the formula (3). A simple mutation is made with probability $p_c! \sim f$ and intensity $p_m! \sim f$, as according to the mechanism of clonal selection, mutation of antibodies with higher affinity is performed with less probability and with less intensity. In this case, the probability p_c determines the fact of applying the formula (3) to an individual. The same probability p_c determines that the DE-mutation $deMutation()$

```
1:  ClonalgDE Algorithm( P, P_C, k_Ab, P_C, P_m, P_DE, F)
2:  G := 0;
3:  Ab^G := initPop(P);
4:  evaluate(Ab^G);
5:  while(!terminationCondition()) do
6:      Ab^C := cloning(Ab^G, P_c);
7:      Ab^M := mutation(Ab^C, p_c, p_m);
8:      Ab^DE := deMutation(Ab^m, 1-p_c, P_DE, F);
9:      evaluate(Ab^DE);
10:     Ab^{G+1} := deSelection(Ab^DE, k_Ab);
11:     G := G + 1;
12: end_while
```

Fig. 4. Pseudo-code of the hybrid clonal selection and differential evolution algorithm

(second phase) is applied to the antibody only in a mutually exclusive order. That is, if a simple mutation is used, then the DE mutation is not used and vice versa. As with the simple mutation, the intensity of the DE mutation is inversely proportional to affinity ($p_{DE}! \sim f$). DE mutation is performed by formulas (4) and (5). A mutated population ($p_{DE}! \sim f$) of clones is $evaluate(Ab^{(DE)})$. The transfer of antibodies to the main population is carried out using DE-selection according to the following formula:

$$\overline{x}_i^{G+1} = \begin{cases} \overline{u}_i^{G+1}, \ if \ f\left(\overline{u}_i^{G+1}\right) \leq f\left(\overline{x}_i^G\right) \\ \overline{x}_i^G, \ else \end{cases}, Ab^{(DE)} \tag{7}$$

where $\overline{u}_i^{G+1} \in Ab^{(DE)}$ – an individual modified by the use of mutation operators.

To maintain a constant size for the main population, DE-selection includes tournaments of size k_{Ab}. The number of tournaments is equal to P.

5 Experiments

To test the effectiveness of the proposed hybrid algorithm, the authors used twelve standard test sequences that are found in the literature as a reference for evaluating the developed heuristic methods [14,38,39]. The length (n), the structure and the minimum energy of the conformation ($E^{\min}$) of each sequence are shown in Table 1. To shorten the record, the hp-sequence in the table contains indices h_i p_i, and $(\ldots)_i$, where i means the number of repetitions of the corresponding symbol or group of symbols. The minimum energy of the conformation is calculated for the convolution of a sequence on a two-dimensional square lattice.

Table 1. Standard tests of the HP model on a two-dimensional square

No	n	Sequence	$(E^{\min})$
1	20	$hphp_2h_2php_2hph_2p_2hph$	-9
2	24	$h_2p_2(hp_2)_6h_2$	-9
3	25	$p_2hp_2(h_2p_4)_3h_2$	-8
4	36	$p_3h_2p_2h_2p_5h_7p_2h_2p_4h_2p_2hp_2$	-14
5	48	$p_2h(p_2h_2)_2p_5h_{10}p_6(h_2p_2)_2hp_2h_5$	-22
6	50	$h_2(ph)_3ph_4p(hp_3)_2hp_4(hp_3)_2hph_4(ph)_3ph_2$	-21
7	60	$p_2h_3ph_8p_3h_{10}php_3h_{12}p_4h_6ph_2php$	-36
8	64	$h_{12}(ph)_2(p_2h_2)_2p_2h(p_2h_2)_2p_2h(p_2h_2)_2p_2(hp)_2h_{12}$	-42
9	20	$h_3p_2(hp)_2hp_2(hp)_2hp_2h$	-10
10	18	$php_2hph_3ph_2ph_5$	-9
11	18	$hphph_3p_3h_4p_2h_2$	-8
12	18	$h_2p_5h_2p_3hp_3hp$	-4

The values of the parameters of the hybrid algorithm, which are not changed throughout all the experiments, are presented in Table 2. One of the criteria for the efficiency of the algorithm is the indicator $N^{(E)}$ that fixes the number of starts of the conformation energy calculation procedure. Performance of the method is higher when the value of indicator is lower. $N^{(E)}$ is convenient to compare with the results of similar studies found in the literature.

In this paper, the experiment is considered successful if the stop condition on the found minimum energy is triggered first. In this case, the value is fixed $N^{(E)}$. If the algorithm stops by the maximum number of generations and the value $E^{\min}$ is not reached, the experiment is considered unsuccessful. The ratio

Table 2. General parameters of the algorithm for all experiments

Parameter name	Parameter value
The coding method of individuals	Real
Number of generations ($G^{\max}$)	300
Selection coefficient (q/P)	1.0
Selection type	Tournament
Terminal condition	Number of generations or the found E^{min}
Probability of simple mutation (p_c)	0.1
Intensity simple mutation (p_m)	0.05
The intensity of DE-mutation (p_{DE})	1.0
Scale factor (F)	0.8
DE-selection	(7) is disabled, tournaments only

of the number of successful experiments to their total number is indicated as k_{SR} which is also used in comparisons as a criterion for the efficiency of the algorithm. Experimental studies are divided into two stages. At the first stage, the algorithm is tested on twelve sequences. In each experiment, the authors conducted ten runs of the algorithm. At the second stage, one sequence was selected under number 4, for which a study was conducted on the effect of the size of the main population (P) and the population of clones (P_C) on indicators $N^{(E)}$ and k_{SR}. Here, each experiment was provided with twenty launches of the algorithm.

6   Results and Discussion

This section presents a comparison of the research results of the developed hybrid algorithm with similar results of the immune algorithm (IA) described in [12]. Table 3 contains data for the most successful indicators obtained during testing of IA, which are compared with the corresponding testing indicators of the proposed method.

Table 3. Comparative test results of the proposed algorithm

Protein	IA				CLONALG+DE			
No	$E^{\min}$	k_{SR}	$\overline{N}^E$	E^*	k_{SR}	$\overline{N}^E$	E^*	$P/P_C/k_{Ab}$
1	-9	1.0	20352.4	-9	1.0	2003.5	-9	300/900/5
2	-9	1.0	27923.1	-9	1.0	3865	-9	300/900/5
3	-8	1.0	138035.37	-8	1.0	4136.8	-8	900/1500/5
4	-14	1.0	2032504	-14	1.0	50444.1	-14	3000/9000/5
5	-22	56.67	6403985.3	-22	1.0	49021.1	-22	2000/6500/5
6	-21	1.0	778906.4	-21	0.8	172271.6	-21	6000/21000/10
7	-36	0.0	–	-35	0.0	–	-35	3000/9000/10
8	-42	3.33	8078548	-42	0.6	804132	-42	2000/27000/5
9	-10	1.0	18085.8	-10	1.0	2977	-10	900/1500/5
10	-9	1.0	69210	-9	0.7	59252.7	-9	6000/21000/10
11	-8	1.0	41724.2	-8	1.0	23212.5	-8	3000/9000/5
12	-4	1.0	68289.6	-4	1.0	17490.4	-4	3000/9000/5

Here $\overline{N}^{(E)}$ is the average value of the index $N^{(E)}$ obtained during one experiment; E^* - the minimum value of the energy of the conformation, achieved as a result of the experiment. A comparative analysis allows us to conclude that most of the values of the performance indicator of the proposed algorithm significantly exceed those of the IA. However, experiments with sequences No. 6 and No. 9, in contrast to the IA, were not always completed successfully.

In these tests, the search "stuck" in local optima. Sequences No. 7 and No. 8 are the most complex, have the maximum length among all tests. As in the case of IA, test No. 7 was unsuccessful for the developed algorithm. Here the minimum value of energy $E^* = -35$ is achieved with the known $E^{\min} = -36$. Test No. 8 gave the best results despite the fact that the indicator is $\overline{N}^{(E)}$ 10 times less than that of the IA. Table 4 provides comparative data of the performance of several well-known search methods in relation to the problem in question. In this case, the indicators of the most successful launch $N^{(E)}$ are compared, i.e. when this indicator took the minimum value.

Table 4. The minimum indicators $N^{(E)}$ for a number of search methods

No	GA	EMC	MMA	EDA	IA	CLONALG+DE
1	30492	9374	14621	4510	1925	1253
2	30491	6929	–	–	2479	2235
3	20400	7202	18736	13880	4212	2933
4	301339	12447	208233	113000	43416	32837
5	126547	165791	1155656	53995	37269	27571
6	592887	74613	336763	118000	18919	116956
7	−34/208781	−35/203729	–	−35/473500	−35/161448	−35/57298
8	−37/187393	−39/564809	–	−41/595900	−39/377404	−42/327987

The comparison involves the genetic algorithm (GA) and the Monte Carlo method (MCM) [14], the multi-memetic algorithm (MMA) [38,39], one of the varieties of the probabilistic genetic algorithm (PGA) [40,41]. In these experiments, in most cases, the developed hybrid algorithm produced the best results. Tests No. 4 and No. 6 turned out to be unsuccessful, where the Monte-Carlo method and the IA were the best, respectively. Despite the fact that the last test in the table (No. 8) turned out to be worse $N^{(E)}$ in comparison with the genetic algorithm, the authors found it advantageous because it reached the minimum value of the energy of the conformation. The last column of Table 3 contains the settings that had the greatest impact on the quality of the algorithm during the research. As the authors found out, the number of successful launches significantly depends on the size of the main population and the clone population. In this regard, the paper stipulates a separate experiment with the test sequence No. 4. In this experiment, the effect of population size on the search capabilities of the developed hybrid is investigated (Table 5).

As can be seen from the table, with values $P = 50$ and $p = 300$, the algorithm never reached a minimum of energy ($E^{\min} = -14$) over twenty starts. Moreover, even the value $E^* = -13$ is quite rare. Basically, the algorithm converges to the value $E^* = -12$. As the size of populations increases, the number of successful starts increases. When reaching values $P = 2000$ and $P_C = 7000$, all launches become successful. From this point on, further growth P P_C and leads to an increase in the stability of the algorithm. This is evident from the dynamics of the coefficient of variation c_V calculated for $N^{(E)}$.

Table 5. Investigation of the influence of population size on the quality of the hybrid algorithm, with reference to protein no. 4

P/P_C	$N^{(E)}$	$\overline{N}^{(E)}$	k_{SR}	$\overline{E}^*$	c_V
50/300	–	–	0.0	−12.25	–
300/900	3250	148671	0.25	−12.95	1.25
600/1800	10057	257672.28	0.35	−13.2	1.31
900/3000	9829	166804.9	0.5	−13.5	1.323
1500/4500	20340	120766.44	0.8	−13.8	2.05
2000/7000	20873	36939.65	1.0	−14	0.3
3000/9000	30220	49654.65	1.0	−14	0.23
6000/18000	61307	96748.5	1.0	−14	0.225
12000/27000	70446	123894.05	1.0	−14	0.197

7   Conclusions

Prediction of the tertiary structure of a protein is one of the complex tasks of biology, requiring the development of new efficient computational methods for its solution. One of the ways to increase the accuracy and speed of calculations is the hybridization of existing algorithms. Hybrid methods need to be thoroughly tested to prove their superiority. Paper proposes a hybrid method and an algorithm for solving the protein folding problem based on a combination of an artificial immune system in the form of a clonal selection algorithm and a differential evolution algorithm. In order to test the effectiveness of the proposed approach, experiments were carried out on test hp-sequences, for which convolutions on a two-dimensional square lattice were calculated. Experiments have shown a generally significant increase in productivity (in terms of N^E) the developed hybrid compared with other published methods. However, in a small number of tests, the algorithm produced the worst results, which is associated with a "stuck" search in a local optimum.

In the future, it is planned to continue testing the developed hybrid on other types of models and lattices, as well as to improve the design in order to reduce the likelihood of falling into local optima.

References

1. Dill K, Bromberg S, Yue K, Fiebig K, Yee D, Thomas P, Chan H (1995) Principles of protein folding - a perspective from simple exact models. Protein Sci 4:561–602
2. Shmygelska A, Hoos H (2005) An ant colony optimization algorithm for the 2D and 3D hydrophobic polar protein folding problem. BMC Bioinf 6(30):30–52
3. Greene LH, Lewis TE, Addou SC, Cuff A, Dallman T, Dibley M (2007) The CATH domain structure database: new protocols and classification levels give a more comprehensive resource for expoling evolution. Nucleic Acids Res 35:D291–D297

4. Qian B, Raman S, Das R, Bradley Ph, McCoy A (2007) High-resolution structure prediction and the crystallographic phase problem. Nature 450:259–264
5. Xiang Z (2006) Advances in homology protein structure modeling. Curr Protein Pept Sci 7:217–227
6. Kolinski A, Rotkiewicz P, Ilkowski B, Skolnick J (1999) A method for the improvement of threading-based protein models. Proteins 9(37(4)):592–610
7. Gulyanitskiy LF, Rudyk VA (2010) Simulation of protein coagulation in space. Komp'yuternaya matematika 1:128–137 (in Russian)
8. Berger B, Leighton T (1998) Protein folding in the hydrophobic-hydrophilic (HP) model is NP-complete. In: International conference on research in computational molecular biology, vol 5, no 1. ACM, New York, pp 30–39
9. Crescenzi P, Goldman D, Papadimitriou C, Piccolboni A, Yannakakis M (1998) On the complexity of protein folding. J Comput Biol 5(3):423–465
10. Dill KA (1985) Theory for the folding and stability of globular proteins. Biochemistry 24(6):1501–1509
11. Gulyanitskiy L, Rudyk V (2012) Analysis of prediction algorithms for tertiary protein structure based on the ant colony optimization method. V.M. Glushkov Institute of Cybernetics, ITHEA, Kiev-Sofia, pp 152–159. (in Russian)
12. Cutello V, Niscosia G, Pavone M, Timmis J (2007) An immune algorithm for protein structure prediction on lattice models. IEEE Trans Evol Comput 11(1):101–117
13. Hoque M, Chetty M, Lewis A, Sattar A (2011) Twin removal in genetic algorithms for protein structure prediction using low-resolution model. IEEE/ACM Trans Comput Biol Bioinform 8(1):234–245
14. Unger R, Moult J (1993) Genetic algorithms for protein folding simulations. J Mol Biol 231(1):75–81
15. Unger R (2004) The genetic algorithm approach to protein structure prediction. Appl Evol Comput Chem 110:2697–2699
16. Chira C (2011) A hybrid evolutionary approach to protein structure prediction with lattice models. In: IEEE congress on evolutionary computation, New Orleans, USA, pp 2300–2306
17. Islam M, Chetty M, Murshed M (2011) Novel local improvement techniques in clustered memetic algorithm for protein structure prediction. In: IEEE congress on evolutionary computation, New Orleans, LA, USA, pp 1003–1011
18. Baǎutu A, Luchian H (2010) Protein structure prediction in lattice models with particle swarm optimization. In: Swarm intelligence, vol 6234. Lecture notes in computer science, pp 512–519
19. Santos J, Direguez M (2011) Differential evolution for protein structure prediction using the HP model. In: Foundations on natural and artificial computation, vol 6686. Lecture notes in computer science, pp 323–333
20. Lesh N, Mitzenmacher M, Whitesides S (2003) A complete and effective move set for simplified protein folding. In: Proceedings 7th annual international conference on research in computational molecular biology, Berlin, Germany, April 2003, pp 188–195
21. Zhang JF, Kou SC, Liu JS (2007) Biopolymer structure simulation and optimization via fragment regrowth Monte Carlo. J Chem Phys 2007:126–225
22. Hansmann U, Okamoto Y (1996) Monte Carlo simulations in generalized ensemble: multicanonical algorithm versus simulated tempering. Phys Rev 1996(54):5863–5865

23. Steinhöfel K, Skaliotis A, Albrecht A (2007) Relating time complexity of protein folding simulation to approximations of folding time. Comput Phys Commun 2007(176):165–170
24. Albrecht A, Skaliotis A, Steinhöfel K (2008) Stochastic protein folding simulation in the three-dimensional HP-model. Comput Biol Chem 2008(32):248–255
25. Zhang YD, Wu LC, Wang SH (2013) Solving two-dimensional HP model by firefly algorithm and simplified energy function. Math Probl Eng 2013:398–441
26. Perdomo A, Truncik C, Tubert-Brohman I, Rose G, Aspuru-Guzik A (2008) Construction of model Hamiltonians for adiabatic quantum computation and its application to finding low-energy conformations of lattice protein models. Phys Rev 2008(78):1232–1235
27. Perdomo-Ortiz A, Dickson N, Drew-Brook M, Rose G, Aspuru-Guzik A (2012) Finding low-energy conformations of lattice protein models by quantum annealing. Sci Rep 2:571
28. Lau FF, Dill KA (1989) A lattice statistical mechanics model of the conformational and sequence spaces of proteins. Macromolecules 1989(22):3986–3997
29. Bromberg S, Yue K, Fiebig K, Yee D, Thomas P, Chan H (1995) Principles of protein folding - a perspective from simple exact models. Protein Sci 1995(4):561–602
30. Moreno-Hernández S, Levitt M (2012) Comparative modeling and protein-like features of hydrophobicpolar models on a two-dimensional lattice. Proteins 2012(80):1683
31. Miyazawa S, Jernigan R (1985) Estimation of effective interresidue contact energies from protein crystal structures: quasi-chemical approximation. Macromolecules 1985(18):534–552
32. Miyazawa S, Jernigan R (1996) Residue-residue potentials with a favorable contact pair term and an unfavorable high packing density term, for simulation and threading. J Mol Biol 1996(256):623–644
33. Krasnogor N, Hart W, Smith J, Pelta D (1999) Protein structure prediction with evolutionary algorithms. In: Proceedings of the 1st annual conference on genetic and evolutionary computation, Orlando, FL, July 1999, pp 1596–1601
34. Burnet FM (1976) A modification of Jerne's theory of antibody production using the concept of clonal selection. CA: Cancer J Clin 26(2):119–121
35. De Castro LN, Von Zuben FJ (2002) Learning and optimization using the clonal selection principle. IEEE Trans Evol Comput 6(3):239–251
36. Storn R, Price K (1997) Differential evolution—a simple and efficient heuristic for global optimization over continuous spaces. J Glob Optim 11(4):341–359
37. Storn R, Price K (1996) Minimizing the real function of the ICEC 1996 contest by differential evolution. In: IEEE international conference on evolutionary computation, Nagoya, Japan, May 1996, pp 842–844
38. Krasnogor NC, Blackburne BP, Burke EK, Hirst JD (2002) Multimeme algorithms for protein structure prediction. In: Proceeding of international conference on parallel problem solving from nature (PPSN VII), Granada, Spain, September 2002, pp 769–778
39. Toma L, Toma S (1996) Contact interactions method: a new algorithm for protein folding simulations. Protein Sci 5:147–153

40. Pelta D, Krasnogor N (2004) Multimeme algorithms using fuzzy logic based memes for protein structure prediction. In: Hart WE, Smith JE, Krasnogor N (eds) Recent advances in memetic algorithms. Springer, Berlin, pp 49–64
41. Santana R, Larrañaga P, Lozano J (2004) Protein folding in 2-dimensional lattices with estimation of distribution algorithms. In: Proceeding of the 5th international symposium on biological and medical data analysis, Barcelona, Spain, November 2004, pp 388–398

Low-Contrast Image Segmentation by Using of the Type-2 Fuzzy Clustering Based on the Membership Function Statistical Characteristics

Lyudmila Akhmetshina$^{(\boxtimes)}$ and Artyom Yegorov

Oles Honchar Dnipro National University, Gagarin Ave. 72, Dnipro, Ukraine
akhmlu1@gmail.com

Abstract. Informational capabilities of the method of low-contrast half-tone image segmentation based on the Type-2 Fuzzy Clustering Approach were studied. Uncertainty of the image segmentation is explained by inaccuracy of the source data and decision model. In the article, the existing approaches of the Type-2 Fuzzy Clustering used for modeling inaccuracy of the fuzzy clustering method are described, and influence of the method of their formation on sensitivity of the image segmentation results is considered. In order to transfer to the Type-2 Fuzzy Clustering, the proposed algorithm uses statistical characteristics of the Type-1 Fuzzy membership functions obtained during the iterative process of the basic FCM method. The modified algorithm and experimental data are presented on the example of processing the half-tone medical images, which demonstrate that the image segmentation sensitivity and accuracy significantly depends on the method of the Type-2 Fuzzy Sets generation.

Keywords: Image segmentation · Fuzzy clustering · Type-2 fuzzy

1 Introduction

In practice, transformation of the images is closely connected with the problem of extracting information for the purpose of its better interpretation by people or its use by automated systems. A key step of information processing is its segmentation, which provides distinguishing of areas with homogeneous characteristics, such as brightness, texture, etc., and which can be interpreted as a process of visualization of the clustering result [1].

The stage of image clustering is ambiguous due to many reasons. In many problems, for example, in the sphere of medicine, geology, ecology, etc., there is a priori no information about the availability and location of objects of interest (i.e. of the anomalies) that should be analyzed. In addition, uncertainty is always present in the source data and decision model. Of particular complexity is analysis of low-contrast images with insignificant differences in characteristics of the objects of interest or when the clusters with complex shape overlap each other (for example, a cancer tumor with metastases, mineral layers, etc.).

© Springer Nature Switzerland AG 2020
V. Lytvynenko et al. (Eds.): ISDMCI 2019, AISC 1020, pp. 689–700, 2020.
https://doi.org/10.1007/978-3-030-26474-1_48

2   Review of Literature

The fuzzy clustering method (FCM) allows assessing grade of the object membership in this or that class on the basis of analysis of the Type-1 membership functions (MFT1), which vary in the interval $[0 \div 1]$, and improving segmentation accuracy through taking into account the uncertainty [2]. In recent years, this approach has been successfully used in many applications, and various modifications of the FCM method have appeared, which consider specific features of the source data and the assigned objective [3–5]. In the work [6], informational capabilities of such algorithms are considered in terms of improving reliability and accuracy of the low-contrast image segmentation.

However, the Type-1 membership functions does not consider a concept of "fuzzy grade of membership", as well as cannot describe uncertainties in the fuzzy set and take into account uncertainties of the solution process, the result of which is characterized by clear values [7–9].

Fuzzy sets of the type-2 were introduced by Zadeh L. as a generalized concept of the theory of ordinary fuzzy sets. The corresponding membership functions of the type-2 (MFT2) are determined as a generalized fuzzy set (FS) by introducing fuzzy intervals; such approach correlates with inaccuracy perception by humans [10].

The T2FCM algorithm is a generalized FCM algorithm, which extends the grade of each object membership with the help of the Type-2 membership functions. In the work [5], results of segmentation of the brain MRI images with the help of T2FCM algorithm are presented, which demonstrate better detection of abnormal areas in comparison with the FCM method [11,12].

3   Problem Statement

Purpose of this work was to study method of the type-2 fuzzy set formation in the Fuzzy Clustering method by using statistical characteristics of the type-1 membership functions and its influence on sensitivity of the low-contrast halftone image segmentation.

4   Materials and Methods

The FCM method assumes an iterative calculation of centers of the fuzzy clusters and their Type-1 membership functions for each pixel x_i of the analyzed image by determining its location in the feature space [13]. Use of the algorithm requires a priori specifying of initial number of clusters c, centroids of fuzzy classes v^0, and fuzzification parameter (exponential weight) m.

The following factors of the algorithm uncertainty should be mentioned:

– random specifying of initial values for the centers of desired clusters ensures finding of local minimum, and not the global minimum;

– there is no theoretical justification for choosing a value for m, which determines degree of the solution fuzziness (the greater is m, the more "blurred" becomes the final matrix of the c-partition, and when $m \longrightarrow \infty$, it becomes $1/c$, that is, grade of membership of the objects is the same in all clusters);
– in practice, number of clusters is often unknown a priori; however, this factor significantly affects the clusterization detailing and accuracy of the final result of image segmentation.

Fuzzy sets of the type-2 (FST2) feature fuzzy membership functions and are able to simulate such uncertainties [7,14,15]. In Fig. 1, graphical image of the FST2 is shown, which is characterized by upper and lower boundaries, each of them is determined by the lower (LMF) and upper (UMF) membership functions of the type-1 (are shown by thick solid and dashed lines in Fig. 1a, respectively).

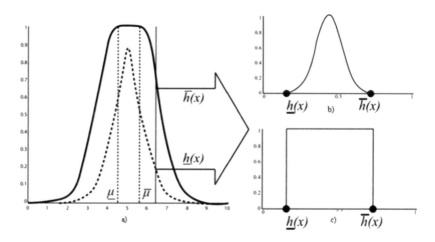

Fig. 1. Type-2 fuzzy set: a is lower $h(x)$ and upper $\overline{h(x)}$ membership functions; b, c are the Gaussian fuzzy membership function type-2 and its interval

The shaded area in Fig. 1a is called a footprint of uncertainty (FOU) and displays uncertainty of the solution made in the range between the upper and the lower boundaries. The distribution form is determined by nature of the problem uncertainty. The membership estimation for each element of the FST2 is a fuzzy set (FS) with the value of the MFT2 in the segment [0, 1].

There are two types of fuzziness of the type-2: generalized and interval. The generalized type-2 defines the upper $\overline{h(x)}$ and lower $h(x)$ values as the spread of the MFT2 values between these boundaries; these values are determined either probabilistically or fuzzy with the help of the MFT1 (Fig. 1b). The interval type-2 is most often used in applications and is a simplified form of the FST2 (Fig. 1c). In this case, the upper $\overline{h(x)}$ and lower $h(x)$ boundaries of membership are clear, and value of the membership function is uniformly distributed along the axis Ox with the MF value for any point in the interval equal to 1 [10].

In the problems of image segmentation, the following approaches to the transition to MFT2 can be singled out with the help of the T2FCM method:

(a) At the step t, iterative calculation of the membership value for the FST2 $a_{k,i}$ on the basis of the Type-1 membership functions method of fuzzy clustering $u_{k,i}$:

$$a^t_{k,i} = u^t_{k,i} - \frac{1 - u^t_{k,i}}{2}. \tag{1}$$

(b) On the basis of difference between the "upper" u^t_h and "lower" h^t_l matrices of the membership functions (are correspond to the upper $\overline{h(x)}$ and lower $\underline{h(x)}$ boundaries). One of the approaches for obtaining these matrices is the use of interval fuzzifier $[m_l, m_h]$, which display different degrees of fuzziness during the formation of Type-1 membership functions. Further, from this set, the maximum is chosen for the elements of the matrix u^t_h and the minimum is chosen for the matrix u^t_h respectively. Determination of interval FST2 with the help of power functions:

$$a^t_{k,i} = (u^t_{k,i})^{1/\alpha} - (u^t_{k,i})^\alpha, \tag{2}$$

where $\alpha \in (1, \infty)$. Usually, for image $\alpha \gg 2$, this parameter is not matter.

The form of the Type-2 membership functions is shown in Fig. 2b, c. These functions is obtained by transformation of the Type-1 membership functions of one of the classes with the help of interval fuzzifier $[3, 2]$ and power functions at $\alpha = 0.5$ and $\alpha = 2$ respectively. The clusterization was performed for the MRI image of brain shown in Fig. 2a as the result of the FCM method applied with $c = 5$.

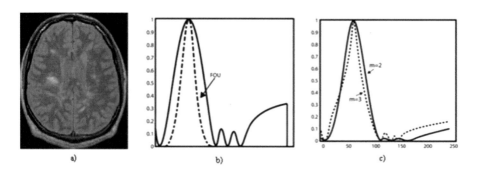

Fig. 2. Type-2 fuzzy set: a - original image; b - power functions: $\cdots \alpha = 0.5$; $-\alpha = 2$; c - is the interval fuzzifier

At updating the cluster center with the help of the MFT2 method, role of patterns with low grade of membership in some cluster is relatively of less importance. Thanks to the MFT2, locations of the cluster centers become more reliable in comparison with the centers obtained by the FCM method of type-1, especially in the presence of noises.

Degree of the MFT2 influence on the result of image segmentation depends both on the method of its formation and on the fuzzy clustering algorithm.

In this work, an algorithm of the FCM dynamic fuzzy clustering [6] was used. The FCM advantages include additional stopping criteria, control based on fuzziness indicator and Xie-Bieni indicator in the training process. These features reduce the necessity of "retraining" (increasing of the formed partition fuzziness).

The algorithm includes the following steps:

1. Initialization of c, v^0 and m.
2. Calculation of current values of the membership function u^t by the formula:

$$u^t_{k,i} = \sum_{L=1}^{c} [\frac{D_{i,k}}{D_{i,L}}]^{\frac{-2}{m-1}} \left(\begin{matrix} \forall k \in \{1,...,c\}, \\ \forall i \in \{1,...,n\} \end{matrix} \right), \tag{3}$$

where t is current step, c is number of fuzzy clusters, n is number of copies of the source data, u is matrix with the size $[c \times n]$, where each of the rows is a function of membership in the k-th center of all vectors of the source data described by the matrix X; the matrix D is calculated by the formula:

$$D^t_{i,k} = \sqrt{(X_i - v^t_k)^T A(X_i - v^t_k)}, \ \forall i \in \{1,...,n\}, \ \forall k \in \{1,...,c\}, \tag{4}$$

where v is matrix of centers of fuzzy clusters, and $A = I$.

3. Formation of a new matrix a^t (MFT2). For this case, various approaches can be used, though the method of matrix a^t formation has a significant impact on the final result of clustering.
4. The matrix v^t is formed by the formula:

$$v^t_{k,j} = \frac{\sum_{i=1}^{n}(u^t_{k,j})^m \cdot X_{i,j}}{\sum_{i=1}^{n}(u^t_{k,i})^m}, \ \forall k \in \{1,...,c\}, \ \forall j \in \{1,...,q\}, \tag{5}$$

where q is number of informative features of each vector of the source data. This matrix will be used at the beginning of the next iteration.

5. Values for Δ^t_v are calculated as average for the matrix of distances between the centers v^t and v^{t-1}; and criteria V^t_{xb} (the Xie-Bieni indicator) and V^t_{fz} (fuzziness indicator) are calculated by the formulas:

$$V^t_{xb} = \frac{\sum_{k=1}^{c} \sum_{i=1}^{n}(a^t_{k,i})^m \cdot \sum_{j=1}^{q}(X_{i,j} - v^t_{k,j})^2}{n \cdot (d^e_{min})^2}, \tag{6}$$

$$V^t_{fz} = \frac{\sum_{k=1}^{c} \sum_{i=1}^{n}(a^t_{k,i})^2}{n}, \tag{7}$$

where d^e_{min} is the minimum Euclidean distance between the centers of fuzzy clusters. Decrease of the Xie-Bieni indicator and increase of the fuzziness indicator characterize improvement of quality of fuzzy partitioning into groups.

6. If the condition:

$$C^t_{fz} \geq C^{max}_{fz} \tag{8}$$

is satisfied, and the value C^t_{fz} is calculated by the following formula for obtaining the final result:

$$C^t_{fz} = \frac{V^t_{fz}}{V^t_{xb}} \tag{9}$$

where C^{max}_{fz} is maximal coefficient among the coefficients C^t_{fz} obtained in the process of training, then the following values are stored: $\Delta^{max}_v = \Delta^t_v$, $C^{max}_{fz} = C^t_{fz}$, $u^{max} = a^t$ and $v^{max} = v^t$.

7. If the following conditions are not satisfied:

$$\Delta^t_v < \varepsilon \quad or \quad (|V^t_{xb} - V^{t-1}_{xb}| < \varepsilon \text{ and } |V^t_{fz} - V^{t-1}_{fz}| < \varepsilon), \tag{10}$$

where V^{t-1}_{xb} and V^{t-1}_{fz} are the Xie-Bieni indicator and fuzziness indicator from the previous iterations, respectively, then it is necessary to return to the step 2.

8. If the conditions

$$C^t_{fz} < C^t_{fz} \quad and \quad (\Delta^t_v > \Delta^{max}_v \text{ or } (\Delta^t_v < \Delta^{max}_v \text{ and } p_{\Delta_v} > p_c)) \tag{11}$$

are satisfied, then the coefficients p_{Δ_v} and p_c are determined by the expressions:

$$p_c = \frac{|C^t_{fz} - C^{max}_{fz}|}{max(C^t_{fz}, C^{max}_{fz})} \cdot \frac{1}{C^{max}_{fz} - C^{min}_{fz}}, \tag{12}$$

$$p_{\Delta_v} = \frac{|\Delta^t_v - \Delta^{max}_v|}{max(\Delta^t_v, \Delta^{max}_v)} \cdot \frac{1}{(\Delta^{max}_v)' - \Delta^{min}_v}, \tag{13}$$

where C^{min}_{fz} and Δ^{min}_v are minimal values of the parameters C^t_{fz} and Δ^t_v respectively, and $(\Delta^{max}_v)'$ is maximal value of the criterion Δ^t_v, then the process returns to the stored values of the matrices of fuzzy membership function u^{max} and centers of fuzzy clusters v^{max}, which are exactly the result of the training.

Thus, in order to calculate the matrix a^t at the 3-rd step of the above mentioned algorithm, the following approaches were used:

(a) On the basis of the formula (1). Direct application of the formula (1) leads to formation of values for the matrix a^t in the segment $[0,1]$. In order to improve image of the matrix a^t in the segment $[0,1]$ the following relationship was used:

$$a^t_{k,i} = \overline{u^t_k} + (u^t_{k,i} - (1 - u^t_{k,i})/2) \cdot (1 - \overline{u^t_k}), \tag{14}$$

where $\overline{u^t_k}$ is the average value for each row of the matrix u^t. However, it should be noted that use of the formula (14) also does not ensure complete displaying of the matrix a^t in the segment $[0, 1]$. In order to improve brightness and contrast of the image after the segmentation, the resulting matrix a^t can be subjected to the following transformation:

$$a^t_{k,i} = |a^t_{k,i}|^{1-sgn(a^t_{k,i})|a^t_{k,i}|^{1-a^t_{k,i}}}. \tag{15}$$

(b) On the basis of the matrix a^t formed as a difference between the "upper" u^t_h and "lower" u^t_l matrices of the membership functions.

When using this approach on the basis of formula (2), statistical characteristics of the matrix u^t were taken into account in the following way:

$$a^t_{k,i} = (u^t_{k,i})^{\overline{u^t_k}+1/\alpha} - (u^t_{k,i})^{\overline{u^t_k}+\alpha}. \tag{16}$$

Taking into account the relaying on the experimental studies, it is recommended to use $\alpha \approx 2$. As in the previous case, in order to improve brightness and contrast of the image, the following transformation can be applied to the resulting matrix a^t:

$$a^t_{k,i} = (a^t_{k,i})^{1-a^t_{k,i}}. \tag{17}$$

Further, it was proposed to use function $y = x^{1-kx}$ while transiting to the MFT2. With values $x \in [0,1]$, this function provides a proportional increase/decrease (depending on the sign of the coefficient k) of the MFT1 value, and also ensures that $y \in [0,1]$. Thus, the matrix a^t was calculated by the following formulas:

$$a^t_{k,i} = (\Delta u^t_{k,i})^{1.5-\overline{\Delta u^t_k}}, \tag{18}$$

$$\Delta u^t_{k,i} = (u^t_h)_{k,i} - (u^t_l)_{k,i}, \tag{19}$$

where $\overline{\Delta u^t_k}$ is average value for each row of the matrix Δu^t, and values for the matrices u^t_h and u^t_l are calculated in the following way:

$$(u^t_h)_{k,i} = (u^t_{k,i})^{1-u^t_{k,i}}^{1-u^t_{k,i}}, \tag{20}$$

$$(u^t_l)_{k,i} = (u^t_{k,i})^{1+u^t_{k,i}}^{1-u^t_{k,i}}. \tag{21}$$

5   Results and Discussion

The proposed algorithm was applied for segmentation of various half-tone images, examples of which are the images shown in Figs. 3 and 4.

In Fig. 3a, a tomogram of the brain is shown, which has been made in order to diagnose the presence of hematoma and to determine area of its influence. According to the histogram of the original image (Fig. 3b), this image cannot be classified as low-contrast one. However, it is difficult to diagnose the hematoma in the original image because of its location in the low-contrast area of the image (see fragment in the rectangle), and area of its influence cannot be determined visually. Introduction of radiographic contrast substance (Fig. 3c) makes

696   L. Akhmetshina and A. Yegorov

it possible to distinguish the hematoma more clearly, but it does not help to determine area of its influence. In Fig. 4, the T1 spin-lattice relaxation of the NMR (nuclear magnetic resonance) examination of the brain region is shown, from the histogram of which (Fig. 4b) it follows that this image is dark. There is a low-contrast area in the low left corner of the picture.

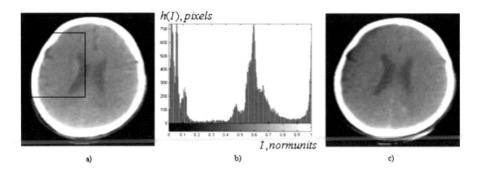

Fig. 3. X-ray tomogram of the brain: *a* - initial half-tone image (204 × 201); *b* - its histogram; *c* - the results after introduction of radiographic contrast substance

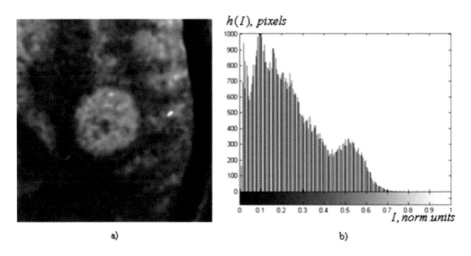

Fig. 4. The T1 spin-lattice relaxation of the NMR examination of the brain: *a* - initial image (256 × 256); *b* - its histogram

Clusterization was performed with the following control parameters: $c = 6$, $m = 2$, $\alpha = 2$. Visualization of the fuzzy clustering results was carried out by way of their comparison with initial data [6]. The results of segmentation of the image showed in Fig. 3a are presented in Fig. 5. Although the segmentation was carried out without transition to the MFT2 (Fig. 5a), it allows distinguishing, to a certain extent, the hematoma, however, it does not allow identifying area of its influence.

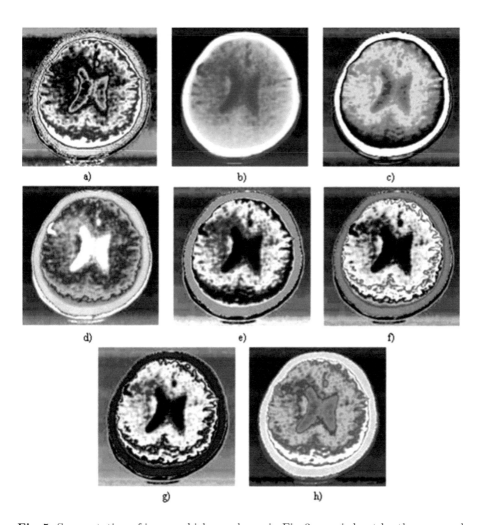

Fig. 5. Segmentation of image which are shown in Fig. 3a carried out by the proposed method: a - without performing step 3; with calculation of the matrix a^t by the formulas: b - (1), c - (14), d - (15), e - (2), f - (16), g - (17), h - (18)

The use of MFT2 improves determining both objects of the interest. The use of the MFT2 statistical characteristics (see Fig. 5c, d, f, g, h) allows distinguishing more clearly both hematoma and structure of the area of its influence in comparison with traditional approaches based on the formulas (1) and (2) (Fig. 5b, e). It should be further noted that use of the formulas (15) and (17)

(Fig. 5d, g) allows to obtain a higher level of brightness and contrast in the resulting image, which leads to a slightly more accurate identification of area of the hematoma influence. The use of formula (18) (Fig. 5h) provides the best accuracy among the considered results in identifying the area of the hematoma influence with good detailing of the image as a whole.

The results of segmentation of the image which are shown in Fig. 4a are presented in Fig. 6. The segmentation without calculation by the Type-2 membership functions (Fig. 6a) leads, on the one hand, to excessive detailing of the result, though, on the other hand, it does not ensure high-performance displaying of the image structure.

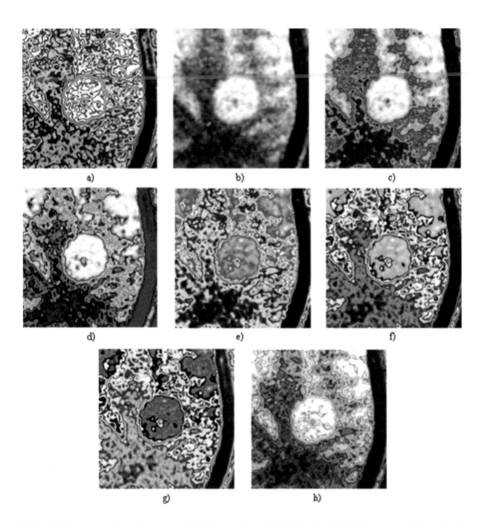

Fig. 6. Segmentation of image which are shown in Fig. 4a with the proposed method: a - without performing step 3; with calculation of the matrix a^t by the formulas: b - (1), c - (14), d - (15), e – (2), f - (16), g - (17), h - (18)

The use of the Type-2 membership functions provides high quality of the image structure displaying without obtaining the excessive detailing. In this case, as well as for the previous image, calculation of the Type-2 membership functions with taking into account statistical characteristics of the membership function matrix (Fig. 6c, d, e, f, g, respectively) provides better detailing than with calculation of the Type-2 membership functions with basic relations(1) and (2) (see Fig. 6b, d). Similarly to the previous case, use of formulas (15) and (17) improves brightness and contrast (Fig. 6d, g), and use of formula (18) more accurately reflects structure of the region of interest and the image as a whole.

During the experiments performing when transition to the Type-2 membership functions was applied, a decrease in the number of the training iterations was also noted. It should be further mentioned that, at calculation of the Type-2 membership functions by the standard approaches, and despite obtaining of acceptable level of detailing, there was some instability in the training process (in particular, a sharp increase of the Xie-Bieni indicator values was observed), which made it necessary to return to intermediate training results (which is realized by the 8th step of the proposed algorithm). Application of formula (18) used for transition to the Type-2 membership functions did not lead to similar problems during the training.

6   Conclusions

On the basis of analysis of the obtained experimental results the following conclusions have been made:

- segmentation of low-contrast images with the help of dynamic fuzzy clustering algorithm and with use of the type-2 membership functions allows to achieve a higher degree of detailing in comparison with the basic algorithm;
- the method of transition to the Type-2 membership functions has a significant effect on the segmentation sensitivity;
- at transition to the Type-2 membership functions, use of the Type-1 membership functions statistical characteristics allows improving the segmentation sensitivity.
- the proposed in this paper method for transition to the Type-2 membership functions based on formula (18) provides a higher level of detailing.

References

1. Pratt WK (2001) Digital image processing. Wiley, New York
2. Bezdek JC, Keller J, Krishnapuram R, Pal NR (1999) Fuzzy models and algorithms for pattern recognition and image processing. Handbooks of fuzzy sets series. Kluwer Academic Publisher, Boston
3. Deepali A, Kumar RT (2013) Fuzzy clustering algorithms for effective medical image segmentation. Intell Syst Appl 11:55–61. https://doi.org/10.5815/ijisa.2013.11.06

4. Colliot O, Camara O, Bloch I (2006) Integration of fuzzy spatial relations in deformable models - application to brain MRI segmentation. Pattern Recogn 39(8):1401–1414
5. Chen WJ, Giger ML, Bick U (2006) A fuzzy c-means (FCM)-based approach for computerized segmentation of breast lesions in dynamic contrast enhanced MRI images. Acad Radiol 13(1):63–72
6. Yegorov A, Akhmetshina L (2015) Optimizatsiya yarkosti izobrazheniy na osnove neyro-fazzi tekhnologiy. Monografiya. Izd. Lambert
7. Castillo O, Melin P (2008) Type-2 fuzzy logic: theory and applications. Springer, Heidelberg
8. Wu H, Wu Y, Luo J (2009) An interval type-2 fuzzy rough set model for attribute reduction. In: IEEE transactions on fuzzy systems. IEEE Press, pp 301–315. https://doi.org/10.1109/TFUZZ.2009.2013458
9. Mendel JM, John R (2002) Type 2 fuzzy sets made simple. IEEE Trans Fuzzy Syst 10(2):117–127
10. Mendel JM et al (2006) Interval type 2 fuzzy logic systems made simple. IEEE Trans Fuzzy Syst 14(6):808–821
11. Zhang WB, Liu WJ (2007) Fuzzy clustering for rule extraction of interval type-2 fuzzy logic system. In: Proceedings of 46th IEEE conference on decision and control, pp 5318–5322
12. Hwang C, Rhee F (2004) An interval type-2 fuzzy C spherical shells algorithm. In: Proceedings of international conference on fuzzy system, vol 2, pp 1117–1122
13. Chi Z (1998) Fuzzy algorithms: with applications to image processing and pattern. Word Scientific, Singapore
14. Zhang WB et al (2007) Rules extraction of interval type-2 fuzzy logic system based on fuzzy c-means clustering. In: 4th international conference on fuzzy systems and knowledge discovery, vol 2, pp 256–260
15. Maity S, Sil J (2009) Color image segmentation using type-2 fuzzy sets. Int J Comput Electr Eng 1(3):1793–8163

Binary Classification of Fractal Time Series by Machine Learning Methods

Lyudmyla Kirichenko$^{(\boxtimes)}$, Tamara Radivilova , and Vitalii Bulakh

Kharkiv National University of Radioelectronics, Kharkiv, Ukraine
lyudmyla.kirichenko@nure.ua

Abstract. The paper considers the binary classification of time series based on their fractal properties by machine learning. This approach is applied to the realizations of normal and attacked network traffic, which allows to detect DDoS-attacks. A comparative analysis of the results of the classification by the random forest and neural network - fully connected multi-layer perceptron is carried out. The statistical, fractal and recurrence characteristics calculated from each time series were used as features for classification. The analysis showed that both methods provide highly accurate of classification and can be used to detect attacks in intrusion detection systems.

Keywords: Fractal time series · Random forest · Neural network · Machine learning classification · Hurst exponent · Recurrence plot

1 Introduction

Many informational stochastic processes generate time series with self-similar (fractal) properties. One of the most famous examples of self-similarity is the realizations of infocommunication traffic, the fractal properties of which cause overloads in networks and data loss [1,2]. The results of the fractal analysis of time series are widely used in telecommunication and information systems to prevent network overload during the transmission of self-similar data traffic and network attacks [2,3]. In recent years, fractal analysis has been used to detect DoS and DDoS attacks, which allow bringing to failure almost any infocommunication system [4,5].

Research has demonstrated that one of the characteristic signs of a DDoS attack is a change of fractal properties of traffic under the attack action [6,7]. Recognition and classification of fractal series, including the attacked traffic, most often is done by time series analysis of fractal characteristics [1,8–10]. However, in recent years, for the analysis and classification of the time series, the methods of machine learning are used [11–15].

The features used to train the classifier model have a great influence on the quality of the results. Search and selection of features in the source data is an important step in learning. Some of the new, previously unused classification

© Springer Nature Switzerland AG 2020
V. Lytvynenko et al. (Eds.): ISDMCI 2019, AISC 1020, pp. 701–711, 2020.
https://doi.org/10.1007/978-3-030-26474-1_49

features of time series can be obtained using nonlinear dynamics methods, in particular, recurrence analysis [16, 17].

The purpose of this article is a comparative analysis of the detection of DDoS attacks by machine learning methods based on the fractal properties of network traffic realizations.

2   Classification of Time Series Based on Their Fractal Properties

Continuous time stochastic process $X(t)$ is called self-similar if it preserves its distribution laws at time scaling [18]:

$$Law\{X(at)\} = Law\{a^H X(t)\}, a > 0 \tag{1}$$

The value a^H shows how the quantitative characteristics of the process $X(t)$ change. The value H is called the Hurst exponent. It is a measure of self-similarity. Simultaneously H is a measure of long-range dependence of $X(t)$.

If the process is represented by a time series $X_t = (X_1, X_2, ...)$, the concept of an aggregate series is used to the definition of self-similarity. Aggregate series $X^{(m)} = \left\{X_1^{(m)}, X_2^{(m)}, ...\right\}$ is the initial series averaged over blocks of length m, where

$$X_t^{(m)} = \frac{1}{m}(X_{tm-m+1} + ... + X_{tm}), m, t \in N. \tag{2}$$

The time series X_t is self-similar with the parameter H, if

$$Law\left\{m^{1-H} X_t^{(m)}\right\} = Law\{X_t\}, \tag{3}$$

that X_t does not change its distribution laws after averaging over blocks of length m. The moments of the self-similar random process have scaling

$$E\left[|X(t)|^q\right] \propto t^{qH}, \tag{4}$$

where q is a real number. If the value of q is equal 2 we obtain the relation

$$E\left[|X(t)|^2\right] \propto t^{2H}, \tag{5}$$

which is the basis of most of the time series estimation methods for the Hurst exponent. To estimate the Hurst parameter H, many methods have been proposed [9, 18–20]. All of them have a sufficiently large error, especially with a small length of the series.

Unlike self-similar processes (1), multifractal processes have a more flexible scaling behavior [18, 21]:

$$Law\{X(at)\} = Law\{M(a) \cdot X(t)\}, a > 0, \tag{6}$$

where $M(a)$ is the independent of $X(t)$ random function. In the case of a self-similar process $M(a) = a^H$. Multifractal processes have the following scaling behavior of moment characteristics:

$$E\left[|X(t)|^q\right] \propto t^{qh(q)}, \tag{7}$$

where q is a real number, $h(q)$ is a generalized Hurst exponent, for which $h(2) = H$. The self-similar process is a monofractal process. For a monofractal process, the generalized Hurst exponent is constant: $h(q) = H$.

There are many time series estimation methods of $h(q)$ for small values of parameter q (as a rule $|q| \leqslant 10$) that also have a significant error for short time series [18].

Usually, the classification of time series by fractal properties is based on the evaluation of their fractal characteristics. In the works [22,23] it was proposed to classify the time series according to fractal properties, i.e. the values of Hurst exponent using, machine learning, in particular, classifiers based on decision trees. The use of statistical and multifractal characteristics as features was showed a significant increase in the classification accuracy in comparison with traditional classification methods. So, when binary classifying multifractal time series, we have the following values of the average probability of correctly determining the class [23], that are shown in Table 1.

Table 1. The average probability of correctly determining the class at binary classification

Length time series	Probability	
	Random forest	Time series estimate H
512	0.94	0.75
4096	0.96	0.78

In this case, the classified time series had a Hurst exponent in the range (0.5, 0.8) for one class and (0.8, 0.95) for another. The following values were presented as features to the classifier input: mean, maximum value, standard deviation, a median of time series, standard deviation, the mean and standard deviation of the generalized Hurst exponent $H(q)$ and others. However, it should be noted that the investigated time series had strong multifractal properties.

Research of monofractal time series and series with weakly expressed multifractal properties (when the values of $h(q)$ vary in a small range) was conducted and showed that the set of features described above is not sufficient for correct classification.

Therefore, it was proposed to use as features for classification the time series recurrence characteristics.

3   Time Series Classification Using Recurrence Plots

The recurrence plot is a projection of a m - dimensional pseudo-phase space onto a plane. Let the point x_i corresponds to the point of the pseudo-phase trajectory $X(t) = [x(t), x(t + \tau), ..., x(t + m\tau)]$ describing the dynamic system that represented by the time series $x(t)$ in m-dimensional space at the time moment $t = i$ for $i = 1, ..., N$. Then the recurrence plot RP is an array of points, where a non-zero element with coordinates (i, j) corresponds to the case when the distance between x_j and x_i less ϵ:

$$RP_{i,j} = \Theta(\epsilon - \|x_i - x_j\|), x_i, x_j \in R^m, i, j = 1, ...N \qquad (8)$$

where ϵ is the size of the neighborhood of the point x_i, $\|x_i - x_j\|$ is the distance between the points, $\Theta(\cdot)$ is the Heaviside function [17].

Analysis of the topological structure of the recurrence plot allows us to classify the observed processes: homogeneous processes with independent random values; processes with slowly changing parameters; periodic or oscillating processes, etc. [24,25]. Numerical analysis of recurrence plots allows calculating quantitative measures of the complexity of the recurrence plots structures, such as the measures of recurrence, determinism, entropy, and others [26,27]. Among the many quantitative recurrence measures are the following: the measure of recurrence RR shows the density of recurrence points; measure of determinism Det is a characteristic of the predictability of the process behavior; the average length of the diagonal lines L is the average time during which two sections of the trajectory pass close to one another; the measure LAM characterizes states when the system moves along a phase trajectory stops or moves very slowly.

In [28,29], it was shown that the quantitative recurrence characteristics of self-similar time series significantly depend on the Hurst exponent H (see Table 2).

Table 2. Quantitative recurrence characteristics

H	RR	Det	L	LAM
0,6	0,027	0,684	3,866	0,824
0,7	0,035	0,752	8,371	0,880
0,8	0,044	0,858	15,959	0,910
0,9	0,049	0,937	23,292	0,937

In [30], it was shown that the classification based on statistical and fractal characteristics for time series with weak multifractal properties is not effective and it was proposed to use recurrence characteristics as features. Classification of monofractal time series based on recurrence characteristics using the random forest method showed the average probability of determining the class from 0.8 to 0.95, depending on the classification conditions.

Thus, to classify fractal time series with different multifractal properties, it would be expedient to use a set of statistical, multifractal and recurrence features.

4   Simulation of Attacked Traffic

To conduct an experiment with the detection of DDoS attacks, it is necessary to have a training and test samples containing the realizations of normal and attacked traffic. In this case, it is advisable to use model traffic realizations. In [31], a model of multifractal traffic which is generated on the basis of fractal Brownian motion was proposed.

The fractal Brownian motion (fBm) is the most well-known and simple model of the self-similar process. On the one hand, fBm is a self-similar process with a single scaling parameter, the Hurst exponent H. On the other hand, fBm can be considered as a multifractal process, in which the generalized Hurst exponent is a constant: $h(q) = H$. Increments of fBm (fractal Gaussian noise) are a Gaussian stationary process with a long-term dependence.

The presented model of multifractal traffic is obtained using the exponential transform of fractal Gaussian noise:

$$Y(t) = Exp\left[k * X(t)\right] \qquad (9)$$

where $X(t)$ is the fractal Gaussian noise with the Hurst exponent H, k is the factor determining the coefficient of variation of time series $Y(t)$, that is, the degree of its multifractality. Thus, using (9), it is possible to model multifractal time series $Y(t)$ with given Hurst exponent and multifractal properties.

For the simulation of DDoS attacks, the dataset described in [32] was used. In this dataset, the results of real experiments are presented that allowed to isolate and record in digital form the realizations of typical DDoS attacks of several types. The fractal analysis of the realizations of DDoS attacks, presented in [33], showed that they have strong multifractal properties and large values of the Hurst exponent ($H > 0.8$).

The realization of traffic under the action of DDoS attacks can be represented as the sum of the realization of traffic and the realization of one of the types of attacks. The sum of two fractal realizations has fractal properties, which are determined by the values of the Hurst exponents and the ratio of the corresponding coefficients of variation [34,35].

Thus, in the work, the realizations of attacked traffic were obtained by summation the model traffic with a given H and part of the realization of DDoS attacks from [32]. It is possible to vary the level of attack (the ratio of the average of the attack realization to the average of traffic), the duration of the attack and the time of the attack. A training and test samples for machine learning classification were a set of realizations of normal and attacked traffics.

5   Classification Methods

5.1   Random Forest

The decision tree method is one of the simplest and most effective methods for classifying various data, including time series [36,37]. Decision trees perform the splitting of the objects in accordance with some splitting rules.

The popularity of the classification and regression decision trees is due to many reasons, in particular, the fact that they allow obtaining easily interpretable models, can work with variables of any type and automatically perform the selection of informative features. However, models of decision trees are unstable: even a small change in the training set can lead to significant changes in the tree structure. In this case, it is expedient to use ensembles of models.

Ensemble modeling turned out to be one of the most powerful methods of machine learning. Random forest is one of the ensemble algorithms that solves both regression problems and classifications [38]. It is based on a set of decision trees and averaging the result of their predictions (in the case of solving a regression problem) or making a decision by voting (in the case of classification). An important point of the algorithm is the element of randomness in the creation of each tree. In random forest, in addition to a random selection of training objects, feature selection is also performed randomly.

5.2   Neural Network

To carry out the classification using a neural network, a fully connected multilayer perceptron was chosen. The neural network had ten hidden layers, five of which were fully connected layers with a sigmoidal neuron activation function. To prevent the effect of overfitting, regularization layers were included in the network, one layer after each fully connected layer. The batch normalization [39] was used as a regularization method. This method allows to increase productivity and stabilize the operation of the neural network. Normalization of the input network layer is usually performed by scaling the submitted data on the activation function. Data normalization can be performed in hidden layers of neural networks as well. The batch normalization method has the following advantages: reduces the value of covariance shift; faster convergence of models is achieved; allows each layer of the network to learn more independently of other layers; it becomes possible to use a higher learning rate, since the outputs of the neural network nodes will not have too large or small values; the regularization mechanism is being implemented; models become less sensitive to the initial initialization of the weights.

The Adam method of stochastic optimization (Adaptive Moment Estimation) was chosen as the method for training the neural network [40]. The Adam optimization algorithm is an extension of the stochastic gradient descent method with iterative updating of the network weights based on adaptive estimates of lower-order moments. The advantages of this method: it is simple to implement;

computationally efficient; does not require large memory; works well for tasks that are large in terms of data and parameters.

The developed neural network in total possessed 250 neurons. The vector of features (statistical, fractal and recurrence characteristics) was fed to the network input, and the output was the presence indicator of a DDoS attack in the traffic time series.

6   Description of the Experiment and Results

The classification was carried out by two different methods: a neural network such as multilayer perceptron type and a random forest method. After a series of experiments, the structure of the neural network was chosen, which contained 10 hidden layers with a total of 250 neurons, and 100 regression trees were taken to construct a random forest.

For the classification two samples, training and test were generated. They consisted of traffic realizations, containing and not containing the attacks. Model traffic was built on the basis of the exponential transform (9). The realization of the DDoS-attack was added to the random position of the traffic in the range from 1 to 1/2 of the traffic length.

Statistical, fractal and recurrence characteristics calculated by the time series were features and they were fed to the input of the classifier. The output was the values 1 or 0: the presence or absence of a DDoS attack in the traffic time series.

Training classifiers were conducted for different types of DDoS attacks when changing the following parameters:

- Hurst exponent of model traffic, which varied from 0.5 to 0.9 in increments of 0.05;
- level of attack which was chosen 10, 15, 20 and 25%.

To implement the classification methods, the language Python was used, the libraries of which contain a large number of machine learning methods. Numerical experiments were carried out for model traffic of different lengths and 6 variants of DDoss attacks. Below, the results for time series with a length of 500 values are given, which is sufficient for processing real-time data and one kind of attacks, which is typical.

Table 3 presents the average probabilities of detecting an attack in the case that the attack level is 20%. The value of True Positive corresponds to the probability of correct detection of an attack, the value False Negative corresponds to false detection of normal traffic as attacked, the value of the F-measure means the aggregated test of attack detection. The values of the Hurst exponent are presented in the range $H = 0.65 - 0.9$, which corresponds to the majority of real traffic realizations.

The values of F-measure depending on the Hurst exponent are given in Table 4. The attack level, in this case, is 20%. It should be noted that the probability of attack detecting significantly depends on the value of the Hurst parameter. The highest probability is when the value of the Hurst exponent of normal

Table 3. Values of True Positive and False Negative for neural network and random forest depending on attack level ($H = 0.85$)

Attack level (%)	Probability	Random forest	Neuron network
10	True Positive	0.56	0.48
	False Negative	0.44	0.6
15	True Positive	0.68	0.64
	False Negative	0.2	0.28
20	True Positive	0.72	0.8
	False Negative	0.44	0.2
25	True Positive	0.88	0.84
	False Negative	0.28	0.24

traffic is the most different from the value of the Hurst exponent of attack. Since attacks have high values H, the probability of attack recognizing is maximum for traffic with small values of exponent H. Similar remarks are easy to do when analyzing Table 3.

Table 4. Values of F-measure for neural network and random forest depending on the Hurst exponent

F-measure	Hurst exponent					
	0.65	0.7	0.75	0.8	0.85	0.9
Random forest	0.868	0.824	0.745	0.8	0.609	0.68
Neuron network	0.868	0.842	0.836	0.852	0.8	0.64

7   Conclusion

A comparative analysis of the binary classification of fractal time series by machine learning methods has been carried out. The classification was considered on the example of the detection of DDoS-attacks in traffic realizations. The random forest with regression trees and multilayer perceptron with batch normalization were chosen as classification methods. Traffic realizations with attacks were obtained by summation of simulated multifractal traffic and realizations of DDoS-attack.

The results of the classification have shown that the probability of detecting attack significantly depends on the fractal properties of normal traffic. This is explained by the fact that usually DDoS attack realizations have a great value of the Hurst exponent and the more the fractal properties of normal traffic and attack realizations differ, the easier it is to detect an intrusion.

The probability of detecting an attack depends on the level of attack with respect to the traffic and varies from about 0.6 at 15% to 0.9 at a 20% level.

The random forest method has performed slightly better at recognizing normal traffic than a neural network, but neural networks have great potential for improvement.

In our future research, we intend to concentrate on construction and training of the neural network for detecting different types of attacks. Also, the next stage of our research will be using datasets with real data traffic.

References

1. Shelukhin OI, Smolskiy SM, Osin AV (2007) Self-similar processes in telecommunications. Wiley, New York, pp 183–196. https://doi.org/10.1002/9780470062098
2. Radivilova T, Kirichenko L, Ageiev D, Bulakh V (2020) The methods to improve quality of service by accounting secure parameters. In: Hu Z, Petoukhov S, Dychka I, He M (eds) Advances in computer science for engineering and education II. ICCSEEA 2019. Advances in intelligent systems and computing, vol 938. Springer, Cham, pp 346–355. https://doi.org/10.1007/978-3-030-16621-2_32
3. Daradkeh YI, Kirichenko L, Radivilova T (2018) Development of QoS methods in the information networks with fractal traffic. Int J Electron Telecommun 64(1):27–32. https://doi.org/10.24425/118142
4. Gupta N, Srivastava K, Sharma A (2016) Reducing false positive in intrusion detection system: a survey. Int J Comput Sci Inf Technol 7(3):1600–1603
5. Popa SM, Manea GM (2015) Using traffic self-similarity for network anomalies detection. In: Proceedings of 20-th international conference on control systems and computer science, Bucharest, pp 639–644
6. Kaur G, Saxena V, Gupta J (2017) Detection of TCP targeted high bandwidth attacks using self-similarity. J King Saud Univ Comput Inf Sci. https://doi.org/10.1016/j.jksuci.2017.05.004
7. Deka R, Bhattacharyya D (2016) Self-similarity based DDoS attack detection using Hurst parameter. Secur Commun Netw 9(17):4468–4481. https://doi.org/10.1002/sec.1639
8. Brambila F (2017) Fractal analysis - applications in physics, engineering and technology. IntechOpen. https://doi.org/10.5772/65531
9. Rea W, Oxley L, Reale M, Brown J (2009) Estimators for long range dependence: an empirical study. Electron J Stat
10. Hippenstiel R, El-Kishky H, Radev P (2004) On time-series analysis and signal classification - part I: fractal dimensions. In: Proceedings of 38th asilomar conference on signals systems and computers, Pacific Grove, USA, vol 2, pp 2121–2125. https://doi.org/10.1109/ACSSC.2004.1399541
11. Ledesma-Orozco SE, Ruiz J, García G, Aviña G, Hernández D (2011) Analysis of self-similar data by artificial neural networks. In: Proceedings of the 2011 international conference on networking, sensing and control, Delft, pp 480–485. https://doi.org/10.1109/ICNSC.2011.5874873
12. Kirichenko L, Radivilova T, Bulakh V (2019) Machine learning in classification time series with fractal properties. Data 4(1):5. https://doi.org/10.3390/data4010005
13. Tyralis H, Dimitriadis P, Koutsoyiannis D, O'Connell PE, Tzouka K, Iliopoulou T (2018) On the long-range dependence properties of annual precipitation using a global network of instrumental measurements. Adv Water Resour 111:301–318. https://doi.org/10.1016/j.advwatres.2017.11.010

14. Arjunan SP, Kumar DK, Naik GR (2010) A machine learning based method for classification of fractal features of forearm sEMG using Twin Support vector machines. In: Proceedings of the 2010 annual international conference of the IEEE engineering in medicine and biology, Buenos Aires, pp 4821–4824. https://doi.org/10.1109/IEMBS.2010.5627902

15. Matuszewski J, Sikorska-Łukasiewicz K (2017) Neural network application for emitter identification. In: Proceedings of the 18th international radar symposium (IRS), Prague, Czech Republic, 15 Aug 2017, pp 1–8. https://doi.org/10.23919/IRS.2017.8008202

16. Korus L, Piórek M (2015) Compound method of time series classification. Nonlinear Anal Model Control 20(4):545–560. https://doi.org/10.15388/NA.2015.4.6

17. Eckmann JP, Kamphorst SO, Ruelle D (1987) Recurrence plots of dynamical systems. Europhys Lett 5:973–977

18. Kantelhardt JW (2012) Fractal and multifractal time series. In: Meyers R (ed) Mathematics of complexity and dynamical systems. Springer, New York. https://doi.org/10.1007/978-1-4614-1806-1_30

19. Taqqu M, Teverovsky V, Willinger W (1995) Estimators for long-range dependence: an empirical study. Fractals 3(4):785–798. https://doi.org/10.1142/S0218348X95000692

20. Kirichenko L, Radivilova T, Deineko Z (2011) Comparative analysis for estimating of the hurst exponent for stationary and nonstationary time series. Int J Inf Technol Knowl 5:371–388

21. Riedi RH: Multifractal Processes. https://www.researchgate.net/publication/2839202_Multifractal_Processes. Accessed 27 Mar 2019

22. Bulakh V, Kirichenko L, Radivilova T (2018) Classification of multifractal time series by decision tree methods. In: Proceedings of the 14th international conference on ICT in education, research and industrial applications. Integration, harmonization and knowledge transfer. Volume I: Main conference, vol 2105, Kyiv, Ukraine

23. Bulakh V, Kirichenko L, Radivilova T (2018) Time series classification based on fractal properties. In Proceedings of the 2018 IEEE second international conference on data stream mining and processing (DSMP), Lviv, Ukraine, pp 198–201. https://doi.org/10.1109/DSMP.2018.8478532

24. Marwan N, Wessel N, Meyerfeldt U, Schirdewan A, Kurths J (2002) Recurrence-plots-based measures of complexity and application to heart-rate-variability data. Phys Rev E Stat Nonlin Soft Matter Phys 66(2):026702. https://doi.org/10.1103/PhysRevE.66.026702

25. Marwan N, Romano M, Thiel M, Kurths J (2007) Recurrence plots for the analysis of complex system. Phys Rep 438(5–6):237–329. https://doi.org/10.1016/j.physrep.2006.11.001

26. Zbilut JP, Zaldivar-Comenges J-M, Stozzi F (2002) Recurrence quantification based Liapunov exponent for monitoring divergence in experimental data. Phys Lett A 297(3–4):173–181. https://doi.org/10.1016/S0375-9601(02)00436-X

27. Ngamga EJ, Nandi A, Ramaswamy R, Romano MC, Thiel M, Kurths J (2007) Recurrence analysis of strange nonchaotic dynamics. Phys Rev E 75(3):036222. https://doi.org/10.1103/PhysRevE.75.036222

28. Kirichenko L, Kobitskaya Yu, Habacheva A (2014) Comparative analysis of the complexity of chaotic and stochastic time series. Radio Electron Comput Sci Control 2(31):126–134

29. Kirichenko L, Baranovskyi O, Kobitskaya Y (2016) Recurrence analysis of self-similar and multifractal time series. Inf Content Process 3(1):16–37

30. Kirichenko L, Radivilova T, Bulakh V (2018) Classification of fractal time series using recurrence plots. In: Proceedings of the 2018 international scientific-practical conference problems of infocommunications. Science and Technology (PIC S&T), Kharkiv, Ukraine, pp 719–724. https://doi.org/10.1109/INFOCOMMST.2018.8632010

31. Kirichenko L, Radivilova T, Alghawli AS (2013) Mathematical simulation of self-similar network traffic with aimed parameters. Anale Seria Inform 11(1):17–22

32. Al-kasassbeh M, Al-Naymat G, Al-Hawari E (2016) Towards generating realistic SNMP-MIB dataset for network anomaly detection. Int J Comput Sci Inf Secur 14(9):1162–1185

33. Radivilova T, Kirichenko L, Bulakh V (2019) Detection of DDoS attacks by mashing learning based on fractal properties. In: Security in cervatury the social internet space in context values and hazards, Kharkiv, Ukraine, pp 299–315

34. Ivanisenko I, Kirichenko L, Radivilova T (2015) Investigation of self-similar properties of additive data traffic. In: Proceedings of the 2015 Xth international scientific and technical conference "Computer Sciences and Information Technologies" (CSIT), Lviv, Ukraine, pp 169–171. https://doi.org/10.1109/STC-CSIT.2015.7325459

35. Ivanisenko I, Kirichenko L, Radivilova T (2016) Investigation of multifractal properties of additive data stream. In: Proceedings of the 2016 IEEE first international conference on data stream mining and processing (DSMP), Lviv, Ukraine, pp 305–308. https://doi.org/10.1109/DSMP.2016.7583564

36. Cielen D, Meysman A, Ali M (2016) Introducing data science: big data, machine learning, and more, using python tools. Manning Publications, Shelter Island

37. Quinlan JR (1993) C4.5: Programs for machine learning. Morgan Kaufmann Publishers Inc, San Francisco

38. Breiman L (2001) Random forests. Mach Learn 45(1):5–32. https://doi.org/10.1023/A:1010933404324

39. Ioffe S, Szegedy C (2015) Batch normalization: accelerating deep network training by reducing internal covariate shift. In: Proceedings of the 32nd international conference on machine learning, PMLR, Lille, France, vol 37, pp 448–456. https://arxiv.org/abs/1502.03167

40. Kingma DP, Ba J (2015) Adam: a method for stochastic optimization. In: Proceedings of the 3rd international conference on learning representations (ICLR), San Diego, USA. https://arxiv.org/abs/1412.6980

Author Index

© Springer Nature Switzerland AG 2020
V. Lytvynenko et al. (Eds.): ISDMCI 2019, AISC 1020, pp. 713–715, 2020.
https://doi.org/10.1007/978-3-030-26474-1